MICRO TOTAL ANALYSIS SYSTEMS 2001

Micro Total Analysis Systems 2001

Proceedings of the μTAS 2001 Symposium,
held in Monterey, CA, USA
21–25 October, 2001

edited by

J. Michael Ramsey
Oak Ridge National Laboratory,
Oak Ridge, Tennessee, USA

and

Albert van den Berg
MESA[+] Research Institute,
University of Twente,
Enschede, The Netherlands

KLUWER ACADEMIC PUBLISHERS
DORDRECHT / BOSTON / LONDON

Library of Congress Cataloging-in-Publication Data

ISBN 1-4020-0148-7

Published by Kluwer Academic Publishers,
P.O. Box 17, 3300 AA Dordrecht, The Netherlands.

Sold and distributed in North, Central and South America
by Kluwer Academic Publishers,
101 Philip Drive, Norwell, MA 02061, U.S.A.

In all other countries, sold and distributed
by Kluwer Academic Publishers,
P.O. Box 322, 3300 AH Dordrecht, The Netherlands.

Printed on acid-free paper

Printed in the Netherlands

Program Committee μTAS 2001

Dr. J. Michael Ramsey (Conference Chair)
Oak Ridge National Laboratory

Professor Yoshinobu Baba
University of Tokushima

Professor David Beebe
University of Wisconsin

Professor Albert van den Berg
University of Twente
MESA Research Institute

Professor D. Jed Harrison
University of Alberta

Dr. Stephen C. Jacobson
Oak Ridge National Laboratory

Professor Klavs Jensen
Massachusetts Institute of Technology

Professor Thomas Laurell
Lund Institute of Technology

Professor Andreas Manz
Imperial College of London

Dr. M. Allen Northrup
Microfluidic Systems Inc.

Professor Shuichi Shoji
Waseda University

Dr. Sabeth Verpoorte
University of Neuchâtel

The Program Committee and the Organizers kindly thank the following sponsors:

caliper

COVENTOR

Waters

Sandia National Laboratories

ACLARA BioSciences Inc.

amersham pharmacia biotech

MESA+

MOTOROLA Life Sciences

RS•C

Preface

The Fifth International Conference on Micro Total Analysis Systems, also known as µTAS 2001, will highlight the latest exciting events in the world of miniaturized devices and systems for performing chemical and biochemical experimentation This conference has become mandatory for those of us working in this field as it is indeed helping to define our discipline. We are grateful to the people of the MESA Research Institute of the University of Twente, particularly Piet Bergveld and Albert van den Berg, for starting this meeting in 1994. Their original intention was for the µTAS meeting to be a small informal workshop. This workshop flavor was sustained through the second meeting held in Basel in 1996, but already in 1998 at the third meeting in Banff it was clear that the "workshop" had become a conference with 420 attendees. It was due to this clearly growing interest in microchemical systems that it was decided we should consider gradually moving toward an annual format and prepare for the possibility that the meeting would increase in popularity. Albert van den Berg was still yearning for a workshop at the µTAS 2000 meeting and planned a single session format. Again there was a large increase in submitted abstracts (more than 230 total) and a further increase in attendance. The µTAS steering committee again agreed that we would have to prepare to address the demand the meeting was receiving. Twelve months after the 2000 meeting, nearly 400 submissions were received for the 2001 meeting. The program committee truly had a huge amount of work to do, sifting through all of these abstracts. We were very pleased to find that the quality of submissions remained very high. Having planned for a larger venue, we were able to accommodate 76 submissions as oral presentations and 200 abstracts as poster presentations or approximately a 75% acceptance rate. In addition, we have an excellent slate of invited speakers coming from universities, national laboratories and the commercial sector. I hope everyone will agree that the program committee did an excellent job in selecting abstracts and arranging the program. This is the first of the µTAS programs that uses a parallel session format throughout the week, with plenary sessions for the invited speakers. I encourage the attendees to provide feedback to the steering committee concerning the advantages and disadvantages of the larger parallel session format.

The program contains both anticipated and surprising features. There are a number of papers that discuss improvements upon known strategies and several new developments. Electrokinetically driven chemical separations continue to be a strong theme with DNA remaining a popular sample but with strong competition from proteins. µTAS and lab-on-a-chip devices have their roots in electrophoretic separations and such devices are exhibiting ever greater performance. For example, the efficiency of the separations is improving and some convincing sample processing steps are being incorporated into separations microchips. Moreover, the first true commercial lab-on-a-chip device will be discussed where proteins are electrophoretically sized and fluorescently stained in an integrated post-separation labeling reactor. Previous commercial devices only performed electrophoretic separations of nucleic acids. Other expected developments, but nonetheless exciting, are microfabricated elements for fluid transport, rapid mixing, and dispensing. Electrokinetic instability is being used for the accelerated mixing of materials. This is a phenomena that I believe many labs have observed, and now it is being systematically studied. Areas of growth include cellular manipulations and assays, and coupling of microchip devices to mass spectrometry *via* electrospray and MALDI. A more surprising area of growth for the meeting is the discussion of detection technologies that specifically address problems dealing with the necessarily small probe volumes associated with microfluidics. Surface plasmon resonance detection is being developed for interrogation of chemical arrays and several optical absorption strategies are discussed that attempt to achieve useful detection sensitivities. Interestingly there are also papers discussing miniaturization of traditionally bulky apparatus such as NMR, mass spectrometers and ion mobility spectrometers. We also find that there are several submissions addressing nanofluidic transport, an area that provides many challenges as well as exciting possibilities.

We were again nearly completely successful in obtaining proceedings papers from the presenters. The attendees and others who possess this book will have a lasting snap shot of the meeting. The book is organized so that the oral papers appear in the same order as they unfold in the program chronologically. The posters also appear in the same order as they were presented at the meeting. Thus this proceedings book is an extended program for use during and after the conference. The reader will also find author and keyword indices to help navigate the book.

As with any undertaking of this magnitude, there are many people who deserve credit for its success. Karen Bertani, Catherine Federele and Patty Shaw of the California Separation Science Society (CaSSS) deserve many thanks for their efforts in all aspects of the conference organization. They are also commended for their patience in putting up with me and the rest of the steering committee who were not timid with their suggestions about how to do most everything about the meeting. I would like to particularly thank Jed Harrison and Albert van den Berg for sharing their experiences with previous meetings and providing many helpful suggestions. In addition I would like to thank Steve Jacobson, Chris Culbertson and Rose Ramsey for being sounding boards for many organizational aspects of the conference. I also thank the local arrangements committee who shared their knowledge of the Monterey Bay area and provided many useful comments concerning the technical portion of the meeting. Finally, I want to thank the most important people of all, the technical presenters and participants! I hope that everyone has a wonderful time, both professionally and socially!

J. Michael Ramsey
μTAS 2001 Chair
August 31, 2001

CONTENTS

Day 1

Keynote Address

Day 2

Plenary Session #1 Invited Speakers

DNA Analysis

Fluidic Mixing and Devices

Plenary Session #2 Invited Speakers

Protein Separations

Nanoscale Fluidics

Protein Analysis and Synthesis

Fluid Dispensing

Posters Day 2

Application – Genomics & Proteomics

Application – Separation Science I

Microfluidics – Fluid Dispensing

Micromachining – Microfluidic Components

Nanotechnology – Nanoengineered Materials

Nanotechnology – Nanofluidic Components

Day 3

Plenary Session #3 Invited Speakers

Mass Spectrometry Technology

Novel Micropumps & Fluid Diagnostics

Plenary Session #4 Invited Speakers

Cellular Analysis I

Optical Detection Techniques I

Cellular Analysis II

Optical Detection Techniques II

Posters Day 3

Application – Cellular Analysis

Detection Technologies – Electrochemical Detection

Detection Technologies – Mass Spectrometry

Detection Technologies – Micro-optical Systems

Microfluidics – World-to-chip Interfacing

Micromachining – Micromachining Methods

Micromachining – Micropumps and Valves

Micromachining – Plastics Machining

Day 4

Plenary Session #5 Invited Speakers

Bioassays

Miniature Detection Technologies

Plenary Session #6 Invited Speakers

DNA Diagnostics

Sorting Technology

Drug Discovery

Flow Control

Posters Day 4

Application – Clinical Diagnostics

Application – Drug Discovery

Application – Environmental Assays

Application – Separation Science II

Detection Technologies – Novel Quantification Strategies

Materials – Chemical & Biochemical Arrays

Materials – Interfacial Characterization

Materials – Surface Modification

Microfluidics – Fluid Mechanics & Modeling Tools

Day 5

Plenary Session #7 Invited Speakers

Microfabricated Separations Technology

System Integration

NANOMECHANICAL SYSTEMS: APPLICATIONS IN BIOTECHNOLOGY

Michael Roukes
Condensed Matter Physics, Caltech 114-36, Pasadena, CA 91125 USA
roukes@caltech.edu

Abstract

Nanoelectromechanical systems (NEMS) offer important new prospects for ultrasensitive chemical and biological measurements. In my talk I will describe two specific applications of NEMS that we are pursuing: vacuum-based mass spectrometry and fluid-based chemical force assays. In both of these arenas ultraminiature mechanical devices offer intriguing new functionality the single-molecule limit.

1. Introduction

NEMS are microelectromechanical devices scaled to submicron dimensions. With their reduced size it is possible to attain extremely high fundamental frequencies while simultaneously preserving very high mechanical responsivity (small force constants). For vacuum-based applications this powerful combination of attributes translates directly into high force and mass sensitivity, ultimately at the attonewton and single-Dalton level respectively. In fluidic media, even though the high quality factors attainable with NEMS in vacuum become precipitously damped, their small size and high compliance still yields response at the piconewton level – this is sufficient for sensing the breaking of *individual* hydrogen bonds within a macromolecule.

2. Ultrasensitive NEMS Mass Sensing

We are exploring the application of NEMS to ultrasensitive mass detection in vacuum. Their extremely small active masses and high Q render high frequency NEMS resonators exquisitely sensitive to added mass. Even the simplest initial realizations of NEMS mass sensing appear to offer immense advances to the current state-of-the-art of mass spectrometry. Most notable are enhanced sensitivity, the sensing of absolute mass (rather than the m/Z ratio of an ion), the possibility of realizing electrically-neutral, adsorption-based mass spectrometry of fragile molecules (*i.e.* ones difficult to ionize non-destructively), and the prospect of exceptionally large dynamic range for mass measurements. Most exciting is that advances in NEMS realizable in the near term make feasible mass spectrometry with single-Dalton resolution. This offers the prospect of, in effect, "weighing", and thereby characterizing, *individual* molecules. To provide an initial demonstration of NEMS mass sensing we have carried out studies of the response of nanometer scale VHF mechanical resonators to a gated, ballistic flux of atoms. In these experiments, the fundamental-mode resonance frequency of a doubly-clamped silicon carbide (SiC) beam is tracked continuously by a phase-locked loop

J.M. Ramsey and A. van den Berg (eds.), Micro Total Analysis Systems 2001, 1–3.

(PLL) circuit. These experiments are carried out under UHV conditions within the apparatus that permits us to direct an extremely weak, ballistic, flux of atoms onto the surface of a NEMS device. Using a 32.8 MHz nanomechanical resonator, we have initially demonstrated a noise floor for mass measurements of $\sim 1.46 \times 10^6$ Daltons, for a 2ms averaging time. Devices with fundamental-mode resonances at 72 MHz have provided even better inverse responsivity, yielding values of $\sim 8 \times 10^{-20}$ grams/Hz (5×10^4 Da/Hz). Careful theoretical analysis of mass sensing using NEMS agrees with these initial measurements and enables us to project the requirements for attaining single-Dalton resolution. These are within reach; advances in NEMS transduction within our group have recent enabled us to detect fundamental mechanical resonances at 635 MHz – devices at this frequency should be at the threshold of single-molecule performance.

3. BioNEMS: Biofunctionalized Nanoelectromechanical Systems

New methods of detection have provided greatly increased force sensitivity for mechanical measurements in solution. Many intriguing new classes of studies have become possible, including direct measurements of intermolecular binding forces, of the forces involved in configurational changes of molecules, and of the forces of adhesion. The growing literature of chemical force microscopy (CFM) – which is atomic force microscopy (AFM) with a biofunctionalized probe – shows that this technique can enable measurements of the binding force of interactions ranging from single hydrogen bond, to single receptor-ligand interactions, to single covalent bonds. We have been exploring the prospects attainable with continued miniaturization of these techniques. Our recent work shows that the best properties of CFM are greatly enhanced with further miniaturization of the mechanical probes. These practical realization of these devices, which we term biofunctionalized nanoelectromechanical systems (BioNEMS), offers the prospect of dynamical single-molecule force assays with the very important advantage that they are amenable to large scale integration. Our initial investigations focus upon the force sensitivity of nanometer-scale cantilevers in fluid, in the important regime of low Reynolds number flow. Recent experiments and theoretical analyses indicate the limits to force sensitivity imposed by fluidic fluctuations appear better than 10fN/√Hz in ultraminiature devices now realizable through advanced nanofabrication. In self-sensing BioNEMS devices with integrated piezoresistive transducers, coupled force sensitivity well below 100fN/√Hz appears attainable with bandwidths exceeding 1MHz. This opens many new possibilities for biological force measurements on the sub-microsecond time scale. In addition to providing real prospects for single-molecule force assays, this level of response offers the opportunity to study conformational dynamics with previously unattainable temporal resolution.

Acknowledgements

This work is being carried out at Caltech in close collaboration with Prof. Scott E. Fraser, Prof. Michael C. Cross, Dr. Jerry Solomon, Dr. Kamil Ekinci and graduate students Jessica Arlett and X.M. Henry Huang.

Table 1. Chemical and Biochemical Force Scales

Nature of Interaction	Interaction Force
Chemical Bond	1–10nN
Covalent(C-C, C-O, C-N)	4.0–4.5nN
Covalent (Au-S, Si-C)	1–3nN
Hydrogen Bond	10pN
Receptor/Ligand Interaction	50–250pN
Avidin-Biotin	90–260 pN
Antibody-Antigen	50–300pN
Cadherin-Cadherin	35–55pN
DNA Hybridization	65pN–1.5nN
Unfolding Forces	100–300pN
Protein (Titin) unfolding	150–300pN
Dextran bond twists	100–300pN

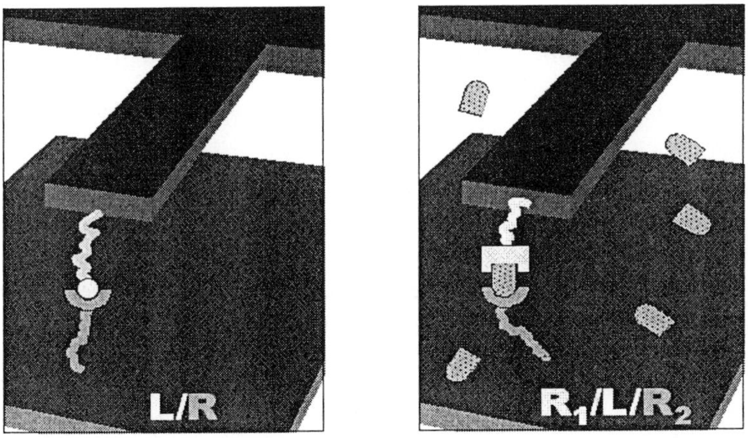

Figure 1. Schematic depiction of two potential applications of BioNEMS that are based upon molecular modification of nanocantilever dynamics. The left panel illustrates the simplest geometry that uses the enhanced temporal response of BioNEMS to track the kinetics of ligand/receptor interactions. This configuration also provides an assay for free ligands or receptors in the surrounding fluid; either component will compete with its tethered counterpart and alter cantilever dynamics. The more complex scheme on the right involves two different binding sites on the same ligand (L) that are recognized by two different tethered receptors (R1, R2). A functional link between the substrate and the cantilever is present only when both receptors are engaged. This offers exceptionally selective molecular recognition for a particular ligand in solution, as both target sites must be present on the same ligand molecule for the cantilever dynamics to be altered.

ENGINEERING ACTIVE INTERFACES BETWEEN CELLS AND ELECTRONICS

Milan Mrksich
The University of Chicago, Department of Chemistry
5735 South Ellis Avenue, Chicago, IL 60637
mmrksich@midway.uchicago.edu

Abstract

This paper introduces a strategy for integrating the functions of mammalian cells with electrodes. The strategy is based on the use of self-assembled monolayers of alkanethiolates on gold that present ligands for the adhesion of cells. The monolayers have been engineered such that applied potentials result in the modulation of ligand activities at the substrate and a corresponding influence over the behavior of attached cells. Early examples have demonstrated electroactive substrates that can turn on the migration of cells and that can release adherent cells. These examples represent a general route to integrating electrical processes with cellular processes.

Keywords: cell adhesion; dynamic substrate; electroactive self-assembled monolayer

1. Introduction

The microelectronics and biotechnology industries are both based on established scientific and engineering platforms, yet these platforms are sufficiently distinct that there has been little cross-talk between these two communities. Microelectronics, for example, uses lithographic methods to assemble inorganic and metallic materials into layered devices, whereas, biology uses self-assembly and enzymatic routes to assemble biopolymers into complex three-dimensional architectures. Further, information in microelectronics is transmitted with electrical currents and potentials whereas biology relies on molecular recognition and enzymatic reactions to transmit information. Technical advances that bridge these differences could enable new classes of hybrid technologies. The integration of cells with microsystems, for example, is currently important for the development of biowarfare agent sensors and may prove useful for adding acutation, power and synthesis capabilities to microsystems. Below we describe a new approach for integrating cellular and electrical processes [1]. We describe an electroactive self-assembled monolayer that can translate electrical signals into biological signals that can then elicit specific behaviors in attached cells.

2. Technical Approach

Figure 1 presents an example of an electroactive monolayer that can regulate cell behavior, in this case by turning on the migration of fibroblast cells that are initially

J.M. Ramsey and A. van den Berg (eds.), Micro Total Analysis Systems 2001, 4–6.

patterned to circular regions of the substrate [2]. The strategy is based on self-assembled monolayers that present an electroactive molecule and tri(ethylene glycol) groups. The former are designed to be switched by applying potentials of ~ 0.5 V to the substrate while the glycol groups are important for preventing unwanted biomolecular interactions at the interface. In Figure 1, the electroactive group is a hydroquinone, which can be switched to the benzoquinone [3], which in turn provides a site for the selective immobilization of peptide ligands. The use of the Arg-Gly-Asp (RGD) peptide provides a surface to which cells can adhere and on which cells can migrate [4].

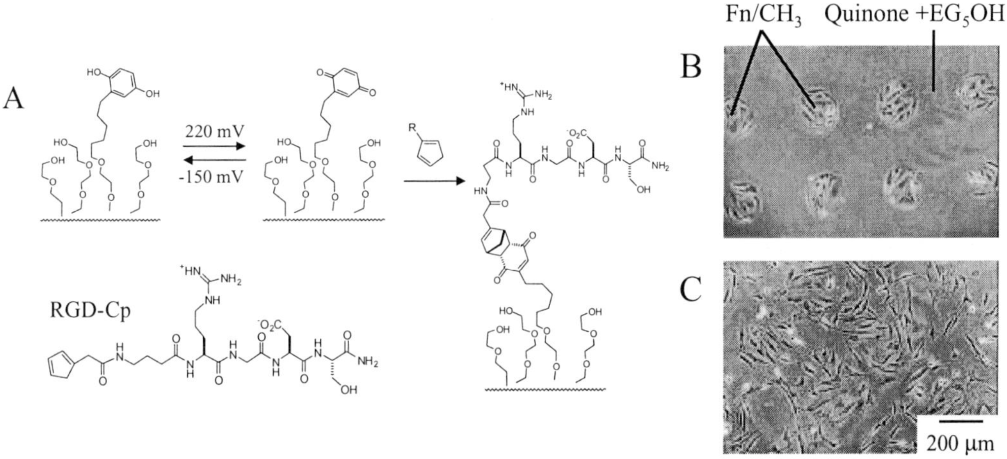

Figure 1. Design of substrates that can turn on cell spreading and growth. (A) Monolayers that present hydroquinone groups mixed with penta(ethylene glycol) groups are inert to cell attachment. Electrochemical oxidation converts the hydroquinone groups to quinone groups, which then undergo Diels-Alder reaction with a diene-Arg-Gly-Asp conjugate (RGD-Cp) to introduce the peptide onto the surface. These modified monolayers support the attachment and spreading of cells. A monolayer was patterned into regions presenting the adhesive protein fibronectin and regions presenting the hydroquinone and glycol groups. Cells attached only to the protein-coated regions (B), but after oxidation of the substrate in the presence of RGD-Cp, cells were able to migrate and then grow to fill the entire monolayer (C).

3. Future Directions

We believe that this and related dynamic substrates will be important for both fundamental studies in cell biology and for fabricating integrated chip-based microsystems. It has already proven important for patterning multiple cell types to a common substrate [5]. We are also demonstrating related strategies for electrical manipulation of cell-substrate interactions. One example uses reduction of a benzoquinone to selectively release ligands [6] and even entire cells [7] from the substrate. These examples emphasize the role for molecular engineering strategies in bridging between biological and electrical processes.

4. Conclusions

The field of cell-based engineering is at a very early stage of development and will require innovations in several technical areas before it becomes technologically significant. This paper presents one strategy for engineering interfaces that can join cells to electrical components and that can translate electrical processes to regulate cell behavior. The development of this and other classes of dynamic interfaces will ultimately provide a basis for integrating cells with electronics and for demonstrating a range of functions that emerge from hybrid devices that contain living and non-living components.

Acknowledgements

We are grateful for support provided by DARPA (N00173-01-1-G010), NSF (BES-9980850) and NIH (GM 63116).

References

1. Dynamic Substrates: Modulating the Behaviors of Attached Cells. M. N. Yousaf and M. Mrksich **2000** In *New Technologies for Life Sciences: A Trends Guide,* (Wilson, E. *et al.*, eds: Elsevier), pp 28-35.
2. Turning on Cell Growth with Electroactive Substrates. M. N. Yousaf, B. T. Houseman and M. Mrksich *Angew. Chem. Int. Ed.*, **2001**, *40*, 1093-1096.
3. Diels Alder Reaction for the Selective Immobilization of Protein to Electroactive Self-Assembled Monolayers. M. N. Yousaf and M. Mrksich *J. Am. Chem. Soc*, **1999**, *121*, 4286-4287.
4. Model Organic Surfaces for Mechanistic Studies of Cell Adhesion. M. Mrksich *Chem. Soc. Rev.*, **2000**, *29*, 267-273.
5. Using Electroactive Substrates to Pattern the Attachment of Two Different Cell Types. M. N. Yousaf, B. T. Houseman and M. Mrksich *Proc. Natl. Acad. Sci. USA*, **2001**, *98*, 5992-5996.
6. Biomolecular Surfaces that Release Ligands Under Electrochemical Control. C. D. Hodneland and M. Mrksich *J. Am. Chem. Soc,* **2000**, *122*, 4235-4236.
7. Electroactive Monolayer Substrates that Selectively Release Adherent Cells. W.-S. Yeo, C. D. Hodneland and M. Mrksich *ChemBioChem*, **2001**, *7/8*, 590-593.

COMMERCIALIZED AND EMERGING LAB-ON-A-CHIP APPLICATIONS

Michael R. Knapp, Anne Kopf-Sill, Robert Dubrow, Andrea Chow, Ring-Ling Chien, Calvin Chow, J. Wallace Parce
Caliper Technologies Corp., 605 Fairchild Dr., Mountain View, CA 94043, USA
(http://www.calipertech.com)

Abstract

Despite considerable progress in lab-on-a-chip technology, few commercial products are available. Now that commercial processing platforms are available, the stage is set for a rapid emergence of new lab-on-a-chip applications. The processing formats currently available embrace both low and high throughput analytical needs. It would be desirable for the basic toolset of microfluidic experimentation to be more widely disseminated.

Keywords: Lab-on-a-chip, microfluidics, commercialization.

1. Introduction

Researchers have been reporting progress in "lab-on-a-chip" technology development for more than a decade. Despite considerable interest in the field, however, few products based on this technology are yet on the market. To understand this discrepancy, it is probably necessary to define the term "lab-on-a-chip" well enough to be able to recognize one when you see one. In our group, the term refers to products that make use of three technological elements: 1) miniaturization, 2) integration, and 3) automation. The first and last of those are self-evident. By "intregation", we mean that two or more steps of an experimental protocol are performed by the same device. Integration can also refer to inclusion of functionality usually performed by two or more different conventional instruments. As a disclaimer, it should be said that the edges of the lab-on-a-chip space with our definition are fuzzy, and reasonable people of good faith can disagree on the categorization of any particular product or design.

Another important aspect is that the majority of lab-on-a-chip investigations have segregated the processing component of the design from the sensing and actuation parts of the system. Thus, in our first commercial products, the fluid, molecules or particles that move through microscopic conduits etched in glass are propelled by forces supplied from a separate appliance. This instrument also performs the task of detecting biochemical or cellular interactions with fluorescence spectroscopy. Thus while the focus of lab-on-a-chip application development has been most often on the chip itself, a principle need in commercialization of lab-on-a-chip applications is the instrument. Finally, the infrastructure for manufacturing chips is not broadly available to applications specialists. Yet these are the scientists who would most clearly define the specific tasks to which this new and remarkable experimentation platform could be most profitably be directed.

J.M. Ramsey and A. van den Berg (eds.), Micro Total Analysis Systems 2001, 7–9.

2. Results and Discussion

The first applications to emerge as lab-on-a-chip products involve sizing and quantitation of nucleic acids. This is not surprising since the early lab-on-a-chip proofs-of-concept involved deployment of capillary electrophoresis methods in glass chips. An example of a commercial protein sizing and quantitation chip together with data produced from a commercial system are given in figure 1.

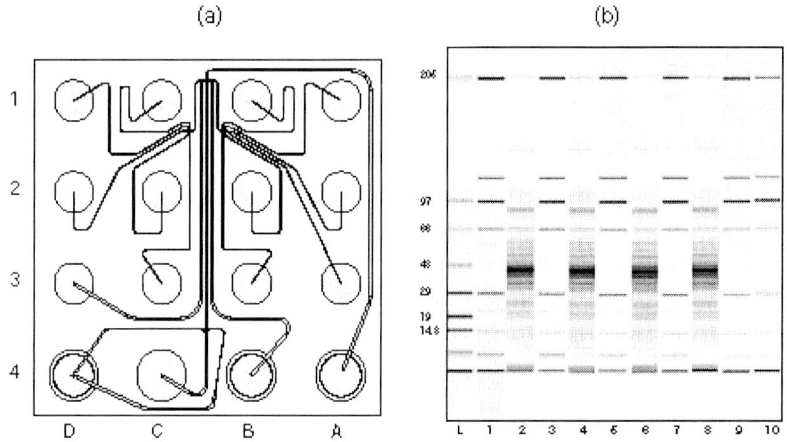

Figure 1. (a) Diagram of a commercial chip used for protein sizing and quantitation. Samples and reagents are placed in reservoir wells. The design includes a quantitative injection site and a destaining site. (b) The "pseudogel" resulting from digital analysis of proteins labeled, separated on the chip and analyzed by fluorescence spectroscopy.

By using only a few hundred picoliters of the sample in the analysis, this application is definitely of miniature scale. It integrates many steps of the conventional process of quantitative protein analysis: electrophoresis, staining, destaining, scanning, comparison to standards, and data analysis. After the samples are put into the chip, all of these steps are performed automatically. A second potentially interesting general category of experimentation suited to lab-on-a-chip technology is high throughput experimentation. This type of experimentation can be expensive. In addition, database-quality information from high throughput experiments has been achievable for only a limited number of experiment types. The lab-on-a-chip format can potentially expand the range of experiments amenable to the "industrialized biology" approach to discovery.

Figure 2 demonstrates that the lab-on-a-chip idea can be made applicable to high throughput experimentation if a solution can be found to the "world-to-chip" interface problem: getting large numbers of different reagents in nanoliter quantities rapidly into the chip. We have engineered capillaries into our chips such that the lumen communicates fluidically with the chip conduits in connections that have virtually no dead volume. Using a continuous flow assay strategy, we have found that the controlled environment of the chip enables production of data having unprecedented precision.

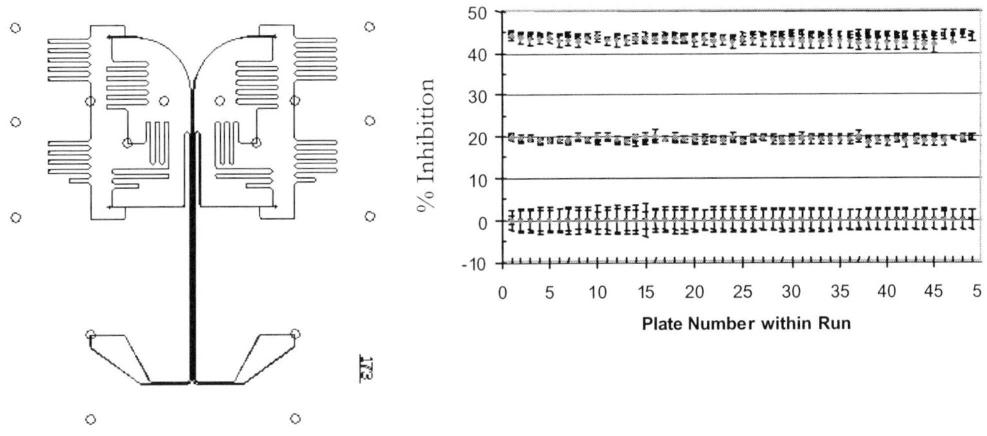

(a) (b)

Figure 2. (a) Diagram of a high throughput chip that processes samples continuously and analyzes them for their ability to inhibit a fluorogenic enzyme reaction. Sipper capillaries attached at four points permit nanoliter quantities of reagents to be imported into the chip from conventional liquid sources such as multiwell plates. (b) Data from a large scale experiment where samples of known potency were analyzed 4 times per plate with respect to 49 plates twice per day over the course of 26 days.

The real promise of the lab-on-a-chip will be realized in collaborations between people developing the technology and those scientists pushing the frontiers of particular fields desperately in need of new technology. Scientists need to begin exploring the microfluidic format. Our approach to this need has been to make available the instrument shown in figure 3 which is a microfluidic application development station. This instrument's main attributes include a flexible optical system, control at all ports over electrokinetic and pressure/vacuum actuation, and a powerful software package for microfluidic control and data acquisition.

Figure 3. A microfludic applications developer station.

3. Conclusions

Lab-on-a-chip technology is demonstrably commercializable. It would now be desirable for a greater community of applications specialists to explore the extraordinary power of this new experimentation format.

ENHANCING THE MICROFLUIDIC TOOLBOX FOR FUNCTIONAL GENOMICS AND RECOMBINANT DNA METHODS

D. Jed Harrison, Ehsan Majid, Said Attiya, Guifeng Jiang
Dept. of Chemistry, University of Alberta, Edmonton, AB, T6G 2G2, Canada

Abstract
Two microfluidic designs are presented which increase the range of techniques available for performing recombinant DNA procedures on-chip: mRNA purification with single strand DNA synthesis, and DNA transfection in cells via electroporation.

Keywords: DNA, Transfection, cDNA library, reverse transcription PCR

Introduction
Microfluidic devices have already demonstrated they can play a major role in genetic analysis and in the future they will play a significant role in microbiology and in recombinant DNA technology. However, there is a need to build up a microfluidics toolbox of functions important to recombinant DNA efforts. As a vehicle for this effort we have focussed on developing an integrated microfluidic device for creating a complementary DNA (c-DNA) library. We have already described magnetic Dynal bead based capture and purification of messenger RNA (mRNA) on-chip from total RNA (TRNA) [1]. Here we describe mRNA capture followed by first strand cDNA synthesis within a microfluidic device. In the final steps of library formation the cDNA must be transfected into cells. A novel integrated electroporation system used for DNA transfection of *E. coli* cells is presented here. While not yet integrated together, these two devices represent the initial and penultimate stages of library preparation, and significantly increase the number of tools available on chip for recombinant DNA work.

Results and Discussion
Figure 1 illustrates the Y-shaped flow channel used for mRNA capture and purification, with a heating element to allow first strand cDNA synthesis. TRNA from *Drosophilia Melanogaster* was treated on-chip to isolate mRNA, followed by on-chip first strand cDNA synthesis (reverse transcription). The beads were then collected from the chip by delivering them to a reservoir using a moving magnet. The low abundance bicoid gene cDNA was amplified off-chip by PCR, and determined using capillary gel

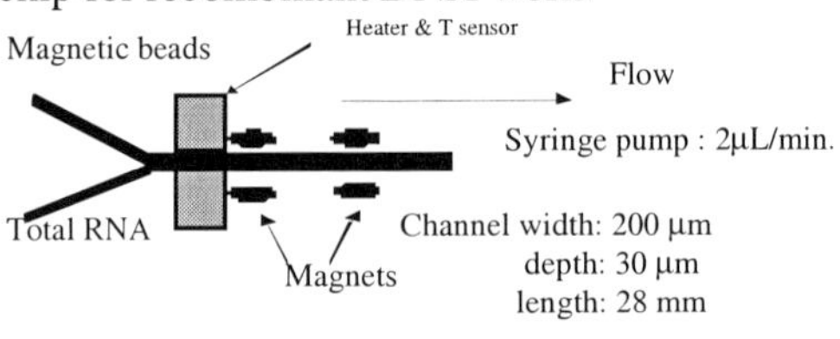

Fig. 1. Y-shaped channel design. TRNA is placed in one arm and magnetic beads are the other. Negative pressure draws materials towards the magnetic trapping zone. Heat for the cDNA synthesis is then applied.

J.M. Ramsey and A. van den Berg (eds.), Micro Total Analysis Systems 2001, 10–12.

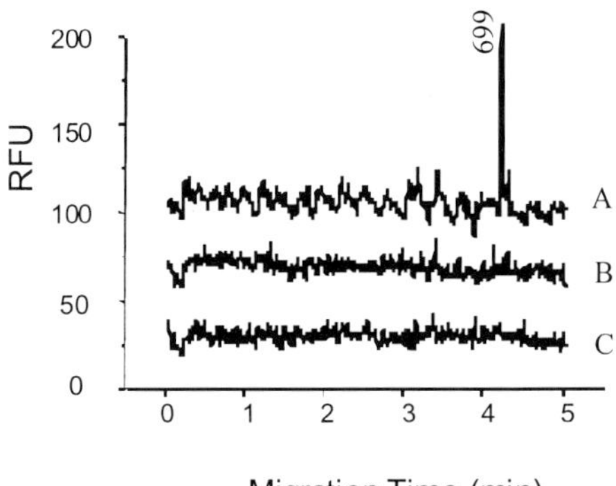

Fig. 2. A) 699 bp PCR product for bicoid gene, with on-chip capture and on-chip synthesis of first strand DNA. Magnets were moved to break up channeling and mix the bed. B) Same experiment without moving magnets C) Negative control of blank sample.

Fig. 3. Top view of magnetic bead bed occupying central region of channel during flow. Channeling of flow along the outside walls (black lines) and across zones of the bed is observed, indicating inefficient exposure of the bed to the flow stream.

electrophoresis, as shown in Fig. 2. Successful synthesis occurred only when the bed was agitated during washing and reaction steps, by 1-5 mm motion of the trapping magnets. Figure 3 illustrates the channeling of flow that occurred in the magnetic bed during solvent flow, causing considerable mass transport inefficiency for the cDNA synthesis step. Agitation of the bed disrupted the channeling. Optimization of the on-chip reverse transcription reaction temperature is illustrated in Figure 4. The results demonstrate these first two key steps of cDNA library formation can be performed on chip. However, the channeling illustrates one of the difficulties of using magnetically trapped beds, in that optimal packing of the bed does not occur. This suggests a mechanical trapping procedure, such as the use of a weir may be advisable in future.

We have also integrated a process for perforating living cells and injecting them with different biochemically active species, commonly known as electroporation. This is the key step in cDNA library construction, prior to plating of the cells for growth. Cells were directed through a straight channel using a syringe pump, and a high voltage pulse was applied across a small segment of the channel, via a set of four cross channels, as illustrated in Figure 5. Weirs were located in the cross channels to prevent solvent and cell flow in these channels. To determine electroporation conditions, *E.Coli* DH5α cells were treated with fluo-

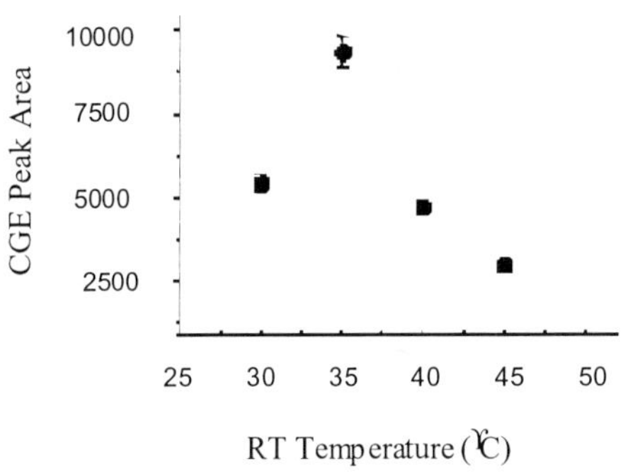

Fig. 4 Optimization of reverse transcription temperature.

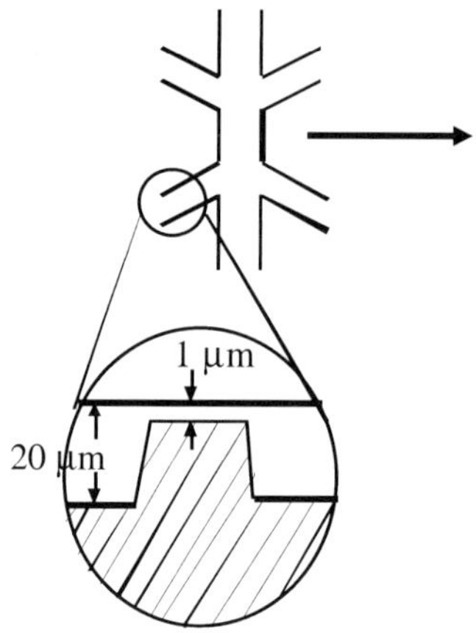

Fig. 5. Blow up of cell transfection structure. Electroporation voltage is applied across the weirs located in the side arms, which restrict bulk flow of solvent into the arms.

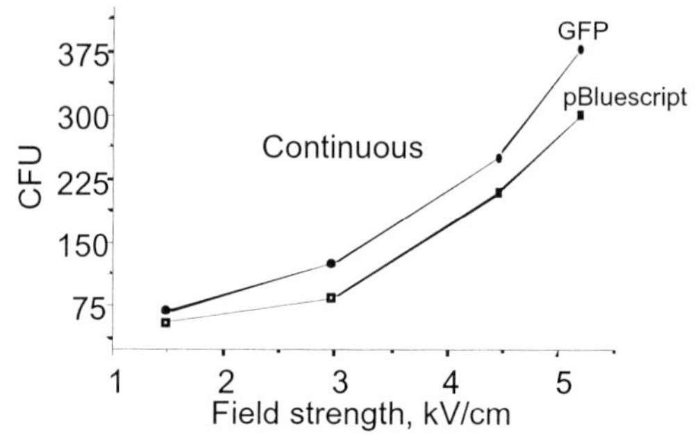

Fig. 6. Colony forming units (CFU or cells) obtained on agar plate after on-chip electroporation at indicated field strength, using green fluorescent protein (GFP) of pBluscript plasmids for transfection.

4 dye then pumped within a stream containing Ca^{2+} through the cross region, with a high voltage continuously applied at the cross. Fluorescence detection downstream identified the formation of a Ca^{2+}-fluo-4 complex within the cells after electroporation. Similar experiments were then performed using ampicillin resistant pbskll (pBluescript) or green fluorescent protein plasmids for DNA transfection. The cells were harvested from the chip, diluted with LB broth, and cultivated on ampicillin loaded agar plates. Cells which had undergone successful transfection survived and grew into new colonies. Figure 6 shows the effect of applied electric field on the number of colonies in a diluted cell mixture harvested from the chip. The yield per µg plasmid and per cell as a function of field is about the same as that obtained in conventional systems [2]. Higher fields are needed for better yields, but could not be applied in the device in Fig 5, so a revised version without weirs was developed, allowing fields of 10 kV/cm. The resulting device gave a higher yield of transfected cells .

Conclusions

The results demonstrate that successful electroporation, transfection and gene expression can be initiated on-chip. Importantly, both of these studies establish that when using particles (cells or beads) it is possible to recapture material introduced on-chip for subsequent experimentation off-chip. The world/chip interface is thus not limited to a one way configuration.

References

1. G. Jiang, D.J. Harrison *Analyst* 2000, 125, 2176-2179.
2. W.J. Dower, J.F. Miller, C.W. Ragsdale *Nucleic Acids Res* 1998, 16, 6127-6145.

(ULTRA)^2-HIGH THROUGHPUT GENETIC ANALYSIS USING MICROFABRICATED CAPILLARY ARRAY ELECTROPHORESIS DEVICES

Charles A. Emrich†, Igor L. Medintz, Huijun Tian, Lorenzo Berti and Richard A. Mathies
Department of Chemistry and †Department of Biophysics, University of California, Berkeley, CA 94720

Abstract

We have developed technologies enabling the next generation in high-throughput electrophoretic genotyping using our rotary confocal scanner. The capabilities of this system are illustrated by (i) a short tandem repeat (STR) DNA genotyping separation performed on a 150-mm diameter, radial 96-lane μCAE microplate in less than 8 minutes, and (ii) the parallel assay of the common H63D mutation in the human *HFE* gene on a 384-lane, 200-mm diameter microplate in only 7 minutes.

Keywords: Capillary Array Electrophoresis, Rotary Confocal Fluorescence Scanner, DNA Genotyping

1. Introduction

The completion of the draft sequence of the human genome heralds a revolution in our understanding of genetic complexity and variation. Higher throughput genotyping assays are now needed to identify polymorphisms and screen for affected individuals. Conventional CAE has brought us far along this path, but a paradigm shift to higher throughput microfabricated-Capillary Array Electrophoresis (μCAE) devices is now at hand. We present here approaches for ultra-high throughput, multi-color, multiplex genotyping assays on 96-lane microplates as well as massively-parallel (ultra)^2-high throughput genotyping on 384-lane microplates.

High speed genotyping was first performed on μCAE devices by Woolley et al. [1] in 1997 using a 12-lane design and a linear confocal scanner, later expanded to accommodate serial loading of up to 48 samples [2]. Subsequently, a 48-lane device employing a galvanometric scanner for detection was used to analyze 96 samples [3]. Introduction of the Berkeley rotary confocal fluorescence scanner together with the 96-lane, radial μCAE design in 1999 [4] provided the paradigm shift necessary to expand to the extremely high device density and design freedom illustrated here.

2. Results and Discussion

Radial 96-lane microplates (see Fig. 1A) have been used to perform a variety of high-speed genotyping experiments such as STR, RFLP, and allele-specific PCR SNP typings of human populations [5, 6]. As an example of the versatility of the μCAE platform we present here high-speed 4-color STR typing on 150-mm diameter, 96-lane μCAE plates. Four-color energy-transfer (ET) cassette-labelled primers were used to produce the STR amplicons. Three of the four colors are used for genotyping while the fourth is reserved for a size standard. The amplicons were rapidly separated (< 8 mins) with single base resolution. The powerful capabilities of this system were demonstrated by genotyping more than 300 loci in a single run [7] on 150-mm diameter devices enabling throughputs of 1 allele every 1.5 seconds (Fig. 1B).

J.M. Ramsey and A. van den Berg (eds.), Micro Total Analysis Systems 2001, 13–15.

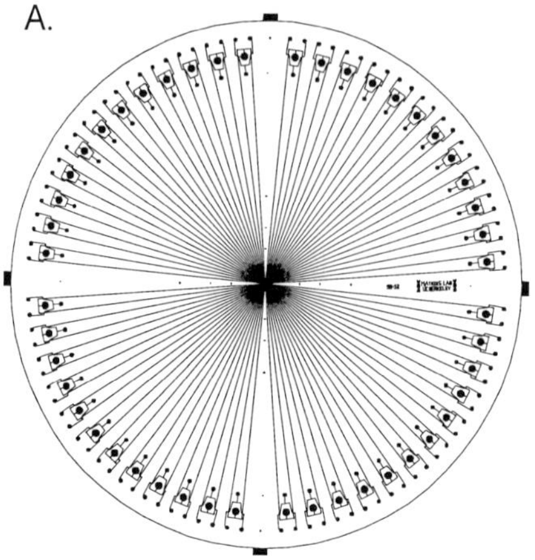

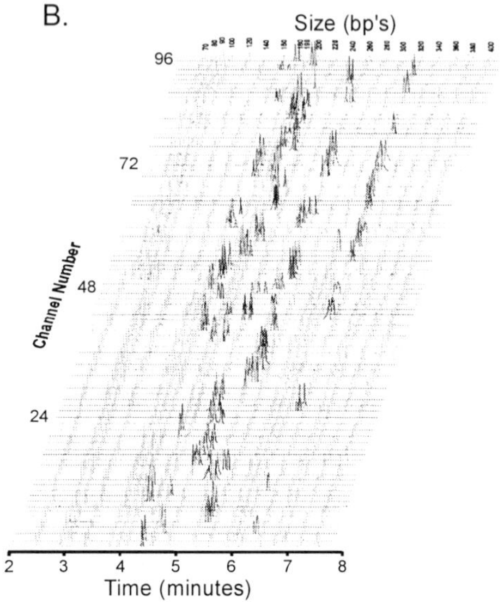

Figure 1. A) Layout of the 150-mm diameter 96-lane chip used for multi-allelic genotyping. The effective separation length is 5.5 cm and the injector is a twin-T design. B) Multiple loci STR samples genotyped on a 96-lane μCAE device. Amplicons were mixed with an ET-400-R ladder (red - shown in grey) in lanes 1-48. A FAM mapmarker (also in grey) was used as the standard in lanes 49-96. Fragment sizes of the standard are listed at the top of the electropherograms and STR amplicons are highlighted in black.

To further demonstrate the power of our radial confocal scanning system and the versatility of the radial format, we have designed, fabricated, and performed separations on a μCAE device with 384 radial lanes (Fig. 2). This massively parallel μCAE device was patterned on 200-mm diameter glass wafers with standard procedures. The larger substrate size allows a higher number of sample wells and longer (8.0 cm) effective separation length. A direct injection scheme is employed, allowing better than 10-bp resolution with representative DNA ladders. The 384-lane μCAE device was used to perform an RFLP analysis of the H63D (187C→G) mutation in the *HFE* gene. This relatively common mutation is associated with the occurrence of Hereditary Hemochromatosis (HHC), an autosomal recessive disease that causes excessive iron intake [8]. All runs on the 384-lane μCAE device were performed using

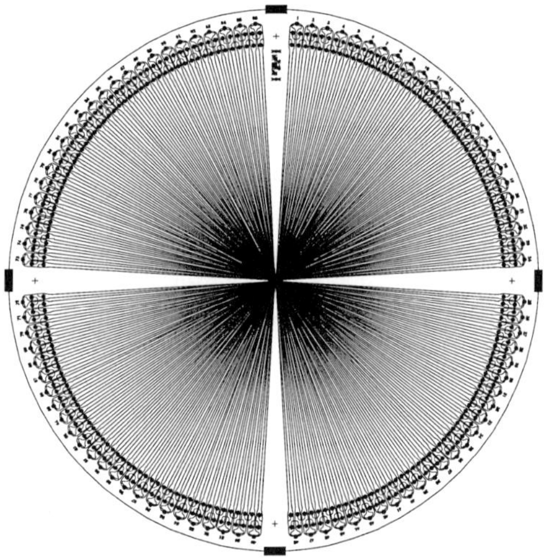

Figure 2. Layout of the 384-lane μCAE device with sample and cathode wells arrayed about the outer edge of the wafer. Each cathode reservoir is shared by 4 adjacent lanes.

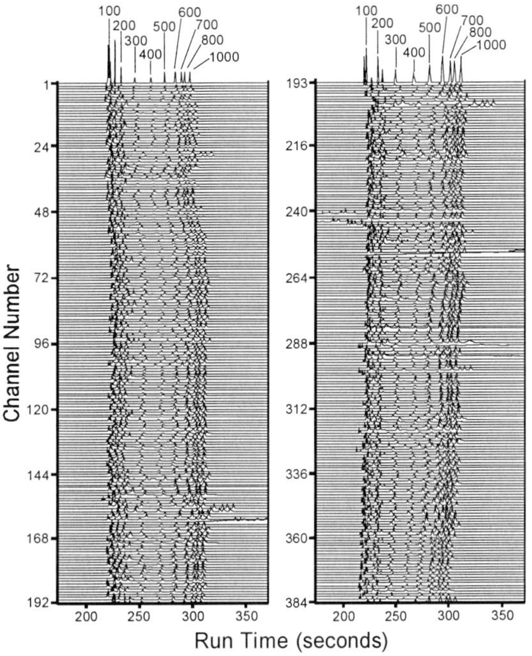

Figure 3. Separation of PCR-RFLP genotyping fragments for the H63D mutation. The mutant allele produces a 246 bp fragment while the wild type allele produces 172 and 74 bp fragments. Samples are sized against a 100 bp ladder.

dsDNA and thiazole orange intercalation dye labelling. The genotyping analysis of 384 samples presented in Fig. 3 was complete in less than 7 minutes with greater than 98% success.

3. Conclusions

We have successfully demonstrated next generation ultra-high throughput genotyping on microfabricated μCAE devices employing the radial chip design and a rotary confocal scanner. To demonstrate the utility of our system a 384-lane device was used to perform fragment sizing at better than 1 sample per second. Functional implementation of either or both of these technologies represents a great leap forward in high-throughput genetic analysis systems.

Acknowledgements
This work was funded by the National Institutes of Health and the U.S. Department of Energy. The authors also acknowledge Amersham-Pharmacia Biotech for support.

References

(1) A. T. Woolley, G. F. Sensabaugh, R. A. Mathies, *Analytical Chemistry* 1997, **69**, 2181-2186.
(2) P. C. Simpson, R. A. Mathies, *SPIE* 1998, **3258**, 170-180.
(3) P. C. Simpson, D. Roach, A. T. Woolley, T. Thorsen, G. F. Sensabaugh, R. A. Mathies, *PNAS* 1998, **95**, 2256-2261.
(4) Y. N. Shi, P. C. Simpson, J. R. Scherer, D. Wexler, C. Skibola, M. T. Smith, R. A. Mathies, *Analytical Chemistry* 1999, **71**, 5354-5361.
(5) I. Medintz, W. W. Wong, G. Sensabaugh, R. A. Mathies, *Electrophoresis* 2000, **21**, 2352-2358.
(6) I. Medintz, W. W. Wong, L. Berti, L. Shiow, J. Tom, J. Scherer, R. A. Mathies, *Genome Research* 2001, **11**, 413-421.
(7) I. Medintz, L. Berti, C. Emrich, J. Tom, J. Scherer, R. A. Mathies, *Clinical Chemistry* 2001, (in press).
(8) H. Zoller, A. Pietrangelo, W. Vogel, G. Weiss, *Lancet* 1999, **353**, 2120-2123.

BIOMEMS-768 DNA Sequencer

Daniel Ehrlich, Aram Adourian, Charles Barr, David Breslau, Scott Buonocore, Robert Burger, Loucinda Carey, Steve Carson, Jeffrey Chiou, Richard Dee, Samantha Desmarais, Sameh El-Difrawy, Robert King, Lance Koutny, Roger Lam, Paul Matsudaira, Luba Mitnik-Gankin, Tom O'Neil, Mark Novotny, Gil Saber, Oscar Salas-Solano, Dieter Schmalzing, Alok Srivastava, Maribel Vazquez

Whitehead Institute for Biomedical Research/MIT, Nine Cambridge Center, Cambridge MA 02142

TEL (617) 258-7283, FAX (617) 258 7226
Ehrlich@wi.mit.edu

Abstract

The Whitehead Institute has developed an automated DNA sequencer that will go into final Genome Center testing during the summer of 2001. The system comprises a total of 768 separation channels distributed over two plates. The working elements are 50-cm x 25-cm, 384-lane microfabricated glass elements, which undergo alternating electrophoresis and regeneration under use with an exchangeable sieving matrix. The microfluidic and sample transfer devices needed to service the plates are integrated into the same compact instrument. Concurrently with the instrument development we have developed all the protocols, materials and surface preparations to achieve matrix-limited separations of single-stranded DNA in the microfabricated plate format.

We will outline some of the most important scientific problems that were overcome in this multiyear project. These include (a) the high-yield microfabrication of zero-defect bonded glass plates at sizes greatly exceeding semiconductor fabrication, (b) the integration of electrodes, reservoirs, and high-pressure seals into these plates, (c) detector optimization for the required very high rate data acquisition and (d) microfluidics issues and optimizations related to channel geometry.

Key Words: DNA Sequencer, BioMEMS

1. Introduction

Practical successes for microelectromechanical systems (MEMS) will require microfabricated devices to solve real world problems better than their classical counterparts. In most chemical assays, data quality is a more important metric than all other factors such as device size, assay speed, or other traditional innate strengths of microfabrication. This is particularly true for genomics applications of DNA sequencing devices. The next most important factor for genome centers is full system integration and automation.

J.M. Ramsey and A. van den Berg (eds.), Micro Total Analysis Systems 2001, 16–18.

The BioMEMS-768 sequencer was designed from the ground up to exceed the current generation of commercial 96-lane capillary array sequencers by providing both longer DNA read-length and an order-of-magnitude increase in raw sequence productivity. Long read length is essential, since genome center costs are (i) proportional to the number of samples run (i.e., longer reads give data for free) and (ii) sequence "assembly" costs (stitching of sequence gaps) are very strongly dependent on read length. The system design is based on our previous characterization of DNA assays in microchip format [References 1-3].

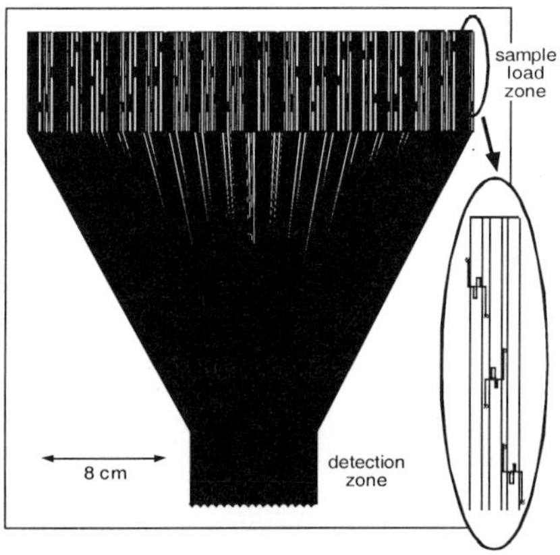

Figure 1. Channel layout for 384 Lane plate. These MEMS parts are fabricated in 50-cm x 25-cm glass plates and are the defining elements of the BioMEMS-768 sequencer.

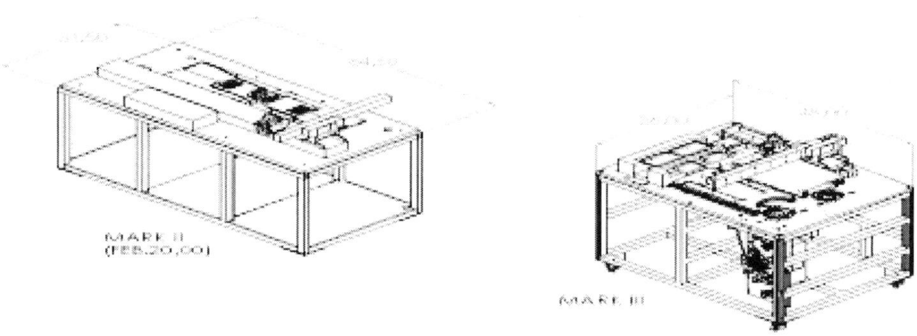

Figure 2. The sequencer has dual spinning epifluorescence microscope detectors, an automated micropipettor, and fully automated microfluidics systems. The system has evolved into a 36-in. by 48-in. footprint.

2. Summary

The BioMEMS-768 system is based on large-format 50- by 25- cm glass MEMS parts (Figure 1). The 50-cm etched channel length is required in order to get the 800+ base read length stipulated by the Whitehead Institute Center for Genome Research. Although fast low-resolution DNA assays are possible on small "chips", when well designed, all chip and capillary systems become limited by the intrinsic properties of the linear polyachrylamide (LPA) sieving matrix. A channel length of 50-cm is required in order to get the most out of the matrix. The choice of glass instead of polymer is mandated by the current status of available materials coatings. Rigorous charge neutralization on the channel wall is required in order to suppress electroosmotic flow and can only be achieved on glass at the moment.

Other features of the system (Figure 2) include a custom designed microfluidic network for pumping of LPA at 700 PSI pressure and for (low pressure) exchange of buffer solutions during the plate refreshing sequence (after every electrophoresis run). Another microfluidics system handles pipetting operations for DNA sample introduction onto the 384-lane plates. We will also summarize DNA sequencing performance of the system (Figures 3, 4).

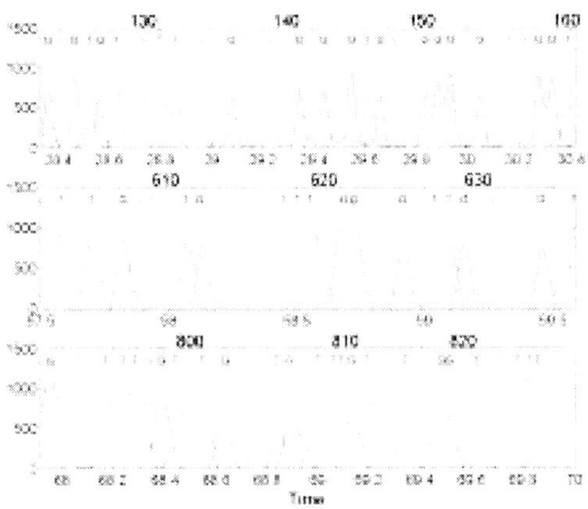

Figure 3. Sequencing data

Figure 4. Data in image format

Acknowledgement

This work was supported by the National Institutes of Health, National Institute of Justice, and the Air Force Office of Scientific Research.

References

1. D. Schmalzing, L. Koutny, A. Adourian, P. Belgrader, P. Matsudaira, and D. Ehrlich, *DNA typing in thirty seconds with a microfabricated device,* Proc. Natl. Acad. Sci. 94, 10273-278 (1997).
2. D. Schmalzing, N. Tsao, L. Koutny, D. Chisholm, A. Srivastava, A. Adourian, L. Linton, P. McEwan, P. Matsudaira, and D. Ehrlich, *Toward real-world sequencing using microdevices,* Genome Research 9, 853-858 (1999)
3. D. Schmalzing, A. Adourian, L. Koutny, L. Ziaugra, P. Matsudaira, and D. Ehrlich, *DNA sequencing on microfabricated electrophoretic devices,* Anal. Chem. 70, 2303-2310 (1998).

PLASTIC MICROFLUIDIC DEVICES FOR DNA SEQUENCING AND PROTEIN SEPARATIONS

Z. H. Fan, W. Tan, H. Tan, X. C. Qiu, T. D. Boone, P. Kao, A. J. Ricco,
ACLARA BioSciences, Inc., 1288 Pear Ave., Mountain View, CA 94043
M. Desmond, S. Bay, K. Hennessy
Applied Biosystems, 850 Lincoln Center Dr., Foster City, CA 94404

Abstract
Plastic microfluidic devices with 18-cm separation length and an offset "double-T" injection system were used to achieve DNA sequencing in 30 minutes with 98%-correct identification to a read length of 650 bases. We recently demonstrated two proteomics-enabling technologies on plastic fluidic devices: isoelectric focusing (IEF) and sodium dodecyl sulfate (SDS)-polyacrylamide gel electrophoresis (PAGE).

Keywords: Plastic microfluidics, DNA sequencing, IEF, protein separations

1. Introduction

Since capillary electrophoresis (CE) was demonstrated in microfabricated devices in the early 1990's [1,2], the "lab-on-a-chip" concept has been explored for many applications, including DNA analysis [3-5] and protein separations [6,7]. The range of device substrate materials now has expanded beyond silicon [8] and glass [1-7] to include plastics [9-11]. Plastic devices offer biochemical compatibility; and a variety of well-developed plastics manufacturing methods enables mass production at low cost, leading to disposable devices that obviate assay-to-assay cross-contamination.

Here we report plastic microfluidic devices fabricated particularly for DNA sequencing and protein separations. Although a first draft of the human genome was recently completed—due in large measure to the utilization of 96-channel CE-array instruments—sequencing still plays a key role in many bioassays. The potential to integrate sample-processing steps, coupled with ease of mass production, makes highly multiplexed microfluidic devices leading candidates for the next generation of sequencers. In the emerging field of proteomics, on-chip IEF and SDS-PAGE may play key roles, offering such advantages as rapid analysis and enhanced separation resolution.

2. Experimental Details

Plastic microfluidic devices were fabricated by ACLARA [9,12]. Photolithographic definition of microchannels on a glass plate or Si wafer is followed by electroplating a metal layer 100s to 1000s of μm thick. This molding tool is mounted on an embossing or molding system to produce plastic cards with the features of the original template structure; thousands to millions of parts can be made using one electroform. Each device is completed by bonding a cover layer using thermal, ultrasonic, or adhesive means. Completed devices have channels 15 – 100 μm deep by 30 – 200 μm wide.

For CE experiments on plastic chips, a "breadboard" workstation including an inverted epifluorescence microscope and in-house-built power supplies is utilized [9,12]. A laser or filtered UV lamp provides excitation, while fluorescent emission is detected by a photomultiplier tube or charge-coupled device. A four-color version of this system

J.M. Ramsey and A. van den Berg (eds.), Micro Total Analysis Systems 2001, 19–21.

for DNA sequencing experiments uses additional detectors to record emission simultaneously at four spectrally resolved wavelengths. Voltages are controlled, and data acquired, by computer.

3. Results and Discussion

Fabrication process optimization. Microfluidic devices were made from plastics including polycarbonates, polyolefins, and acrylics. These materials were used to fabricate a variety of microstructures (some of which are shown in Figure 1) to evaluate manufacturability and performance. Key processing parameters—temperature, pressure, time—in molding, embossing, and lamination contribute to part quality and feature fidelity, and process optimization is required for each material and device.

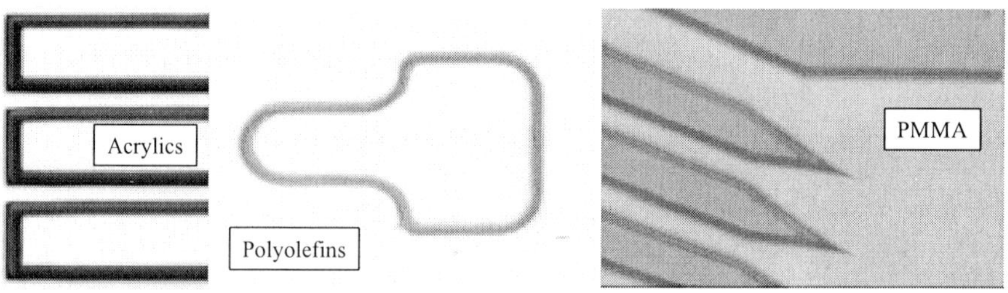

Figure 1. *Photographs of fluidic microstructures in several plastics. Excellent fidelity between master and replicate devices was obtained once the process was optimized.*

DNA sequencing. A variety of multichannel integrated plastic devices, with channel number ranging from 3 to 96, was designed and fabricated. Several sample-injector designs were investigated (Figure 2). Using a 13-channel plastic device with a separation distance of 18 cm, we accomplished 4-color DNA sequencing in ~30 minutes with identification of ~650 bases at better than 98% accuracy (Figure 2).

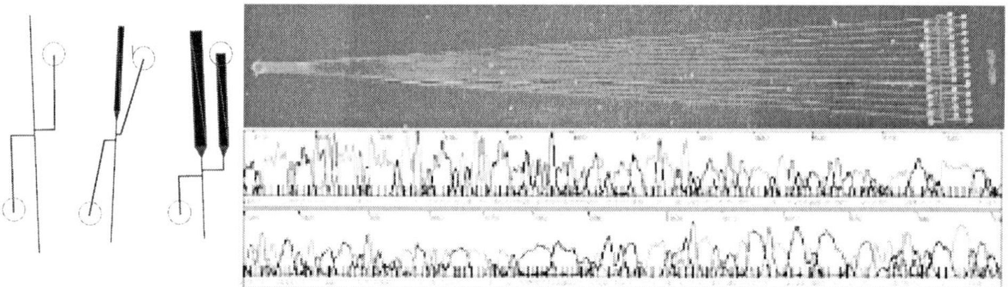

Figure 2. *Clockwise from left: Several injector designs; photo of a 13-channel plastic device; part of a 4-color DNA sequencing electropherogram.*

Protein Separations. "Proteomics" includes such activities as profiling proteins, analyzing protein structures and activities, studying protein interactions, and comparing protein expression levels between samples (e.g., healthy and diseased tissues); two-dimensional (2D) PAGE is widely used to separate proteins for such applications. We are developing a microfluidics-enabled 2D-PAGE approach that offers rapid analysis, increased separation resolution, improved identification accuracy, and ease of automation.

Figure 3 shows IEF of five BioRad pI markers in a plastic microfluidic LabCard™ device. Focusing took place in a 5 cm-long channel using an electric field of 490V/cm, and required as little as ~2 minutes. After focusing, a mobilizer was used to move the focused samples through a detector and the series of fluorescent peaks was recorded (note that the migration time to image the peaks is not, in general, the same as the focusing time). In addition, we recently demonstrated SDS-PAGE protein separations and on-chip labeling, further supporting the feasibility of the 2D-PAGE chip concept.

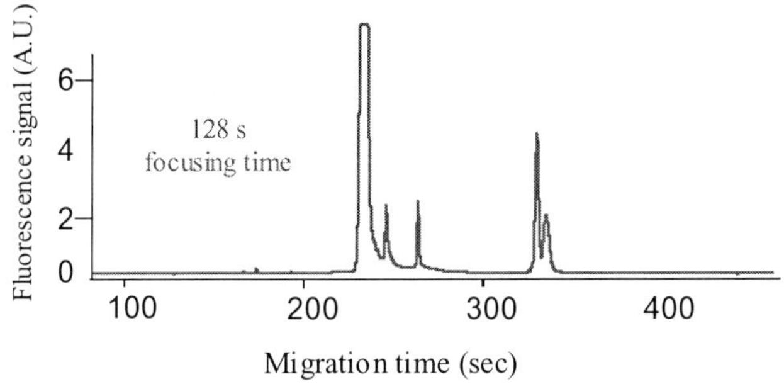

Figure 3. *IEF (separation) of five pI markers (pI = 5.3, 6.4, 7.4, 8.4, 10.4) in a plastic microfluidic device. The markers are labeled with Alexa Fluor 488 (Molecular Probes). Intensity variations probably result from differences in labeling efficiency.*

4. Conclusions

Plastic microfluidic devices combine biochemical compatibility and low-cost disposability with the capability to integrate diverse functions needed for such current applications as multiplexed DNA analysis and proteomics.

Acknowledgements

We thank our colleagues at ACLARA and Applied Biosystems for device prototyping and application development. Partial support for this work was provided by DARPA under the MicroFlumes and BioFlips Programs.

References

1. Manz, A.; Fettinger, J. C.; *et al. Trends in Anal. Chem.* **1991**, *10*, 144-149.
2. Harrison, D. J.; Manz, A.; Fan, Z.; *et al. Anal. Chem.* **1992**, *64*, 1926-1932.
3. Woolley, A. T.; Mathies, R. A. *Proc. Natl. Acad. Sci. USA* **1994**, *91*, 11348-11352.
4. Backhouse, C.; Caamano, M.; Oaks, F.; *et al. Electrophoresis* **2000**, *21*, 150-156.
5. Liu, S.; Ren, H.; Gao, Q.; *et al. Proc. Natl. Acad. Sci. USA* **2000**, *97*, 5369-5374.
6. Kroutchinina, N.; Jacobson, S.C.; *et al. Anal. Chem.* **2001**, *73*, 2669-2674.
7. Bousse, L.; Mouradian, S.; Minalla, A.; *et al. Anal. Chem.* **2001**, *73*; 1207-1212.
8. Burns, M.; Johnson, B.; Brahmasandra, S.; *et al. Science* **1998,** *282,* 484-487.
9. McCormick, R.; Nelson, R.; Alonso, M. G.; *et al. Anal. Chem.* **1997**, *69*, 2626.
10. McDonald, J. C.; Duffy, D. C.; *et al. Electrophoresis* **2000**, *21*, 27-40.
11. Soper, S. A.; Ford, S. M.; Qi, S.; *et al. Anal. Chem.* **2000**, *72*, 643 A-651 A.
12. Sassi, A.; Cruzado, I.; Björnson, T.; *et al. J. Chromatogr. A*, **2000**, *894*, 203-217.

SNP ANALYSIS BY ALLELE-SPECIFIC PYROSEQUENCING EXTENSION IN A MICROMACHINED FILTER-CHAMBER DEVICE

Aman Russom[1,2], Afshin Ahmadian[2], Helene Andersson[1], Wouter van der Wijngaart[1], Joakim Lundeberg[2], Mathias Uhlen[2], Goran Stemme[1], Peter Nilsson[2]

[1]Department of Signals, Sensors and Systems, [2]Department of Biotechnology, Royal Institute of Technology, S-10044 Stockholm, Sweden

aman.russom@s3.kth.se

Abstract

In this work, the principle of allele-specific extension by pyrosequencing reaction was employed to analyze single nucleotide polymorphisms (SNPs) in a micromachined filter-chamber device. Two SNP sites located on chromosome 17p and 9q were selected to evaluate the device. The bead-trapping device enabled analysis of genetic variations by using pyrosequencing chemistry in reaction-chamber volume as small as 12.5 nL. This leads to a 4000 fold reduction of the reagent consumption compared to the present standard volume. The results in this work demonstrate the capability of performing allele-specific extension of SNPs by pyrosequencing chemistry, which opens up the possibilities to perform base-by-base sequencing in this filter-chamber device.

Keywords: SNP, Pyrosequencing, filter—chamber, beads, Microfluidics

1. Introduction

It is believed that single nucleotide polymorphisms are associated with genetic diseases. As the number of identified SNPs increases, there will be an increasing demand for efficient methods to type these sequence variations. A recently developed technique for SNP genotyping is pyrosequencing [1]. The technique relies on incorporation of nucleotides by DNA polymerase and release of PPi, which is converted to ATP and then detectable light. In the standard procedure of pyrosequencing, dNTPs are added iteratively. However, for SNP analysis by allele-specific extension, a mixture of all four nucleotides is added and primers with alternating 3 -ends are utilized. In order to miniaturize pyrosequencing, a microfluidic filter-chamber device for chemical reactions on beads has been designed, manufactured and characterized, see figure 1 [2]. In this study, analysis of all variants of two SNP sites located on chromosome 17p (codon 72 of the p53 gene) and 9q (wiaf 1764) was performed to evaluate the filter-chamber device.

2. Experimental

Two single nucleotide polymorphism positions in the p53 tumor suppressor gene (exon4, codon72), and wiaf 1764 were selected to be analyzed. PCRs were performed to generate ~80 bp fragments. One of the PCR primers was biotinylated at the 5 -end to

J.M. Ramsey and A. van den Berg (eds.), Micro Total Analysis Systems 2001, 22–24.

allow immobilization and single-stranded DNA was obtained. The immobilized strand was divided into two wells and allele-specific primers with alternating 3 -ends were added in each well. Single stranded DNA with annealed primer was captured in the reaction chamber and a pyrosequencing mixture was added. The filter-chamber device was then placed on a CCD camera and the collected data was analyzed.

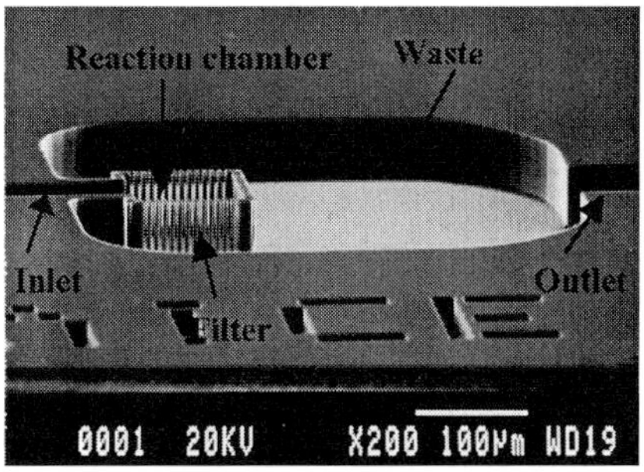

Figure 1. A SEM overview of the micromachined filter-chamber device.

3. Results and discussion

Figure 2 illustrates the priciple of allele-specific extension by pyrosequencing chemistry.

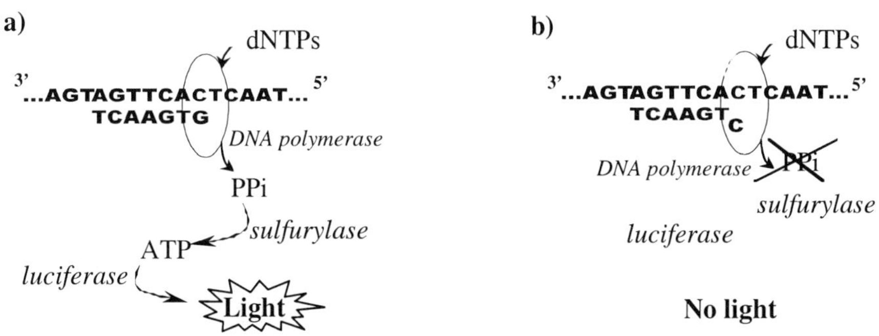

Figure 2. The principle of allele-specific pyrosequencing extension. In the matched primer (a), incorporation event of dNTP s is allowed by the DNA polymerase with release of PPi. The PP is converted to ATP and detectable light is produced by luciferase. If the primer does not match the DNA template (b), the DNA polymerase will not incorporate dNTP s and no light will be produced.

The principle was used to analyze two SNP sites. The results using this method are shown in figure 3 for all three possible variants of the SNPs at codon 72 and wiaf 1764. The coding SNP at codon 72 involves either G or C residues while at wiaf 1764 contains G or T residues. In the homozygoes variants, all the matched primers are extended successively with increasing light signals and are distinguished from the mismatched variants. The mismatched signals are almost constant at a base line level. In the heterozygoes cases, both alleles are being extended producing detectable light. This is expected because both allele-specific primers result in matched configurations.

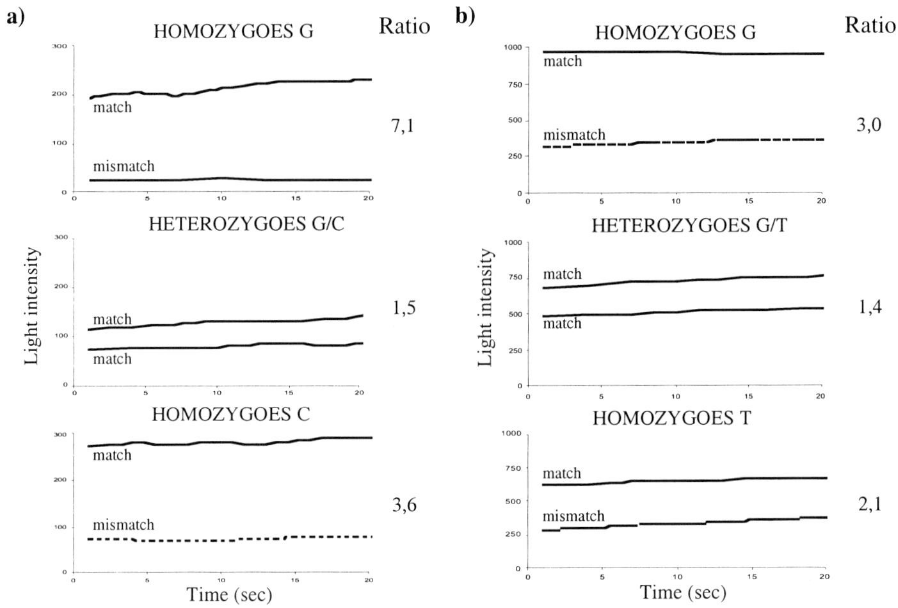

Figure 3. The signals obtained for the three different variants of the coding SNP on the p53 gene codon 72 (a) and wiaf 1764 (b) using pyrosequencing reaction in the micromachined device. The obtained ratio illustrates the discrimination factor in the different situations.

4. Conclusions

The results in this work demonstrate the capability of scoring SNPs by allele-specific extension using pyrosequencing, which opens up the possibilities to perform base-by-base sequencing in the filter-chamber device.

References

1. A. Ahmadian, B. Gharizadeh, A. Gustafsson, F. Sterky, P. Nyr n, M. Uhl n, J. Lundeberg. Single-nucleotide polymorphism analysis by pyrosequencing. *Anal Biochem* 280, 103-110 (2000).

2. H. Andersson, W. van der Wijngaart, P. Enoksson, G. Stemme. Micromachined. Flow-through Filter-Chamber for Chemical Reactions on Beads. Sensors and Actuators B: Chemical, 67, 203-208, (2000).

A RAPID VORTEX MICROMIXER FOR STUDYING HIGH-SPEED CHEMICAL REACTIONS

Sebastian Böhm[1], Ken Greiner[2], Stefan Schlautmann[3], Simon de Vries[4] and Albert van den Berg[3]
[1]Coventor B.V., Amsterdam, The Netherlands
[2]Coventor Inc. Cambridge MA, U.S.A.
[3]MESA+ Research Institute, University of Twente, The Netherlands
[4]Dept. Applied Physics, University of Delft, The Netherlands
(sebastian.bohm@coventor.com)

Abstract

A micromixer for the studying of high-speed chemical reactions is presented. The mixer comprises a circular vortex chamber with 16 tangential inlets for injecting the liquids to be mixed at velocities high enough to induce a swirling flow field. In this paper a design optimisation study is presented indicating the influence of various geometrical variations on mixing performance together with first experimental data on flow visualization by a fluorescent dye.

Keywords: micromixer, vortex, high speed mixing.

1. Introduction

During the evolvement of the μTAS field in last decade, a large number of micromixers were presented [1, 2]. The majority of these micromixers where targeted towards the rapid mixing of compounds for performing reactions in micro chemical systems. Besides enabling on-chip reactions as such, micromixers also play a role in the study of chemical reactions on the microsecond time scale [3]. A requirement of micromixers for this application is that the mixing can be performed typically at a timescale which is much shorter than the timescale in which the reaction takes place.

2. Theory

For rapid mixing, the reduction of diffusion distance is of utmost importance [4]. A decrease of the diffusion distance by a factor of two results in a decrease of mixing time by a factor of 4.

Figure 1 shows the general geometry of the vortex micromixer. From 16 tangential inlets, the liquids to be mixed (A and B) enter at high speed (m/s) the vortex chamber. Due to the high injection speed, in the vortex chamber a rotational flow field is induced which effectively reduces the distance between the lamella. Because of this reduction of diffusion distance, mixing times can be drastically reduced.

J.M. Ramsey and A. van den Berg (eds.), Micro Total Analysis Systems 2001, 25–27.

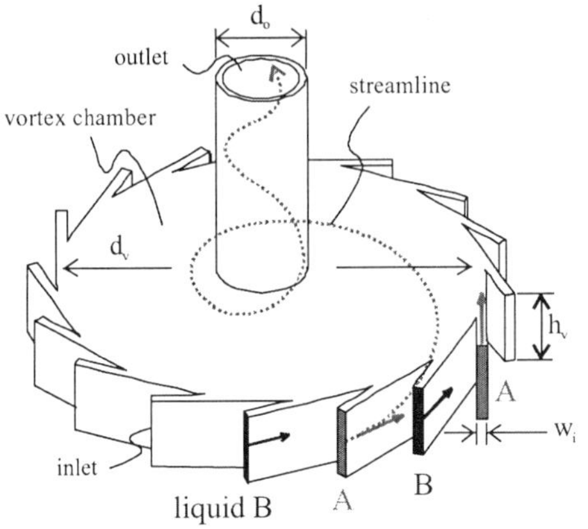

Figure. 1: Geometry of the vortex micromixer.

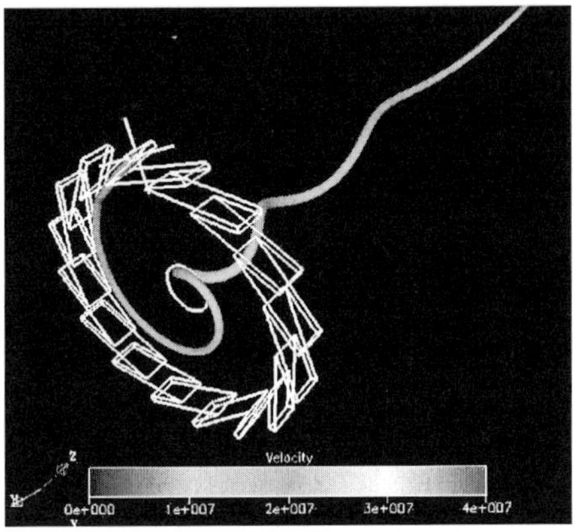

Figure 2: Streamline indicating rotational flow field (velocity in μm/s).

In order to derive an effective design, the influence of key dimensions on mixing speed was studied using Finite Element Modelling tools [4]. As for every full rotation of the liquid within the device the mixing speed increases by a factor of four, the total rotation of a liquid volume passing the device was taken as optimisation parameter. Figure. 2 visualizes this rotational flow field, which extends in the outlet of the mixer (inlet speed 10 m/s, Re max 200, pressure drop around 15 bar).

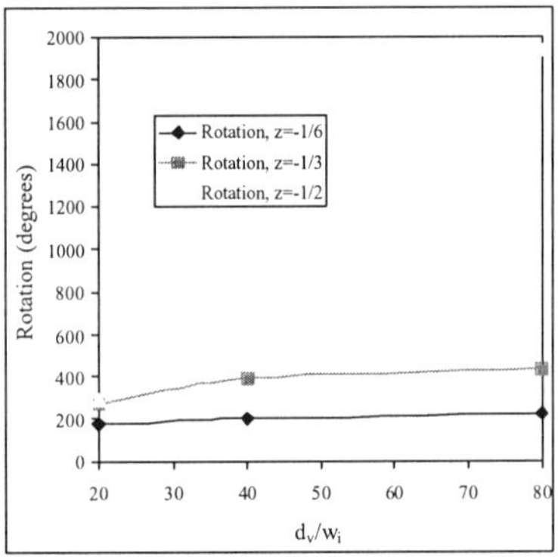

Figure. 3: Total rotation vs. d_v/w_i for volume elements injected at different heights (1/6, 1/3, and 1/2 of h_v).

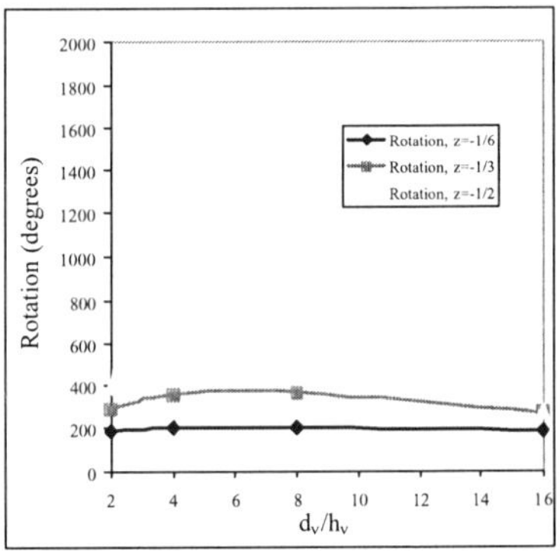

Figure. 4: Total rotation vs. d_v/h_v for volume elements injected at different heights (1/6, 1/3, and 1/2 of h_v).

Figures 3 and 4 show the total rotation of as function of respectively d_v/w_i and d_v/h_v. It can be seen that volume elements entering at different heights undergo different rotations, which is another factor reducing the diffusion distance.

It can be concluded that volume elements entering at ½ h_v undergo the largest rotation, From Figure 3 it can be seen that minimizing w_i is beneficial whilst Figure 4 suggests an optimal ratio d_v/h_v of about 4 to result in a maximum rotation.

In Figure 5 the cross-sectional concentration profile in the vortex chamber is plotted. Clearly the spiral pattern resulting from the rotational flow field can be seen. In this case mixing is completed within 5 ms.

3. Experimental

Devices were made by sandwiching a two step DRIE-etched Si wafer between two powderblasted Pyrex® wafers with through-holes and feed-rings (Figure. 6, insert). First experiments were carried out with demi water (A) and a 0.1 mM fluoresceine solution (B) (Figure. 6, injection pressure 1 bar). Currently high-pressure experiments are in progress.

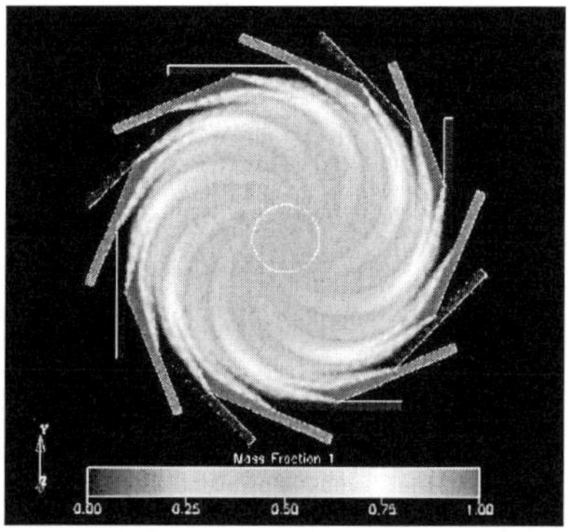

Figure. 5: Cross-section of vortex chamber showing mixing process.

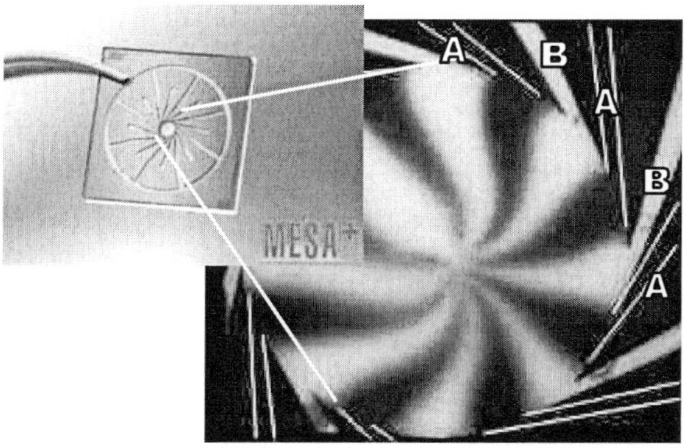

Figure. 6: Back: Mixing of dye at low velocity (video capture). Insert: Photo of realized device.

4. Conclusions

A novel micromixer has been presented together with a detailed design analysis indicating the effects of dimensions on mixing performance. Micromixers were successfully implemented by Dry Reactive Ion Etching and a preliminary test for visualisation was performed.

References

1. M. Elwenspoek, T.S.J. Lammerink, R. Miyake and J.H.J. Fluitman, "Towards integrated microliquid handling systems", *J. Micromech. Microeng.* **4**, pp. 227-245, 1994.
2 F.G. Bessoth, A.J. deMello and A. Manz, 'Microstructure for efficient continuous flow mixing', *Anal. Commun.*, **36**, (1999), pp. 213-215.
3 M.C. Ramachandra Shastry, S.D. Luck and H. Roder, 'A Continuous-Flow Capillary Mixing Method to Monitor Reactions on the Microsecond Time Scale', *Bioph. J.*, **74**, pp. 2714-2721, (1998).
4 Mixing speed inversely depends on time t [s] to overcome distance d [m] by diffusion: $t=d^2/D$ (D=diffusion coefficient [m^2/s]).
5 CoventorWare v. 2001.1: Incompressible liquid, $D=5mm^2/s$.

A NOVEL MICROSTIRRER AND ARRAYS FOR MICROFLUIDIC MIXING

Liang-Hsuan Lu, Kee Suk Ryu, and Chang Liu
Department of Electrical and Computer Engineering
University of Illinois at Urbana-Champaign, Urbana, IL 61801, USA

Abstract

We present a novel active microstirrer and its arrays for micro total analysis systems. These mixers are capable of rapid and on-demand mixing. Fabrication of the microstirrers is described. The mixing efficiency is investigated using experimental means and computer-aided fluid simulation.

Keywords: Microfluidics, active mixer, mixer array, magnetic actuation

1. Introduction

Microscale fluid mixing is crucial and challenging for µTAS. Previous approaches fall into two categories: passive mixing and active mixing. Passive mixing is accomplished by configuring the fluid channels structurally [1-4]. Active mixing methods have been reported using pressure driven flow or electrokinetic (EK) flow [5-6]. Active mixers can be activated on-demand. However, these are generally more complex to package and control. EK-based mixing is conjectured to be dependent on electrochemical properties of fluids. We present for the first time an active mixer employing micro magnetic rotating bar stirrers and arrays.

A micro magnetic bar stirrer is a miniaturized mixer of the kind commonly found in chemistry laboratories. A hub allows the bar to rotate but prevents dislocation from the substrate. When a rotating external magnetic field is applied, the rotor (micro magnetic bar) will magnetize and experience a torque, causing the bar to follow the external magnetic field. A micro stirrer offers the following performance characteristics and advantages. (1) Rapid mixing in a channel can be realized within the characteristic length of a stirrer bar. (2) The mixer array allows effective mixing in a larger reaction chamber for sequential analysis. (3) Magnetic actuation eliminates the need for tether wires. (4) The mixer can be used toward a wide variety of fluids. (5) Mixing is active, adjustable, and on-demand.

2. Microfabrication

Figure 1 shows the fabrication process of a microstirrer. This microstirrer consists of a cap, a hub, and two rotary blades, all made of Permalloy, an electroplated ferromagnetic material. The rotor consists of tips on the bottom to reduce friction. Mixer fabrication is low temperature in nature and can be applied to a variety of substrates. So far, we have

J.M. Ramsey and A. van den Berg (eds.), Micro Total Analysis Systems 2001, 28–30.

demonstrated fabrication on glass and silicon substrates. Micrographs of a single microstirrer and a 5x3 array are shown in Figure 2.

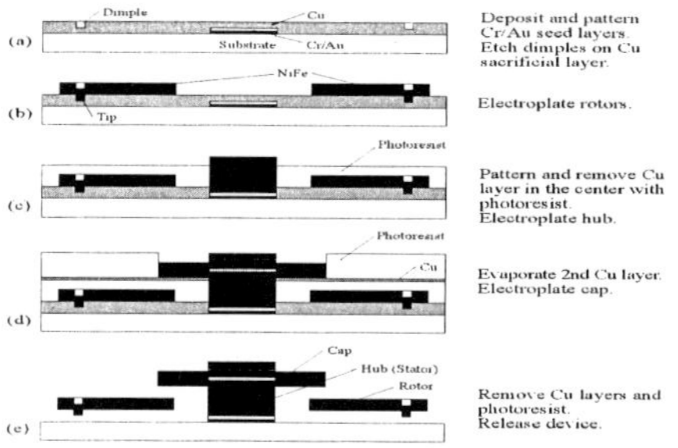

Figure 1. A schematic illustration of the fabrication process.

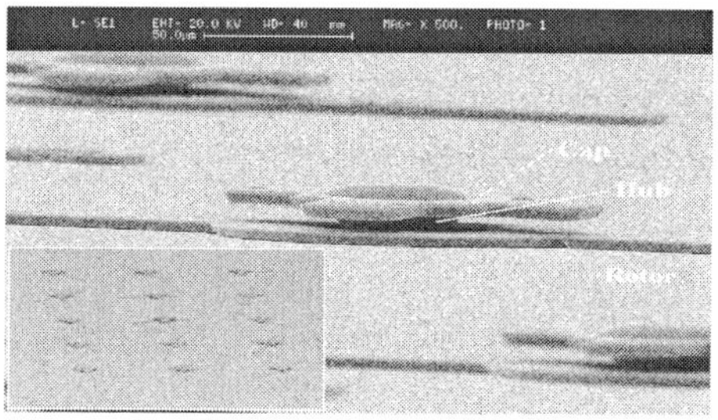

Figure 2. SEM micrograph showing the perspective view of a microstirrer in a 5x3 mixer array (insert).

3. Experimental

Microfluidic devices are formed by bonding PDMS pieces having molded, recessed structures onto glass chips with build-in mixers. When actuated by the external magnetic field of a commercial hotplate/stirrer, the rotating speed of microstirrers can exceed 600 rpm, resulting in fast mixing. Figure 3 shows two colored fluids in a 750-μm wide and 40-μm-tall channel being mixed with a single stirrer rotating at 120 rpm. The pressure-driven flow rate is 0.17μl/min. Scanometric color index profiles measured upstream and downstream of the stirrer have confirmed effective mixing. We find that the mixing performance is influenced by the stirrer speed. Figure 4 demonstrates a microstirrer array mixing at 600 rpm in a reaction chamber with a diameter of 2.5 mm and a height of 40 μm. The mixing causes the color in the chamber to become more uniformly distributed, resulting in a narrow peak in the color histogram. Complete mixing is achieved in 55 seconds as shown in Figure 5.

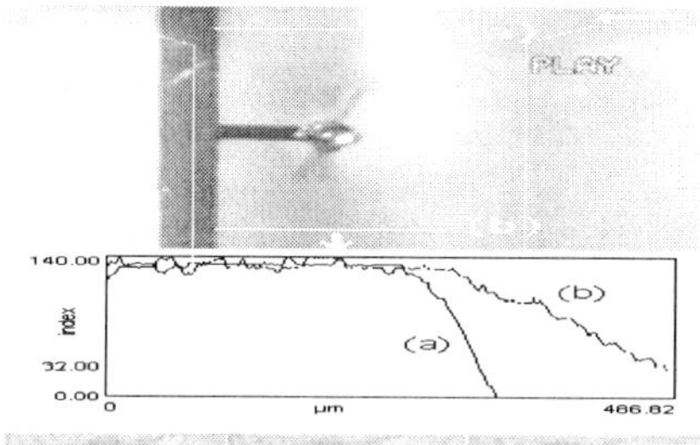

Figure 3. Top: Channel mixing near a T-junction. Lines (a) and (b) are 450 μm away from the mixer hub. Bottom: Scanometric index profiles along lines (a) and (b) show color of fluids averaging out significantly after mixing. (Line (b) represents downstream profile).

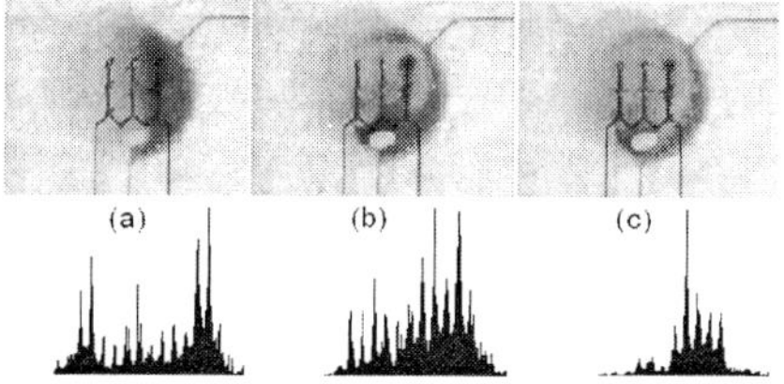

Figure 4. Top: Sequential video frames of mixing by an array (3x3) of stirrers. Results at (a) 0s (b) 30s and (c) 60s. Bottom: Color histogram corresponding to each case illustrated above.

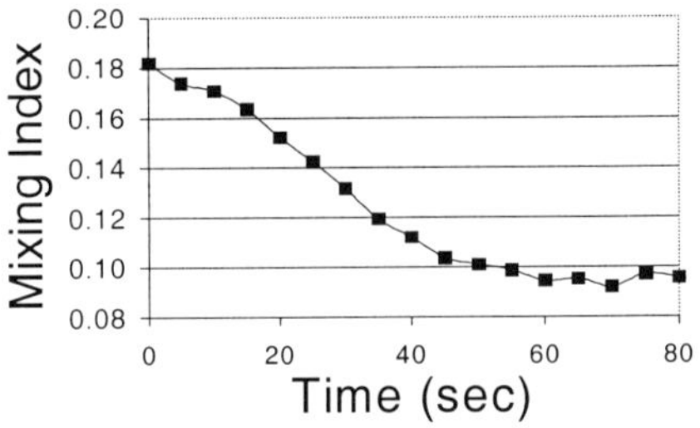

Figure 5. Mixing index (the standard deviation divided by average color index) versus time in the reaction chamber.

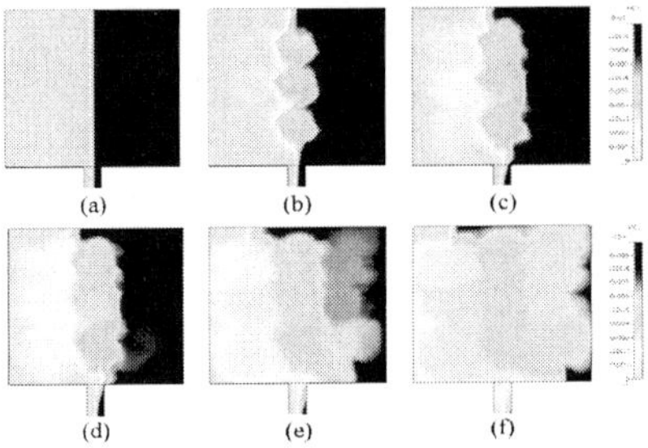

Figure 6. Simulation of array mixing. (a) Chamber is initially filled with two colors. (b)-(f) Mixing after 1s, 5s, 10s, 30s and 60s, respectively.

4. Simulation

A computer aided simulation model is developed based on CFD-ACE to facilitate the development and optimization of microstirrer mixers. Typical results are shown in Figure. 6. No mixing occurs when two fluids are initially brought into a 2000x2000x50 μm square chamber. The mixing results at six time intervals are shown to illustrate successful mixing with a 3x3-microstirrer array each rotating at 600 rpm.

5. Conclusion

We have presented a novel microstirrer and its arrays for microfluid mixing. Experimental results have shown excellent performance of the microstirrers both in continuous channel mixing and sequential chamber mixing. Rapid and on-demand mixing is achieved and validated by computer-aided simulation.

References

1. K. F. Jensen, "The Impact of MEMS on the Chemical and Pharmaceutical Industrials," Proc. IEEE Sensor and Actuator Workshop, (2000) 105-10.
2. J. Branebjerg, et al., "Fast Mixing by Lamination", Proc. IEEE MEMS Workshop, San Diego, (1996) 441-6.
3. T. Fujii, et al., "Development of a Microfabricated Biochemical Workbench -Improving the Mixing Efficiency-," Proc. Micro Total Analysis Systems,(1996) 173-6.
4. R. H. Liu, et al., "Passive Mixing in a Three-Dimensional Serpentine Microchannel", J. Microelectro. Sys., 9(2) (2000) 190-7.
5. R. M. Moroney, et al., "Ultrasonically induced microtransport", Proc. IEEE MEMS Workshop, Amsterdam, The Netherlands, (1991) 277-82.
6. Y.-K. Lee, et al., "Chaotic Mixing in Electrokinetically and Pressure Driven Micro Flow", Proc. IEEE MEMS Workshop, (2001) 483-6.

A NOVEL IN-PLANE PASSIVE MICROMIXER USING COANDA EFFECT

Chien-Chong Hong, Jin-Woo Choi, and Chong H. Ahn
MicroSystems and BioMEMS Lab
University of Cincinnati, P.O. Box 210030, Cincinnati, OH 45221-0030, USA

Abstract

In this paper, an innovative in-plane passive micromixer using the "Coanda effect" has been designed, simulated, fabricated and successfully characterized. Simulation and experimental results of the developed novel micromixer have shown excellent mixing performances over a wide range of flow rate from 1 μl/min to 100 μl/min. The micromixer realized in this work has achieved even better mixing performance at a higher flow rate of 100 μl/min. Due to a simple in-plane structure of the novel micromixer, the mixer can be easily fabricated and integrated with total microfluidic systems.

Keywords: Passive micromixer, Coanda effect, microfluidic, in-plane micromixer

1. Introduction

In microfluidic-based biochemical analysis systems, mixing of the liquid samples is considered as one of the most challenging tasks to achieve an appropriate reaction in a short period of time. Both turbulence and diffusion mixing concepts are available principles in realizing a passive mixer. Most passive micromixers usually rely on diffusion mixing of two or more substreams because low Reynolds number in microfluidic devices usually produces a laminar flow. For this reason, most passive micromixers [1-3] reported in the past few years showed relatively good mixing performances just at low flow rates, while ideal micromixer should have uniform mixing capability independent to flow rates. Although passive mixing at higher flow rate can be achieved in a three-dimensional passive micromixer, integration of the device with microfluidic system is not easy because of complex structure of 3-D micromixer [1]. In this paper, we propose an innovative in-plane passive micromixer using the "Coanda effect". Due to a simple in-plane structure of the novel micromixer, the mixer can be easily fabricated and integrated with micro total analysis systems (μ-TAS).

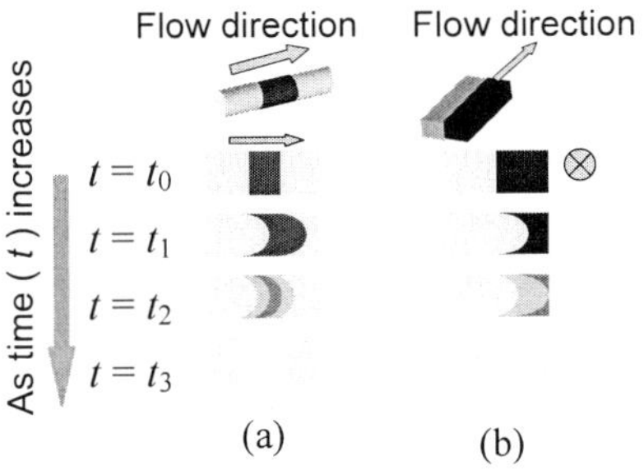

Figure 1. Mixing pattern at different time steps: (a) Taylor dispersion in a capillary tube [4] and (b) the fluidic behavior in our mixer.

2. Principle and Design

The dispersion of two miscible liquids in a laminar-flow capillary tube, along the major axis of the channel was described by Taylor dispersion [4]. From

J.M. Ramsey and A. van den Berg (eds.), Micro Total Analysis Systems 2001, 31–33.

the analysis, axial convection establishes large radial concentration gradients. Obviously, the velocity profile will significantly contribute to the mixing behavior, as shown in Figure 1(a). This concept was applied to this micromixer design. In-Plane microvalves and pumps with similar Tesla structures using the "Coanda effect" have been successfully realized by Jang *et al.* and Duffy *et al.* [5-6], but none of the mixers using the structure have been reported yet. Now this structure was used to produce a Poisseuille flow in the perpendicular direction of the flow as shown in Figure 1(b). Figure 2 shows the principle of a novel passive micromixer using the Coanda effect in the newly modified Tesla structure for our mixing purpose. Fluids will tend to flow nearby angled surface or steep angled surface by the Coanda effect, which is used to guide the fluid to collide.

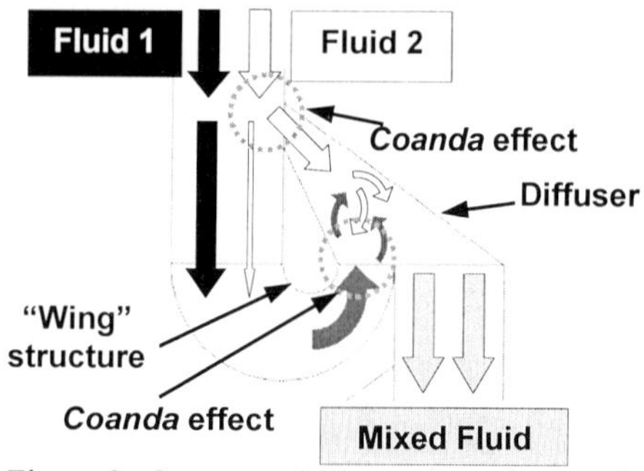

Figure 2. Conceptual illustration of a novel in-plane passive micromixer using the "Coanda" effect.

3. Simulation and Experimental Results

In the micromixer, one of the fluids is divided into two substreams and then merged again with the other fluid from the main channel of the micromixer. Then they will be mixed with the other substream, producing a strong turbulence around the sub-channel of the micromixer. Most mixing occurs in the latter stage according to the fluidic simulation by CFD-ACE+ as shown in Figure 3.

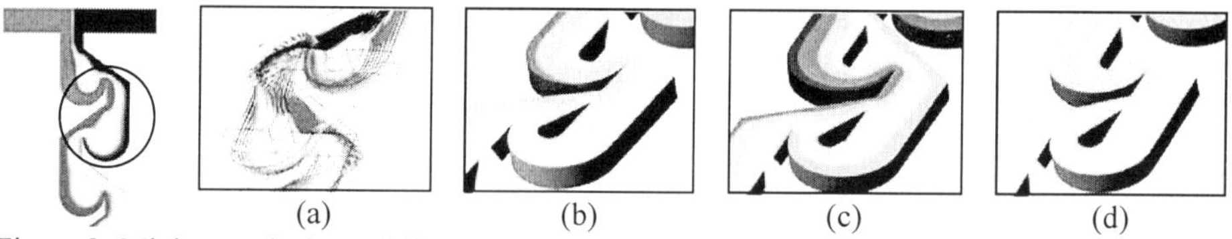

Figure 3. Mixing analysis at different flow rates by CFD-ACE+: (a) velocity field; (b) mixing at 1 μl/min; (c) mixing at 10 μl/min, and (d) mixing at 100 μl/min.

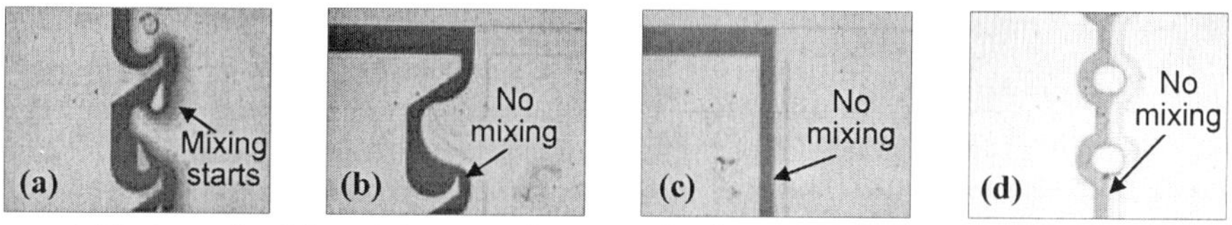

Figure 4. Photograph of the mixing experiments at the flow rate of 50 μl/min: (a) our design; (b) without "wing" structure; (c) T-type mixer, and (d) Ring-type mixer.

The new micromixer was fabricated using PDMS, and blue- and yellow-dyed DI water were injected into the micromixer by syringe pump to characterize its mixing performance. Several passive micromixers (T-type and Ring-type micromixer) with different geometries were compared by fluidic simulation for the same channel dimension (200 μm x 110 μm) as shown in Figure 4, where the micromixer with the Coanda effect shows the most preferable result.

Figure 5 clearly shows mixing achievement at the different positions at a flow rate of 50 μl/min. While the mixing performance of T-type micromixer sharply falls as the flow rate increases, the new micromixer realized in this work achieves fairly uniform mixing performance at the wide variation of flow rate as shown in Figure 6. For a higher flow rate than 50 μl/min, mixing starts at even early stage after the first mixer cell.

4. Conclusions

The new passive micromixer presented in this work has achieved an outstanding mixing performance, which is almost comparable to an ideal passive mixer. The passive mixer realized in this work will be immediately useful for most micro total analysis systems, which require in-plane passive mixers.

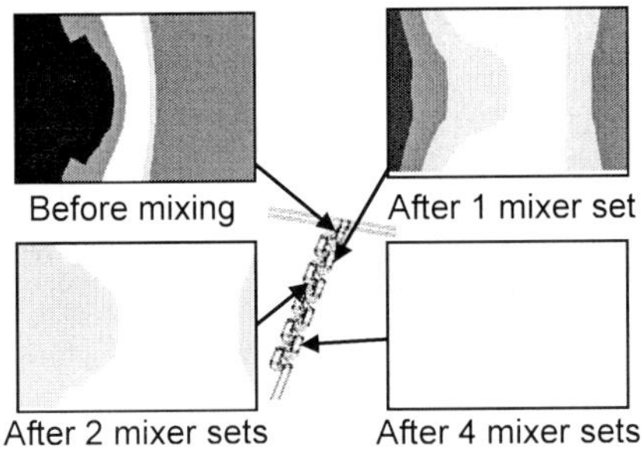

Figure 5. Cross-sections of mixing pattern in the micromixer at the flow rate of 50 μl/min. After 4 mixer sets, it shows that two fluids have been completely mixed.

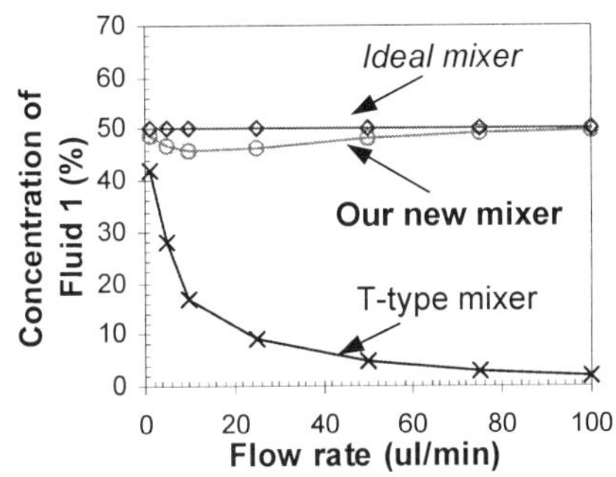

Figure 6. Comparison of mixing performance for T-type mixer and a novel micromixer using Coanda effect at different flow rates.

Acknowledgements

This research was fully supported by a DARPA grant under contract AF F30602-00-1-0569 from the *BioFlips* program, DoD, USA.

References

1. W. Ehrfeld, V. Hessel and H. Lowe, "Microreactors-New Technology for Modern Chemistry,"New York: Wiley-Vch 2000
2. F. G. Bessoth, A. J. deMello and A. Manz, "Microstructure for Efficient Continuous Flow Mixing," Anal. Commun., Vol. 36, pp. 213-215, 1999.
3. M. A. Stremler, M. G. Olsen, B. H. Jo, R. J. Adrian, H. Aref and D. J. Beebe, "Chaotic Mixing in Microfluidic Systems," Tech. Digest of Solid-State Sen. & Act. Workshop, pp. 187-190, 2000.
4. R. F. Probstein, "Physicochemical Hydrodynamics: an Introduction," New York: John Wiley & Sons, 1994
5. L.S. Jang, C. J. Morris, N. R. Sharma, R. L. Bardell and F. K. Forster, "Transport of Partical-Laden Fluids Through Fixed-Valve Micropumps," Proc. of MEMS-Vol. 1, ASME, pp. 503-509, 1999.
6. D. C. Duffy, O. J. Achueller and S. T. Brittain and G. M. Whitesides., "Rapid Prototyping of Microfluidic Switches in Poly (dimethyl siloxane) And Their Actuation by Electro-Osmotic Flow," Journal of Micromechanics and Microengineering, Vol. 9, pp. 211-217, 1999.

ELECTROKINETIC INSTABILITY MICROMIXERS

M. H. Oddy, J. G. Santiago, and J. C. Mikkelsen
Department of Mechanical Engineering, Stanford University
Bldg. 500, Room 500Y, Stanford, CA 94305 USA
oddy@stanford.edu

Abstract

We have observed that microchannel liquid flows subjected to sinusoidally alternating electric fields exhibit, under certain electric field strengths and geometries, a flow instability capable of rapidly stirring micro- and nanoliter volume solutions. We present the design, fabrication, and characterization of two micromixers: a prototype device Micromixer I and a more robust, 2nd generation device Micromixer II that utilize an electrokinetically induced instability as the stirring mechanism and may be easily integrated into many microfluidic chip designs.

Keywords: Micromixing, Electrokinetic Instability, AC Electrokinetics

1. Introduction

Rapid mixing of macromolecular solutions in bioanalytical microsystems is often difficult due to the characteristically low Reynolds number of microflows and the low diffusion coefficients of the molecules. Many biochemical sensing techniques such as immunoassays and hybridization analyses require the rapid, homogeneous mixing of macromolecular solutions, such as DNA or globular proteins. We have used an electrokinetic instability (EKI) to efficiently stir microflows with Reynolds numbers based on channel depth and RMS velocity as low as 1.5.

2. Theory

For typical microfluidic device buffer chemistries, applying an alternating (AC) electric field to a microchannel flow causes the electric double layer (EDL) to oscillate with the same frequency as the applied electric field [1]. Viscous shear causes the bulk fluid to oscillate with diminished amplitude and a phase lag. The effects of the oscillatory EDL on the bulk fluid may be modeled with a Helmholtz-Smoluchowski slip velocity given by, $U_{wall} = (\varepsilon\zeta|E|/\mu)\exp(i\omega t)$, provided that the Debye length is small relative to typical channel dimensions. The velocity field shown in Figure 1 is determined from the Navier-Stokes equations,

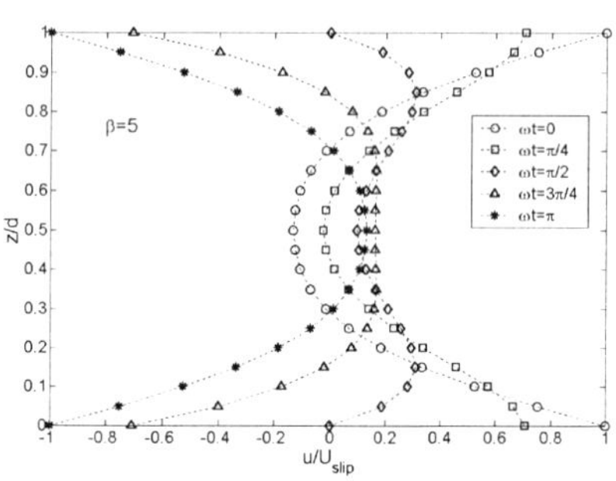

Figure 1. Theoretically predicted velocity profiles of electroosmotically oscillating flow between two infinite flat planes for the dynamically stable base state, $\beta = 5$.

J.M. Ramsey and A. van den Berg (eds.), Micro Total Analysis Systems 2001, 34–36.

similar to Stokes 2nd problem [2]. β is the ratio of d/δ, where d is the channel depth and δ is the momentum diffusion length scale $(2\nu/\omega)$^0.5.

3. Experimental

The micromixer designs (Figure 2) comprise simple, two-dimensional layouts capable of stirring two fluid streams continuously or intermittently (in a stopped flow mode). The experimental setup and conditions for both micromixers are shown in Figure 3 and listed in Table 1. For Micromixer II, porous, dielectric frits (0.5 μm pore size) were integrated into the AC side ports to mechanically isolate the electrodes from the mixing chamber fluid, yet provide an electrically conductive path between the electrodes and the fluid.

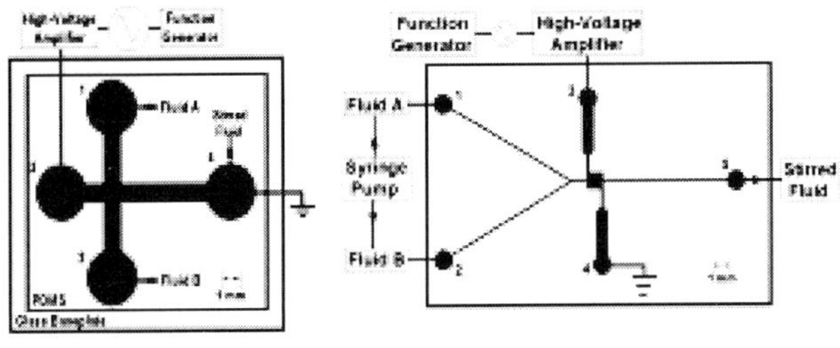

(a) (b)

Figure 2. Dyed and undyed buffer streams are fed through ports 1 and 2. Platinum electrodes inserted into the AC side channel ports 3 and 4 induce EKI stirring by conducting a high voltage sine wave from a function generator and a high voltage amplifier. (a) Micromixer I constructed from PDMS/Glass (300 μm deep). Stirring occurs along the entire channel stretching from ports 3 to 4. (b) Micromixer II constructed from wet-etched Borofloat (100μm deep). Stirring is effectively confined to the square mixing chamber.

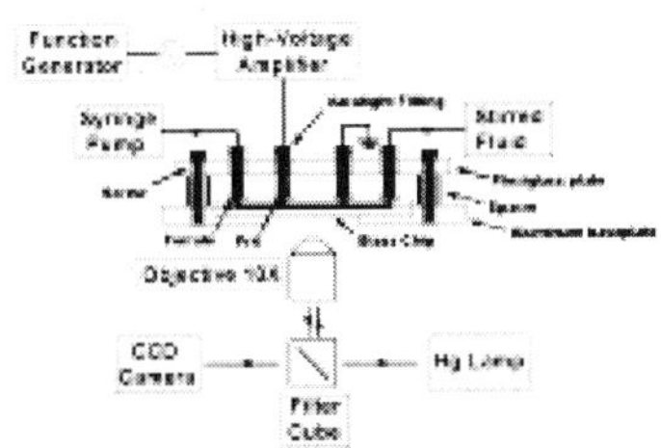

Figure 3. Experimental setup for Micromixer II, shown with optical system and AC excitation sources. Epifluorescent microscopes and cooled 12-bit CCD cameras were used to image and record fluid stirring. Micromixer I used a similar setup.

4. Results and discussion

Time stamped images obtained from both micromixers are shown in Figure 4. The EKI rapidly deforms the initially distinct dyed/undyed interface upon application of the AC field (3rd frame for both sets of experiments). Spatial intensity profiles, provided by the epifluorescence images (Figure 4), represent the spatio-temporal integration of fluorophores within discrete fluid

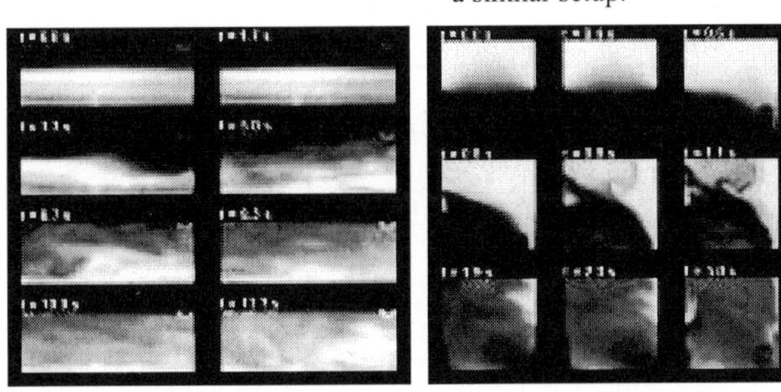

(a) (b)

Figure 4. (a) 4X objective images of EKI stirring within Micromixer I of 2MDalton Dextran/Fluorescein conjugate (10 μM) in a 2 mM Borate buffer. The electric field strength and frequency were 120 V/mm and 10 Hz. (b) 10X objective images from the square mixing chamber of Micromixer II. 20 μM Fluorescein was used for the dye and the background buffer was a 5 mM HEPES buffer. The estimated field strength within the chamber was 300 V/mm with a frequency of 5 Hz.

measurement volumes or voxels, and over the exposure time of the image. In our optical setup, the voxel dimensions correspond to the size of the binned pixels and the depth of the mixer or 10.7 x 10.7 x 100 μm in the object plane. We have initially characterized the mixer performance of Micromixer II by computing ensemble-averaged probability density functions (PDF's) of the instantaneous, spatial fluorescence intensity fields. The PDF's (Figure 6) provide a description of the distribution of voxel-averaged intensity values and is one measure of the mixing dynamics of the entire micromixer region.

5. Conclusions

A rapid stirring technique has been developed and demonstrated in two micromixing devices. The EKI stirring technique has the potential of improving the functionality of bioanalytical, microfluidic devices. These micromixers require no moving parts and have a small footprint. Ensemble-averaged PDF's were used to assess the micromixer performance using a Fluorescein dilution experiment. The characteristic diffusion time of Fluorescein across the mixer half-width (0.5 mm) of 500 s may be compared to the EKI "time-to-mix" of 3 s. Experimental observation has shown that parameters influencing the instability include geometry, electric field strength, and frequency.

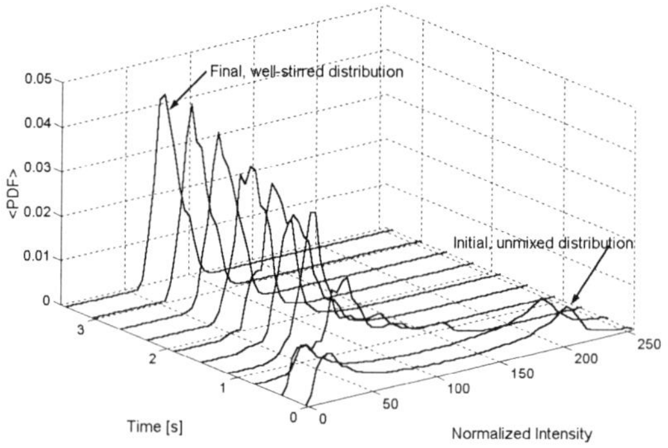

Figure 6. Ensemble-averaged temporal evolution of voxel-averaged spatial intensity PDF's. Each ensemble consists of nine realizations. The initially bimodal PDF is indicative of the unmixed initial state of the fluid composed of mostly unmixed dyed and undyed streams. After 3 s, the mixer has stirred the two liquid streams to within the scale of the voxels and the PDF is unimodal.

Acknowledgements

The authors would like to acknowledge the support of a Department of Defense Graduate Fellowship, Agilent Technologies, and the Defense Advanced Research Projects Agency contract F33615-98-1-2853 with Drs. Abe Lee and Michael Krihak as contract monitors.

References

1. J. Santiago, "Electroosmotic flows in microchannels with finite inertial and pressure forces," *Anal. Chem.,* **73** (#10), (2001), pp. 2353-2365.
2. M. Oddy, J. Santiago, and J. Mikkelsen, "Electrokinetic instability micromixing," submitted Anal. Chem. (2001).
3. F. Bessoth, A. de Mello, and A. Manz, "Microstructure for efficient continuous flow mixing," *Anal.Commun.,* **36**, (1999), pp. 213-215.

USING MICROFLUIDIC SYSTEMS AS ANALOG DEVICES FOR SOLVING COMPUTATIONAL PROBLEMS

Darwin R. Reyes[1], George M. Whitesides[2], Andreas Manz[1]
[1]Astra Zeneca/SmithKline Beecham Centre for Analytical Sciences, Imperial College, London SW7 2AY, UK
[2]Department of Chemistry and Chemical Biology, Harvard University, Cambridge, MA, USA
d.reyes@ic.ac.uk

Abstract

An analog computational approach using a microfluidic chip to solve shortest path problems is presented. The computational process is based on the lighting up of a glow discharge between two points in a graphical representation on a microfluidic device. This work shows 1) the solution and visible display for the shortest path between two points in mazes and a map and 2) the small computational time difference between problems of varying sizes and complexity.

Keywords: Analog computing, microfluidic device, shortest path problem

Introduction

Digital computing has been used successfully in many applications, but its use is still limited by both the time required to devise and code an efficient algorithm, and the increased computational time required to solve larger instances of a problem. An analog approach simulating a physical system could address this limitation.

Recently, microfluidic devices have been used to solve combinatorial optimization problems [1,2], based on coding an algorithm that fit the properties of the microfluidic device. Here we present an analog computational approach to solve the shortest path problem. For this purpose a microfluidic device for computation based on the lighting up of a glow discharge in a system of interconnected micro channels is used.

Experimental

Several problems of different size were posed. First, graphical representations of the problems were generated. Then, the graphical representations were fabricated on glass (soda-lime) chips using standard microfabrication procedures [3]. Starting and ending points for the problems were selected and the computation between these two point or landmarks was executed. The width and depth of the channels were ca. 250μm and 100μm, respectively. Capillaries of 250 μm i.d., glued onto the chip, were used to introduce helium gas. Low pressure was obtained using a high vacuum pump (see Fig. 1). The outlet capillary was connected to an external pressure sensor. Tungsten wires of 25 μm were inserted in the appropriate channels to designate the starting and ending points. A high voltage (HV) power supply provided the breakdown voltages used to perform the desired computation.

J.M. Ramsey and A. van den Berg (eds.), Micro Total Analysis Systems 2001, 37–39.

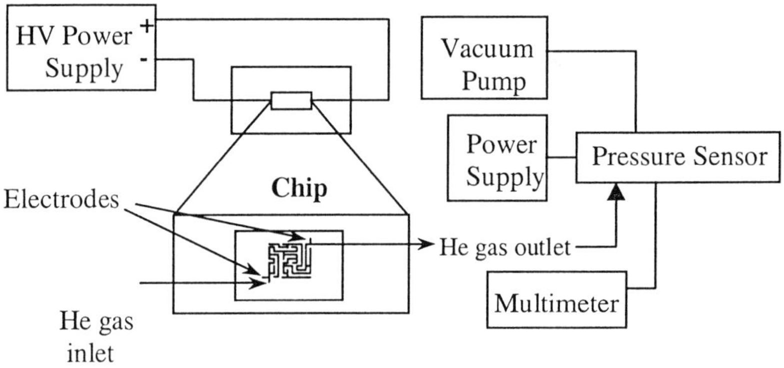

Figure 1. Schematic of the setup.

Results and Discussions

Two problems of increasing size were posed, maze searches and the computation of the shortest path. To solve these problems, a voltage higher than the breakdown voltage was applied to the graphical representations and a glow discharge was generated providing the solutions to each of these. Figure 2A shows a schematic of a maze and 2 B the two possible paths. The solution given by the glow discharge in this maze, as is observed in fig. 2C, was the shortest path. The longest path was not chosen.

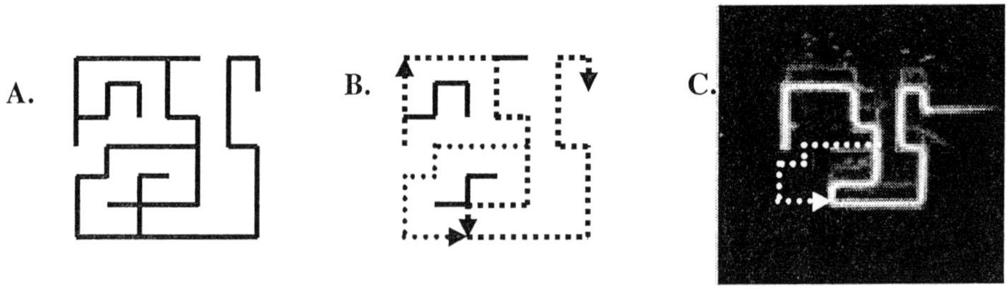

Figure 2. A) Diagram of a maze with two possible paths. The shortest path shown (■) is 34.7 mm long and the longest path (•) is 39.4 mm. B) Photo of the glow discharge showing the solution of the problem, the shortest path. The longest path on the photo was drawn (•). The voltage applied was 21 kV and the pressure was 89 torr.

A graphical representation of an interconnected network, with increased size, was also studied. When the voltage was applied between two landmarks the shortest path or the next shortest paths (k-shortest paths) were observed, depending on the voltage applied. Figure 3 shows a graphical representation of a segment of a map of London. In figure 3A the shortest path from Trafalgar Square to Victoria Station is observed (highlighted) when a potential of 15.75 kV was applied between these two landmarks. In figure 3B multiple solutions (1-3 in increasing length) of the route from Trafalgar to Victoria Station are observed when the potential was raised from 15.7 to 18.9 kV. The pressure of both experiments was 160 torr.

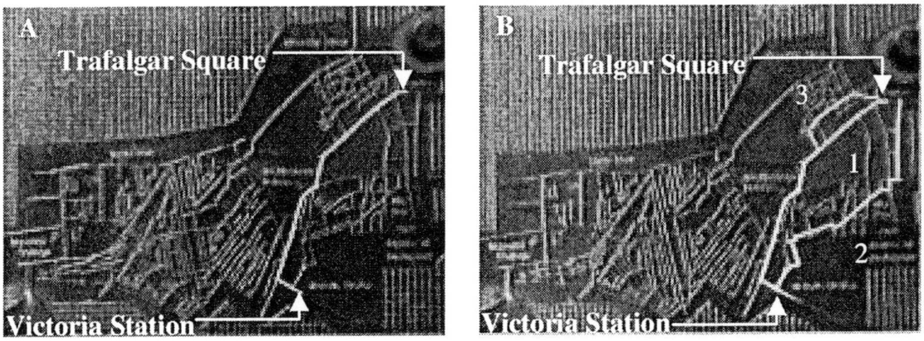

Figure 3. London map on a chip.

The problem solving times were measured at several applied potentials for three graphical representations which vary in size and complexity. Figure 4 shows the computational time for a straight line, a maze, and two landmarks in the London map.

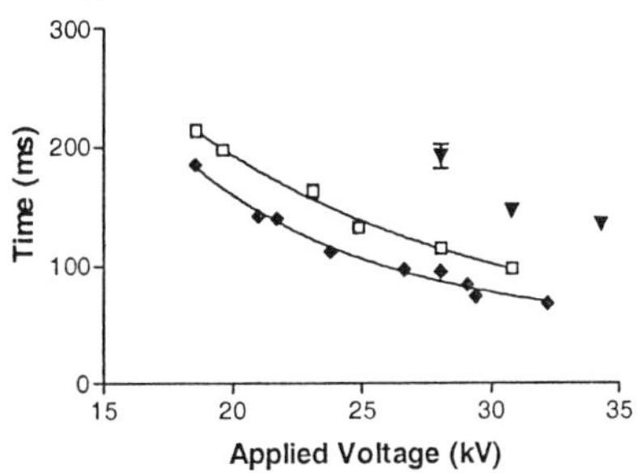

Figure 4. Problem solving time measured as a function of the applied voltage for a straight line (□), a maze (◆) and two landmarks in the London map (▼).

Conclusion

This analog computer proved to be useful finding the shortest path in a maze and the shortest and the next shortest paths (k-shortest paths) in a map. The difference in response time with increasing size and complexity of the problems studied was found to be insignificant and the solutions of the problems were obtained in less than 250 ms. We consider this approach as a promising starting point for the use of microfluidic systems to solve new computational applications, e. g. NP-hard problems.

References

1. D. T. Chiu et al., *Proc.Natl. Acad. Sci. U. S. A.* ***98***, 2961 (2001).
2. J. S. McCaskill, *Biosystems* ***59***, 125 (2001).
3. D. J. Harrison, et al., *Analytical Chemistry* ***64***, 1926 (1992).

Acknowledgements

This material is based upon work supported by the NSF under Grant No. INT-0000462.

MOLECULAR SEPARATIONS IN NANOFABRICATED SYSTEMS

H.G. Craighead
Applied & Engineering Physics and the Nanobiotechnology Center, Cornell University, Ithaca, NY 14853, USA

Abstract
Nanofabrication allows the creation of ultra-small fluid systems in a variety of material systems. These small-scale controlled geometries can be integrated with microfluidic systems to provide for molecular separation or signal-to-noise improvements in molecular detection.

Keywords: Single molecule detection, nanofabrication

1. Introduction
High-resolution lithography or controlled etching processes can be used to make fluid control systems with dimensions of tens of nanometers. Most conventional chemical separation systems use porous materials with pore sizes down to the nanometer range. By engineering systems with greater geometrical control and integrating these into larger-scale microfluidics new functionality can be incorporated in fabricated analytical systems.

2. Nanofabrication Approaches
A variety of approaches are evolving for fabricating fluid systems with nanoscale constrictions and objects. We have explored methods for incorporating nanoscale engineered inclusions in microfluidic systems.[1] These include etching variable depth fluid channels and sacrificial layer removal approaches with structures defined by high-resolution lithography. New materials systems have been investigated for the sacrificial layers including polymeric systems that can be directly patterned by electron or ultra violet light exposure.[2,3]

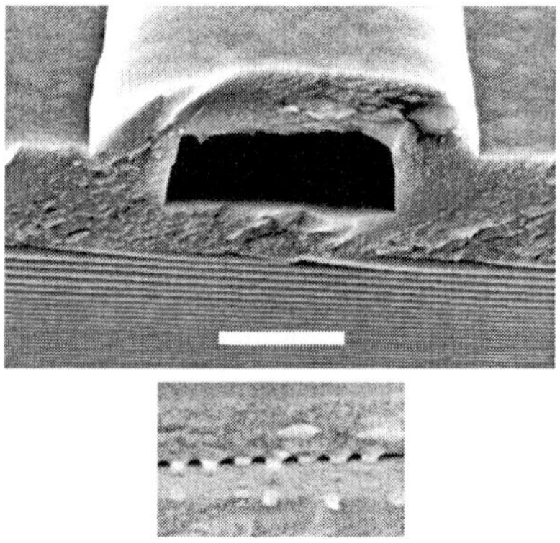

Figure 1. Scanning electron micrograph of a silicon nitride tube created by chemical etching of a silicon sacrificial layer using the approach of reference 2. Both images are on the same scale with the white bar representing 0.5 μm.

3. Molecular Sorting Devices
Controlled geometries enable experimentation with new approaches to molecular sorting. These approaches include separation based on electrophoretic mobility differences in gel-free nanostructured channels.[4-6] Similar fabrication approaches have been used to create devices for sorting based on diffusion rate differences.[7]

J.M. Ramsey and A. van den Berg (eds.), Micro Total Analysis Systems 2001, 40–41.

4. Single Molecule Detection

Nanostructures can be used to create systems where the interaction volume for optical excitation of fluid systems is defined by the physical dimensions of the fluid containing nanostructure rather than optical diffraction. This has been done in both narrow flow channels and in aperture defined optical systems.[8,9]

5. Conclusions

The technology for fluid handling in very small dimensions is expanding with new material systems and fabrication approaches. There are beginning to be explored for use in a variety of experiments in molecular separation and analysis. Some of these approaches may ultimately prove useful for analytical systems.

Acknowledgements

This work was done in collaboration with M. Foquet, J. Han, C. Harnett, J. Henion, J. Kameoka, A. Lopez, A. Turner, S. W. Turner, and G. Coates. Funding was provided by NSF, Dept. of Energy, DARPA, NIH and NYSTAR.

References

1. H. G. Craighead, *Nanoelectromechanical Systems*, Science, 290, 1532 (2000).
2. C. K. Harnett, G. W. Coates, and H. G. Craighead, *Heat-Depolymerizable Polycarbonates as electron Beam Patternable Sacrificial Layers for Nanofluidics*, (in press).
3. S. Turner, A. M. Perez, A. Lopez and H. G. Craighead, *Monolithic Nanofluid Sieving Structures for DNA Manipulation*, J. Vac. Sci. Technol B, 16 3835-40 (1998).
4. J. Han and H. G. Craighead, *Entropic Trapping and Sieving of Long DNA Molecules in a Nanofluidic Channel*, J. Vac. Technol. A., 17 2142-2147 (1999).
5. J. Han, S. W. Turner and H. G. Craighead, *Entropic trapping and escape of long DNA molecules at a submicron size constriction*, Phys. Rev. Lett., 83, 1688-91 (1999).
6. Han and H. G. Craighead, Science, *Separation of Long DNA Molecules in a Microfabricated Entropic Trap Array,* J. 288, 1026-1029, (2000).
7. Chai-Fu Chou, O. Bakajin, S.W. P. Turner, T. A. Duke, S. S. Chan, E.C. Cox, H.G. Craighead and R. H. Austin, *Sorting by Diffusion: An Asymmetric Obstacle Course for Continuous Molecular Separation*, PNAS, 96, 13762-5 (1999).
8. M. E. Foquet, J. Han, A. Lopez, W. Wright, and H. G. Craighead, *Fabrication of Microcapillaries and Waveguides for Single Molecule Detection,* Proc. SPIE, 3258, 141-147 (1998).
9. M. Levene, D. Larson, J. Korlach, M. Foquet, S. W. Turner, H. G. Craighead, and W. W. Webb, *Nanometer Constrained Observation volumes for High Concentration Fluorescence Correlation Spectroscopy and Single Molecule Dynamics*, Biophys. J., 78, 402A, Part 2 (2000).

MICROARRAY IMMUNOASSAYS IN THE MICROFLUIDIC TRIAGE® PROTEIN CHIP

Kenneth F. Buechler[1] Paul McPherson[2] Joseph Anderberg[3] Kevin Nakamura[4] Stephen Lesefko[5] and Barry Noar[6]
Biosite Inc., 11030 Roselle Street, San Diego, CA 92121, USA

Abstract

The Triage® protein chip is a tool for the simultaneous measurement of up to 100 different proteins by immunoassay. The immunoassays are performed in a microfluidic plastic chip and results are achieved in 15 minutes with picomolar sensitivities directly from several drops of whole blood or plasma. Microfluidic fluid flow is controlled passively in the protein chip by the surface architecture and surface hydrophobicity in the microcapillaries. The immunoassays utilize high affinity antibodies and a near infrared fluorescent label, which is read by a portable, battery powered fluorometer. The products aid emergency room physicians in the rapid diagnosis of myocardial infarction, congestive heart failure, and drug overdose.

Keywords: Protein Chip, Immunoassay, Microarray, Microcapillaries

1. Introduction

Immunoassays have classically been performed on robotic instruments. With the advent of point of care testing, where the testing of whole blood samples in emergency rooms of hospitals has become important for optimum patient care, there has been a need for rapid and simple immunoassays. This is particularly true for aiding emergency room physicians in the rapid diagnosis of myocardial infarction, congestive heart failure and drugs of abuse overdose. The Triage® protein chip has met a market need as a quantitative point of care immunoassay.

2. Description

Triage® Protein Chip

The protein chip is comprised of a plastic base and lid, and a blood filter. The base and the lid are ultrasonically joined to create a series of precise capillaries through which the sample flows. A defined surface architecture of different textures in the base and the lid control the movement of sample through the microcapillaries.

Protein chip utilization

The test procedure involves addition of several drops of sample to the protein chip, insertion of the chip in the Triage® Meter and reading the results from the display screen. The quantitative results are generated within approximately 15 minutes after addition of the sample to the protein chip.

J.M. Ramsey and A. van den Berg (eds.), Micro Total Analysis Systems 2001, 42–44.

Microcapillary flow

The immunoassays are performed in the microcapillaries of the protein chip. Fluid flow is controlled passively in the microcapillaries by the texture and architecture of the microcapillaries. Figure 1 shows a skeleton diagram of the top view of the Triage® protein chip. The blood sample is added at the sample addition zone. The sample enters a filter where red blood cells and plasma are separated by lateral flow in the filter. The plasma exits the filter and enters into a high capillarity region termed the sample reaction barrier. The sample reaction barrier helps form the volume of the reaction chamber. In the walls of the sample reaction barrier are grooves that promote the capillary flow of the sample into the reaction chamber and in doing so, the fluid moves from higher capillarity of the sample reaction barrier to the lower capillarity of the reaction chamber. The sample enters and fills the reaction chamber and the fluid flow is delayed at the area of the time gate. The flow delays so that the analytes in the sample can react with the fluorescent antibody conjugates in the reaction chamber to form a reaction mixture. The time gate substantially delays fluid flow through the protein chip because the surface of the time gate is hydrophobic. Proteins from the sample bind to the hydrophobic surface and make it hydrophilic. The rate of binding of the proteins and the relative hydrophobicity of the surface dictate the incubation time of the reaction mixture in the reaction chamber. This time is generally around 2 minutes, but can be varied by changing the relative hydrophobicity of the surface. The reaction mixture then flows from the time gate into the protein array. The change of the surface of the time gate to a hydrophilic surface allows an unimpeded flow of sample down the protein array. On the surface of the protein array are antibodies immobilized in discrete zones (detection zone antibodies). Up to 100 discrete zones have been immobilized on the protein chip. The analyte bound to the fluorescent antibody conjugate binds to the respective zone as the reaction mixture flows through the protein array. The excess plasma that was added to the protein chip acts as a wash to remove unbound fluorescent conjugates from the protein array. The wash solution fills the perimeter of the protein chip in the capillary of the used reagent reservoir. The fluorescence of the analyte zones on the protein array is directly proportional to the analyte concentrations.

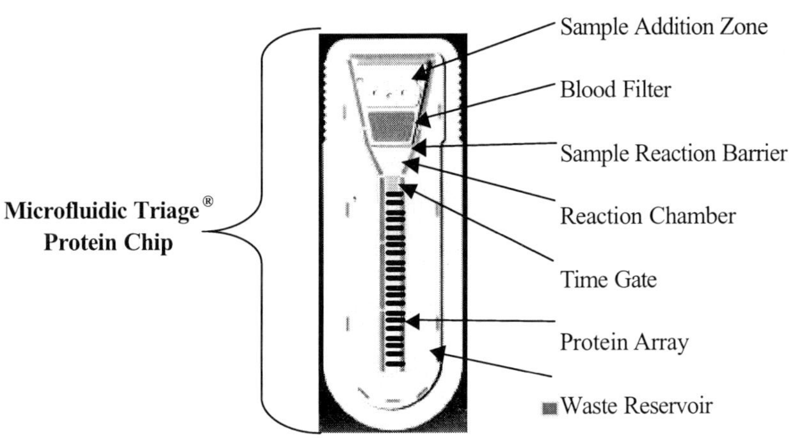

Figure 1.

The surface texture and the relatively high capillarity of the sample reaction barrier assure that sample will predominantly fill the capillary uniformly prior to the sample entering the reaction chamber. The texture in the reaction chamber promotes uniform drying of the assay reagents during manufacture and aids in the filling of the reaction chamber with sample. The grooves at the time gate direct the fluid flow front to be essentially perpendicular to the direction of fluid flow. This minimizes the formation of air bubbles in the time gate capillary and promotes consistent time gate times. The texture of the protein array and the used reagent reservoir is designed to promote uniform fluid flow through the microcapillary where the fluorescent latex is captured at discrete zones on the protein array.

Triage® Meter

The Triage® Meter is a portable, battery powered fluorometer. The fluorescent label utilizes fluorescence energy transfer to achieve a 90nm Stokes shift. The fluorescent dyes are derivatives of phthalocyanine dyes. The excitation wavelength of 670nm of the donor dye coincides with the output of the laser diode in the instrument. The detector is a silicon photodiode that measures the 760nm emission light of the acceptor dye with a quantum efficiency of about 85%. A motor in the meter drives the protein chip under the optic block for the measurement of fluorescence at the discrete assay zones. The fluorescence measurement is converted into an electrical signal that is then transformed into an analyte concentration using the calibration curve stored in the meter memory. The calibration curves and the meter software are downloaded into the instrument memory using a small code chip (EE prom). The assay results are displayed on a LCD, printed, or uploaded to a computer.

3. Conclusions

The Triage® protein chip functions as a microfluidic immunoassay device for performing quantitative immunoassays for the point of care market. The system, comprised of a portable meter and a protein chip, meets the needs of emergency room physicians as an aid in the diagnosis of myocardial infarction, congestive heart failure and drugs of abuse overdose. The Triage® protein chip passively performs simultaneous immunoassasys from whole blood and uses a textured surface to control fluid flow in microcapillaries. The Triage® Meter, which is a portable fluorometer, quantifies the concentration of the analytes in the blood sample.

A LAB-ON-A-CHIP SYSTEM FOR AUTOMATED PROTEIN SIZING

Luc Bousse, Stephane Mouradian, Wally Ausserer, and Robert Dubrow
Caliper Technologies Corp., 605 Fairchild Drive, Mountain View, CA 94043

Abstract

We describe a lab-on-a-chip system capable of carrying out all the steps neeeded for a complete assay, namely SDS protein sizing. Nanoliter-level sample access is followed by on-chip dilution, marker addition, dye staining, separation, destaining, and detection.

Keywords: Lab on a chip, protein separation, automated sample access

1. Introduction

The promise of Lab-on-a-Chip technology is to automate on a microscale all the steps necessary to perform a complex assay, including all the liquid handling operations. The desired result is walk-away automation, namely a system that analyzes many samples, stored in a convenient format such as microtiter plates, without user intervention.

We describe here a system that accomplishes all of these goals for one of the most commonly performed biological assays, SDS protein sizing. The current method for performing this assay on slab gels is slow (4 to 12 hours) and entirely manual. One of the key requirements of a true Lab-on-a-Chip system is to automate sample access at the nanoliter scale, as opposed to most electrophoretic microchips used today.

2. System Description

The protein sizing assay described here automates and integrates on one chip the following steps, starting with an array of denatured protein samples on a microtiter plate.

1. Nanoliter-scale sample access using a short (25 mm) fused-silica capillary integrated with the chip, as described in [1]. This is done by applying a certain amount of vacuum on one of the wells.
2. On-chip sample dilution and addition of size markers.
3. Electrophoretic injection of a sample plug into the separation gel.
4. Non-covalently labeling the sample plug with a fluorescent dye.
5. Separation by size in the separation gel, which is a polymeric sieving matrix containing 9 mM SDS.

J.M. Ramsey and A. van den Berg (eds.), Micro Total Analysis Systems 2001, 45–47.

6. SDS dilution, which is similar to destaining of a conventional gel. This step increases signal amplitude and reduces the background fluorescence caused by dye/SDS complexes (see Ref. [2]).
7. Fluorescent detection, using a red laser.
8. Data interpretation by software (peak areas, sizes, and concentration).

Each of these steps has an equivalent in the conventional method, which must be carried out manually and individually.

3. Chip Description

The chip design that carries out all of these steps is shown in Figure 1. The channels shown in blue are vacuum driven, and sip sample through the capillary, dilute it at the point where it enters the chip, and fill the remainder of the channel to the vacuum well (steps 1 to 2). The channels in red are much shallower and filled with a sieving matrix, so that electrophoretic movement in those channels can occur without hindrance from the hydrodynamic flow and pressure in the blue channels. At the injection cross-section, a sample plug is created and immediately dye-labeled as it enters the separation column (steps 3 and 4). After separation (step 5), at the destain intersection, dilution from the destain wells without dye or SDS occurs, typically in a 10:1 ratio (step 6). A short distance away from the destain intersection, the fluorescent signal is detected (step 7)

4. Results and Discussion

Figure 2 shows an example of the separation of a protein ladder, with high signal-to noise ratio, and high separation efficiency (over 10^5 theoretical plates), in less than 50s. We have run this assay automatically on all the wells of a microtiter plate, with results similar to what is shown here for a single well. A portion of such a data set displayed in a gel-like format can be seen in Figure 3. The first sample contains the protein ladder, and the next 12 samples are replicates of a mixture of BSA and carbonic anhydrase.

Acknowledgements

The authors thank Mac McReynolds and Abdel Minalla for their assistance.

References

1. A. Chow *et al.,, High-Throughput Screening on Microchips*, Proc. Micro Total Analysis Systems 2000, Kluwer Academic Publisher, Dordrecht, The Netherlands, pp. 489-492, (2000).
2. L. Bousse, S. Mouradian, A. Minalla, H. Yee, K. Williams, and R. Dubrow, *Protein sizing on a microchip*, Anal. Chem., **73** (2001) 1207-1212.

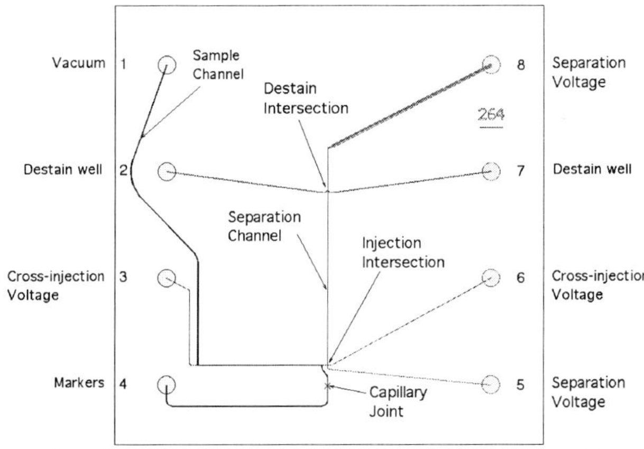

Figure 1. Schematic of chip used for the automated SDS protein sizing assay. The capillary is 25 mm long, and enters the chip at right angles at the indicated location.

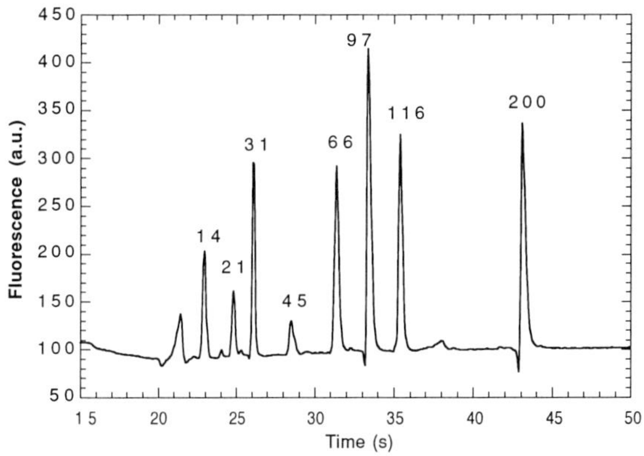

Figure 2. Example of a separation of a BioRad protein ladder in this chip.

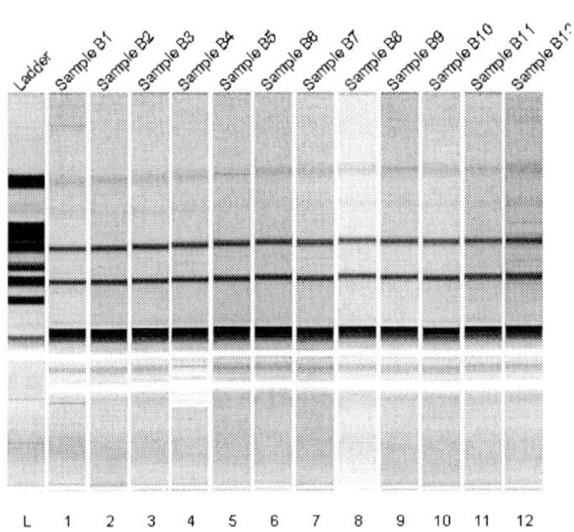

Figure 3. Gel-view display of 12 samples from a microtiter plate.

CHROMATOGRAPHIC SEPARATION OF PROTEINS ON A PDMS-POLYMER CHIP BY PRESSURE FLOW

Minoru Seki[1, #], Masumi Yamada[1], Ryutaro Ezaki[1], Ryusuke Aoyama[1] and Jong Wook Hong[2]

[1]Deparment of Chemistry and Biotechnology, Graduate School of Engineering, The University of Tokyo, 7-3-1 Hongo, Bunkyo-ku, Tokyo 113-8656, Japan

[2]Institute of Industrial Science, University of Tokyo, 4-6-1 Komaba, Meguro-ku, Tokyo 153-8505, Japan, (#e-mail: mseki@chembio.t.u-tokyo.ac.jp)

Abstract

We have developed an ion-exchange chromatography system on a polydimethylsiloxane (PDMS) microchip. PDMS is a useful material for its optical transparency and its simple fabrication and sealing method. We have devised the simple fabrication step for a three-dimensional microfluidic channel packed with mono-sized beads for anion-exchange, that was made by molding PDMS against two masters. The system enables the efficient separation of a mixture of proteins by ion-exchange chromatography without the effects of electrokinetic interaction, so that it is suitable for most analytes, even polar compounds.

Keywords: chromatography, protein-separation, polymer-chip, pressure-flow

1. Introduction

Although capillary chromatography on a microchip has been studied by many researchers, in the most studies the chip substrates were limited to quartz or glass, which needs time- and labour-consuming fabrication and binding processes. Moreover, these studies mostly used electro-osmotic flow, so that the applications of chromatographic separation were suitable for analyses of neutral or less polar compounds. Therefore, we have developed a PDMS-polymer microchip for chromatographic separation of proteins using pressure-driven flows.

2. Experimental

Chip fabrication

A layout of the basic chip design is shown in Figure 1. The basic chip consists of a cross channel with 200 μm in width and 180 μm in depth for sample introduction and chromatographic separation, and four ports for sample and buffer introduction and the waste outlets. At the end of the separation channel, the frit structure was fabricated to keep the ion-exchange beads in the channel. The commercial anion-exchange beads (Source 30Q, Pharmacia) with 30 μm in diameter were packed in the channel forming a 3-mm packed bed. The chip is fabricated mainly using PDMS (polydimethylsiloxane)

J.M. Ramsey and A. van den Berg (eds.), Micro Total Analysis Systems 2001, 48–50.

polymer by the replica molding method as shown in Figure 2. A PDMS plate with a main cross channel with a short gap for the frit structure was permanently bonded with the other PDMS plate with four access holes and a short channel with a 17-μm depth that covers the frit structure. The ion exchange beads could be physically retained by the narrow microchannel structure with a 17-μm height, which is much smaller than the beads diameter.

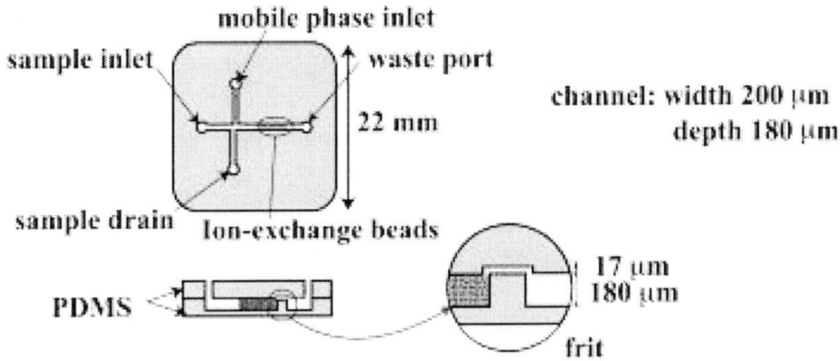

Figure 1. Layout of the PDMS microchip for chromatographic protein separation.

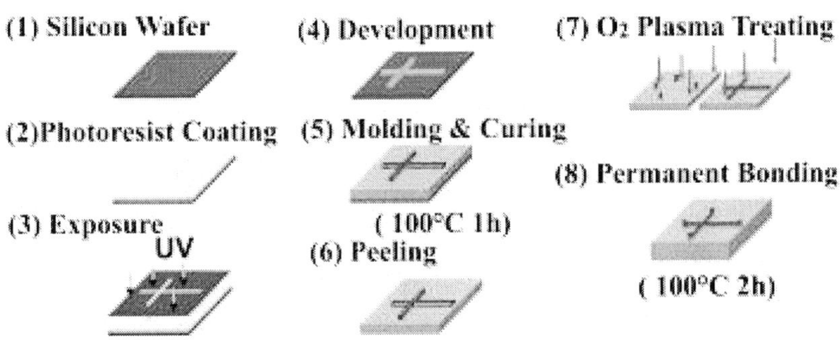

Figure 2. The fabrication process of the PDMS microchip.

Sample injection and separation

In our study, the sample solution and a mobile phase were driven only by a pressure flow with syringe pumps. The sample solution was injected by the gating method, in which the sample drain port was closed for a short instance and reopened.

Surface treatment

PDMS has a relatively hydrophobic surface with a water contact angle of about 110 degree, which affects the adsorption property of proteins on a channel wall of the PDMS chip. The surface of the channel wall was hydrophilized by a concentrated HCl solution to prevent the non-specific adsorption of proteins on the wall. Both contact angle of water on the PDMS and the protein adsorption amount on the surface were decreased with the increasing time for HCl treatment. Sixty minutes was enough for reducing the protein adsorption amount less than a glass surface.

3. Results and discussion

Protein separation on a PDMS chromatography system

A mixture of two proteins, BSA and human IgG, was successfully separated using the PDMS-based polymer chip by a pressure-driven flow. Protein concentrations were determined by the fluorescence of FITC-protein conjugates using a laser spectroscopy. Adsorbed proteins were eluted by a Tris/HCl buffer (pH. 8.0) containing NaCl of which concentration was changed from 0.05 to 1.2 M. In this microchip chromatography, the theoretical plate number per unit column length was much larger than that in a conventional HPLC system.

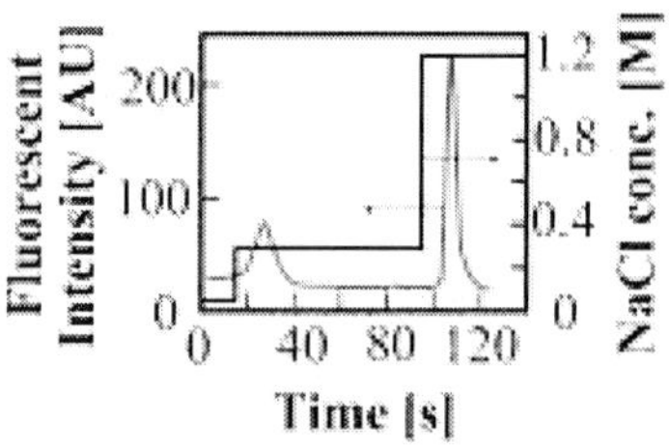

Figure 3. Result of chromatographic separation of the mixture of proteins.

4. Conclusions

Chromatographic protein separation was successfully conducted on a PDMS-polymer microchip with pressure-driven flow. This means the large varieties of conventional HPLC separation mode should be applied directly to the on-chip chromatographic separation of proteins. It is necessary to increase an accuracy of injected sample volume and to decrease the dead volume in a microchannel for reducing the effects of diffusion and distribution of injected sample.

Acknowledgements

This work was supported in part by Grants for Scientific Research on Priority Areas (A) (10145104 & 13025216) from the Ministry of Education, Culture, Sports, Science and Technology in Japan.

References

1. J. W. Hong. T. Fujii, M. Seki, T. Yamamoto and I. Endo, "Integration of gene amplification and capillary gel electrophoresis on a PDMS-glass hybrid microchip," *Electrophoresis*, 22, pp. 328-333, (2001).
2. M. Seki, R. Ezaki, J. W. Hong, T. Fujii and I. Endo, *Ion-Exchange Chromatography on a PDMS Microchip*, Proc. 14th Intl. Symp. on Microscale Separation & Anal. (*HPCE 2001*), p.120 (2001).

MINIATURIZED ISOELECTRIC FOCUSING (μIEF) AS A COMPONENT OF A MULTI-DIMENSIONAL MICROFLUIDIC SYSTEM

A. E. Herr[1], J. I. Molho[1], R. Bharadwaj[1], J. C. Mikkelsen[1], J. G. Santiago[1], T. W. Kenny[1], D. A. Borkholder[2], M. A. Northrup[2]
[1]Department of Mechanical Engineering, Stanford University, CA 94304,USA
[2]Cepheid,1190 Borregas Ave, Sunnyvale, CA USA
aeh@mems.stanford.edu

Abstract

Combining several one-dimensional miniaturized separation techniques in series has the potential to be an exceptionally comprehensive analysis approach. Sample species can be separated and identified based on a battery of characteristics not available from any single technique. μIEF is attractive as a first dimension of such a multi-dimensional system due to the equilibrium nature of the final separation, rapid separation capabilities (1min), concentrating qualities (100x), and high resolution (ΔpI = 0.005). As a step towards integration of μIEF with a second dimension (e.g., miniaturized capillary electrophoresis), this work investigates the robustness of μIEF as an initial dimension and explores means to reliably extract a purified, concentrated sample zone from a complete μIEF assay.

Keywords: isoelectric focusing, multi-dimensional separations, full-field fluorescence detection, electrokinetic extraction

1. Introduction

Sample complexity often necessitates either a reduction in the number of sample components or high separation resolution[1]. This high-resolution goal is applicable to separation techniques regardless of implementation (slab, capillary, or chip-based). Coupling of multiple separation techniques can increase the overall system resolution[1]. This coupling is often referred to as a *multi- or n-dimensional* separation.

Single- and multi-dimensional techniques have been demonstrated in microfluidic devices[2, 3]. Previously, we have demonstrated IEF on a microchip platform (μIEF) in conjunction with a full-field imaging approach[4]. This detection scheme allows for real-time fluorescence data acquisition and, thus, analysis of fluid transport. In this work, we employ this transport information to design and implement a post-μIEF, electrokinetic sample extraction and injection scheme. This scheme shows promise for the sequential coupling of μIEF with other separation mechanisms.

2. Theory

In IEF, a separation channel is filled with a mixture of ampholytes and sample. Application of an axial electric field to such a channel results in the formation of an axial pH gradient and the migration of proteins their respective isoelectric points.

J.M. Ramsey and A. van den Berg (eds.), Micro Total Analysis Systems 2001, 51–53.

A chip-based coupling of IEF and CE would provide sequential, independent, selective displacement mechanisms for each dimension potentially resulting in high resolving power[1]. A chip-based approach would allow for ease of channel coupling and enhanced geometric complexity.

3. Experimental

Imaging System. Full-field imaging of the time-varying fluorescence signal present in the channels was accomplished using a standard epi-fluorescent microscope set-up (Olympus). A 100W mercury lamp was used as the excitation source, while a cooled charge-coupled device (CCD) was employed for detection (Roper Scientific).

Microchip Device. Our microchip platform consisted of embossed acrylic (PMMA) substrates (ACLARA). A t-shaped channel layout was used (short leg: 22mm, long leg: 25.4mm). Channels were 200μm wide and 40μm deep. All channels were filled with sample solution. A custom fixture enabled fluidic porting, as well as electrokinetic and hydraulic control of the system.

Extraction. For sample extraction, electrokinetic flow was used to mobilize the initial μIEF assay past an extraction junction, initiate sample extraction, and further manipulate post-μIEF sample zones. This focusing and extraction system is attractive for miniaturized systems, as no mechanical on-chip actuation is necessary.

4. Results and discussion

The width of the focused zone (50μm at E = 800V/cm, 600μm at E = 100V/cm) and the speed of the initial focusing (1 min) agree well with both steady-state theory and observations from our previous 1D miniaturized system (Figure 1).

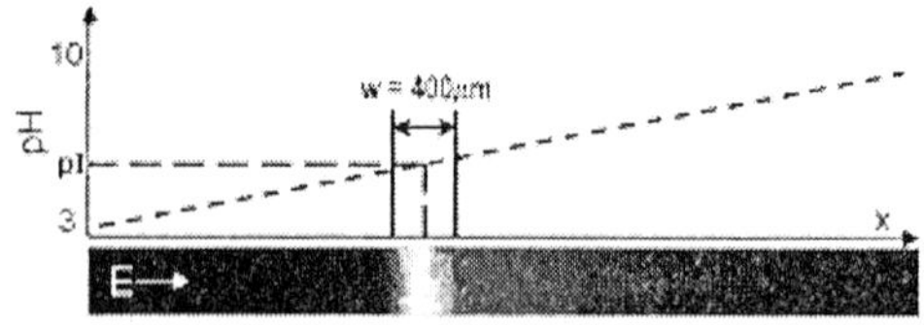

Figure 1. Focusing of GFP in μIEF. Protein that is initially distributed in the column migrates to the pI location. L = 6cm, w = 400μm results in n ~ 150. 190nM GFP, E = 500V/cm.

Electroosmotic flow (EOF) mobilized the zone past the extraction junction. For our experimental conditions, the EOF shows evidence of both suppressed velocities, as compared to EOF in PMMA channels filled with simple buffer solutions and a marked decrease over time, as would be qualitatively expected due to mobilization of the pH gradient.

To assess the robustness of μIEF, we introduced additional sample mass to a post-μIEF protein zone. Figure 2 shows a time sequence of the injection, general refocusing in the μIEF channel, and the two sample plugs finally merging at their common pI. These observations suggest that our implementation of μIEF may be sufficiently insensitive to the fluidic perturbations necessary for sample handling in a 2D system.

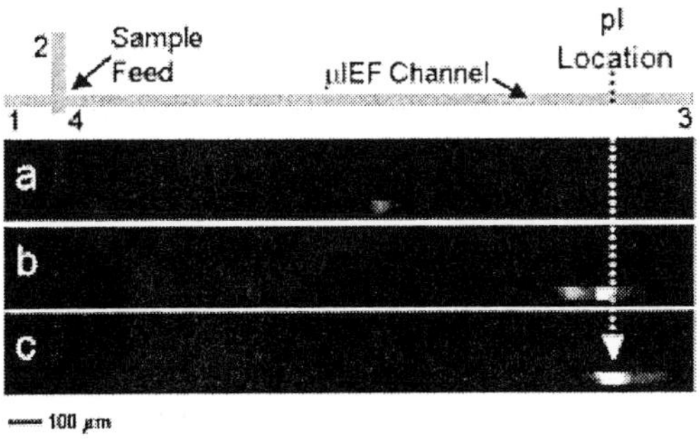

Figure 2. Robustness of µIEF. a. Δt=13s, pressure-driven flow has introduced sample mass from the Sample Feed channel. E_{13}=800V/cm causing the added sample mass to move towards its pI. **b.** Δt=21s, the added sample mass and pre-focused band each continue to move towards their common pI (pI ~ 5.6) **c.** Δt=25s, the zones join.

Electrokinetic extraction of a concentrated zone (concentrated ~130x to ~2.5µM) resulted in a small volume (2nL aliquot) of a single sample available for further analysis (Figure 3). Electrokinetic injection, while not optimised with pull-back, reduced the extraction-induced width broadening to 10% of that observed with pressure-driven extraction.

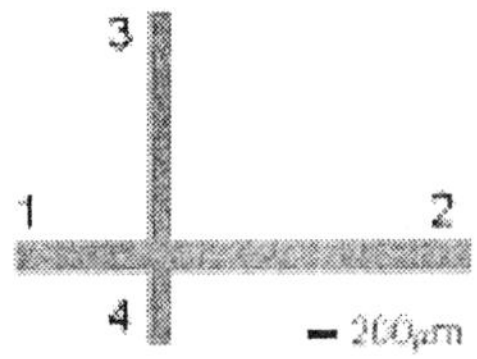

Figure 3. Electrokinetic extraction of focused µIEF zone. a. Sample mobilization U_{mob}= 30µm/s, E_{12} = 235V/cm. **b. Extraction initiation** E_{34} = 310 V/cm. **c.–d. Non-optimised zone extraction** $U_{extract}$ = 450µm/s Δt_c = 13s, Δt_d = 14s. Collection: 2x magnification (NA = 0.08) and 0.3 demagnification.

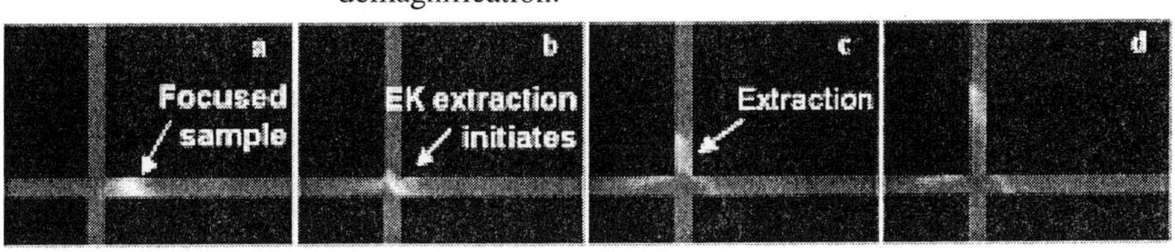

5. Conclusions

A full-field imaging technique was used to enable electrokinetic extraction of sample from a chip-based µIEF assay. Such post-µIEF electrokinetic manipulation suggests sequential coupling of this high resolution, quasi-steady state technique to a second technique (e.g., chip-based electrophoresis) may be feasible. We believe this coupling holds promise for the development of fast, low complexity, ultra-high resolution, chip-based multi-dimensional separation schemes involving µIEF. We are currently integrating µIEF in a chip-based multi-dimensional system.

References

(1) Giddings, J. C. *Anal Chem* **1984**, *56*, 1258-&.

(2) Colyer, C. L., *et al. Electrophoresis ;* **1997**, *18*, 1733-1741.

(3) Rocklin, R. D., *et al. Anal Chem* **2000**, *72*, 5244-5249.

(4) Herr, A. E., *et al.*, Enschede, The Netherlands, May 14-18, 2000 2000; Kluwer Academic Publishers; 367-370.

A QUANTITATIVE NANOSCALE COULTER COUNTER

O. A. Saleh and L. L. Sohn
Dept. of Physics, Princeton University, Princeton, NJ 08544
(oasaleh@princeton.edu)

Abstract

We have fabricated a microchip Coulter counter on a quartz substrate, and have used it to detect individual colloidal particles with a sensitivity proportional to each particle s size. We demonstrate the ability of this device to distinguish colloids whose diameters differ by less than 10%. Furthermore, the device is sensitive to individual molecules of λ-phage DNA, which is a first step towards the application of this technique to the size-based differentiation of single biological macromolecules.

Keywords: On-chip Coulter counter, artificial nanopore, resistive sensing

1. Introduction

We have developed a chip-based device able to detect and differentiate polydisperse solutions of colloids [1], and to detect individual molecules of λ-phage DNA. Our all-electronic detection technique is based on the Coulter method of particle sensing [2], and is comparable to, but simpler than, current optical methods, since little sample preparation (e.g. fluorescent dyes) is needed. Coulter counters typically consist of two reservoirs of particle-laden solution separated by a membrane, and connected by a single pore through that membrane (Fig. 1a). An individual particle passing through the pore causes a change in the pore s electrical resistance; by sensing the magnitude of this resistance change, a Coulter counter is able to measure the size of the particle. By

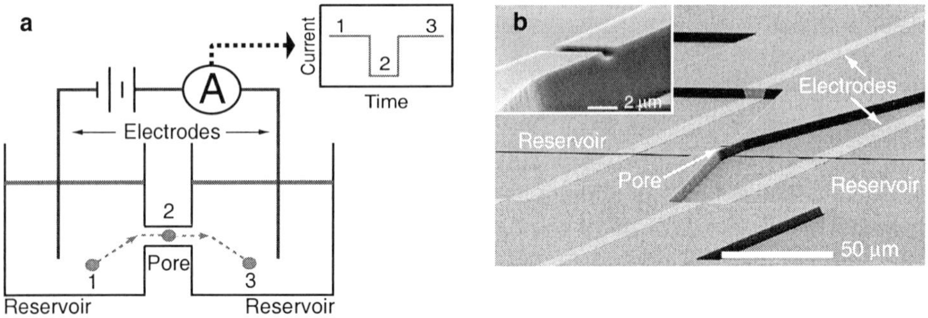

Figure 1. (a) Schematic of a Coulter counter. A particle passing through the pore causes a transient decrease in the electrical current. (b) Our on-chip Coulter counter, which has a pore and reservoirs etched into a quartz substrate.

J.M. Ramsey and A. van den Berg (eds.), Micro Total Analysis Systems 2001, 54–56.

fabricating a Coulter counter on a chip, we are able to dramatically decrease the pore dimensions, thus allowing the measurement of nanoscale particles, most notably biological molecules.

2. Experimental

We have fabricated and tested two types of on-chip Coulter counters using a variety of micro- and nano-fabrication techniques. The first type (Fig. 1b) consists of a quartz substrate into which the reservoirs and pore have been etched as channels. The device is completed by patterning Ti/Pt electrodes in the reservoirs, and sealing the channels with a PDMS-coated coverslip. The second type (not shown) consists of a glass substrate that supports only the Ti/Pt electrodes, and a molded PDMS reservoir and pore. The device is completed by aligning and sealing the PDMS mold to the electrodes. For both types of devices, the pore diameter can be adjusted from <100 nm to 1 μm, and the pore length from 1 to 10 μm.

3. Results and discussion

Fig. 2a shows the normalized current recorded vs. time on a device containing a solution of colloids of four different diameters; as can be seen, the size of the current pulse clearly depends on the size of the colloid passing through the pore. This is further emphasized in Fig. 2b, which is a histogram of over 3000 pulse heights recorded in this

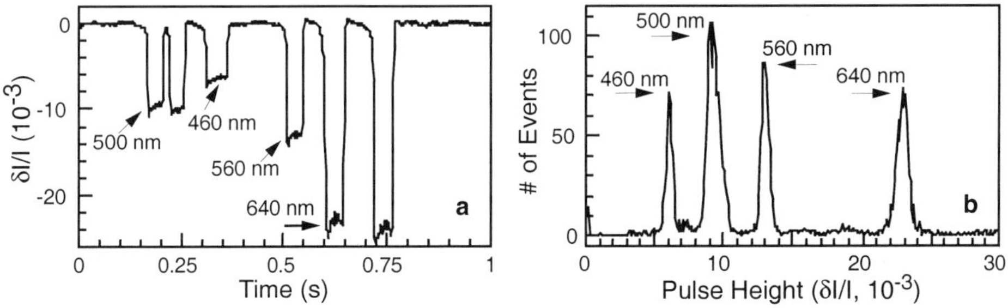

Figure 2. (a) Relative changes in current vs. time measured using a pore of size 9.5 μm by 1.2 μm^2 on a polydisperse colloidal solution. Each downward pulse represents a single colloid (diameter labeled) passing through the pore, and the height of each pulse reflects the colloid s diameter. (b) Histogram of ~3000 pulse heights measured using the same solution and pore as (a). From this histogram, we estimate the device s resolution to be –10 nm for a ~500 nm diameter colloid.

experiment. Using only the measured pore size, upper and lower bounds on the expected pulse height as a function of colloid diameter can be calculated [3,4]. In Fig. 3,

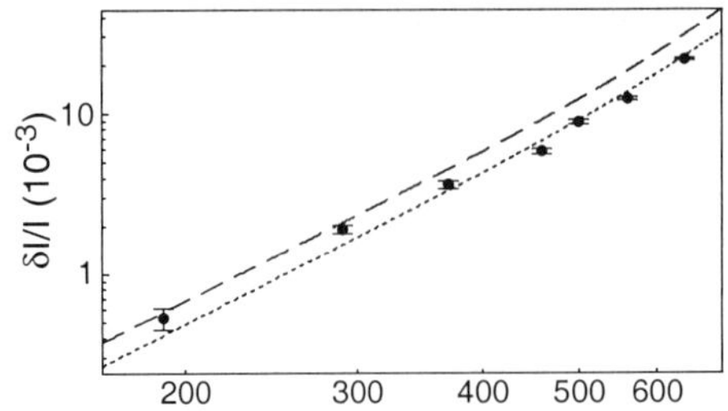

Figure 3. Measured pulse heights for several different colloid diameters (filled circles) plotted with pulse heights predicted by upper and lower bound theories [3,4].

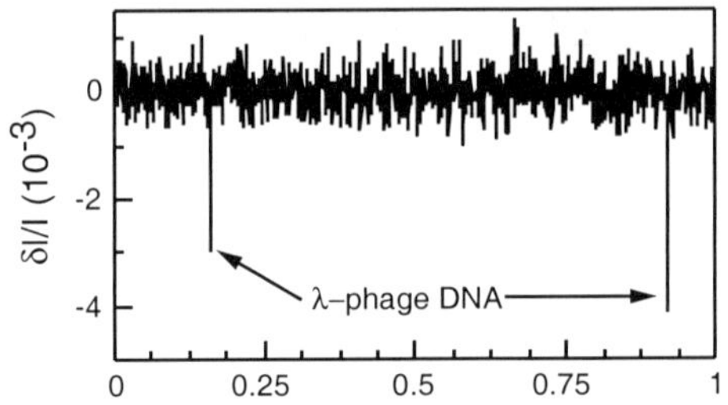

Figure 4. Relative current measured vs. time for a pore of size 4 μm by 0.16 μm^2 on a solution of λ-phage DNA. Each downward spike corresponds to a single DNA molecule passing through the pore.

we show that there is excellent agreement between the calculated and measured pulse heights.

Fig. 4 shows data taken using a smaller pore on a solution of λ-phage DNA. As each DNA molecule passes through the pore, it partially blocks the electrical current in the same manner as the colloids, causing the downward spikes shown here. This is a significant result, as it indicates the possibility of using our device to characterize solutions of DNA, perhaps through rough measurements of size (in a manner similar to the colloid data) or concentration.

4. Conclusions

We have demonstrated the successful fabrication and quantitative testing of an on-chip Coulter counter. Furthermore, we have shown that the device is sensitive to large DNA molecules. We expect that further miniaturization of the pore will increase the device s sensitivity to macromolecules, thus enabling applications to size-based fractionation of DNA or large proteins.

Acknowledgements

O. A. S. acknowledges support from the Fannie and John Hertz Foundation. This work was supported in part by DARPA and an NSF CAREER award.

References

1. O. A. Saleh and L. L. Sohn, Submitted to Rev. Sci. Instrum.
2. W. H. Coulter, U. S. Patent No. 2,656,508, issued 20 Oct. 1953
3. R. W. DeBlois and C. P. Bean, Rev. Sci. Instrum. **41** (7), 909 (1970)
4. E. C. Gregg and K. D. Steidley, Biophys. J. **5**, 393 (1965)

ELECTROKINETIC TRANSPORT THROUGH NANOMETER DEEP CHANNELS

Stephen C. Jacobson, Jean Pierre Alarie, and J. Michael Ramsey
Oak Ridge National Laboratory, Oak Ridge, Tennessee 37831-6142 USA

Abstract

Electroosmotic flow was evaluated in fluidic devices with channel depths of 98 nm, 300 nm, and 10.4 μm. At each channel depth electric double layer thicknesses of 21.5, 6.80, 2.15, and 0.78 nm were examined. For the 21.5 nm double layer thickness, a 35% decrease in the electroosmotic mobility was observed with the 98 nm deep channel relative to the 300 nm and 10.4 μm channel depths. These results were in good agreement with theoretical predictions.

Keywords: nanochannel, electrokinetic transport, double layer thickness

1. Introduction

A common method for moving materials through the channels of microfabricated fluidic devices (microchips) is electrokinetic transport. When the channel dimensions are large, e.g. 10-100 μm, the electric double layer thickness is negligible compared to the channel dimensions, and the flow profile is planar normal to the axis of flow. However, when channel dimensions begin to approach the double layer thickness, flow rates are expected to decrease with cross-sectional area in a non-linear fashion [1, 2]. To test this theory, we fabricated devices with channels that were confined to the nanometer range in one dimension (depth), measured the electroosmotic mobilities over a range of double layer thicknesses, and compared the results to theory.

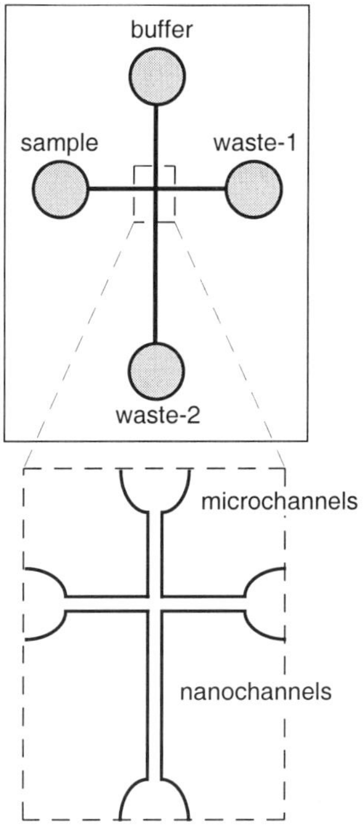

Figure 1. Schematic of the chip with nanochannels.

2. Experimental

Figure 1 shows a schematic of the fluidic chip with nano- and microchannels. The channels were fabricated using standard photolithographic and wet chemical etching techniques. The substrates with nanometer deep channels were initially etched to ≈100 or 300 nm deep. The nanochannels including

J.M. Ramsey and A. van den Berg (eds.), Micro Total Analysis Systems 2001, 57–59.

the cross intersection were then coated with photoresist to prevent further etching, and the uncoated portions of the channels were etched to 5 μm deep. The 10.4 μm deep channels were etched in a single step. Cover plates were then thermally bonded to the substrates

The double layer thickness was controlled by changing the buffer concentration. Sodium tetraborate buffers of 0.20, 2.0, 20, and 150 mM with 50% (v/v) methanol were used and corresponded to calculated double layer thicknesses (1/κ) of 21.5, 6.80, 2.15, and 0.78 nm, respectively [3]. The electro-osmotic mobilities were measured by injecting a neutral dye, rhodamine B, and monitoring the arrival time 1 mm downstream from the injection cross by confocal fluorescence detection.

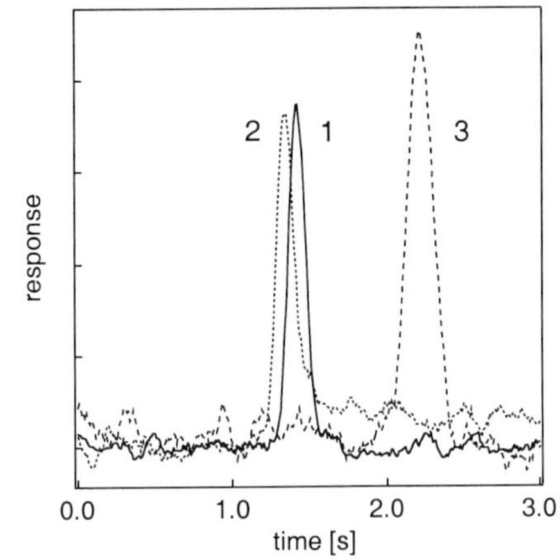

Figure 2. Temporal profiles measured with the 98 nm deep channel for (1) 21.5, (2) 6.80, and (3) 2.15 nm double layer thicknesses. The electric field strength

3. Results and Discussion

The electroosmotic mobilities were measured in three microchips with channel depths of 98 nm, 300 nm, and 10.4 μm. The 10.4 μm channel served as a control because the double layer thickness was assumed to be negligible relative to the channel depth. Figure 2 shows a series of temporal profiles obtained with the 98 nm deep channels for double layer thicknesses of 21.5, 6.80, and 2.15 nm. The electroosmotic mobility was expected to increase with increasing double layer thickness (decreasing buffer concentration). However, when the double layer thickness extends significantly into the channel, the electroosmotic flow decreases. This is seen in Figure 2 where profile 1 eluted after profile 2, and the 21.5 nm double layer was 44% of the 49 nm half-depth of the channel. Figure 3 shows the variation of the electroosmotic mobilities with the reciprocal of the double layer thickness (κ). For double layer thicknesses ≤ 6.80 nm, the electroosmotic mobility was independent of channel depth. For the 21.5 nm double layer thickness, the electroosmotic mobility in the 98 nm deep channels was 35% lower than the electroosmotic mobility in the 300 nm and 10.4 μm deep channels.

Equation 1 was used to calculate the electroosmotic mobility for different channel depths and double layer thicknesses [2].

$$\frac{u_{eo}}{E}(y) = \frac{\varepsilon\varepsilon_0\zeta}{\eta}\left[1 - \frac{\cosh[\kappa(h-y)]}{\cosh(\kappa h)}\right] \tag{1}$$

where u_{eo}, E, h, y, ζ, η, ε, and ε_0 are the electroosmotic velocity, electric field strength, 1/2 channel height, distance from channel wall, zeta potential, viscosity, dielectric constant, and permittivity of vacuum, respectively. The zeta potentials at different double layer thicknesses were calculated from a fit to the experimental values obtained using the 10.4 μm deep channel. The dielectric constant was 59.24 [4], and the viscosity 168 mPa·s. The calculated electroosmotic mobilities are plotted in Figure 3 as the solid curve.

These experiments showed reduced electroosmotic mobilities in fluidic channels where the double layer thickness was a significant fraction of the channel depth. The results compared well with calculated mobilities from Equation 1. However, theory predicted that a slightly greater reduction in the mobility should be observed.

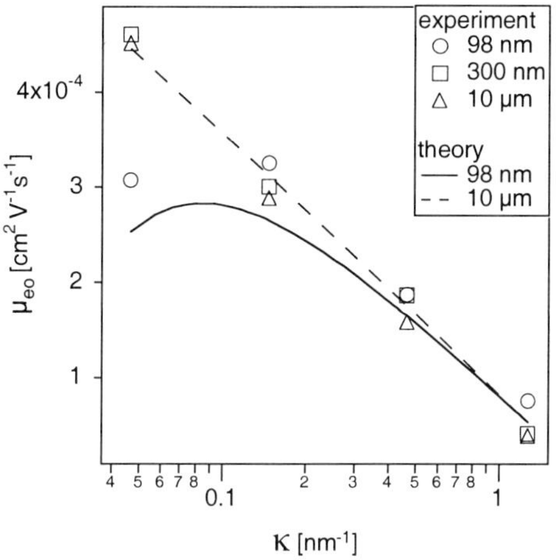

Figure 3. Variation of electroosmotic mobility (μ_{eo}) with the reciprocal of the double layer thickness (κ). Equation 1 was used for the theoretical calculations.

Acknowledgement

Oak Ridge National Laboratory is managed and operated by UT-Battelle, LLC for the U.S. Department of Energy under contract DE-AC05-00OR22725.

References

1. C.L. Rice, R. Whitehead, *J. Phys. Chem.* 1965, *69*, 417-424.
2. S. Levine, J.R. Marriott, K. Robinson, *J. Chem. Soc. Farad. Trans. 2* 1975, *71*, 1-11.
3. A.J. Bard, L.R. Faulkner, *Electrochemical Methods*, John Wiley and Sons, New York, 1980.
4. C. Wohlfarth, O. Madelung, *Static Dielectric Constants of Pure Liquids and Binary Liquid Mixtures*, Springer-Verlag, New York, 1991.

Three-Dimensional Fluidic Architectures Using Nanofluidic Diodes to Control Transport between Microfluidic Channels in Microelectromechanical Devices

Tzu-Chi Kuo, Donald M. Cannon, Jr., Wenju Feng, Mark A. Shannon, Jonathan V. Sweedler, and Paul W. Bohn

Department of Chemistry, Department of Mechanical Engineering, and Beckman Institute for Advanced Science and Technology, University of Illinois at Urbana-Champaign, 600 S. Mathews Ave., Urbana, Illinois 61801.

Abstract

Nuclear track-etched membranes are used as gateable interconnects between vertically separated microfluidic channels to achieve three-dimensional fluidic architectures. Transport through the membranes may be electrically gated to control the transfer of material between channels in an externally controllable fashion. Because unidirectional flow may be gated on or off with the application of the appropriate bias conditions, these structures are likened to electronic diodes.

Keywords: molecular gates, interconnects, nanofluidics, nuclear-track etched membrane

Introduction

Microfluidic devices occupy an increasingly significant position in chemical and biochemical sensing, molecular separations, drug delivery and other forefront technologies. As powerful as currently available μTAS technologies have become, however, they are primarily two-dimensional. Extension of fluidic devices to three-dimensions opens new vistas for applications and parallels the massively three-dimensional architectures characteristic of electronic devices. Externally controllable interconnects, employing nuclear track-etched polycarbonate membranes with nanometer diameter pores, are described that produce hybrid three-dimensional fluidic architectures.

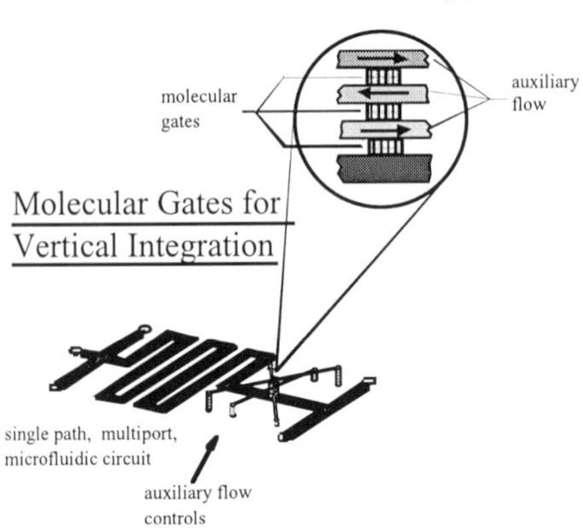

Figure 1: Schematic diagram of a microfluidic structure with three stacked fluidic layers and associated molecular gates. By applying appropriate biases, analytes can be directed through each of the gate structures.

J.M. Ramsey and A. van den Berg (eds.), Micro Total Analysis Systems 2001, 60–62.

Experimental

Crossed microfluidic channels are fabricated in polydimethylsiloxane, PDMS, in planes above and below a 5 μm-thick nanoporous membrane, *cf.* Fig. 1. The membrane is a nuclear track-etched polycarbonate film with a nearly monodisperse distribution of cylindrical channels with diameter 15 nm $< d <$ 220 nm and pore densities 1×10^8 cm^{-2} $< N_p < 6 \times 10^8$ cm^{-2}. Channels are fabricated using standard rapid prototyping protocols for PDMS as explained in ref. 1. Channels are fabricated either by sandwiching the membrane between two channels manufactured using standard rapid prototyping protocols for PDMS as explained in reference 1, or chemically bonding the membrane to silica channels as described in ref. 2. Channels are 100-μm wide and 30 μm deep. Assembly is accomplished by placing a 10 mm x 1 mm section of membrane centered on the bottom PDMS channel, then placing the top PDMS channel at 90° to the bottom channel within the area of the membrane.

Results and Discussion

Using nanofluidic structures to connect microfluidic channels allows a variety of flow control concepts to be implemented, leading to hybrid fluidic architectures of considerable power and versatility. The key distinguishing characteristic feature of nanofluidic channels is that fluid flow occurs in structures of the same size as physical parameters that govern the flow, *viz.* Fig. 2. Ion migration, electroosmosis and diffusion all contribute to transport under applied bias. Diffusion dominates under field-free conditions with surface hydrophobicity controlling solvent access to the nanochannels, and hence the magnitude of transport by diffusion. In low ionic strength solutions (μ ~ 10 mM), electroosmosis dominates transport when the membranes are biased, and the charge state of the surface determines the direction of flow. At high ionic strength (μ ~ 1 M) ion migration dominates in hydrophobic membranes, and diffusion is controlling in hydrophilic membranes. The magnitude and polarity of the interior surface charge is controlled by surface functionality and exhibits a large impact on molecular transport. Transport can be fine-tuned by adjusting pH under low ionic strength conditions in either type of membrane. Increasing the

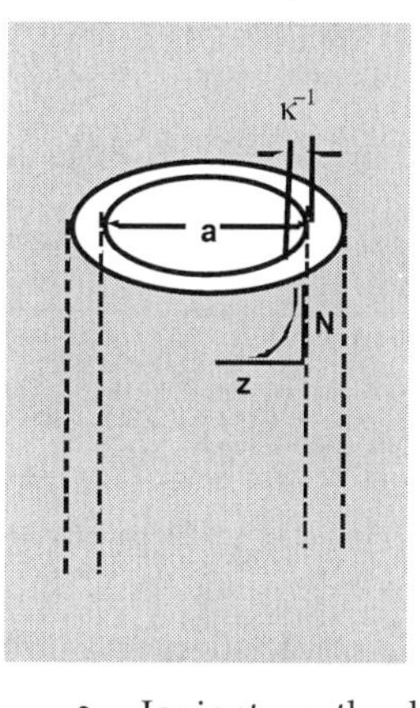

Electroosmotic Flow at Nanometer Dimensions

$$\kappa = \sqrt{\frac{8\pi e^2 \sum_i n_i z_i^2}{\varepsilon kT}}$$

- Ionic strength adjusts κa
- At κa << 1 electroosmotic flow dominates
- At κa >> 1 ion migration dominates

Figure 2: Schematic representation of a single nanopore in a molecular gate with dimensions on the order of 15 - 220 nm.

surface charge density, σ_s, enhances the mobile counterion concentration, increasing the electroosmotically-driven flux. Comparisons of behavior under different conditions are understood by reference to the product, κa, of the inverse Debye length, κ, and the pore diameter, a, *cf.* Fig. 2.

Furthermore, the separations capacity factor, k', governed by the surface-to-volume ratio, can be quite large. For example, k' increases by ~120 when a 200 nm i.d. nanopore with a 10 nm thick coating is compared with a 20 μm i.d. wall-coated open tubular column with the same coating. These nanofluidic interconnects can be thought of as fluidic diodes, albeit with a much richer array of parameters to control biasing, where forward/reverse bias is controlled by applied potential, surface charge density (pH controllable), ionic strength, and even by the characteristics of the fluidic network in which the interconnect is placed. Fig. 3 demonstrates an example of diode-like behavior achieved with a 220 nm PVPF membrane using fluorescein as the model analyte. The magnitude of the forward-bias current is governed by the surface charge density on interior walls of the nanofluidic channel.[3]

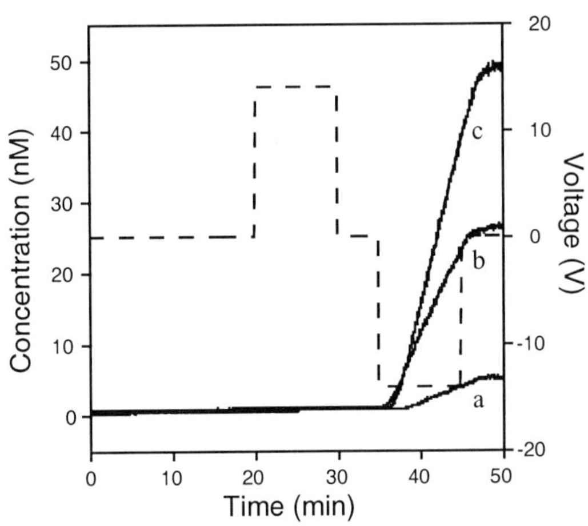

Figure 3. Field-induced permeation of Fl^{2-} in a 220 nm pore diameter PVPF membrane in 5 mM PBS at pH 7 (a), 8 (b), and 11 (c). The applied dc offset potential program is referenced to the right ordinate. Fl^{2-} concentration in the receiving compartment is referenced to the left ordinate.

Conclusions

The use of nuclear-track etched membranes makes possible the intelligent control of fluidic transport between vertically stacked microfluidic layers and thus constitutes an important enabling step in the drive to fabricate functional three-dimensional fluidic architectures.

References

1. J. C. McDonald, *et al.*, *Electrophoresis* **21**, 27-40 (2000).
2. M.A. Shannon *et al.*, 2001 μTAS Abstracts (2001).
3. T. -C. Kuo, L. A. Sloan, J. V. Sweedler, P. W. Bohn, *Langmuir*, submitted (2001).

SYSTEM DESIGN OF TWO DIMENSIONAL MICROCHIP SEPARATION DEVICES

John C. Harley[1], Richard F. Day[1], John R. Gilbert[1], Manish Deshpande[1] J. Michael Ramsey[2] and Stephen C. Jacobson[2]
[1]Coventor, 1820 Gateway Dr., Ste. 330, San Mateo, CA 94404
[2]Oak Ridge National Laboratory, P. O. Box 2008, Oak Ridge, TN, 37831-6142

Abstract

Two dimensional (2D) separation systems have been used extensively for the analysis of complex protein and peptide mixtures because of the increased peak capacity they provide relative to one-dimensional separations [1,2,3,4,5]. The effects of miniaturization of various separation schemes have been reported by Manz et al [6]. In general, miniaturization has been shown to provide high separation efficiencies and a convenient means of manipulating extremely small sample volumes. Recently, microscale 2D systems using micellar electrokinetic chromatography (MEKC) with capillary electrophoresis (CE) and open channel electrochromatography (OCEC) with CE have been reported [7,8]. In this paper, we will demonstrate the use of CAD software for predictive design of 2D microchips, using both detailed simulation for optimization of electrophoretic injections and turn designs, as well as reduced order modeling to better understand system performance trade-offs. Modeling results will be compared against experimental observations by Ramsey et al [7, 8].

Keywords: CAD, Separation, System Design, Two Dimensional

1. Introduction

Giddings [9] developed the concept of peak capacity, n_c, as a means of comparing the maximum number of components that could be resolved in a separation. Provided the separation modalities are orthogonal in a 2D system, the peak capacity of the system is approximately the product of the peak capacities of the individual separations: $n_c = n_{c1} \times n_{c2}$, where n_{c1} and n_{c2} are the peak capacities of the first and second dimensions, respectively. This multiplicative nature of the peak capacities of the individual separation modes can provide very large overall system peak capacities. However, in order to realize large n_c values and maintain system resolution, the effective design of 2D separation systems involves optimization of a number of interdependent parameters such as the relative separation times of the individual dimensions, the area or "footprint" of the device, the overall analysis time, and the process of transferal of analyte from the first to the second dimension.

While Murphy et al [10] have investigated the effect of peak broadening due to finite volume sampling from the first dimension into the second dimension, to the authors' knowledge, no comprehensive analysis of system level design trade-offs for 2D microchips has been presented. Optimization of system peak capacity can best be achieved through a design analysis of the 2D system and its key components as described herein.

J.M. Ramsey and A. van den Berg (eds.), Micro Total Analysis Systems 2001, 63–65.

2. Methodology

The design methodology described here begins with a reduced order model of the desired system. This model for the 2D system of [7] (layout shown in Figure 1) is used to examine design trade-offs, such as length of the first dimension vs. the second dimension, the effect of bends and injections, and injection duration and rate for each column.

For each species in a separation, a separation mobility, ν, can be defined. ν is the property of a species that responds to the driving modality of the separation in question. For instance, for CE, ν equals the electrophoretic mobility of the species, and for MEKC, ν is related to the hydrophobicity of the species. In a given sample, there will be a range of separation mobilities and diffusivities, *D,* of the subcomponents of the sample. Those species with the highest ν and lowest *D* will elute first and with the smallest peak widths, while those with the lowest ν and highest *D* will elute last and with the broadest peaks. Thus, in the ν-*D* space of the analytes, there is generally a region of analytic interest, for which the separation device must be able to resolve all subcomponents (Figure 2). For a 2D system, this is further complicated by the fact that the separation mobilities will certainly be different in the two dimensions, and diffusivities may also be different, so there are in fact two separate ν-*D* spaces to consider. Full experimental optimization of the device over the required intersection of the 2 ν-*D* spaces, while maintaining the minimal system footprint and analysis time, is nearly impossible due to the large parameter space involved. However, numerical tools, especially system simulation tools, can be effectively employed for this purpose.

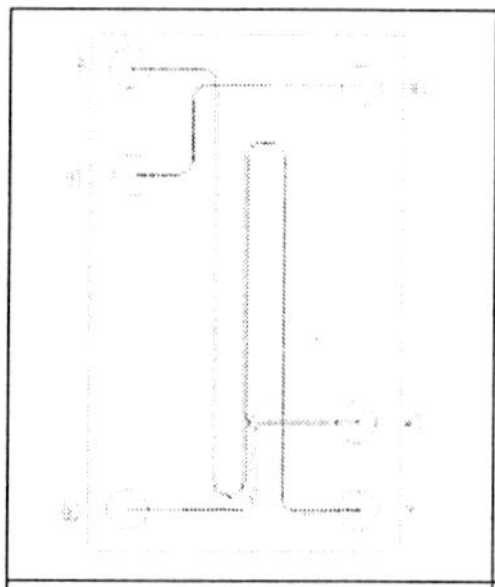

Figure 1: Layout of Rocklin, Ramsey, and Ramsey MEKC/CE 2D system [7]

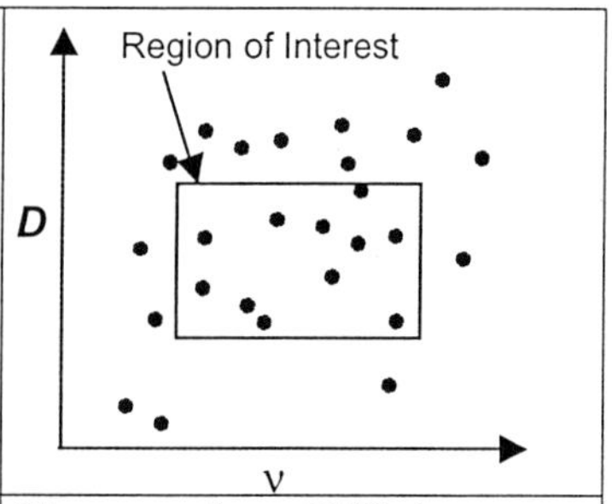

Figure 2: Diffusion vs. separation mobility plot for subspecies in an injected sample, showing a region of interest.

3. Results

Figure 3 shows a separation performance plot for the 2D system of [7]. Here, a desired minimum separation in ν/*D* is chosen (here ν/*D* = 7). Figure 3 then indicates that for a mobility to diffusivity ratio greater than 60, the system is no longer able to resolve two peaks that are separated by less than 7 in ν/*D*.

From this analysis, it is clear that the injection process from the first to

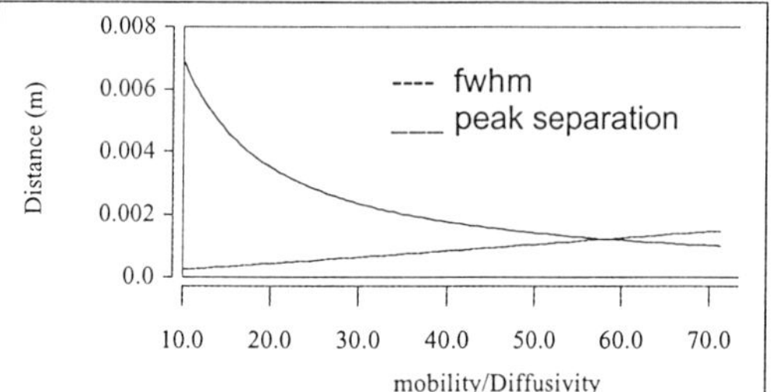

Figure 3: Crossover plot showing peak width (fwhm) and peak separation distance plotted versus separation mobility/diffusivity

second dimension in [7] is sub-optimal. A new approach is suggested where the electric field strength in both dimensions is lowered during the injection. This is modeled in detail to examine the resulting injection profile, and then the effect on system performance is determined using system modeling. Figure 5 indicates that making this change improves the system performance to the point that peaks separated by a ν/*D* ratio of 7 can now be resolved as distinct peaks for ν/*D* ratios of 70 or more.

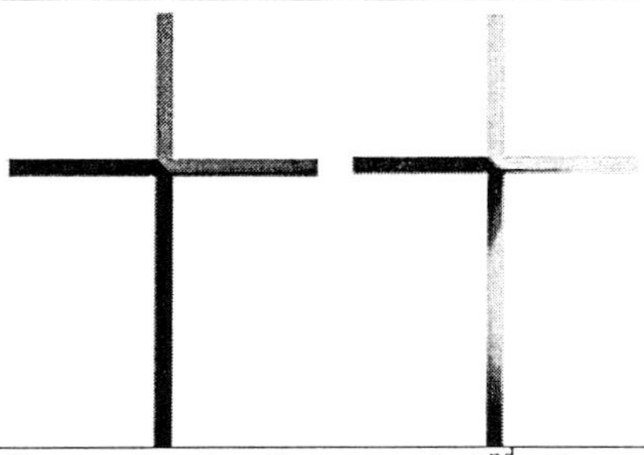

Figure 4a, b: Suggested 2nd dimension injection sequence. Field strength in both columns is reduced during injection to minimize injected peak width. (Note: images rotated 90 degrees from Figure 1)

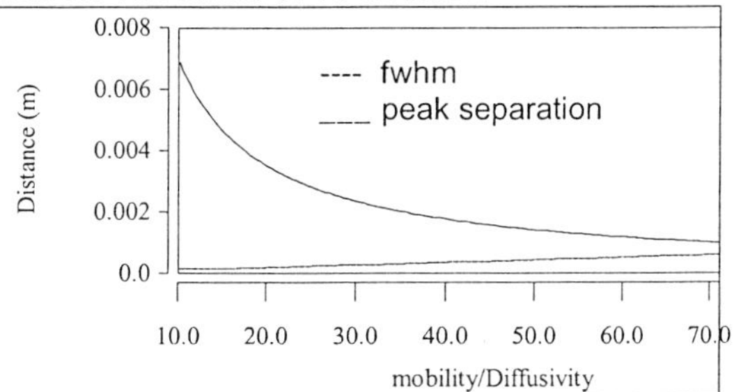

Figure 5: Separation distance and peak width (fwhm) vs separation mobility/Diffusivity. Gain in switch performance (Figure 4) allows for separation of peaks separated by ν/*D*=7 for ν/*D* ratios as large as 70.

4. Conclusion

System modeling is shown to be an effective means of design of 2D separation systems. Because of the interaction between dimensions that occurs, design of these systems without benefit of modeling results in systems that are not fully optimized. Here, the example of improving second dimension injection is presented. The effects on both separate dimensions, as well as the overall result on the system, are then observed. This approach will also yield the required footprint and analysis time of a device in order to obtain a given resolution and peak capacity.

References

1 Giddings, J. C.: *Anal Chem* **56** (1984), 1258A-1270A.

2 Liu, Y., and Sweedler, J. V.: *Anal Chem* **68** (1996) 3928-3933

3 Hooker, T. F., Jeffery, D. J., Jorgenson, J. W.: *High Performance Capillary Electrophoresis*, Khaledi, M. G. Ed., **146** (1998), John Wiley & Sons.

4 Opiteck, G. J., Jorgenson, J. W., Anderegg, R. J.: *Anal Chem.* **69** (1997), 2283-2291.

5 O'Farrell, P. H. *J Biol. Chem.* **250** (1975), 4007-4021.

6 Manz, A., Harrison, D. J., Verpoorte, E., Widmer, H. M.: *Adv. Chromatogr.* **33:1** (1993)

7 Rocklin, R. D., Ramsey, R. S., Ramsey, J. M., *Anal. Chem.* **72** (2000), 5244-5249.

8 Gottschlich, N., Jacobson, S. C., Culbertson, C. T., Ramsey, J. M.: Anal. Chem. **73** (2001), 2669-2674

9 Giddings, J. C.: *Anal Chem* **39** (1967) 1027-1028

10 Murphy, R. E., Schure, M. R., Foley, J. P., *Anal. Chem.* **70** (1998) 1585-1594

CHEMICAL IC CHIP FAMILY FOR ON-CHIP CELL-FREE PROTEIN SYNTHESIS FROM DNA

Koji Ikuta Kota Ikeda Atsushi Takahashi Shoji Maruo
Department of Micro System Engineering, Graduate School of Engineering, Nagoya University, JAPAN (e-mail: ikuta@mech.nagoya-u.ac.jp)

Abstract
We have already demonstrated world's first "In-chip cell-free protein synthesis from DNA" by using the Biochemical IC chips [1]. In this paper, we newly developed high-performance Biochemical IC chips for cell-free protein synthesis. These new Biochemical IC chips gave 17-fold increase in protein production compared to that of previous chips. Our successful results provide new and strong method demanded from post-genome research to find the relationship between genes and proteins. User-assemble versatile Biochemical IC chips enable us to reduce experimental through put time, the cost of expensive biochemical reagent and apparatus for various kinds of protein. DNA originated cell-free protein synthesis is one of the biggest key technology for not only life science but also medicine and drug discovery.

Keywords: Biochemical IC chips, Cell-free protein synthesis, Microstereolithography

1. Introduction

We have been developing versatile micro fluidic device named "Biochemical IC Family" for these years [1-4]. Various kinds of Biochemical IC chips containing multiple micro fluidic devices such as reactors, concentrators, one-way valves, micro pumps and micro switching valves have been developed successfully. Our original 3D micro fabrication process named "Hybrid IH process" [3] was utilized to hybridize various micro parts (ultrafiltration membrane, silicon rubber, shape memory alloy actuator etc.) with UV-polymer structure. Leak-free and transparent modules are made. Moreover, total chemical system can be constructed by stacking Biochemical IC chips and basic performance as generally usable micro chemical device was demonstrated. Recently we have succeeded in realizing "In-chip cell-free protein synthesis from DNA" by using Biochemical IC chips [1].

To increase both the productivity and the reliability of protein production, in this paper, we have improved the Biochemical IC chips for cell-free protein synthesis. These Biochemical IC chips should become practical microdevices for post-genome research to find the relationship between genes and proteins.

2. Biochemical IC chips for cell-free protein synthesis

In the cell-free protein synthesis system proposed by Spirin et al. [5], protein synthesis can be continued for more than 20 hours because of continuous supplement of energy sources such as ATP and GTP, while batch cell-free protein synthesis usually stops after only 1-2 hours. We intend to achieve continuous in-chip cell-free protein synthesis by using Biochemical IC chips.

Fig.1 shows a schematic design of the improved Biochemical IC chip-set for cell-free protein synthesis. The lower one is the reservoir chip with three micro reservoirs. The second one is the one-way valve chip to prevent inverse flow during experiments. The upper one is the reactor chip with a check valve and an ultrafiltration membrane to synthesize protein. During protein synthesis process, cell extract and a solution included in DNA, amino acids, ATP, etc. are mixed in the reactor, and then assay reagent is added to measure the amount of resultant protein.

J.M. Ramsey and A. van den Berg (eds.), Micro Total Analysis Systems 2001, 66–68.

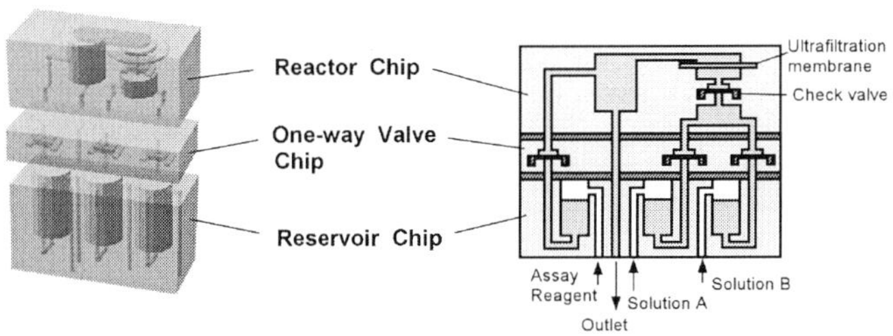

Fig. 1 Schematic design Biochemical IC chips for cell-free protein synthesis from DNA

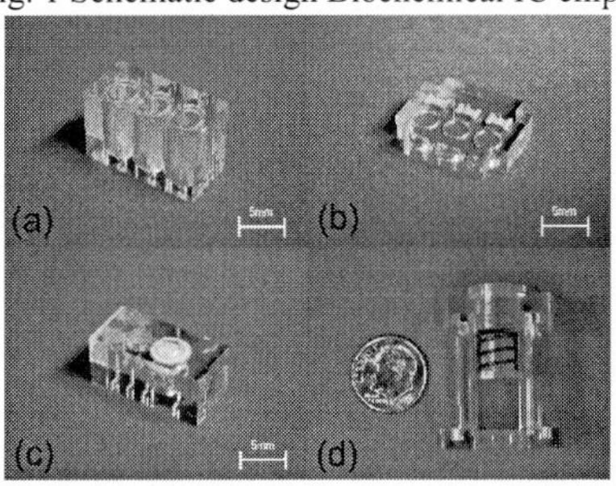

Fig. 2 Biochemical IC chips fabricated by Hybrid IH process
(a) Reservoir chip (b) Valve chip (c) Reactor chip (d) Holder unit

Fig. 3 Biochemical IC chips
in a holder unit

The major improvements of the Biochemical IC chips are as follows:

1) The reactor chip with a built-in check valve was developed.
The check valve included in the reactor prevents solutions for protein synthesis flowing out, so that the productivity and reliability of protein synthesis can be drastically improved.

2) The check valves were changed cantilever type into fixed-beam type.
In the previous Biochemical IC chips, since cantilever-type check valves were used, pressure loss is very low. However, sealing is not enough tight to prevent back flow of solutions. By contrast, the fixed-beam type enables tighter sealing compared to cantilever type, so that precise flow control is realized. The reliability and the durability of the Biochemical IC chips are improved.

3) The layout of the Biochemical IC chips were modified for easy handling and monitoring.
The reservoir chip is set at the bottom, whereas that was set at the top in the previous Biochemical IC chips. This modification enables easy assembly of Biochemical IC chips, because the leakage of solutions between the reservoir chip and the valve chip during the setting is eliminated. In addition, each elements such as a reactor, reservoirs and valves are arranged on the same vertical plane. This allows us to observe the chemical reactions inside the Biochemical IC chips from the outside.

Fig.2 shows the prototypes of Biochemical IC chips fabricated by hybrid IH process and a holder unit. The valve chip contains three check valves made from silicone rubber. These silicone rubber films were formed with polymeric mold fabricated by microstereolithography. In the reactor chip, ultrafiltration membrane (YM100) is used for the separation of synthesized protein. For experiments, these three chips are set in the holder unit as shown in Fig.3. These chips are coupled by spring force without leakage.

3. Experimental verification

Fig.4 illustrates a total experimental setup for DNA originated cell-free protein synthesis by using our Biochemical IC chips. The solution A (DNA, S30 Premix, Amino acid complete, SW)

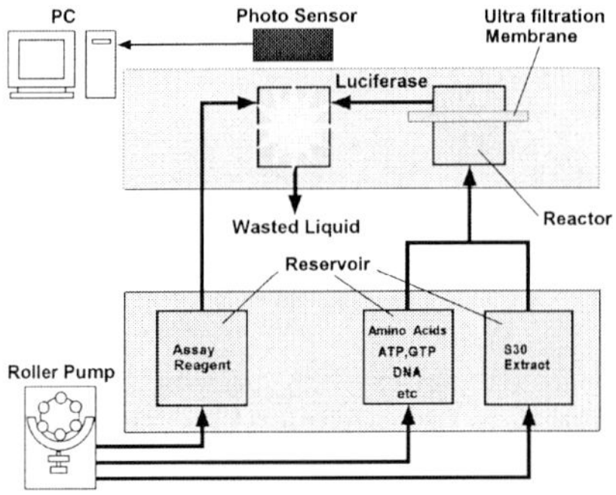

Fig. 4 Experimental setup for cell-free protein synthesis by using Biochemical IC chips

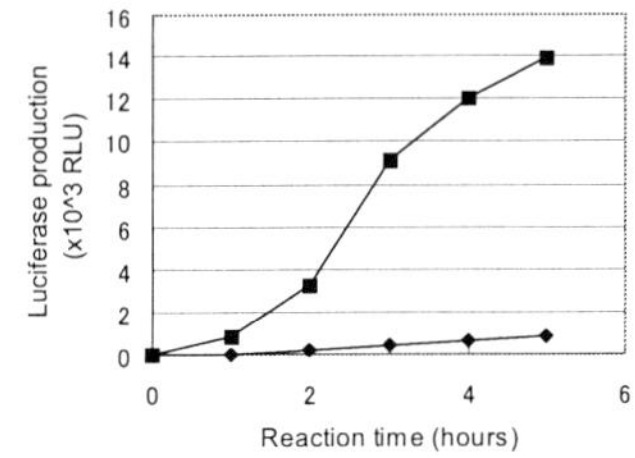

Fig. 5 Comparison between the reactor chip with built-in check valve (■) and the reactor without check valve(◆) in protein production

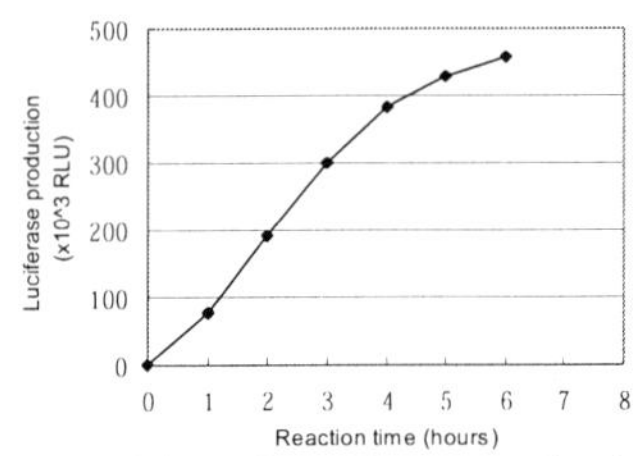

Fig.6 DNA originated Luciferase production with high reliability

and solution B (S30 Extract) are mixed in the reactor and synthesized Luciferase is filtered through an ultrafiltration membrane (YM100). An assay reagent is mixed in the second reactor (left side) and luminescence caused by the biochemical reaction between Luciferase and Luciferin is generated. By measuring the luminescence, we can monitor the production of protein in real time manner.

Fig. 5 shows an experimental result to verify the validity of the reactor chip with the built-in check valve. After a 5-hours run of the protein synthesis process, the reactor with the built-in check valve produced a 17-fold amount of Luciferase compared to a valveless reactor. These results indicate that the check valve is useful to produce sufficient protein synthesis reaction without flowing out.

Moreover, we succeeded to synthesize Luciferase with high reliability. Fig. 6 shows a typical experimental result of cell-free Luciferase synthesis. Although it seems slow reaction rate in comparison with normal chemical reaction, it is much faster then conventional protein synthesis process by using colon bacillus. Moreover, our biochemical IC chip approach drastically miniaturize both total experiment setup and required chemical solutions such as expensive DNA and bio-reagents.

4. Conclusions

We have developed high-performance Biochemical IC chips to synthesize protein from DNA without living cells. The above "palm-top" protein synthesis devices are demanded from not only post-genome researchers but also many life scientists and doctors to make an order-made examination and drugs in the near future. Moreover, our approach is useful for implantable artificial organs.

References

1. K. Ikuta, A. Takahashi and S. Maruo, Proc. of MEMS 2001, 455-458 (2001).
2. K. Ikuta, T. Ogata, M. Tsuboi and S. Kojima, Proc. of MEMS'96, 301-306 (1996).
3. K. Ikuta, K. Ikuta, S. Maruo, T. Fujisawa and A. Yamada, Proc. of MEMS99, 376-381 (1999).
4. K. Ikuta, T. Hasegawa, T. Adachi and S. Maruo, Proc. of MEMS 2000, 739-744 (2000).
5. A. S. Spirin, V. I. Baranov, L. A. Ryabova, S. Y. Ovodov and Y. B. Alakhov, Science **242**, 1162-1164 (1988).

CELL-FREE PROTEIN SYNTHESIS IN PDMS-BASED PARALLEL MICROREACTORS

T. Yamamoto[1], T. Nojima[2], and T. Fujii[1]
[1]Institute of Industrial Science, Univ. of Tokyo,
4-6-1 Komaba Meguro-ku Tokyo,153-8558, JAPAN
[2]Advanced Research Institute for Science and Engineering, Waseda University,
3-4-1-55-S508 Okubo Shinjuku-ku Tokyo, 169-8555, JAPAN
(tfujii@iis.u-tokyo.ac.jp)

Abstract

Cell-free protein synthesis is one of the most useful techniques to amplify proteins and to characterize the mechanisms of gene expression. The synethesis reactions are performed in parallel microreactors fabricated on a PDMS-glass hybrid microchip. To achieve the favorable reaction condition heat control devices are integrated on the glass substrate of the hybrid structure. It is demonstrated that fluorescent proteins, GFP and BFP, can be successfully synthesized in the present microreactors.

Keywords: Cell-free protein synthesis, Microreactor, PDMS

1. Introduction

Cell-free protein synthesis[1] is one of the most useful techniques in biochemical operations for characterization of proteins and elucidation of the control mechanism of gene expression. In the preparation step of the proteome analysis, for example, specific proteins can be amplified by this technique without using any living cells like *E. coli*. The reaction of the protein synthesis (translation of the gene) is conducted by many kinds of molecular machinery, such as ribosomes, tRNAs, solubleprotein factors, etc., as shown in Fig.1. Since the mechanism is relatively complicated, the activity of the reaction, especially when it is in vitro, is strongly dependent on ambient temperature (Fig.2). Instrumentation of this operation has been first attempted by Spirin, et al. [2] in a continuous flow reactor with 1mL reaction volume. This principle has been commercialized as RTS (Rapid Translation System)[3], in which conventional plastic tubes with 1mL volume are used.

In order to realize cell-free protein synthesis operation as a part of the chip-based biochemical analysis systems, we have already tried to conduct the reacion on silicon-glass hybrid microchips[4, 5]. Here, we present newly designed parallel microreactors with temperature control capability. Considering the compatibility, the reactors are fabricated on a PDMS-glass hybrid microchip, which is the same format as other components

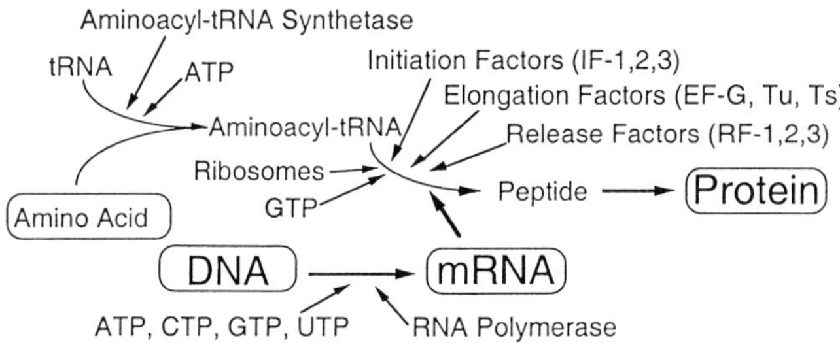

Figure 1 Schematic of the protein synthesis reacion

J.M. Ramsey and A. van den Berg (eds.), Micro Total Analysis Systems 2001, 69–71.

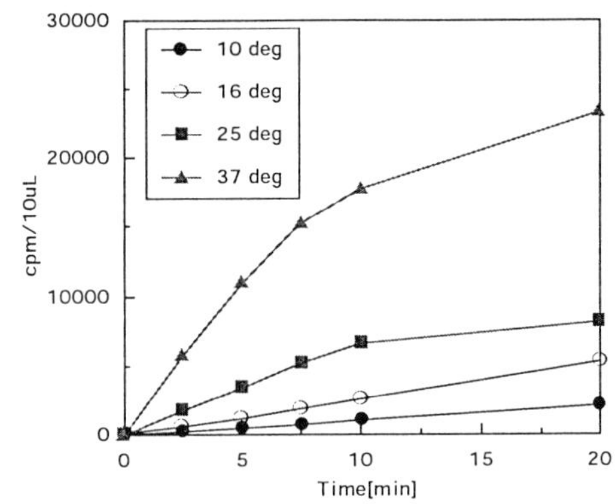

Figure 2 Temperature dependency of cell-free translation: the activity of poly-U dependent polyphenyl alanine synthesis is measured by RI Labelled phenylalanine.

developed in our group, such as CGE (capillary gel electrophoresis) chip[6], PCR-CGE chip[7], small volume liquid handling chip[8], etc. It is shown that GFP (Green Fluorescent Protein) and BFP (Blue Fluorescent Protein), which are widely used as tracer molecules in the field of cell biology, can be successfully synthesized in the present reactors.

2. Design and Fabrication

Figure 3 shows the design of the parallel reactor. Two layers of ITO electrodes are deposited on a glass substrate for heating(lower layer) and temperature measurement (upper layer). The reaction chamber with 1μL volume and the microchannel for sample introduction are formed on a PDMS microchip by replica molding process[6]. The PDMS chip is put on the glass substrate to have a complete structure (Fig.4). In the current layout, seven reactors are integrated on a 2 x 3 inches hybrid microchip.

3. Experimental

The performance of the temperature control device is evaluated by making a feedback control loop using the resisitant measurement of the upper layer electrode. The maximum

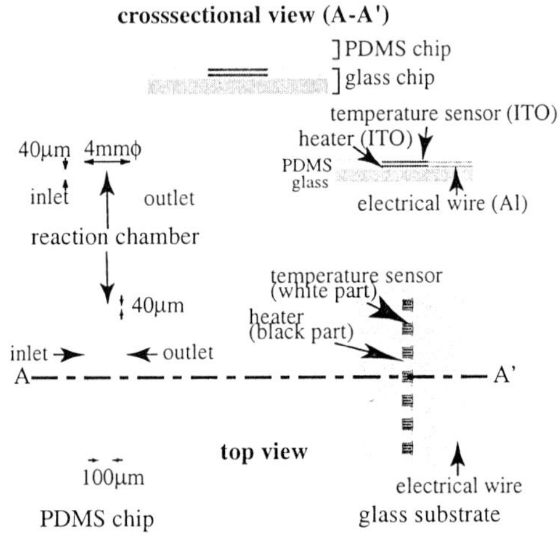

Figure 3 Design of the Parallel Microreactor

Figure 4 Photo of the Parallel Microreactor

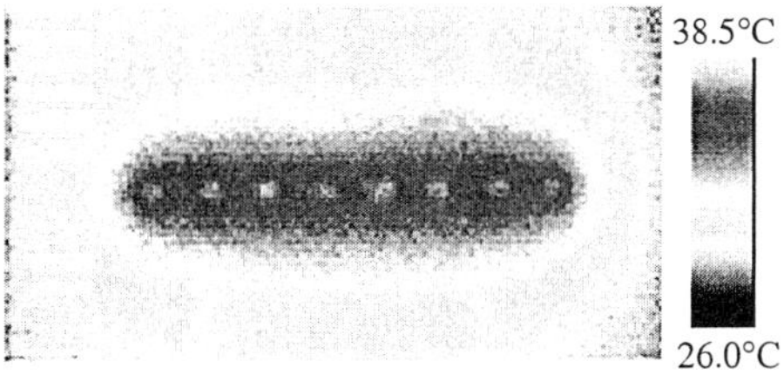

Figure 5 Thermograph of the Microreactors

heating rate is 50°C/sec and the stability is ±0.1°C. Figure 5 shows the image taken by a thermography when the temperature control is turned on to keep the reaction chamber at 37°C. GFPuv gene and BFP gene are successfully translated thourgh cell-free protein synthesis in the present parallel microreactors. Fluorescence of the GFP and BFP could be observed in each corresponding reaction chamber (Fig.6).

4. Conclusion

It is demonstrated that protein synthesis can be conducted in the present parallel microreactor with heat control capability. This leads to the integration of the new operation, protein synthesis / amplification into chip-based biochemical analysis systems.

References

[1] Nirenberg,M.W., and Matthaei,J.H.,The dependence of cell-free protein synthesis in E.coli upon naturally occuring or synthetic polyribonucleotides, Proc. Natl. Acad. Sci., Vol. 47 (1961) pp.1588-1602

[2] Spirin, A.S., et al.,A continuous cell-free translation system capable of producing polypeptides in high yield", Science, Vo.242 (1988) pp.1162-1164

[3] http://biochem.roche.com/RTS/

[4] Fujii, T., et al., A microfabricated reactor for cell-free protein synthesis, Proc. EMBS'97, Chicago (1997) pp.2572-2574

[5] Nojima, T., et al.,A microfabricated reactor for cell-free protein synthesis (2nd report: translation of natural mRNAs),Proc. EMBS'98, Hong Kong (1998)pp.2978-2980

[6] Hong, J.W., Hosokawa, K., Fujii, T., Seki, M., and Endo, I., An inexpensive PDMS (polydimethylsiloxane) microchip for capillary gel electrophoresis, Proc. Transducers '99, Sendai, Japan (1999) pp.760-763

[7] Hong, J.-W., Fujii, T., Seki, M., Yamamoto, T., and Endo, I., Integration of gene amplification and capillary gel electrohpresis on a polydimethylsiloxane-glass hybrid microchip, electrophoresis, Vol.22, No.2 (2001) pp.328-333

[8] Hosokawa, K., Fujii, T., and Endo, I.,Handling of picoliter liquid samples in a polydimethylsiloxane (PDMS) -based microfluidic device, Anal. Chem., Vol. 71, No. 20 (1999) pp. 4781-4785

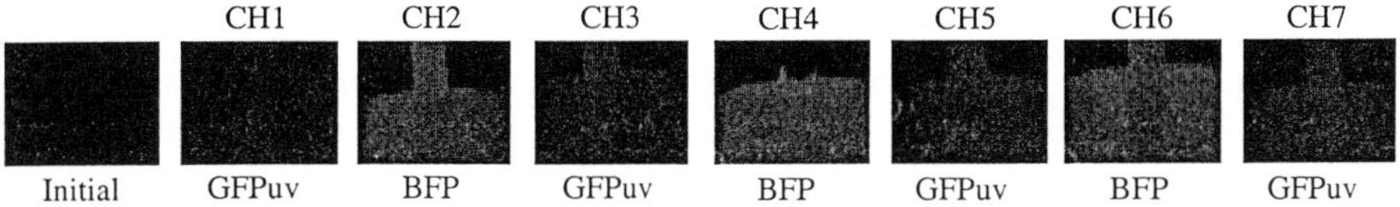

Figure 6 Results of the cell-free protein synthesis in the parallel microreactors: GFPuv gene (508nm emission) and BFP gene (440nm) are translated.

A PROGRAMMABLE DIELECTROPHORETIC FLUID PROCESSOR FOR DROPLET-BASED CHEMISTRY

Jody Vykoukal, Jon A. Schwartz, Frederick F. Becker and Peter R. C. Gascoyne
The University of Texas M. D. Anderson Cancer Center
Department of Molecular Pathology 1515 Holcombe Boulevard Houston, Texas 77030

Abstract We demonstrate the use of dielectrophoresis (DEP) in a programmable microsystem to drive the formation and handling of small (10 to 1000 nL) droplets of sample and reagent on a channel-less, planar reaction surface. DEP-controlled metering and mixing of reagent droplets is shown in a proof-of-principle protein concentration assay. The electronically reconfigurable droplet processor described here is expected to provide functionality for many chemical, biological, and diagnostic applications.

Keywords: dielectrophoresis, channel-less, fluid dispensing, fluid handling

1. Introduction In principle, many microfluidic systems could be based on a reconfigurable architecture that would enable a single device to be programmed to accomplish a wide range of synthesis and analysis tasks, in much the same way that a digital microprocessor is programmed to meet multiple needs in electronics. Microfluidic devices that employ pumps, valves, or other means to direct fluids though networks of dedicated channels are intrinsically hardwired and unable to achieve such diverse functionality. To surmount these limitations, dielectrophoresis can be applied to the metering and handling of reagents in a programmable, channel-less microsystem.

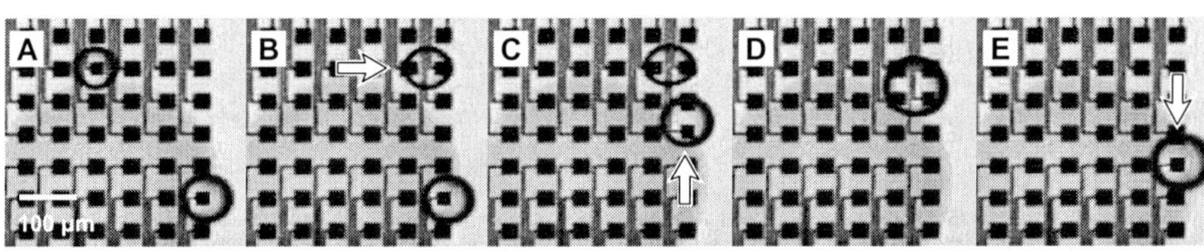

Fig. 1 Droplets are manipulated in two-dimensions on the surface of the DEP fluid processor. **(A)** Two droplets are held in discrete DEP energy minima that surround the energized electrodes in an array of 30 µm x 30 µm gold electrode pads. **(B)** A 240 pl droplet is moved along the x-axis of the array to a new location. **(C)** A 340 pl droplet is moved along the y-axis while the smaller droplet is held in place. **(D)** The droplets spontaneously fuse when they are brought into contact. **(E)** The resultant droplet is shuttled to a new location on the microarray.

2. Theory From the perspective of droplet handling, dielectrophoresis (DEP) is the tendency of a fluid of high dielectric permittivity to migrate in an electrical field gradient to the region of maximum field strength [1-3]. In our programmable fluid processor, DEP forces are generated by a two-dimensional array of individually addressable microelectrodes on the bottom surface of the device. Sample and reagent droplets are shuttled and interacted in unit steps along arbitrarily chosen paths anywhere on an X-Y plane under computer control without the need for channels, valves or any off-chip pumping mechanism (Fig. 1). This scheme allows multiple droplets from

J.M. Ramsey and A. van den Berg (eds.), Micro Total Analysis Systems 2001, 72–74.

different sources to be moved independently along intersecting pathways with droplet interactions limited to instances where two or more are explicitly brought into contact. The only requirement for manipulation of the droplets is that their dielectric permittivity be higher than that of their suspending medium, a condition satisfied for droplets of polar liquids, such as water, suspended in a nonpolar liquid (or gas). Because an electrical field creates the motive force, no contact or interaction with the reaction surface or electrodes is needed, and the DEP fluid-handling force is independent of the droplet net charge, pH, conductivity, solute or particulate content.

Our general-purpose microfluidic platform comprises not only a means of manipulating, combining and collecting fluid droplets, but also a method for introducing metered, 0.01 to 10 nanoliter aliquots of reagents from multiple reservoirs or other sources on demand to form droplets of desired sizes. Our injection scheme is based on the application of dielectric forces for building droplets of controlled volume from pressurized reservoirs adjacent to our reaction surface. When the reservoir pressure is low, the fluid at the injector orifice assumes a flattened shape (Fig. 2A). As the reservoir pressure is increased, a nascent aliquot of increasing volume, but decreasing sphere radius, is formed. The internal pressure of the aliquot reaches a maximum (Fig. 2B) when the radius of the droplet is equal to the radius of the injector orifice and the droplet is hemispherical in shape. Further increase of the droplet volume is associated with an increased droplet radius and decreased internal pressure. In dielectrophoretic-injection, a hydrostatic pressure less than that required to overcome the maximum hold-off is applied to the fluid in the reservoir. The application of an AC electrical field between the droplet and a nearby electrode provides a supplemental *DEP pressure* (Fig. 2C) to overcome the hold-off condition, triggering aliquot growth and introducing an additional lateral DEP force component (Fig. 2D) to detach the aliquot from the injector aperture at a well-defined volume. Once they leave the orifice, fluid aliquots spontaneously coalesce at the collecting electrode, allowing droplets to be built *in situ* from metered units and then moved to other regions of the reaction surface.

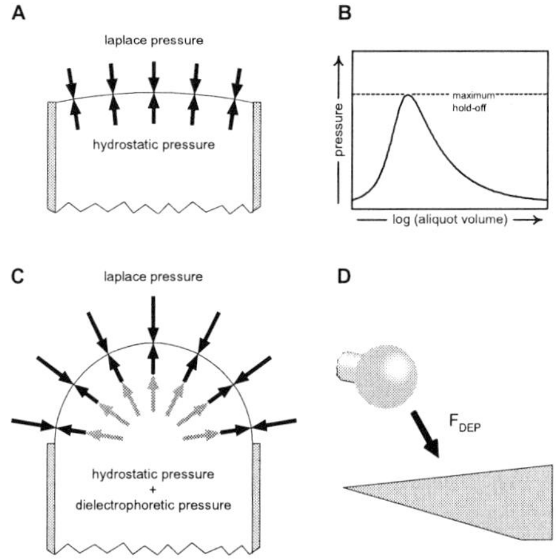

Fig. 2 A combination of hydrostatic, interfacial and dielectrophoretic forces are used to control injection of metered aliquots of reagents and samples into the droplet processor (*see main text for details*).

3. Results and Discussion In our experiments, droplets ranging from 10 to 1000 nL were built on the reaction surface by combining 0.01 to 10 nL aliquots from one or more injectors. This allowed sets of droplets containing different reagent concentrations to be formed as needed. The process of building droplets through sequential addition of aliquots is shown in Fig. 3, which demonstrates injection of an

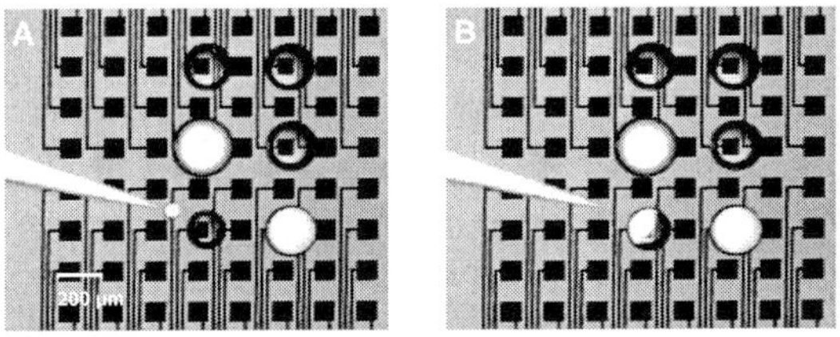

Fig. 3 Building droplets by DEP reagent metering. Six aqueous droplets, viewed using a combination of epifluorescence and backlighting, are shown **(A)** A 100 pl aliquot is shown in mid-flight as it is drawn from the injector toward a 4.2 nl droplet under the influence of an applied DEP force. The diffusion front of the injected aliquot can be seen **(B)** advancing across the droplet with which it has fused.

aliquot of fluorescein-containing solution into a droplet of water that had been built previously using aliquots from a different injector. Spontaneous mixing of an injected aliquot with a droplet occurs by diffusion in only a few milliseconds as the observed rate of diffusion is >100 μm/sec. This eliminates the need for explicit fluid mixing components. For illustrative purposes, the injector tip in Fig. 3 has been moved close to the droplet into which fluorescein reagent is being injected. More typically, droplets are built from injectors at the periphery of the electrode array and then shuttled, as required, to other sites on the reaction surface by DEP manipulation where they may be observed or interacted with droplets containing other reagents or samples.

To demonstrate the utility of the programmable droplet processor, a protein concentration assay was performed. Nine droplets of similar volume, each containing the fluorescent protein indicator OPA (o-pthalaldehyde), were built from a single DEP injector. An additional injector was used to meter between 1 and 128 aliquots of a bovine serum albumin solution into each of the protein indicator droplets. The protein diffused rapidly after injection into the droplets and reacted with the indicator to form a fluorescent product. Droplets were arrayed on the reaction surface and their fluorescence was measured by image analysis. The protein concentration in the various droplets (Fig. 4) ranged from 0 (bottom-right) to 229 (top-left) μg/ml, and the total fluorescence of each droplet was proportional to its protein content.

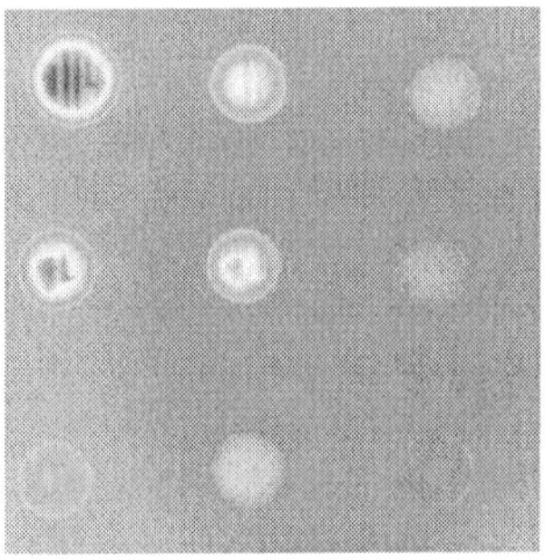

Fig. 4 The fluorescent OPA derivative of bovine serum albumin as an indicator of protein concentration in a droplet-based microassay.

Acknowledgements We are grateful to P. Krulevitch and J. Hamilton at LLNL for electrode fabrication. Supported by DARPA under contract N66001-97-C-8608 and grant agreement DAAD19-00-1-0515 from the Army Research Office.

References

1. Pohl, H.A. Dielectrophoresis. Cambridge University Press, New York (1978).
2. Gascoyne, P.R.C., Huang, Y., Pethig, R., Vykoukal, J. & Becker, F. F. Dielectrophoretic separation of mammalian cells studied by computerized image analysis. *Meas. Sci. Technol.* 3, 439-445 (1992).
3. Jones, T.B., Gunji, M., Washizu, M. & Feldman, M.J. Dielectrophoretic liquid actuation and nanodroplet formation. J. Appl. Phys. 89, 1441 (2001).

COMPOUND MICRODISPENSING

Johan Nilsson, Jonas Bergkvist, Simon Ekström, Lars Wallman and Thomas Laurell
Department of Electrical Measurements, Lund Institute of Technology, P.O. Box 118, SE-221 00 Lund, Sweden
(E-mail: johan.nilsson@elmat.lth.se)

Abstract

A flow-through piezo-electric microdispenser is presented in which the ejected droplets are covered with a secondary layer of liquid thereby decreasing the contamination of the nozzle exit by the primary liquid. The working principle was verified using two different peptides in the primary and secondary liquids. The collected droplets were analyzed using a MALDI-TOF MS-system.

Keywords: Flow-through, piezoelectric dispensing, compound microdispenser.

1. Introduction

Flow-through piezoelectric microdispensing is a technique that has proven useful in several applications ranging from liquid handling in an ultrasonic levitation system for living cell response studies [1] to sample handling in MALDI-TOF mass spectrometry analysis for protein identification in the area of proteomics [2]. The main advantages of the flow-through dispenser is the possibility to dispense samples from flowing liquids, the non-contact way of sample supply, the small size of the droplets, typically 40 µm and the high droplet ejection frequency, up to 5 kHz. The small droplet size and high droplet rate enables on-spot enrichment of analytes which is crucial to achieve high sensitivity, not to the least in MALDI-TOF MS.

Although the flow-through dispenser works well in the applications, we have found shortcomings especially when it comes to dispense liquids with high concentrations of solutes, like different matrices in the MALDI applications. The tendency of forming crystals at the nozzle makes it very difficult to run the dispenser unattended for a long time. However, by improving the design and adding a second layer of "clean" liquid covering the primary nozzle it may be possible to overcome this problem.

2. Experimental

Figure 1 shows the design of the compound microdispenser. The device consists of three silicon layers forming the flow channels and the nozzles. The overall size is 20 mm x 6 mm x 0.75 mm with inner channel volume of 0.4 µl and outer channel volume of 16 µl. The sizes of the inner and outer nozzles are 40 µm x 40 µm and 80 µm x 80 µm, respectively. The silicon structures are fabricated using wet etching and pn-etch

J.M. Ramsey and A. van den Berg (eds.), Micro Total Analysis Systems 2001, 75–77.

stop technique [3]. A piezo-electric multilayer element is used for actuation. Upon droplet ejection the liquid in the inner chamber will pass the second liquid layer which will cover the inner liquid so no contamination of the outer nozzle will occur by the inner liquid, fig. 2.

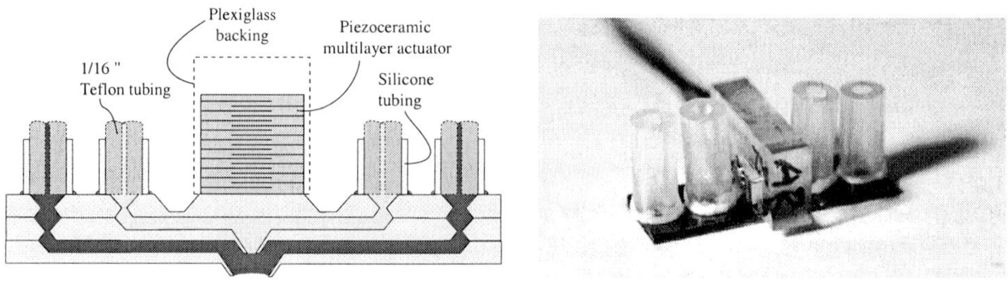

Figure 1. Design of the compound microdispenser.

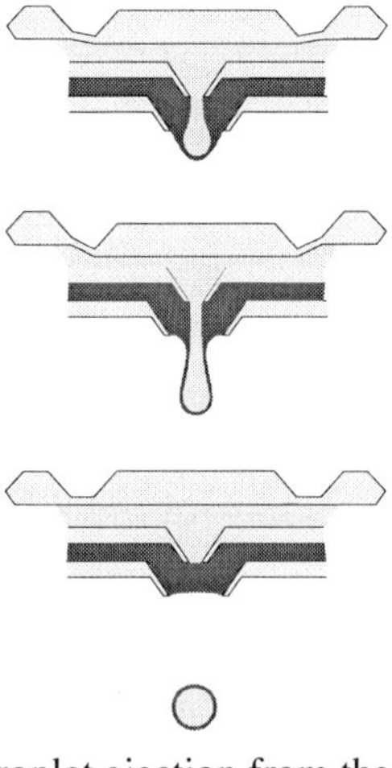

Figure 2. Principle for the droplet ejection from the compound microdispenser.

The function of the compound microdispenser was tested using two different peptides in the inner and outer channel, Angiotensin I in the inner and Angiotensin II in the outer channel. The compound droplets were ejected onto a target plate, which was analyzed in a MALDI-TOF MS-system in order to verify that both liquids were present in the ejected droplets.

4. Results and discussion

Figure 3 shows the MALDI-TOF MS spectrum obtained from the liquid in the collected droplets. Both peptides are present in the output indicating that liquids from both the inner and outer channel are ejected simultaneously in the droplets. By keeping the secondary liquid clean, for instance by only using solvent, contamination on the outer nozzle exit can be kept to a minimum. This will prevent crystallization and also reduce cross-contamination if different liquids are supplied in the primary channel. The flow-through design of channels enables a continuous flow in the outer channel that will, for example, wash away analytes that may diffuse through the inner nozzle before droplet ejection. It is also possible to use different liquids in the channels that will be mixed in the ejected droplets.

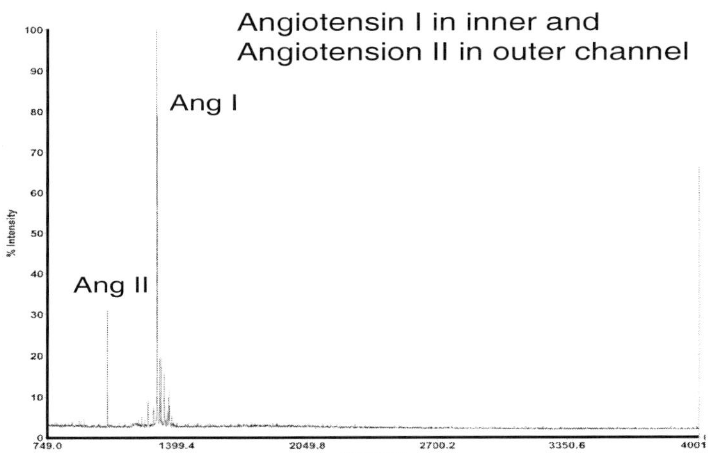

Figure 3. MALDI spectrum obtained from a run with two different peptides in the liquids in the inner and outer channels.

References

1. S. Santesson, A. Martinsson, E. Degerman, T. Johansson, J. Nilsson and S. Nilsson. *Airborne cell analysis.* Analytical Chemistry, 2000, 72, 3412-3418.
2. Ekström S., Ericsson D., Önnerfjord P., Bengtsson M., Nilsson J., Marko-Varga G. and Laurell T. *Signal amplification using "spot-on-a-chip" technology for the identification of proteins via MALDI-TOF MS.* Analytical Chemistry, 2001, 73, 214-219.
3. Thomas Laurell, Lars Wallman and Johan Nilsson. *Design and development of a silicon microfabricated flow-through cell for on-line picoliter sample handling.* J. Micromech. Microeng. 9 (1999) 369-376.

AN AIR-DRIVEN FLUIDIC MULTIPLEXER INTEGRATED WITH MICRODISPENSERS

Aniruddha Puntambekar[1], Jin-Woo Choi[1], Chong H. Ahn[1], Sura Kim[2], Sami Bayyuk[2] and Vinod Makhijani[2]
[1]*Microsystems and BioMEMS Laboratory*
Department of Electrical and Computer Engineering and Computer Science
University of Cincinnati, P.O. Box 210030, Cincinnati, OH 45221-0030, USA
[2]CFD Research Corporation, Huntsville, AL 35805, USA

Abstract

In this work, we have successfully designed, fabricated and characterized an air driven, discrete volume multiplexer with integrated microdispenser. Using sPROMs (*structurally programmable microfluidic systems*) concepts we have developed discrete volume generating microdispensers that can dispense a volume of 280 *nL*. Using low-pressure air-driven actuation, the dispensed volume is then precisely multiplexed into four volumes of 70 *nL* each. The multiplexer with the integrated microdispenser delivers 70 *nL* at each of the four outputs with an accuracy of 99%.

Keywords: air driven, microdispenser, fluidic multiplexer, discrete volume handling

1. Introduction

Microfluidic biochemical analysis systems, also referred as μTAS or Lab-On-A-Chip devices, have made great progress in the recent years. Most of the benefits offered by μTAS devices are derived from their ability to handle extremely small volumes of samples and reagents [1]. Hence, it is critical to develop a reliable and simple technique to manipulate fluid volumes in the *nL* – μ*L* range.

Recently, we have demonstrated a passive microfluidic control approach based on sPROMs technology, realizing continuous flow multiplexers and discrete volume microdispensers using sPROMs [2,3]. By integrating the microdispenser with the continuous flow multiplexer, we can realize a more powerful tool for discrete volume manipulation using air actuation. The devices developed in this work offer immediate solutions to the challenging problem of on-chip sample preparation and manipulation.

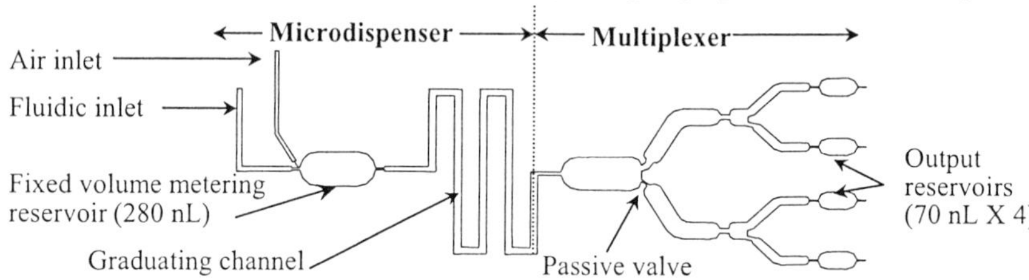

Figure 1. Schematic sketch of air driven multiplexer integrated with microdispenser.

The microdispenser uses a fixed volume reservoir to define the dispensing volume and passive valves to contain the fluid within the reservoir. Air pressure from the air inlet causes a split in the liquid column just before the reservoir and dispenses an

J.M. Ramsey and A. van den Berg (eds.), Micro Total Analysis Systems 2001, 78–80.

exact amount of liquid. The liquid is then pushed towards the multiplexer by continued air driving. The geometry and physical characteristics of the substrate surface cause the fluid to travel sequentially down the multiplexer resulting in simultaneous fluid division. Detailed principles of operation for the multiplexer and dispenser are listed in references 2 and 3, respectively.

2. Design and Fabrication

The geometry and position of the air inlet plays a crucial role in the liquid column splitting process. Since the accuracy of the splitting process affects the accuracy of the dispensed volume, detailed simulations were conducted using CFD-ACE+ software to investigate the splitting process and optimize microdispenser performance.

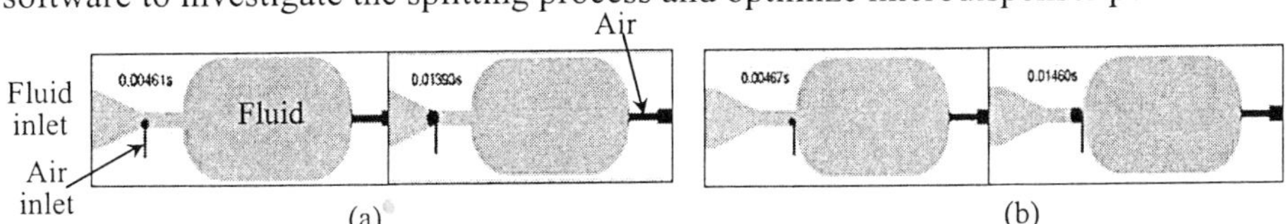

Figure 2. Simulation results: (a) air inlet located upstream of reservoir inlet and (b) air inlet located close to reservoir inlet.

Figure 2 shows a section of Figure 1, specifically the air inlet at the start of the fixed-volume reservoir. As shown in Figure 2, the position of the air inlet determines the location of the liquid column split. The mode of the air bubble growth and position of the split can be accurately predicted using the simulation model. Using the simulation results for optimal position of the air inlet, a fluidic multiplexer has been designed and fabricated as shown in Figure 3, where a volume of 280 *nL* is multiplexed into four volumes of 70 *nL* each.

The designed devices were fabricated using a UV-LIGA fabrication process with SU-8 photoresist. After UV-LIGA lithography the final devices were fabricated in PDMS using casting techniques.

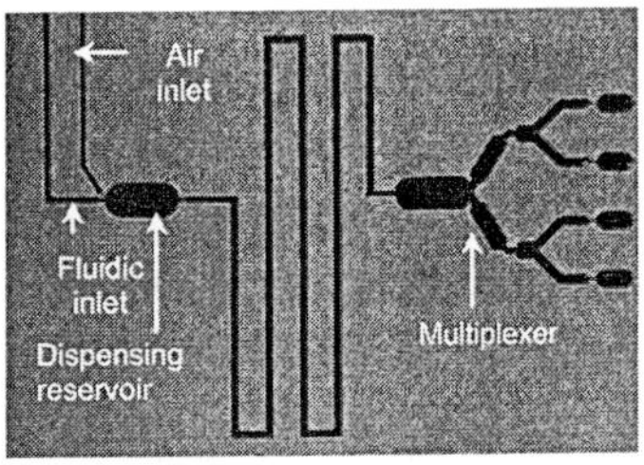

Figure 3. Microphotograph of fabricated device.

3. Experimental Results

Figure 4 shows the sequence of operation for the microdispenser.

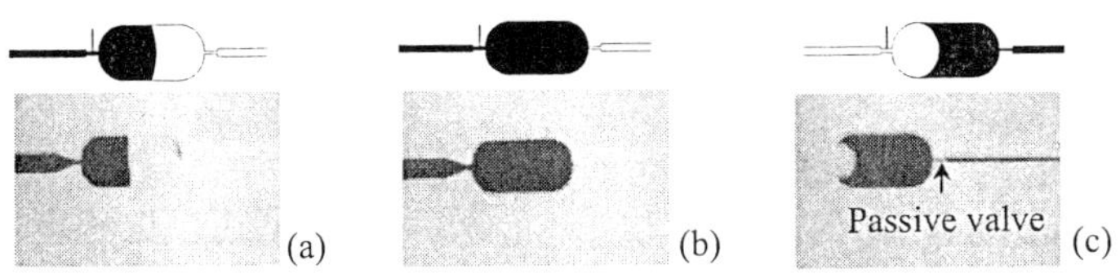

Figure 4. Microphotographs showing operation of microdispenser: (a) filling sequence; (b) fluid locked in the reservoir by passive valve; and (c) split in liquid column by air injection and subsequent dispensing.

Figure 4(c) clearly shows the liquid column split and simultaneously fluid being ejected to the right. Figure 5 shows the sequence of operation of the multiplexer after the dispensed volume enters via the graduating channel.

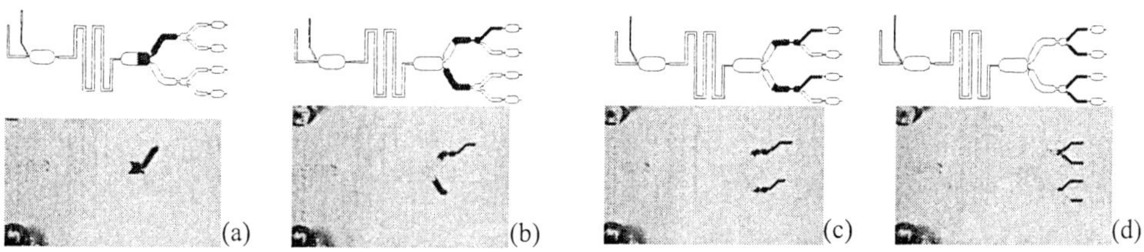

Figure 5. Microphotographs showing operation of sequential multiplexer: (a) first level division; (b) second level division; (c) continued multiplexing; and (d) end of sequential multiplexing sequence.

For characterization the fluidic volume was measured at two locations. First, the output volume of the microdispenser without the subsequent multiplexer was measured as shown in Figure 6(a). The dispensed volume was drawn into a microcapillary and the weight of the dispensed volume was measured to calculate the volume. Next, the discrete volume at each of the four outputs of the multiplexer with integrated microdispenser was measured as shown in Figure 6(b). Using the above data the accuracy of the microdispenser is 99.4% and for the multiplexed output it is 99%. The value of multiplexed output accuracy is limited by the resolution of the measurement system (0.1 μg which corresponds to 0.1 nL for water) and not device performance.

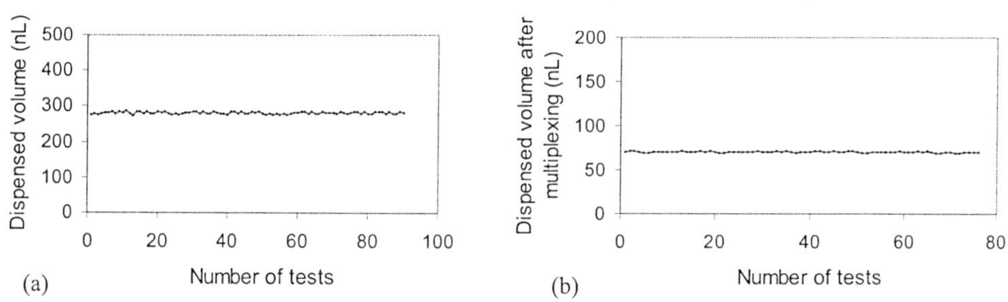

Figure 6: Volume measurement results for: (a) microdispenser and (b) dispensed volume at multiplexer output.

4. Conclusions

An innovative air driven, discrete volume multiplexing system integrated with microdispenser has been proposed and successfully demonstrated in this work. Accuracy of the order of 99% has been achieved for fluid dispensing and multiplexing. The integration of an on-chip microdispenser offers significant prospects for on-chip sample loading, preparation and manipulation that would greatly enhance the performance of biochemical analysis systems.

Acknowledgements

This research was fully supported by a DARPA grant under contract AF F30602-00-1-0569 from the *BioFlips* program, DoD, USA.

Reference:

1) P. Merel, "The 'lab-on-a-chip' concept", Biofutur, Vol. 1999, Issue 193, pp. 38-41, 1999.
2) C. Ahn, A. Puntambekar, S. Lee, and H. Cho, "Structurally Programmable Microfluidic Systems", Proc. of the μTAS, pp. 205-208, Enschede, The Netherlands, 2000.
3) A. Puntambekar, H. Cho, C. Hong, J. Choi, C. Ahn, S. Kim and V. Makhijani, "A New Fixed-Volume Metering Microdispenser Module Based on sPROMs Technology", Proc. Transducers'01, pp. 1240-1243, Munich, Germany, 2001.

OPTIMISED POROUS SILICON μIMERs FOR HIGH-SPEED ON-LINE PROTEIN DIGESTION

Martin Bengtsson[1], Simon Ekström[1], György Marko-Varga[2] and Thomas Laurell[1]
[1] Dept. Electrical Measurements, Lund University ,Sweden
[2] AstraZeneca, Molecular Science, Sweden

Abstract

In order to achieve a homogeneous porous layer on a μIMER in porous silicon, different geometries and anodisation conditions are studied. It is clearly shown that an a homogeneous porous depth is possible to achieve and that it improves the catalytic performance of the μIMER.

Keywords: Porous silicon, IMER, microreactor, enzyme, carrier matrix

Introduction

Immobilised enzyme reactors (IMER) are a commonly used tool for biospecific detection. Our group has developed a porous silicon microstructured parallel channel enzyme reactor. The catalytic performance of such reactors is strongly dependent of the porous silicon morphology for enzyme immobilisation. For high aspect ratio structured silicon, such as the μIMER, however, it is difficult to achieve a homogeneous porous depth all over the structure. In order to achieve such a layer, and thus maximise the μIMER efficiency, μIMERs with different geometry were fabricated and anodised under different comditions. The catalytic performance of the different μIMER geometries were then investigated with respect to the pore morphology.

Experimental

The microreactors were fabricated [1] , with a wall thickness and channel width of 25μm, 50μm or 75μm in different combinations. Then the walls were made porous by anodising in an HF-ethanol solution at a constant current, 10mA/cm^2 or 50mA/cm^2, for anodisation times ranging from 5 to 50 minutes, during backside illumination. The porous depth at different heights, was estimated from SEM photographs. Enzyme was coupled onto the porous matrix following a three step (silanisation / glutaraldehyde / enzyme) procedure [2]. The catalytic performance of the reactors were evaluated for two enzyme categories using a) a fast diffusing low molecular metabolite: glucose and immobilised glucose oxidase measured in a colourimetric assay and b) large slow diffusing protein, β-casein and immobilised trypsin, where the trypsin were used to digest the protein in an on-line format and the digest was subsequently analysed by MALDI-TOF MS.

J.M. Ramsey and A. van den Berg (eds.), Micro Total Analysis Systems 2001, 81–82.

Results and discussion

The reactors with thicker walls (75 μm), showed a porous profile with equal porous depth over the full height of the wall, figure 1, even for anodisation times as long as 50 minutes and pore depths as deep as half the wall thickness, 35μm. The highest turn-over rate, both for GOx and trypsin, is shown for the 75μm wall reactor anodised at 50mA/cm^2 for 10 minutes. In spite of the differences in enzyme and molecule size of the substrate the combined pore size and porous depth gives an advantageous porous matrix for both enzymes. The mass spectra resulting from the MALDI-TOF MS analysis of the digest from the best μIMER is presented in figure 2. After database search, using PeptiIdent, ten peaks could be assigned to β-casein. Successful protein identification was demonstrated after only 12 sec. residence time in the μIMER with 75 μm channel wall, 25 μm channel width, anodised at 50 mA/cm^2 for 10 minutes.

Figure 1. Homogeneous porous layer on high aspect ratio channel walls. μIMER anodised at 10mA/cm^2 for 50min

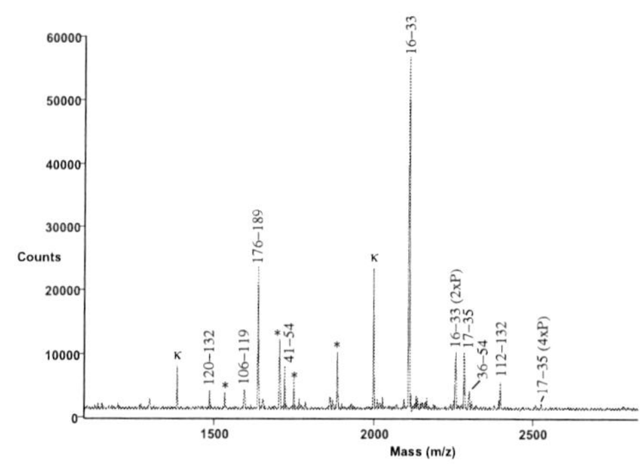

Figure 2. Mass spectra of β-casein digested in the μIMER, immobilised at 50mA/cm^2 for 10min

Conclusions

The importance of a correctly designed μIMER to fully utilise the benefit of integrating porous silicon in chemical microsystems as a surface enlarging carrier matrix for enzymes has clearly been demonstrated. By tailoring the μIMER channel cross-section geometry, homogeneous PS layers were obtained on high aspect ratio silicon microstructures, e.g. the μIMER, yielding improved biocatalytic performance. In the case of the protein digestion application the outcome of the optimisation work is crucial, enabling high-speed online protein digestion.

References

1. Drott, J., Lindstrom, K., Rosengren, L., Laurell, T. (1997) Porous silicon as the carrier matrix in microstructured enzyme reactors yielding high enzyme activities. *Journal of Micromechanics and Microengineering* **7**, 14-23.
2. Weetall, H.H. (1976) Covalent couplings methods for inorganic Support Materials. In *Methods in enzymology*, Volume 44 (K. Mosbach, ed) Academic Press, N.Y. 134-148.

A DISPOSABLE CELL-BASED BIOCHIP FOR DETECTION OF HORMONES AND DRUGS

Tauseef R. Butt[1], Hiep Tran[1], Jin-Woo Choi[2], Hyoung J. Cho[2], and Chong H. Ahn[2]
[1]LifeSensors Inc., 271 Great Valley Pkwy, Malvern, PA 19355, USA
[2]University of Cincinnati, PO Box 210030, Cincinnati, OH 45221-0030, USA

Abstract

This paper presents development and characterization of a disposable cell-based biochip for detection of hormone dependent responses of human nuclear receptors. The disposable cell-based biochip has chamber arrays for immobilize cell-based sensors and microfluidic channels for hormone and substrate injection. A prototype plastic microarray has been designed, fabricated, and validated for hormone dependent, cell-based, human nuclear receptor function. Depending upon the receptor employed, these sensors can determine predisposition to type II diabetes and cardiovascular diseases.

Keywords: Cell-based biochip, yeast cell, disposable biochip, hormone

1. Introduction

Human genome sequencing has uncovered several novel receptors that may be important diagnostic and therapeutic targets [1]. It is not possible to develop ultra-sensitive *in vitro* methods to detect ligand-mediated functions of receptors, and cell-based approaches are required. The disposable cell-based biochip has chamber arrays for immobilize cell-based sensors and microfluidic channels for hormone and substrate injection as shown in Figure 1. The prototype cell-based biochip employs human estrogen receptor (ER) or thyroid hormone receptor β (TRβ) engineered yeast cells. The cell-based sensor is dormant, until a drug or a hormone is channeled. The yeast cells that are immobilized in microarray chambers were responsive to injected estrogen or thyroid hormone by changing color or fluorescing thus demonstrating feasibility of using cell-based sensors as biochips.

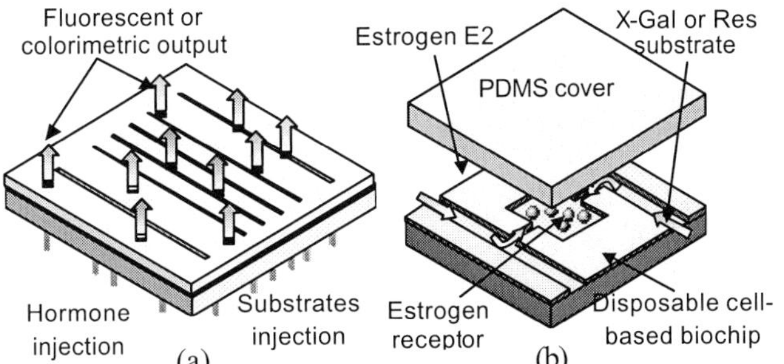

Figure 1. Conceptual illustration of a disposable cell-based biochip: (a) array type cell-based biochip and (b) magnified view of a cell chamber array.

2. Cell-Based Biochip and Experiments

To develop a system that will house the cells and amenable to drug and hormone response, a 3 x 3 cell immobilizing chamber arrays and microfluidic channels on a chip have been designed and fabricated as shown in Figure 2. UV-LIGA processing was

J.M. Ramsey and A. van den Berg (eds.), Micro Total Analysis Systems 2001, 83–84.

employed to fabricate a precisely patterned injection mold-disk and cast master for use of a plastic injection molding technique and PDMS casting technique.

1 μL of human estrogen receptor yeast engineered cells, where the β-galactosidase reporter gene was under the control of 17 β estradiol (E2) were embedded in the chamber along with 1 – 1000 nM hormone. After few hours resorufin-β-D-galactopyranoside (Res) substrate was added and the chip was allowed to develop. The substrate fluoresces after cleavage with β-galactosidase. The signal was detected by placing the chip in fluorescence detector as shown in Table 1. The data shows that hormone mediated reporter responses were easily detected on the chip. The data also shows that the chip material is compatible with the cell growth and the material used in the development of the chip did not inhibit cellular signal transduction pathways. These hormone dependent responses observed in plastic chip were completely consistent with the one obtained with reporter assays performed with a traditional assay format [2].

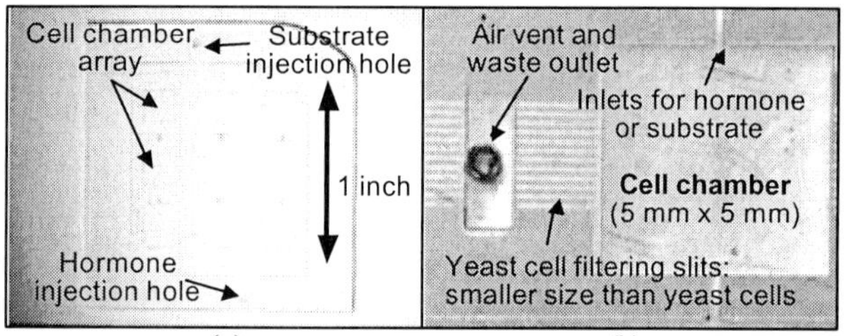

(a) (b)

Figure 2. Fabricated cell-based biochip: (a) 9 x 9 array of cell-based biochip using replaceable mold-disk technique for injection molding and (b) microphotograph of a cell chamber of the microarray cell-based biochip.

Table 1. Hormone mediated sensor responses on the developed biochip. The data is shown as fluorescence units. DMSO is used as a control.

Res. 1/500	**Res. 1/50**	**X-gal**	
0.0000	0.0000	0.0000	**DMSO**
0.0168	0.1183	0.0000	**200nM E2**
0.0153	0.1175	0.0000	**20nM E2**

3. Conclusions

A prototype disposable cell-based biochip has been successfully developed and tested for detection of hormone dependent responses of human nuclear receptors. Detection of agents in human bodily fluids, that regulate receptor function, in microfluidic and disposable plastic chips will have wide application in health care. Capability of the developed biochip can be expanded to number of other receptors, to genomics/proteomics applications, and to clinical diagnostics and/or drug discovery.

References

1. T. R. Butt and D. J. Chen, "Reconstruction of Mammalian Nuclear Receptor Function in Saccharomyces Cerevisiae," J. Davies and C. Harshberger (Eds.), *Manual of Industrial Microbiology and Biotechnology*, 2nd Edition, Section V, American Society for Microbiology, Washington, DC, pp. 527-538, 1999.
2. M. Anafi, Y.-F. Yang, N. A. Berlev, S. L. Berger, T. R. Butt, and P. G. Walfish, "GCN5 and ADA Adapter Proteins Regulate T3/GRIP Coactivator Dependent Gene Activation by Human Thyroid Hormone Receptor," *Mol. Endo.* Vol. 14, pp. 718-732, 2000.

PROTEIN CHIPS BASED ON RECOMBINANT ANTIBODY FRAGMENTS

Simon Ekström[1], Carl A.K. Borrebaeck[2], Ann Christin Malmborg Hager[2] Johan Nilsson[1], György Marko-Varga[3], Thomas Laurell[1]
[1]Department Electrical Measurements, Lund University, Sweden
[2]Department of Immunotechnology, Lund University, Sweden
[3]Department of Analytical Chemistry, Lund University, Sweden

Abstract

This report demonstrates how functionalised protein chips can be made by using recombinant scFv antibody fragments, all based on the same single framework scaffold [1], in combination with piezoelectric dispensing onto microfabricated silicon chips followed by mass spectrometric readout.

Keywords: protein chips, recombinant antibodies, matrix-assisted laser desorption/ionization time-of-flight mass spectrometry (MALDI-TOF MS), microdispensing.

1. Introduction

The ability to perform protein expression analysis in microstructure formats is currently a major objective in proteomics and engages a number of investigators and biotech companies. In analogy with the DNA microarrays the protein arrays, or so called protein chips, could offer a possibility to develop a rapid method to probe an entire proteome.

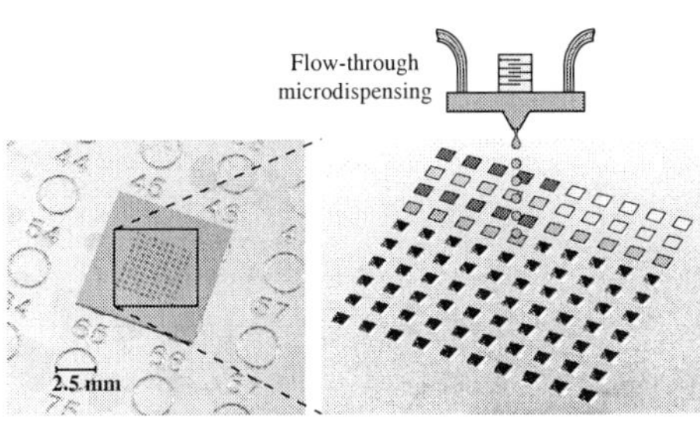

Figure 1. Nanovial antibody array.

2. Experimental

The antibody fragments and the antigens were both deposited onto the nanovial array chip using the in-house developed piezoelectric flow-through dispensing technology [2] in array format of 300 x 300 x 50 µm nanovials with a density of 270 spots/cm^2, Figure 1. The following procedure was applied to run the assay: 1) formation of thin polymer layer on the microchip; 2) microdispense the scFv antibody fragment into the vial; 3) washing of chip surface, repeated 4 times 4) microdispense the antigens into the vial; 5) washing of chip surface, repeated 4 times 6) apply the MALDI-matrix by microdispensing; 7) analyze with MALDI-TOF MS.

J.M. Ramsey and A. van den Berg (eds.), Micro Total Analysis Systems 2001, 85–86.

3. Results and discussion

The scFv fragment, specific for choleratoxin (CT-17), was spotted onto these chips and allowed to interact with a mixture of different antigens The antigen solution contained the choleratoxin subunit (11 kDa) at 50 nM and three different background proteins (MW 12-17 kDa) at a level of 500 nM. After microdispensing of this protein mixture into the antibody activated nanovials (20 fmol antigen solution/spot), the mass spectroscopy analysis was performed and the resulting MALDI spectra are shown in Figure 2. The present report demonstrates the possibilities to design protein chips based on specially constructed, small recombinant antibody fragments, using microstructure surfaces with biocompatible characteristics resulting in highly sensitive detection of antigen in the 600 attomole range.

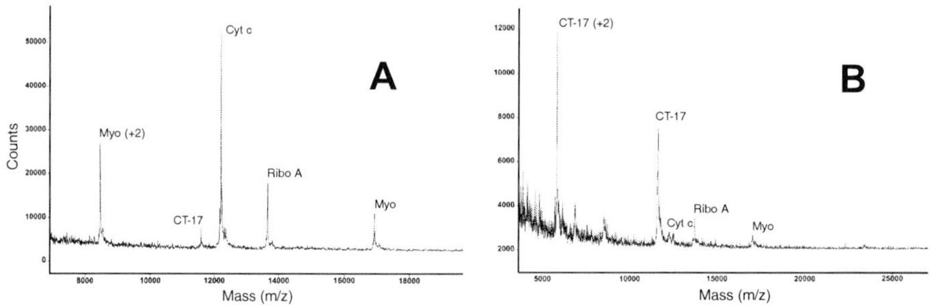

Figure 2. The MALDI spectrum obtained before washing the chips (Fig. 2A) revealed the presence of background proteins in high yields, whereas the specific choleratoxin antigen signal was weak. However, after the washing step (Fig. 2B) a significant enhancement (40-50 times) in MS signal ratio between choleratoxin and the unrelated proteins could be observed.

4. Conclusions

The combination of a highly selective affinity interaction and the high sensitivity obtained by MALDI-TOF MS provides unique data features, where both biorecognition and structural information are obtained simultaneously in one assay cycle.

5. Acknowledgements

The single chain Fv antibody fragments from the n-CoDeR library were kindly supplied by Bo Jansson, **BioInvent Therapeutics AB**, Lund. This investigation was supported by the National Research Council of Engineering Sciences (TFR).

References

1. Borrebaeck, C.A.K. Antibodies in diagnostics – from immunoassays to protein chips *Immunol. Today* **21**, 379-382 (2000).
2. Laurell, T., Wallman, L. & Nilsson, J. Design and development of a silicon microfabricated flow-through cell for on-line picolitre sample handling. *J. Micromech. Microeng.* **9**, 369-376 (1999).

DISPENSER-AIDED NANODIGESTION FOR RAPID PROTEIN IDENTIFICATION IN MALDI-TOF NANOVIAL ARRAYS

David Ericsson[a], Simon Ekström[b], Johan Nilsson[b], Jonas Bergquist[c], György Marko-Varga[d], and Thomas Laurell[b]
[a]Dept. Analytical Chemistry, Lund University, P.O. Box 124, 221 00 Lund, Sweden
[b]Dept. Electrical Measurements, Lund Institute of Technology, P.O. Box 118, 221 00 Lund, Sweden
[c]Dept. of Analytical Chemistry, Uppsala University, P.O. Box 531, 751 21 Uppsala, Sweden
[d]AstraZeneca R&D, Molecular Sciences, 221 87 Lund, Sweden
E-mail: David.Ericsson@analykem.lu.se

Abstract
A technique for protein identification through enzymatic digestion in nanovial (nL) arrays and analysis with MALDI-TOF MS is presented. Through micro-dispensing of low-level protein solutions (10 nM-500 nM) and simultaneous evaporation the protein is concentrated up to 300 times in-vial, which increases Michaelis-Menten kinetics, resulting in identification after 0.2-10min digestions.

Keywords: Matrix-assisted laser desorption ionization- mass spectrometry, Protein identification, Nanovial array, Silicon flow-through micro-dispenser

1. Introduction

Protein identification through peptide mass fingerprinting by proteolytic cleavage and analysis with Matrix-assisted laser desorption/ionization–mass spectrometry (MALDI-MS) has proven to be a fast and highly automatable MS-technique. In order to detect and identify as many proteins as possible, the detection volumes should be kept down to enable a high degree of sample preconcentration, beneficial for reduced sample loss, improved Michaelis-Menten kinetics and enhanced detectability. Silicon vials as a target plate for MALDI-MS has been used earlier[1,2] as well as the use of micro-dispensing for MALDI-MS sample preparation[2] but our group has more thoroughly penetrated these topics[3,4].

2. Theory

Rapid and low-level enzymatic catalysis is enabled by the large surface-to-volume ratio in the nL-vials and the dispenser-aided sample enrichment feature. By dispensing of the protein sample and simultaneous evaporation, the protein concentration [S] increases, which gives a higher reaction velocity (V) according to the Michaelis-Menten equation.

$$V = V_{max} [S] / ([S] + K_M) \qquad (1)$$

J.M. Ramsey and A. van den Berg (eds.), Micro Total Analysis Systems 2001, 87–88.

3. Experimental

The microdispenser ejects 100-pL droplets (insert B Fig. 1) and it has an inlet to nozzle volume of 250 nL. The chip-based nanovials has the dimensions 300x300x20 μm (1.8 nL) (insert A Fig.1) placed in arrays of 10x10 per chip. Both the microdispenser and the nanovials were made by anisotropic wet-etching of <100>-oriented monocrystalline silicon. A motor-driven x-y stage controls the dispensing into different vials.

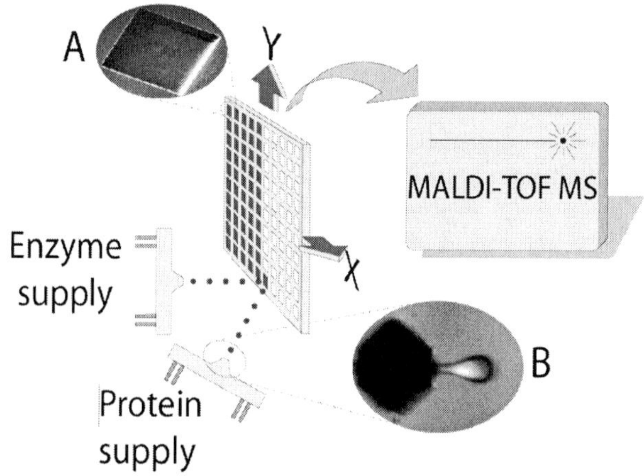

Figure 1. Nanodigestion platform.

4. Results

The fast catalysis of in-vial digestion was demonstrated by the analysis of hemoglobin (10μM). Peptides from the α- (5 peptides, covering 43% of the amino acid sequence) and β-subunit (4 peptides, 31% of aa seq.) of hemoglobin were achieved after only 30-sec digestion, resulting in unambiguous identification of both subunits by database matching. Low-concentration digestion and identification was exemplified by a 10-nM solution of lysozyme C, which was catalyzed for 5 min, resulting in 7 peptides corresponding to 46% of the amino acid sequence of lysozyme C.

5. Discussion

The well-defined dimensions in the nanovial array enable both automated dispenser sample handling and MALDI-TOF analysis. Miniaturized nanovial arrays promote that hundreds or thousands of samples can be analyzed in the same lot and be stored in a high-density form for later re-analysis.

References

1. S. Jespersen, et al., *Rapid Commun. Mass Spectrom.,* pp. 581-584, *8*, (1994).
2. D. P. Little, et al., *Anal. Chem.,* pp. 4540-4546, *69*, (1997).
3. S. Ekström, P. Önnerfjord, J. Nilsson, M. Bengtsson, T. Laurell, G. Marko-Varga, *Anal. Chem.,* pp. 286-293, *72*, (2000).
4. S. Ekström, D. Ericsson, P. Önnerfjord, M. Bengtsson, J. Nilsson, G. Marko-Varga, T. Laurell, *Anal. Chem.,* pp. 214-219, *73*, (2001).

MICROMANIPULATION OF A SINGLE DNA MOLECULE USING LASER ASSEMBLING OF MICROPARTICLES

Ken Hirano[1], Yukiko Matsuzawa[2], Yoshinobu Baba[1], and Akira Mizuno[2]
[1]Department of Medicinal Chemistry, The University of Tokushima
1-78 Sho-machi, Tokushima City, Tokushima 770-0044, Japan
[2]Department of Ecological Engineering, Toyohashi University of Technology
1-1 Hibarigaoka, Tempaku-cho, Toyohashi, Aichi 441-8580, Japan

Abstract
We have developed a manipulation method of a single DNA molecule for oncoming analysis targetting in a single DNA on a chip. Using laser assembling of beads, it was possible to manipulate a single DNA molecule without any modification of it. Moreover, the properties of laser assembling of microparticles was confirmed.

Keywords: DNA, single molecule, laser assembling, laser trapping

1. Introduction

Novel micromanipulation technique of the single DNA molecule should be developed towards a single DNA analysis on a chip [1]. Laser trapping technique is the most widely recognized as a technology for single DNA molecule manipulation, but it is not applicable to the manipulation of native DNA molecule without some modification, such as DNA-bead complex [2]. A single particle such as a bead and a cell can be trapped easily for manipulation by laser trapping technique [3], whereas a random-coiled DNA molecule cannot be trapped.

In this study, we propose novel micromanipulation technique for native DNA molecule with laser-assembling method. Figure 1 shows a concept of DNA manipulation using laser assembling of bead. This technique is based on laser trapping to assemble microparticles at laser focal point. This assembling process is effective in catching a DNA molecule as a long polymer by gathering it with beads. A feature of laser assembling of beads is investigated. This method allows us to manipulate efficiently native DNA molecule without special modification such as DNA-bead complex. Furthermore, the method may have potential to manipulate long biopolymers such as chromosomal DNA molecule and actin filament for catching and separating inside channels on a chip.

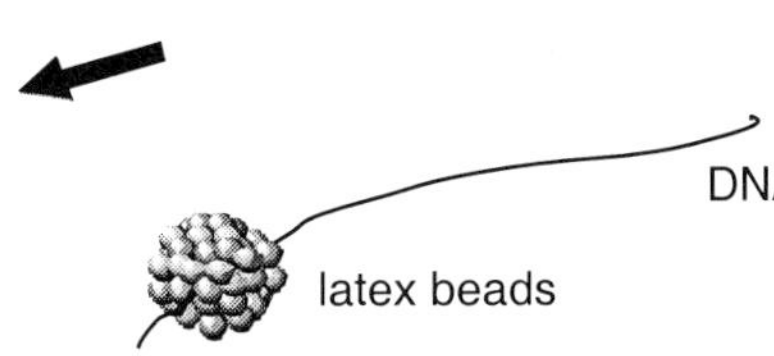

Fig. 1. Concept of manipulation of a single DNA molecule.

2. Properties of laser assembling

Trapping capacities against microparticle size, laser power and moving velocity as temporary flow velocity, are confirmed. Figure 2 shows number of laser trapped beads depended

J.M. Ramsey and A. van den Berg (eds.), Micro Total Analysis Systems 2001, 89–90.

upon bead size. The trapped beads increases with decreasing particle size, and latex bead in 0.2μm diameter is peak of the number. A laser power also affects number of trapped beads (Fig. 3). The number increases depended upon laser power. Moreover, the number is inversely proportional to moving velocity of assembled beads by drag force of solution (Fig. 4).

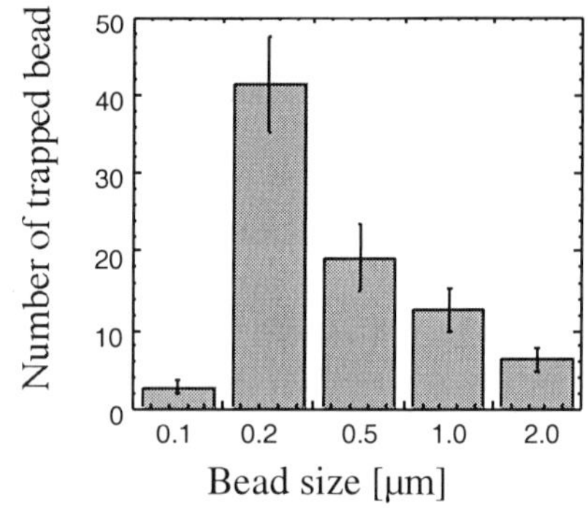

Fig. 2. Bead number vs bead size.

3. Micromanipulation of a single DNA molecule

Figure 5 shows snapshots of manipulation of a single DNA molecule by laser assembling of 0.2 μm latex beads. The laser assembling was carried out at 400mW by Nd:YAG laser, 1064nm in wavelength. The laser beam turned on over DNA molecule at which time, a single T4 phage DNA (166kbp) was grasped (Fig. 5a). When a microscope stage was moved, the held DNA molecule was manipulate following stage moved direction to upward (Fig. 5b) and leftward (Fig. 5c-d). At turned off the laser beam during manipulation, the DNA molecule could be released from trapped point immediately. This method provides simple way for manipulation of a coiled state DNA molecule.

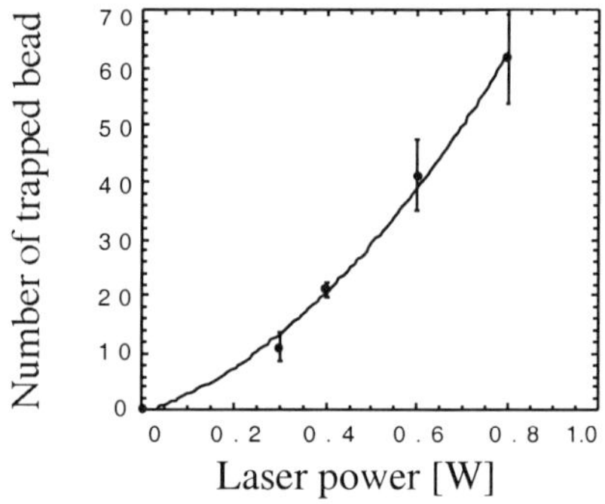

Fig. 3: Bead number vs laser power.

References

[1] K. Hirano, *et al.*, *Proceedings of the μTAS2000 Symposium*, 439-442, 2000.

[2] T. T. Perkins, *et al.*, *Science*, **264**, 822, 1994; Y. Arai, *et al.*, *Nature*, **399**, 446, 1999.

[3] A. R. Wheeler, *et al.*, *Proceedings of the μTAS2000 Symposium*, 25-28, 2000.

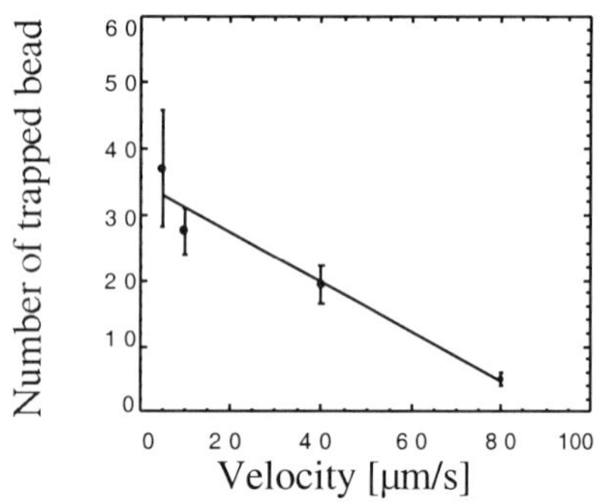

Fig. 4. Bead number vs moving velocity.

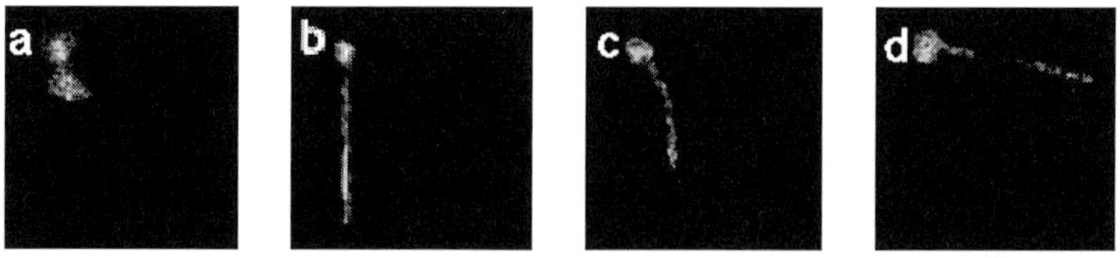

Fig. 5. Manipulation of a single T4 phage DNA by laser-assembling method.

QUANTITATIVE BIO-AFFINITY ASSAYS AT NANOLITER SCALE, PARALLEL ANALYSIS OF CRUDE PROTEIN MIXTURES

Mats Inganäs, Gunnar Ekstrand, Johan Engström, Ann Eckersten, Helene Dérand and Per Andersson
Gyros AB, Uppsala, Sweden

Keywords: integrated microfluidics, bio-affinity assay, crude sample protein analysis, monoclonal antibodies

1. Introduction

The traditional approach to identify/analyze proteins in a proteomics perspective is to separate crude samples of proteins by 2-D electrophoresis prior to further processing and identification of individual proteins by, for example, MALDI–MS. The process is cumbersome and time consuming and allows only limited parallel processing of samples. In this work we describe an alternative approach to quantitatively determine the presence of multiple proteins in crude mixtures in parallel.

2. Results and Discussion

The assay employs specially designed microfluidic reactions structures in a CD format. Each structure has a specific functionality necessary for the quantitative analysis of proteins (Figure 1). One key function is the capturing bed that has to fulfil requirements such as appropriate ligand density, suitable flow properties and compatibility with fluorescent in situ detection.

We have used human myoglobin as a model system to optimize the bio-affinity assay. Monoclonal antibodies against two different epitopes on myoglobin were used to construct a flow-based bio-affinity assay. Liquid movement within the reaction structure is driven by centrifugal force when spinning the CD. The assay principle is based on the interaction between the ligand in excess and the analyte in the crude mixture, as liquid flows slowly through the capturing bed in a controlled fashion. The overall process is controlled by timely sample and reagent flow over the bed (5-15 nl) at flow rates (~1 nl/s). Flow rates must be compatible with the diffusion properties and reaction kinetics needed to achieve efficient analyte capture. Bound analyte is subsequently detected using a fluorescent second reagent in excess.
Achievements so far indicate that parallel analyses can be performed within 10-15 min, using 50-200 nl of sample and generating a tentative dynamic range of myoglobin of 1-100 nM (Figure 2). Further optimization is in progress.

The assay procedure offers a promising alternative for rapid, parallel analysis of multiple proteins in crude samples by combining bio-affinity principles and highly integrated microfluidics at nanoliter scale working with femtomolar amounts.

J.M. Ramsey and A. van den Berg (eds.), Micro Total Analysis Systems 2001, 91–92.

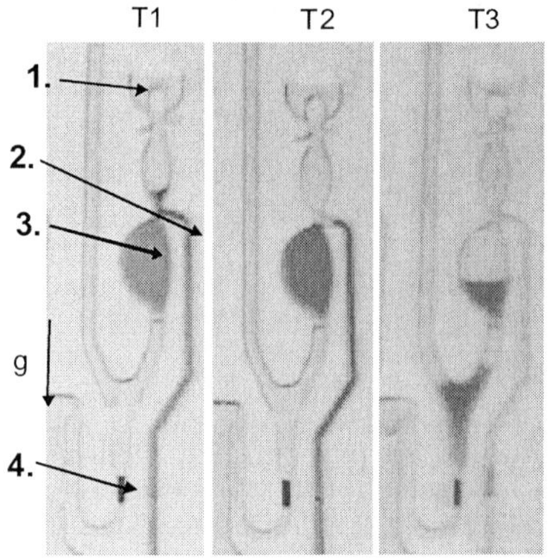

Figure 1. Microfluidic reactions structures in a CD format.
The structure has a specific functionality necessary for the quantitative analysis of proteins: 1. sample inlet; 2. overflow channel, 3. volume defining container (60 nl); 4. particle capturing bed. T1= overflow channel activated, 2 seconds at 2000 rpm; T2=volume defined in container (3), 3 seconds at 2000 rpm; T3= loading the sample by increasing the rpm, 1 second at 3000rpm. The direction of g-force is labelled "g".

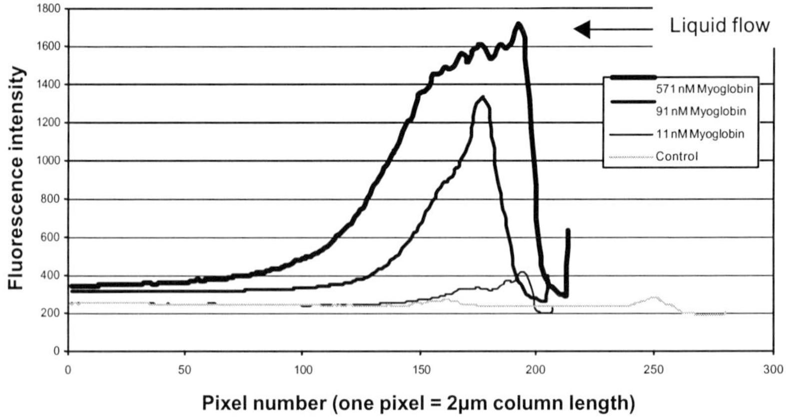

Figure 2. Dose-response relation between myoglobin concentration and fluorescence intensity.
A capture bed of approximately 400 µm length (~10 nl), prepared by covalently coupling of anti-myoglobin monoclonal antibody to NHS-activated Superose™ 12 HR, was packed into the reaction structure. Samples containing myoglobin at various concentrations were simultaneously processed in four reaction structures equipped with a 200 nl volume definition structure. Liquid flow was controlled by spinning the disc to generate sample flow along the longitudinal axis of the capture bed as indicated by the arrow. Bound myoglobin was detected over a one second time period using an ALEXA-labelled monoclonal antibody.

NOVEL METHOD FOR STRETCHING OF MEGA BASE PAIR DNA IN A MICROSPACE TOWARD A SINGLE CHROMOSOMAL DNA ANALYSIS

Noritada Kaji, Masanori Ueda and Yoshinobu Baba
Department of Medicinal Chemistry, Faculty of Pharmaceutical Sciences, The University of Tokushima, CREST, Japan Science and Technology Corporation (JST), Tokushima 770-8505, Japan

Abstract

Here we propose novel and fundamental method toward realizing a single chromosomal DNA analysis for future human chromosomal sequencing. Long DNA sized over a few kilo-basepair, especially chromosomal DNA, exhibit random-coiled conformation in a solution. To obtain information on genome directly, stretched conformation is more suitable than random-coiled one. In this paper, we propose a novel method for stretching of chromosomal DNA in a miniaturized agarose gel applying a low frequency AC electric field.

Keywords: Chromosomal DNA, Direct Observation, AC electric field, agarose gel

1. Introduction

Human genome sequencing has been almost completed [1], but determination of the genome sequence of many other organisms is still required, since comparative genomics allows biologists to identify conserved functional features and recognize new innovations in specific lineages. To realize the full promise of comparative genomics, however, it needs to become simple and inexpensive to sequence the genome of any organism. Microfabricated chip technology is one of the promising technology for future sequencing and more revolutionary method, single-molecule sequencing, has not yet been fully developed, but hold great promise and deserve strong encouragement [1].

2. Experimental

The direct observation of stretching process of a phage DNA (T4 DNA; 165.6 kbp) and a chromosomal DNA (*Saccharomyces cerevisiae*; 0.225 – 2.2 Mbp) was carried out using fluorescence microscope equipped with miniaturized gel electrophoresis set (Fig. 1). After sample DNA were injected into working gel applying low steady electric field, we investigated effective condition for stretching DNA.

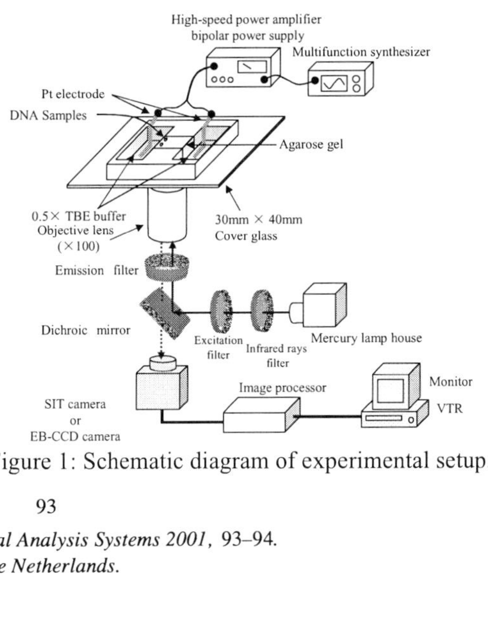

Figure 1: Schematic diagram of experimental setup.

J.M. Ramsey and A. van den Berg (eds.), Micro Total Analysis Systems 2001, 93–94.

3. Results and Discussion

We found applying 10-Hz sinusoidal AC electric field is extremely effective to stretch single DNA molecule in a miniaturized agarose gel as shown in Fig. 2B, whereas all DNA molecules are random-coiled state without applying electric field (Fig. 2A). We examined several conditions, including the frequency, the concentration of gel, field strength, and different DNA fragments. The frequency is the most efficient parameter for the stretching of DNA molecule, whereas other parameters are less effective for the stretching.

Optimized condition was successfully applied to stretching of *Saccharomyces cerevisiae* chromosomal DNA as shown in Fig. 3. The 10-Hz sinusoidal AC electric field is remarkably suitable to stretch even over 300-μm long DNA molecule corresponding to 2.2 Mega base pairs. The method reported here is promising to contribute to more rapid and inexpensive single genome analysis for future chromosomal sequencing.

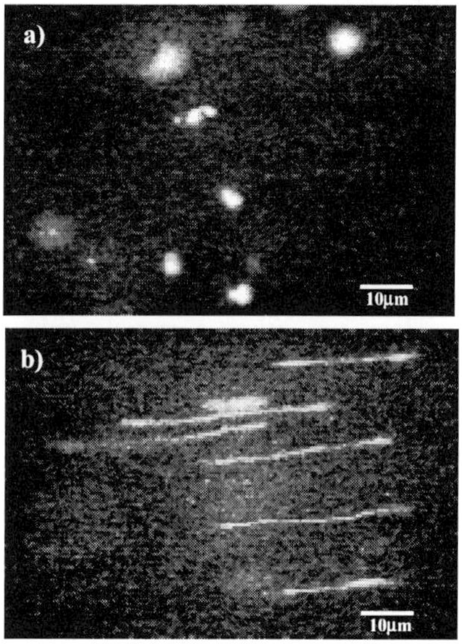

Figure 2: Fluorescence images of T4 DNA in 1% agarose gel. A) In the absent of external electric field. B) Under sinusoidal AC electric field (200V/cm, 9Hz). The electric field was applied in the horizontal direction.

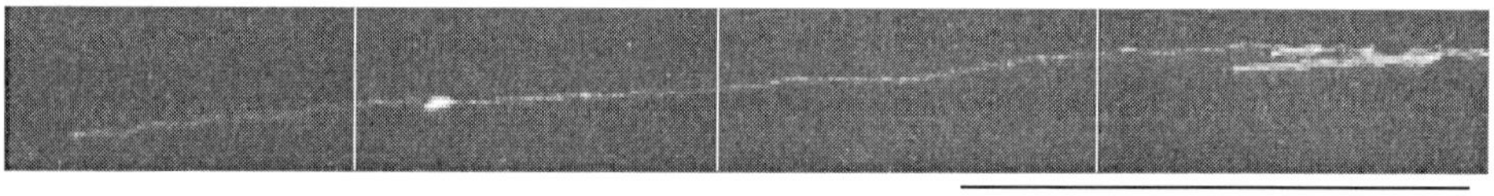

Figure 3: Fluorescence image of *Saccharomyces cerevisiae* chromosomal DNA, 2.2 Mbp fragments, under AC field ($200V_{pp}$/cm, 10Hz) in 1% agarose gel. Coil and bent parts never solute in spite of applying AC field more than few minutes. This image was recovered from four images to obtain well-focused one.

4. Conclusion

The investigation presented here suggests that the most important factor to stretch chromosomal DNA in agarose gel is the frequency of AC electric field. The method reported here provides a novel method for stretching of human chromosomal DNA and contribute to realizing single genome analysis.

Reference

1. International Humane Genome Sequencing Consortium, Initial sequencing and analysis of the human genome, *Nature*, 409, 860-921 (2001).

ANTIBODY LIBRARY SCREENING IN THROUGH-HOLE MICROARRAYS

Tanya S. Kanigan[1], Colin J. H. Brenan[1], Tara Heitner[1], Robert Hess[1], Arrin Katz[1], Elen Ortenberg[1], Karl Yoder[1], James D. Marks[2]
[1]BioTrove Inc., Cambridge, Massachusetts
[2]University of California, San Francisco, California

Keywords: microarray, biochip, phage display, antibody, screening

Introduction

Through application of recombinant DNA technology monoclonal human antibody libraries of over 10^9 antibodies are stored as plasmids in bacterial (*E. coli*) colonies. We have developed ultra high throughput methods suitable for screening these libraries after only one round of panning in precisely constructed, high density arrays of micro-channels, or bottomless wells.

Through-Hole Microarray Technology

This technology, which we refer to as the Living Chip,™ may be thought of as a dense array of test-tubes that are easily filled and washed. Currently we are working with microarrays containing up to 10,000 microchannels, each holding ~50 nl of sample. Aqueous solutions may be instantaneously loaded into the hydrophilic channels by dipping. Once loaded, the solutions are retained by means of surface tension. By aligning and stacking two chips, mixing of reagents may be performed on all micro-channels simultaneously (Fig. 1), thus accomplishing massively parallel initiation of reactions. Inhomogeneous assays, such as ELISA, may be performed by immobilizing compounds inside the channels and taking advantage of the flow-through nature of the chips. Reactions are monitored in parallel by a CCD-based fluorescent imaging system.

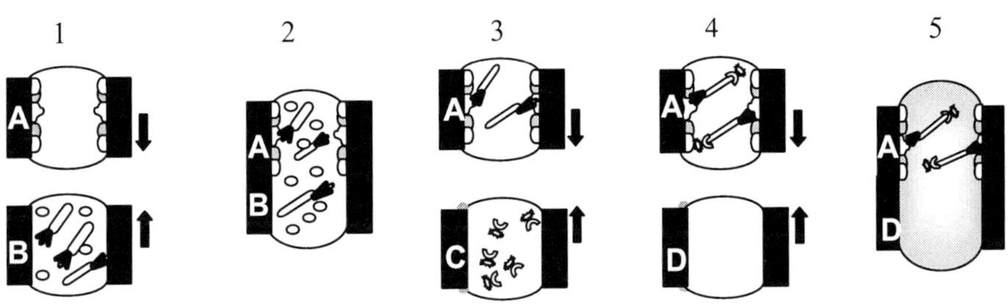

Fig. 1. The sequence of microarray stacking operations used to detect the binding of a phage displayed antibody to a complimentary antigen, shown for a single channel of a through-hole microarray.

J.M. Ramsey and A. van den Berg (eds.), Micro Total Analysis Systems 2001, 95–96.

Antibody Capture and Detection

As a proof of concept we have demonstrated the expression of phage in *E. coli*, and the selective binding of the phage-displayed antibody to a known antigen in a 100-channel microarray. Referring to fig. 1: Antigen was immobilized in the channels of Chip A. Non-selective binding was blocked with a solution of BSA. Chip B was dip-loaded with a solution of *E.. coli* expressing phage (1). The two chips were stacked and incubated to allow for binding of antibody-antigen interaction (2). After incubation the chips were washed and separated. Chip A was then stacked with Chip C containing α-M13-HRP, which binds to a protein expressed on the phage outer membrane (3). Again the stack was incubated, rinsed and separated. Chip A was then stacked with Chip D filled with solution of amplex-red substrate. The chip was then imaged with a CCD-based fluorescent imaging system to detect the conversion of the substrate by the enzyme bound to the phage (Fig. 2). A significant increase in signal (4:1) was detected in channels containing antibody expressing phage compared to the control, a phage-displayed antibody that exhibits non-specific binding.

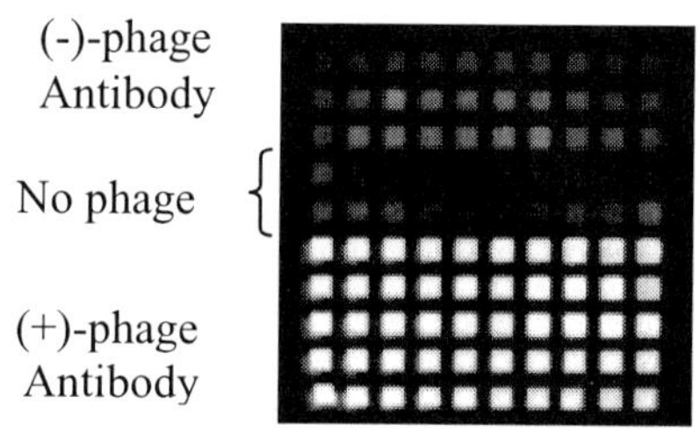

Fig. 2. ELISA detection of selective binding of phage-displayed antibody to a known antigen in a 10 x 10 through-hole microarray. This image was collected with a 1s camera integration time.

Summary

We have demonstrated a number of basic operations critical to screening antibody libraries in our unique Living Chip™ microassay platform. These operations include growth of bacterial colonies, expression of phage, antibody immobilization in the through-holes, blocking, washing, dip-loading of solutions, incubation and fluorescent detection.

Acknowledgements

This work was performed under the support of the U.S. Department of Commerce, National Institute of Standards and Technology, Advanced Technology Program, Cooperative Agreement Number 70NANB1H3003.

IN VITRO PROTEIN SYNTHESIS ON A HIGH-ACCUMULATED MICROCHAMBER CHIP

Takeshi Kinpara, Yuji Murakami, Yasutaka Morita, Kenji Yokoyama and Eiichi Tamiya
The School of Materials Science, Japan Advanced Institute of Science and Technology, 1-1 Asahidai, Tatsunokuchi Ishikawa 923-1292 JAPAN
(kinpara@jaist.ac.jp)

Abstract

The high-accumulated microchamber was fabricated with polydimethylsiloxane (PDMS). A thin PDMS sheet having pores through the sheet was bonded with slide grass using the oxygen plasma treatment. The volume of a microchamber was about 150 pL. The *in vitro* protein synthesis was performed on the chip.

Keywords: *in vitro* protein synthesis, PDMS, microchamber, protein chip

1. Introduction

It is an important technique to develop a protein library chip from a DNA library for post-genome analysis. In the proteome analysis it is principal to detect the protein-protein interaction or other compounds interacted with protein. The protein library chip is useful for these methods. The accumulated protein chip will allow becoming the high throughput chip. We were succeeded to make a microchamber for micro PCR [1] and perform one cell PCR on the chip. In this report, we fabricated a microchamber chip from polydimethylsiloxane (PDMS) and performed the *in vitro* protein synthesis on the chip.

2. Experimental

Fabrication of PDMS chamber

First, the template was molded on the silicon plate by using the SU-8, a thick photoresist. The SU-8 template had about 20 μm height. PDMS was poured onto the template and a transparency film was placed over the prepolymer mixture. The template/prepolymer/film stack was then clamped. The clamped PDMS sandwich was cured for 1.5 h at 70°C. The PDMS thin layer was cleaned with diluted HCl and was extended in methanol or acetone. The extended sheet was put on a slide glass and treated with oxygen plasma to bond the sheet and the grass. One well size was about 100 μm x 100 μm and the depth was about 15 μm. The volume of one well was about 150 pL. The chip size was 4 cm^2 and there were 10,000 wells on the chip

On chip *in vitro* protein synthesis

J.M. Ramsey and A. van den Berg (eds.), Micro Total Analysis Systems 2001, 97–98.

The microchamber made into PDMS was used for *in vitro* protein synthesis. This chamber had hydrophobic surface (PDMS) and hydrophilic bottom (grass). This structure prevented the contamination over the wells. Ribosome system based on the lysate from *Escherichia coli* (RTS-500 kit from Roche) was used. Green fluorescence protein (GFP) gene was expressed in the chamber. The microchamber chip was covered with cover glass and sealed with manicure to prevent the drying. 0.1% BSA was added to synthesis solution as a blocking agent against the absorption.

3. Results and discussion

The chamber structure having hydrophobic surface and hydrophilic bottom prevented the contamination over the wells. Thus, the protein synthesis solution poured onto the chip was only staid in the wells.

In this experiment, the GFP expression was used as a marker. Figure 1 showed the expression of GFP gene after 0 hour (A) and 1 hour (B). GFP expression was observed with fluorescent intensity. Fluorescence intensity was arisen in 1 hour and then little change was observed till 12 hours after starting of protein synthesis.

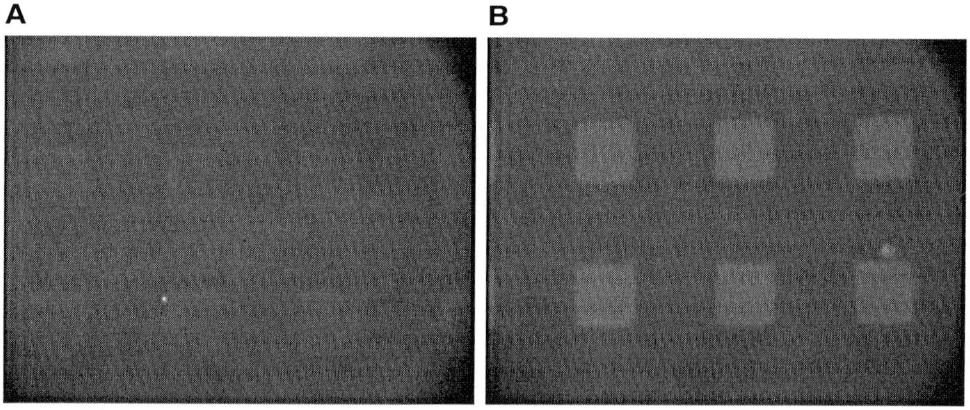

Figure 1. Fluorescence images of GFP expressed on the PDMS chip.
A: after 0 hour. B: after 1hour.

4. Conclusions

We made a PDMS chamber having hydrophobic surface and hydrophilic bottom. This chamber enable to retain the solution in the wells. *In vitro* protein synthesis on this chip was successful and the GFP expression was detectable within 1 hour.

References

1. H. Nagai, Y. Murakami, Y. Morita, K. Yokoyama and E. Tamiya, *Development of a microchamber array for picoliter PCR*, Anal. Chem., **73 (5)**, 1043-1047, (2001).

A MICROFABRICATED HYBRID DEVICE FOR DNA SEQUENCING

Sharong Liu, Jianzhong Zhang, Hongji Ren, Jianbiao Zheng, and Hanghui Liu
Molecular Dynamics/Amersham Pharmacia Biotech, Sunnyvale, CA 94086

Abstract

Round channels of different diameters have been microfabricated on chips to connect capillaries. Semicircular grooves are obtained using masks with very narrow line-width and a standard isotropic photolithograohic etching process. Round channels are formed when two etched wafers are face-to-face aligned and bonded (Figure 1). To extend straight separation channel on electrophoresis chips for DNA sequencing, we have created a hybrid device that couples a microfabricated twin-T injector on a microchip to a separation capillary (Figure 2). A two-mask fabrication process is used to make round channels of two different diameters, one matches the i.d. and the other the o.d. of the joining capillary. The capillary is inserted into a microchip through the larger channels, allowing integration of capillary bores to the rest of the channels on the chip. The dead-volume of the connection is estimated about 5 pico-liters. To demonstrate the efficiency of this hybrid device, we have performed four-color DNA sequencing on it. Using a 200-μm twin-T injector coupled with a capillary of 20-cm effective separation distance, readlengths of more than 800 bases have been obtained at an accuracy of 98.5% in 56 min. At higher separation field strength, the separation time has reduced to 20 min to separate 700 bases at 98.5% base-calling accuracy. Performance of a 16-channel hybrid device with 30-cm long effective separation distance demonstrated robust DNA sequencing over 800 bases at 98.5% in less than 1 hour.

Keywords: Microfabrication, Round Channel, DNA Sequencing

1. Introduction

Microfabricated electrophoresis chips for DNA sequencing have many advantages due to their unique injection schemes [1]. First, a uniform signal intensity profile is usually obtained over a wide range of fragment sizes. Secondly, a chip injector introduces narrow sample plugs into a separation channel. Thirdly, DNA template can be removed using a chip-injector [1]. Finally, a chip-injector requires a very small volume of sample [2]. However, it is challenging for microfabricated electrophoresis chips to achieve very long readlength DNA sequencing, due to limited straight channel lengths. In order to attack this problem, large plates were employed to fabricate half-meter-long separation channels [3,4]. Fabrication of this huge chip is challenging and its yield is low.

We present here a method of making chip-injector arrays of round channels on a chip and separation capillaries are coupled with these chip-injector arrays. Based on this

J.M. Ramsey and A. van den Berg (eds.), Micro Total Analysis Systems 2001, 99–100.

hybrid device, we have demonstrated the practicability and productability of microfabricated devices.

2. Experimental

Figure 1 presents a schematic scheme of the fabrication process of making round channels. Figure 2 presents a few images of the real device. In Figure 2a, the fat channel

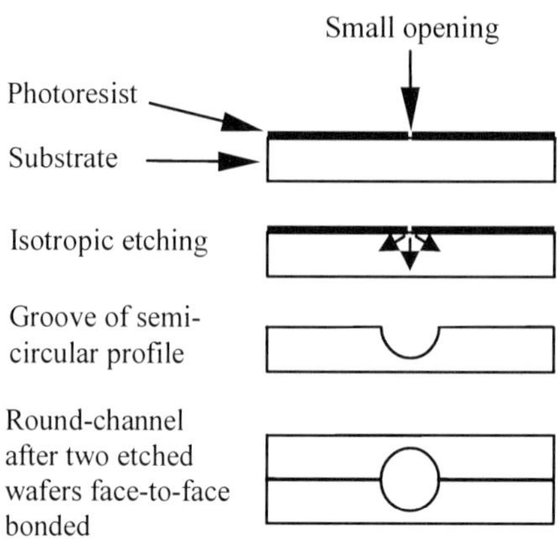

Figure 1. A schematic diagram for making round channels on microchips

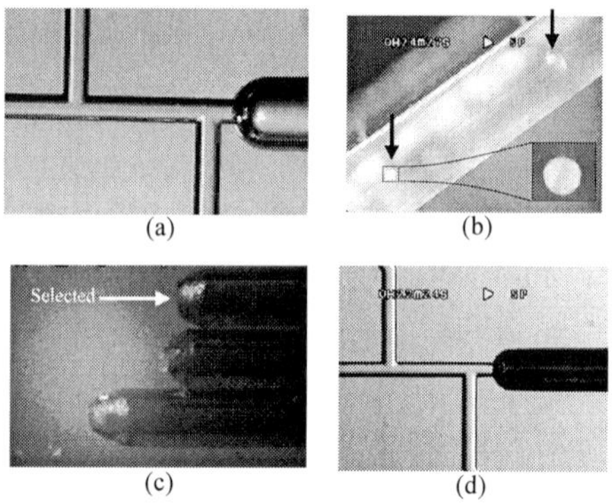

Figure 2. Images of real devices
(a) a 400 µm twin-T injector;
(b) cross section profile of a round channel;
(c) ground capillary tips;
(d) a twin-T injector incorporated with a capillary

on the right-hand side has a diameter of ca. 200 µm to host a 200-µm-o.d. capillary. All other channels have a diameter of ca. 75 µm. Figure 2b presents round profiles. Three capillary tips ground at different angles are presented in Figure 2c. In this experiment, the top one was used. The ground angle of this tip is about 30°. Figure 2d presents the injector region after the capillary is connected. From this figure, we can see that the o.d. and i.d of the capillary matched the channels pretty well. Table 1 summarizes sequencing data using a 16-channel hybrid device for 25 runs.

Table 1. Summary of sequencing conditions and results on a 16-channel hybrid device

Conditions		Results	
No. of channels:	16	No. of runs:	25
Chip-injector size:	250 µm	No. of separations:	400 (25 x 16 lanes)
Effective capillary length:	30 cm	Average readlength	
Total capillary length:	50 cm	(300 base cut-off):	824 bases
Separation field strength:	140 V/cm	Average success rate:	99%
Injection time:	90 s	Average run time:	60 min
Injection field strength:	140 V/cm		
Sieving matrix:	3% LPA		
Separation temperature:	65 °C		
Sample:	M13		
Scanning frequency:	2.7 Hz		

A NOVEL, HYDROPHILIC, ADSORPTIVE POLYMERIC COATING FOR ELECTROPHORETIC SEPARATION OF DNA AND PROTEINS IN SILICA MICROCHANNELS

M. N. Albarghouthi[1], B. A. Buchholz[1], J. P. Huiberts[2], T. M. Stein[2], A. E. Barron[1]

[1]Dept. of Chemical Engineering, Northwestern University, Evanston, IL, USA
[2]BioWhittaker Molecular Applications, Walkersville, MD, USA

Abstract

A novel, replaceable polymer matrix based on a new *N*-substituted acrylamide monomer Duramide^{TM} has been synthesized for application in DNA separations by microchannel electrophoresis. The monomer was found to be more hydrophilic than both acrylamide and *N,N*-dimethylacrylamide (DMA). Polymers were synthesized by free radical polymerization in aqueous solution and characterized by tandem gel permeation chromatography – multi-angle laser light scattering. PolyDuramide^{TM} exhibits good capillary-coating ability *via* adsorption from aqueous solution, efficiently suppressing electroosmotic flow (EOF) in a manner comparable to that of poly-*N,N*-dimethylacrylamide (PDMA). Under DNA sequencing conditions, adsorptive polyDuramide^{TM} coatings proved to be stable and to maintain negligible EOF for over 600 hours of electrophoresis. Resolution of DNA sequencing fragments, particularly fragments > 500 bases, in the novel matrix generally improves with increasing polymer concentration and decreasing electric field strength. Using polyDuramide both as a separation matrix and as a dynamic coating in bare fused silica capillaries, the matrix can resolve over 620 bases of contiguous DNA sequence within about 3 hours. Within a glass chip, the same results could likely be achieved in ~30 minutes. These results demonstrate the potential of the novel matrix for use in DNA separation, as well as proteins, by microchannel electrophoresis.

Keywords: capillary electrophoresis, chip, silica wall coating, polymer

1. Introduction

Hydrophilic polymer matrices are needed for the electrophoretic separation of DNA in glass and plastic microchannels to improve resolution and increase sequencing read length. Silica microchannels require polymeric coatings to suppress EOF. Preparation of covalently coated capillaries requires a difficult-to-control, tedious procedure, which limits reproducibility and increases channel cost. Here we report a novel hydrophilic polymer, polyDuramide^{TM}, that can be used as a DNA sequencing matrix in microchannel electrophoresis, as well as an adsorptive microchannel wall coating, efficiently suppressing EOF and minimizing band-broadening due to analyte adsorption.

2. Experimental

Monomer hydrophilicity was assessed by measuring the monomer partitioning

J.M. Ramsey and A. van den Berg (eds.), Micro Total Analysis Systems 2001, 101–102.

coefficient in water/1-octanol system. Polymers were synthesized in aqueous solution using V-50 initiator. EOF in uncoated and polymer-coated capillary was measured. Stability of the polyDuramide coating was assessed by measuring variation of EOF with electrophoresis time in a capillary filled with DNA sequencing buffer containing 0.1% polyDuramide™. DNA sequencing was carried out on MegaBACE 1000™ (Molecular Dynamics, Sunnyvale, CA) in polyacrylamide (LPA) covalently coated and in polyDuramide™ adsorptively coated capillaries. Testing in chips is work in progress.

3. Results and discussion

Monomer partitioning in water/1-octanol shows that Duramide™ monomer is more hydrophilic than acrylamide and DMA (Table 1). PolyDuramide™ also shows a remarkable ability to suppress EOF in fused silica capillaries (Table 2). The polymeric wall coating is stable for at least 600 hours of electrophoresis time in conditions similar to those employed in DNA sequencing.

Table 1. Partition coefficient of monomers in water/1-octanol

Monomer	Partition coefficient
Duramide	0.11 ± 0.02
Acrylamide	0.17 ± 0.03
DMA	0.50 ± 0.02

Table 2. Comparison of mobility of EOF, μ_{eo}, in polymer-coated capillary

Coating polymer	$\mu_{eo} \times 10^8$ (m^2/Vs)
No polymer	4.177 ± 0.226
PDMA	0.075 ± 0.027
PolyDuramide	0.070 ± 0.011

Using LPA-coated capillaries, the read length produced by polyDuramide matrices, as well as the run times, increase with increasing polymer concentration and decreasing electric field strength (Table 3). In uncoated capillaries, a polyDuramide matrix resolves 620 DNA bases at 98.5% accuracy in about 3 hours of run time. This read length is comparable to that produced by commercially available DNA sequencing instruments. Further improvements of read length may be achieved by using higher temperatures and optimizing base-calling software for use with this novel matrix.

Table 3: Read lengths and run times obtained in polyDuramide matrices (5.2 MDa) under different electric field strengths and polymer concentrations.

READ LENGTHS (bases)	Field strength (V/cm)			
PolyDuramide conc. (%)	140	117	94	70
4	445	550	560	620
5	570	620	660	700
6	620	680	735	750

RUN TIMES (min)	Field strength (V/cm)			
PolyDuramide conc. (%)	140	117	94	70
4	65	83	167	219
5	93	155	168	224
6	109	185	245	304

LATERALLY ASYMMETRIC DIFFUSION ARRAY WITH OUT-OF-PLANE SAMPLE INJECTION FOR CONTINUOUS SORTING OF DNA MOLECULES

Mario Cabodi[1], Yi-Fan Chen[2], Stephen Turner[2], Harold Craighead[2]
[1]Physics Department, Cornell University, Ithaca NY
[2]School of Applied And Engineering Physics, Cornell University, Ithaca NY

Abstract:

A new injection scheme is proposed and tested that allows continuous injection of sample in a molecular filtration device. This scheme can be used in many other types of devices.

Keywords: DNA Fractionation, Brownian Motion Rectification, Laser Micromachining, Microfluidics, Out-of-plane Sample Injection

1. Introduction:

A DNA fractionation method based on rectification of Brownian motion was first proposed by Duke and Austin [1] and Ertas [2] in 1997, and subsequently demonstrated to work by Chou et al. [3] in 1999. A two-dimensional array of obstacles can be used to rectify the Brownian motion of particles (in this case DNA molecules) so that they diffuse preferentially in one direction, perpendicularly to the direction of the applied field. This type of device has not yet been used for actual fractionation of DNA molecules, due to difficulties in injection of the DNA sample. Here we show a new injection strategy that provides a well-defined, narrow and continuous stream of molecules into the separation channel.
This injection method should enable this already proven separation technique to be used in a working device, as well as be applied to other microfluidic systems.

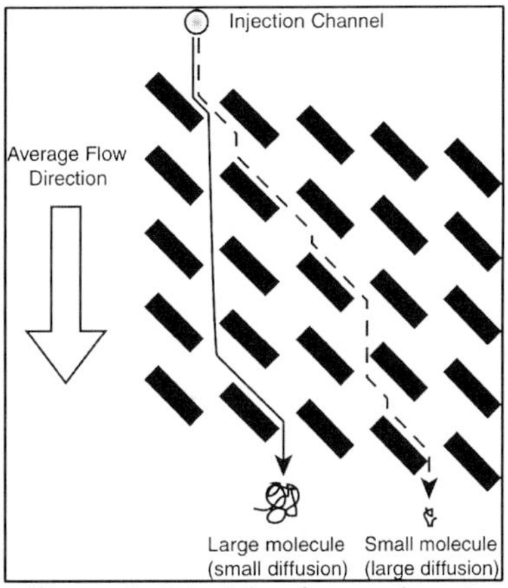

Figure 1: Principle of separation. An average flow is maintained in the main channel direction. When encountering an obstacle, smaller molecules (larger diffusion constant) will tend to be deflected to the right, while large molecules will tend to go straight.

2. Fabrication:

Conventional silicon etching techniques were used to create the 2D obstacle array. The channel depth was 20 microns and the obstacle size was 12x6 microns. An excimer

J.M. Ramsey and A. van den Berg (eds.), Micro Total Analysis Systems 2001, 103–104.

laser was used to create a small vertical (through-wafer) channel for sample injection. The device was then oxidized and anodically bonded to a Pyrex® glass coverslide and PDMS reservoirs were glued to the backside.

3. Out-of-plane sample injection scheme:

Both hydrostatic pressure and electrophoresis can be used to optimize the injection conditions. With the correct hydrostatic pressure conditions, the sample can be kept from spreading into the separation channel. Electrophoresis can then be used to inject molecules in a controlled fashion. This method is superior to in-plane injection schemes because it avoids the broadening of the sample plug when injecting from a narrow channel into a larger separation channel.

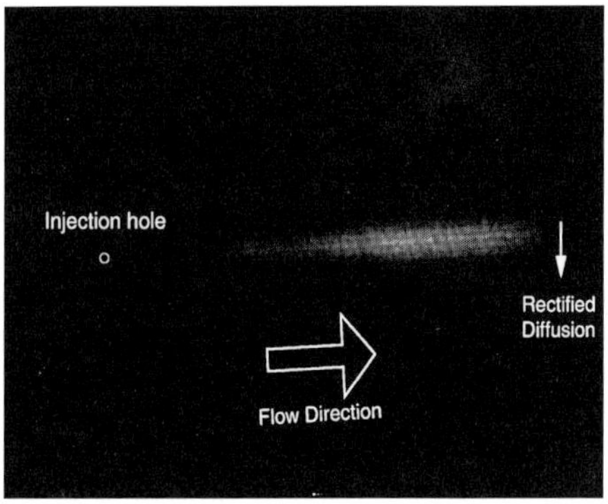

Figure 2: A photograph of the DNA molecules downstream of the injection port. The molecules can be seen to diffuse preferentially in one direction.

4. Experimental:

A sample containing a mixture of T7 and T2 DNA (39 kbp and 164 kbp respectively) was injected through the vertical channel, while a uniform field was maintained in the channel direction by applying a voltage to the reservoir electrodes. Individual DNA molecules were tracked and observed to diffuse laterally as they traveled down the device. The amount of lateral diffusion was consistent with the observed molecule size. Further downstream this lateral diffusion would result in the separation of the two species of molecules.

5. Conclusions:

We successfully tested a novel injection scheme (Out-of-plane Sample Injection) for continuous sample injection into a laterally asymmetric diffusion array. We expect this scheme to work with many other types of microfluidic devices.

Acknowledgements:
This project was supported by the National Institute of Health and the Department of Energy. All the fabrication work was carried out at the Cornell Nanofabrication Facility.

References:

[1] Duke, T. A. J. & Austin, R. H., *Phys. Rev. Lett.* **80,** pp.1552–1555. (1998)
[2] Ertas, D., *Phys. Rev. Lett.* **80,** pp.1548–1551. (1998)
[3] Chou C.F., Bakajin O., Turner S.W.P., Duke T.A.J., Chan S.S., Cox E.C., Craighead H.G., Austin R.H., **PNAS** 96 (24): pp.13762-13765

MAPPING OF pH GRADIENTS IN MICROFLUIDIC ELECTROKINETIC DEVICES

Catherine R. Cabrera[1], Paul Yager[1]
[1]Department of Bioengineering, University of Washington, Seattle, Washington

Abstract

As part of a larger effort to perform isoelectric focusing and electrophoresis in a microfluidic device [1,2], image analysis techniques that permit non-invasive quantitative measurement of pH in a microfluidic channel have been developed [3]. By using two independent color channels, the ratio of the two forms of the indicator dye can be measured; this ratio is then correlated to pH. Methods for accounting for illumination variation have also been developed and shown to reduce scatter in the data. Results of this technique are compared to model predictions of dye concentration and pH gradients and found to be in good agreement.

Keywords: isoelectric focusing, electrophoresis, image analysis, pH measurement

1. Introduction

In sealed microfluidic devices, pH cannot be measured with standard invasive techniques. By quantifying the relationship between color and local pH for a non-fluorescent pH indicator dye, the pH in a microfluidic device can be non-invasively resolved both spatially and temporally.

2. Theory

The pH indicator dyes used in this technique change color upon protonation/deprotonation. The optical absorption at a given wavelength can be modeled by modifying Beer's law to account for absorbance contributions from multiple species (yellow and purple in the case of bromocresol purple):

$$OD_\lambda = \varepsilon_y C_y d + \varepsilon_p C_p d \qquad (1)$$

For a 3-chip CCD camera with log-linear response, the log of pixel value (PV) for each channel can be substituted for OD into Equation 1, resulting in the following equation:

$$\mathrm{Log(PV)} \propto \varepsilon_y C_y d + \varepsilon_p C_p d = m_y C_y + m_p C_p \qquad (2)$$

Two independent channels (typically green and blue) provide sufficient data to determine the pH at a given pixel position. The constants m_y and m_p are determined for each channel empirically from a calibration curve.

J.M. Ramsey and A. van den Berg (eds.), Micro Total Analysis Systems 2001, 105–106.

3. Experimental

The microfluidic electrochemical flow cell is constructed of layers of laminate sandwiching two pieces of palladium, which act as both the electrodes and two of the channel walls. For details of construction and use, see Reference 2.

The device is filled with a buffered solution of colorimetric pH indicator dye, such as bromocresol purple, and a voltage (~2.1V) is applied. Images are taken of the channel with an RGB cooled CCD camera coupled to the binocular output of a microscope. The images are processed to reduce the effects of illumination variation both within each image and among the set of images.

4. Results and discussion

The results of the image analysis are superimposed on model predictions of pH gradient formation [3] in Figure 1. The two data sets compare well at early times and at later times. The measurable pH range is typically ~1.5 pH units to either side of the pKa of the indicator dye, which for bromocresol purple is 6.3. The spatial resolution of this method is ~10 μm, determined by the degree of magnification of the channel and the resolution of the camera itself.

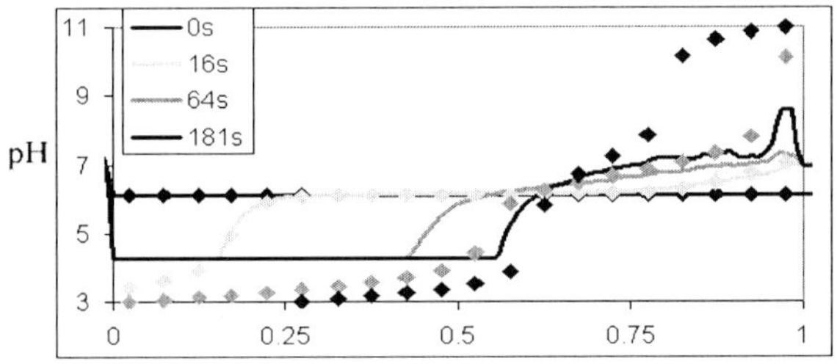

Figure 1. Comparison of model predictions (diamonds) to experimental data (lines).

5. Conclusions

A non-invasive quantitative method of pH gradient measurement has been applied successfully to a microfluidic device. Work is underway to develop calibration curves for a range of pH indicator dyes, thus expanding the sensitive range of the technique.

References

1. C.R. Cabrera, P. Yager, *Continuous concentration of bacteria in a microfluidic flow cell using electrokinetic techniques*, Electrophoresis, 22, pp. 355-362, (2001).
2. K. Macounova, C.R. Cabrera, P. Yager, *Concentration and separation of proteins in microfluidic channels on the basis of transverse IEF*, Analytical. Chemistry, 73, pp. 1627-1633, (2001).
3. C.R. Cabrera, B. Finlayson, P. Yager, *Formation of natural pH gradients in a microfluidic device under flow conditions. Model and experimental validation*, Analytical Chemistry, 73, pp. 658-666, (2001).

DIFFUSION BASED PROTEIN SEPARATION IN MICROFLUIDIC DEVICES

J. Gaudioso[1], D. E. Miller[1], S. W. P Turner[1], T. Jeoh[2], L. P. Walker[2] and H. G. Craighead[1]
[1]Nanobiotechnology Center and Department of Applied and Engineering Physics, Cornell University, Ithaca, NY 14853, USA
[2]Department of Agricultural and Biological Engineering, Cornell University, Ithaca, NY 14853, USA

Abstract
Preliminary results show separation of Cel9A and Cel5A cellulases in a diffusion based microfluidic device and they are in agreement with the model that we present. These proteins are difficult to separate with conventional separation techniques.

Keywords: on-chip separation, electrophoresis, chromatography, microfluidics

1. Introduction
Conventional chromatography and electrophoretic techniques can be miniaturized and, by operating on more biologically and chemically relevant lengthscales, new separation schemes can be devised. [1] We have developed prototypes for microfabricated diffusion based separation techniques (Fig. 1) that we are modeling through simulations and comparing to experimental results. Though these devices should be broadly applicable for separating many types of protein systems, we are studying enzymes that hydrolyze cellulose - cellulases.

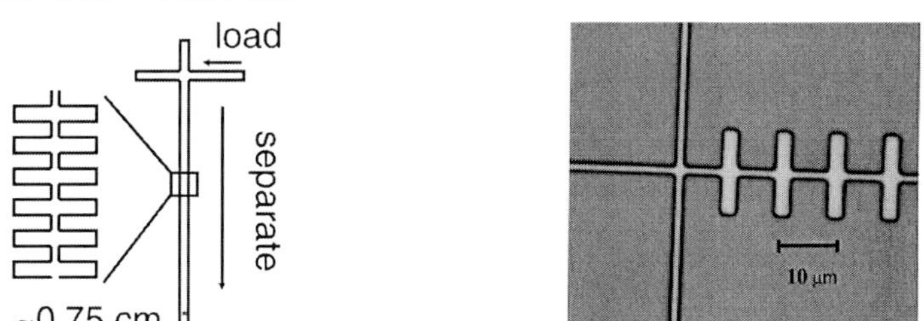

Figure 1. Schematic drawing (not to scale) and an optical micrograph of a prototype column for diffusion based protein separations.

2. Experimental
The diffusion well device is fabricated in a silicon wafer with standard photolithography and etching techniques at the Cornell Nanofabrication Facility. The channels are two µm wide and the wells are ten µm deep and have four µm wide openings. The cellulases are fluorescently labeled and an inverted Olympus IX70 microscope is used to monitor the fluorescence as the proteins move through the device.

J.M. Ramsey and A. van den Berg (eds.), Micro Total Analysis Systems 2001, 107–108.

3. Results and discussion

In this diffusion well prototype, the faster diffusing proteins have a higher likelihood of entering the wells and being retarded relative to the slower diffusing proteins. Thus the slower diffusing protein should elute from the column first. In preliminary experiments, we demonstrate separation of the Cel9A and Cel5A cellulases. Figure 2 shows the fluorescence intensity as a function of time at one point along the separation channel. The first peak is currently assigned as the slower diffusing Cel9A cellulase and the second peak is assigned as the Cel5A cellulase. These results are consistent with the model but further experiments are ongoing to verify the cause of separation.

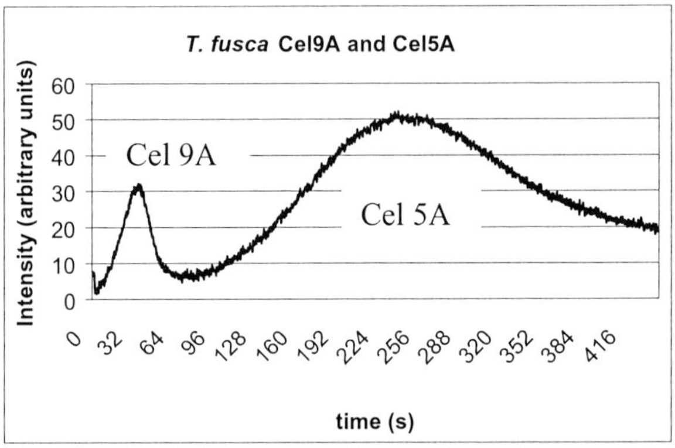

Figure 2. Observed separation of a cellulase binary mixture in the diffusion well device.

4. Conclusions

Mixtures of the *Thermobifida fusca* Cel6B, Cel9A, and Cel5A cellulases have proven difficult to separate by more conventional means [2] and this is an important hurdle to overcome in the process of understanding the molecular mechanisms of enzymatic cellulose hydrolysis to fermentable sugars. Thus our preliminary separation results illustrate some of the advantages of working on the microscale.

Acknowledgements

Funding from the DARPA Center for Biochemical Optoelectronic Microsystems and use of the Cornell Nanofabrication Facility is gratefully acknowledged.

References

1. J.P.Kutter, "Current developments in electrophoretic and chromatographic separation methods on microfabricated devices," Trends Anal. Chem., Vol.19, p352, (2000).
2. D.L.Watson, "Synergism in Binary Mixtures of Thermomonospora fusca Cellulases E3, E4, and E5 on BMCC and Avicel. Influence of Cellulase Mole Fraction and Concentration in Agricultural and Biological Engineering," (1997) Cornell University, Ithaca, NY.

DYNAMIC LABELING OF PROTEIN-SODIUM DODECYL SULFATE (SDS) COMPLEXES FOR LASER INDUCED FLUORESCENCE (LIF) DETECTION ON MICROCHIPS

B.C. Giordano[1], A.J. Couch[1], H. Ahmadzadeh[1], L.J. Jin[1], J.P. Landers[1, 2]
[1]Department of Chemistry, University of Virginia, Virginia, USA
[2]Department of Pathology, University of Virginia, Virginia, USA

Abstract
In order for protein separations to be performed on microchips, it is often necessary to covalently label the proteins prior to separation with fluorescent tags for laser-induced fluorescence (LIF) detection. This is necessary due to the difficulties in implementing UV detection on microchips. These tagging techniques are often labor-intensive and time-consuming. We present a method for dynamic labeling of protein-SDS complexes for LIF detection on microchips for the purpose of protein sizing.

Keywords: labeling, proteins, size separations, microchips

1. Introduction

While capillary electrophoretic separations of DNA and proteins are equally common, DNA separations are the predominant choice for microchip electrophoresis. This can be attributed to the ease of implementation of laser induced fluorescence detection (LIF) on microchips through the use of fluorescent dyes that dynamically intercalate into dsDNA. A similar mechanism for dynamic labeling protein-SDS complexes with a dye suitable for LIF detection during the electrophoretic separation has been determined and is presented here.

2. Theory

A commercial fluorescent dye, NanoOrange, is included in the run buffer during a protein-SDS size-based separation. This commercially-available dye is used for quantifying proteins in free solution. Upon binding to proteins or protein-SDS complexes, the dye undergoes a conformational change resulting in fluorescent emission (590 nm max) with excitation using a 488 nm light source. Several examples of CZE separations using NanoOrange for LIF detection have recently been published [1, 2]. However, in these works, NanoOrange binds to the native proteins, changing the electrophoretic mobility of the protein. It is necessary in CZE separation to utilize postcolumn labeling [1] or to address the mobility differences observed when NanoOrange is included in separation buffer [2]. Mobility differences are not observed when protein-SDS complexes are separated in the presence of NanoOrange.

3. Experimental

Electrophoretic separation conditions were first optimized on capillaries then translated to microchips. We show the results from evaluation of a series of buffer and

J.M. Ramsey and A. van den Berg (eds.), Micro Total Analysis Systems 2001, 109–110.

polymer systems suitable for on-column dynamic labeling with this dye for both capillary and microchip separations. The major advantages of this method of labeling include low background fluorescence, high signal-to-noise ratio, and ease of use (simply add the dye to the run buffer) without on-chip manipulation.

4. Results and discussion

The separation in Figure 1A resolves 8 proteins ranging in size from 14 to 200 kDa with UV detection implemented. Figure 1B and 1C demonstrates that NanoOrange can simply be added to CE-SDS separation buffer to allow for LIF detection. Comparision of Figure 1A and 1B demonstrates that NanoOrange can bind to protein-SDS complexes without significantly changing mobility.

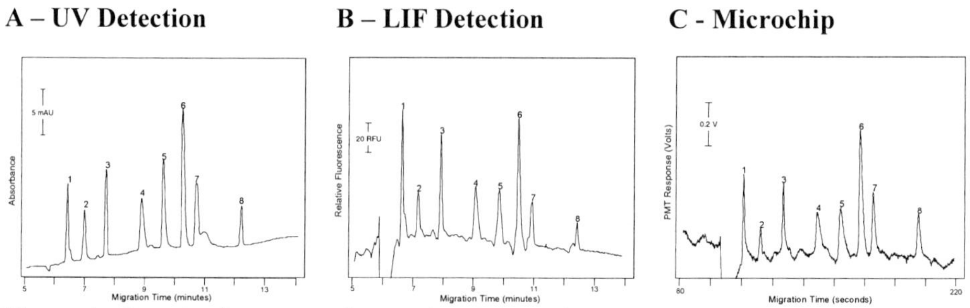

Figure 1. Protein-size separations using Bio-Rad CE-SDS separation buffer. Peak identification: 1 – Lysozyme (14.4 kDa), 2 – Trypsin inhibitor (21.5 kDa), 3 – Carbonic anhydrase (31.0 kDa), 4 – Ovalbumin (45.0 kDa), 5 – Serum albumin (66.2 kDa), 6 – Phosphorylase B (97.0 kDa), 7 – B-galactosidase (116 kDa), 8 – Myosin (200 kDa). A – Separation in a 27 cm (20 cm effective length) capillary, 50 micron i.d. using UV detection (214 nm). B – Separation as in A with buffer containing 0.2% NanoOrange with LIF detection (488 nm excitation). C – Separation performed in 7 cm microchip with buffer containing 0.2% NanoOrange.

5. Conclusions

We demonstrate that simple addition of a fluorescent dye to the run buffer of a CE-SDS separation allows for LIF detection on both the capillary and microchip platforms without any microchip manipulation.

References

1. Y. Liu, R.S. Foote, S.C. Jacobson, R.S. Ramsey and J.M. Ramsey, *Electrophoretic Separations of Proteins on a Microchip with Noncovalent, Postcolumn Labeling*, Anal. Chem. 72, 4606-4613, (2000).
2. M.D. Harvey, V. Bablekis, P.R. Banks and C.D. Skinner, *Utilization of the Non-covalent Fluorescent Dye, NanoOrange, as a Potential Clinical Diagnostic Tool: Nanomolar Human Serum Albumin Quantitation,* J. Chromatogr. B. 754, 345-356, (2001).

PARTICLE FOCUSING AND SEPARATION USING DIELECTROPHORESIS IN A MICROFLUIDIC DEVICE.

David Holmes and Hywel Morgan
Bioelectronics Research Centre, Department of Electronics and Electrical Engineering, University of Glasgow, Glasgow, G12 8LT, U.K.
(email: d.holmes@elec.gla.ac.uk, hywel@elec.gla.ac.uk)

Abstract

Separation of rare or unwanted bio-particles from heterogeneous mixtures is of great interest both for science and medicine. The ability to purify blood products of virus, isolate rare cell types for diagnostic purposes, obtain pure cell samples for further study are all important problems. We have developed a microfluidic device capable of addressing these problems. The device uses dielectrophoretic force to manipulate particles as they flow along a channel, with the desired subset of particles being retained within the channel.

Keywords: dielectrophoresis, particle separation, electrokinetic focusing, rare cells

1. Introduction

Dielectrophoresis (DEP) has been used increasingly in the field of particle separation. The technique has been shown to be useful for the manipulation of viruses, bacteria, cells as well as other biological entities [1]. Here we describe a microfluidic system capable of isolating individual particle types from a heterogeneous population, based upon the difference in their dielectric properties. The system can be used for the depletion of unwanted particles from a fluid (e.g. virus from blood plasma), and also for the continuous concentration of rare particles from large sample volumes.

2. Experimental

A fluidic chamber was fabricated, 100μm high, 10mm wide and 50mm long, in SU-8 on glass. Arrays of interdigitated microelectrodes were patterned onto the top and bottom glass surfaces. The initial top and bottom electrode arrays were energised such that all particles underwent negative DEP and were focused into a tight band 5-10um thick midway between the top and bottom of the channel. The second array of electrodes (only on the bottom wall of the chamber) were energised so particles of interest were pulled down and held at the electrode surface while other particles were transported through the device.

Figure 1. Separation device showing focusing and separation electrodes along with fluidic and electrical connections.

J.M. Ramsey and A. van den Berg (eds.), Micro Total Analysis Systems 2001, 111–112.

3. Computer simulation

The 2D analytical solution of the electric field above an array of interdigitated microelectrodes was used to calculate the DEP force experienced by a particle at any point in space [2]. The equations of motion for the system were solved numerically. Knowledge of the electrical properties of a particle and the suspending medium allow the trajectories of different particles to be predicted and the optimum experimental conditions chosen.

4. Results and discussion

A 20:1 mix of human peripheral blood mononuclear cells (PBMCs) and 6μm latex beads suspended in low conductivity sugar solution showed almost complete removal of PBMCs when passed through the device (see figure 2). The initial focusing of the particle stream results in banding of particles along the length of the separation electrode. The position of the band depends on the flow rate, applied voltage, particle and medium properties and electrode size. The various particle types follow different trajectories as they flow through the chamber. The system also shows removal of virus-sized particles from a protein rich medium similar to removal of virus from blood plasma for purification.

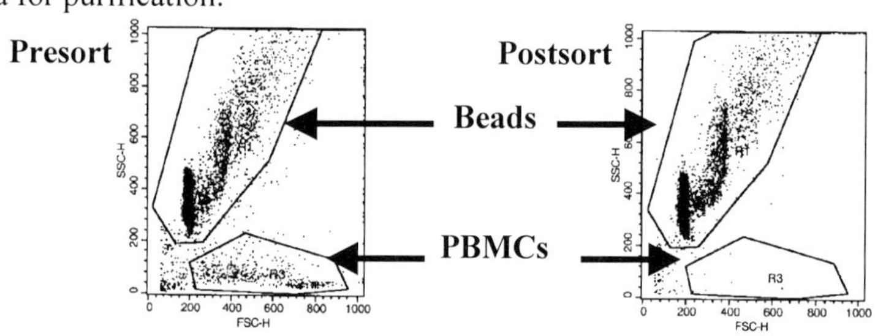

Figure 2. Separation of PBMCs from 6μm latex beads.

5. Conclusions

A method of separating bio-particles using DEP has been demonstrated. The system described can be used for the continuous isolation and concentration of rare particles from large samples. The initial focusing of the particle stream leads to banding of particles along the separation electrode, resulting in a novel form of on-chip separation.

Acknowledgements

DH would like to thank the EPSRC for funding and Dr M. Thomas for FACS use.

Reference

1. Pethig R. Dielectrophoresis: using inhomogeneous AC electrical fields to separate and manipulate cells. *Crit Rev Biotech* 1996, 16(4):331-348.
2. Morgan H., Izquierdo A.G., Bakewell D.J., Green N.G. and Ramos A. The dielectrophoretic and travelling wave forces for interdigitated electrode arrays I: Analytical solution using Fourier series J.Phys.D: Appl. Phys. 2000.

SEPARATION AND COLLECTION OF A SPECIFIED DNA FRAGMENT BY CHIP-BASED CE SYSTEM

Jong Wook Hong[1], Hisashi Hagiwara[2,3], Teruo Fujii*[1], Hiroaki Machida[2], Mika Inoue[4], Minoru Seki[5], and Isao Endo[6]
[1]Institute of Industrial Science, the University of Tokyo, Tokyo 153-8505, Japan (*e-mail: tfujii@iis.u-tokyo.ac.jp), [2]Bio-Equipment System Group, Hitachi Electronics Engineering Co., Ltd., [3]Human Genome Center, Institute of Medical Science, the University of Tokyo, [4]Aida Engineering Ltd., [5]Department of Chemistry and Biotechnology, the University of Tokyo, [6]Department of Applied Biochemistry, Utsunomia University

Abstract

Collection of a specified DNA fragment following electrophoretic separation on a chip has been firstly reported. The chip consists of two cross channels and six ports. A specified DNA fragment from 10 different-sized bands was collected successfully by controlling applied voltages.

Keywords: Chip-based CE, Separation and Collection, DNA fragment

1. Introduction

Chip-based CE systems for DNA samples have been widely investigated but most of them are just for separation of the sample. Although some trials to integrate PCR with chip-based CE system have been reported, collection of a specified DNA fragment following electrophoretic separation on a chip has not been reported. Collection of a specified DNA sample would expand applications of a microCE system, for example, amplification of specified fragment after fast separation. Here, we first report the on-chip operation of band collection of a separated DNA fragment.

2. Experimental

Chip fabrication & Imaging System

Lay out of the chip is shown in Figure 1. The chip is fabricated with PDMS (polydimethylsiloxane) by replica molding method reported previously [1]. The chip consists of two cross channels for sample introduction and separated band collection, and six ports for sample, buffer, and electrode introduction. Fluorescence of DNA samples was detected by an inverted fluorescence microscope and an ICCD camera. The image was recorded by a video recorder. Electrical voltage control is critical to collect specified DNA fragment, especially for two cross channels on the chip. We have developed a power supply system having 6 channels for electrical voltage control and current measurement. The applied voltage is summarized in Table 1.

J.M. Ramsey and A. van den Berg (eds.), Micro Total Analysis Systems 2001, 113–114.

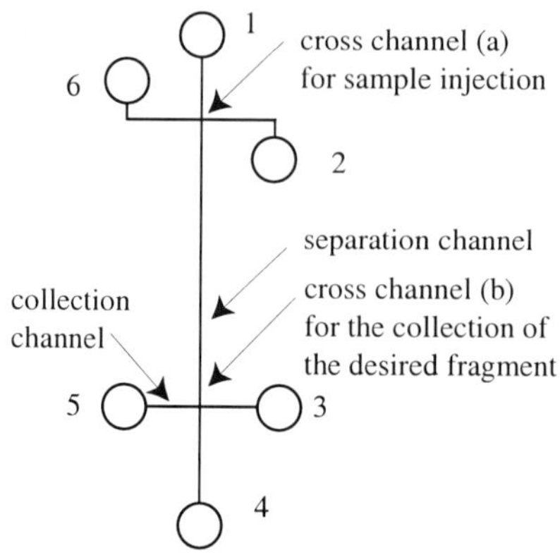

Fig. 1. Microfabricated PDMS chip for separation and collection of a specified DNA fragment.

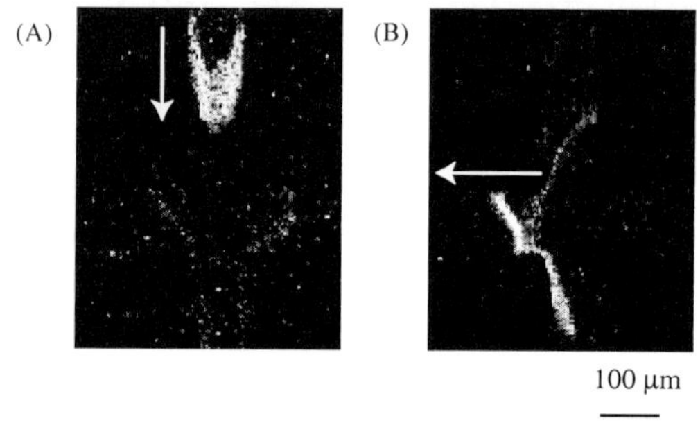

Fig. 2. Images of electrophoretic separation of specified DNA fragment from 10 different DNA bands (A), and collection of the specified fragment (B). Pictures were taken at the cross channel (b).

Table 1 Applying volatage (V) conditions for sample introduction (S1), sample separation (S2), and collection of a spcified sample (S3) on a microchip.

Port	1	2	3	4	5	6
S1	G	350	G	G	G	G
S2	G	130	G	750	G	130
S3	G	130	G	130	750	130

G is ground.

Separation and collection of a specified DNA fragment

Commercial 100 bp DNA size standard (ladder) was purchased from Bio-Rad Co. The sample was stained by SYBR-Green I prior to sample loading. Specified DNA fragment from 10 different DNA bands was collected by controlling the voltage among the channels.

3. Results and discussion

On-chip operation of band collection of a separated DNA fragment was successfully conducted by accurate and fast control of the electrical voltage among 6 ports (Fig. 2). On going research is the optimization of this process and the amplification of specified DNA fragment by PCR as the downstream operation.

Reference

1. J. W. Hong, T. Fujii, M. Seki, T. Yamamoto, and I. Endo, Integration of gene amplification and capillary gel electrophoresis on a PDMS-glass hybrid microchip, *Electrophoresis*, 22 (2), pp. 328-333, 2001

Purification of DNA Sequencing Products with a Model Compound in a High-Throughput Microfluidic Format, the ACAPELLA-5K

Ling-Sheng Jang, Mohan S. Saini, Mark R. Holl, and Deirdre R. Meldrum
Department of Electrical Engineering, Genomation Laboratory
University of Washington, Seattle, Washington, USA

Abstract
The development of a procedure based on streptavidin magnetic beads for purification of sequencing reaction products in microfluidic glass capillary channels is presented. Recovery of the DNA, up to 80%, was achieved using big dye terminator reagents in 2 μl reaction volume.

Keywords: Microfluidics, Sample preparation, DNA purification

1. Introduction

The purification of DNA sequencing products for high-throughput capillary array electrophoresis (CAE) sequencer analysis is a significant process bottleneck. The ACAPELLA-5K (A5K) automated microfluidic sample processing system can prepare and thermal cycle 5000 DNA templates per 8-hour day in 2 μL reaction volume [1]. Purification is required to isolate the sequencing reaction DNA fragments from excess dye terminators, salts, proteins and nucleotides. In this purification protocol, labeled oligo is used as a model compound for the DNA fragments. The labeled oligos are present in a biotinylated complex (Fig. 1). The streptavidin magnetic beads capture the biotinylated DNA complex. Miniature rare earth magnets are used to separate the magnetic beads from the carrier solution. Quantitative validation of the protocol was performed using a fluorescein labeled model compound.

2. Method

Sequencing Reaction Preparation: The sequencing reaction was prepared using big dye terminator reagents (Applied Biosystems, Foster City, CA) combined with our proprietary biotinylation reagent and throroughly mixed using the A5K.
Purification Protocol: Streptavidin magnetic beads (Seradyn, Indianapolis, IN, 0.8 μm, 0.5 μL, 10 mg/ml) were added to each capillary (length 55 mm; O.D. 0.84 mm; I.D. 340 μm; volume 5 μL). Thorough mixing of the beads with the solution was accomplished using the A5K piezo-mixer technology for 3 minutes [2]. The beads were magnetically separated from the solution and washed with rinse buffer (Fig. 2). Beads were subsequently captured again and washed with 70% ethanol. Residual ethanol was evaporated by vacuum. The DNA was eluted from the beads with 2 μL of 80 % formamide. The recovery of the label was calculated from its fluorescence against standards on Packard FluoroCount Fluorometer. The mixing efficiency depends on the

J.M. Ramsey and A. van den Berg (eds.), Micro Total Analysis Systems 2001, 115–116.

total throw distance and frequency. The total throw distance and frequency used were respectively, 16 mm and 3 Hz.

3. Conclusions

We have demonstrated the feasibility of a procedure to purify DNA using streptavidin magnetic beads, with a model label compound achieving up to 80% recovery. The protocol does not require extreme conditions; pH, denaturants like guanidium chloride, or heat to elute the DNA from the streptavidin magnetic beads, common in other commercially available protocols. This protocol, when fully automated, will result in significant cost savings of reagent, labor and time.

Acknowledgements

This research was supported by a NIH grant, Grant #HG02215-01.

References

1. D. R. Meldrum, H. T. Evensen, W. H. Pence, S. E. Moody, D. L. Cunningham, and P. J. Wiktor, *Genome Res*, vol. 10, pp. 95-104, 2000.
2. H. T. Evensen, D. R. Meldrum, and D. L. Cunningham, *Review of Scientific Instruments*, vol. 69, pp. 519-526, 1998.

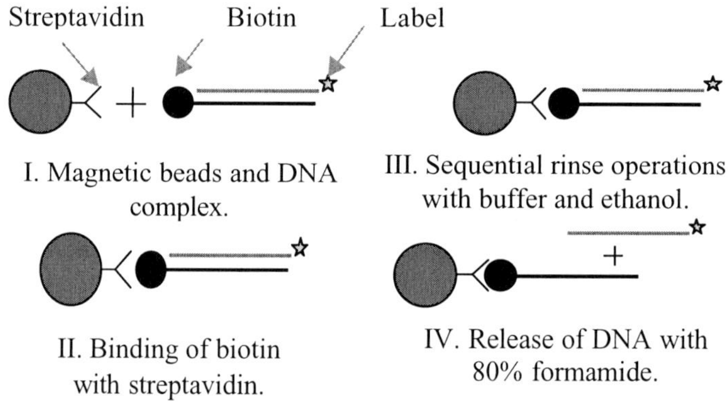

Figure 1: Biochemistry protocol.

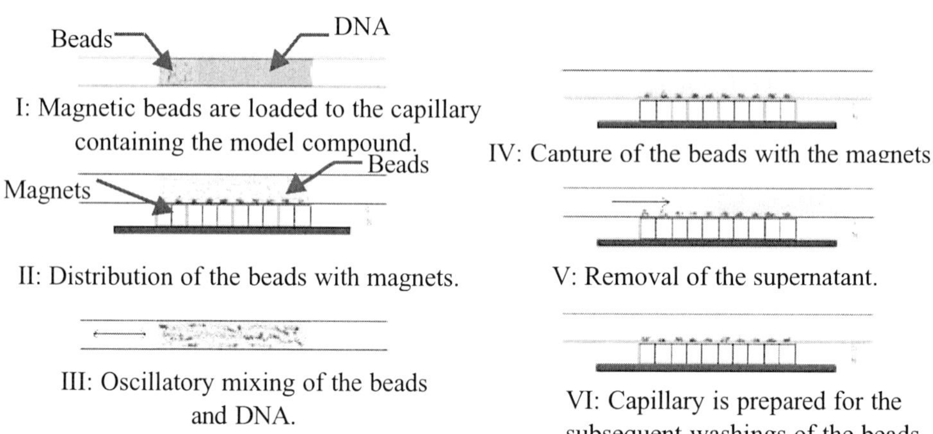

Figure 2: Procedure for the purification of a model compound.

INTEGRATED PCR-CE SYSTEM FOR DNA ANALYSIS TO THE SINGLE MOLECULE LIMIT

Eric T. Lagally[1] and Richard A. Mathies[1,2]
[1]UCB/UCSF Joint Bioengineering Graduate Group and
[2]Department of Chemistry, University of California, Berkeley, CA 94720

Abstract

We have developed an integrated PCR-CE system for DNA amplification and analysis from nanoliter volumes with single molecule sensitivity. In addition, thermal cycling elements have been microfabricated directly onto the device for improved thermal cycling efficiency. PCR-CE data demonstrating sex determination from human genomic DNA are presented.

Keywords: PCR, capillary electrophoresis, single molecule detection, RTD

1. Introduction

Our early work on microfabricated integrated genomic analysis systems included the development of the first integrated PCR reactor with a capillary electrophoresis (CE) system [1]. More recently, we devised a monolithic system for conducting PCR directly connected to a CE microchannel for product separation and analysis [2]. Samples are loaded precisely into a 200 nL PCR reactor using 50 nL dead volume valves and hydrophobic vents. The device takes 30 seconds/cycle and exhibits 10° C/second heating and cooling rates, representing a vast improvement over conventional thermal cycling systems. This PCR-CE device has recently demonstrated successful detection of a PCR product amplified from a single DNA template molecule, bringing this technology to the limiting molecular sensitivity of the PCR reaction [3].

2. Integrated thermal cycling elements

Recent improvements in our PCR-CE device include the fabrication of integrated resistance temperature detectors (RTDs) and heaters shown in Figure 1. Platinum RTDs are fabricated inside the PCR chambers and platinum heaters with gold leads are fabricated on the backside of the device. These integrated components have resulted in more accurate temperature measurement and more efficient heating of the reactor than previously possible. New heating and cooling rates are up to 20° C/second. PCR-CE analyses of sex determination from human genomic DNA [4] are presented in Figure 1.

3. Conclusion

These integrated nanoliter analysis systems can be easily configured to perform parallel amplification and separation reactions on a single device, enabling high throughput while greatly reducing reagent costs and eliminating manual steps. High-sensitivity analyses, such as comparative gene expression studies from individual cells, should also be possible. These results demonstrate a key step in the development of an

J.M. Ramsey and A. van den Berg (eds.), Micro Total Analysis Systems 2001, 117–118.

integrated microfluidic system or microprocessor that performs complete genetic analyses at sub-microliter volumes.

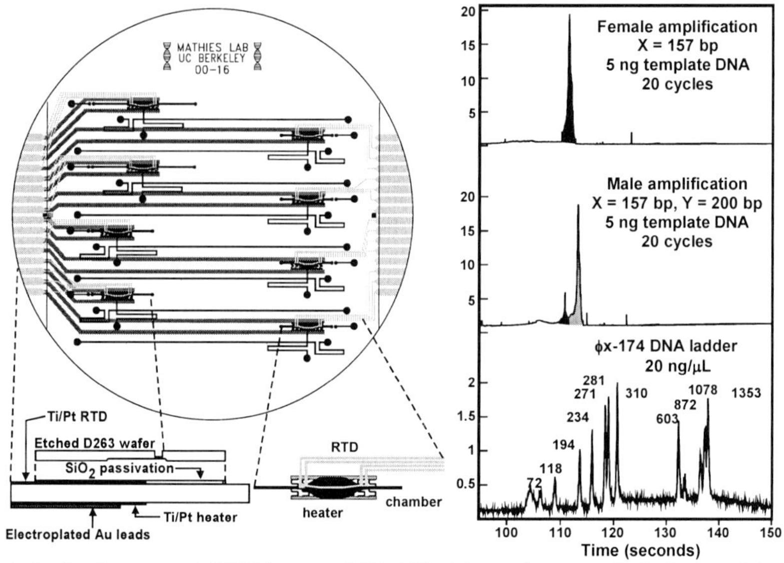

Figure 1. Left: Integrated RTD/heater PCR-CE chip with expanded views of the device cross-section and the chamber. Right: Series of centromeric alphoid repeat analyses from human genomic DNA performed with the PCR-CE chip. Total analysis time was 15 minutes.

Acknowledgements

This research was supported by the Director, Office of Science, Office of Biological and Environmental Research of the U.S. Department of Energy under Contract DE-FG-91ER61125. E.T.L. gratefully acknowledges the support of a Whitaker Foundation Predoctoral fellowship.

References:

1. A. T. Woolley, D. Hadley, P. Landre, A. J. deMello, R. A. Mathies and M. A. Northrup, *Functional Integration of PCR Amplification and Capillary Electrophoresis in a Microfabricated DNA Analysis Device*, Anal. Chem., 68, pp. 4081-4086, (1996).
2. E. T. Lagally, P. C. Simpson and R. A. Mathies, *Monolithic Integrated Microfluidic DNA Amplification and Capillary Electrophoresis System*, Sens. Actuat. B, 63, pp. 138-146, (2000).
3. E. T. Lagally, I. Medintz and R. A. Mathies, *Single-Molecule DNA Amplification and Analysis in an Integrated Microfluidic Device*, Anal. Chem., 73, pp. 565-570, (2001).
4. D. Neeser and S. Liechtigallati, *Sex Determination of Forensic Samples by Simultaneous PCR Amplification of α-Satellite DNA from Both the X and Y Chromosomes*, J. Forensic Sci., 40, pp. 239-241, (1995).

MICROFABRICATED POLYCARBONATE CE DEVICES FOR DNA ANALYSIS

Yingjie Liu[1], Dale Ganser[1], Alan Schneider[1], Robin Liu[1], Piotr Grodzinski[1], Natasha Kroutchinina[2]
1.Physical Sciences Research Laboratories, Motorola Labs, Motorola, Inc.
2. Motorola Biochip Systems
7700 S. River Parkway, MD-ML34, Tempe, AZ, 85284

Abstract
The microchip CE devices were fabricated in polycarbonate (PC) material by compression molding. Aqueous fluid transport inside the plastic CE devices was enhanced by UV irradiation treatment of the PC surfaces prior to thermal bonding. CE DNA separation was demonstrated in these PC CE devices.

Keywords: polycarbonate, capillary electrophoresis, DNA

1. Introduction

Polymer substrates are promising alternatives for building complex MEMS systems (1). Techniques such as plasma etching, laser ablation, imprinting, and injection molding are typically applicable to fabrication of devices in plastic materials. In this paper we will present the fabrication of polycarbonate microchip CE devices for DNA analysis.

2. Experimental

Polycarbonate CE devices were fabricated from PC pellets by compression molding. CE microchannels were enclosed using thermal bonding. PC surface was treated with UV irradiation before bonding to increase surface hydrophilicity.

3. Results and discussion

Because of strong hydrophobicity, filling PC CE channel networks with aqueous solution is very difficult. We have developed a method to increase PC surface hydrophilicity by UV irradiation, using exposure to UV light. When exposed to UV light in air, the surface of PC polymer is oxidised, which leads to the introduction of functional groups, and contributes to the increase of surface hydrophilicity (2). Figure 1 shows the contact angle between deionized water and UV exposed PC surfaces. A larger contact

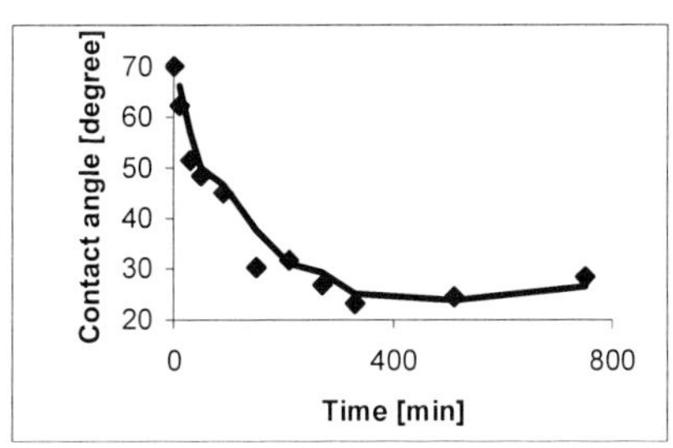

Figure 1. Effect of UV irradiation on contact angle between deionized water and UV irradiated polycarbonate wafer.

J.M. Ramsey and A. van den Berg (eds.), Micro Total Analysis Systems 2001, 119–120.

angle is associated with higher hydrophobicity.

The EOF flow velocities of different pH buffers were measured in native and UV treated PC microchannels (Figure 2). EOF in UV treated channels is higher than that in native PC channels under the same applied electric field and buffer condition. EOF mobility in native PC channels is independent of pH at the two-measured pH valves of 7 and 9.

Figure 3 shows the separation of DNA100 bp marker on PC CE chips. All 11 fragments of the marker were very well resolved. The CE chips used in this demonstration were four weeks old, and no signs of chip performance deterioration were observed. This means that the UV irradiation modified PC surface is stable under normal storage conditions

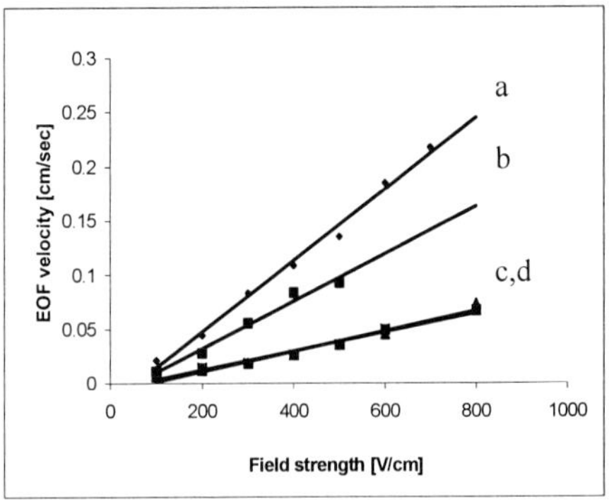

Figure 2. Effects of applied electric field strength and buffer pH on EOF of UV treated and native PC microchannels. a. pH: 9; channel wall was UV treated for 3 hours. b. pH: 7; channel wall was UV treated for 3 hours. c. pH: 9; native channel wall. d. pH: 7; native channel wall. Each data point in the plots is the average of three individual measurements.

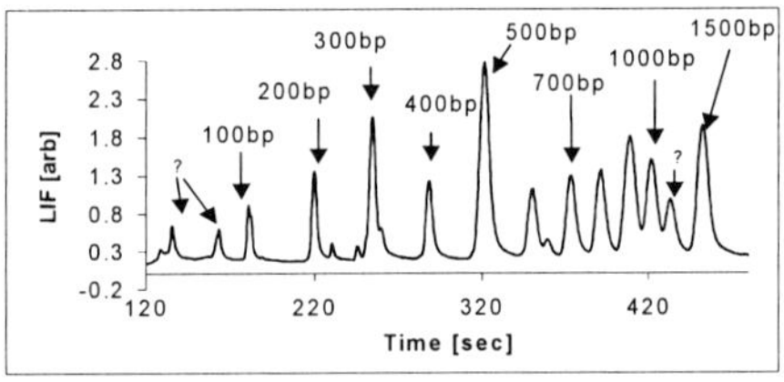

Figure 3. DNA separation on PC CE devices. Sample: 100bp ladder. Sieving matrix: 2% HEC. Electric field strength: 160 V/cm. Separation length: 2.8 cm

4. Conclusions

We have demonstrated the feasibility of CE DNA separation in compression molded PC microdevices. Surface treatment by UV irradiation increases PC surface hydrophilicity. Thermal bonding was performed after this surface treatment, and aqueous fluid transport inside the treated PC microchannels was seen to improve.

References

(1) Soper, S. A.; Ford, S. M.; Qi, S.; McCarley, R. L.; Kelly, K.; Murphy, M. C. *Anal. Chem.* 2000, *72*, 642A-651A.

(2) Scott, G. *Polymer Degradation and Stabilization*; Elsevier Applied Science: London, 1990.

THE IMPACT OF ELECTRIC FIELD DEPENDENCE OF DIFFUSION COEFFICIENTS ON MICROCHIP SEPARATIONS

Abdel Minalla and Luc Bousse
Caliper Technologies Corp., 605 Fairchild Drive, Mountain View, CA 94043

Abstract

In this work, we describe a method to measure diffusion coefficients for single-stranded and double-stranded DNA as a function of the separation field. We found that diffusivity is strongly field-dependent with or without the presence of a sieving matrix.

Keywords: Microchip separations, DNA separation, diffusion coefficient

1. Introduction

Study of the scaling of the speed and resolution of microchip separations with sample plug size [1] shows that there is an optimal separation length given by $L_0 = \frac{\mu E w^2}{24D}$, where w is the size of the initial sample plug. When the separation length is significantly greater than L_0, separations become diffusion-limited, and the microchip format loses the advantage of being able to define very small sample plugs. However, in order to apply this scaling law, it is necessary to know the diffusion coefficient, which is usually not accurately known.

2. Measurement of Diffusion Coefficients on Microchips

One method to measure a diffusion coefficient D in a capillary is to create a small plug of material, and observe how it spreads out as a function of time. When the detection point is fixed, time can only be varied by changing the electric field E. This will produce unreliable data if D is also a function of the field. In a microchip, however, it is easy to vary the detection point at a constant field. This method allows the measurement of D as a function of E. We have verified that for simple fluorophores such as fluorescein and rhodamine both methods give the same results and that D does not depend on E.

3. Measurement setup

To measure D, we use a glass microfluidic chip with channels 10 by 30 μm in cross-section, fabricated as reported previously [2]. The chip was mounted on a fluorescence microscope and connected to an eight-channel high-voltage power supply. Data was collected either by a photomultiplier tube, or by an intensified CCD camera. A simple cross device was used to generate plugs, whose size was tracked at various locations along the separation channel as they move at constant E. To measure diffusion at zero field, a plug was created, moved into the channel, and then allowed to diffuse there for a variable amount of time without moving. The plug was then moved past the detector location to measure its width. In all cases, D is determined from the slope of σ_x^2 as a function of the diffusion time.

4. Results for DNA

Results were obtained for both single-stranded and double-stranded DNA, initially in a sieving matrix (4% polydimethylacrylamide). Figure 1 shows the data for a 500 bp fragment detected with an intercalating dye, and a 21 base covalently labeled oligo. Diffusivity increases strongly with the electric field, especially for dsDNA. At zero field, D for dsDNA is very low (around 1×10^{-8} cm^2/s), but it increases by an order of magnitude by the time a

J.M. Ramsey and A. van den Berg (eds.), Micro Total Analysis Systems 2001, 121–122.

field of 200 V/cm is reached. The data for ssDNA is similar, but the field dependence is lower. To investigate whether this strong field dependence of D is due to DNA-sieving matrix interactions, we repeated the experiments without the sieving matrix. Somewhat surprisingly, we found that the field dependence was still present, but that all diffusivities were 3 to 4 times higher. In order to confirm these results, we took CCD images of moving or stationary ssDNA plugs. Figures 2 to 4 show these results on the same scale, and it can be seen that the plug size of ssDNA at 60 V/cm increases much faster than at zero field.

5. Conclusions

The most important conclusion is that increasing the electric field can significantly harm DNA separations by increasing dispersion. That means an optimum field must exist for each type of separation in a given sieving matrix. The data of Figure 1 will allow us to determine that optimum, both for dsDNA and ssDNA oligos.

References

[1] L. Bousse *et al.,* Annu. Rev. Biophys. Biomol. Struct. 29:155-181 (2000)
[2] L. Bousse *et al.,* μTAS 1998 Workshop Digest, p. 271.

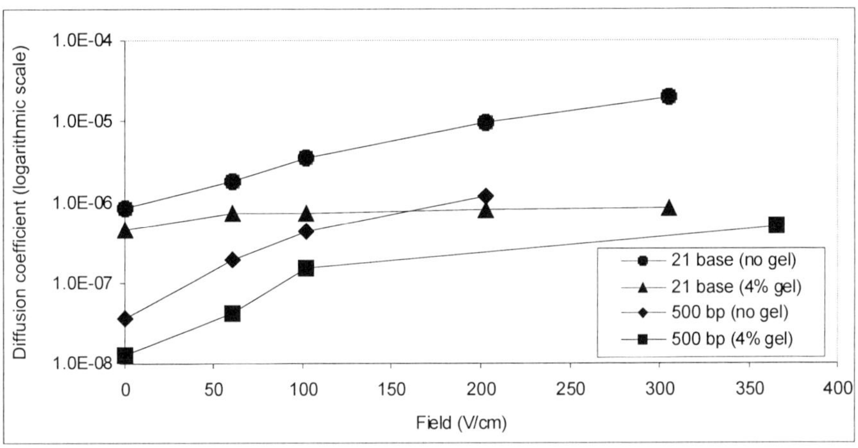

Figure 1: Diffusion coefficient of DNA as a function of the electric field.

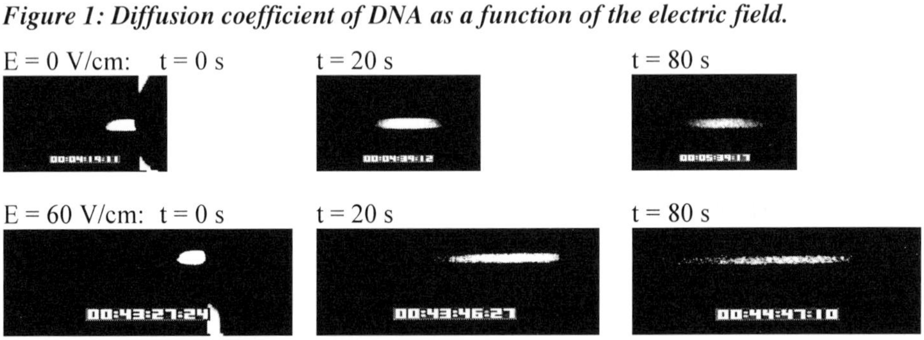

Scale for all pictures: 400 μm

Figures 2: Dispersion of ssDNA (50 bases) as a function of time in gel-free medium.

RAPID SAMPLE CLEANUP MICROCHIP FOR PROTEIN ANALYSIS BY ELECTROSPRAY IONIZATION MASS SPECTROMETRY

Young Chan Kim, Kyung Won Ro, Nokyoung Park, and Jong Hoon Hahn*
Department of Chemistry, Pohang University of Science and Technology,
San 31 Hyoja-Dong, Pohang, 790-784, South Korea
(email: hahn@postech.ac.kr)

Abstract

We have developed a simple, rapid sample cleanup method for the electrospray ionization (ESI) mass spectrometry of proteins by using a plastic microchip. Diffusion-based cleanup method between laminar flows offers effective removal of salts in about 1-second, and greatly enhanced signal-to-noise ratios of ESI mass spectrum.

Keywords: sample cleanup microchip, protein analysis, ESI mass spectrometry

1. Introduction

The operating paradigm for proteome analysis has quickly become the combination of two-dimensional gel electrophoresis with mass spectrometry [1]. The electrospray ionization (ESI) is one of the methods for mass spectrometric analysis in proteomics. The ionization efficiency of ESI process is, however, greatly reduced by the presence of salts in protein samples. Conventionally, dialysis has been employed in desalting from protein samples prior to ESI mass spectrometric analysis. However, it is very time consuming and has difficulties in handling small amounts of samples. We have developed a simple, rapid sample cleanup method for ESI mass spectrometry of proteins by using plastic microchip

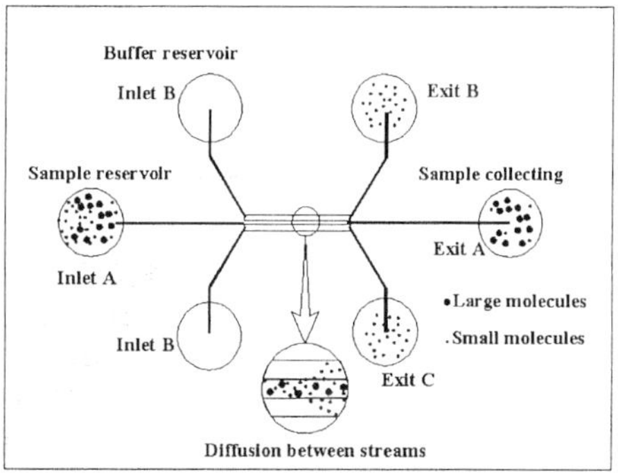

Figure 1. The concept of diffusion-based sample cleanup. A sample solution with small and large molecules enters the micro chip from the inlet A, and the buffer solutions serve as sheath flows enter from two inlets B All streams flow laminarly one another after merging at the central channel. Small molecules such as salts can diffuse out from the central stream but large molecules such as proteins do not. The three paralleled flows are then split up into three outlet channels. At the exit A, the sample with a reduced concentration of small molecules is collected.

2. Diffusion-Based Cleanup Microchip

The microfluidic device designed for this method is featured by diffusion-based rapid extraction between laminar flows [2]. In

J.M. Ramsey and A. van den Berg (eds.), Micro Total Analysis Systems 2001, 123–124.

this microchip, the sample laminar flow is sheathed by two buffer laminar flows for washing out salts in the protein sample as shown in figure 1. This microchip consists of a poly(dimethylsiloxane) (PDMS) plate, with channels molded on its surface, sealed against the flat surface of a slide glass. The cleanup is demonstrated in both off- and on-line modes, and the schematic of on-line cleanup system is shown in figure 2.

3. 1-Second Sample Cleanup

About 1-second diffusion-based cleanup of horse heart myoglobin with 500 mM NaCl gives an ESI mass spectrum as good as one obtained after 2-hour dialysis (figure 3). It should be also noted that only 0.5 micro liters of the sample, injected into the cleanup chip, is sufficient to obtain a mass spectrum with an excellent signal-to-noise ratio.

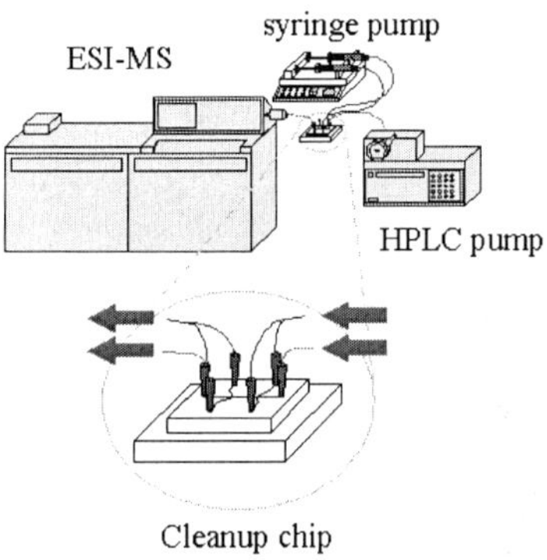

Figure 2. Schematic of on-line mode cleanup chip system.

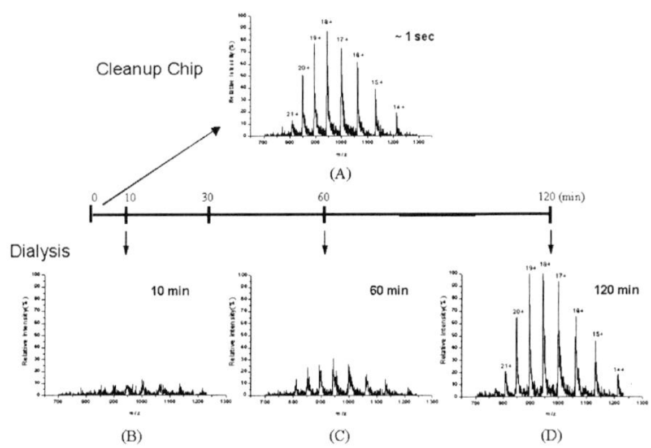

Figure 3. (A) ESI-mass spectrum of 1 mg/mL horse heart myoglobin in 500 mM NaCl, 100 mM Tris, and 10 mM EDTA after ~1 second off-line mode cleanup. (B) ESI-mass spectrum of the myoglobin sample after 10 min dialysis using a 1200 MWCO dialysis membrane tube. (C) After 60 min dialysis. (D) After 120 min dialysis.

4. Conclusions

We have developed a diffusion-based sample cleanup method for ESI mass spectrometry of proteins in a microchip. This approach is expected to greatly facilitate the high-throughput proteome analysis using ESI mass spectrometry.

References

1. H. W. Lahm, H. Langen, "Mass Spectrometry: A Tool for the Identification of Proteins Separated by Gels", *Electrophoresis*, **21**, (2000), 2105-2114.
2. B. H. Weigl, P. Yarger, "Microfluidic Diffusion-Based Separation and Detection", *Science*, **283**, (1999), 346-348.

FREE FLOW ELECTROPHORESIS MODULE FOR RAPID SAMPLE PREPARATION

Etsuo Shinohara[1], Nobuyoshi Tajima[1], Jun Funazaki[1], Koichi Tashiro[2], Shuichi Shoji[2], Yuichi Utsumi[3] and Tadashi Hattori[3]

[1] Genome Medical Business Project, Olympus Optical Co., Ltd., Tokyo, 192-8512 Japan
[2]School of Science and Engineering, Waseda University, Tokyo, 169-8555 Japan
[3]Laboratory of Advanced Science and Technology for Industry, Himeji Institute of Technology, Hyogo, 178-1205 Japan

Abstract
Newly designed FFE modules and a control system were developed. The separation time during the sample flow from the inlet to the outlets was one second when a mixture of genomic DNA, plasmid and albmin was used as a sample. Very high-speed separation and collection system was realized by the microfabricated FFE module.

Keywords: microfabrication, free flow electrophoresis, sample preparation

1. Introduction

Progress of molecular biology in recent years has contributed greatly to accelerate Genomics or Proteomics, and further progress in these fields would be expected from now on. In order to simplify the sample preparation procedure, we have been developed a free flow electrophoresis (FFE) module by means of microfabrication technology [1,2]. In this presentation, newly developed FFE modules and a control system for the module are described.

2. Experimental

The size of the module was 66 X 70 mm, in which 47 X 40mm separation bed with 30 μm deep flow gap was fabricated. Seven inlets of buffer solution, three inlets of sample and 19 outlets were formed on the module. Two platinum electrodes were formed on the both edges parallel to its flow direction (Fig.1). Automatic liquid handling system and silicon resin connector arrays for the FFE module were also developed to use the module easily.

The flow in the separation bed showed very stable laminar flow and homogeneous velocity distribution over wide range (Fig. 2), because of a wide but very narrow structure.

J.M. Ramsey and A. van den Berg (eds.), Micro Total Analysis Systems 2001, 125–126.

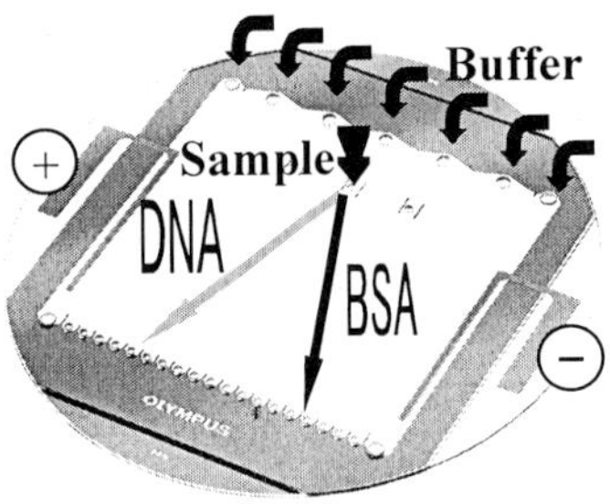

Fig.1 Free Flow Electrophoresis Module
Whole size:ϕ100 x 2mm
Size of separation bed:
56.5(W) x 35(L) x <u>0.03</u>(D)mm

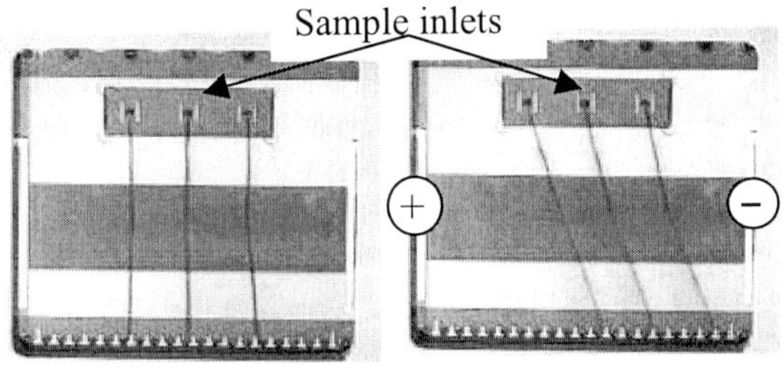

applied voltage: 0V 5kV

Fig.2 Test flow
In order to visualize the sample flow, Xylene cyanol dye was used as a sample.

2. Results and discussion

To investigate its function a mixture of genomic DNA, plasmid and albmin was applied to the FFE with 4 kV at a flow rate of 40 mm/sec. The genomic DNA and plasmid were recovered from outlet No. 4 and albmin from No.3. The recovery of the injected total volume of genomic DNA from outlet No.4 was about 83 % and the separation ratio was 96 percent (Fig.3). In this condition, the separation time during the sample flow from the inlet to the outlets was one second. Very high-speed separation and collection system was realized by the microfabricated FFE module.

3. Conclusions

FFE module based on a combination of microfluidics effects and electrophoresis was demonstrated to be a very powerful tool for sample preparation. Furthermore, its application would be expand in giant molecule separation such as organelles or small cells.

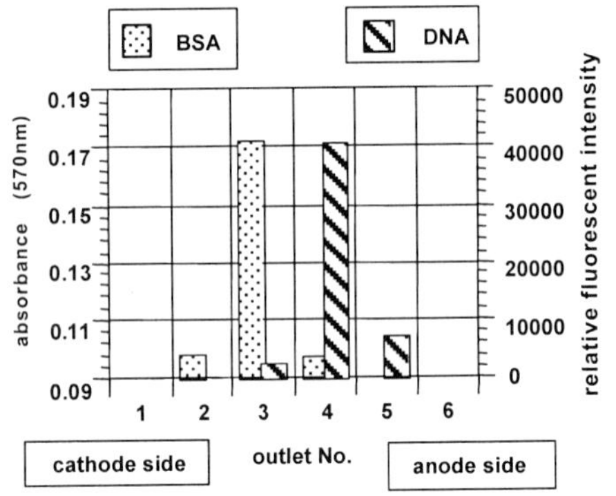

Fig.3 Fractions recovered from outlets
- mixture;DNA: 50mg/ml,BSA: 250mg/ml
- buffer: Gly-Tris (20mM) pH8+ 10% EtOH
- applied voltage: 4kV

Acknowledgment
This work was supported by New Energy and Industrial Technology Development Organization. This work was partly supported by the Grant-in-Aid for COE of Waseda University and for Scientific Basic Research (B) No.12450167.

References

1. E.Shinohara, N.Tajima, H.Suzuki and K.Kano, Proc. of μTAS'98, 319-323 (1998)
2. N.Tajima, H.Suzuki, K.Kano and E.Shinohara, Proc. of the 22nd International Symposium on Capillary Chromatography, PI 414 (1999)

Isoelectric Focusing Using Electrokinetically-generated Pressure Mobilization

Victoria A. VanderNoot[1], Gary Hux[2], Joseph Schoeniger[1] & Tim Shepodd[1]
[1]Sandia National Labs, P.O. Box 969, Livermore, CA 94551-0969, USA
[2]Eksigent Technologies, Livermore, CA 94501, USA

Abstract

One potential barrier to the miniaturization of isoelectric focusing has been the incorporation of mobilization strategies that are compatible with micro-scale devices. This work demonstrates the feasibility of using capillary sized electrokinetic pumps made from porous polymer monoliths for the mobilization of focused protein bands.

Keywords: electrokinetic pump, porous polymer monolith, isoelectric focusing

1. Introduction

Research into the transition of isoelectric focusing (IEF) to chip-based platforms has been pursued by several research groups [1, 2]. Our approach is to employ micro-scale electrokinetic (EK) pumps to mobilize focused proteins past a detector. In a chip format EK pumps have the added benefit of being able to function in fluid handling.

Porous polymer monoliths (PPM) as separation supports in applications such as capillary electrochromatography have been demonstrated [3]. The same types of PPMs can be used as small and versatile EK pumps [4]. EK pumps prepared from positively charged monolith are suitable for operation at low pH (i.e. anode electrolyte or "anolyte" in IEF) where the surface functional groups are ionised.

2. Experimental

Monolith preparation: Capillaries were treated with N-[3-(trimethoxysilyl)propyl] N'-(4-vinylbenzyl)ethylenediamine hydrochloride overnight to reverse EOF. The treated capillaries were flushed with a degassed solution made up of 2:1 solvents {60:20:20v% acetonitrile:ethanol:5 mM sodium phosphate buffer pH6.8} to monomers {30:33:35:2 v% ethylene glycol dimethacrylate:butyl methacrylate:tetrahydrofurfuryl methacrylate: [2-(Methacryloyloxyethyl]trimethyl ammonium methyl sulfate} containing 0.5 w% of 2,2'-azobis-isobutyronitrile. After polymerising overnight at 65°C, the monoliths were purged with anolyte and trimmed to length (~3cm).

Isoelectric Focusing: The experimental set-up is illustrated in Figure 1. An isolated or "floating" power supply was used to generate the mobilization voltages without impacting the focusing voltages. Commercial reagents (Beckman cIEF kit) were used. The ampholyte/pI marker mixture was flushed through the capillary and allowed to focus for ~5 minutes at 15 kV prior to initiating EK mobilization.

3. Results and discussion

The performance of the EK pumps was observed and as expected for EOF-generated flow, the flow rate was a linear function of applied voltage and higher anolyte concentrations (5-20 mM HCl) yielded slower flow rates. The floating power supply

J.M. Ramsey and A. van den Berg (eds.), Micro Total Analysis Systems 2001, 127–128.

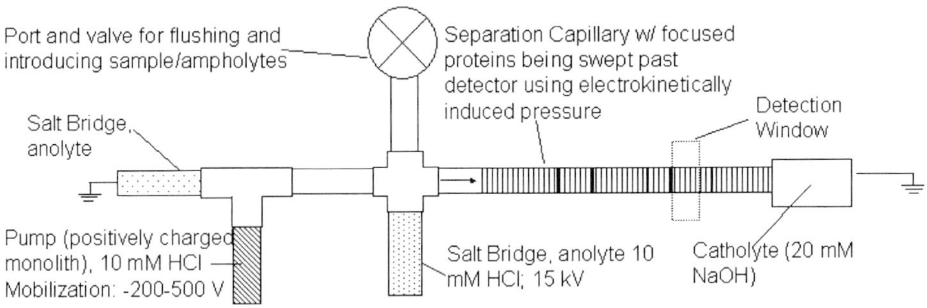

Figure 1. Schematic Design used for Isoelectric Focusing With Electrokinetically-generated Pressure Mobilization

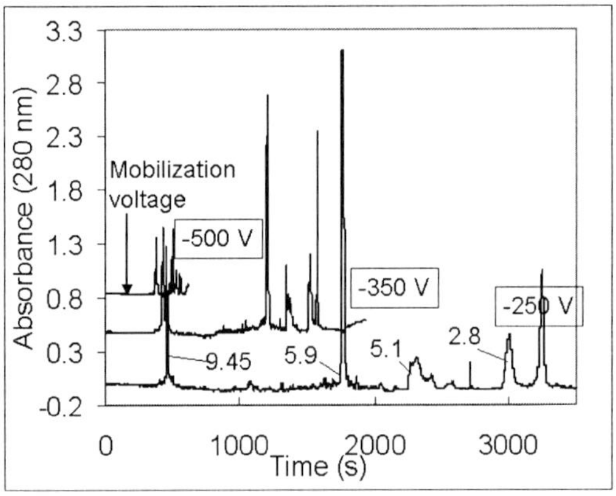

Figure 2. Isoelectric Focusing of pI Markers with a Variety of Mobilization Voltages

showed good isolation of the focusing and mobilization voltages (shown by unperturbed focusing current after mobilization). Figure 3 shows the IEF traces for pI markers with three different mobilization voltages. Although the slowest mobilization gave the best resolution and linearity of pH gradient, even the trace mobilized at 514 V gave a linearity greater than R^2=0.95 and was complete in ~10 minutes.

This work demonstrates an effective application for EK pumps and a successful first step in the transition of IEF to μTAS.

This work was sponsored by the U.S. DOE under contract DE-AC04-94AL85000.

References

1. K. Macounova et al., *Generation of natural pH gradients in microfluidics channels for use in isoelectric focusing*, Analytical Chemistry, 72, 3745-3751, (2000).

2. Q. L. Mao and J. Pawliszyn, *Demonstration of isoelectric focusing on an etched quartz chip with UV absorption imaging detection*, Analyst, 124, 637-641, (1999).

3. F. Svec and J.M.J Fréchet, *Molded rigid monolithic porous polymers*, Ind. Eng. Chem. Res., 38, 34-48, (1999).

4. P. Paul et al, *Electrokinetic Pump Application in Micro-Total Analysis Systems Mechanical Actuation to HPLC*, Proc. Micro Total Analysis Systems 2000, Kluwer Academic Publisher, Dordrecht, The Netherlands, pp. 583-590, (2000).

CONTROL OF TRANSITION BETWEEN ISOTACHOPHORESIS AND ZONE ELECTROPHORESIS BY BASE-MEDIATED DE-STACKING

W.N. Vreeland[1,2], S.J. Williams[1], A.P. Sassi[1], M.T. Cronin[1]
[1]ACLARA Biosciences, Inc., Mountain View, CA, USA
[2]Northwestern University, Dept. of Chemical Engineering, Evanston, IL, USA

Abstract

Key features of microchannel electrophoresis are speed and efficiency. These result from the ability to control accurately the volume of the injected sample plug with cross-channel injectors and from the short separation lengths afforded by microfabricated devices. However, narrow cross-channel injector geometries limit the total amount of sample injected into the separation channel, and therefore limit the detection sensitivity for dilute samples. Here, we present a tandem isotachophoresis-zone electrophoresis (ITP-ZE) method that allows the amount of analyte injected to be decoupled from the effective injection width. This method relies upon the use of an electrophoresis buffer system that functions initially in an ITP mode and, after a controlled amount current has passed through the system, operates in a ZE mode. The transition between the two modes of electrophoresis can be controlled by adjusting the concentration of components of the buffer. This method has been applied to separations of ACLARA's eTag™ reporter molecules to enable detection of concentrations up to two orders of magnitude lower than those detectable in standard ZE in microchannels. The tandem ITP-ZE mode gives separations of eTag™ reporter molecules with resolutions and analysis times equivalent to ZE alone.

Keywords: Isotachophoresis, zone electrophoresis, stacking

1. Introduction

In ITP, a discontinuous, inhomogeneous electrolyte system is used to create step changes in the electric field gradient along the separation axis. During electrophoresis in aqueous systems, electrolysis at the electrodes generates H^+ and OH^- ions. In the case of ITP of anionic analytes, the production of OH^- ions at the cathode can disrupt the step changes in the electric field potential gradient and must be buffered sufficiently to maintain stability of the ITP conditions. If the production of OH^- ions is too great, the capacity of the buffering agent will become saturated, eliminating the step changes in the electric field gradient. ITP behavior will then be lost, and the system will transition to ZE.

During tandem ITP-ZE, the boundary of OH^- ions titrating with the cationic buffering agent moves toward the anode, eventually passing through and de-stacking the sample zone. At this point ITP of the analytes is lost, and the analytes begin to migrate by ZE

J.M. Ramsey and A. van den Berg (eds.), Micro Total Analysis Systems 2001, 129–130.

and are separated. Thus, the duration of ITP can be controlled through the concentration of buffering ions in the electrolyte system.

2. Results

As seen in Fig 1, the effective peak width of an analyte decreases as the concentration of buffering agent (Tris) increases. This effect is due to higher concentrations of buffering agent quenching electrolytically produced OH^- for a longer time in comparison to lower buffer concentrations. As a result, higher buffer concentrations allow ITP to persist for a longer fraction of the total separation time which concentrates the sample into a narrower volume, thus producing a narrower peak.

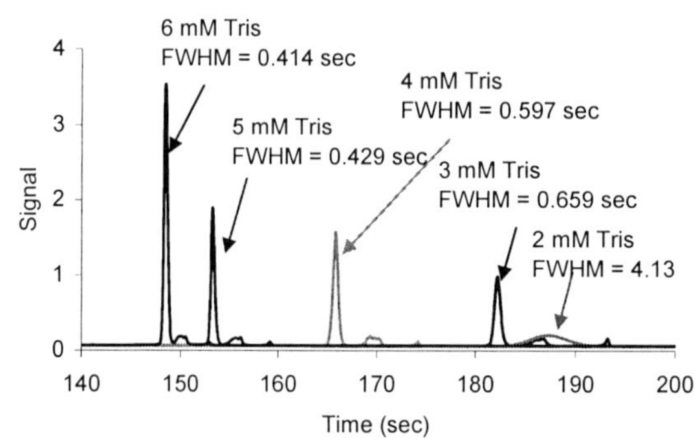

Fig 1. Tandem ITP-ZE separation of eTag™ reporters in buffers of increasing cationic buffering agent concentration.

When this technique is used in microfluidic chips with relatively long injector geometries (up to 3 mm in this work) to analyze dilute samples, electrophoresis functions initially in an ITP "concentrating" mode, during which the dilute analytes are concentrated into a single, narrow band (~30 μm), and then transitions to a ZE mode during which the analytes are separated. This tandem ITP-ZE technique gives separations that are of equivalent efficiency and with markedly more intense peaks when compared to those given by pure ZE, as seen in Fig. 2. Tandem ITP-ZE allows for facile separation and detection of analytes at concentrations well below the limit of detection accessible with more traditional injection techniques.

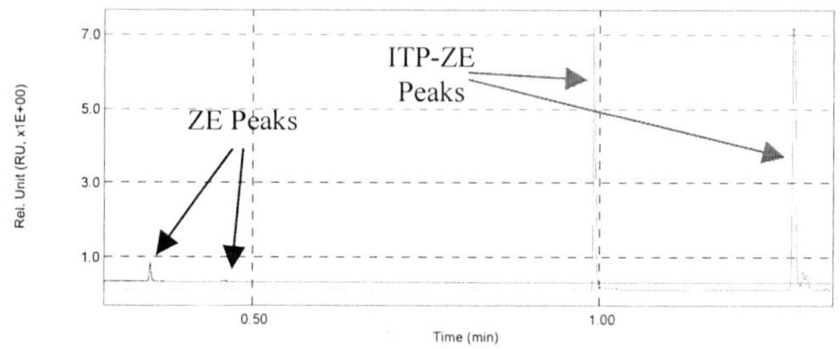

Fig 2. Comparison of the separation of two eTag™ reporters using standard ZE and tandem ITP-ZE. In both separations, the sample was at the same concentration.

ELECTROOSMOTICALLY INDUCED HYDRAULIC PUMPING ON MICROCHIPS

Jean Pierre Alarie, Stephen C. Jacobson, B. Scott Broyles, Timothy E. McKnight, Christopher T. Culbertson, and J. Michael Ramsey
Oak Ridge National Laboratory, Oak Ridge, Tennessee 37831-6142 USA

Abstract

We report the design and testing of microchips with salt-bridge junctions that were capable of inducing hydraulic pumping electroosmotically. The salt-bridge junction was a thin gap or channel (< 100 nm deep) connecting two closely spaced microchannels. When an electric potential was applied across the junction, the electroosmotic flow was lower in the thin channel relative to the microchannels. This generates a pressure at the junction that in turn produces hydraulic flow.

Keywords: electroosmotically induced hydraulic pumping, salt-bridge junction

1. Introduction

The ability to control the electric potential at any point within a microfabricated fluidic network enhances the operability of the device. An early approach was to use thin metal films for electrical contacts, but due to electrolysis, bubbles were generated at the electrode and in turn disconnected the electrode from the buffer solutions. We are investigating several techniques for forming intrachannel electrical connections that circumvent electrolysis issues. These salt-bridge junctions conduct electrical current by small ion transport but minimize or inhibit bulk fluid transport. In this paper, hydraulic pumping in the field-free channel was induced electroosmotically by applying a potential between the side 1 and waste 2 reservoirs and across the salt-bridge junction (Figure 1).

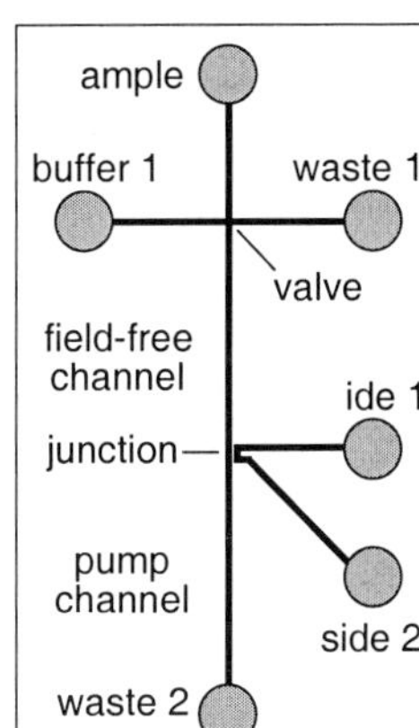

Figure 1. Schematic of microchip with a salt-bridge junction.

2. Experimental

For the microchip in Figure 1, the side 1 and pump channels were fabricated in close proximity to each other. For example, the initial spacing between these channels on the photomask was 25 μm. If the channels were etched 11 μm deep, the distance between the tops of the channels was 3 μm due to isotropic etching of the substrate (Figure 2). This distance is referred to as the length of the salt-bridge junction. The cover plate was then thermally bonded to the substrate at 500°C. Due to incomplete bonding over short

J.M. Ramsey and A. van den Berg (eds.), Micro Total Analysis Systems 2001, 131–132.

distances, e.g. 3 μm, a thin gap or channel exists between the side 1 and pump channels and the cover plate. This junction allowed electric current to pass but exhibited reduced electro-osmotic flow relative to the microchannels. If electrical conduction across the junction was not observed, a high potential, e.g., 8 kV, was applied between the side 1 and waste 2 reservoirs to break down the bonded region.

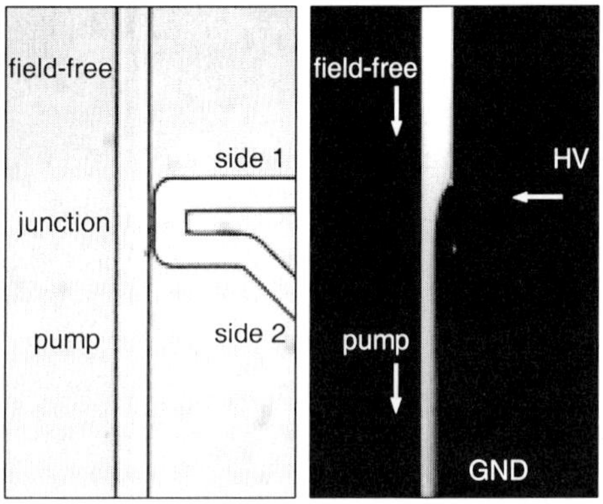

Figure 2. White light image of salt-bridge junction and fluorescence image of flow at junction. Side 1 reservoir is maintained at 2 kV (HV) and waste 2 reservoir is grounded (GND). Arrows depict direction of flow.

3. Results and Discussion

Figure 2 shows a fluorescence image of fluid flow at the junction with a potential applied between the side 1 and waste 2 reservoirs. The electro-osmotic flow in the pump channel was greater than the electroosmotic flow across the junction. Consequently, a negative pressure was generated at the junction, and sample was drawn from the field-free channel into the pump channel. This can be seen in Figure 2 where the left side of the pump channel was filled with a fluorescent dye (bright) and right side was filled with buffer (dark). By drawing a front of fluorescent dye through the field-free and pump channels, the pumping velocities were measured (Figure 3). The ratios of the flow rates in the field-free and pump channels (pumping efficiency) were 0.43 and 0.63 for the 1 and 3 μm wide bridges, respectively.

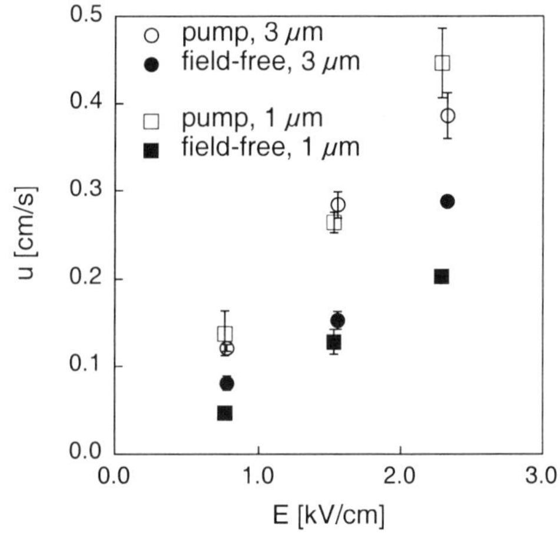

Figure 3. Variation of velocity (u) with field strength (E) in the pump and field-free channels for 1 and 3 μm wide junctions.

Acknowledgement

Oak Ridge National Laboratory is managed and operated by UT-Battelle, LLC for the U.S. Department of Energy under contract DE-AC05-00OR22725.

3D-FLOW CONFINEMENT FOR EFFICIENT SAMPLE DELIVERY – APPLICATION TO IMMUNOASSAYS ON PLANAR OPTICAL WAVEGUIDES

Oliver Hofmann[1], Guy Voirin[2], Philippe Niedermann[2], Andreas Manz[1]

[1]AstraZeneca/SmithKline Beecham Centre for Analytical Science, Imperial College of Science, Technology & Medicine, Exhibition Road, London SW7 2AY, UK
[2]CSEM, Rue Jaquet-Droz 1, CH-2007 Neuchatel, Switzerland

Keywords: 3D-microchip, flow confinement, sample delivery, immunoassays

Abstract

In this paper a microchip-based "flow confinement" method for sample delivery is validated for the first time by applying it to immunoassays on optical waveguides. With binding ligands in the sample flow and receptors immobilized on the waveguide it results in faster binding.

1. Introduction

For flow confinement, a sample flow is joined with a perpendicular make-up flow. The laminar flow conditions typical in microchannels preclude turbulent mixing. The make-up flow "confines" the reagent into a thin layer above the sensing area and increases its velocity. This benefits diffusion limited processes such as immunoassays. While shallower microchannels could be used for physical confinement these are often prone to clogging and expensive to fabricate. Other approaches to deliver small sample volumes include the use of a closed loop and oscillating sample flow, which are based on multiple sample passing. However, progressive depletion renders these "recycling methods" inefficient for low analyte concentrations.

2. Experimental

To implement flow confinement in a planar sheet geometry, a three-dimensional microchannel network is required. A novel modular fabrication method based on molded polydimethylsiloxane (PDMS) building blocks with multiple side microstructuring was developed. For assembly of the microfluidic chip and interfacing to the waveguide, the adhesive surface properties of the elastomer were used (1). To evaluate the benefits of flow confinement, the microfluidic chip was applied to a high-affinity, excessive reagent immunoassay with rabbit IgG antibodies immobilised on the waveguide surface and labelled anti-rabbit IgG as sample flow (Figure 1a). The binding was monitored via evanescent field detection with a single grating used for in-coupling of excitation light and out-coupling of fluorescence light resulting from molecules on the waveguide surface (Figure 1b).

J.M. Ramsey and A. van den Berg (eds.), Micro Total Analysis Systems 2001, 133–134.

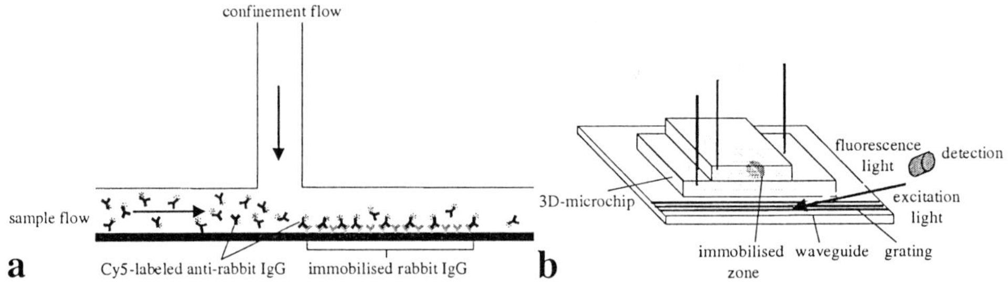

Figure 1. a. Model assay for flow confinement. b. Measurement principle.

3. Results and discussion

When flow confinement was applied to the model immunoassay, completion times could be reduced from 1 hour for static drop conditions to 13 minutes. Reference experiments with sample filling the whole channel flowing at the corresponding total flow rates indicate that this is mainly due to the velocity increase induced by the confinement. Compared to this mode, however, sample volume consumption is much reduced by the flow confinement (Figure 2).

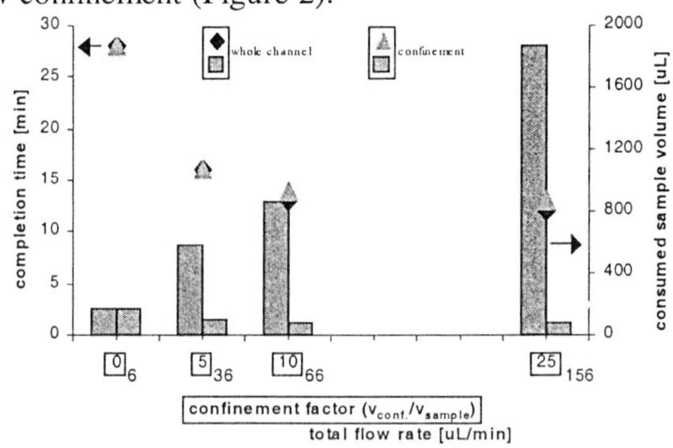

Figure 2. Comparison of assay completion times and consumed sample volumes for whole channel and confinement experiments. The sample flow rate was 6 μL/min for the confinement experiments and 6 μL/min + 6 μL/min × confinement factor (total flow rate) for the whole channel experiments.

Assays were allowed to reach equilibrium in order to ensure comparability between runs (sensorgrams normalised on plateau value). For typical analytical applications, however, this is not required and the enhanced mass transport obtained with flow confinement should enable very short analysis times. The main limitation of flow confinement is diffusional loss of analyte into the confining layer which will be discussed elsewhere (2).

References

(1) Hofmann, O.; Niedermann, P.; Manz, A. *submitted to Lab on a Chip,* 2001.
(2) Hofmann, O.; Voirin, G.; Niedermann, P.; Manz, A. *in preparation,* 2001.

CONTROLLED PIPETTING OF LIQUIDS IN THE SUB-µL-RANGE BASED ON THE LEVEL DETECTION INSIDE A CAPILLARY

T. Lisec, S. Mühlmann, S. Grünzig, B. Wagner
Fraunhofer-Institute for Silicon Technology,
Fraunhoferstraße 1, D-25524 Itzehoe, Germany

Abstract

This paper presents a new pipetting technique allowing parallel transfer of different liquid volumes in the µl- and nl-range by individually working units. Each single unit contains a level detection sensor, an actuator and electronics. Resulting from the sensing principle an extensive real-time control of the whole pipetting procedure is possible.

Keywords: level detection, fluid transfer

1. Working principle

Main part of the so called sensor pipette shown in Fig. 1 is a level detection sensor. This is a silicon chip with an periodical arrangement of electrode pairs. The electrodes are located inside a capillary tube formed by the silicon chip and a cover. Impedance measurements allow to assign a phase boundary between air and a conducting driving liquid within the sensing area to only one of the electrode pairs, see Fig. 2. Consequently, with a suitable actuator the monitored phase boundary can be positioned at any electrode pair. Since the driving liquid acts like a piston the displaced sample volume in the tip is proportional to the distance the phase boundary is moved. In such a way aspiration as well as dispensing of fluids can be realized.

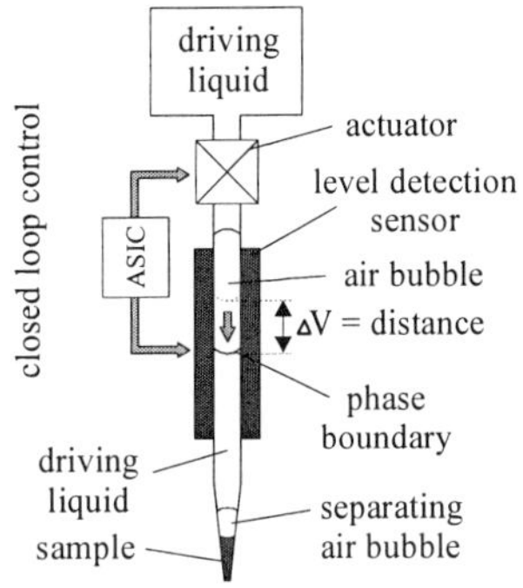

Fig. 1 Cross-sectional view of the sensor pipette.

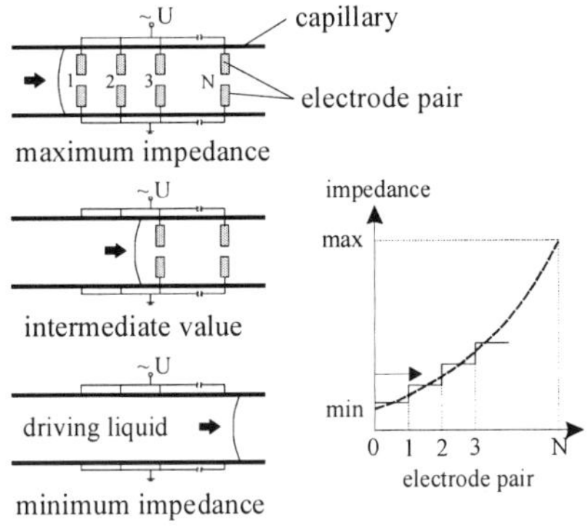

Fig. 2 Principle of phase boundary (level) detection

Volume range and volume step are defined by the number and the pitch of the electrode pairs and the cross-section of the capillary. Drift effects due to temperature or conductivity changes in the driving liquid are eliminated by an in-situ calibration of maximum and minimum impedance values before any pipetting cycle.

J.M. Ramsey and A. van den Berg (eds.), Micro Total Analysis Systems 2001, 135–136.

2. Experiment and discussion

Resulting from the working principle the sensor pipette shows some unique features. Failures like tip clogging or sample losses during robotic handling will be detected. The properties of the sample liquid (for example the viscosity) have no influence on the pipetting accuracy. Utilizing the capillary effect the presence of a liquid in a well can be monitored before aspiration or dispensing. Since the dead volume of the pipette is reduced to the air bubble in the tip precise transfer of volumes in the nl-range is possible. At the same time sample losses are nearly zero.

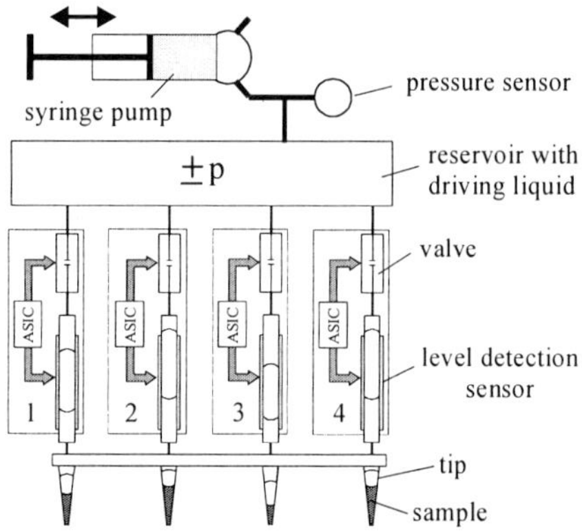

Fig. 3 4-channel pipettor with valves and a syringe pump for pressure generation.

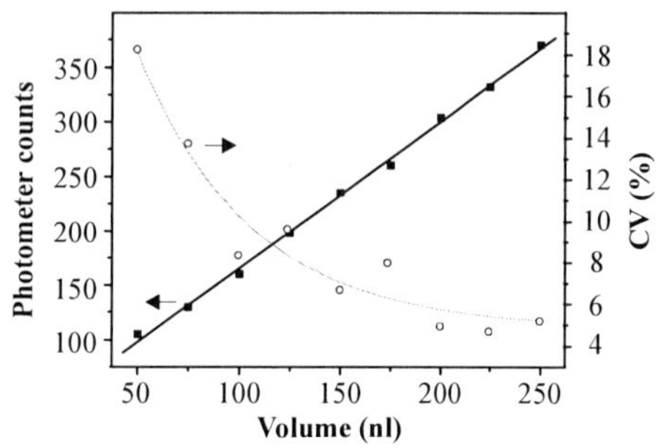

Fig. 4 Pipetting of nl-volumes of aqueous dye solution in a wet microtiter-plate using a level detection sensor with 5 μl total volume and 25 nl step and a valve as actuator.

Fig. 3 shows the scheme of an automatic 4-channel pipettor based on miniature valves as actuators and a syringe pump for pressure generation. One channel including actuator and electronics can fit within an area of 9×9 mm^2. Fig. 4 presents the pipetting results obtained for one single channel equipped with a sensor with 25 nl step (5 μl total volume) and a steel tip with 0.3 mm inner diameter. Aqueous 4-nitrophenole solution was dispensed in the wells of a standard 96-microtiterplate prefilled with 200 μl buffer solution. The CV´s are final values including photometer reading and mixing errors. They depends on the sensor type, the tip and the actuator. To achieve a good CV the monitored phase boundary must be displaced for a minimum number of steps since the sensor works quasi-digital. The sensor volume can easily be scaled down to 1 μl with a step of 5 nl. Theoretically volumes down to some tenth of nl could be transferred with high accuracy. At the same time the orifice of the tip must be small enough. For disposable plastics similar results has been obtained as for steel tips. Although, for very low volumes valves may be unsuitable as actuators. Other concepts must be tested, for example electrolytic or thermal bubble generation or piezoceramic actuators.

3. Conclusions

A new pipetting technique based on a level detection sensor has been demonstrated. An array of sensor pipettes will be able to aspirate and dispense different liquid volumes in parallel independent from each other allowing at the same time a real-time control of the whole pipetting procedure.

DROPLET FORMATION IN A MICROCHANNEL ON PMMA PLATE

Takasi Nisisako, Toru Torii, Toshiro Higuchi
Department of Precision Engineering, The University of Tokyo, Japan
(nisisako@intellect.pe.u-tokyo.ac.jp)

Abstract

A method for generating droplets in a microchannel network has been studied. Using oil as continuous phase and water as dispersed phase, nanoliter-sized water droplets are generated in the oil flow at a T-junction.

Keywords: microchannel, T-junction, microcapsule, emulsification

1. Introduction

Over the past decade, there has been an increased interest in chemical reactions and their analyses in microfabricated devices. For successful analyses in such devices, accurate metering of liquid samples is necessary [1]. This paper is intended to report a method for preparing regular-sized discrete droplets in a microchannel network.

2. Fabrication

A T-junction structure was fabricated on a polymetyl metacrylate (PMMA) plate, using a 100 μm diameter end mill. The channels are 100 μm deep: the narrower channel for water flow is 100 μm wide and the other for oil flow is 500 μm wide. The cross section of these channels is rectangular. Adhesive tape is used for enclosing those channels.

3. Experimental

Water and high oleic sunflower oil (triolein, 80%) are driven into channels respectively by syringe pumps. Semi-conductor pressure sensors are embedded in the plate to investigate conditions for droplet formation. Using a microscope and a high-speed video camera, the motion of droplet formation at the T-junction is observed (Fig. 1).

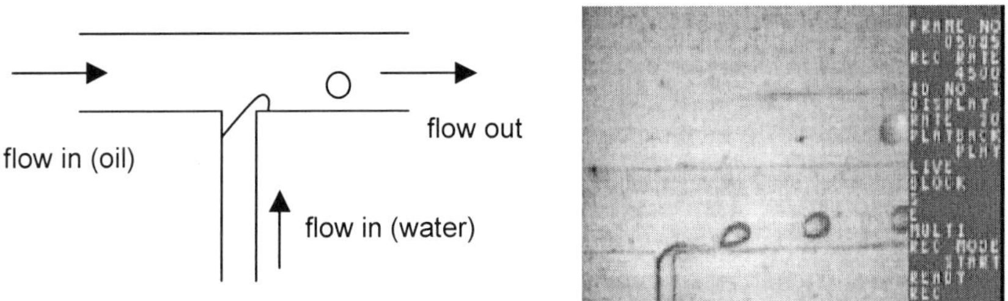

Figure 1. Schematic of droplet formation at a T-junction (left) and an image by a high-speed video camera (right)

J.M. Ramsey and A. van den Berg (eds.), Micro Total Analysis Systems 2001, 137–138.

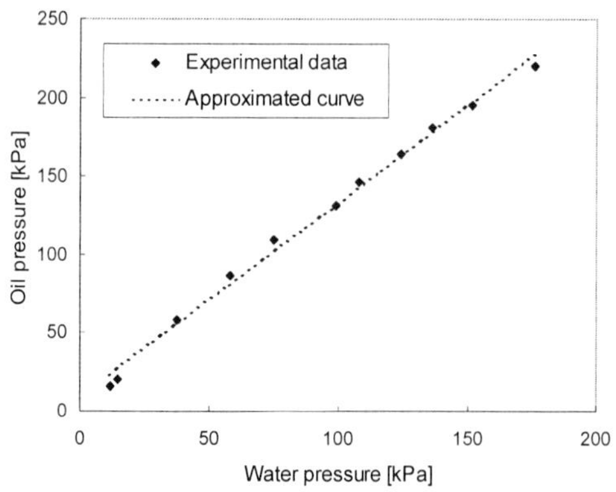

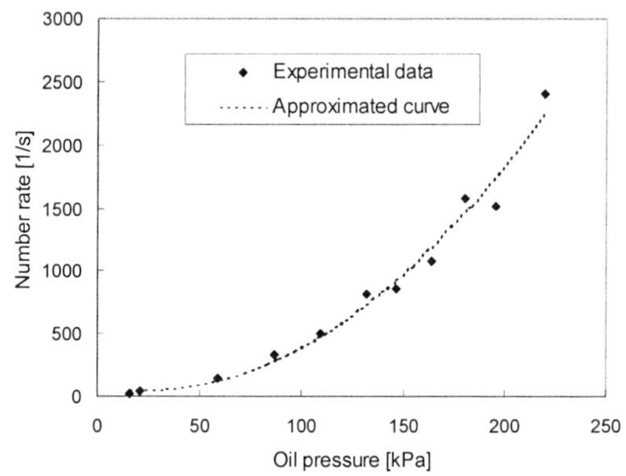

Figure 2. Water pressure versus oil pressure when the diameter of water droplets is approximately 100 μ m (left), and effect of continuous phase (oil) pressure on droplet production rate (right).

4. Results and discussion

Regular-sized water-in-oil droplets were generated at the T-junction. The droplet size was controllable by changing flow rates of oil and water. The minimum diameter of droplets was approximately 100 μ m. Oil pressure, water pressure, and droplet production rate were measured when the diameter of droplets was minimized. Linearity was found between oil pressure and water pressure (Fig. 2, left). The production rate is estimated to be proportional to the square of oil pressure (Fig. 2, right). Assuming that the production rate obeys this law, this method will be effective in the case of preparing a number of nanoliter-sized discrete samples in a short time.

5. Conclusion

In the present work a method for generating water-in-oil droplets in a microchannel network has been studied. Further improvement of this technique will enable us to obtain oil-in-water droplets and microcapsules as well in microfabricated devices. Besides, this method will be applicable, not only to liquid metering, but also to emulsification, drug delivery system (DDS), and so on.

References

1. K. Handique, D. T. Burke, et al, *Nanoliter liquid metering in microchannels using hydrophobic patterns*, Anal. Chem. 72, pp. 4100-4109, (2000).

DISPERSION CONTROL IN NANO-CHANNEL SYSTEMS BY LOCALIZED ZETA-POTENTIAL VARIATION

R. Qiao and N. R. Aluru
Beckman Institute for Advanced Science and Technology
University of Illinois at Urbana-Champaign, Urbana, IL 61801

Abstract

A novel technique for controlling turn-induced dispersion in nano-fluidic channels through localized ζ-potential variation is proposed. Algorithms were developed to search for the optimal distribution of ζ-potential near the turn. Results for an L-turn show that the technique can lower the dispersion induced by the 90° turn significantly.

Keywords: Dispersion control, nano-fluidic channels, zeta-potential, optimization

1. Introduction

Turns and bends, such as an L-turn or a U-turn, in micro and nano-fluidic channels can induce significant dispersion leading to a low efficiency electrophoretic separation process. Turn-induced dispersion can be lowered by optimizing the geometry of the turns [1,2]. In this paper, we propose a new approach for controlling turn-induced dispersion in nano-fluidic channels. The approach is based on the idea of locally controlling the ζ-potential at turns and bends.

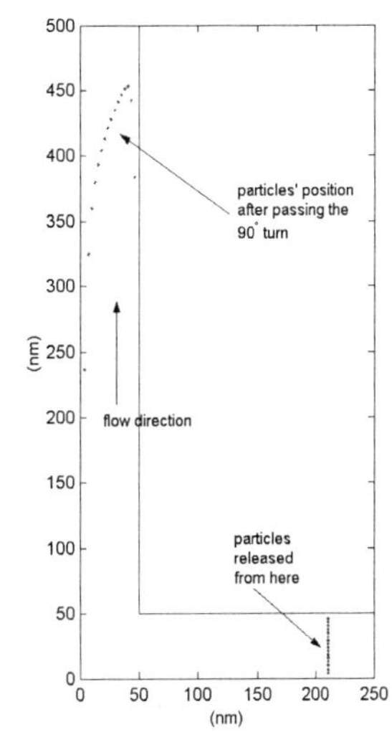

Figure 1. Dispersion in L-turn without control

2. Numerical Model and Control Strategy

The dispersion of species in nano-diameter channels can be caused by both advection and diffusion. For simplicity, we consider the dispersion caused by advection only and dispersion is calculated by tracking the particles released from the upstream of the turn. The flow field is governed by the Stokes equations coupled with the Poisson-Boltzmann equation for the ζ-potential and the Laplace equation for the applied potential as described in [3]. This is a more general numerical model compared to the previous approaches [1,2] where the velocity is assumed to be proportional to the gradient of the applied potential field.

Figure 1 shows the dispersion of particles as they go through a 90° turn. The dispersion can be reduced by applying a two-stage control strategy shown in Figure 2.

3. Optimization Algorithm and Results

A fixed-step optimization algorithm is used to search for the optimal ζ-potential distribution shown in Figure 2. The search is done in two steps. The objective function for the first control step is given by:

$$F_1 = |y_1 - y_n| \qquad (1)$$

J.M. Ramsey and A. van den Berg (eds.), Micro Total Analysis Systems 2001, 139–140.

where y_1 is the position of the first particle, which is also the particle closest to the top or inner wall, and y_n is the position of the nth particle, which is also the particle closest to the bottom or outer wall. The objective function for the second control step is given by:

$$F_2 = \sum_{i=r}^{s} \frac{|y_i - \bar{y}|^2}{s - r} \qquad (2)$$

In the second objective function, only the particles outside of the velocity boundary layer are included. The indices r and s refer to the first and last particle outside of the boundary layer, respectively. For the first stage control, we assume that the ζ-potential distribution varies linearly along the two patches (patches OA and OB in Figure 2). The optimization parameters are the length of each patch and the zeta-potential variation on each patch. For the second stage control, we assume that the ζ-potential is constant and length of the control region is equal to the channel width. Hence, there is only one optimization parameter, which is the ζ-potential value.

Optimal ζ-potential distribution has been found using the algorithm described above and the new dispersion results are shown in Figure 3. It can be observed that the dispersion is significantly lowered compared to the dispersion without control, which is shown in Figure 1.

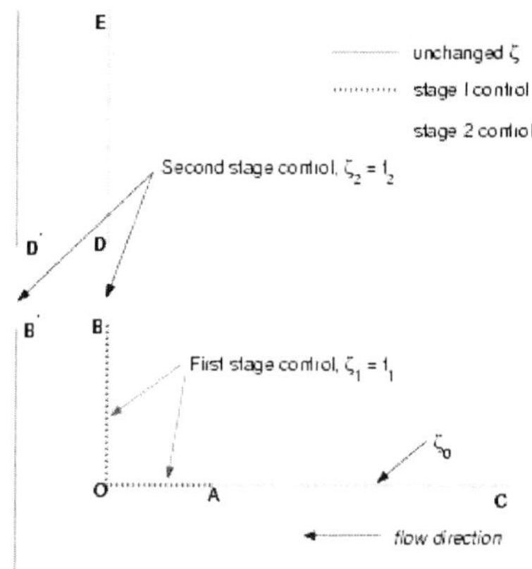

Figure 2. Sketch of control strategy used for an L-turn

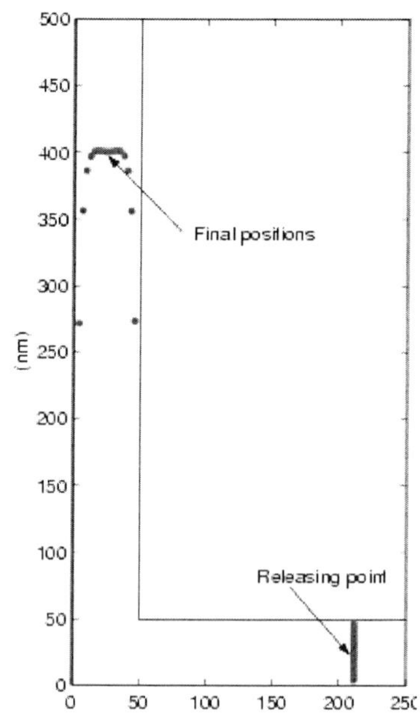

Figure 3. Dispersion in L-Turn after control algorithm is applied.

4. Conclusion

The results presented in this paper indicate that the flow control by local ζ-potential changes is a promising new approach. Even though results are shown for nano channels, the approach can be extended in a trivial manner for micro channels.

References

1. S. K. Griffiths and R. H. Nilson, *Low dispersion turns and junctions for microchannel systems*. Ana. Chem., Vol.73(2), pp272-278.
2. J. I. Molho et. al., *Optimization of turn geometries for microchip electrophoresis*, Ana. Chem., 73(6), pp1350-1360.
3. M. Mitchell, R. Qiao, N.R. Aluru, *Meshless Analysis of Steady-State Electroosmotic Transport*, Jour. of Microelectromechanical Systems, Vol. 9(4), pp 435-449.

MICROFLUIDIC MIXER EMPLOYING TEMPORALLY-INTERLEAVED LIQUID SLUGS AND PARABOLIC FLOW

M.C.Tracey[1], T.I.Cox[2], J.B.Davis[1].
[1]STRC, University of Hertfordshire, Hertfordshire AL10 9AB, UK ,
[2] Sensors and Electronics Division, QinetiQ, Malvern, Worcs WR14 3PS, UK

Keywords Mixing, laminar flow, diffusion, variable ratio mixing

Abstract

Efficient mixing is achieved by alternately introducing slugs of liquids into a microfluidic channel and driving them along the channel under pressure-driven flow.

1. Introduction

Rapid mixing of liquids is valuable for efficient, on-chip, chemical synthesis and analysis. Previous approaches, e.g. [1], have sought to form domains of the two liquids with a high interfacial area to volume ratio. Mixing by diffusion then occurs rapidly. We present a new technique to achieve such favourable ratios by introducing time-interleaved, pressure-driven, slugs to a mixing channel.

2. Experimental

Chips were fabricated by precision-milling PMMA laminates which were then sealed by diffusion bonding. The extent of mixing of liquids coloured with blue (A) and yellow (B) dyes was investigated by image processing of the colour images (the accompanying graphics are monochrome representions in which blue is dark, yellow is light and the resulting green is grey). Figure1 details a mixing channel arranged as a serpentine structure, 100 mm long, 1000 μm wide and 200 μm deep with 20 μL volume. Liquids A and B, are injected, via miniature, conventional valves via two injectors, 100 μm width and 30 μm depth, and length chosen to provide an appropriate delivery rate. Currently, the valve switching rate is limited to 1Hz.

3. Results

Figure 1 shows the case of both valves continuously open to allow equal flows of A and B displaying the classic, 'striated' laminar flow pattern of 'A' and 'B'. Figure 2 shows the corresponding images for the two input valves alternately opened to alternately inject equal slugs of A and B. In the top left, the two slugs can be clearly seen. As the fluid is driven along the channel, the contrast between the slugs rapidly disappears due to efficient mixing. Figure 3 shows a concentration profile along the mixing channel. The rapid progress of mixing is evident as the slugs proceed along the channel.

J.M. Ramsey and A. van den Berg (eds.), Micro Total Analysis Systems 2001, 141–142.

4. Discussion

Consider the flow channel, shown in Figure 4 at time zero with interleaved slugs of A and B. CFD simulation confirms that, under pressure driven flow, the characteristic parabolic velocity-profile develops. Figure 5 shows the mixing process schematically and the high interfacial area to volume ratio, giving rapid diffusional mixing. This technique has the added advantage that the ratio of A to B can be controlled by the relative opening times of the input valves.

5. Acknowledgements

This work was partly carried out as part of the Technology Group 07 of the UK MoD Corporate Research Programme.

6. References

1 F. G. Bessoth, et al,, Anal. Commun., **36**, pp. 213 – 215 (1996).

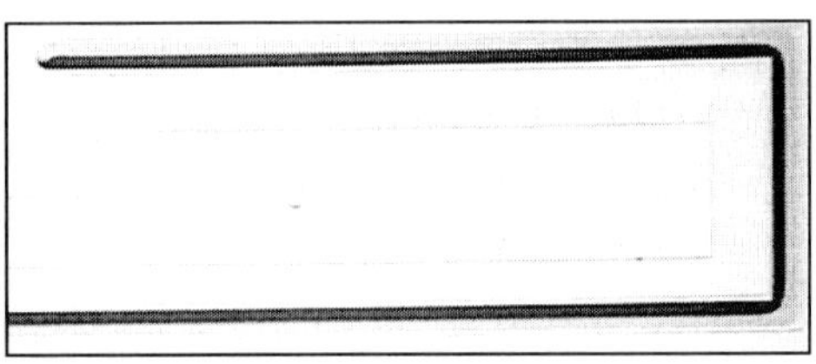

Fig. 1 Both injectors open

Fig.2. Injectors operated in antiphase.

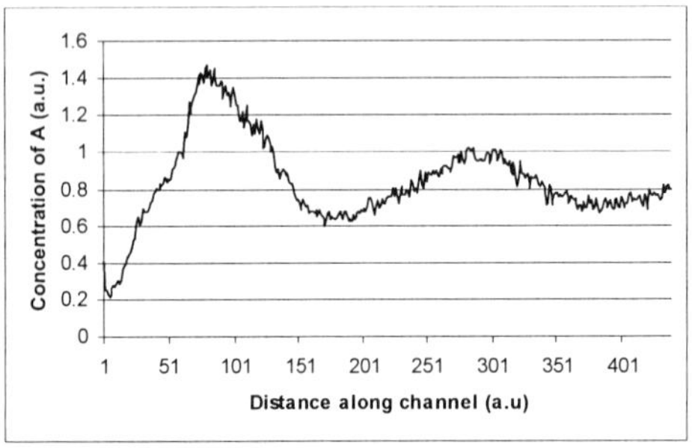

Fig.3 Concentration of A along the channel

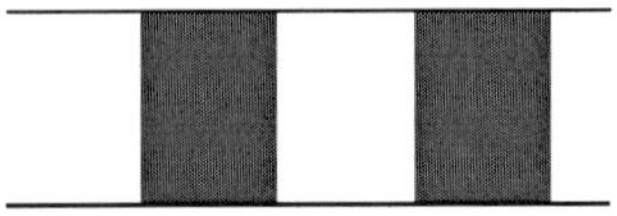

Fig.4 Interleaved slugs of A and B

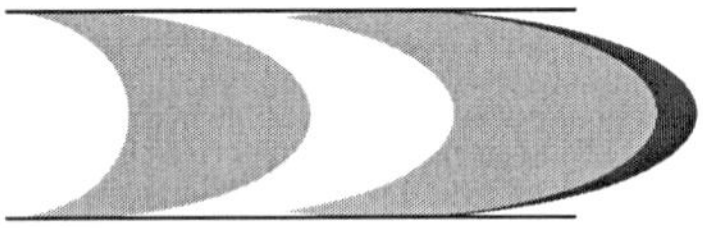

Fig.5. Mixing under pressure driven flow

A DYNAMIC SPOTTING METHOD FOR SPLIT-PIN BASED MICROARRAYS

Jun Zeng, Manish Deshpande, Heng-Chuan Kan and John R. Gilbert
Coventor Inc., 625 Mount Auburn Street, Cambridge MA 02138, USA
jun.zeng@coventor.com

Abstract

The authors propose a non-contact spotting concept for split-pin based microarrays utilising dynamic control of the trajectory of the split-pin. Numerical simulation demonstrates that this novel method not only avoids the necessity of the pin tip striking the surface of the substrate , but also offers a new mechanism to realize spot-volume-on-demand, as well as enhance the uniformity of sample spots.

Keywords: liquid dispensing, split pin printing, microarray, CAD, microfluidics

1. Introduction

Microarray technology was first introduced by the Brown and Davis groups [1][2][3] to deposit pre-synthesized oligonucleotides or PCR products onto solid substrates. The substrates are then processed and analyzed in parallel. This powerful technology enables the simultaneous analysis of thousands of sequences of DNA, and has found wide use in genomics (for instance, [4]), and more recently, in proteomics [3].

Pin based liquid sample deposition (spotting) is widely employed due to its low cost, simplicity and robustness[1]. Figure 1 shows a split-pin fabricated by the Brown group. The pin is first dipped into the sample solution and a small volume of fluid is transferred into the slot. The pin is then guided to "tap" the substrate leaving a spot of the sample.

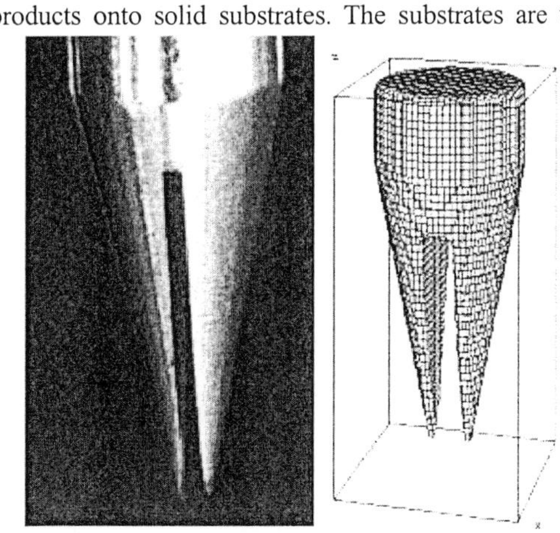

Fig. 1:At left is a split-pin fabricated by Brown group. A fine slot is machined into the end of the pin to accommodate liquid sample (courtesy professor Patrick O. Brown, Stanford University). At right shows solid structure and meshes used in numerical simulations.

The tapping force required to deposit a sample spot on the substrate is not insignificant [1]. Such direct contact can lead to mechanical damage to the pin tips. In addition, the sample droplet volume is affected by the speed at which the pin strikes the surface (for example,2cm/sec [1]), which depends on the local mechanical contact property of the substrate - it can therefore be a major source of droplet volume disparity.

J.M. Ramsey and A. van den Berg (eds.), Micro Total Analysis Systems 2001, 143–144.

2. A Non-Contact Pin Printing Concept: Dynamic Control of Pin Trajectory

A split-pin based non-contact spotting method is described here, that uses precise control of the pin trajectory. The displacement of the pin in the spotting sequence is shown in Figure 2 using numerical simulation [5][6]. The sample spot is deposited on the surface without tapping, by triggering the required instability to break the droplet through the acceleration of the pin away from the substrate. This approach overcomes the disadvantages of the conventional approach described above. This spotting approach enhances the uniformity of volume of sample spots. Furthermore, it offers a new mechanism to actively control the spot volume, or *spotting-volume-on-demand*. This can be accomplished by adjusting the pin's pulling velocity – spot sizes from 1-1000 pL can be deposited as shown in Figure 3.

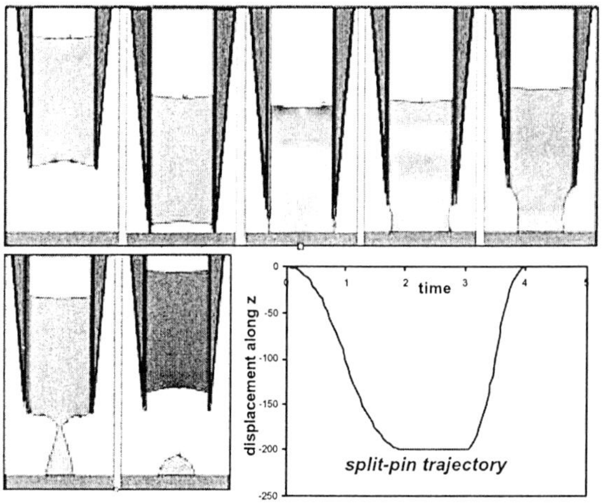

Fig. 2: A time sequence of liquid sample spotting. Motion of the split-pin is defined as shown trajectory. Frames are corresponding to time 0, 1.8, 3.3, 3.5, 3.6, 3.8, 4.0 milliseconds. Minimum distance between pin tip and substrate is 2 μm.

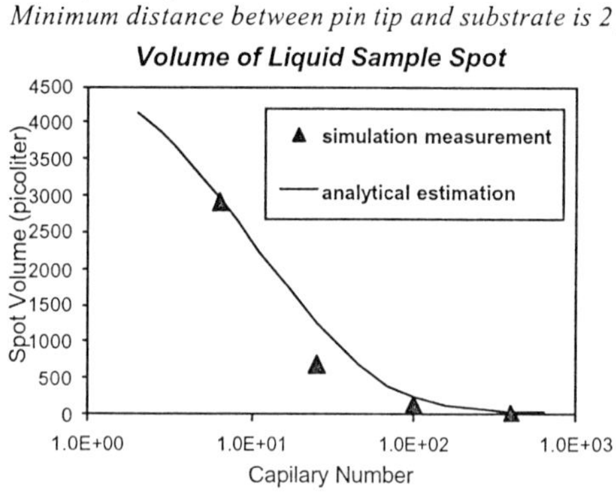

Fig. 3: Adjusting pin's pulling velocity can actively control the spotted liquid volume. Measured results are from numerical simulation, which agree with analytical estimation.

The results in Figure 3 are compared with the analytical estimate, where V is droplet sample volume,

$$V = \frac{\pi}{3}\tan\theta\left(\frac{a\cdot\cos\theta}{\cos\theta + \sin\theta + \sin(4\sqrt{Ca} - \theta)}\right)^3$$

capillary number $Ca = 2\sigma/(\rho u^2 a)$, θ static contact angle, a dimension of the slot, , σ surface tension coefficient, u average pin pulling velocity,. The agreement between numerical and analytical measurements is very good, as the figure shows.

Acknowledgements

This work was funded in part by DARPA (Grant no. Award No. DAAD10-00-1-0515).

References

[1] Rose, D., in "Microarray Biochip Technology", edited by Mark Schena, Eaton Publishing, 2000

[2] DeRisi, J. L., Iyer, V. R. and Brown, P. O., Science 278, pp.680-686, 1997

[3] Tseng, F. G., et al, ASME MEMS 2000, Vol. 2, pp.659-663

[4] Genetic MicroSystems, GMS 417 Arrayer User Guide, 1999

[5] Zeng, J., et al, μTAS 2000, pp. 579-582

[6] Zeng, J., et al, ASME MEMS 2000, Vol. 2, pp.581-587

EXPANDABLE MICROSPHERES: A NOVEL TOOL IN MICROFLUIDICS

Helene Andersson, Patrick Griss, Göran Stemme
Royal Institute of Technology
Department of Signals, Sensors and Systems, Microsystem Technology

Abstract
In this study a novel component for microfluidics is introduced. Expandable microspheres have been studied for their application in microfluidics. Two methods for selective immobilization of expandable microspheres on silicon, including patterning by photolithography and self-assembly based on surface-chemistry has been shown. After the immobilization step the microspheres were expanded thermally. The expansion is irreversible and the volume of the microspheres increases more than 60 times. Patterns of microspheres with features as small as 15 μm have successfully been generated by photolithography. By using self-assembly the microspheres can conveniently be immobilized in a monolayers.

Keywords: expandable microspheres, actuator, self-assembly, surface enlargement

1. Introduction
In this study a novel tool for microfluidics is introduced. Expandable microspheres (Expancel, SE) are small spherical plastic particles, which consist of a polymer shell encapsulating a hydrocarbon liquid. When the microspheres are heated the liquid transforms into a gas, which increases its pressure. Together with softening of the thermoplastic shell this result in a dramatic irreversible increase in the volume of the microspheres, see figure 1. When fully expanded, the volume of the microspheres increases more than 60 times. Expandable microspheres are today used as blowing agents in many areas such as printing inks, paper, textiles, polyurethanes, and PVC-plastics [1].

A prerequisite for using expandable microspheres as a tool in microfluidics is the ability to selectively place them in a predefined area of a device. In this study, two methods for selective immobilization of expandable microspheres have been studied for their application in MEMS, including patterning by photolithography and microcontact printing combined with self-assembly based on surface-chemistry. After the immobilization step the microspheres were expanded. The expansion can be used to displace a liquid volume, generate a pressure or enlarge the internal reactive surface of a microfluidic system.

2. Experimental
Photoresist patterning: Two types of expandable microspheres were used in this study, Expancel DU (Dry, Unexpanded) 820 and 091. Two different microspheres/photoresist mixtures were used for each microsphere type: 3.7 w% and 6.4 w%. The microspheres were deposited on the surface of silicon wafers by a process based on a positive photoresist [2]. *Surface chemistry patterning:* 4-(Chlorosulfonyl)benzoic acid was printed on the silicon substrate and transformed to an anhydride according to [3]. A

J.M. Ramsey and A. van den Berg (eds.), Micro Total Analysis Systems 2001, 145–146.

drop of an aqueous slurry of microspheres was added to the substrate after microcontact printing and the chip was incubated for 1 h. The chip was then washed in water. *Microspheres expansion:* Patterned microspheres, by using photoresist or surface chemistry, were expanded by using a hot plate (50-120 °C).

3. Results and Discussion

Figure 2 (a) shows selectively deposited expandable microspheres (before expansion) on a silicon wafer, using a photoresist technique developed in this study. Figure 2 (b) shows the same pattern after expansion. The diameters of the unexpanded microspheres are 10-16 μm. The shown pattern is a test pattern consisting of 100 μm wide parallel lines with vertical 15 μm wide lines with a pitch of 100 μm. The density of deposited microspheres, *i.e.* the number of microspheres per area, can be influenced by varying the amount of microspheres per photoresist volume or by varying the spin speed. The actuation is irreversible, *i.e.* the expandable microspheres maintain their expanded volume/pressure when cooled down again.

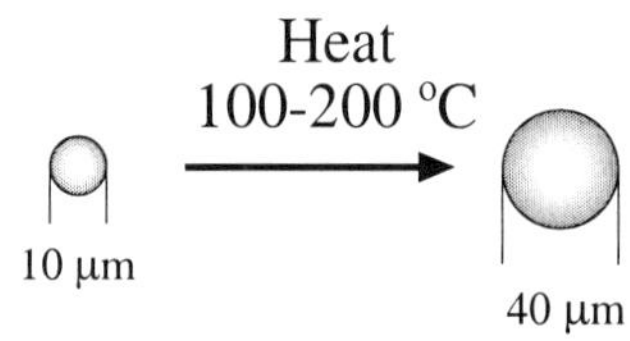

Fig.1. A schematic of the expansion principle of the expandable microspheres.

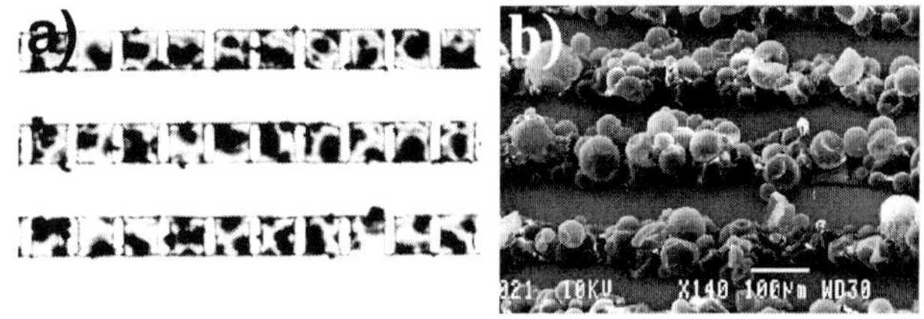

Fig. 2. Photos showing selectively deposited (using photoresist) expandable microspheres (a) before and (b) after expansion.

In many biomedical applications it is not possible to have patches of photoresist in the final device due to its toxicity. Therefore, an alternative way of patterning the expandable microspheres was investigated. Hydroxy-functionalized beads have previously been patterned and immobilized on silicon substrates by means of surface chemistry [3]. The beads were thought to be covalently bound to the substrate via ester bond formation between the anhydride-functionalized surface and the hydroxy groups on the beads. The same method was successfully applied for immobilization of the expandable microspheres.

Future applications of the expandable microspheres can be to replace fluidic components such as one-shot valves or micropumps. The expandable microspheres might also be used to coat the internal walls of microfluidic channels in order to enlarge the reactive surface area. It should also be possible to chemically modify the surface groups of the microspheres in a variety of ways for screening and analysis within the fields of biochemistry and organic chemistry.

References

[1] Expancel microspheres: An introduction, Expancel, SE, [2] H. Andersson, P. Griss, G. Stemme, Expandable Microspheres: A novel Tool in Microfluidics, Submitted to Sensors and Actuators B, [3] H. Andersson, C. Jönsson, C. Moberg, G. Stemme, Self-assembled and self-sorted chemically active beads for analytical and biochemical screening, *Talanta*, In press.

MINIATURIZED FLOW-THROUGH MICRODISPENSER WITH PIEZOCERAMIC TRIPOD ACTUATION

Jonas Bergkvist[1], Tobias Rosqvist[2], Stefan Johansson[2], Johan Nilsson[1] and Thomas Laurell[1]
[1] Lund Institute of Technology, P.O. Box 118, SE-221 00 Lund, Sweden
[2] Uppsala University, P.O. Box 534, SE-751 21 Uppsala, Sweden

Abstract

The development of a piezoceramic tripod actuated silicon flow-through microdispenser is described. The design is improved by replacing the previously employed piezoelectric bimorph rod with an in-house developed multilayer piezo actuator. Properties relevant for chemical analysis systems are significantly improved by further miniaturizing of the channel structure. To conform with this compact version the tripod actuator design is utilized. The work demonstrates the flow-through microdispenser potential of high speed sample handling and crucial improvements in usefulness for chemical analysis systems.

Keywords: droplet dispenser, miniaturization, piezoceramic tripod

1. Introduction

The possibility of making smaller and more stable dispensers, as they thereby become more useful components in chemical analysis work, is demonstrated.

2. Silicon microstructure design and assembly

Two separate silicon wafer chips are engaged in the total unit build up. Connecting those two structures by gluing introduces a number of unwanted uncertainties to the design. To reduce such effects silicon direct bonding (SDB) is utilized instead. This results in improved quality as the internal volume variations in principal vanishes. Figure 1 shows the total unit build-up with the actuating piezo element, interfacing silicone tubes, the upper chip with a p-n etch stop defined membrane and an integrated push-bar, the lower chip with a flow-through channel and the protruding nozzle. For practical reasons the nozzle is depicted in more than ten times the magnification of the other components.

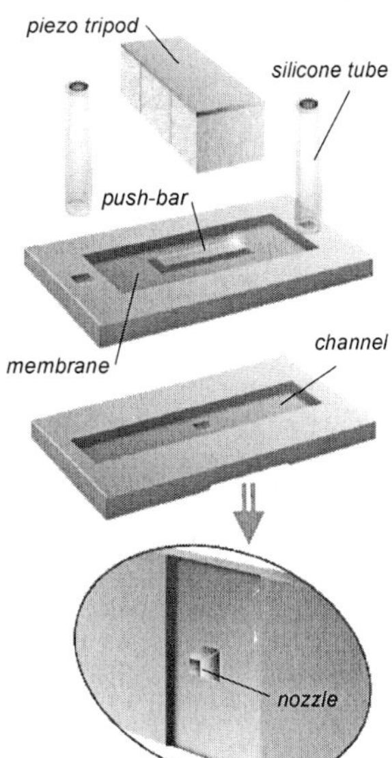

Figure 1, Schematic sketch of dispenser build-up.

J.M. Ramsey and A. van den Berg (eds.), Micro Total Analysis Systems 2001, 147–148.

3. Piezoceramic design and actuation

The piezoactuators are manufactured using a multilayer wet casting and screen-printing technology as illustrated in figure 2a. The multilayered green body is cut into microelements that are shaped into the tripod design, figure 2b, using a dicing saw. Electrode layers are then contacted at the ends using silver paste, as seen in figure 2c.

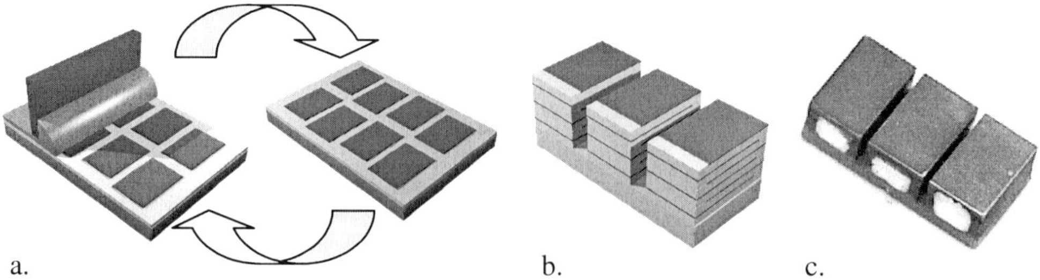

a. b. c.

Figure 2, Casting of multilayered piezo (a). Sketch (b) and photo (c) of tripod actuator.

4. Results and discussion

Reduced dispenser dimensions enables the use of smaller sample volumes which many times is a critical issue in the field of life science research, e.g. proteomics, where only single microliters of precious analyte may be available. Also, the throughput of the device increases since sample switching and washing is faster. Separation performance in chromatography systems increases as the dimensions are reduced [1]. This fact implies the necessity to further miniaturize the flow-through microdispenser when used in microfractionating applications [2]. Miniaturizing the dispenser increases its natural frequencies, which in turn allows a higher maximum operating frequency of the device. Successful dispensing has been accomplished at frequencies up to 8 kHz, figure 3.

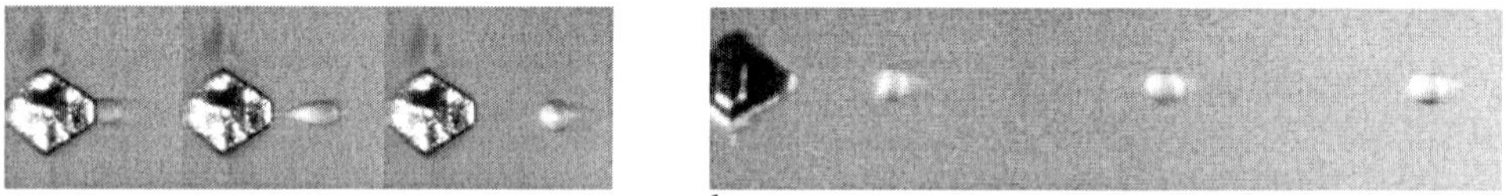

a. b.

Figure 3, Droplet generation at the nozzle (a) and 5 kHz high speed dispensing (b).

References

1. Manz A., Harrison J. D., Verpoorte E. and Widmer H.M., Planar chips technology for miniaturisation of separation systems: A developing perspective in chemical monitoring, in Advances in Chromatography, vol. 33, eds. Brown P. R and Grushka E., Marcel Dekker, Inc.
2. Miliotis T., Kjellström S., Nilsson J., Laurell T., Edholm L-E. and Marko-Varga G., Capillary liquid chromatography interfaced to MALDI-TOF MS using an on-line coupled piezoelectric flow-through microdispenser, J. Mass Spectrom. 35, 369-377 (2000)

Semiflexible Chains of Magnetic Particles

Sibani Lisa Biswal, Alice P. Gast
Department of Chemical Engineering, Stanford University

Abstract
We report the synthesis of semiflexible chains of colloidal particles through ligand binding of polymeric spacers. We create permanently-linked, magneto-responsive chains. We measure the flexural rigidity of these structures using mechanical measurements performed with optical traps and by monitoring the thermal fluctuations of free chains.

Keywords: paramagnetic particles, optical trapping, fluctuations, flexural rigidity

1. Introduction

There is much interest in the use of paramagnetic particles in fluidic microchips to form magneto-responsive structures. Paramagnetic particles form reversible chains by acquiring dipole moments in the presence of an external magnetic field. By physically crosslinking paramagnetic colloidal particles with a spacer, a flexible permanently-linked chain can be formed. The flexibility of the chain is a function of the length of the spacer molecule between the microspheres. These linked chains remain responsive to a magnetic field, allowing their orientation to be controlled by directing the field. When the field is removed, the chains relax and follow the fluid flow profile.

2. Flexural Rigidity Theory

The flexural rigidity or bending stiffness describes the resistance of a chain to bending forces. According to linear elasticity theory, the force per unit length acting on a thin flexible rod is described by the following equation:

$$F = EI\frac{d^4y}{dx^4} \tag{1}$$

where EI is the flexural rigidity and the coordinate x is along the rod and y is orthogonal to it. Using the boundary conditions with equation (1) in a chosen frame of reference the shape of the chain can be described in terms of the force acting upon it and L, the total length of the chain:

$$y = \frac{F}{6EI}\left(2L^3 - 3L^2x + x^3\right) \tag{2}$$

By knowing the deflection and the magnitude of the force acting on the bent chain, the flexural rigidity can be determined.

3. Experimental

Crosslinking Paramagnetic Microspheres

We link the paramagnetic microspheres using streptavidin-biotin binding. Streptavidin is a tetrameric protein with a high affinity for the molecule biotin. The microspheres used are 0.85 microns in diameter polystyrene beads containing 55% (weight) magnetite, coated with streptavidin. These microspheres are placed in a channel and a magnetic field is applied across the channel causing the particles to form chains. A solution of polymeric linkers of bis-biotin-polyethylene glycol (MW=3400) is flowed

J.M. Ramsey and A. van den Berg (eds.), Micro Total Analysis Systems 2001, 149–150.

into the channel in the presence of the magnetic field. This method is successful in creating flexible and linked paramagnetic chains. We attach 3 micron polystyrene microspheres at the ends of the chains to serve as "handles" for the optical trapping experiments since the magnetic particles are difficult to trap directly.

Video Microscopy and Optical Trapping

Observations were made using a 63x/1.30 objective (Zeiss). Images were recorded with an S-VHS recorder and the shape of the chain was digitized using Scion Image software. A sequence of images displaying the chains thermal fluctuations is shown in Figure 1.

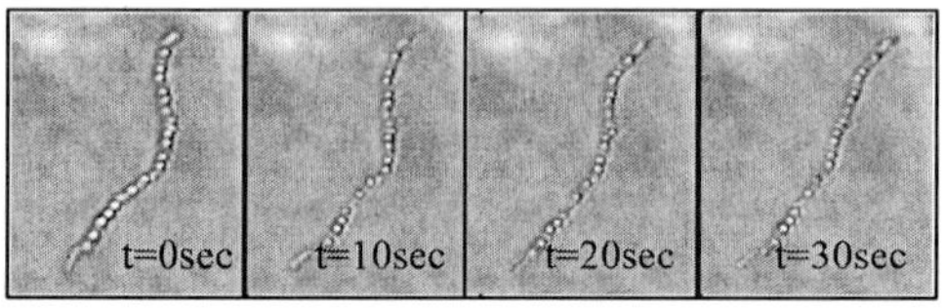

Figure 1. Video images of thermal fluctuations of a chain

Optical trapping is a powerful tool that can be used to study the mechanical properties of these permanently linked chains. It uses highly focused laser light to trap and manipulate microscopic objects. The optical tweezers are formed by coupling an expanded beam of a CW solid state diode pumped laser (λ=532nm) into the objective. The optical trapping experiments allow us to manipulate these chains to study their flexibility and deformation. We calibrate the trapping force using Stokes' flow past trapped spheres. From this calibration, the force on a handle can be determined from the deflection in the trap.

4. Results and Discussion

We performed two types of experiments to determine the flexural rigidity of the chains. In the first experiment, we analyze the thermally driven fluctuations in the shape of the chain to determine its flexural rigidity[1]. Measurements on chains of various lengths provided the flexural rigidity on the order of 10^{-20} Nm^2.

The next experiment involves bending a chain with the optical traps. We bend the chain by manipulating optically trapped handle particles attached to the chain.

We found the flexural rigidity from the force and the shape of the chain following equation (2). The flexural rigidity found from the bending experiments corresponded to that found from the thermal fluctuations: EI on the order of 10^{-20} Nm^2.

5. Conclusions

We have synthesized and measured the bending stiffness of permanently-linked paramagnetic microspheres. By understanding the dynamics of this new class of materials, a magnetically controlled microstructure can be created.

References

1. Gittes, F., Mickey, B., Nettleton, J. and Howard, J. (1993). Flexural rigidity of microtubules and actin filaments measured from thermal fluctuations in shape. *J Cell Biology*. 120, 923-934.

A Miniaturized Cyclic PCR Device

C.F. Chou[1], R. Changrani[1], P. Roberts[1], D. Sadler[1], S. Lin[2], A. Mulholland[2], N. Swami[2], R. Terbrueggen[2] and F. Zenhausern[1]
[1]CTRL, Solid State Research Center, Motorola Labs, 7700 S. River Pkwy, Tempe, AZ 85284
[2]Clinical Micro Sensors, Motorola Life Sciences Division, 757 S. Raymond Ave., Pasadena, CA 91105, USA
(Frederic.Zenhausern@motorola.com)

Keywords: PCR, microreactor, ceramics

1. Introduction

DNA amplification with polymerase chain reaction (PCR) is often performed in a batch process using a bench-top thermal cycler or in a micromachined PCR chamber. Recent effort has shown a continuous flow through PCR system can be realized by a time-space conversion in the PCR system [1]. Devices of this kind open up new areas of application in medical diagnostics, such as the online amplification and monitoring of a specific gene. Here we take advantage of the vertical integration capability of multi-layer ceramic technology and demonstrate a miniaturized cyclic PCR device. A cyclic device promises much smaller footprint and ease of integration into a micro total analysis system.

2. Device fabrication

The design (Fig. 1) essentially comprises of a serpentine channel with different cross-sectional areas in different zones to provide adequate residence time for the melting, annealing, and extension reaction to take place. Air-gaps were built in the device to thermally isolate the individual reaction zones and prevent cross talking. Electric heaters below each reactor provide the necessary energy input for carrying out the biological reactions. The device was fabricated in ceramic substrate (Du Pont T2000 tapes) using multi-layer low-temperature cofired lamination process. The heaters were screen-printed using Ag/Pd paste (30 mOhm/square) and the reactor chambers were defined by mechanical punching or laser cutting in different layers of ceramic tapes to form microfluidic channels. Once the functionality of each layer is defined, all layers are then laminated at 1000 psi and cofired at a final temperature of 875°C. The complete three-dimensional momentum and heat transport equations for the entire system was simulated using CFDRC ACE+.™ The system of equations accounted for the heat transport within the ceramic monolith and the momentum and heat transport within the liquid flowing in the microchannels (Fig. 2A).

J.M. Ramsey and A. van den Berg (eds.), Micro Total Analysis Systems 2001, 151–152.

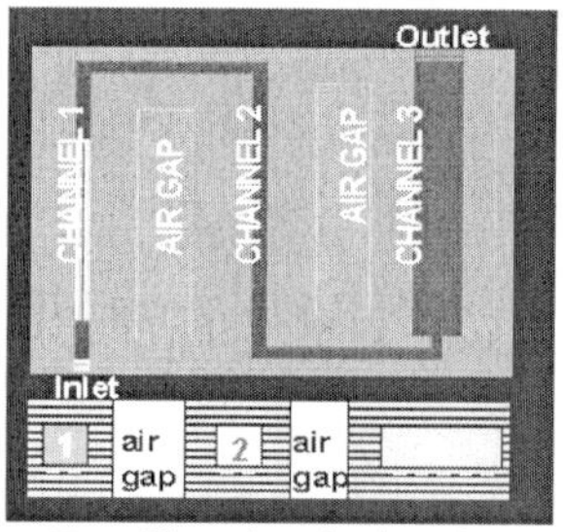

Fig. 1 Top view (top) and cross- Section (bottom) of the cyclic PCR chip.

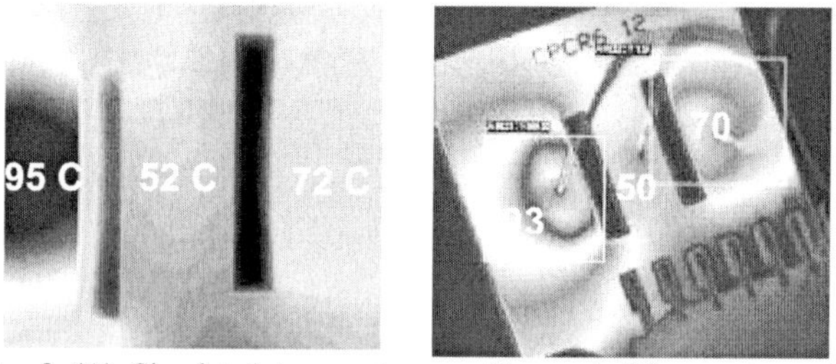

Fig. 2 (A) Simulated temperature profiles and (B) experimentally validated infrared image of the surface temperature profiles of the three reactors. Reactors 1, 2, and 3 are measured at 93, 50, and 70°C, respectively. The surface temperatures are, in general, 2-3°C lower than that in the fluidic channels.

3. Experimental

Temperature control of the three reaction zones is done with three temperature controllers each using a pre-calibrated thermal couple (TC) inserted into a 250 μm via which is in direct contact with the PCR solution. The three temperature zones are set at 95, 50, 72°C, with power input of 350, 1, 250 mW, corresponding to melting, annealing, and extension temperature for the PCR reaction, respectively. The temperature stability is ± 0.2 °C. An infrared image validates the temperature profile of the three reaction zones (Fig. 2B). The sample manipulation is achieved using a peristaltic pump connected to the inlet and outlet of the cyclic PCR device. DNA amplification using Hfe-H fragment (209 bp) as template was demonstrated after 20 cycles of recirculation (Fig. 3). The PCR product was also used with a DNA bioelectronic sensor chip for a genotyping experiment.

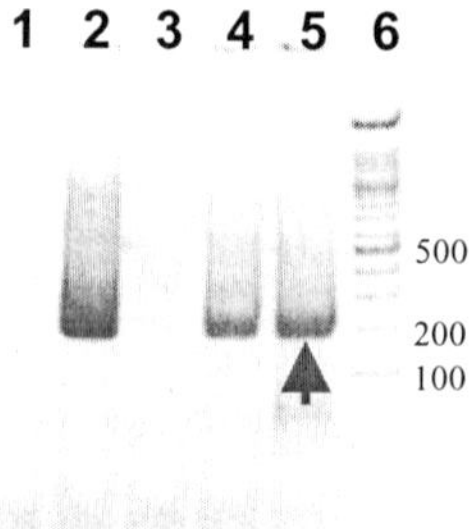

Fig. 3 DNA amplification (20 cycles) from a commercial thermal cycler compared with that from a cyclic PCR device. The template used was Hfe-H fragment (209 bp). Lane 1-4 was performed in a commercial thermal cycler (MJ Research, MA): 1, 3 (2, 4)- negative (positive) controls with different Taq concentration. Lane 5: PCR product from our cyclic PCR device. Different enzyme concentration used to investigate the surface area/volume ratio in our cyclic PCR device as compared to the batch process. Lane 6: 100 bp ladder.

Reference

1. M.U. Kopp, A.J. de Mello, A. Manz, Chemical Amplification: Continuous-Flow PCR on a Chip, Science 280, 1998, p. 1046-1048.

INTERLOCKING MECHANICAL AND MICROFLUIDIC INTERCONNECTIONS FABRICATED BY DEEP REACTIVE ION ETCHING

Bonnie L. Gray*, Scott D. Collins, Rosemary L. Smith
Micro Instruments and Systems Laboratory, University of California at Davis

Abstract

Deep reactive ion etching of bulk silicon is employed to fabricate vertical interlocking structures with which high-density fluidic interconnections are provided between two microchannel containing substrates. Components and boards can be reversibly connected/disconnected. Fluid flow between component and board channels is demonstrated using these structures and sealed using several materials and methods.

Keywords: microfluidic interconnect, packaging, deep reactive ion etching

Introduction

Many applications in μTAS can benefit from a breadboard approach to the integration of microfluidic components onto one substrate. A number of microfluidic instrument board approaches have been developed (e.g., [1,2]). Silicon based processes allow a substantial reduction in the overall size of a microfluidic instrument board as well as an increased level of integration complexity over other materials [3]. One problem previous approaches have in common is the use of permanent bonding methods for component-to-board attachment. A reversible breadboarding method of assembling silicon based fluidic microsystems would enable flexible, rapid prototyping and testing of microfluidic components and systems.

Approach and Methods

The approach presented here uses DRIE (deep reactive ion etching) of silicon to fabricate mechanically interlocking structures with which high-density reversible fluidic connections are possible. The geometric versatility and simplicity of DRIE makes it attractive for the fabrication of these structures, but other methods, such as LIGA, could also be implemented. Figure 1 shows interlocking hole and notched cylinder pairs that accomplish mechanical and fluidic interconnect in the same structure.

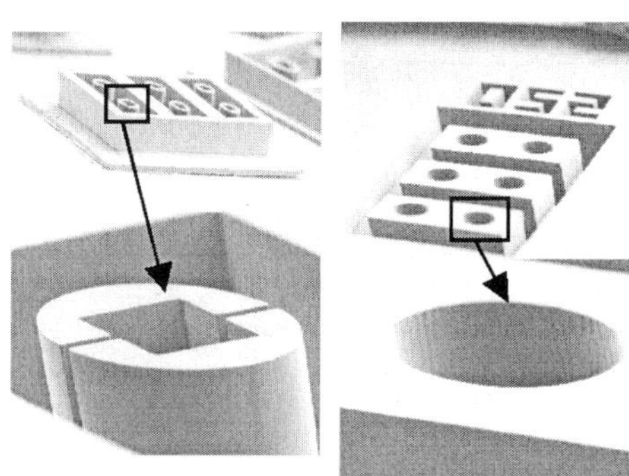

Figure 1. SEM photographs of 250-μm-diameter notched cylinders and holes. Each part has six fluidic connections.

Each cylinder is "notched" so that it will compress slightly when inserted into the

* currently at the Electronic Instrumentation Laboratory, Delft University of Technology, Mekelweg 4, 2628 CD Delft, The Netherlands, b.l.gray@its.tudelft.nl.

J.M. Ramsey and A. van den Berg (eds.), Micro Total Analysis Systems 2001, 153–154.

mating hole structure which is a few microns smaller in diameter. The friction between the contact surfaces, and the restoring forces of the beams that form the cylinder, hold the cylinder in the hole. Fluidic interconnect is accomplished through the aligned hole and cylinder bore. Gasketing of cylinder sidewalls and the spaces between cylinders enables an increase in pressure range, and is accomplished by using photopatterned PDMS O-rings [4], low-temperature melting wax or vapor-deposited Parylene™. Figure 2 shows a close-up of a 250-μm-diameter notched cylinder compressed into a 247-μm-diameter board hole. The theoretical compressive force on the cylinder is estimated to be 3N.

Experimental Testing and Results

Isolation and cross-talk between interconnects were tested using a syringe pump with the board and component system shown in Figure 3. The test channels on the component are each connected via two cylinder/hole pairs to channels on the microfluidic board, demonstrating the use of the structures for component-to-board attachment. Tygon™ tubing attaches to hexagonal ports [4] that connect to the board channels. Each channel was filled with water containing one of three differently colored dyes. No leakage or cross-talk between the interconnects is evident.

Figure 2. Close-up of one notched cylinder inserted into a board hole.

Pressure versus flow rate for a single channel system gasketed with wax was measured using a syringe pump and pressure gauge. The results are shown in Figure 4. The interconnects were leak-tight up to a pressure of 70 kPa, which corresponded to a flow rate of 1.95 ml/min.

Figure 3. Photograph of interconnect isolation testing. The inset shows a close-up of the three component test channels.

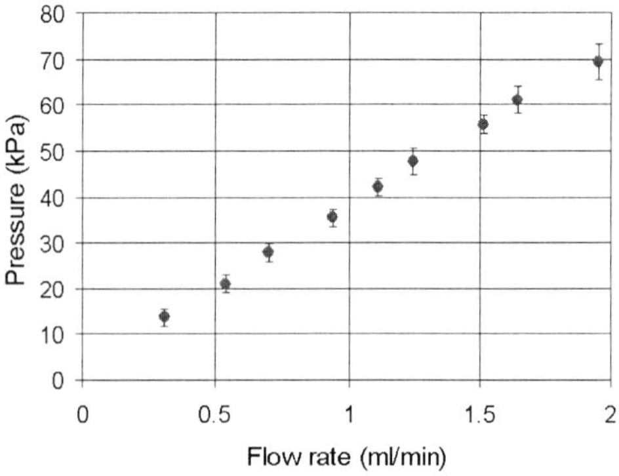

Figure 4. Pressure versus flow rate for the center 392-μm-diameter notched cylinder and 390-μm-diameter hole pair shown in Figure 3.

References

1. C. G. J. Schabmueller, et. al., *J. Micromech. Microeng.*, v9 (1999), pp. 176-179.
2. T. S. J. Lammerink, et. al., *IEEE Proc. 9th Int. Workshop on Micro Electro Mechanical Syst., (MEMS '96)*, San Diego, Feb. 1996, pp. 389-394.
3. B. L. Gray, et al., *Sensors and Actuators A*, v77 (1999), pp. 57-65.
4. C. Gonzalez, et. al., *Sensors and Actuators A*, v66 (1998), pp. 315-332.

POLYMER MICRO NEEDLES WITH THROUGH-GOING CAPILLARIES

A.E. Guber[1], H. Dittrich[1], M . Heckele[1], D. Herrmann[1], A. Muslija[1], W. Pfleging[2] and Th. Schaller[3]
Forschungszentrum Karlsruhe, P.O. Box 3640, 76021 Karlsruhe, Germany
[1]Institut für Mikrostrukturtechnik, [2]Institut für Materialforschung-I,
[3]Institut für Mikroverfahrenstechnik
e-mail address: andreas.guber@imt.fzk.de

Abstract

Removal or exact transfer of minimum substance volumes from reservoirs or micro fluidic systems (e.g. lab-on-a-chip systems) may be accomplished by using miniaturized tips with integrated through-going capillaries. A method to fabricate polymer micro needles with through-going capillaries by double-sided hot embossing has been developed.

Keywords: Polymer micro needles, capillaries, micro machining, double-sided hot embossing

1. Introduction

Different applications in biomedical engineering like the application of drugs, life sciences technologies or micro nozzles require the use of disposable polymer micro needles with through-going capillaries.

2. Micro technologies

For this purpose, double-sided hot embossing of polymer material has now been applied successfully for the first time [1]. 1 mm high, miniaturized needles were fabricated from PSU with through-going micro channels of variable diameter from 100 μm down to 15 μm only. The micro fabrication of micro tips with through-going capillaries on thin plastic substrates requires two metal mold inserts adapted to each other (Fig. 1). Several mold-making techniques available at the Forschungszentrum Karlsruhe have been employed. The first mold insert is made of nickel and the microstructure is a 1 mm deep conic hole. This mold insert is manufactured by a combination of mechanical micro machining with a subsequent electroforming process. First, a metallic pin is made of a brass substrate and then transferred by electroforming into a nickel mold insert with a conic hole. On the opposite mold insert there is a thin needle of 1.1 mm in height (bottom diameter: 0.1 mm and top diameter: 0.01 mm), which is ground from a tungsten carbide rod and inserted into a brass body. Prior to double-sided melding, both mold inserts are adjusted precisely in the embossing machine. The metallic pin is aligned relative to the center of the conic hole. During hot embossing, the two mold inserts are moved towards each other.

J.M. Ramsey and A. van den Berg (eds.), Micro Total Analysis Systems 2001, 155–156.

3. Results and conclusions

After demolding, polymer substrates of 0.3 mm thickness and 1 mm high micro tips are obtained. To realize the through-going holes the residual layer due to the minimal distance of the mold inserts, which covers the top of the micro tip has to be removed. By applying precise horizontal cuts or laser micro machining (Figs. 2 and 3). Laser drilling of high aspect ratios up to 20 is performed with Excimer laser radiation, with accuracies in the μm-range and with hole diameters of about 10 μm [2,3]. After this, the substrate with the integrated micro tip is bonded to a separately manufactured micro fluidic channel plate from the rear using special assembly and bonding technologies, thus providing for a tight sealing. A special test facility has been set up to measure the fluidic penetrability of the micro tips. For this purpose, one of the two openings located on the micro fluidic channel plate is connected with a pressure sensor, while the other opening is closed. The micro tip with the bonded channel plate is put onto an adapter. Using a vacuum pump, pressure is reduced from 1 bar to about 0.5 bar within a period of about 30 seconds. Thus, principal penetrability of the micro capillaries in the micro tip is demonstrated. For the first time, we thus succeeded in producing by double-sided hot embossing very filigree three-dimensional micro capillary structures for various applications in the life sciences sector.

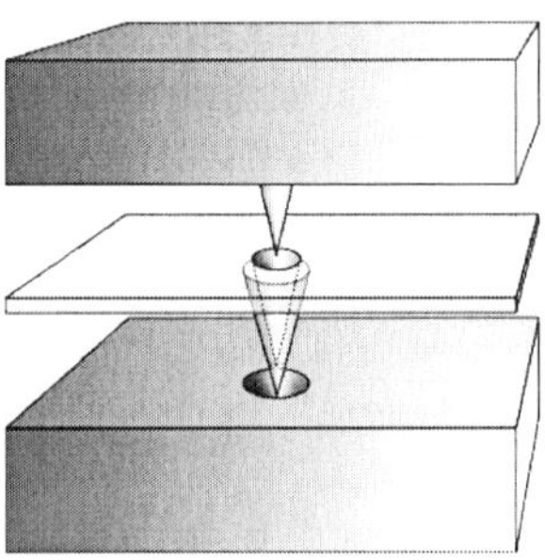

Figure 1. Double-sided hot embossing for the replication of polymer micro tips using a mold insert with a pin (top) and a mold insert with a conic hole (bottom).

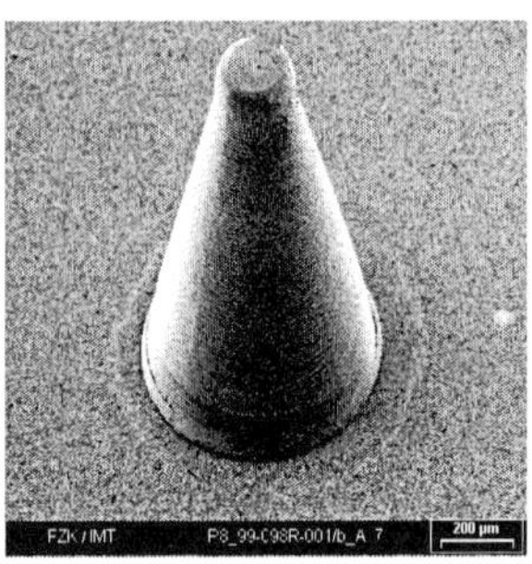

Figure 2. Replicated hollow micro needle made of PSU with a through-going capillary.

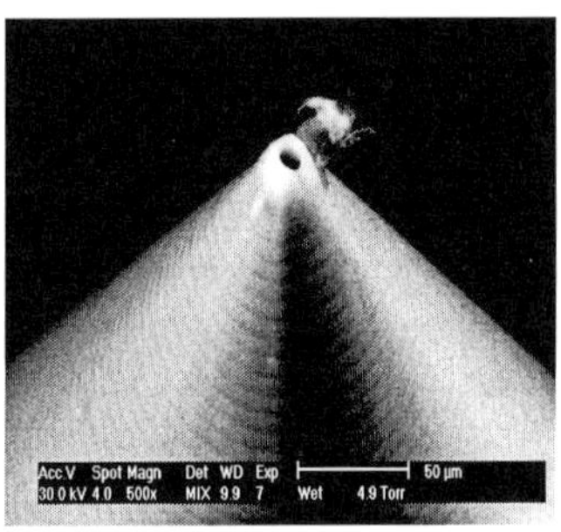

Figure 3. Top part of the micro needle with a laser drilled internal capillary (∅=10 μm).

References

1. M. Heckele, H. Dittrich, A. Guber, A. Muslija and T. Schaller, *Double-sided hot embossing of microstructures*, Proc. HARMST 2001, pp. 137-138, (2001).
2. W. Pfleging, *Laser material processing for applications in medical and microsystem technologies*, LaserOpto, 31, No. 5, pp. 54-57, (1999).
3. M. Torge and W. Pfleging, *Laser assisted patterning of polymers and CE-Chip production with motorized mask*, Proc. HARMST 2001, pp. 107-108, (2001).

Hydrophobic Sealing for Integrated Microactuators in Biofluidic Systems

Ki-Hun Jeong, Sebaek Oh, and Luke P. Lee
Berkeley Sensor and Actuator Center, University of California, Berkeley, CA, USA

Abstract

Selective hydrophobic sealing for integrated microactuators in biofludic system is described in this paper. A selectively deposited hydrophobic polymer on the surface of a microchannel keeps a fluid from entering into the channel. Through this hydrophobic channel, an electrostatic microactuators can be coupled to a hydrophilic channel. Integrated microactuatos in biofluidic systems can be used for sinlge molecule detections and manipulations.

Keywords: hydrophobic sealing, microactuator, and biofludic device

1. Introduction

Advanced microfluidic integrated devices with microactuators, microvalves, or micropumps require an effective sealing mechanism for fluid leakage. The role of nano and microscale actuators is important for single cellular and molecular manipulation. However, they have not yet been realized in a microfluidic environment due to electrical short through ionized fluid leakage. In this paper, an effective hydrophobic sealing method for an integrated microactuator within a biofluidic system is demonstrated.

2. Design

Hydrophobic sealing is achieved by depositing a hydrophobic polymer on a channel coupled with a hydrophilic channel. The unfavorable capillary forces due to the hydrophobic surface keep a fluid from entering a hydrophobic channel coupled with a microactuator as illustrated in Fig.1. Threshold flow rates above which the fluid advances across the interface between a hydrophobic and a hydrophilic channel can be estimated by combining the Hele-Shaw equation and surface tension equation with the different channel geometric parameters and contact angles as shown in Eq.1 and Fig.2[1].

3. Fabrication

As a test structure of hydrophobic sealing, two types of microchannels were fabricated. A Teflon®-like polymer using C_4F_8 was selectively patterned on hydrophobic channels by lift off process after deep etching and oxygen plasma on single crystal silicon. An integrated biofludic device which has microactuators coupled with a microfluidic channel was fabricated by single crystal silicon reactive etching and metallization process. The lift off process was also used to selectively deposit hydrophobic polymer on the released structures.

J.M. Ramsey and A. van den Berg (eds.), Micro Total Analysis Systems 2001, 157–158.

$$Q = \frac{\sigma \cos \theta_c}{6\mu L} w_m h^2 \left(1 + \frac{h}{w_c}\right) \qquad [1]$$

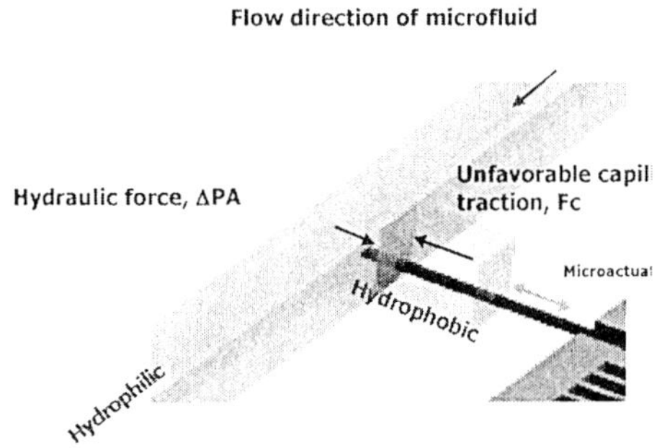

Fig.1 Schematics of hydrophobic sealing

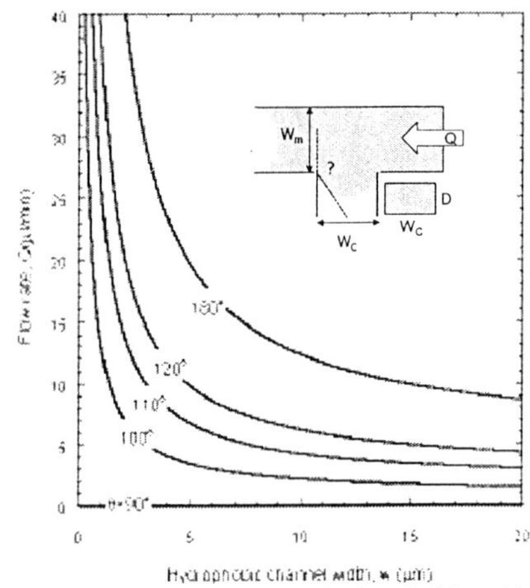

Fig. 2 Threshold flow rate vs. hydrophobic width

4. Results and discussion

Hydrophobic sealing due to unfavorable capillary forces is shown in Fig.3. It is shown that hydrophobic sealing is not dependant on the geometry of a hydrophobic channel under no pressure. Fig.4 shows microactuator can be coupled to a microfludic channel. Liquid is also blocked at a probe tip where the hydrophobic surface starts.

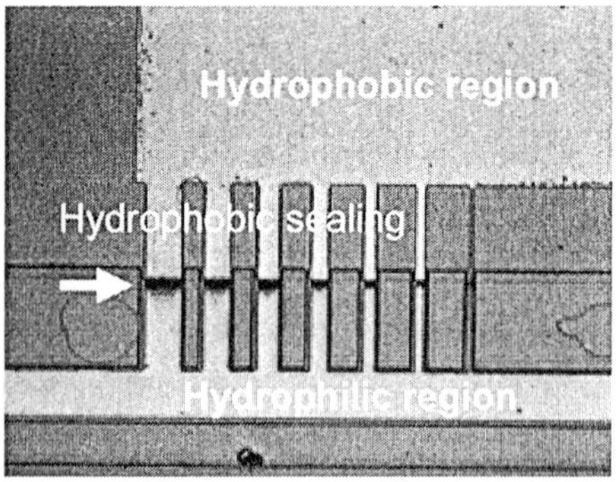

Fig.3 Test structure of hydrophobic sealing

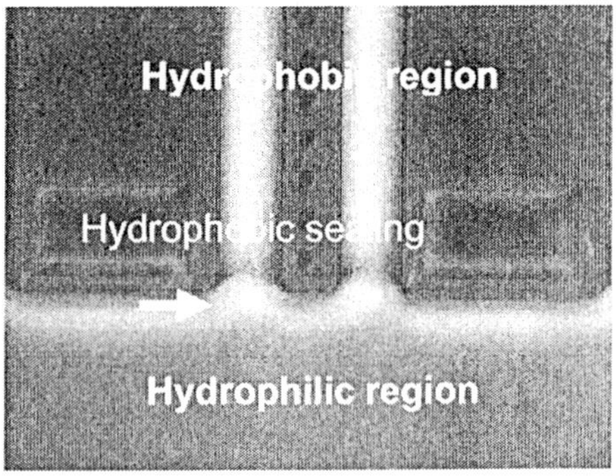

Fig.4 Microactuator coupled to a microfludic channel using hydrophobic sealing

5. Conclusions

We have demonstrated the possibility of incorporating microactuators with a microfludic channel, using hydrophobic sealing. An integrated biofluidic system for a single molecule manipulation system is in progress as a further study.

References

1. Helene Andersson, and et al., Hydrophobic valves of plasma deposited octafluorocyclobutane in DRIE channels, Sensors and actuators B, vol.75, pp.136-141, 2001

METHODOLOGY FOR JUNCTION DILUTION COMPENSATION PATTERN AND EMBEDDED ELECTRODE IN CE SEPARATOR

Y.W.Jeong[1], S.Y.Kim[1], S.Chung[2], S.J.Paik[1], Y.S.Han[3], J.K.Chang[1], D.Dan.Cho[1], and D.S.Chung[3], and K.Chun[1]

[1] School of Electrical Engineering, [2] School of Mechanical and Aerospace Engineering, [3] Dept. of Chemistry, Seoul National University, Seoul, Korea

Abstract

Cyclic capillary electrophoresis has advantages in scaling down of size and low operation voltage, as we described before. [1] However, injected sample plug has been diluted because it met so many junctions in circular channel. We estimated that it is a critical point in a cyclic CE. So we simulated the junction dilution problem and designed the dilution compensation pattern. In fabrication, we developed embedded electrode technology for removing spacer material, which is conventionally used in bonding with electrode. We chose two method of quartz bonding. One is thermal direct bonding, and the other is CYTOP (teflon-like fluorocarbon polymer) bonding.

Keywords: Capillary electrophoresis, diffusion, embedded electrode, cyclic

1. Introduction and theory of dilution

At corner junction, there are so many parameters for affecting the fluidic control. We approached the junction dilution toward fringing field of junction. [fig. 1] Its phenomenon of dilution is simulated as Flumecad^R (Coventor Co.). The designed compensation methodology as result of simulation is that the dilution is controlled by W1/W2 channel width ratio. Actually, as W1/W2 ratio is bigger, the dilution is weaker. [fig. 2] But, if W1/W2 ratio is bigger than 5, the minute pattern is difficult to fabricate and the injection of buffer by just capillary force is difficult because the junction acts like a nozzle. Our optimum W1/W2 ratio is 3.

2. Embedded electrode methodology & fabrication

For ICP dry etching of quartz, the charge must be shielded by aluminum layer. It can make the roughness of etched surface minimize and accelerate the etch rate of quartz. The channel and electrode part is dry etched by the charge shielding method. We made a hole of a 500 μm depth reservoir by a sandblaster with a etch mask of 100 μm dry PR, and we sputtered 1 μm Ti/Au on the electrode part, which is etched as 3 μm. After 3 μm CMP process, Au is polished away except only electrode part. [fig. 3] The intermediate layer (silicon dioxide or CYTOP), which is used for insulating layer can be used optionally. In these methodologies, it doesn't need any spacing material at bonding process. [fig. 4] Its technology can be applied to dielectrophoresis and electrode integration of CE separator.

J.M. Ramsey and A. van den Berg (eds.), Micro Total Analysis Systems 2001, 159–160.

3. Experimental

In opposition to thermal direct bonding, CYTOP bonding process makes the property of the channel hydrophobically. So, methanol must be mixed in buffer solution to inject the buffer solution in channel by just capillary force. The fluidic property of the system is investigated using dyestuff.

Acknowledgements

Support for this work has been provided by the Frontier program and ERC of ministry of education, Korea.

References

1.N. Burggraf, A. Martz, C. S. Effenhauser, E. Verpoorte, N. F. de Rooij, and H. M. Widmer, "Synchronized Cyclic Capillary Electrophoresis – A Novel Approach to Ion Separations in Solution," Journal of High Resolution Chromatography, vol. 16, pp. 594-596,(1993)

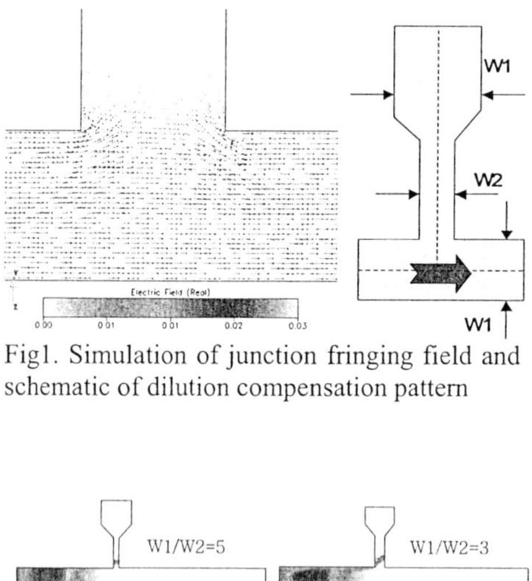

Fig1. Simulation of junction fringing field and schematic of dilution compensation pattern

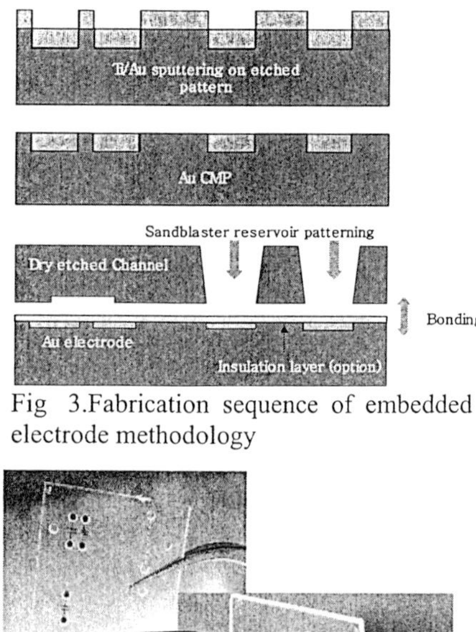

Fig 3.Fabrication sequence of embedded electrode methodology

Fig 2. Simulation results of dilution compensation patterns

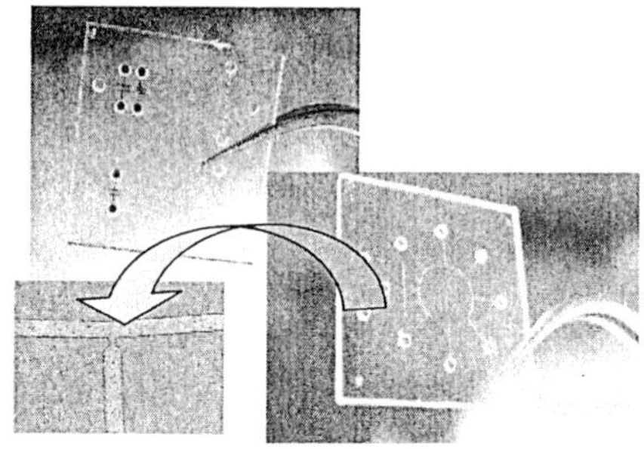

Fig 4. Fabrication result of cyclic CE separator and embedded electrode

FABRICATION OF A GLASS MICROCHIP WITH A THREE-DIMENSIONAL CHANNEL NETWORK AND ITS APPLICATION TO A SINGLE-CHIP COMBINATORIAL SYNTHETIC REACTOR

Y. Kikutani[1], H. Hisamoto[1,2], M. Tokeshi[1], T. Kitamori[1,2]
[1]Kanagawa Academy of Science and Technology, Kawasaki, Japan
[2]The University of Tokyo, Department of Applied Chemistry, Tokyo, Japan

Abstract

With two-level crossing structures constructed by utilizing three layers of glass plates, a three-dimensional channel network that makes micro combinatorial synthesis possible was successfully integrated on a single glass microchip.

Keywords: **Two-level crossing, 3-D microchannel network, multi-layered glass microchip, combinatorial chemistry**

1. Introduction

Glass is widely used as a versatile material for microchips, and the most popular method in making a glass chip is carving microchannels on a flat glass surface by photolithography/wet etching followed by bonding a cover plate to make tunnel structures. Only two-dimensional patterns can be made by this method. Here we present a simple expansion of this conventional fabrication method by utilizing three layers of glass plates, which makes a two-level crossing of microchannels possible. We applied this new fabrication technique to a single-chip micro combinatorial synthetic reactor.

2. Experimental

As a simple model, we designed a 2 x 2 combinatorial synthetic reactor. Channel layout in the reactor is schematically presented in Fig. 1. Each reagent stream introduced from the four inlet ports forks into two branches. The branched streams are joined in a way that reagents of 2-member libraries {A, B} and {C, D} are mixed to produce AC, AD, BC, BD. For this channel layout to be integrated in a single chip, two-level crossings of channels are geometrically inevitable. To achieve the requirement, we used three layers of glass plates. Microchannels (width: 240 μm, depth: 60 μm) were fabricated on the upper and the bottom plates (30 x 70 x

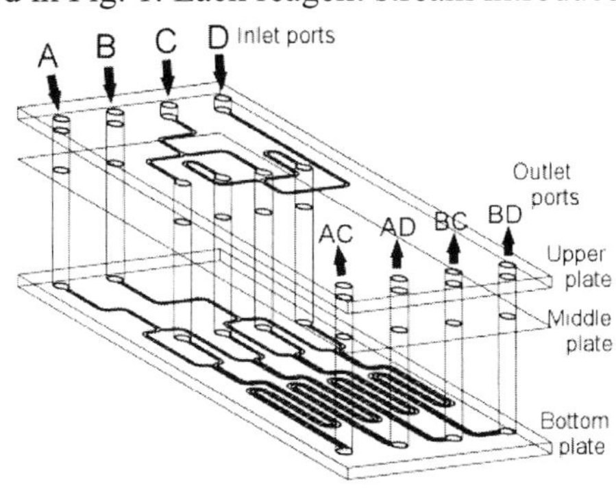

Figure 1. Channel layout inside the glass microchip for 2x2 combinatorial synthesis

J.M. Ramsey and A. van den Berg (eds.), Micro Total Analysis Systems 2001, 161–162.

0.7 mm) by the same method used in conventional glass microchip fabrication. Between the etched surfaces of the upper and the bottom plates, a glass middle plate (30 x 70 x 0.2 mm) with holes pierced at channel paths was placed, and the three plates were thermally bonded at 650 °C, 5 h, under pressure. A photograph of the combinatorial synthetic chip is shown in Fig. 2.

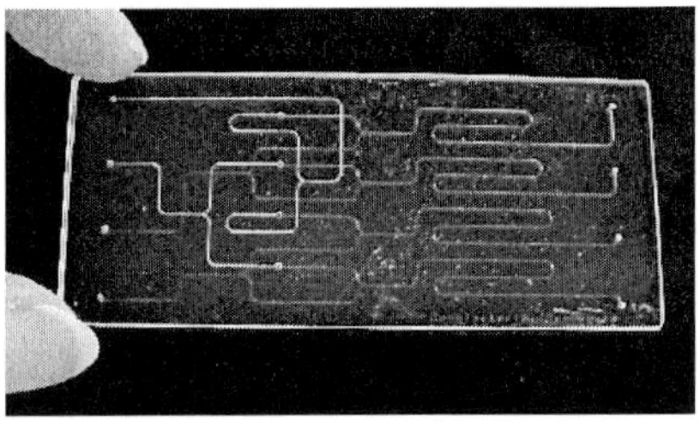

Figure 2. Glass microchip for 2x2 combinatorial synthesis

To illustrate applicability of the three-dimensional channel network, combinatorial synthesis of amide compounds between two amines (1 x 10^{-2} M / 0.1 M NaOH solutions of A: DL-1-phenyl- ethylamine, B: 4-amino-1-benzylpiperidine) and two acid chlorides (1 x 10^{-2} M / ethylacetate solutions of C: 3,5-dinirobenzoyl chloride, D: 3-nitrobenzoyl chloride) was carried out.

3. Results and Discussion

With only a two-dimensional channel pattern, it is impossible to mix four different reagents in four different combinations. In our system, TLC spots of products from the four outlet ports had different Rf values (Fig. 3), and it was confirmed that they coincided those of corresponding target compounds. That means we could successfully construct two-level crossings inside a glass microchip by the simple method. It is significant for further development of the microchip technology because, in principle, any three-dimensional channel layout can be built if a two-level crossing structure can be fabricated. Further integration of chemical processing in microchips, especially of high-throughput systems for industrial applications, requires complicated three-dimensional channel layout which allows multi-channels to work in parallel.

A microspace in a microchannel has characteristics such as short diffusion length and small heat capacity. By utilizing these properties with a microreactor, efficient chemical synthesis can be carried out [1,2]. Whereas a conventional automated combinatorial synthetic system in which robots manipulate ordinary-sized vessels needs milliliters of reagents, micro combinatorial synthesis needs as small amount of reagents as in μ-TAS. Recent development of μ-TAS enables assay in the nanoliter scale. Integration of an assay system on combinatorial synthesis chip is expected to give a highly efficient screening system.

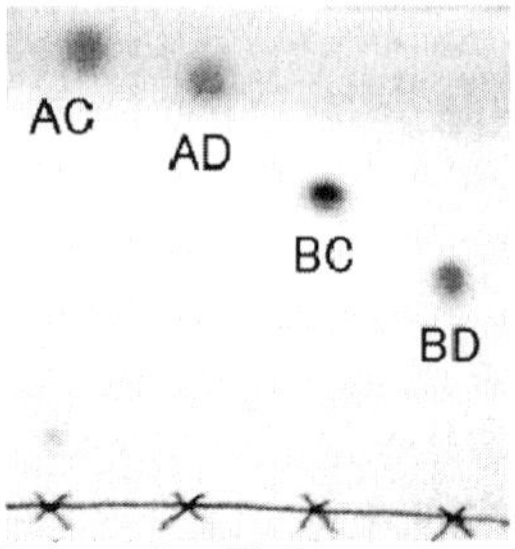

Figure 3. TLC spots of products from four outlet ports of the combinatorial synthetic reactor chip

References

1. W. Ehrfeld, V. Hessel, H. Lowe, Microreactors, John Wiley & Sons, 2000
2. H. Hisamoto, M. Tokeshi, T. Kitamori, Program of 14th Symposium on Microscale Separation and Analysis, HPCE 2001, P0811Ta, p.117

PLASTIC IN-LINE CHAOTIC MICROMIXER FOR BIOLOGICAL APPLICATIONS

Robin H. Liu^, Michael Ward^, Justin Bonanno^, Dale Ganser^, Mahesh Athavale*, Piotr Grodzinski^
^Microfluidics Laboratory, Motorola Labs, Motorola Inc., Tempe, AZ
*CFD Research Corp., Huntsville, AL

Abstract
A plastic 3D "L-shaped" serpentine micromixer is developed to enhance mixing of biological samples. Both numerical simulation and experiments show such a device has high mixing efficiency and low shear strain field.

Keywords: Micromixer, chaotic advection, cell capture

1. Introduction

Rapid in-line mixing is essential for microfluidic systems in biological applications. Mixing in microfluidics is typically dominated by diffusion that can be very inefficient. A "C-shaped" 3D serpentine microchannel based on chaotic advection has been demonstrated to achieve effective mixing by increasing the interfacial surface area between initially distinct fluid regions so that diffusion can complete the mixing process in a reasonable amount of time [1]. However, application of this mixer in biological processes has not been reported. In this paper, we report the development of a plastic "L-shaped" serpentine mixer and its use in biological sample preparation.

2. Theory and Design

Chaotic advection is a laminar stirring phenomena [2] that has been verified in steady laminar flow through a sequence of three-dimensional channel geometries (such as the "L-shaped" serpentine design shown in Fig.1) resulting enhanced transverse and longitudinal stirring thus rapid mixing due to secondary flow instability (at Re>1). Mixing process in the "L-shaped" channel is simulated using CFD by adding different reactants in the two incoming streams form a dye product (Fig.2). At Re > 20, about three serpentine blocks provide >90% mixing. Numerical results of the degree of mixing at different flowrates and at different locations in the mixer are shown in Fig.3. Numerical studies on shear stress fields and hemolysis index values are also performed.

3. Experimental Results and Discussion

Two micromixers are fabricated: one using 3D PDMS microfabrication and the other one using plastic (polycarbonate) micromolding technique. Fluidic dye experiments (Fig.4) demonstrate the ability of flow in the "L-shaped" channel to mix faster and more uniformly than pure molecular diffusion and the "C-shaped" mixer for small Re (Re>1). At Re>30, the mixing curve turns towards an exponential relationship suggesting the

J.M. Ramsey and A. van den Berg (eds.), Micro Total Analysis Systems 2001, 163–164.

occurrence of chaotic advection in the channel (Fig.5). Mixing rate increases at the higher Re are consistent with mixing being dominated by chaotic advection. The flow stirring results in twisted and stretched interfaces between streams giving diffusion a larger area to work, thus increasing the overall mixing rate. Bacteria capture by mixing blood/bacteria samples with magnetic beads is performed in the "L-shaped" micromixer. Results show high capture efficiency (~ 99%) was achieved within a short of time (0.15 sec). UV absorbance measurement shows blood cells and bacteria were intact after the mixing process indicating that the mixer has low shear strain field.

Fig. 1 Schematic of a 3D "L-shaped" micromixer.

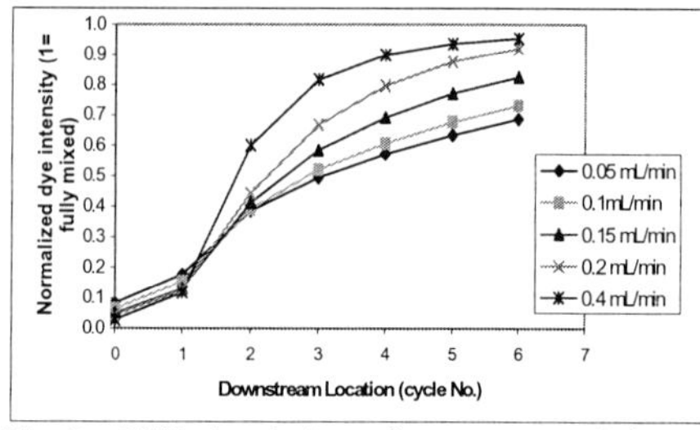

Fig. 3: CFD simulation of percentage mixing versus downstream location in the 3D L-shaped mixer.

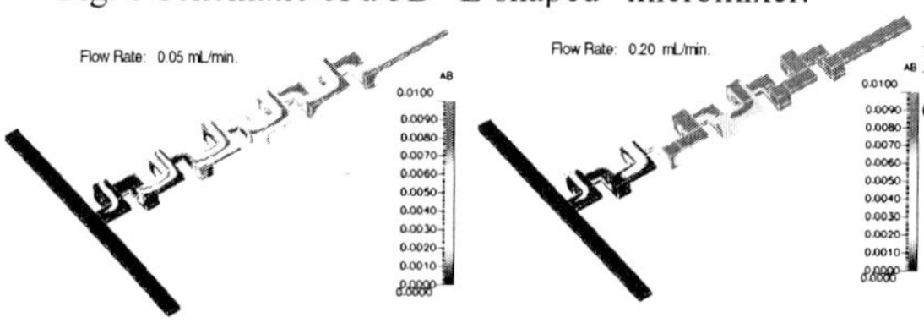

Fig. 2: CFD simulation of mixing at different flowrates

Fig. 4: Video prints of the flows (Two inlet streams, NaOH and phenolphthalein solution, flow from right to left) in the micromixer. Streams produce a red dye when contact.

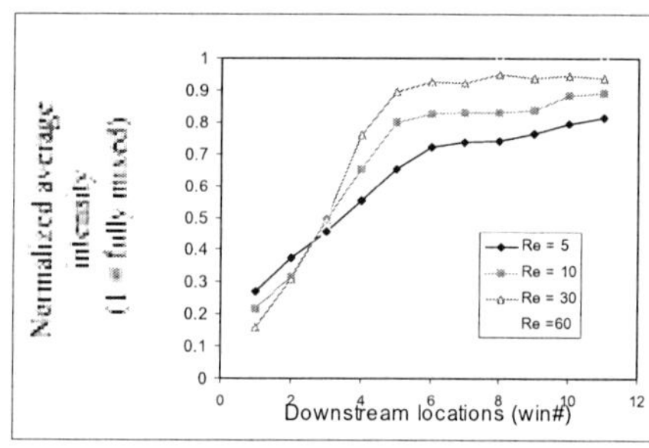

Fig. 5: Plot of percentage mixing versus downstream location in the serpentine mixer.

4. Conclusions

A plastic in-line chaotic micromixer with high mixing efficiency and fluid throughput is demonstrated. This device is particularly useful for bio-sample processing due to low shear strain field.

Acknowledgements

This work is sponsored by DARPA contract #MDA972-01-3-0001.

References:

1. Liu, R. H., et al., "A Passive Micromixer: 3-D C-shaped Serpentine Microchannel", *JMEMS*, Vol. 9, No. 2, June, 2000, pp190-197.
2. Jones, S.W., et al. , "Chaotic Advection by Laminar flow in a Twisted Pipe", *J. Fluid Mech,* 1989. **209**: p. 335-357.

MICROSYRINGE ARRAYS IN POLY(DIMETHYLE SILOXANE)

Mariam N. Maghribi[1, 2], Peter Krulevitch[1] and Julie K. Hamilton[1]
[1]Lawrence Livermore National Laboratory, 7000 East Ave. L-223 Livermore, CA 94550, USA
[2]University of California Davis, One Shields Ave. Biomedical Engineering Davis, CA 95616, USA
(maghribi1@llnl.gov)

Abstract

In this paper we discuss the development of integrated microsyringe arrays, fabricated using a rapid low-cost process that is easily adapted for many microfluidics systems. A key innovation to the development of the microsyringes is a process for producing microchannels in poly(dimethyl siloxane) (PDMS) with perfectly round cross-sections. The flexible PDMS conforms to an oversized spherical metal piston inserted into the channel, forming a leak-proof seal. This is the reverse of conventional syringes, which use a rigid vessel and compliant plunger.

Keywords: Circular channels, poly(dimethyl siloxane), PDMS, drug delivery

1. Introduction

Microsyringes are useful for several fluid control functions, including pumping, valving, and metering. Addressable thermopneumatic actuation sources [1,2] are integrated with metal pistons within PDMS microchannels. The pneumatic fluid is heated creating sufficient pressure to drive the piston within the microchannel, delivering the required reagent, as illustrated in Figure 1. The microsyringes are an attractive alternative to existing methods for micro-scale fluid control, such as piezoelectric-actuated pumps, electro-osmotic flow, and external pressure sources. These approaches suffer from several drawbacks: they require peripheral energy sources to generate large pressures and offer limited fluid control capability. The ultimate objective of this research is to integrate these fully self-contained microsyringe arrays in a transdermal drug delivery system.

2. Experimental

Ranges of channel and piston diameters were characterized using a conventional syringe pump to determine leak-proof ratios. Piston actuation pressures were measured using a Brass Bourdon Tube pressure gauge connected via Teflon tubing between the syringe pump and microsyringe.

PDMS surface treatment in oxygen plasma provides a temporary hydrophilic characteristic that facilitates the process of heater bonding, pneumatic fluid loading and channel sealing. Since surface effects dominate fluid flow on the microscale, a process was developed to selectively modify the surface. The thermopneumatic actuation reservoir contained fluid with a high coefficient of expansion (Fluorinert 77 from 3M) activated by a resistive heater that is driven using a battery-powered microprocessor as illustrated in Figure 1.

3. Results and Discussion

J.M. Ramsey and A. van den Berg (eds.), Micro Total Analysis Systems 2001, 165–166.

A crucial aspect of the microsyringe is the circular cross-sectional geometry, because it minimizes friction and provides a uniform, leak-proof seal with a spherical or cylindrical piston. We have developed a novel technique for rapidly fabricating circular, smooth channels in PDMS, as seen in Figure 2. Existing fabrication methods of circular channels are expensive, time consuming, and encounter numerous difficulties including alignment problems and etching defects. We are capable of producing channels as small as 30μm in diameter, but our current microsyringes range from 254-508 μm in diameter due to piston size limitation. These microsyringes can be adapted to meet desired volumetric needs from nano-liters to micro-liters.

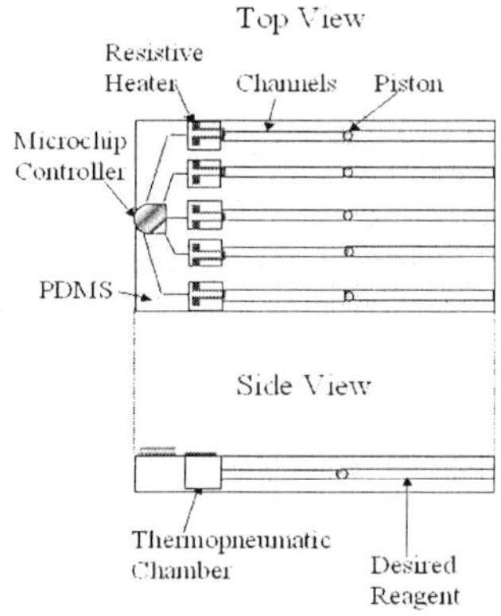

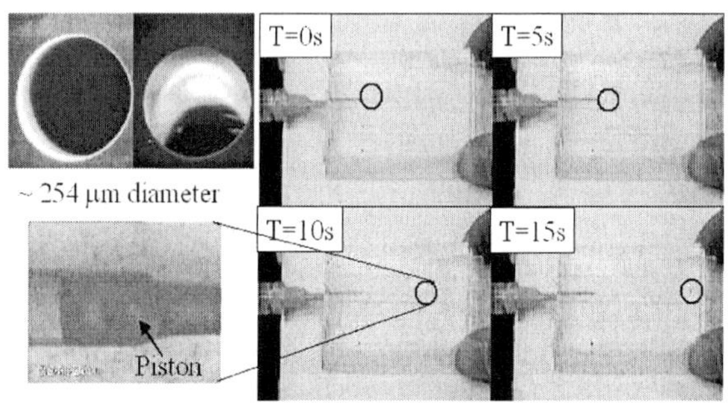

Fig. 1. Integrated Microsyringes ***Fig. 2. SEM of Microsyringe & time sequence***

The initial prototype of the microsyringe arrays is shown in Figure 2, which also illustrates a controlled time sequence movement of the piston within the PDMS microchannel. Optimal channel diameter was found through multiple iterations to be ~60-65% of the piston diameter to achieve a leak proof seal without obstructing piston movement within the channel. The pressure requirement to initiate piston movement was between 65-100 psi, varying with microsyringe diameter.

4. Conclusion

The present analysis indicates the fabrication feasibility of a programmable, control release drug delivery device. It is expected that the performance of this device will significantly improve transdermal drug delivery over current technology. These microsyringes also offer the capability of truly integrated fluid handling for miniature and field-portable micro total analysis systems.

Acknowledgements

We gratefully acknowledge Bill Benett and Paul Stratton for their advice and assistance. This work was conducted under the auspices of the U.S. Department of Energy by Lawrence Livermore National Laboratory, contract number W-7405-ENG-48.

References

1. Zdeblick, Mark J. et. al. *Thermpneumatically actuated microvalves and integrated electro-fluidic circuits*. Solid-State and Actuator workshop June 13-16 1994. Hilton Head, South Carolina, pp.251-255,(1994).
2. Henning, Albert K. *Microfluidic MEMS*. IEEE, pp.471-486, (1998).

A MEMS BODY FLUID FLOW SENSOR

Ellis Meng[1], Sascha Gassmann[2], and Yu-Chong Tai[1]
[1]California Institute of Technology, Pasadena, CA 91126, USA
[2]ETH Zurich, Institute of Mechanics, CH-8092 Zurich, Switzerland

Abstract
To achieve *in vitro* flow rate measurements of biological fluids in such tasks as hematological studies and urinalysis, a MEMS flow sensor has been developed. Flow sensing is achieved by measuring the forced convective heat transfer from a thermal sensing element to the fluid. Currently, fluid flow down to 10 μl/min can be detected.

Keywords: Flow sensor, thermal sensing, King's Law

1. Introduction
Heat transfer is the most promising flow sensing principle for measuring very low flow rates (< 1 ml/min). Several thermal anemometer type sensors have been introduced in previous work [1-4]. Many use polysilicon thermistors as heating and temperature sensitive elements. Here, a metallic resistive sensing element is placed on a channel wall to sense flow rate. When operated in constant current mode, the convective heat transfer from this element to the fluid can be measured and correlated to flow rate.

2. Sensor Design and Fabrication
Sensors (Fig.s 1-2) consist of platinum resistors on a parylene membrane over a bulk micromachined silicon channel (1 mm × .5 mm × 8 mm). This structure prevents resistor contact with the fluid and possible issues with corrosion. Platinum is chosen as the sensor material for its stability, accuracy, and high temperature coefficient of resistivity (TCR). Additional packaging is performed to form fluidic connections using micromachined fluidic couplers [5].

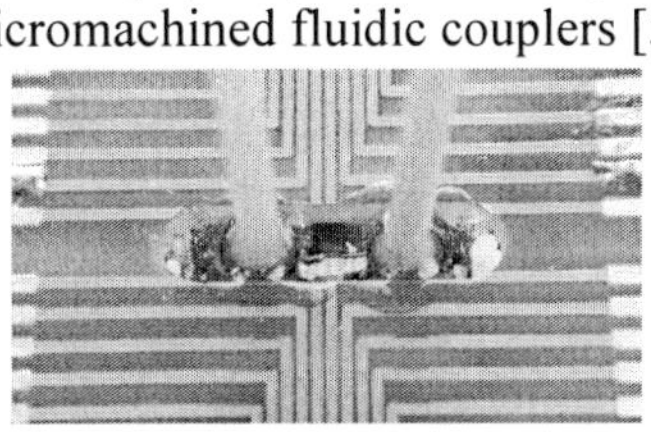

Fig. 1 Flow Channel & Fluidic Connections

Fig. 2 Sensor Close-Up

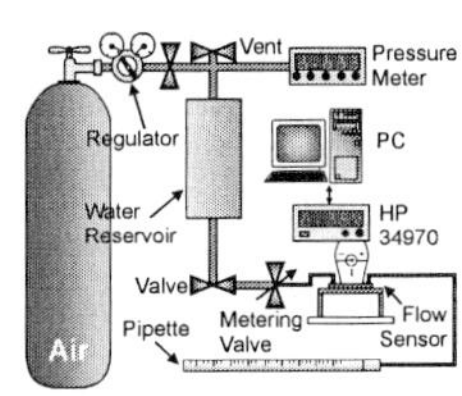

Fig. 3 Flow Testing Setup

3. Experimental
To determine flow rate, the sensor voltage is recorded under constant current (30 mA) while fluid is forced through the device using compressed air. The flow rate is adjusted by a metering valve and calibrated using a stopwatch and precision pipette (Fig. 3).

4. Results and discussion
Temperature calibration and transient behavior of the device under water flow and no flow conditions are shown in Fig.s 4-6. As expected, the device responds faster to

J.M. Ramsey and A. van den Berg (eds.), Micro Total Analysis Systems 2001, 167–168.

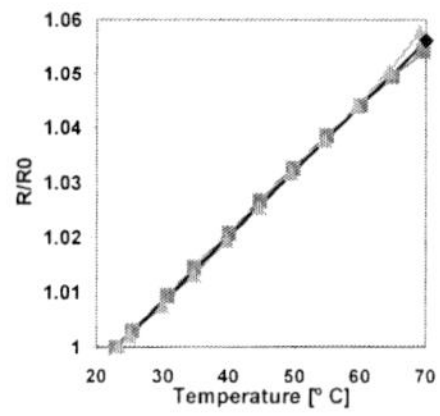

Fig. 4 Temperature Calibration

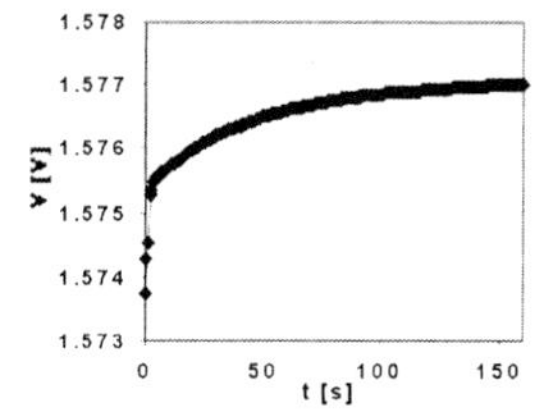

Fig. 5 Transient Response to No Flow

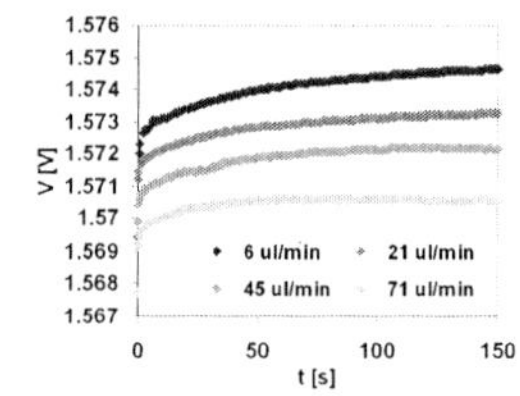

Fig. 6 Transient Response to Water Flow

higher flow rates. The transient response has two associated time constants, the first being less than 1 s and the second ranges from 10 to 60 s. These increase as flow rate decreases. The sensor response to flow rate was adjusted to remove the effects of ambient temperature fluctuations is shown in Fig. 7. Power consumption and overheat ratio were 36 mW and 1.9%, respectively. The sensor can resolve up to 10 μl/min flow. Commercial devices with such resolution are not currently available. These results are in agreement with behavior predicted by King's Law, $H(\nu) = A + B\,\nu^{1/2}$, where A = 0.8559, B = -0.878849, and n = 0.51 (Fig. 8).

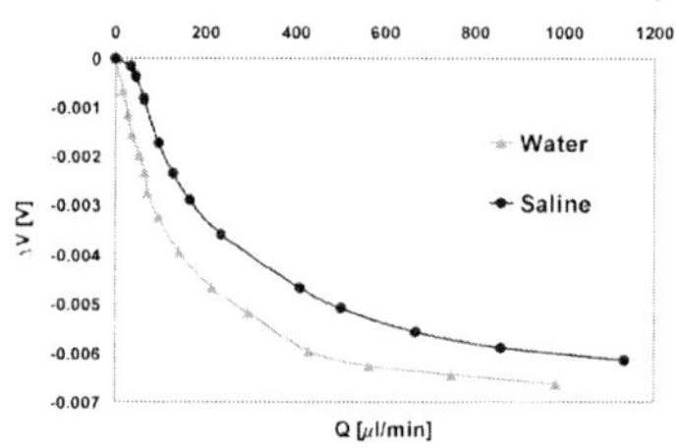

Fig. 7 Sensor Response to Flow

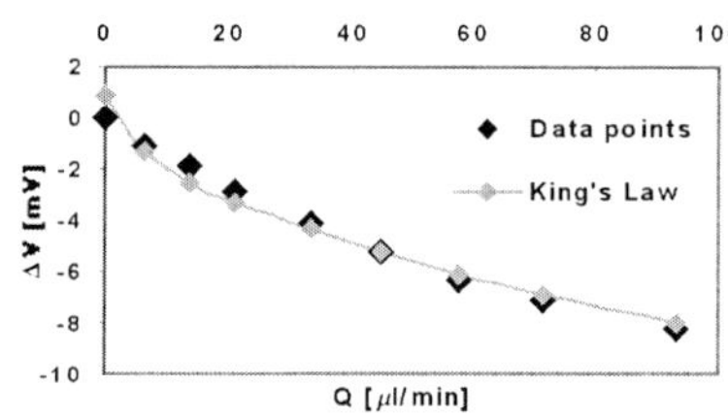

Fig. 8 King's Law vs. Experimental Results

5. Conclusion

A MEMS fluid flow sensor capable of detecting 10 μl/min flows has been demonstrated. Future work will include testing with different biological fluids, e.g. blood and urine, and various detergents. As many biological fluids contain particulate matter, the effect these have on measurements will also be examined.

Acknowledgements

The authors would like to thank Trevor Roper for help with processing and the NSF Center for Neuromorphic Systems Engineering and IRIS, Inc. for funding.

References

1. Yang, C. and H. Soeberg, Sensors and Actuators A, **33**: pp. 143-153, (1992).
2. Lammerink, T.S.J., *et al.*, Sensors and Actuators A, **37-38**: pp. 45-50, (1993).
3. Nguyen, N.T. and R. Kiehnscherf, Sensors and Actuators A, **49**: pp. 17-20, (1995).
4. Wu, S., *et al.* MEMS 2000, Miyazaki, Japan, (2000).
5. Meng, E., *et al.* Micro Total Analysis Systems 2000, Enschede, The Netherlands, pp. 41-44, (2000).

CONSTRUCTION AND APPLICATION OF RESPONSIVE MICROSTRUCTURES: FIBERS, TUBES AND SENSORS

G. Mensing and D. J. Beebe

Dept. of Biomedical Engineering, Univ. of Wisconsin-Madison

Abstract

We report the construction and application of microstructures consisting of responsive hydrogel material. Micron-sized fibers and tubes have been fabricated using liquid phase photopolymerization. Applications of the responsive fibers include a dipstick pH sensor with a built in readout. The tubes show a fast response to changes in pH. Flow through the responsive tubes has been demonstrated.

Keywords: hydrogels, sensors, fibers

1. Introduction

In previous work [1,2], we have incorporated responsive hydrogels inside microchannels to form valves and sensors. These materials consist of a pH sensitive hydrogel made of acrylic acid and HEMA, both monomers; EeGDMA, a crosslinker; and DMPA, a photoinitiator. Here we explore the use of responsive fibers and tubes both within a microenvironment and as independent fluid components.

2. Experimental

Micron-sized responsive fibers (~ 150 to 200 μm dia.) have been fabricated (see Fig. 1). A core material (nylon thread or other thin fiber) can also be introduced before polymerization and later removed to form a tube. These structures can be used alone or reintroduced into a microenvironment and used to detect the presence of a stimulus.

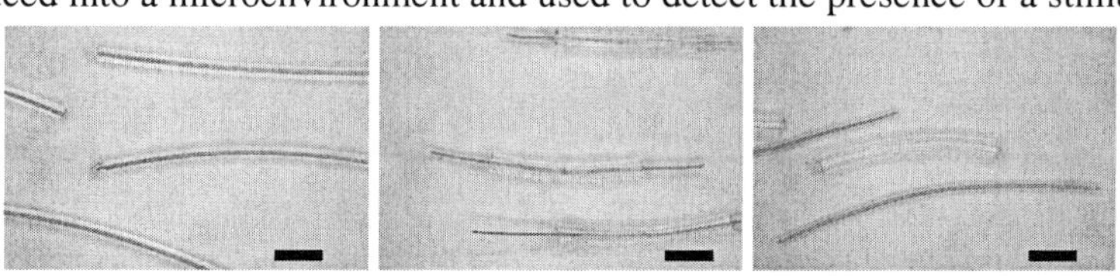

(a) (b) (c)

Figure 2. (a) Responsive hydrogel fibers are polymerized inside stretched poly-ethyl vinyl acetate tubing. (b) The tubing is heated and returns to the original diameter. (c) The responsive fibers are pulled from the tubing. Scale bars = 1 mm.

3. Results and Discussion

An example of using the fibers in a microenvironment is a dipstick sensor (figure 2a). The capillary contains pieces of responsive hydrogel fiber positioned between different thicknesses of glass filter paper plugs. The capillary tube is placed in a high pH solution, and the fluid wicks through the filter material by capillary action. In the high pH solution, the fibers expand and push the smaller filter paper plug upward (figure 2b). The design provides for a simple "dipstick" test with easily interpreted results (figure 2c).

J.M. Ramsey and A. van den Berg (eds.), Micro Total Analysis Systems 2001, 169–170.

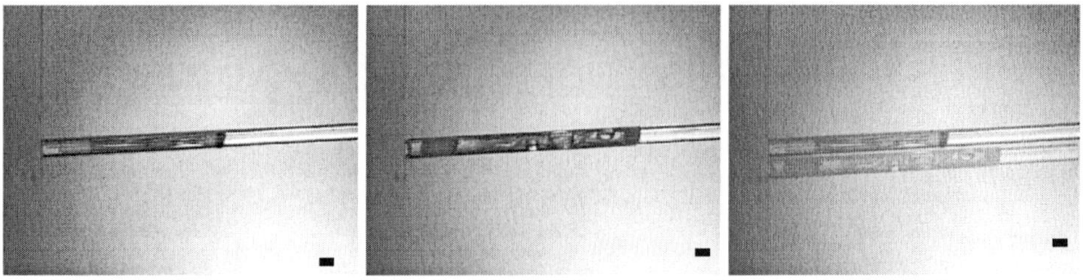
(a) (b) (c)

Figure 2. Responsive fibers are inserted between glass filter paper plugs (a) inside a capillary tube, and exposed to a high pH solution (b) that shows expansion of the fibers. A comparison of the two images (c) shows the difference between the two states. Scale bars are approximately 1mm.

Using the techniques described above, responsive tubes that sense and respond to local stimuli have been fabricated. When the tube is unconstrained and exposed to a high pH solution, expansion occurs in all directions, as shown in Figure 3a-b. The tubes can also be connected to pulled pipettes and encased in a silicon elastomer that constrains the tube and adds stability. Fluid flow through the responsive tube has been demonstrated and is shown in Figure 3c.

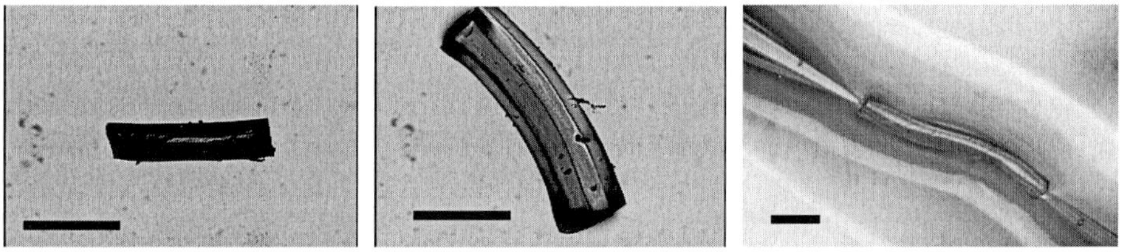
(a) (b) (c)

Figure 3. A dyed responsive tube shown in the contracted state (a) and the expanded state (b). A clear responsive tube (c) attached to pulled pipettes and flushed with a red dye solution. Scale bars = 500 μm.

4. Conclusion

We report the fabrication of microstructures made outside a microenvironment that both sense and respond to local stimuli. Advantages of a dipstick device made with responsive fibers are small volume sample, ease of use, and built in readout.

Acknowledgement

We gratefully acknowledge the support from DARPA-MTO under grant number F30602—00-1-0570 (Program manager: Dr. Michael Krihak).

References

1. D. J. Beebe, J. S. Moore, J. M. Bauer, Q. Yu, R. Liu, C. Devadoss, B. Jo, "Functional hydrogel structures for autonomous flow control inside microfluidic channels", *Nature*, Vol. 404, pp. 588-590, 2000.
2. Q. Yu, J. M. Bauer, J. S. Moore, D. J. Beebe, "A biomimetic hydrogel check valve for fluid control", *Appl. Phys. Lett.*, **78**, (17) 2589-2591.

FLEXIBLE POLYIMIDE-BASED MICROCHANNELS WITH EMBEDDED MICROELECTRODES

Stefan Metz, Shady Gawad, and Philippe Renaud
Swiss Federal Institute of Technology Lausanne, EPFL DMT-IMS,
1015 Lausanne, Switzerland

Abstract:
This paper describes the fabrication of flexible, polyimide-based microchannels with embedded microelectrodes. The structures combine a flexible microfluidic system with microelectrodes inside or outside the channels. Microchannels were realized by an adapted lamination technique and combined with a polyimide-platinum-polyimide sandwich structure.

Keywords: Microchannel, Microelectrode, Polyimide

1. Introduction

Polymer-based microfluidic devices have gained increased interest for "lab-on-a-chip" and μTAS microdevices. For a wide range of applications it is desirable to additionally embed microelectrodes inside fluidic networks for flow control, electrochemical analysis, or particle handling.

2. Fabrication

The modified lamination technique is based on the adaptation of cure cycles [1] and the use of swelling agents [2] to improve polyimide/polyimide adhesion. For the release of final structures, a layer of chrome/aluminum (500 nm/500 nm) is evaporated on silicon substrates. On the aluminum surface a 5 to 20 μm thick layer of photosensitive polyimide precursor (PI-2732, DuPont) is applied, photostructured and cured. Subsequently, a layer of titanium/platinum (50 nm/200 nm) is sputtered on the substrate and structured by dry etching (Fig. 1a). A second layer of photosensitive polyimide defines the channel geometries (Fig. 1b).

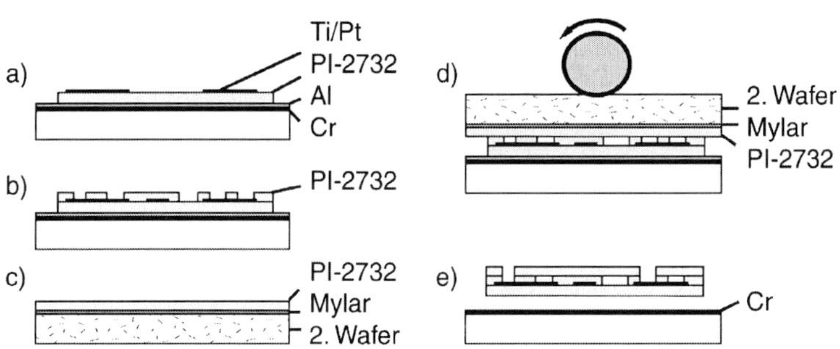

Figure 1 : Process steps for the fabrication of flexible polyimide-based micro-channels with embedded electrodes.

J.M. Ramsey and A. van den Berg (eds.), Micro Total Analysis Systems 2001, 171–172.

On a second carrier substrate a thin Mylar™ sheet is made to adhere temporarily by injecting water between the foil and the wafer. The foil is spin coated with a 5 to 20 μm thick layer of photosensitive polyimide (Fig. 1c). Following a softbake step, the second substrate is bonded to the open channel structures of the first wafer (Fig. 1d). After removing the second carrier wafer and the Mylar foil, the top layer of polyimide is photostructured and the structures are cured. The final devices are dissociated from the fabrication support by anodic metal dissolution of the sacrificial aluminum layer (Fig. 1e).

3. Results

Flexible microchannels with embedded microelectrodes have been fabricated in a variety of shapes and dimensions. The channel heights are 15 to 30 μm, widths range from 50 to 500 μm and lengths are of up to 5 cm. Electrode sites inside the channel ranged from $50 \times 50\ \mu m^2$ to $200 \times 200\ \mu m^2$ (Fig. 2). The channels were operated at pressures up to 19 bar and flow rates up to 10 ml/min without leakage for several hours.

Figure 2 : Cross-sectional view of a microchannel (height 20 μm, width 100 μm) with embedded micro-electrode at the bottom of the channel (indicated by right arrow) and three insulated metallization lines (indicated by three left arrows). No interface is visible where the top layer has been laminated. -
(Figure reproduced by permission of The Royal Society of Chemistry)

4. Conclusion

The fabrication of polyimide-based microfluidic channels with integrated electrodes has been described. By combining a modified lamination technique with a polyimide/metallization/polyimide process, versatile devices can be fabricated which meet the demands for "lab-on-a-chip" and μTAS applications.

References:

[1] H. R. Brown, A. C. M. Yang, T. P. Russell, W. Volksen, and E. J. Kramer, "Diffusion and Self-Adhesion of the Polyimide Pmda-Oda," *Polymer*, vol. 29, pp. 1807-1811, 1988.

[2] K. L. Saenger, H. M. Tong, and R. D. Haynes, "Improved Polyimide Polyimide Adhesion Via Swelling Agent Enhanced Interdiffusion," *Journal of Polymer Science Part C-Polymer Letters*, vol. 27, pp. 235-237, 1989.

Autonomous Colorimetric Readout for Microfluidic Devices

Jaisree Moorthy, David J. Beebe
Department of Biomedical Engineering, University of Wisconsin at Madison, WI 53706.

Abstract
Autonomous readouts that work on colorimetric principles were fabricated inside a microchannel by immobilizing dyes (congo red and phenolphthalein) in a hydrogel matrix. The response time of the readout varied between 20 s and 45 s depending on the dye. To test for reusability, the phenolphthalein containing gels were recycled between acidic and basic solutions at different flow rates (0.1, 0.25 and 0.5 mL/min).

Keywords: Combinatorial, in-situ polymerization, pH sensitive

1. Introduction

Various types of sensitive detection techniques (e.g. fluorescence spectroscopy) have been developed in microfluidics. These are typically associated with complex instrumentation and power requirements making it difficult to perform tests outside a laboratory environment. Here we demonstrate a 'readout' prepared by immobilizing ion sensitive dyes in a hydrogel matrix. This approach is similar to Mirzabekov's DNA-oligonucleotide microchips [1] but, without the requirement of covalent tethers. Also, the readouts are patterned inside the microchannel by liquid phase photopolymerization [2], making fabrication easy and allowing flexibility in integration with other microfluidic system components (valves, filters, etc.). The readout operation is based on colorimetric principles, requires no power or leads and the high volume support of the gel enables perception by the naked eye (unlike surface immobilized dyes which typically require optical/electronic detection due to low intensities). Two pH sensitive dyes – congo red (CR) and phenolphthalein (PP) were studied. CR changes color from blue to red at around 3 – 5 pH while PP becomes pink from colorless at around 8 – 9 pH.

2. Experimental

The pre-gel mixture for each readout type was prepared by mixing the dye powder (10% by wt) in HEMA-based prepolymer mixture. After thorough mixing, the pre-gel solution was injected into a straight channel (30 mm long × 1 mm wide × 0.13 mm deep) and the readout structures were polymerized and patterned by illuminating with UV light (Spectroline source) through a mask. The channels were then flushed with methanol and water and dried by baking at 45 °C for 8 hours.

3. Results and Discussions

The time required for the readout to completely change color at high and low pH solutions varies between 20 s and 45 s. PP gels had a faster time response while the CR gels required a longer time to achieve color change. The response time at different flow

J.M. Ramsey and A. van den Berg (eds.), Micro Total Analysis Systems 2001, 173–174.

rates was tested and was found to be similar, confirming that diffusion is the main transport process. The PP readouts were cycled between acid (pH 2) and base (pH 12.5) to test for reusability, at flow rates of 0.1, 0.25, and 0.5 mL/min. The normalized intensity of the dye is shown in Figure 1. The trend indicates that flow rate is not the main variable that influences the reusability of the readout. Several parameters may affect the readout performance including polymerization exposure (time, intensity), dye/gel compatibility, dye absorption and proximity of the structures. The *in-situ* fabrication facilitates the creation of combinatorial readouts. A combinatorial (pH sensor) readout (Figure 2) was prepared by polymerizing PP and CR gels serially.

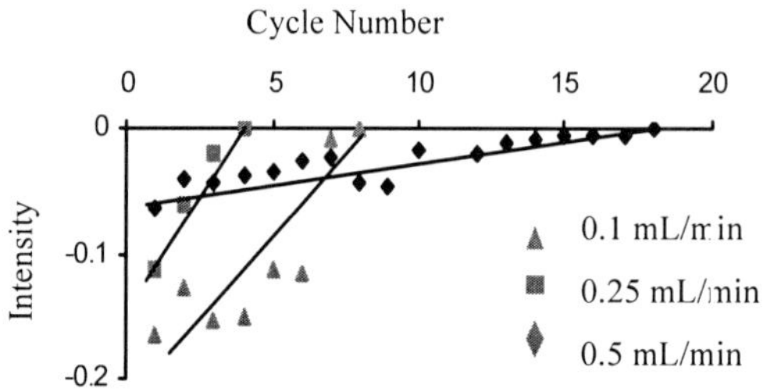

Figure 1. Change in intensity of the PP-gel for different flow rates. The maximum value (0) in the graph indicates background color *i.e.* no dye present.

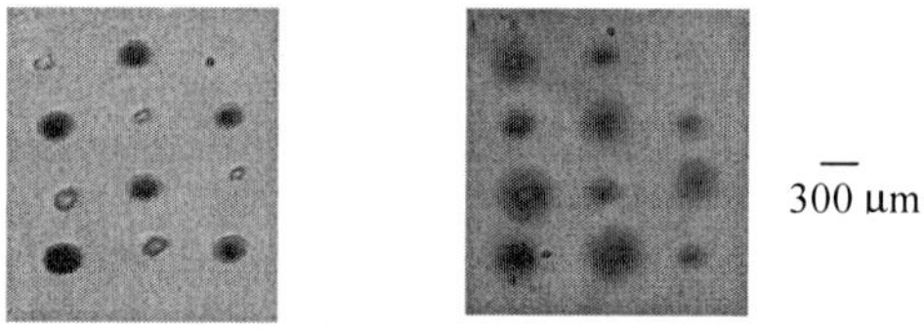

Figure 2. A combinatorial readout (pH sensor) having alternate patterns of congo red and phenolphthalein dyes immobilized in hydrogel matrix in acid and base.

4. Conclusions

Readouts were fabricated inside microchannels and tested. Observations suggest that depending on the end use (disposable, reusable) the readout structures can be appropriately designed. The wide availability of dyes sensitive to both chemical and biological agents will allow extensions of this idea to many applications.

Acknowledgement

This work was supported under a grant from DARPA-MTO (#F30602-00-1-0570) (Program manager: Dr. Michael Krihak). Thanks to Prof. Jeff Moore's group at UIUC for helpful discussions.

Reference

1. D. Proudnikov, E. Timofeev, A. Mirzabekov, *Immobilization of DNA in polyacrylamide gel for the manufacture of DNA and DNA-oligonucleotide Microchips*, Anal. Biochem., Vol. 259, 1998, p.34.
2. D. J. Beebe, J. S. Moore, J. M. Bauer, Q. Yu, R. H. Liu, C. Devadoss, B.- H. Jo, *Functional Hydrogel Structures for Autonomous Flow Control Inside Microfluidic Channels*, Nature, Vol. 404, 2000, p. 588.

ON-CHIP BYPASS STRUCTURE FOR SAMPLE SEGMENT DIVISION AND DILUTION

Yuji Murakami, Takashi Kanekiyo, Takeshi Kinpara, Eiichi Tamiya
The School of Materials Science, Japan Advanced Institute of Science and Technology, 1-1 Asahidai, Tatsunokuchi, Ishikawa, 923-1292 JAPAN
(yuji@jaist.ac.jp)

Abstract

This paper describes theoretical aspects and experimental results of an on-chip structure to divide and dilute a sample segment automatically without the addition of peripherals. The structure is a simple addition of a bypass to a main channel of a flow system. It splits a sample segment into two at a desired ratio without sample broadening.

Keywords: EOF, dilution series, flow visualization, capillary electrophoresis

1. Introduction

An on-chip capillary and its complicated pattern are a key structure of μTAS devices. Complicated structures have been successfully applied to μTAS devices such as double-T injector, mixers, synchronous CE, and dilutors. However, the structures usually require additional pumping mechanism such as electric power supply for electroosmotic flow (EOF) and syringe pump for backpressure pumping. Immunoassay requires the preparation of dilution series of samples. FIA realizes dilution by using broadening of a sample segment in a long pipe, but it cannot split the segment.

2. Theory

We propose the use of a simple bypass structure to make dilution series of sample. When a sample segment comes to the structure, it diverges to the main capillary and the branch (Figure 1). Then the segment from the main channel reaches a mixing point of the structure prior to the segment from the branch. Thus, the segments are mixed with running buffer. Dilution is completed by the diffusion in the narrow capillary downstream of the structure. According to theoretical review on EOF in bypass structure omitting axial diffusion of sample segment, the lengths of the divided segments are the same as the original segment. Parameters, the lengths of both channel, l_i, and cross-sectional area of the channels, S_i, can determine characteristics such as residual time, t_i, and flow-rate, f_i, in each capillary. Ratios of these characteristics are derived as follows:

$$t_{main} : t_{bypass} = l_{main}^2 : l_{bypass}^2$$

$$f_{main} : f_{bypass} = l_{bypass} S_{main} : l_{main} S_{bypass}$$

J.M. Ramsey and A. van den Berg (eds.), Micro Total Analysis Systems 2001, 175–176.

3. Experimental

A typical on-chip CE pattern with a double-T injector was fabricated on glass or PDMS chip. The capillary also has a twice-longer bypass structure with the same cross-sectional area. Fluorescein was injected as a test sample to visualize flow characteristics. Whole capillary was observed by CCD camera system, and CE-LIF system observed fluorescein segment at a downstream of the bypass structure

4. Results and discussion

The CCD camera on a fluorescence microscope recorded the movement of fluorescent sample in the dilution structure. Ratio of the time to pass through the shorter main capillary and the longer bypass capillary was 1 : 3.9, which is in good agreement with the theoretical value, 1:4.

CE-LIF method gave an electropherogram of the fluorescent sample with two peaks, which simply shows successful separation and dilution from a single sample segment. Half height widths of both peak was the same as estimated. However, the ratio of peak heights and the ratio of peak area were 3.0:1 and 3.6:1, respectively. These ratios should be inversely proportional to the ratio of length, and then be 2:1. This theoretical estimate deviates considerably from the actual observation. The reason might be unwanted axial broadening of the sample segment such as leading and tailing at the corners of the structure. This was confirmed with capillaries of different depth.

Acknowledgements

This work is partially supported by Encouragement of Young Scientists (A) of Grants-in-Aid for Scientific Research, and Research for the Future Program of JSPS.

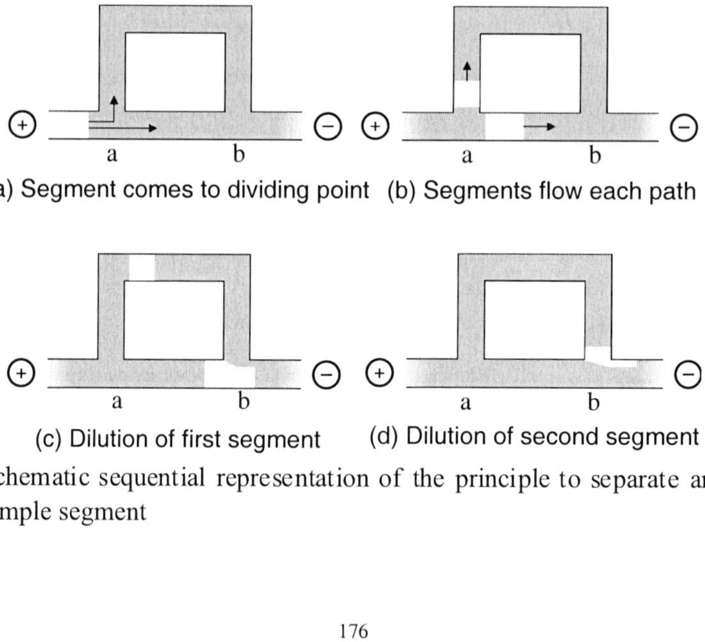

Figures 1 Schematic sequential representation of the principle to separate and dilute a sample segment

DEVELOPMENT OF A MICROMACHINED FLUIDIC STRUCTURE FOR A BIOLOGICAL AND CHEMICAL SENSOR ARRAY

Young-Soo Sohn[1], Adrian P. Goodey[2], Eric V. Anslyn[2], John T. McDevitt[2], Jason B. Shear[2], and Dean P. Neikirk[1]
[1]Department of Electrical and Computer Engineering
[2]Department of Chemistry and Biochemistry
The University of Texas at Austin, Austin, TX 78712

Abstract

The development of a micromachined fluidic structure for a chip-based sensor array composed of polymeric microspheres for the rapid characterization of analytes in solution [1] is presented. One of the key parts of the proposed structure is a passive fluid introduction system driven only by capillary force. This simple means of fluid introduction should allow the fabrication of a compact, portable, and potentially low-cost device.

Keywords: Sensor array, capillary

1. Introduction

One significant challenge for μTAS is the integration of many microfluidic components in a manner that does not interfere with the fluid handling characteristics of the individual components. In addition, systems with no moving components are clearly useful for disposable biochips.

2. Device Design

In our sensor array platform transduction occurs via optical changes to receptor and indicator molecules that are attached to the polymeric microspheres selectively arranged in micromachined cavities localized on silicon wafers [2]. Figure 1 shows a cross sectional view of the system. The microfluidic device consists of a cover glass layer, micromachined silicon, and glass substrate. The cover glass confines the microspheres in their individual cavities while still allowing optical and fluidic access. Furthermore, the hydrophilic surfaces of the cover glass and deposited oxide on the silicon assist the sample fluid flow through the top channel. Sample introduction is driven only by capillary forces. The fabrication process of the microfluidic

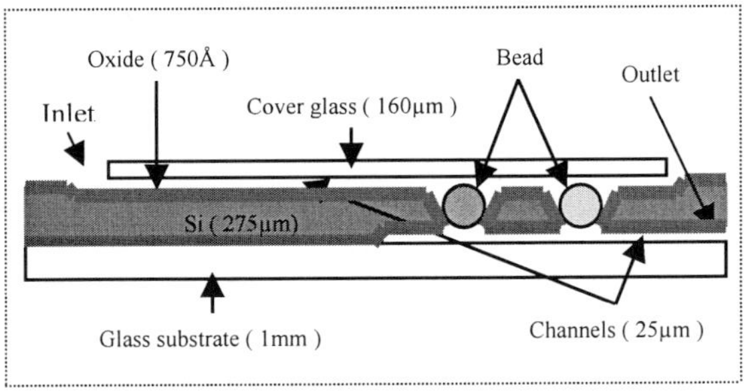

Figure 1

J.M. Ramsey and A. van den Berg (eds.), Micro Total Analysis Systems 2001, 177–178.

device has been selected to protect receptors and/or indicator molecules that may be sensitive to the normal processes used in microfluidic chip fabrication.

3. Experimental Characterization

Optical absorbance changes of alizarin complexone derivatized microspheres as a pH 1 HCl/water solution displaces the air in the array has been studied to characterize time domain fluid flow into the microsphere storage cavities. The flow randomly passes through the cavities with a maximum variation in the initiation time of about 7 second, which may be caused by slightly different surface effects across the micromachined cavities. Typical red, green, and blue (RGB) intensities of the alizarin microspheres in pH 1 HCl solution are shown in Figure 2. Comparison of RGB data and visual images shows that the reaction starts when the fluid first touches the microsphere.

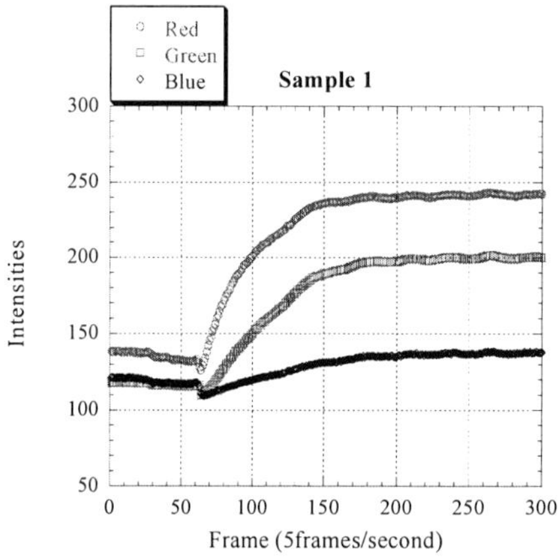

Figure 2

The contact angle between the DI-water and virgin silicon dioxide was about 0°. However, the measured flow velocity in our device was consistent with a contact angle of about 20° for water flowing through the top channel. This suggests that the silicon dioxide layer was modified during device construction, becoming somewhat less hydrophilic.

4. Conclusion

A microfluidic structure for a biochemical sensor array has been proposed and tested in this work. With the microsphere array detection methodology, this simple integrated structure may have a high potential as a disposable and portable biochip.

Acknowledgements

Support for this project was provided by the National Institute of Health and the Army Research Office MURI program (contract number DAAD19-99-1-0207).

References

1 J. J. Lavigne, S. Savoy, M. B. Clevenger, J. E. Ritchie, B. McDoniel, S.-J. Yoo, E. V. Anslyn, J. T. McDevitt, J. B. Shear, and D. Neikirk, "Solution-Based Analysis of Multiple Analytes by a Sensor Array: Toward the Development of an "Electronic Tongue"," *Journal of the American Chemical Society*, vol. 120, pp. 6429-6430, 1998.

2 Adrian Goodey, et al., "Development of Multianalyte Sensor Arrays Composed of Chemically Derivatized Polymeric Microspheres Localized in Micromachined Cavities," *Journal of American Chemical Society*, vol. 123, no. 11, pp. 2559-2570, 2001.

IN-CHANNEL FLOW SENSOR USING DRAG FORCES

Shankar Radhakrishnan, Harun Solak, Amit Lal
*Sonic*MEMS Laboratory
Department of Electrical and Computer Engineering,
University of Wisconsin-Madison

Abstract
We present an analytical model, simulated results and experimental verification for a scalable microchannel-embedded cantilever flow sensor that utilizes drag forces to produce moments in the embedded beams.

Keywords: embedded channel, drag force, flow sensor, microfluidic

1. Introduction

The flow sensor we present addresses the need to be integrated in microfluidic channels for closed-loop control of flow. Also, we can use cover structures for the channels made of a large variety of materials, such as PDMS. This is not the case for most membrane-type pressure sensors formed on walls of the microfluidic channels, where the microfluidic cover requires machining to integrate the sensor. Previously reported in-channel flow sensors utilize lift forces for sensing gas flow for moderately high Reynolds numbers [1].

2. Analytical model

Figure 1

From Stokes equation for flow

$$\nabla P = -\mu \Delta^2 U$$

the pressure gradient in the channel on the two sides of the beam can be written as

$$\frac{\partial P_1}{\partial x} = -\frac{Q_1(x)\mu}{(a+mx)^2 b^2 F(a/2b)}$$

$$\frac{\partial P_2}{\partial x} = -\frac{Q_2(x)\mu}{(a+mx)^2 b^2 F(a/2b)}$$

Here, Q1 and Q2 are the volumetric flows in the two parts of the channel, F(a/2b) is the leakage factor

The net force acting on the beam is given by

$$F_{net} = \frac{4bP_0}{m}\left[a - \frac{1}{mL}\left(a^2 - \frac{m^2L^2}{4}\right)\log\left(\frac{a+\frac{mL}{2}}{a-\frac{mL}{2}}\right)\right]$$

For small sensor dimensions and low Reynolds number flow viscous forces dominate inertial forces. This flow is governed by time-independent Stokes equation. As shown below, a beam placed

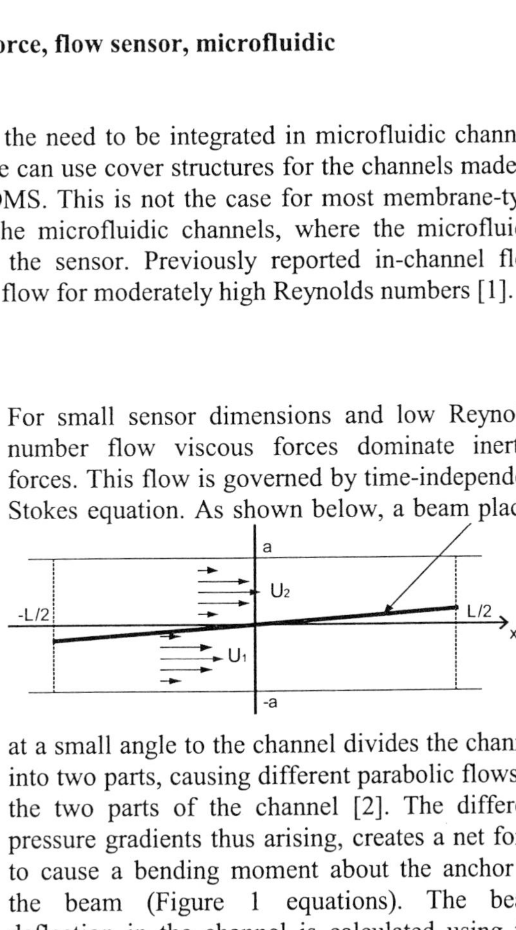

at a small angle to the channel divides the channel into two parts, causing different parabolic flows in the two parts of the channel [2]. The different pressure gradients thus arising, creates a net force to cause a bending moment about the anchor of the beam (Figure 1 equations). The beam deflection in the channel is calculated using the stiffness of the beam and the bending moment in the beam about its anchor. In the ideal case the beam height is equal to the channel height, causing no leakage of fluid between the

J.M. Ramsey and A. van den Berg (eds.), Micro Total Analysis Systems 2001, 179–180.

two parts of the channel. In the non-ideal case, the leakage flow is accounted by flow between parallel plates [3].

3. Simulations, Fabrication and Testing:

We simulated the flow sensor described above using a CFD code by CFD Research Corporation. The simulations were done for different beam lengths (l) and widths (w) for silicon and SU-8 beam materials (Figure 2). The beam deflections were measured in plastic devices fabricated by stereolithography (SLA at Becton Dickinson shown in Figure 3). The SLA beams were 500 μ thick, 160 μ wide and had lengths ranging from 2.3 mm to 13.9 mm. The design of the channel and beams were such that the deflections can be optically measured, differentially or absolutely, using internal or external photodiodes.

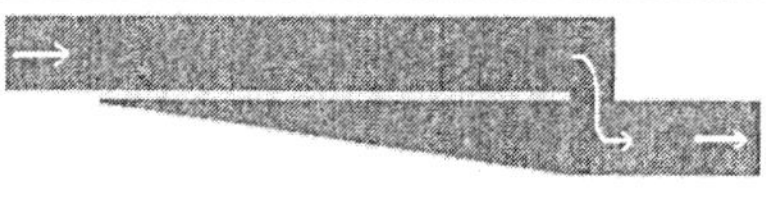

Figure 2 Channel structure

Figure 3 SLA fabricated device

4. Results and Conclusion

Figure 4 shows measured values of deflections of a beam of dimensions 7mm X 160 μ and compares with the simulated values of deflections for the same beam and channel dimensions. According to the analytical model, the sensitivity of the device is proportional to $(l/w)^3$ which agrees with the simulated and measured data shown in Figure 5. These results confirm the working of the flow-sensor.

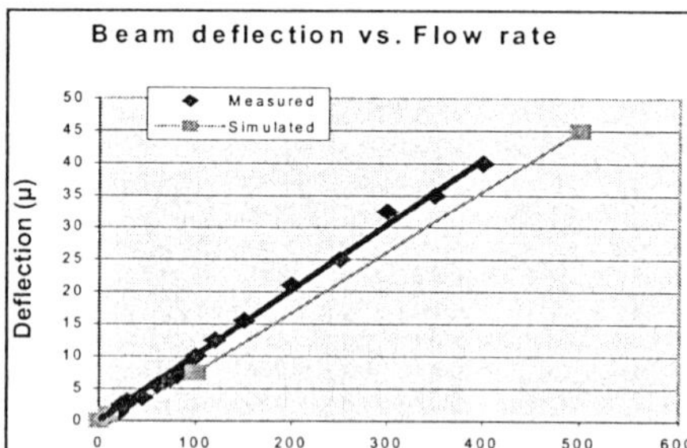

Figure 4

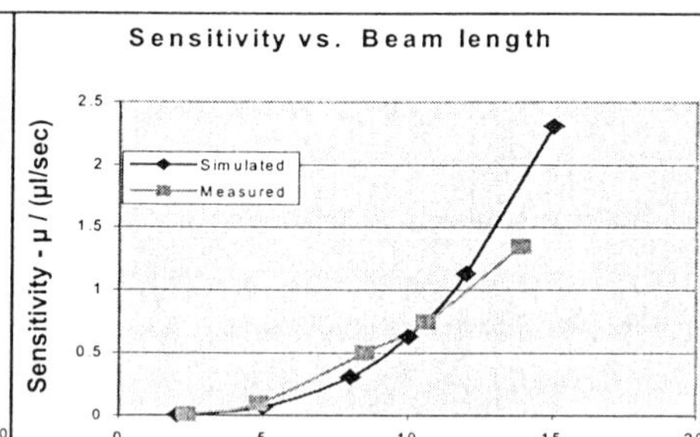

Figure 5

References

1. Svedin, Niklas et al. "A New Silicon gas flow sensor based on lift force" JMEMS, Vol. 7 No. 3, 1998, pp. 303-308.
2. Longwell, Paul A. "Mechanics of fluid flow" New York, McGraw-Hill, 1996 Description: xv, 433 p. illus. 24 cm.
3. Jeong, Jae-Tack, "Two dimensional Stokes flow through a slit in a vertical plate on a plane wall", Journal of the Physical society of Japan, Vol. 67, No. 12, Dec 1998, pp. 4074-4079.

Acknowledgements

This work was funded by DARPA/MTO contract # F30602-00-2-0572.

ON-LINE CHEMILUMINESCENCE DETECTION OF BIOPROCESSES USING POLYMER-BASED MICROCHIPS WITH IMMOBILIZED ENZYMES

W. Rong and J. P. Kutter
MIC-Mikroelektronik Centret, DTU, Building 345 east,
DK-2800 Lyngby, Denmark
Email: rw@mic.dtu.dk

Abstract
A compact analysis system with polymer-based chips was developed. The polymer-based microchips were microfabricated in PMMA by hot embossing. Based on the principle of covalent coupling, enzymes were immobilized on the inner wall of polymer-based microchips. The polymer-based microchips with immobilized enzymes were used successfully for on-line monitoring of bioprocesses. The analysis method was based on enzymatic reaction and chemiluminescence detection.

Keywords: polymer-based microchip, bioprocesses, immobilization of enzymes, chemiluminescence detection

1. Introduction

In recent years there has been a growing interest and activity in the area of miniaturization of analytical systems (μ-TAS), due to the many potential advantages those systems can offer. A majority of the published papers present microfabricated devices based on silicon, planar glass or quartz[1-4]. But, because of some advantageous properties of polymers and, in particular, the low fabrication costs, silicon, glass and quartz are more and more replaced by polymers. Despite these increased research activities, to our knowledge, this is the first description of a polymer-based microchip with immobilized enzymes for on-line chemiluminescence detection of bioprocesses.

2. Experimental

In this study, a polymer-based microchip was microfabricated in PMMA by hot embossing. The design of the chip included an enzyme reactor, a mixer and a chemiluminescence detector (see Fig. 1).

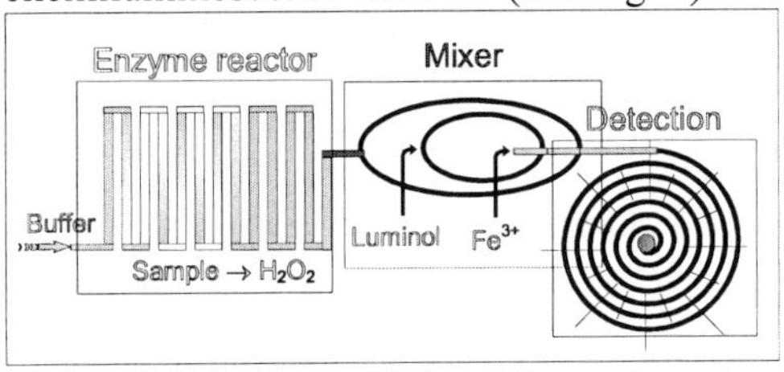

Figure 1. Schematic of the polymer-based microchip with integrated enzyme reactor, mixer and chemiluminescence detector.

J.M. Ramsey and A. van den Berg (eds.), Micro Total Analysis Systems 2001, 181–182.

The enzyme reactor was prepared using an in-house optimized immobilization method based on covalent coupling and glutaraldehyde chemistry. The enzymes were immobilized on the inner wall of a meander structure. After reaction of the substrates with the immobilized enzymes the formed H_2O_2 was mixed with luminol and Fe^{3+} (catalyst) from both sides in a simple diffusion-based mixer structure. Finally, emitted light with a maximum intensity at 430 nm was collected by photodiodes while the reaction between H_2O_2 and luminol took place in the spiral of the chemiluminescence detector.

3. Results and discussion

Two different immobilization procedures were tested for GOD. Yielding different methods with respect to achievable signal levels and the available linear range. The immobilized enzyme displayed an increased stability over enzymes in free solution, lasting for 2-3 weeks. The immobilization methods can be used for immobilization of glucose oxidase (GOD), hexose oxidase, galactose oxidase and lactate oxidase (LOD) for measurements of glucose, saccharides, galactose and lactic acid during bioprocesses, respectively.

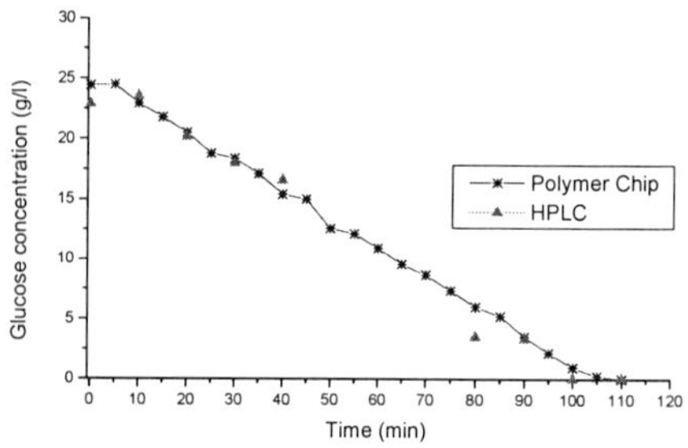

Figure 2. On-line measurements of the glucose concentration during yeast bioprocesses using a polymer-based microchip. Off-line measurements by HPLC are shown for comparison.

The described polymer microchip was used for on-line monitoring of glucose during yeast fermentation. The results are depicted in Fig. 2 and compared to a few off-line HPLC measurements. A good agreement between the measurements from the polymer microchip and the more cumbersome and time-consuming HPLC measurements was achieved.

Reference

1. T. Laurell, J. Drott, L. Rosengen, K. Lindstrom, Sensors and Actuators B 31(1996)161

2. R. Steinku, C. Dumschat, C. Sundermeier, H. Hinkers, R. Renneberg, K. Cammann, M. Knoll, Biosensors & Bioelectronics 11(1996) 187.

3. G. M. Greenway, L. J. Nelstrop, S. N. Port, Analytica Chimica Acta 405 (2000) 43.

4. X. J. Huang, Q. S. Pu, Z. L. Fang, Analyst 126(2001)281.

MANUFACTURING ISSUES OF POLYMER MICROFLUIDIC DEVICES WITH INTEGRATED ELECTRODES FOR ELECTROCHEMICAL DETECTION OF HEAVY METALS IN ENVIRONMENTAL SAMPLES

Wilfried Röpke[1], Alan O'Neill[2], Oliver Rötting[1], John Murrihy[3], Mila Pravda[3], Holger Becker[1]
[1] Mildendo GmbH, Göschwitzer Str 40, 07745 Jena, Germany,
wilfried.roepke@jenoptik.com
[2] NMRC, Cork, Ireland. [3] Chemistry Department, UCC, Cork, Ireland

Abstract
Polymer microfluidic devices with integrated electrodes have been produced. In doing so, we have considered two aspects: The functionality of the microfluidic system and of the electrodes and the possibility to scale up the manufacturing by batch processing.

Keywords: manufacturing, polymer, capillary electrophoresis, electrochemical detection

1. Introduction

Capillary electrophoresis on polymer microfluidic chips [1] with integrated electrodes for separation and detection [2] offers the potential for hand-held μ-TAS. Electrochemical detection conveys the biological/ chemical information directly as an electronic signal. For successful product development of this type of μ-TAS, the price per information unit and the reliability are critical. The commercial goal of this work results in the development of a cost effective, rapid, portable analysis system for the measurement of bioavailable toxic metal or organic contaminants.

2. Chip design and electrochemical detection of heavy metals.

The chip's microfluidic layout (fig.1) consists of a separation channel, a t-injection point formed by the intersection of a sample channel with the separation channel, a pair of sample stream focussing channels and a preconcentration sample reservoir.
Electrochemical detection of heavy metals is permitted by the use of two sensing electrodes located in the separation channel beyond the high voltage electrode. The sensing electrode assembly consists of two gold electrodes with an interstitial distance of 50μm (fig 3).

3. Manufacturing

The process flow for the manufacturing of polymer microfluidic devices features master fabrication, replication processes for the microstructuring of the polymer, hole drilling, metallisation, bonding, dicing and quality control. Advanced silicon etching has been used for master microfabrication. Replication of the master has been achieved using hot embossing.

J.M. Ramsey and A. van den Berg (eds.), Micro Total Analysis Systems 2001, 183–184.

For the creation of microelectrodes on polymer two technologies can be employed. A metal gas phase is deposited through a microstructured mask onto a polymer sheet. Alternatively, a second method uses a polymer sheet coated with a metal film, which is structured by laser ablation or etching to uncover the electrodes. The first technology has been optimised in terms of adhesion forces of the metal-polymer-interface and the geometry of the electrodes (fig 2, 3).
In the subsequent bonding process the polymer sheet is bonded onto the microstructured substrate. In this bonding process damage to the channel geometry, cracks in the electrodes or impairment of electrode surface characteristics has been avoided to allow defined microfluidic behaviour of the channel system and proper electrochemical behaviour. All process steps have been designed to allow for scaling-up the manufacturing to larger series.(fig. 4).

Acknowledgements
This work was supported by the EU-commission under contract EVK1-CT-1999-00008.

References
[1] Gwo-Bin Lee et al., Sensors and Actuators B 3708, 2001, 1-7
[2] Adam T. Woolley et al., Anal. Chem. 1998, 70, 684-688

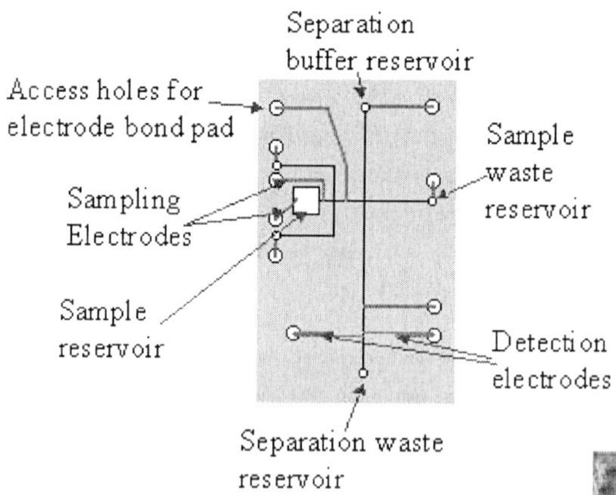

Figure 1. Chip design (O´Neill, NMRC)

Figure 2. Integrated Au-electrodes on microchannel in PMMA

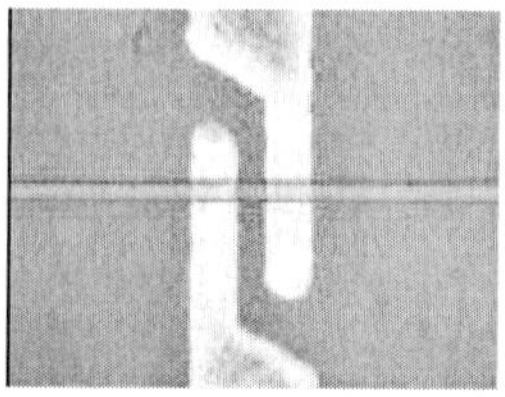
Figure 3. Au-electrodes for on-chip electrochemical detection, gap 50µm

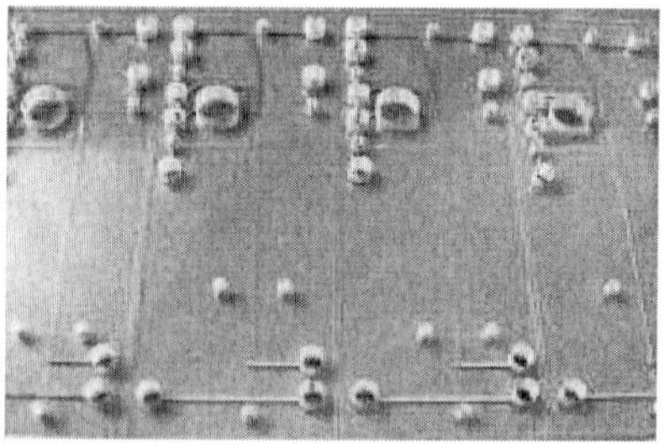
Figure 4. Batch manufacturing of polymer chips with integrated Au-electrodes for separation and electrochemical detection.

AN ACOUSTIC VORTEX PARTICLE CONCENTRATOR

Abhijit Sathaye, Amit Lal
*Sonic*MEMS Laboratory
Department of Electrical and Computer Engineering
University of Wisconsin - Madison

Abstract
In this paper, we report on a novel 'particle capturing' MEMS device for microfluid sample collection. When the resonance modes of the center-anchored surface micromachined polysilicon disk actuator inside the fluid channel are excited, time averaged acoustic force generates pressure gradients along the disk periphery. These pressure gradients are used to obtain fluid vortices by optimum design of poly-disk substrate gap and channel height. The vortices trap particles traveling in the channel due to pressure driven flow. In many ways this vortex capture device is biomimetic – it is like the single-celled organism called *Vorticella*, which collects food by inducing local vortices [1]. This particle concentrator paradigm gives us the tight control of the particle capture location so that we can integrate optical or acoustic sensors to measure the change in captured particle density over time.

Key Words : Concentrator, Acoustic streaming, Vortex, Microfluidics

Introduction
We have recently reported on a frequency addressable microfluidic actuator array in which frequency addressable mixing and pumping was demonstrated by placing the actuators inside the fluid channel [2]. The paper gave a broad framework to further investigate the effects of five length scales namely acoustic wavelength, shear viscous depth, structure (poly disk) substrate gap, evanescent decay length and channel height on the microfluid effects observed. Here, we report on optimization of shear viscous depth and channel height and its use as a particle concentrator for collecting and accumulating beads inside a fluid channel. This strategy eliminates the need for interconnects inside the microfluidic channel.

Theory
After release and critical point drying, the silicon die containing polysilicon micromachines and SU-8 walls was capped with a glass cap and adhesively bonded to a piezoelectric PZT (lead-zirconate-titanate oxide) plate. The ultrasonic vibrations applied to the PZT/Si laminate (<10Vpp) are coupled to the surface micromachines via inertial drive and elastic deformation at the anchor. With the disk actuator under resonance, there is shear viscous coupling between the edges of the disk and the fluid. The shear viscous depth is $\sqrt{v/\omega}$ where v is the length scale over which the shear viscous force is coupled to the fluid and ω□ is

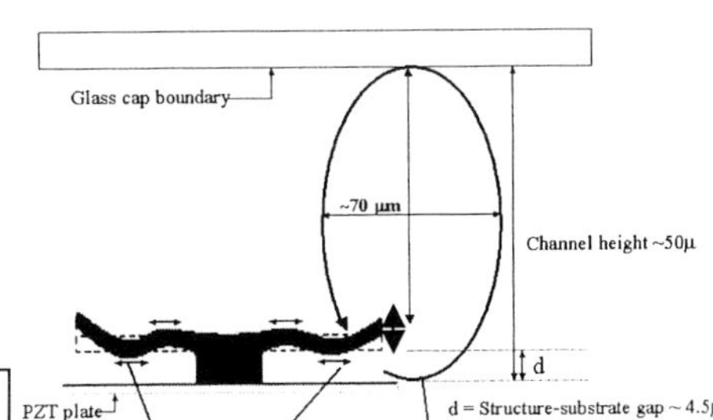

Figure 2. The flexural resonance of hub and corresponding vortex size limited by the channel height.

J.M. Ramsey and A. van den Berg (eds.), Micro Total Analysis Systems 2001, 185–186.

the frequency. This shear viscous depth varies from 2μm at 100kHz to 0.2μm at 10 Mhz for water. Therefore by having structure-substrate gap (4.5μm) and channel height (50 μm) of the same order as of the shear viscous depth, large pressure gradients near the plate edge can be obtained that result in shear-wave vortices. These fluid vortices are found to capture particles passing by the disk. The graph shows the dependence of number of particles trapped with changing frequency. Actuator resonance was observed at 629kHz. The bead density goes to zero when there is no resonance for the actuator (Figure 3). The shear-wave vortex formed near the edge is limited by the channel height. Hence by having a channel with a height gradient, one can have different size fluid vortices to capture particles of different sizes. Also, it was found that when the syringe-pump controlled flow rate increases beyond a critical threshold value, the pressure gradients near the disk periphery are no longer strong enough to trap particles.

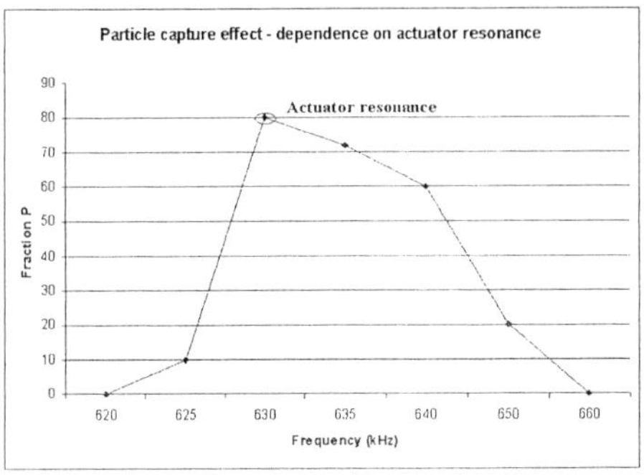

Figure 3. Fraction P = # of beads passing by the disk / # of beads captured inside the peripheral vortices.

Results

Figure 4 shows a schematic of a particle capture device with ports for fluid entry. The fluid vortices were observed with latex beads (1 - 10-micron diameter). Use of polygon-shaped hubs with vertices (Figure 5) could further localize the vortices because of higher acoustic filed gradients at the sharp corners resulting in vortex localization.

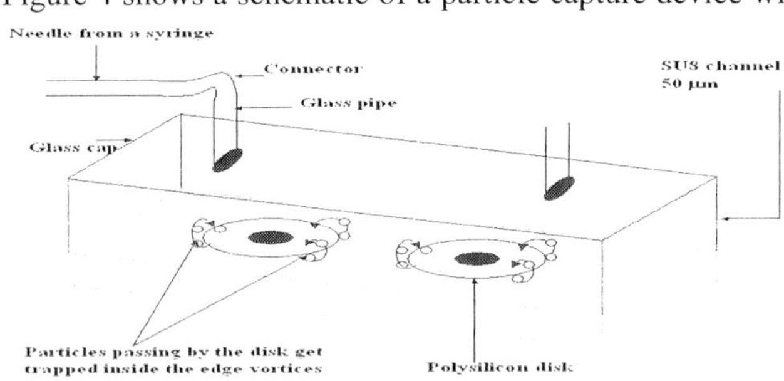

Figure 4. Schematic of the particle capture device.

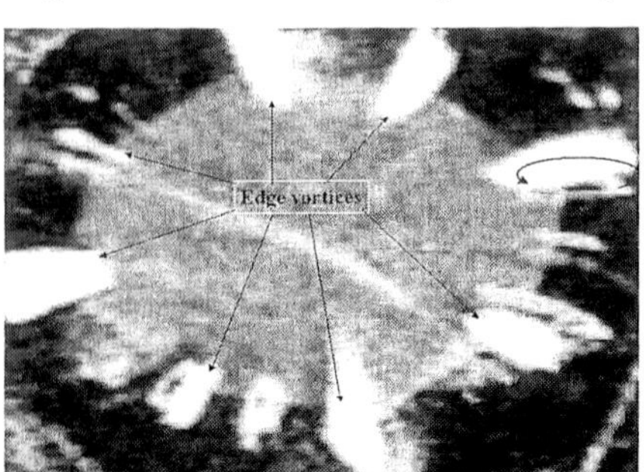

Figure 5. An octagon shaped hub (circum-radius = 400 mm) with preferential vortex formation at the corners due to higher acoustic filed gradient.

References

1. Vogel, S., Life in Moving Fluids – The physical biology of flow, Princeton University Press, 1994.
2. Sathaye, A., Kaajakari, V., Lal, A., "Frequency Addressable Ultrasonic Microfluidic Actuator Array," Transducers'01, Munich, Germany, June 10-14, 2001.

Acknowledgement

This work was funded by DARPA/MTO contract #F30602-00-2-0572.

A SILICON MICROFLUIDIC MULTIPLEXER USING FIELD EFFECT FLOW CONTROL

Nathan J. Sniadecki[1], Pen-Cheng Wang[2], Cheng S. Lee[2], Don L. DeVoe[1]
[1]Department of Mechanical Engineering and Institute for Systems Research,
[2]Department of Chemistry and Biochemisty, University of Maryland, College Park, MD 20742, USA, (email: nsniadec@glue.umd.edu)

Abstract

The design, fabrication, and characterization of a silicon-based microfluidic multiplexer employing field-effect flow control is described. Mulitplexer fabrication uses a combination of selective doping, oxidation, and deep reactive ion etching to form a microchannel network with integrated flow control gates. By modulating the bias of each gate, the zeta potential may be modified to control the electroosmotic flow (EOF) within each branch of the network. Experimentally, the control slope was found to be within reasonable agreement with the theoretical prediction.

Keywords: Field Effect Flow Control, Microfluidic Multiplexer, Electroosmosis

1. Introduction

Field-effect flow control offers the potential for independent fluid flow control within interconnected braches of a microfluidic network. The technique is based on modulation of the zeta potential using an applied gate voltage (V_g) in a voltage divider circuit. It has been shown by C. S. Lee et al. [1] that application of a radial electric field can change the magnitude and direction of the EOF within a fused silica capillary tube. Schasfoort et al. [2] recently extended the concept to silicon nitride microchannels. The presented work demonstrates field-effect flow control in a microfluidic circuit with silicon dioxide microchannel walls. For each flow control gate, the change in the zeta potential ($\Delta\varsigma$) is

$$\Delta\varsigma = \frac{C_d + C_w}{C_w} V_g \qquad (1)$$

where C_d and C_w are the capacitances of the electrostatic diffuse layer and the gate dielectric, respectively.

2. Fabrication

Figure 1 shows the layout of a microfluidic multiplexer. The flow control gates were defined by selectively doping an n-type silicon wafer with boron to form 13 μm deep p-wells. The *pn* junctions enable the voltage of each flow control gate to be biased independently. Deep reactive ion etching was used to form 10 μm deep, 20 μm wide channels, followed by dry oxidation to a thickness of 0.4 μm to form an insulating SiO_2 layer. The SiO_2 at the p-type regions was locally etched for electrical contacts to the gates. The channels were sealed with a layer of polydimethylsiloxane (PDMS) that contained the fluid reservoirs.

J.M. Ramsey and A. van den Berg (eds.), Micro Total Analysis Systems 2001, 187–188.

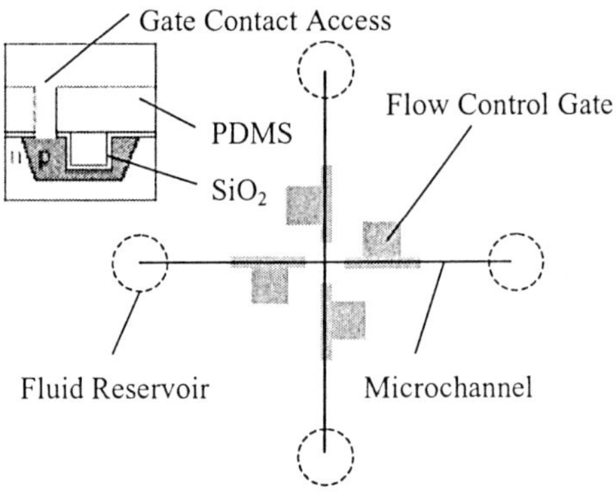

Figure 1. Top and cross-sectional views of microfluidic multiplexer.

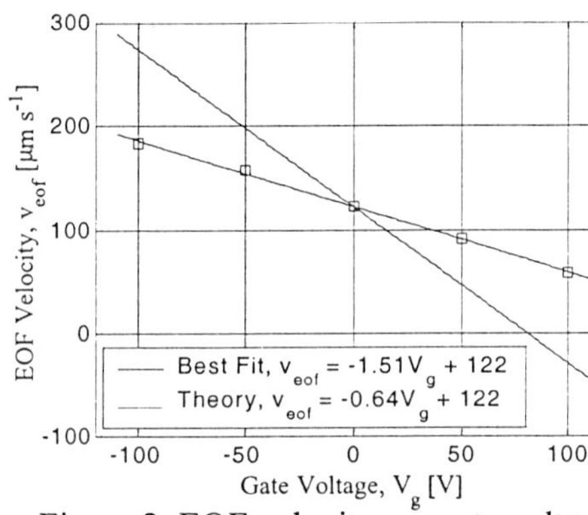

Figure 2. EOF velocity vs. gate voltage for ε-ACA buffer.

3. Results and Discussion

The flow control was characterized by current monitoring as a 19 mM buffer solution replaced a 20 mM buffer solution in a single channel branch. The slope of the EOF mobility versus the applied gate voltage was found to be -4.4×10^{-3} $cm^2/(V^2\ s)$ for phosphate buffer solution at pH 5. In order to minimize the absorbion of the buffer ions into the SiO_2, epsilon-amino-n-caproic acid (ε-ACA) buffer solution was used. For this larger buffer ion, the slope was -7.2×10^{-3} $cm^2/(V^2\ s)$ and -15.5×10^{-3} $cm^2/(V^2\ s)$ for pH 4 and pH 3, respectively. These flow control experiments under various buffer conditions were found to exhibit similar linearity as well as an inverse relationship between the control slope and the pH level.

An example of the flow control measurements is shown in Figure 2 for ε-ACA buffer solution at pH 4 under an electric field of 93 V/cm. The flow control under these conditions exhibits an excellent linearity between applied gate voltage and the resulting flow velocity. Assuming the permittivity and viscosity of a phosphate buffer solution for the ε-ACA buffer solution, the theoretical EOF velocity from Eqn. 1 is also plotted with the experimental slope 57% lower than theory predicts. This error is reasonable given the assumptions inherent in Eqn. 1. In addition, incorporating into the model the measured properties of ε-ACA and the specific hybrid boundrary conditions of the composite PDMS and SiO_2 microchannels would provide more accurate modeling.

Acknowledgements

This work was supported by a National Institutes of Health grant #01521573.

References

1. Lee, C. S., Blanchard, W. C., Wu, C.-T., "Direct Control of the Electro-osmosis in Capillary Zone Electrophoresis by using and External Electric Field." Anal. Chem., 58, p. 1550-1552, (1990)
2. Schasfoort, R. B. M., Schlautmann, S., Hendrikse, J., van den Berg, A., "Field-Effect Flow Control for Microfabricated Fluidic Networks." Science, Vol. 286, p. 942-945, (1999)

MICRO CAPSULE BASED CHEMICAL REACTOR USING AC ELECTRIC FIELD

Tomohiro Taniguchi, Toru Torii, Toshiro Higuchi
Department of Precision Engineering, The University of Tokyo, Japan
(e-mail: tomo@intellect.pe.u-tokyo.ac.jp)

Abstract

Reagent and samples were put in micro capsules which were surrounded by chemically stable medium. They were manipulated independently by electric field, collided with each other, and mixed. Chemical reaction was caused in mixed capsules.

Keywords: micro capsule, electric field, chemical reactor

1. Introduction

It is expected that μTAS application using this Micro Capsule based Reactor method decreases reagent volume and reaction time requisite for chemical analysis extremely. Further, since reagent samples are manipulated using electric field as micro capsules, micro fluidic device, such as pumps and valves are not necessary, and the medium prevents evaporation and contamination. In this study, behavior of micro capsules surrounded by medium was investigated. Further, manipulated micro capsules were collided with each other.

2. Experimental

Micro capsules were manipulated on the electric panel device. A programmable, high voltage power supply is connected to six phases electrodes of the device (figure 1). The supply sequence from the power supply to the six phases electrodes(a,b,c,d,e,f,g) for micro capsules transportation is shown in figure 2. The parallel electrodes of the device have an average pitch length of 400μm, a width of 150μm.An insulating cover-film was 90μm in thickness.

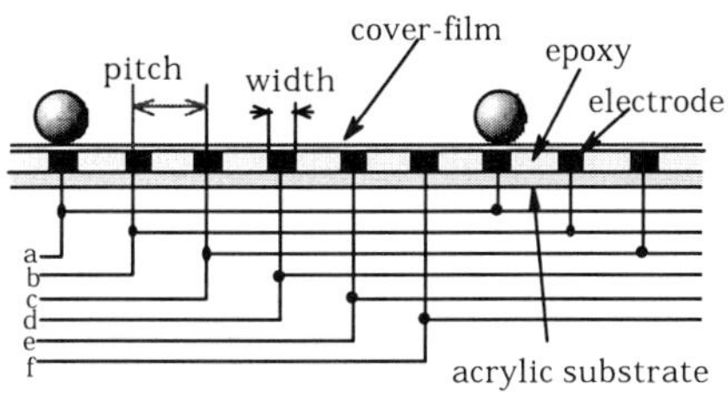

Figure 1: Cross-sectional view of the panel device with six-phase electrodes.

transportation →

1 cycle (rows 1–6)

	a	b	c	d	e	f
1	+	+	0	—	—	0
2	0	+	+	0	—	—
3	—	0	+	+	0	—
4	—	—	0	+	+	0
5	0	—	—	0	+	+
6	+	0	—	—	0	+
7	+	+	0	—	—	0

Figure 2: The supply sequence from the power supply to the six-phases

J.M. Ramsey and A. van den Berg (eds.), Micro Total Analysis Systems 2001, 189–190.

electrodes.

3. Results and discussion

Figure 3 shows a velocity of H_2O micro capsules transported by electric field. These micro capsules were surrounded by medium(oil) under the conditions shown in Table 1. This indicates that micro capsules with 0.6~0.8mm diameters showed synchronized movement with the power supply sequence (frequency = 1Hz). Figure 4 shows micro capsules fusion of phenolphthalein solution and NaOH solution(2mol/L) surrounded by medium. Fusion caused chemical reaction, and mixed micro capsules changed into red color.

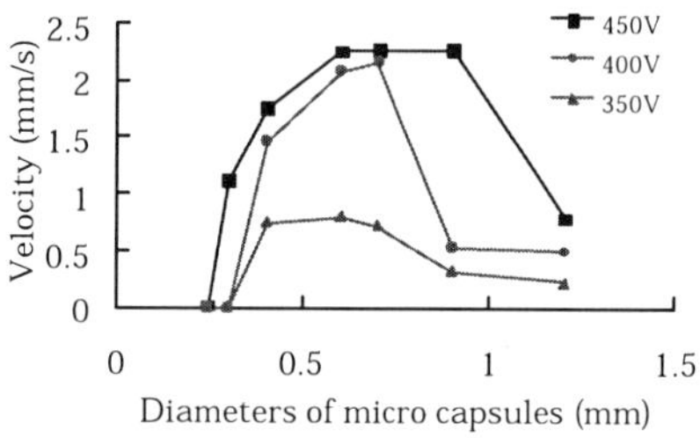

Table 1: Experimental conditions

	density	viscosity
medium	Oil : 0.9 g/cm^3	60 mPa・s
micro capsule	H_2O : 1.0 g/cm^3	1.0 mPa・s

Figure 3: A velocity of H_2O micro capsules transported on the panel device. A frequency of power supply sequence was 1Hz.

Figure 4: Micro capsules fusion of phenolphthalein solution and NaOH solution in the medium(oil).

5. Conclusions

A method for μTAS application was described, where reagent and samples were put in micro capsules which were surrounded by chemically stable medium. In this system, micro capsules on the panel device were manipulated by electric field, and fusion caused chemical reaction. These results suggest that μTAS application using this method is very useful.

FABRICATION AND PERFORMANCE OF ULTRASONIC MICRO-DEGASSING DEVICE

Zhen Yang[1,2], Sohei Matsumoto[1] and Ryutaro Maeda[1]*
[1]Institute of Mechanical System Engineering, AIST, Namiki 1-2, Tsukuba, Japan 305-8564 *(maeda-ryutaro@aist.go.jp)
[2]NEDO,Higashi-Ikebukuro, Tokyo, Japan 170-6028

Abstract

The thermo behavior and quantitative performance evaluation of the first micro-degassing device were reported. The maximum temperature increase of the operating device was 36 °C and that of water flowing through the operating device was 24 °C. 38% of dissolved oxygen was removed after flowing through the operating device and the removing rate of dissolved oxygen increased while the input power increased.

Keywords: Degassing, Ultrasound, Microfabrication

Degassing process is a pretreatment step to samples for chemical analysis. It is also applied to medical treatment, for example, dialysis. Generally, chemical degassing, gas-exchange method and vacuum enhanced diffusion method are unsuitable for μTAS applications. Degassing using ultrasound is an alternative technique. Dissolved gases are actively driven out by means of the cavitation phenomena induced by ultrasound. We use this principle to miniaturize degassing unit.

The device was fabricated using conventional photolithographic technology. The structure, working principle of the device were shown in our previous report. The entire degassing process, from bubble generation to venting, was also shown [1].

The temperature distribution of our device was monitored using a IR thermal imaging camera. Thermocouples were inserted into the inlet and outlet of the device respectively to monitor the temperature of water. The temperatures were measured after 5 min continuously ultrasonic irradiation. The experiment was taken under 16 °C room temperature and the result was shown in Fig. 1b.

The dissolved oxygen concentration was measured using a miniature oxygen electrode. The forward scan current at –885 mV of the second voltammogram was used as the oxygen electrode's response. The oxygen electrode has no built-in temperature compensation feature. Therefore, an additional stainless pipe was inserted between the outlet of the degasser and oxygen electrode. With the help of forced air-cooling, the temperature of the degassed water recovered to room temperature. The results were shown in Fig. 2.

We reported the primary results on MEMS01 [1]. The quantitative evaluations of the device were reported here. The maximum temperature increase of the operating device was 36 °C (Fig. 1b) and that of water flowing through the operating device was 24 °C. 38% (driving at 100 V) of dissolved oxygen was removed after flowing through the operating device and the removing rate of dissolved oxygen increased while the input power increased (Fig. 2). The device would be able to be driven by batteries. This device exhibits great potential for the integration for microchannels, micropumps,

J.M. Ramsey and A. van den Berg (eds.), Micro Total Analysis Systems 2001, 191–192.

micromixers and other components. This is a significant step toward portable integrated systems. The full text of this paper will be reported elsewhere [2].

References

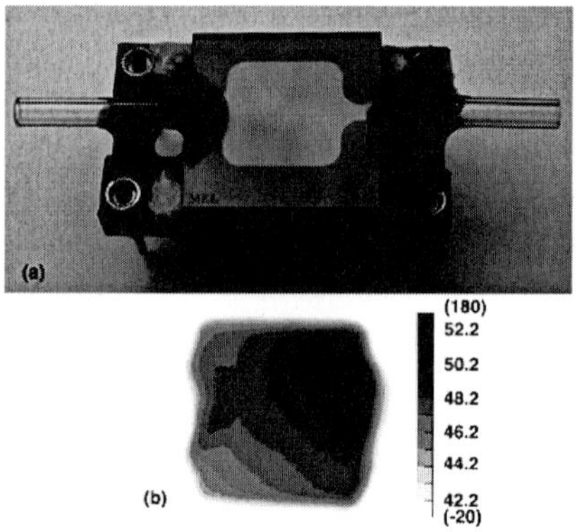

Fig. 1. (a) Photograph of the degassing device. The die size is 10 mm x 10 mm x 0.9 mm. (b) The temperature distribution on the surface of the operating device. The highest temperature was 52 °C and the lowest temperature was 42 °C on the chip.

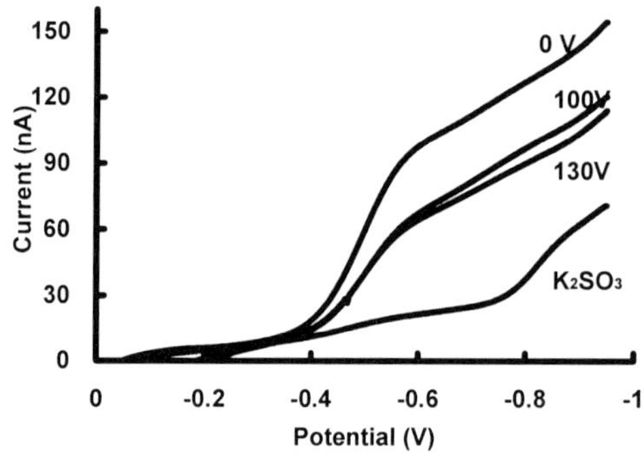

Fig. 2. The forward scans of the second voltammogram from the miniature oxygen electrode under various conditions. Water was used for the evaluations and PZT was excited by 49 kHz square waves. The electrode gave the highest response when the degassing device was off (0 V). Dissolved oxygen decreased when the device was actuated at 100 V. The electrode showed lower response at 130 V than that at 100 V. Thus the more energy input into the degassing device, the more dissolved oxygen was removed. The electrode showed the lowest response to 2% K2SO3 solution for the dissolved oxygen concentration is 0.

1. Z. Yang and R. Maeda, Ultrasonic micro-degassing device, MEMS 2001, Interlaken, Swiss. pp. 471-474, (2001).
2. Z. Yang, S. Matsumoto and R. Maeda, A prototype of ultrasonic micro-degassing device for portable dialysis system, Sens. & Actuaters A, accepted.

FUNCTIONAL AND EFFICIENT ELECTRODE-INTEGRATED MICROFLUIDIC PLASTIC DEVICES

Mingqi Zhao[1], Antonio J. Ricco[1], Uyen Nguyen[1], Richard M. Crooks[2], and Qun Zhu[1]

[1]ACLARA BioSciences, Inc., 1288 Pear Ave., Mountain View, CA 94043
[2]Department of Chemistry, Texas A&M University, College Station, TX 77842

Abstract

Several methods have been developed to deposit and integrate electrodes onto plastic microchips; their application for biochemical analyses is described. The deposited electrodes are compatible with polymer substrates and various bioassay components. These electrodes can be used as on-chip heaters, driving electrodes for capillary electrophoresis (CE), and electrochemical detectors.

Keywords: Integrated electrodes, plastic microfluidic devices, heater, capillary electrophoresis, high-throughput analysis.

1. Introduction

Disposable plastic microfluidic devices offer unique advantages for microanalytical systems [1-3]. For on-chip sample preparation and chip-based CE, integrated electrodes on plastic devices enhance functionality and automation while minimizing carry-over and resultant cross-contamination relative to external, reusable wire electrodes.

2. Methods, Results, and Discussion

Plastic microfluidic devices were fabricated from acrylic, polycarbonate, polystyrene, or polyolefin using published procedures [1,3]. Integrated electrodes were formed on the rigid plastic substrate or the flexible thin-film channel-cover layer *via* each of 3 methods: physical vapor deposition plus conventional photolithography; electroless deposition; and screen printing of commercial conductive ink. Devices were completed by securing the plastic film to the microchip to close the channels, using either thermal lamination or pressure-sensitive adhesive (Figure 1).

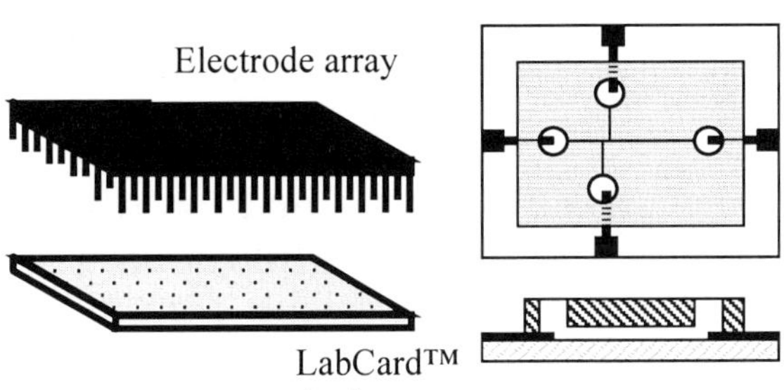

Figure 1. *"Ordinary" and electrode-integrated chips.* ***At left:*** *electrodes on a PC card are placed in the wells to drive CE separations.* ***At right:*** *top and side views of electrodes patterned on the bottoms of reservoirs in plastic chips to serve as driving electrodes for CE separations.*

For nucleic acid amplification and identification applications, we patterned heaters on plastic devices consisting of meandering or straight traces < 25 μm wide; they

J.M. Ramsey and A. van den Berg (eds.), Micro Total Analysis Systems 2001, 193–194.

respond rapidly to the application and removal of power. Importantly, the heater traces can be used simultaneously to probe temperature by monitoring electrical resistance.

We integrated electrodes onto plastic devices to drive chip-based CE for high-throughput drug screening and genetic analysis applications. ACLARA has devised a strategy wherein CE-separable fluorescent eTag™ reporters are selectively released in proportion to the concentration of each nucleic-acid-sequence target during a multiplexed enzyme-based amplification process; this enables high-throughput, highly multiplexed gene-expression, genotyping, and other nucleic acid assays with multiple targets. Figure 2 shows the electrophoretic separation of 13 such eTag reporters using integrated electrodes on plastic microchips. All peaks are well resolved on both chips, but the FWHM and resolution for some probes are somewhat improved using the electrode-integrated chips.

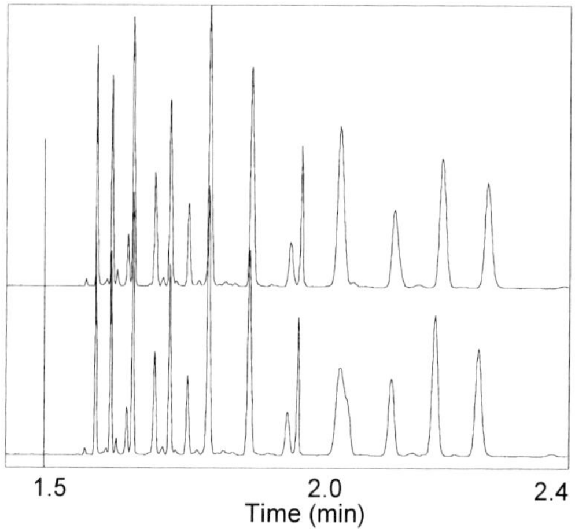

Figure 2*. Separation of 13 fluorescent eTag™ reporters using electrode-integrated plastic chip. The electropherogram at bottom is for a control experiment, where Pt wires were used as driving electrodes. A mixture of 13 probes with concentrations in the range of 4 to 40 nM was added to the sample well. The electric field strength for the separation was 400 V/cm.*

Integrated electrodes sustain high electric fields and associated electrolysis with no observed degradation during CE separations. For example, an electrode-integrated plastic microchip has been used many times for high-resolution separation of single-stranded DNA fragments in sequencing applications, even when each run requires over one hour. We acknowledge DARPA BioFlips in supporting some of this work.

References

1. R. M. McCormick, R. J. Nelson, M. G. Alonso-Amigo, D. J. Benvegnu, and H. H. Hooper, "Microchannel Electrophoretic Separations of DNA in Injection-Molded Plastic Substrates", *Anal. Chem.*, *69*, 2626 (1997).

2. C. S. Effenhauser, G. J. M. Bruin, and A. Paulus, "Integrated Chip-Based Capillary Electrophoresis", *Electrophoresis*, *18*, 2203 (1997).

3. M. Zhao, T. D. Boone, X. C. Qiu, and A. J. Ricco, "Distribution And Mixing Of Reagents on Multichannel Plastic Chips", *Technical Digest Of The 2000 Solid-State Sensor and Actuator Workshop*, Transducers Research Foundation, Cleveland (2000); pp. 183-186.

DESIGN AND FABRICATION OF MULTIPLE NANOFLUIDIC INTERCONNECTS FOR THREE-DIMENSIONAL MICROANALYTICAL SYSTEMS

Mark G. Spencer[1], Bruce R. Flachsbart[1], Thaddeus Yasunaga[1], Tzu-Chi Kuo[2,3], Jonathan V. Sweedler[2,3], Paul W. Bohn[2,3], and Mark A. Shannon[1,3]

[1] Department of Mechanical Engineering, [2] Department of Chemistry, and [3] Beckman Institute for Advanced Science and Technology, University of Illinois at Urbana-Champaign, 1206 W. Green Street, Urbana, Illinois 61801.

(E-mail address for M.A. Shannon: mas1@uiuc.edu)

Abstract

The design and fabrication is described of multi-layered interconnects using nanopore membranes and microfluidic channels to form a three-dimensional microanalytical system.

Keywords: molecular gates, interconnects, nanofluidic, nanopore membranes

Introduction

Nanofluidic devices offer unusual properties for chemical and biochemical sensing, molecular separations, drug delivery and other forefront technologies, due to the affinity of molecular species within nanometer-sized pores to wall interactions and field mediated transport [1]. When the solution Debye length is of the order of the channel dimensions in the nanopores, fluidic transfer may be controlled through applied bias, polarity, density of the immobile nanopore surface charge, and the impedance of the nanopores relative to the microfluidic channels, giving rise to externally controllable gates for molecules in aqueous solutions. While nanofluidic components are often difficult to access and control, it has recently been demonstrated at UIUC that 5-10 μm thick membranes with 10 to 200 nm diameter pores can act as "molecular gates" for digitally manipulating the flow of molecules through a three-dimensional microanalytical system. With this discovery, the goal is to fabricate microanalytical systems that combine 3-D microfluidic circuits driven by electrokinetic flows with nanofluidic interconnects, as depicted in Fig. 1(a). The basic design employs layered nanopore membranes with cross channels to access the gate regions. The inset shows a single four-level microchannel circuit, which has

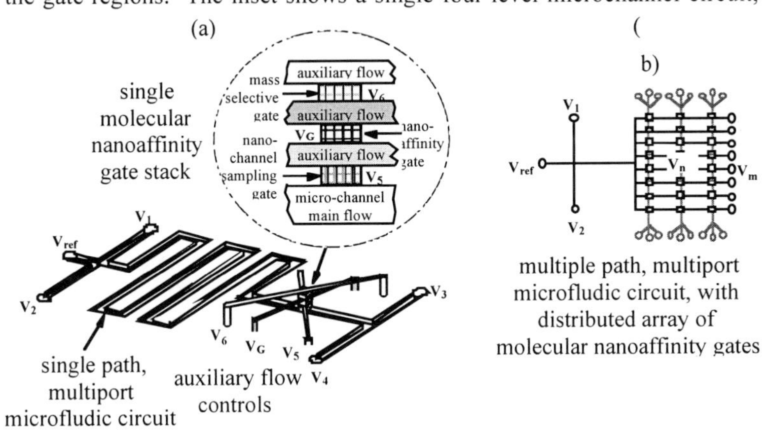

Figure 1. (a) Single system consists of capillary electrophoresis separation, a sampling gate, the nanoaffinity gate, and a mass selective gate. Each auxiliary flow is assigned a specialized fluid handling, separation or sensing task. (b) The single gate region can be expanded to n x m gates to conduct combinatorial manipulations and assays on a large number of molecules.

J.M. Ramsey and A. van den Berg (eds.), Micro Total Analysis Systems 2001, 195–196.

been separated by three nanopore membrane interconnects. The objective is to expand this capability into a distributed array of molecular gates to perform complex operations on a multitude of species, as shown in Fig. 1(b).

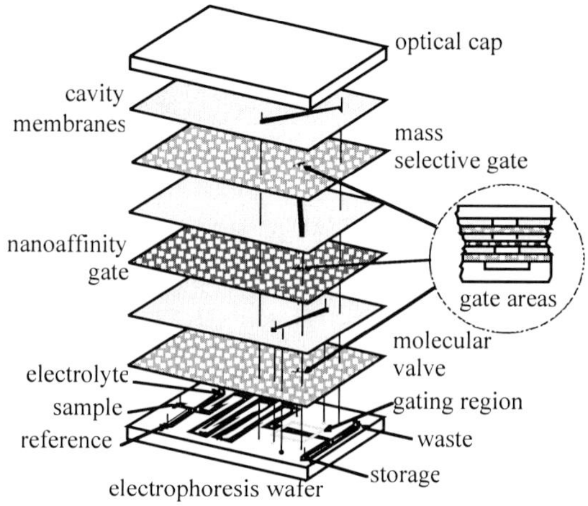

Figure 2. The concept in Fig. 1 is realized by using layered patterned membranes bonded together used to access the crossed-channel gate regions.

Results and Discussion

To achieve the 3-D functionality, we have developed a design and fabrication technology that utilizes a stack of patterned polycarbonate (PC) membranes, bonded together on the base electrophoresis wafer, and capped with a glass to provide optical access, as depicted in Fig. 2. Each layer is separately patterned and processed to achieve the desired functionality, and then assembled together to form the 3-D micro- to nanofluidic system. While microfluidic systems allow access to the nanoscale, and there has been rapid advancements in fabricating porous membrane structures and integrating polymers into microanalytical systems[2], the layered membrane design poses several fabrication challenges.

The key challenges in fabricating layered nanopore PC membranes interleaved with solid patterned PC membranes are to: (i) pattern and transfer the processed labile membranes, (ii) bond the membranes to the substrate and optical cap and to each other without plugging the nanopores, and (iii) achieve sufficient bond strength to seal and maintain fluid integrity in the presence of large wicking forces. The fabrication processes to be presented include: (i) the transfer process developed utilizing a carrier with the "just right" adhesive energy so that the membranes can be separately processed; (ii) the self-assembled monolayer hetero- and homobifunctional linkers with the appropriate reactive end groups to adhere glass and PC to themselves and each other; and (iii) the pressure and temperature: controlled deformation required to achieve nanometer planarity between the layers in order to achieve the desired bond strength. Fig. 3 shows a picture of the first stack that utilized the nanopore membranes on glass. Up to 9 layer stacks have been fabricated using the methodology developed, allowing 3-D nanofluidic interconnect architectures to be employed for microanalytical systems.

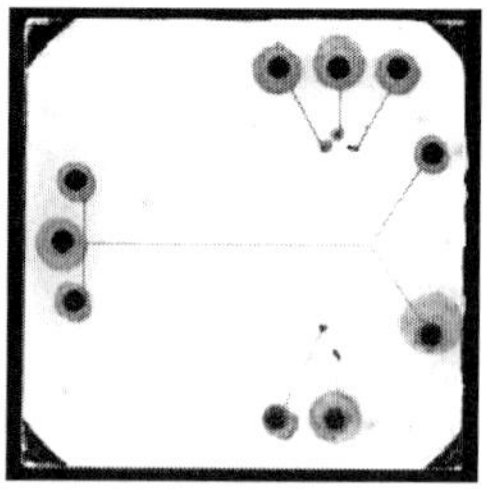

Figure 3. The 3-D design in Fig. 2 made with BOE etched glass microfluidic channels, shown with dye. The nanopore membrane is bonded to a glass slide.

Acknowledgements

The authors wish to thank the DARPA MTO BioFlips program, contract number F30602-00-2-0567, for support of this work.

References

1. Kemery, P. J.; Steehler, J. K.; Bohn, P. W. "Electric Field Mediated Transport in Nanometer Diameter Channels," *Langmuir* **1998**, *14*, 2884-2889.
2. Khandurina, J.; Jacobson, S. C.; Waters, L. C.; Foote, R. S.; Ramsey, J. M. "Microfabricated Porous Membrane Structure for Sample Concentration and Electrophoretic Analysis," *Analyt. Chem.* **1999**, *71*, 1815-1819.

NANOBARCODES^{TM} IDENTIFICATION TAGS IN PROTEOMICS AND SMALL MOLECULE PROFILING

Daojing Wang, Lin He, R. Griffith Freeman, Remy Cromer, Michael J. Natan & Rajendra Singh

SurroMed, Inc., 2375 Garcia Ave., Mountain View, CA 94043, USA

Abstract

We describe novel applications of Nanobarcodes^{TM} Identification Tags (NBCs), nanometer-scale metallic bar codes in proteomics and small molecule profiling. Synthesis, detection and surface derivatization of these nanoparticles are presented. Solid-phase nanoextraction (SPNE) of analytes from biological samples and its unique application in MALDI-TOF mass spectrometry are detailed.

Keywords: Nanobarcodes, proteomics, SPNE, mass spectrometry, MALDI-TOF

1. Introduction

One of the major challenges in the post human genome era is high throughput profiling of proteins and small molecules in biological samples. As a result, there has been a recent paradigm shift in bioanalytical chemistry towards microscale total analytical systems (μTAS). The push for miniaturization and multiplexing, in turn has fueled rapid developments in microfluidics separation chip and protein microarrays. Nevertheless, protein microarrays suffer from slow diffusion of analytes to the surface and limit the dynamic range of analyte detection. Microfluidics systems are plagued with nonspecific binding problems. Solution arrays such as fluorescence-labeled bead-based systems may circumvent these problems but are limited by available spectrally distinguishable fluorophores. Herein we describe a new platform based on nanometer-scale cylindrical metal (e.g. Ag, Au Pd, Ni, Co, Cu and Pt) particles encoded with distinctive stripes (Figure 1) for solid phase nanoextraction.

2. Experimental

NBC synthesis and detection. NBCs were fabricated via template-directed electrochemical deposition [1]. Desired stripping pattern was achieved by sequential electrochemical reduction of metal ion solutions into membrane pores. Optical microscopy was used to identify the striping patterns via differential reflectivity of adjacent metals. NBC length and shape were monitored with SEM.

NBC derivatization. The surface of NBCs were derivatized with thiols such as mercaptoundecanoic acid ($HS(CH_2)_{10}CO_2H$) to generate a self-assembled monolayer (SAM) (Figure 1). Antibodies or other amine-containing reagents were attached to the surface by EDC/S-NHS chemistry.

SPNE. 25 µl of normal human plasma fraction (MW< 10 KD) was mixed with differently derivatized NBCs (3 x 10^6 particles) at desired buffer conditions for 1 hour. After centrifugation and washing, NBCs were suspended in H_2O (2 µl). An aliquot (0.5 µl) was mixed with α-cyano 4-hydroxy cinnamic acid matrix (0.5 µl, 10mg/ml) for MALDI-TOF analysis.

J.M. Ramsey and A. van den Berg (eds.), Micro Total Analysis Systems 2001, 197–198.

MALDI-TOF. MALDI-TOF spectra were obtained on a Voyager DE-PRO (Perseptive Biosystems Inc., Framingham, MA). The instrument was equipped with a linear flight tube of 1.1 m and a delayed extraction ion source and employed a nitrogen laser (λ=337 nm).

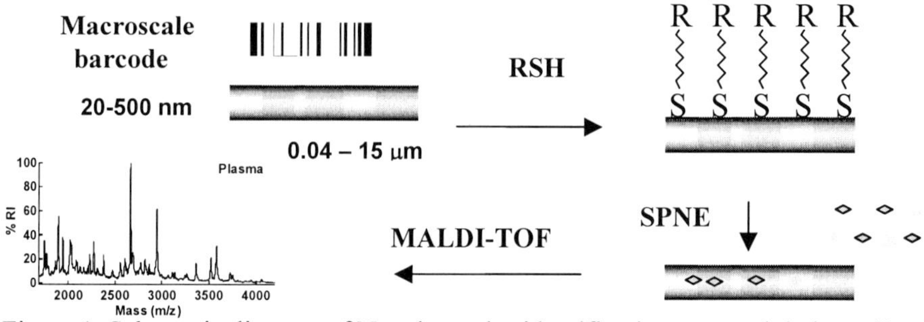

Figure 1. Schematic diagram of Nanobarcodes identification tags and their applications in SPNE and MALDI-TOF. Macroscale barcode is shown on top as a comparison.

3. Results and discussion

Solid phase nanoextraction (SPNE) is a powerful sample preparation technique in which target analytes are extracted and concentrated onto support surfaces. We have demonstrated that NBCs can serve as SPNE media to capture and concentrate protein and small molecules. Surface selectivity and specificity was achieved with well-developed Ag/Au derivatization chemistries, ranging from thiol self-assembled monolayer to coupling of antibody and other proteins. High multiplexing is feasible with barcoding and different surface chemistries. Miniaturization is achieved due to the small size of each NBC. Applications with both nonspecific and specific binding were demonstrated. The former includes extraction from Cyt C trypsin digests and human plasma and the latter includes immunoassays and streptavidin-biotin system.

Mass spectrometry, and especially MALDI-TOF, has proven extremely powerful in proteomics. Nevertheless, MALDI has limitations in terms of sample preparation and detection sensitivity. We have demonstrated the unique application of NBC in MALDI-TOF. MALDI detection limit is increased over polystyrene beads due to the inherent "nano-concentrating" advantages of metallic NBCs resulting from the cylindrical shape of NBCs (with surface area augmentation by a factor of π), as well as strong local heating properties.

4. Conclusion

We have developed a new NBC-MALDI platform for the profiling of proteins and small molecules in biological fluids. This platform can be extended to genomic applications.

References

1. S. R. Nicewarner-Pena, L. He, D. Wang, R. G. Freeman, B. D. Reiss, D. J. Pena, I.D. Walton, R. Cromer, R. Singh, C. D. Keating and M. J. Natan, *Submicron Barcodes*, Science, Submitted (2001).

CHARACTERIZATION OF MOLECULAR TRANSPORT WITHIN GATEABLE NANOFLUIDIC INTERCONNECTS FOR THREE-DIMENSIONAL MICROFLUIDIC SYSTEMS

Donald M. Cannon Jr., Tzu-Chi Kuo, Wenju Feng, Mark A. Shannon, Jonathan V. Sweedler and Paul W. Bohn

Department of Chemistry, Department of Mechanical Engineering, and Beckman Institute for Advanced Science and Technology, University of Illinois at Urbana-Champaign, Urbana, Illinois 61801

Abstract

Novel molecular gate membranes with nanometer diameter pores are described as active interconnects for producing three-dimensional microfluidic systems. Molecular transport studies target the physical and chemical characteristics that influence this highly controllable transport mechanism.

Keywords: molecular gates, interconnects, nanofluidic, nuclear track-etched membranes, three-dimensional

1. Introduction

Microelectromechanical systems have targeted a wide range of applications for fluid processing of molecules within micrometer-scale structures. Three-dimensional microfluidic designs are increasingly being pursued as an added dimension of complexity required for further integration of sophisticated designs and functions [1-3]. Nuclear track-etched membranes, with nanometer diameter pores that support flow across a relatively large (~5 μm) membrane thickness are described as novel "molecular gates" for producing three-dimensional microfluidic devices (fig. 1). En route to understanding the relevant phenomena surrounding fluid dynamics and chemical interactions within the nanometer-scale structures, investigations have targeted the characterization of molecular transport through these pores [4]. Comparing the differing interconnect characteristics using various probe molecules demonstrates the ability to control molecular transport through the molecular gate. The characterization of molecular transport within nanometer pores is essential for an understanding of these externally controllable active interconnects. This ability for intelligent molecular transport is a critical development for further application of three-dimensional microfluidic systems.

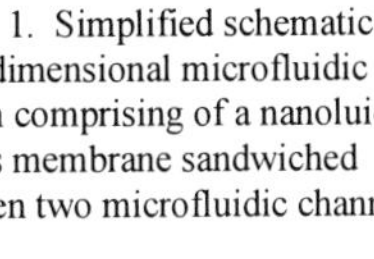

Figure 1. Simplified schematic of three-dimensional microfluidic system comprising of a nanoluidic porous membrane sandwiched between two microfluidic channels.

2. Experimental

The chip design used in these experiments consists of a 5-μm thick nanofluidic membrane sandwiched between two crossed polydimethylsiloxane (PDMS) micro-

J.M. Ramsey and A. van den Berg (eds.), Micro Total Analysis Systems 2001, 199–200.

fluidic channels. The nuclear track-etched polycarbonate membrane has a relatively monodisperse distribution of cylindrical pores with diameters of 15 nm $\leq d \leq$ 220 nm and densities of 1 x 10^8 cm^{-2} $\leq N_p \leq$ 6 x 10^8 cm^{-2}. Standard prototyping protocols for PDMS are used tailored for three-dimensional constructs [5]. Channels are 100-µm wide and 30-µm deep. The crossed microfluidic channels spatially define the transport region associated with the molecular gate. Transport is monitored with laser induced fluorescence detection or fluorescence microscopy in either the source or the receiving microfluidic channel at several positions relative to the nanofluidic-gating region.

3. Results and Discussion

By controlling the chemical and physical properties of the individual gates, molecular transport can be manipulated, providing a highly selective and controllable transport mechanism. Initial efforts have focused on the electric bias control to achieve fast, clean transport of small-molecule analytes of varying polarities. Comparison of the bias controls for 200-nm vs. 15-nm pore diameter membranes shows the direction of transport is reversed. This behavior is attributed to an increased resistance to pressure driven flow as well as an increased voltage drop across the membrane, thereby enhancing the contribution of electroosmotic flow within these smaller pores. Mass-independent gating using the 200-nm pore size membranes is demonstrated with a series of FITC-labeled dextrans up to 2 MDa. Reproducible transport profiles (fig. 2) are composed of a rapid electrophoretic transfer of these larger molecules across the nanofluidic membrane with a slower diminishing component after biasing, corresponding to analyte movement away from the detection zone within the receiving channel. Molecular volume dependence on the gating characteristics of these interconnects is currently being interrogated using a series of protein mass markers. These results demonstrate intelligent molecular transport via externally controllable interconnects, a critical development for further application of three-dimensional microfluidic systems.

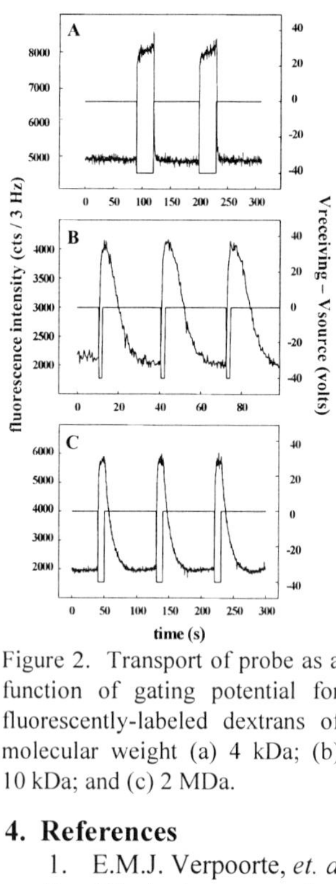

Figure 2. Transport of probe as a function of gating potential for fluorescently-labeled dextrans of molecular weight (a) 4 kDa; (b) 10 kDa; and (c) 2 MDa.

4. References

1. E.M.J. Verpoorte, *et. al.*, *Micromech. Microeng.*, **1994**, *4*, 246-256.
2. J.R. Anderson, *et. al.*, *Anal. Chem.*, **2000**, *72*, 3158-3164.
3. S. Dharmatilleke, H.T. Henderson, *Micro-Electro-Mech. Syst.*, **2000**, *2*, 413-418.
4. T.-C. Kuo, L.A. Sloan, J.V. Sweedler, P.W. Bohn, *Langmuir*, submitted (2001).
5. B.-H. Jo, L.M. Van Lerberghe, K.M. Motsegood, D.J. Beebe, *J. Microelectromech. Syst.*, **2000**, *9*, 76-81.

NANOIMPRINT FABRICATION OF DNA ELECTROPHORESIS CHIPS

V. Studer[1], P. Youinou[1], A. Pépin[1], A. Lebib[1], Y. Chen[1] and Y. Baba[2]
[1]LPN / CNRS, 196 avenue Henri-Ravéra, 92222 Bagneux, FRANCE
[2]Department of Medicinal Chemistry, Univ. of Tokushima, 770-8505, JAPAN

Abstract

We have used nanoimprint-based techniques [1] to fabricate capillary electrophoresis chips with an in-channel array of nanopillars for DNA separation, in both silicon dioxide and plastic substrates.

Keywords: nano-embossing, nanoimprint, microfluidics, DNA, electrophoresis

1. Introduction

The Human Genome Project and human polymorphism analysis have led to much interest being dedicated to the development of microfabrication technology for efficient DNA and protein separation. The structure we study here is a standard cross-shaped electrophoresis channel of 50 μm width ended with 500 μm diameter reservoirs. The sieve, or nanostructured portion of the separation channel, is 7 mm long and consists of a 320 nm period hexagonal array of 150 nm diameter pillars. The channel depth is typically 10 μm outside the array, and either 150 nm or 500 nm inside the array.

2. Experimental

a) 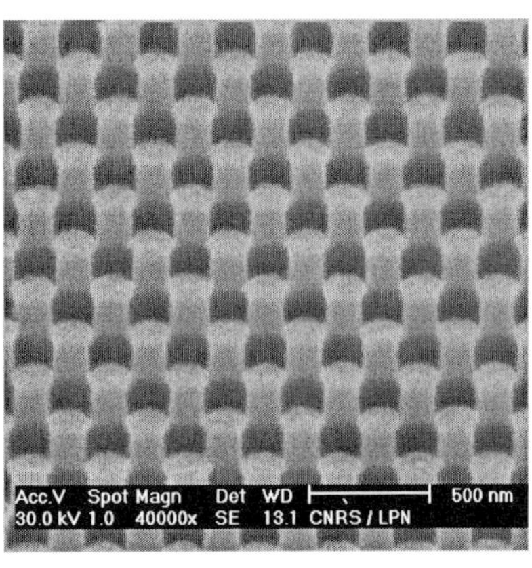

b) 

Figure 1. SEM micrographs of an array of nanopillars of diameter 150 nm fabricated inside a microfluidic channel, a) on a silicon dioxide substrate b) on a PMMA sheet.

J.M. Ramsey and A. van den Berg (eds.), Micro Total Analysis Systems 2001, 201–202.

Two different fabrication techniques were investigated. First, tri-layer-based nanoimprint lithography [1] was used to create the structure in a silicon dioxide substrate. The tri-layer sequence consisted of a thick thermostable PMGI resist layer at the bottom, an intermediate thin layer of Ge and a top thermoplastic resist layer of either polymethylmetacrylate (PMMA) or hyper-branched polymer (hybrane). Nanoimprint temperature (above glass transition temperature) was 200ºC and 150ºC for PMMA and hybrane respectively, with typical pressures around 50 bar. In the second approach, the whole structure was created by one-step nano-embossing of PMMA pellets, carried out at 180ºC with pressures around 50 bar. The imprinting molds made of Si were obtained through conventional nanofabrication technologies [1]. A top cover part made of PMMA and containing the microfluidic connections to the reservoirs was then glued to the fabricated chip, by either thermal bonding between two PMMA sheets or UV-curable glue for PMMA on SiO_2/Si.

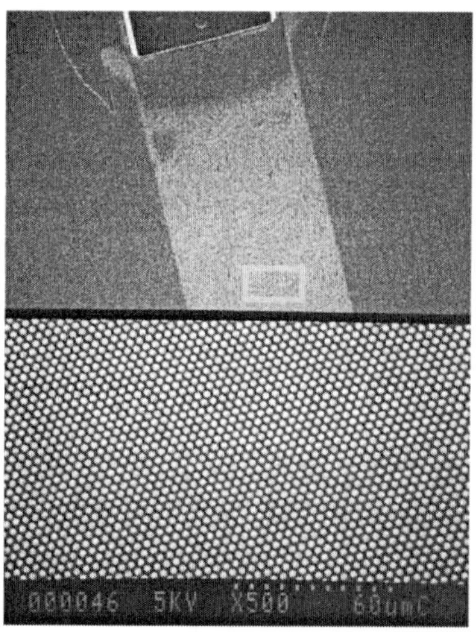

Figure 2. SEM micrograph of one-step nano-embossing of PMMA showing a 10 µm-deep channel connected to a 150 nm-deep nanopillar array.

3. Results

A portion of the highly homogeneous 500 nm-deep nanopillar array obtained in SiO_2/Si is shown in Figure 1(a). The 150 nm-deep array obtained in nano-embossed PMMA is shown in Figure 1(b). Special molds containing both a shallow nanostructured array and deeper channels were fabricated. Figure 2 shows the result of one-step nano-embossing in PMMA using these molds.

4. Conclusions

The technological approach we have been developing is both low-cost and flexible and should allow mass production of micro- and nano-fluidic devices. The efficiency of our device for DNA separation still needs further experiments, which will be carried out in the next step of this work.

References

1. Y. Chen and A. Pépin, *Nanofabrication: conventional and nonconventional methods,* Electrophoresis, 22, pp.187-207, (20001).

MOLECULAR DYNAMICS MODELING OF THE MOLECULAR TRANSPORT THROUGH A NANPORE

S. T. Cui[1,2] and H. D. Cochran[2,1]
[1]Department of Chemical Engineering, University of Tennessee, USA
[2]Chemical Technology Division, Oak Ridge National Laboratory, USA

Keywords: Molecular simulation, nanoscale pore, molecular transport, fluid flow

Introduction There is growing interest in and potential important applications of fluidic devices with nanoscale cross-sectional dimensions [1], where the size of molecules (e.g., polyelectrolytes) and the electrical double layer can be comparable to the pore dimension, and interfacial interaction between the channel wall and fluid may have important impact on the fluid and solute molecule transport. The transport of fluids and samples has traditionally been analyzed using continuum models based on the Poisson-Boltzmann equation and the Navier-Stokes equation [2]. Studies have also been carried out using primitive (charged hard sphere) models for ions in which the solvent is treated as a continuum [3]. The surface charges are treated as being uniformly distributed on the surface. These treatments are appropriate when the cross-section of the pore is much larger than the size of the molecules. The neglect of excluded volume of the solvent, ion hydration, and the discreteness of the surface charges, raises many questions about the ability of such methods to predict the distribution of the ions inside nanoscale pores, however. The predicted flow behavior of the fluid under an electric field and hence sample transport in nanoscale pores, under such assumptions, may be unreliable.

Molecular Modeling We attempt to study the behavior of aqueous solution containing Na^+ and Cl^- ions in a cylindrical pore with properties similar to silica glass using a molecular dynamics method. The pore diameter ranges from 2 to 7 nm. The SPC/E model for water is used. For Na^+ and Cl^-, we chose the models that have been used in a number of recent publications [4]. The interaction between the fluid molecules and the surface is modeled by a Steele 10-4-3 potential [5], with Lennard-Jones parameters for the wall: ε_w = 1.913 kJ mol^{-1}, and σ_w = 3.0 Å. The cross interaction parameters between the wall and water oxygen, Na^+, and Cl^- are determined using Lorentz-Berthelot combining rules. Surface electric charges are represented by discrete charges equal to that of an electron. All interactions are truncated at 11.08 Å. Details of the model Lennard-Jones parameters are listed in Table I.

molecules	σ/Å	ε/kJ mol^{-1}	molecules	σ/Å	ε/kJ mol^{-1}
O-O	3.166	0.6502	Na^+-Na^+	2.73	0.3576
Na^+-O	2.876	0.5216	Cl^--Cl^-	4.86	0.1679
Cl^--O	3.785	0.5216	Na^+-Cl^-	3.87	0.1706

Results and Discussion Fig. 1 shows ionic distribution and average water polarization for the 5 nm diameter cylindrical pore at salt concentration of 80 mM and surface charge density of 0.0484 C/m^2, where ρ and μ are the density and dipole moment of water molecules. One notable feature in the ion distribution, beside the strong Na^+ peak due to adsorption to the negative surface charge, is a region near the surface that exhibits a depletion of ions. It shows that for the

J.M. Ramsey and A. van den Berg (eds.), Micro Total Analysis Systems 2001, 203–204.

nanoscale pore diameter studied here, the ions tend to stay away from the wall of the cylinder. The phenomena can be understood by the polarization of water media and the ion hydration. Near the wall, the water molecules are more polarized, and it would be energetically unfavorable for the ions to be hydrated. In addition, close to the center of the cylinder, the ions would be able to maximize the number of interactions with water molecules on geometric considerations.

Based on the ion distribution in Fig. 1, we numerically solved the momentum equation of motion to obtain a velocity profile across the cylinder for fluid flow under an externally applied axial electric field of 1000 volt/cm using a viscosity 1 cP for water, as plotted in Fig. 2. The profile is seen to be close to a parabolic function (dashed line).

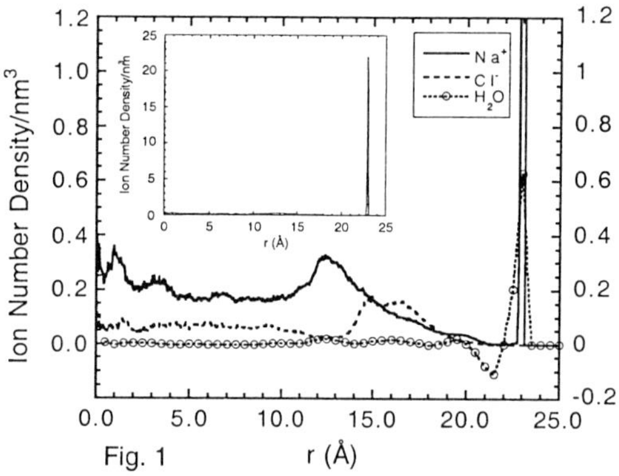

Fig. 1

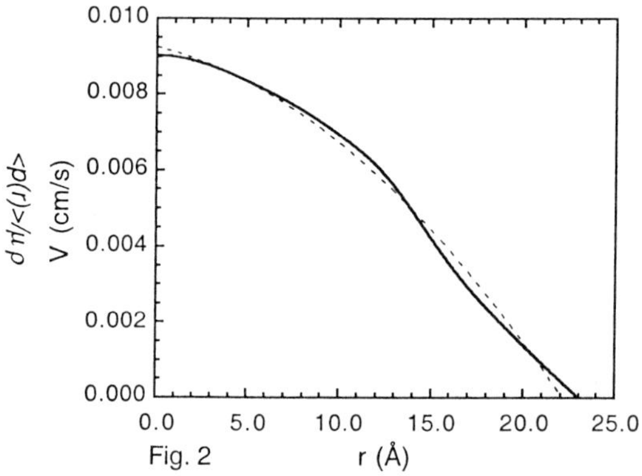

Fig. 2

Conclusions Results obtained in this study suggest the importance of solvent hydration of charged species in very narrow channel and necessitate the explicit treatment of surface charges. Such study provides a more realistic picture for the fluid flow in nanoscale channels. By studying the fluid characteristics at approximately 1-10 nm, we aim to obtain asymptotic behavior for channels on the scale ~ 100 nm.

Acknowledgments The authors acknowledge many helpful discussions with Mike Ramsey and other members of his group. This work is supported by Laboratory-Directed R&D funding at ORNL and NSF funding at UT under grant number BES-0103140. ORNL is operated for the U.S. Department of Energy by UT-Battelle, LLC, under contract number DE-AC05-00OR22725.

References

1. J. Han and H. G. Craighead, *Science* **288**, 1026 (2000).
2. C. L. Rice and R. Whitehead, *J. Phys. Chem.* **69**, 4017 (1965).
3. B. Hribar et al., J. Phys. Chem. B **104**, 11522 (2000).
4. H. J. C. Berendsen et al., J. Phys. Chem. **91**, 6259 (1987); R. M. Lynden-Bell and J. C. Rasaiah, J. Chem. Phys. **105**, 9266 (1996); E. Spohr, Electrochimica Acta **44**, 1697 (1999).
5. W. A. Steele, Surface Sci. **36**, 317 (1973); *The Interaction of Gases with Solid Surfaces* (Oxford: Pergamon Press, 1974).

METERING AND MIXING OF NANOLITER LIQUID IN THE MICROCHANNEL NETWORKS DRIVEN BY FLUOROCARBON SURFACES AND PNEUMATIC CONTROL

Sang-Ho Lee and Yong-Kweon Kim
School of Electrical Engineering and Computer Science, Seoul National University, Kwanak P.O. Box 34, Seoul 151-742, Korea

Abstract

A Mixing and metering of nanoliter liquid was performed in the polydimsiloxane (PDMS)-microchannel. We could meter a water droplet as a precise nanoliter volume of 20 nL (S.D. = 0.24) using the external air pressure and selectively modified surfaces.

Keywords: Discrete flow control, PDMS, microchannels, hydrophobic surface

1. Introduction

In the micro TAS, the sample preparation could involve various fluidic operations such as metering, mixing and reaction as a precise volume. This study reports the nanoliter fluidic handling strategies. It is based on the discrete droplet control using the only external air pressure and selectively patterned hydrophobic fluorocarbon (FC) surfaces.

2. Theory

From the equation (1) (width w, height h, H_2O surface tension γ), we designed the liquid flow direction in the microchannel networks. Positive pressure indicates capillary filling, but negative pressure indicates a repellent force at the solid-liquid-gas interface [1,2].

$$P = \gamma\cos\theta_{PDMS}\left(\frac{2h+w}{hw}\right) + \gamma\cos\theta_{SiO_2\ or\ FC}\left(\frac{w}{hw}\right) \qquad (1)$$

3. Experimental

The microchannel networks were fabricated by bonding of PDMS microchannels on a substrate with the patterned hydrophobic FC surface. FC film was prepared by spin coating using the mixture of 3M's Fluorad^{TM}. FC film was patterned using the lift-off process in an ultrasonic bath. The surface properties of FC film are characterized by contact angle analysis and atomic force microscopy (AFM) observation.

4. Results and discussion

Water contact angles of FC films were at a range of $110 \sim 118^{o}$ higher than solid bulk Teflon and their surface energies were calculated to be $10 \sim 14$ dyne/cm.

In a microchannel 1, the injected liquid flow stopped when it met a hydrophobic FC film (white color) (Figure 1(a)). We could freely meter a droplet as a desired volume by external pressure control (Figure 1(c)). The liquid driven by the external pressure came to

J.M. Ramsey and A. van den Berg (eds.), Micro Total Analysis Systems 2001, 205–206.

be split at the boundary between hydrophilic surface and hydrophobic surface (Figure 1(d)). We could obtain a nanoliter droplet of around 20 nL (S.D.= 0.42) over the five times metering (Figure 1(e)). Figure 2 shows the some mixing performance in the microchannel networks.

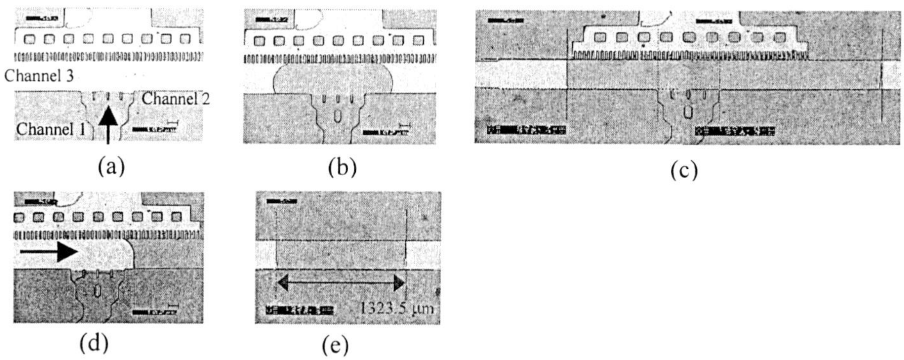

Figure 1. Micrographs of the fluidic behavior manipulated by the external air pressure and FC surface inside PDMS microchannel network (arrow: external pressure direction)

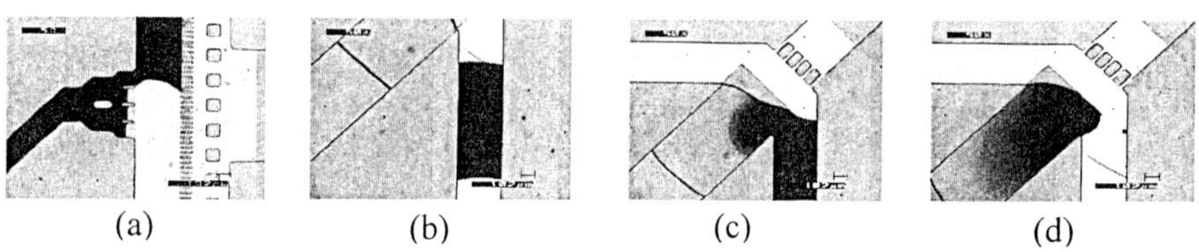

Figure 2. Metering and mixing of two liquids, DI water and blue ink, inside PDMS microchannel networks.

5. Conclusions

We have successfully manipulated the nanoliter liquid droplets using the external pressure and the combination of hydrophobic and hydrophilic surface. Two nanoliter liquid droplets were metered as the desired volume in the two metering regions respectively. Metered two liquid droplets were mixed in the mixing region.

Acknowledgements

This paper was supported by the Nano Bioelectronics and Systems Research Center of Seoul National University.

References

1. K. Hosokawa, T. Fujii, and I. Endo, *Droplet-based Nano/Picoliter Mixer Using Hydrophobic Microcapillary Vent*, Proc. MEMS '99, pp. 388-393, (1999).
2. K. Handique, D. T. Burke, C. H. Mastrangelo and M. A. Burns, *Nanoliter Liquid Metering in Microchannels Using Hydrophobic Patterns*, Anal. Chem., Vol. 72, pp. 4100-4109, (2000).

INTEGRATED MICRO- AND NANOFLUIDICS: SILICON REVISITED

Albert van den Berg
MESA+ Institute, University of Twente, The Netherlands
a.vandenberg@el.utwente.nl

Abstract
Several examples of microfabricated structures in silicon for microanalysis will be given, demonstrating the advantages of the availability of a wide variety of technologies. The monocrystalline, anisotropic nature, the precise dry-etching techniques and sophisticated thin-film surface micromachining are amongst the major advantages. Different micro- and nanofluidic structures for chemical analysis and synthesis are presented and discussed.

Keywords: microfluidics, nanofluidics, silicon, surface micromachining

1. Introduction

The controlled manipulation of liquids on the micron-scale using mechanical actuation, electrical or magnetic fields, pressure or surface tension has been one of the crucial elements in the rapid developments of Miniaturized Total Analysis Systems (µTAS) over the past decade. Although commercially successful µTAS products are often realised in plastics, glass or ceramics [1], in many cases the first demonstration of microfluidic/analytical device functionality was done in silicon. The classical '*µTAS-avant-la-lettre'* example, the micro gas chromatograph by Terry et al. [2], the first prototype of a micro-PCR device by Northrup et al. [3], the first attempts to make an LC-on-chip (4), the first FIA-on a chip (5) and the first controlled pore-size microfilters [6] were all made using silicon microtechnology. There are several reasons for this; most importantly, silicon is an extremely well characterized material with an incredibly large 'toolbox' full of technologies developed by tens of thousands of researchers in the electronic industry. This fact implies that it is virtually always possible to fabricate a certain geometrical structure very precisely in silicon. Secondary reasons for using silicon are its mechanical properties (hard, strong) and its easy coverage by a chemically well-known and relatively inert layer of silicon dioxide. An additional advantage in the use of silicon lies in the straightforward realization of nanometer size structures, giving new challenges for exploration of nanofluidic phenomena. In the following several new examples of demonstrated functionality using silicon will be given, and a concept for realization of so-called surface-micromachined nanofluidic systems will be proposed.

2. Examples of functional demonstrators in silicon

A first example of a device where very precise micromachining in silicon is essential for the device functionality is the so-called hydrodynamic chromatography (HDC) chip. In this device, essentially consisting of a 10 cm long, 1 mm wide and 1 µm high separation

J.M. Ramsey and A. van den Berg (eds.), Micro Total Analysis Systems 2001, 207–209.

channel, small particles and large molecules are separated in a parabolic (pressure driven) flow profile [7]. For a good quality of the separation, a very small variation in the channel height [see fig.1a] is mandatory, and using sacrificial layer techniques a height variation of less than 0.5% over the width of the channel can be realised. This precise control of the separation channel height has resulted in a successful separation of 20 and 80 nm polystyrene beads within 2 minutes.

A second example use is made of the crystalline nature of silicon to create an optical absorbance cell with a long (10 mm) path length. Such a long path length makes it possible to obtain the required detection limit for detection of the reaction product, indophenol blue, that is formed out of the measured ammonia sample in a miniaturised flow injection analysis (FIA) setup {see p.xxx}. In order to reflect a vertically oriented light beam into the horizontal flow channel, a 45-degree reflecting sidewall is needed, which can be made using silicon <110> planes, which can be etched free using a KOH-IPA solution. In figure 1a the schematics of the setup are shown, figure 1b show the realised detection cell. The realised cell resulted in the required detection limit for ammonia of 0.1 mM.

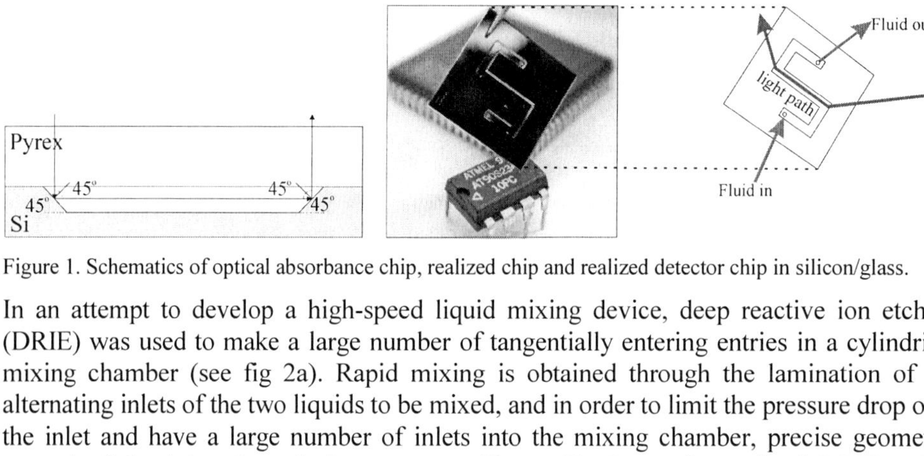

Figure 1. Schematics of optical absorbance chip, realized chip and realized detector chip in silicon/glass.

In an attempt to develop a high-speed liquid mixing device, deep reactive ion etching (DRIE) was used to make a large number of tangentially entering entries in a cylindrical mixing chamber (see fig 2a). Rapid mixing is obtained through the lamination of the alternating inlets of the two liquids to be mixed, and in order to limit the pressure drop over the inlet and have a large number of inlets into the mixing chamber, precise geometric control of the inlet channels is necessary. Figure 2b shows the realised DRIE-etched silicon micromixer chip. Simulations of the device have indicated that indeed sub-millisecond mixing should be possible.

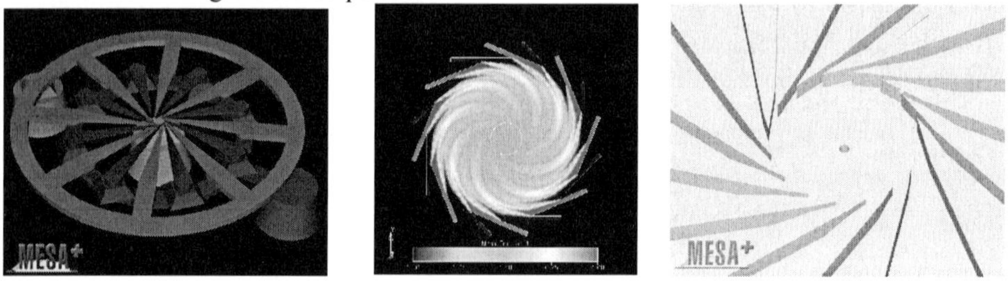

Figure 2. (fltr) Design, simulation and realization of an ultra fast tangential mixer chip in silicon.

The anisotropic nature of silicon is once more utilized for the fabrication of sharp microneedles, used for painless) blood sampling. Using a combination anisotropic KOH etching and DRIE etching we have been able to fabricate a large array of such sharp needles (see figure 3a). The <111>-crystal plane sidewall of the needle ensures extreme sharpness, as is illustrated in an enlarged picture (figure 3b).

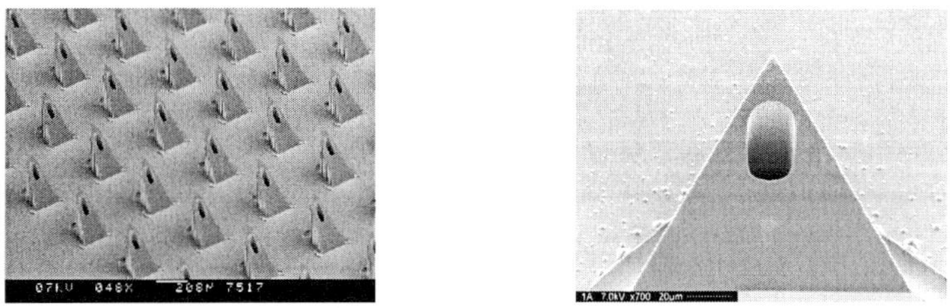

Figure 3. Array of hollow silicon microneedles (left) and one microneedle in detail (right).

A very promising direction is to further scale down the dimensions of the fluid channels to (in one direction) below 1 μm, and enter the area of *nanofluidics*. This is the domain where surface effects start to totally dominate the liquid behavior, and ultimately a liquid can no longer be considered as a continuum but as an ensemble of individual molecules. Bulk properties of the liquid (such as viscosity, dielectric constant) and dissolved molecules and ions are no longer valid, and new phenomena that could be used for separating and detecting species appear. A very versatile technique to fabricate channels with nanometer dimensions (in the vertical direction) is a combination of sacrificial layer techniques and surface micromachining [8]. Using this technique we have been able to make channel crossings, flow sensors, and perforated membranes, and surface micromachined Integrated Fluidic Systems (IFS) can be made. As an example of the potential of this concept we realised a pneumatically actuated, capillary pressure driven micropump capable of delivering picoliter amounts of liquid in a controlled way (fig 4). Optimal exploitation of the large role of surface effects will make the realization of complex micro- and nanofluidic networks possible with interesting applications such as fast fluidic computing and ultra low volume high throughput experimentation and analysis.

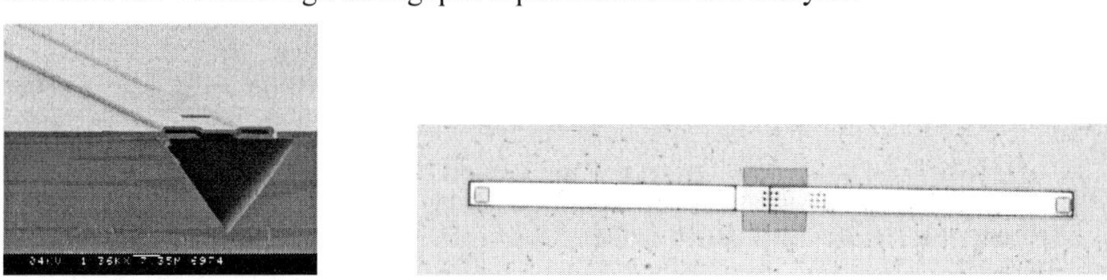

Figure 4. Illustration of IFS technology (left) and optical micrograph of picoliter bubblepump (right).

References

1. see examples in: Proc. Micro Total Analysis Systems 2000, Kluwer Ac., The Netherlands, (2000).
2. S. Terry et al., IEEE Trans Electron. Dev., ED-26, (1979), 1880-1886.
3. A. Northrup et al., Proc. Transducers '93, Yokohama, Japan, (1993), 924-926.
4. A. Manz et al., Sensors and Actuators B1(1-6), (1990), 249-255.
5. B.H. van der Schoot et al., Sensors and Actuators B4(3-4), (1991), 239-241.
6. C. van Rijn et al., Journal of Microelectromechanical Systems, Vol. 6, No. 1, March 1997, 48-54.
7. R. Tijssen et al., Anal. Chem. 58, (1986), 3036-3044
8. J.W. Berenschot et al., Proc. Transducers 2001, Munich, Germany, (2001), 624-627.

MICROCHIP ION TRAP MASS SPECTROMETRY

William B. Whitten, Jeremy Moxom, Peter T. A. Reilly, and J. Michael Ramsey
Chemical and Analytical Sciences Division
Oak Ridge National Laboratory
Oak Ridge, TN 37831-6142
whittenwb@ornl.gov

Abstract
We have been exploring the use of cylindrical ion traps of 1-mm diameter and smaller for mass spectrometry. These miniature ion traps have dimensions that are comparable to the thickness of the substrates used in microfluidic devices and their simple structure is amenable to microfabrication techniques. Operating voltages are lower than for conventional ion traps while RF frequency and working pressure are higher, making miniaturized electronics and vacuum systems possible as well. Mass resolution comparable to or better than for laboratory ion trap mass spectrometers has been demonstrated. The number of ions that can be stored in traps of equal energy depth scales as the radius of the cavity rather than the volume. This property makes two-dimensional arrays of small ion traps attractive for increased dynamic range. Several examples of miniature ion traps, methods of sample ionization, mass analysis, and ion detection will be described.

Keywords: Mass spectrometry, ion trap

1. Introduction

In the more than 10 years since ion trap mass spectrometers were commercially introduced as mass selective detectors for gas chromatographs, they have been shown to have applications as field instruments as well as in the laboratory. Several instrument characteristics have made ion trap mass spectrometers especially powerful for analysis of trace molecular species. First, they can operate at a pressure of up to 10^{-3} torr, several orders of magnitude higher than other mass spectrometers. This results in a proportionally higher sensitivity because more analyte molecules will be present within the trap at a given time. Secondly, the ion trap is a very efficient device for tandem mass spectrometry, or MS/MS. In this process, ions of a particular mass are isolated in the trap by ejecting all other ions, then the remaining ions of interest are collisionally dissociated under conditions where the fragment ions can be trapped and mass analyzed. MS/MS can also be achieved through ion-molecule reactions. The resulting information helps to positively identify the ions present even when the original sample was a complex mixture.

We have recently been exploring the capabilities of micro ion traps of millimeter to submillimeter inner dimension for mass spectrometry applications.[1,2] The micro ion

J.M. Ramsey and A. van den Berg (eds.), Micro Total Analysis Systems 2001, 210–212.

traps are functionally equivalent to their laboratory-sized relatives but operate at higher frequency and lower voltage, so that the electronics package can be considerably smaller in size. Vacuum requirements are reduced because of the smaller chamber size. Mass resolution is comparable to or better than that obtained with the laboratory systems, the micro traps operate at the same or higher pressure, and single ions can be detected. Sensitivity for trace constituents in the atmosphere under various sampling conditions has yet to be determined. We have shown that a two-dimensional array of equivalent ion traps can be operated in parallel for enhanced storage capacity (sensitivity). While the traps studied so far have been assembled from conventionally machined parts, the dimensions and form are amenable to microfabrication for higher precision and eventual integration with the ion source and detector.

2. Experimental

The ion trap electrodes were fabricated as described in Ref. 2 except that the outer diameter was 12.7 mm instead of 25.4 mm so that the assembly would fit in a 1.33-inch mini-Conflat vacuum cube. A rhenium filament served as an electron emitter for sample ionization and a fragment of a microchannel plate was used as an ion detector. In the usual fashion,[3] ions generated by the electron beam were trapped, the range of m/z of the trapped ions depending on the RF and DC voltages applied to the trap electrodes. After an integration time of typically 20 ms, the voltages on the ion trap were scanned so that ions were ejected to the detector at different times depending on their mass to yield a mass spectrum of the initially trapped ions. Mass spectra were obtained by scanning either the RF voltage or a DC voltage applied between the ring electrode and endcaps. Helium was used as a buffer gas at pressures from 10^{-2} torr to greater than 10^{-1} torr. Samples studied included xenon gas and various volatile organic compounds.

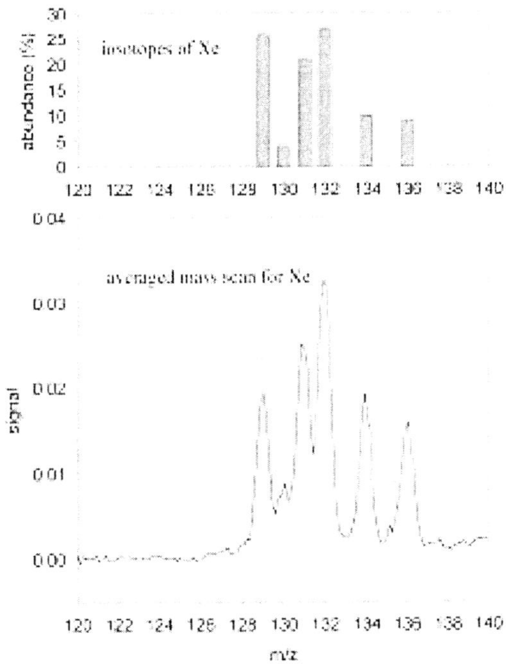

Figure 1. Mass spectrum of xenon isotopes.

3. Results and discussion

An example of a mass spectrum of xenon isotopes is shown in Fig. 1. This spectrum was obtained by decreasing the RF voltage on the ring electrode (DC voltage = 0) so that the ions would selectively be ejected when a nonlinear resonance occurred at $\beta = 0.5$. The spectrum was averaged over several hundred scans, so that isotope ratios could be determined with higher precision than for single scans.

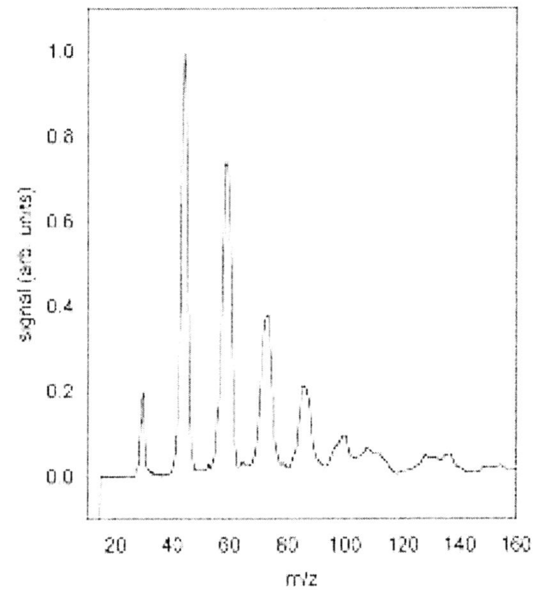

Figure 2. Mass spectrum of dodecane

It was also possible to obtain mass spectra by scanning a DC voltage on the ring electrode while the RF voltage was of constant amplitude. This scan is electronically simpler to implement than the previous scan. A mass spectrum of dodecane obtained in this way is shown in Fig. 2. The various peaks correspond to fragment ions with different numbers of carbon atoms with unresolved attached hydrogen atoms.

Acknowledgements

This research was sponsored by Office of Research and Development, U.S. Department of Energy, under contract DE-AC05-00OR22725 with Oak Ridge National Laboratory, managed and operated by UT-Battelle, LLC.

References

1. O. Kornienko, P. T. A. Reilly, W. B. Whitten, and J. M. Ramsey, *Rapid Comm. Mass Spectrom.*, **13**, 50 (1999).
2. O. Kornienko, P. T. A. Reilly, W. B. Whitten, and J. M. Ramsey, *Rev. Sci. Instrum.* **70**, 3907-3909 (1999).
3. Practical Aspects of Ion Trap Mass Spectrometry, vol. I, R. E. March and J. F. J. Todd, Eds., CRC Press, Boca Raton, 1995.

LOW-COST DISPOSABLE POLYMERIC HIGH-DENSITY NANOVIAL TARGET PLATES FOR MALDI-TOF MASS SPECTROMETRY

Simon Ekström[1], Göran Helldin[2], Johan Nilsson[1]
György Marko-Varga[3] and Thomas Laurell[1]
[1]Department Electrical Measurements, Lund University, Sweden
[2]MicroPlast AB, Box 99, S-532 22 Skara , Sweden
[3]Department of Analytical Chemistry, Lund University, Sweden

Abstract

In order to meet the expected demand for mass spectrometry throughput as a result of the current efforts to map the human proteome, this report presents a new concept for high-throughput protein identification by MALDI-TOF MS peptide mapping using low-cost disposable polymeric high-density nanovial MALDI target plates.

Keywords: nanovial, phosphoprotein, matrix-assisted laser desorption/ionization time-of-flight mass spectrometry (MALDI-TOF MS), proteome analysis

1. Introduction

The nanovial array format has proven to be an attractive form of high-desity MALDI target that has the potential of meeting the on-coming efforts in miniaturizing upstream protein preprocessing methodology. For increased functionality in proteomics work the nanovial array format as such provides a perfect reaction vessel for performing additional processing steps of the sample, at high reaction rates, prior to the MALDI MS-analysis.

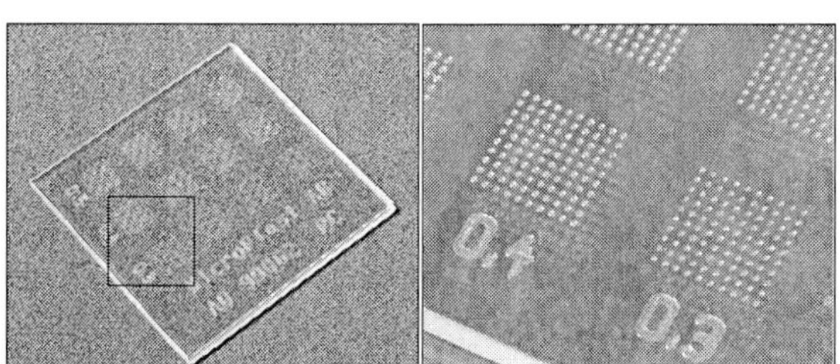

Figure 1. The disposable polymeric high-density MALDI target.

2. Experimental

By means of microfabrication technology precision engineered nanovial arrays was fabricated in polymer substrates such as polymethylmethacrylate (PMMA) and polycarbonate (PC), Figure 1. The target plate fabrication processes investigated were precision micromilling, cold embossing and injection moulding.

J.M. Ramsey and A. van den Berg (eds.), Micro Total Analysis Systems 2001, 213–215.

Investigated nanovial dimensions were 300, 400 or 500 µm, with array densities of 165 nanovials/cm^2. Deposition of samples into the polymeric high-density target plates can be accomplished either by conventional sample deposition using small dimensional pipettes, capillaries, or by using piezoelectric dispensing technology as described earlier by our group [1-3].

3. Results and discussion

The nanovial format allows for "in-vial" sample manipulations, such as digestion of intact proteins. By performing an in-vial dephosphorylation of a nanovial fractionated peptide separation potentially interesting phosphorylated peptides could be identified by the loss of 80 Da. Figure 3 shows an in-vial dephosphorylation of 10 fmol β-Casein by microdispensing of alkaline phosphatase that enabled rapid and sensitive identification of a phosphorylated peptide.

The mass spectra obtained from these polymeric targets was also used in high sensitivity applications, such as peptide map identification of human fibroblast proteins separated by 2D-gel electrophoresis, Figure 4.

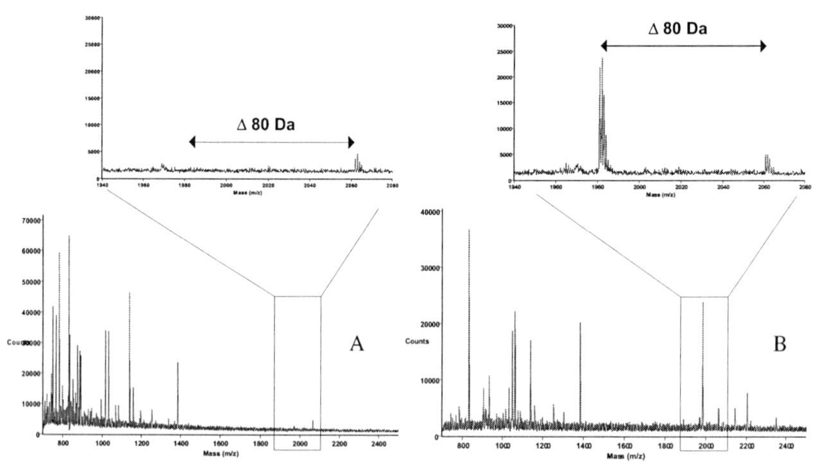

Figure 3. Analysis of phosphoprotein. A) Mass spectra obtained from the untreated β-Casein and B) mass spectra obtained from a nanovial containing < 10 fmol β-Casein after an in-vial dephosphorylation by microdispensing of alkaline phosphatase (0,1 unit/µL in 50 nM NH_4HCO_3, pH 7,8) at 6 Hz for 60 seconds (corresponding to 36 nL alkaline phosphatase). During this treatmeant the target plate was heated to 40° C. The deposited enzyme was then allowed to dry for 5 minutes followed by microdispensing of 10 nL matrix. The insert of the peak in B) shows the dephosphorylated peptide, generated by the loss of a phospate group (-80 Da) through the enzymatic treatment. The same insert for the peptide peak in A) exhibits only the phosphorylated state.

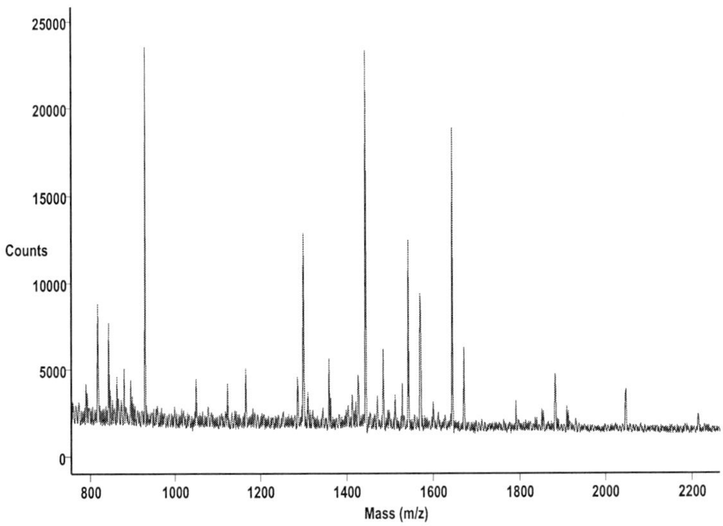

Figure 4. Protein identification by peptide mass fingerprinting. The protein was identified as kinesin (Accession No. Q07866).

4. Conclusions

The presented polymer target plates was proven to generate similar high-quality peptide map fingerprint MALDI-data as standard re-usable gold, or stainless steel target plates. The disposable concept also minimizes the risk for contamination and provides a clean surface for the matrix crystallization process. The use of in-vial dephosphorylation of a phosphoprotein as outlined in this report enables phospho-position identification. As earlier reported by our group [2] on-line capillary liquid chromatography is easily interfaced with nanovial fractionation. The use of an automated in-vial dephosphorylation step on a nanovial fractionated peptide separation would thus enable rapid identification of potentially interesting phosphorylated peptides by the loss of 80 Da.

References

1. Ekström S., Önnerfjord P., Nilsson J., Bengtsson M., Laurell T. and Marko-Varga G., Anal. Chem., 72, (2000) 286-293
2. Miliotis T., Kjellström S., Nilsson J., Laurell T., Edholm L-E. and Marko-Varga G., , J. Mass Spectrom. 35 (2000) 369-377
3. Ekström S., Ericsson D., Önnerfjord P., Bengtsson M., Nilsson J., Marko-Varga G. and Laurell T., Anal. Chem., 73 (2000) 214-219

Integrated Sample Preparation and MALDI MS on a disc

Anders Palm, Susanne R Wallenborg, Magnus Gustafsson, Anders Hedström, Ebru Togan-Tekin, Per Andersson
Gyros AB, Uppsala Science Park, SE-751 83 Uppsala, Sweden

Abstract
A microfluidic platform based on a CD format has been developed for parallel sample preparation prior to matrix assisted laser desorption ionization MS (MALDI MS). Samples are desalted and concentrated on a solid phase and eluted onto a target area defined within the CD. Sensitivities in the high amol range have been achieved for MS analysis of in-solution tryptic digests.

Keywords Sample preparation, MALDI MS, protein analysis, centrifugal force, CD

1. Introduction

The demand for a reliable sample pre-treatment method for low abundance proteins prior to identification by mass spectrometry is well recognized. A number of concentration techniques have been exploited [1,2]. To date, an analytical system for routine preparation of protein digests at concentrations approaching the detection limit for mass spectrometry, is missing. Microfluidic systems on-line with MS would be an intriguing approach for this purpose. However, most MS couplings to chip-based devices have been based on electrospray ionization (ESI), probably due to the compatibility of ESI with the continuous flow of liquid used in most chip-based systems which are hydrodynamically or electroosmotically driven. However, the CD format utilizes discrete volumes and is thus well suited for integration to MALDI MS.

This work will demonstrate a CD based microfluidic technique for rapid sample purification and concentration of peptides in a tryptic digest, followed by protein identification using MALDI MS. The sample preparation procedure, including clean-up and concentration, is based on the interaction between the peptides and a solid phase chromatographic medium followed by elution into a defined MALDI target area within the CD. This technique enables routine concentration/desalting of protein digests close to the detection limit of commercial mass spectrometers.

2. Experimental

The CD master was manufactured by deep reactive ion etching and converted to a mould insert by a nickel electroplating process. Injection moulding was used to manufacture CDs in various polymeric materials such as polycarbonate, zeonex and zeonor. Prior to use the CDs were functionalized using the following steps; (1) patterning with gold using a conventional two-layer sputtering process, (2) treatment in an oxygen plasma to create a hydrophilic surface, (3) deposition of hydrophobic breaks and (4) heat-bonding of a laminate material (DuPont, UK). The microstructures were packed with 15 μm SOURCE RPC (Amersham Pharmacia Biotech, Uppsala, Sweden).

J.M. Ramsey and A. van den Berg (eds.), Micro Total Analysis Systems 2001, 216–218.

Peptides from the packed bed were eluted using 50% acetonitrile containing 4-hydroxy-α-cyano cinnimatic acid below saturation. Figure 1 shows the experimental process including particle packing and equilibration of the capturing material, sample application, washing of the packed bed and elution of peptides.

All fluidic experiments were performed using the instrumentation and software developed at Gyros AB (Uppsala, Sweden) including devices for handling microtiter plate and high precision spinning and robotics for liquid transfer. Prior to MS analysis the CD was cut to fit the ion-source of commercial mass spectrometers.

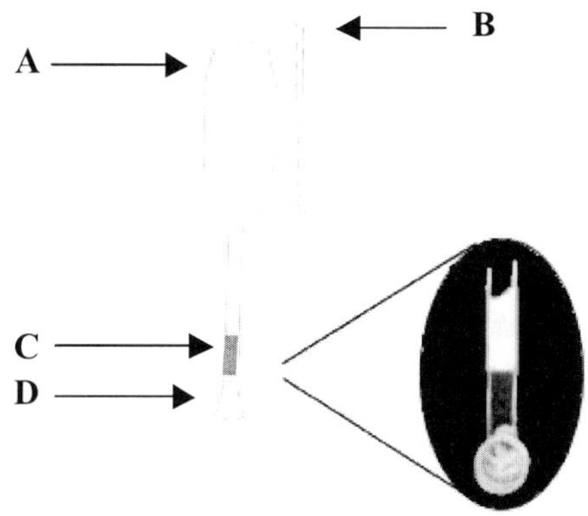

Figure 1. Microfluidic structure for integrated sample preparation and MALDI MS. The microstructures are enclosed by a foil containing access holes for fluid introduction and sample presentation (shaded areas). (A) Port used for addition of packing material, sample application and introduction of conditioning-, washing-, and elution liquids, (B) Air-vent, (C) chromatographic column and (D) MALDI MS target area.

3. Results and discussion

The CD microfluidic technique, employing centrifugal force as the fluid driving mechanism, is well suited for handling large number of discrete nL to μL samples in parallel [3]. Since this microfluidic system allows the use of discrete volumes, it is well suited for MALDI applications which require the evaporation of solvent to generate co-crystallized sample and matrix.

Each MALDI CD contained 48 to 96 microfluidic structures. Each structure was packed with reversed phase chromatographic beads to give a large number of parallel separation columns with a volume of 1-20 nL (figure 2). Each column was capable of processing a large sample volume (0.5-5 μL) as required in proteomic applications to achieve high sensitivity. The liquid flow within the CD was controlled through a combination of hydrophobic breaks deposited in the microchannels and geometric valves, the latter being based on single or double U-bend structures [4]. As peptide elution requires the use of solvents containing organic solvents with low surface tension properties, methods to control the flow of such fluids within the microstructures were evaluated. For aqueous solutions, volumes of 36 nl ± 0.3%, were defined within the CD

itself (confidence interval of 95%, n=92). By varying the rotational speed of the CD, the peptide adsorption to and desorption from the solid phase columns could be controlled.
Figure 2 shows the MALDI MS spectra obtained before and after sample preparation in the CD. Clearly the sample clean-up using the CD microfluidic technology allowed detection of the digested protein while the unprocessed sample could not be analyzed.

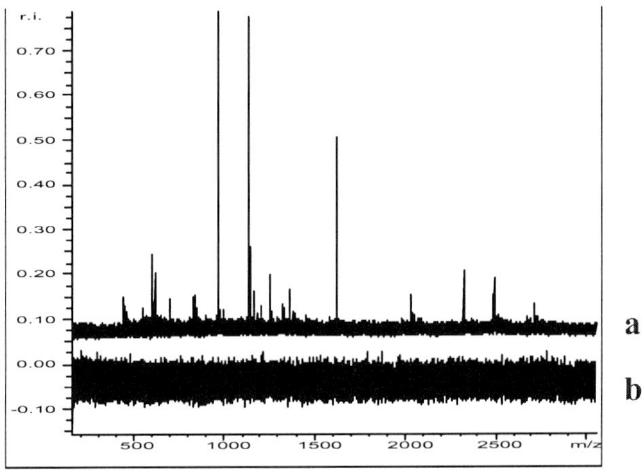

Figure 2. Comparison of MALDI MS spectra obtained for a trypsin digest of ADH (alcohol dehydrogenase) on the CD after clean-up(a) and a stainless steel target (b).

5. Conclusions

The CD-based sample preparation presented here makes it possible to accurately identify proteins at low concentrations and to process many samples in parallel. Further steps will be included in the same CD to increase the functionality.

Acknowledgements

Prof. Hans Jörnvall, Prof. Tomas Bergman, Dr Carina Palmberg and Dr. Daniel Hirschberg at Karolinska Institute for sample preparation and helpful discussions. Dr Martin Schuerenberg at Bruker Daltonik, Germany for fruitful collaboration.

References

[1] Schuerenberg, M., Luebbert, C., Eickhoff, H., Kalkum, M., Lehrach, H. and Nordhoff, E. (2000) *Anal Chem* **72**, 3436-42.

[2] Ekström, S., Onnerfjord, P., Nilsson, J., Bengtsson, M., Laurell, T. and Marko-Varga, G. (2000) *Anal Chem* **72**, 286-93.

[3] A. Eckersten *et al.* Proc. mTAS 2000, Kluwer Academic Publisher, Dordrecht, The Netherlands, pp 521-524, (2000)

[4] G. Ekstrand *et al.* Proc. mTAS 2000, Kluwer Academic Publisher, Dordrecht, The Netherlands, pp 521-524, (2000)

MICROCHIP INTEGRATED ANALYSIS SYSTEM FOR ELECTROSPRAY MASS SPECTROMETRIC ANALYSIS OF COMPLEX PEPTIDE MIXTURES

Iulia M. Lazar and Barry L. Karger
Barnett Institute/Northeastern University, 360 Huntington Avenue, Boston, MA 02115, USA

Abstract

A comprehensive characterization of expressed proteins in a cell lysate or organism, today represents a high priority goal requiring high-throughput instrumentation and rapid analysis protocols. In this presentation, we report on the development of new microfabricated analytical platforms interfaced to electrospray ionization mass spectrometry (ESI-MS) that facilitate on-line analysis of peptide samples. In particular, liquid phase separation elements, microdigestion units, pumping units and electrospray elements were fabricated on a unique microchip platform, and the over-all performance of the integrated system evaluated.

Keywords: microchips, ESI-MS, peptides

1. Introduction

The analysis of protein expression and function for a given cell, tissue or microorganism must provide answers to questions such as: the identity of the present proteins, the level of their expression, the nature and site of posttranslational modifications, the nature and localization of interactions with other proteins, and the specific functions associated with particular proteins. Typically, a protein mix analysis protocol involves a 2D-gel electrophoresis separation step of the protein extract, enzymatic digestion of the separated protein spots, and MS or liquid chromatography (LC)-MS of the generated peptide fragments. Alternative protocols involving batch labeling and digestion of the whole protein mix followed by restrictive affinity isolation and enrichment of labeled peptides, fractionation, and LC of fractionated peptides, were developed.

The final stage of MS detection involves the examination of hundreds of peptide fractions and a parallel processing approach would be highly desirable. Microfabricated devices with integrated operational elements qualify as ideal candidates for inexpensive, disposable, and automation amenable instrumentation, capable of parallel sample processing. Our research focused on developing novel microfabricated devices that integrate sample processing elements necessary for the analysis of peptide or protein mixtures. A fully integrated microanalysis system consisting of a micropump, injection element, separation channel, digestion chamber and ESI-interface element was sought (**Figure 1**). In the case of our prior microfabricated devices that performed infusion or capillary electrophoresis experiments, fluid flows generated by electrical fields were stabilized with the aid of pressure gradients (gas pressures, vacuum and syringe

J.M. Ramsey and A. van den Berg (eds.), Micro Total Analysis Systems 2001, 219–221.

pumping). In this research, we have given priority to stand-alone configurations that did not necessitate external instrument assistance.

2. Experimental

Microchip devices were fabricated in-house from glass using photolithography and wet chemical etching technologies. Fluid flows, electrically or pressure driven, were visualized and optimized with a Nikon epi-fluorescence microscope. Electrospray from the chip was generated by inserting nanospray emitters (5-8 μm i.d. at the tip) in the microchip body. Mass spectra were acquired with a Mariner (Applied Biosystems) or Q-Tof (Micromass) instrument at acquisition rates of 0.5-2 spectra /s. Peptide samples were generated by co-digesting a mixture of various proteins with trypsin in electrosprayable buffers (ammonium bicarbonate).

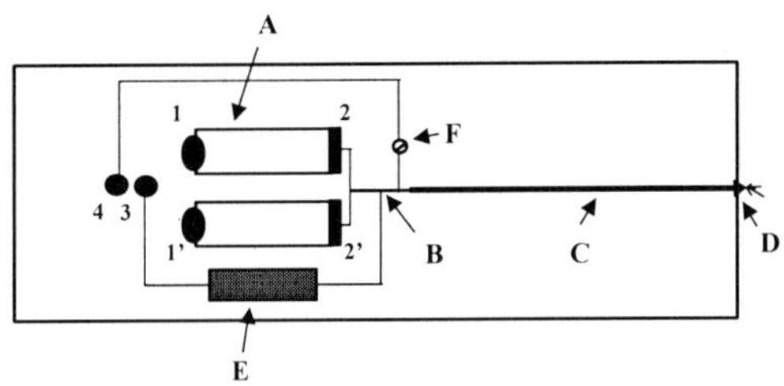

Figure 1. Microchip integrated analysis system. (A) pumping unit; (B) injection element; (C) infusion or separation channel; (D) ESI interface; (E) digestion unit; (F) sample waste valve; 1, 1', 2, 2' 3, 4 – reservoirs.

3. Results and Discussion

The presently developed microfabricated analysis system provides an efficient approach to deliver peptide samples for mass spectrometric analysis. Peptide mixtures generated off or on the chip, are separated in a short channel (3-5 cm) prior to ESI-MS detection. The separation channel can be utilized to perform LC, capillary electrochromatography and capillary electrophoresis, or simple infusion experiments; however, emphasis was placed on developing a micro-LC configuration. Open or packed channel configurations were investigated. Stand-alone operation of the micro-LC system was achieved by integrating a pumping unit on the microfabricated platform. An open channel micropump based on electroosmotic pumping principles [1] was constructed and evaluated. It consists of multiple open channels with small diameters, that all deliver electroosmotic flow (EOF) into a larger channel (the separation channel). The ratio of the forward flow to the electroosmotic flow (F/F_{eof}) generated in the pump will depend on the number of pumping channels and on two ratios, d_1/d_2 and L_1/L_2 (d_1/L_1 and d_2/L_2 are the diameter and length of the pump and separation channel, respectively). Micropumps consisting of 4, 25, and 100 pumping channels were evaluated (**Figure 2**). Flow rates up to 100-150 nL/min and eluent gradients were generated, and it was estimated that pressures as high as 100-150 psi could be produced with certain pump configurations. Even though packed channel electroosmotic pumping systems can generate pressures of thousands of psi [2], the advantages of the open channel configurations are numerous: straightforward fabrication, trouble free operation, and stable fluid delivery without bubble formation as often occurs in packed channels. Generated pressures are sufficient for most operations performed on the chip.

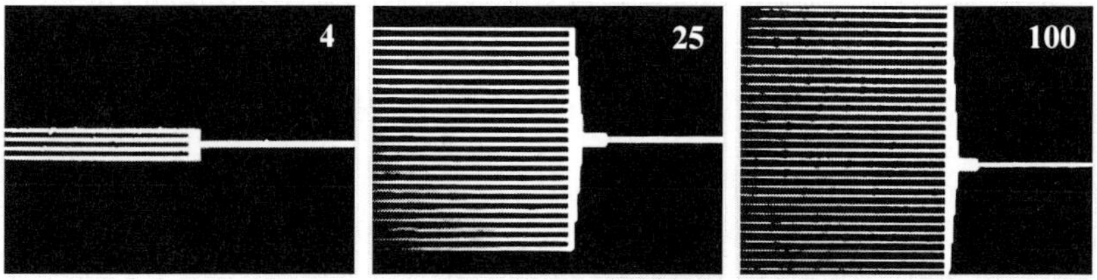

Figure 2. Photographs of micropumps with 4, 25 and 100 pumping channels.

In addition, the on-line integration and performance of a protein microdigestion chamber was investigated. Peptide mixtures containing over a 100 components (**Figure 3**) were generated by simultaneous digestion of a mixture of proteins and were used to test the over-all performance of the microchip device. Total sample consumption for stable operation of the microfluidic device was a few pmoles (approximately 5 microliters), and MS detection of a few fmol of protein was accomplished.

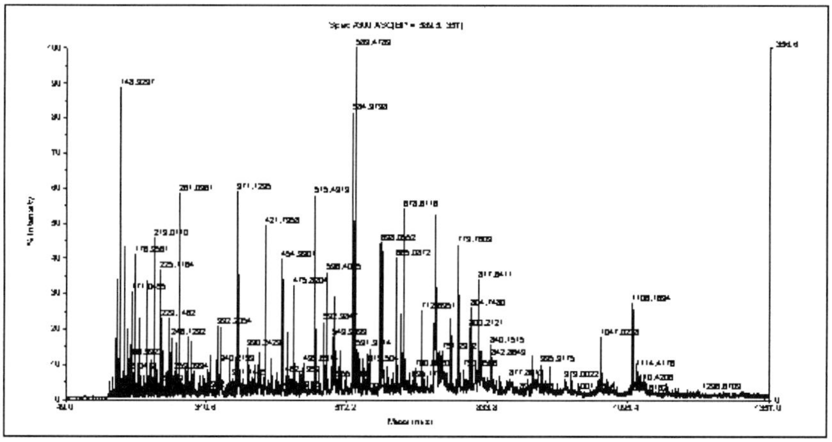

Figure 3. Mass spectrum of a protein mixture digest (approximately 1 μM each).

Acknowledgment

This work was supported by the John Hatsopolous Scholar Fund.

References

1. Pretorius, V.; Hopkins, B. J.; Schieke, J. D. *J. Chromatogr.* **1974**, *99*, 23.
2. Paul, P. H.; Arnold, D. W.; Rakestraw, D. J. *Proceedings of the Micro Total Analysis Systems '98 Workshop*, p. 49-52. Banff, Canada, 13-16 October, **1998**.

ON-CHIP POST-COLUMN DERIVATISATION REACTIONS IN CAPILLARY LIQUID CHROMATOGRAPHY - MASS SPECTROMETRY

Valerie Spikmans[1], Stephen J. Lane[2], Bill Leavens[3], Andreas Manz[1] Norman W. Smith[1].

1. Zeneca / SmithKline Beecham Centre for Analytical Sciences, Department of Chemistry, Imperial College of Science, Technology and Medicine, South Kensington, London, SW7 2AY, UK.
2. GlaxoSmithKline, Analytical Sciences, Gunnels Wood Road, Stevenage, Herts, SG1 2NY, UK.
3. GlaxoSmithKline, DMPK, Park Road, Ware, Herts, SG12 0DP, UK

Abstract

A micromixer chip used previously for organic reaction chemistries [1] is applied for post-column derivatisation in μ-HPLC hyphenated to a Time-of-Flight mass spectrometer (TOFMS).

Keywords: Micromixer chip, post-column derivatisation, mass spectrometry, micro-high performance liquid chromatography (μ-HPLC)

1. Introduction

For the first time a microchip has been successfully introduced for post-column derivatisation in μ-HPLC hyphenated to a time-of-flight mass spectrometer (TOFMS)[2]. Reactions took place in about 15 seconds. In this μ-HPLC-UV-chip-MS set-up, the analytes are separated using μ-HPLC and subsequently derivatised on-chip and detected.

The optimal flow rates in μ-HPLC and MS are compatible with flow rates for the chip and instead of MS detection, UV(/MS) and fluorescence(/MS) detection are applicable, demonstrating the flexibility of the system. The system is robust and easily fully automated.

2. Experimental

Figure 1 shows a silicon-machined micromixing device, which has been developed in our laboratory and is based on the principle of distributive mixing between two inlet flows [3]. This platform allows rapid mixing of reagents and, coupled to the MS, fast analysis of the reagent/product mixture [4]. The integration of the micromixer chip into the system is shown in Figure 2. The outlet of the μ-HPLC column is connected to one inlet and the reagents for the reaction are continuously infused into the other inlet of the chip.

J.M. Ramsey and A. van den Berg (eds.), Micro Total Analysis Systems 2001, 222–223.

Figure 1: Bottom view of the silicon-machined micromixing device, where (a) and (b) are the inlets and (c) is the outlet of the chip. [3]

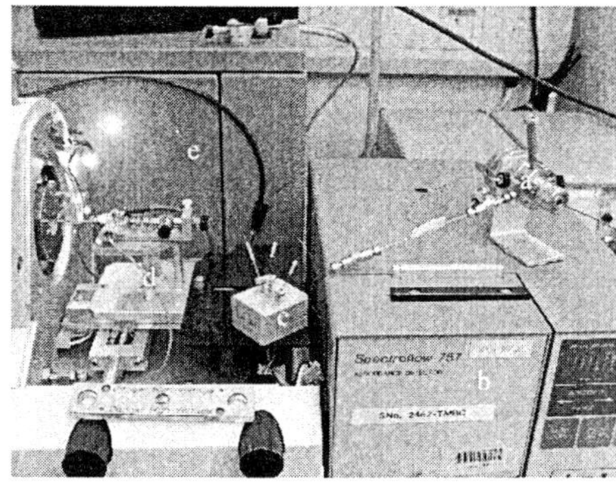

Figure 2: μ-HPLC-UV-chip-MS set-up. (a) injector and μ-HPLC column, (b) UV-Vis detector, (c) micromixer chip in clamp, (d) electrospray interface and (e) Mariner TOF mass spectrometer.

3. Results and discussion

Examples of μ-HPLC separations of amines with their products after derivatisation will be shown during the presentation. An enhancement in speed and product formation is achievable by changing the conditions and set-up of the system. These future ideas will also be presented.

4. Conclusions

The micromixer chip has been successfully introduced for post-column derivatisation reactions in μ-HPLC. The distributive mixing of the chip allows for very fast mixing of reagents and therefore fast chemical reactions, even coupled to a separation technique.

Acknowledgements

The authors would like to thank Vas Nittis from GlaxoSmithKline for the development of the micromixer clamp and Applied Biosytems/MDS Sciex and GlaxoSmithKline for funding towards the Mariner TOF-MS. The authors would also like to thank Zeneca and SmithKline Beecham for funding the analytical centre at Imperial College. This work was supported by a PhD scholarship from GlaxoSmithKline Stevenage.

References

1. M.C. Mitchell, V. Spikmans, A.J. de Mello, *Analyst*, 2001, **126**, 24
2. V. Spikmans, S.J. Lane, B. Leavens, A. Manz and N.W. Smith, *Rapid Commun Mass Spectrom., in press.*
3. F. G. Bessoth, A. J. de Mello and A. Manz, *Anal. Commun.*, 1999, **36**, 213
4. M.C. Mitchell, V. Spikmans, A.Manz and A.J. de Mello, *J. Chem. Soc. Perk. T. 1*, 2001, **5**, 514

A MICRO MASS SPECTROMETER

G. Petzold, P. Siebert, J. Müller
Technische Universität Hamburg-Harburg,
Department of Micro-Systems-Technology
Eissendorfer Str. 42, D-21071 Hamburg, Germany

Abstract

This paper presents the concept, the processing and the simulated and measured characteristics of a Miniaturized Mass Spectrometer (MMS), with dimensions of only a few cm^3 approximately.
The MMS based on a micromachined ion source [1], which delivers an high ion current density in comparison with customary ion sources. Therefore even a mass separator can be miniaturized.
Its fabrication is based on techniques used in micro-system processing and in particular anisotropic etching, thin film deposition, electroplating, and anodic bonding.
In the presented design of the MMS the resolution is in the range of 10 to 20 at a sensitivity of several tens of ppm.

Keywords: Micro Mass Spectrometer, Microwave Plasma Ion Source, UV-LIGA

1. Introduction

Presently available mass spectrometers mostly use relative bulky mass separators. Large dimensions of several tens of centimeter length demands low pressure of 10^{-4} mbar or less to prevent ion collision, which requires a two-stage pump system at least. Thus the pumping system accounts for a considerable share of the overall size and costs.
The above mentioned disadvantages can be greatly reduced by using a mass spectrometer fabricated as a micro-system. The small separator length of 1.5 mm facilitates the mass separation at a pressure of about 1-10 Pa, hence requirements for the vacuum system are tremendously reduced.
The use of standard integrated circuits for the generation of the deflection voltages – even for high field strength due to the small dimensions of a micro-system – greatly enhances portability.
These advantages allow the use of micro mass spectrometers for many existing and potential applications, e.g. process monitoring in microelectronics, aerospace, and optical coating industries as well as in pollution monitoring even though they do not fulfil the characteristic data of the macroscopic mass spectrometers.

2. Design and working principle

As shown in fig. 1 the mass spectrometer consists of three parts which are fabricated separately on different substrates: The plasma electron source, the ion source including an electron optic and the ion optic and the separator.
The electrons are first extracted from the plasma, focused in the electron optic into a small region in the ionisation chamber and ionize the measurand gas by impact. The

J.M. Ramsey and A. van den Berg (eds.), Micro Total Analysis Systems 2001, 224–226.

ionisation chamber is free of electric fields except for the small ion extraction potential introduced by a small slit of 50 μm in the chamber. Therefore the ion beam meets the requirements of the mass separator for nearly mono energetic ions. After extraction from the ionisation chamber the ions are focused by the ion optic.

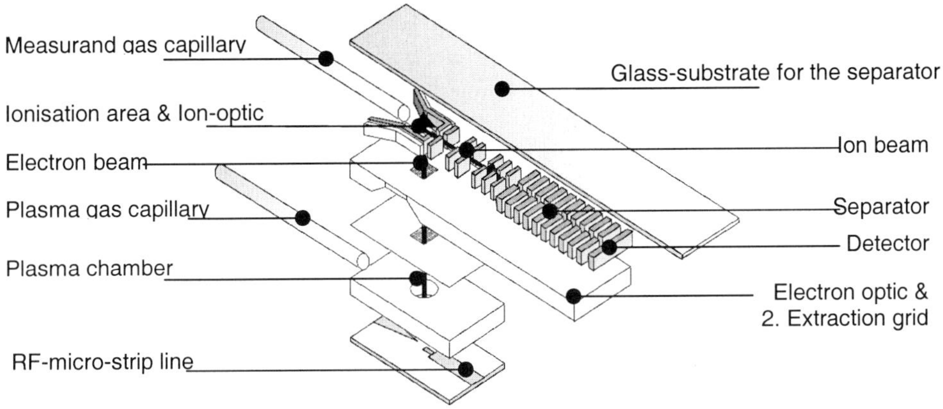

Figure 1: Schematic of the Micro Mass Spectrometer and its sub-units

Since high field strengths are attained at low voltages of about only 10 V at these dimensions specially suited separation signals – even more complex and of higher frequency – can be readily achieved by the use of standard IC's [2].

The mass filter is velocity selective combining the principles of time-of-flight and trajectory stability. A dipole travels along the channel which changes its polarity from one electrode pair to the next.

Ions of the same speed like the travelling dipole are guided through the separator on a sinusoidal trajectory (see fig. 2). Ions faster or slower than the dipole reach static field areas and are neutralized at the electrodes.

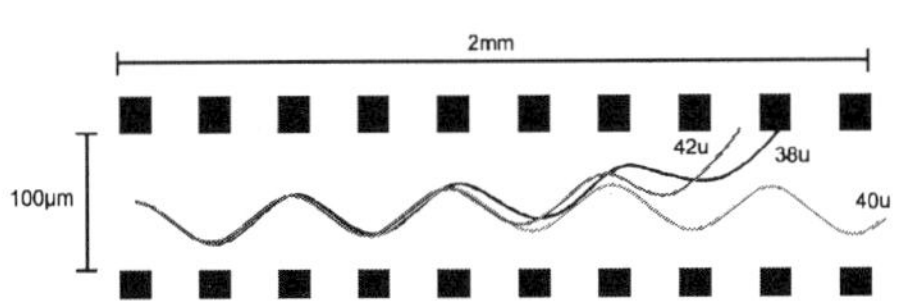

Figure 2: Simulation of mass-resolution with SimIonV6.0.

3. Manufacturing of the Mass Separator

To fabricate such a separator, a rather new process technology known as the 'UV-LIGA' [1,3] is used. These structures are electroplated in an Epon SU-8 form. In fact different shapes and heights of micro structures can readily be achieved by this technology. In contrast to deep etched silicon structures the parasitic capacitance's are kept low since the whole structure is built of metal while glass is used as the substrate material.

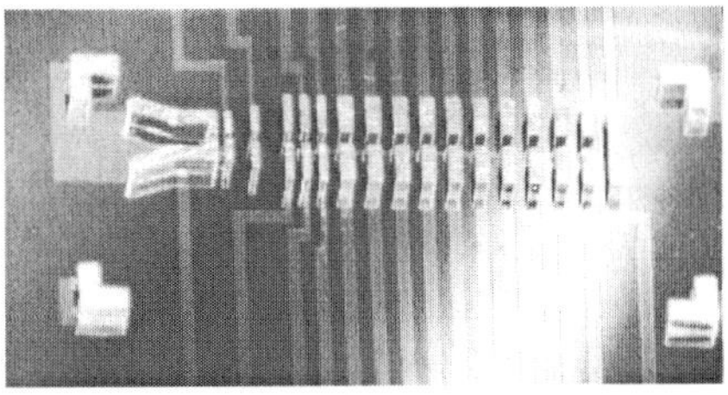

Figure 3: Photo of the mass separator

The ionisation chamber has an inner diameter of 200 μm, the electrodes of the ion optic have a width of 50 μm while the adjacent and opposite electrodes of the separator have a distance of 100 μm. The height of all electrodes is 150 μm.
A photo of the separator is shown in fig. 5.

4. Simulations and measurements of the performance

As already mentioned the MMS based on the electron beam ion source [1]. The extractable ion current is about 3 μA. The Simulation on the ion optic shows an efficiency of 60 % which corresponds to the measured overall ion current on the detector.
Measurements of the separation performance have not been carried out as yet. The resolution (mass/Δmass) of the micro mass spectrometer depends on the amplitude of the travelling dipole field. For an amplitude of 10 V the resolution is about 20 which decrease down to 10 for an amplitude of 2 V.
The same counts for the transmission of the separator. With a rising potential the transmission decrease, i.e. 10 % for 2 V and 1 % for 10 V amplitude voltage.
Based on this simulated data the detection limit of the presented micro mass spectrometer design could be estimated as follows: With a lower limit of a detectable ion current of 1 pA – since a simple electrode and an integrator circuit is used as detector – and a transmission of 1% up to 10 % the sensitivity of the micro mass spectrometer is about 100 or 10 ppm, respectively.

5. Conclusions

A concept for a micro mass spectrometer and its realization in micro-system technology is presented in this paper. The specific advantages of miniaturization as generation of high field strength at low voltage and operation of the mass spectrometer under fine vacuum are exploited.
The resolution of the presented design of a micro mass spectrometer is in the range of m/Δm = 10 to 20 while the detection limit is in the order of several tens of ppm. Since single channel plate channels have dimensions in the range of microns, the integration seems feasible, and thus the detection limit can further be enhanced.
Due to the small size and the low power and gas consumption the MMS is perfectly suitable for any self sufficient mobile analysis systems, e.g. pollution monitoring, emission control or online leak detection of gas pipelines.

Acknowledgements:
This work was supported by the European Union.

References

[1] G. Petzold, P. Siebert, J. Müller: *A micromachined electron beam ion source*, Proc. μTAS 2000, Kluwer Academic Publisher, Dordrecht, The Netherlands, pp. 171-174

[2] P. Siebert, G. Petzold, J. Müller: *Surface microstructure / miniature mass spectrometer,* Applied Physics A, 1998.

[3] H. Lorenz et al: *A low cost negative resist for MEMS,* J. Micromech. Microeng. 7, p. 121-124, 1997.

MOBILE FLOW CONTROL ELEMENTS FOR HIGH-PRESSURE MICRO-ANALYTICAL SYSTEMS FABRICATED USING *IN-SITU* POLYMERIZATION

J. E. Rehm[*†], T. J. Shepodd, E. F. Hasselbrink[**]
Sandia National Laboratories, Livermore, CA, USA

[*] Eksigent Technologies, 2021 Las Positas Court Suite 161, Livermore, CA, USA
[**] Dept. Mechanical Engineering, University of Michigan, MI USA
[†]jerehm@eksigent.com

Abstract

A method for rapidly fabricating a family of robust fluid control elements in microfluidic channels is presented. The polymer devices are lithographically defined *in situ* in glass microfluidic channels in a few seconds on a benchtop. The devices are capable of controlling fluid flow in microchannels at pressures exceeding 5000 psi (340 bar) and can be actuated in milliseconds. In this work we demonstrate chip-based devices, including a piston, check-valve, and a 10 nanoliter pipette.

Keywords: microvalve, microfluidic control, microfluidics, polymer, valve

1. Introduction

Integration and advanced functionality of μTAS have been fundamentally limited by the inability to incorporate fluid control devices with microfluidic channels. The problem is difficult because of the range of requirements for a practical microdevice: it may be required to provide excellent sealing against very high pressures or be compatible with electroosmotic flow (*e.g.*, nonconductive); at the same time, it must be relatively inexpensive and process-compatible. Although significant effort has been placed in developing microscale versions of traditional valves from an array of materials [1], recent work in 'soft' valves [2,3] appears promising. Here we present a simple and inexpensive means for making a new class of "soft" microscale flow control devices.

2. Fabrication

Devices are photo-polymerized inside an existing glass or silica microfluidic channel, where a monomer-solvent-initiator mixture has been introduced (Fig. 1). A typical formulation used in the examples presented here is as follows: 1,3 butanediol diacrylate and trifluoroethylacrylate (mixed 1:1) as the monomer (80%), 2-methoxyethanol, 1,4 dioxane, and 5 mM TRIS buffer (mixed 4:4:1) as the solvent (20%), and 2,2'-azobisisobutyronitrile (0.5 wt.%) as the photoinitiator. Cross-linking is initiated within a selected region of microchannel by UV exposure through a chrome mask (10-45 seconds from a 355 nm Nd:YAG laser at 1 mJ cm^{-2}/pulse, 10 Hz), with about 3μm resolution. Resolution is limited by the competition between radical diffusion and initiation/recombination reactions.

J.M. Ramsey and A. van den Berg (eds.), Micro Total Analysis Systems 2001, 227–229.

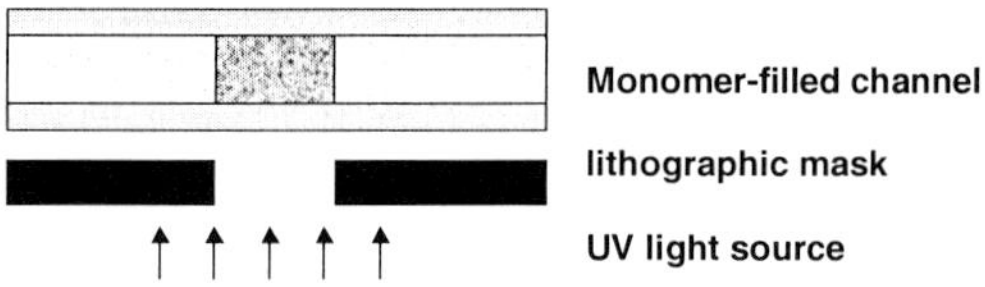

Figure 1. Channels filled with liquid monomer are exposed to a UV light source through a lithographic mask. Photo-polymerization is initiated in the exposed region.

The polymerized element does *not* adhere to the channel walls, creating a small "piston" that can be mobilized by applying pressure. Modifying the initial mixture allows the behavior and functionality of the device to be tailored to a specific application, as even modest variations in formulation can have dramatic effects on the polymer structure and its mechanical properties. In general, this method allows us to produce a polymeric monolith that conforms to the shape of the microchannel and does not bond to surrounding structures with excellent repeatability.

3. Experimental Results

Various combinations of polymer and channel geometry lead to a wide variety of devices. For example, Figure 2 shows a variable-area piston, which successfully seals against 10 bar of air pressure. Figure 3 shows a check-valve configuration, which only allows unidirectional flow due to the location of the piston relative to a bypass channel.

These closed-cell structures have high compressive strength, yet are flexible enough to negotiate 90° turns without damage. Compressive tests indicate most elements have a modulus of elasticity similar to a common rubber stopper (0.01-0.1 GN/m^2). Initial pressure and leakage tests have been performed in silica capillaries and glass chips in a configuration similar to that shown in Fig. 3. Maximum pressures exceed 340 bar, and actuation times are less than 1 video frame (33 msec). Actuation pressure can be less than 1 psi, depending strongly on UV exposure during polymerization as shown in Fig. 4. The leakage rates of these devices vary between 0-50 pL/min over a wide range of pressures (0-

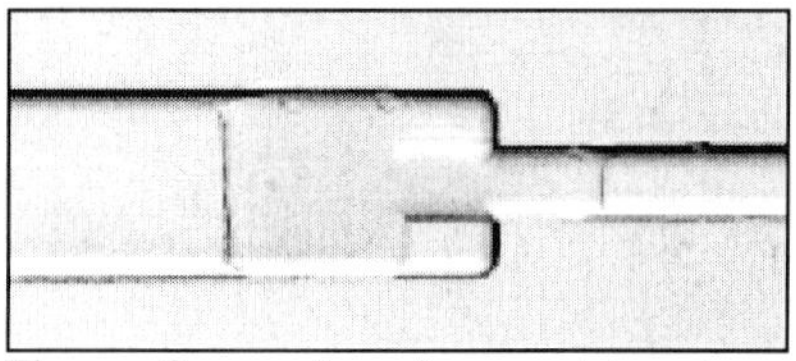

Figure 2. A polymer piston in a microfluidic channel (150 μm wide, narrowing to 50 μm wide, 25 μm. deep). The piston is displaced slightly to the left of its sealing surface. Pistons have been demonstrated to seal against high pressures in liquid and gas environments.

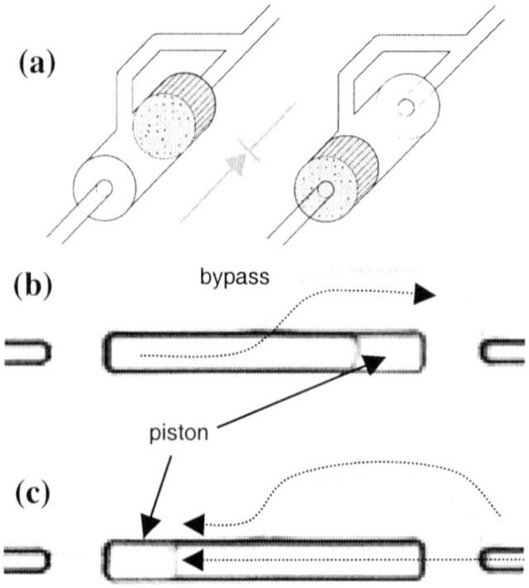

Figure 3. On-chip check valve. (a) Schematic of valve architecture. Both glass wafer surfaces are HF etched to produce interconnecting channels of concentric cylinders when bonded. (b) Flow from left-to-right bypasses the piston seated against the right stop. (c) Flow in opposite direction is prevented when the piston seals against the left stop, 'checking' the flow. Cylinder diameter is 100 μm with 25 μm weirs. Flow direction is indicated by dashed lines.

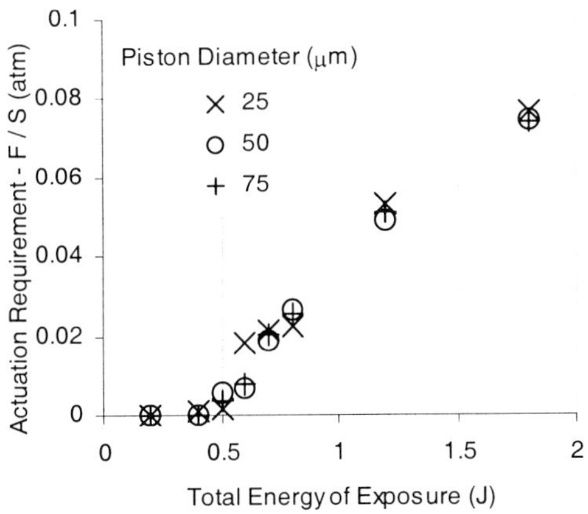

Figure 4. Actuation requirements (force divided by surface area in contact with channel wall) for polymer pistons as a function of initiation energy. Element length is 1 mm.

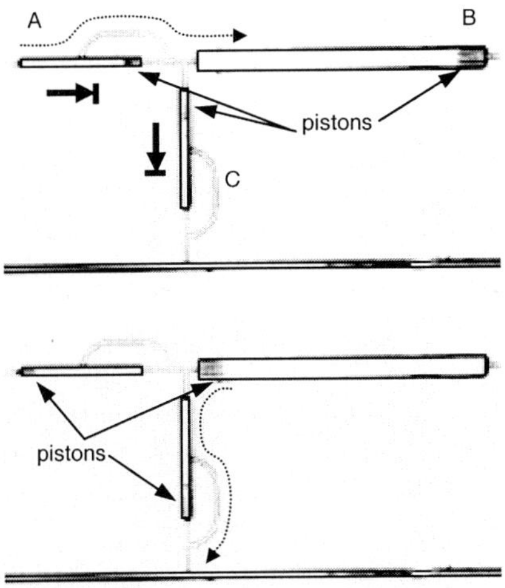

Figure 5. A 10 nanoliter pipette for aliquoting sample into a channel. Dashed lines indicate fluid flow direction. *Top:* Suction is applied at inlet B and the piston in cylinder B moves to the right (while drawing sample from inlet A through open check valve) until the stop is reached. Valve C remains closed. *Bottom:* Pressure is applied at inlet B; A closes, C opens, and the fluid in cylinder B is injected into the channel.

300 bar). As the pressure increases, the sealing behavior improves (leakage decreases), as the polymer deforms to compensate for imperfections in the sealing surface.

Complex functions can be performed with a sequence of dependent elements as shown in Fig. 5, where a piston combined with two check-valves form a pipette, used to aliquot a 10 nanoliter sample volume into a running LC column. A wide range of pipette sample volumes (picoliter to microliter) can be defined by chip geometry.

5. Conclusions

A method of fabricating a family of fluid control elements has been developed that enables the development of highly integrated high-pressure microfluidic systems. These devices are process-compatible with existing chip architectures that allow paths for optical access. The devices are fabricated rapidly and inexpensively, outside of a clean-room, in a single post-microfabrication step.

References

1. Shoji, S., and Esashi, M., Microflow Devices and Systems, *J. Micromech. Microeng.* 4(4), 1994, 157-171.
2. Beebe, D. J., Functional hydrogel structures for autonomous flow control inside microfluidic channels, *Nature* 404, 2000, 588-589.
3. Unger et al., Monolithic microfabricated valves and pumps by multilayer soft lithography, *Science* 288, 2000, 113-116.

LOW-VOLTAGE ELECTROOSMOSIS PUMP AND ITS APPLICATION TO ON-CHIP LINEAR STEPPING PNEUMATIC PRESSURE SOURCE

Yuzuru Takamura, Hiroyuki Onoda, Hiromichi Inokuchi, Sakuichiro Adachi, Akio Oki, and Yasuhiro Horiike
Dept. Material Science, School of Engineering, The Univ. of Tokyo
7-3-1 Hongo, Bunkyo-ku, Tokyo 113-8656, Japan

takamura@micro.mm.t.u-tokyo.ac.jp

Abstract

Low-voltage EOF pumps which consist of narrow gap channels and cascade configuration were microfabricated on quartz chips. A pressure of 500 mmH_2O (~5000Pa) was obtained at 40V using a 120nm gap single stage pump, and it was confirmed that the pressure increase with cascade stage numbers without accumulating voltage. For an example of applications of these pumps, a linear stepping actuator was demonstrated as a pressure source of stand-alone pneumatic-driven microfluidics chips.

Keywords: Electroosmosis flow, low voltage pump, salt bridge, photopolymerize gel, actuator

1. Introduction

Electroosmosis flow (EOF) is widely used for liquid transportation on microfluidics devices. It requires, however, hazardous high voltage, such as a few kV, which prevents not only easy handling but also 1) direct drive by normal IC level signal, 2) co-fabrication on a Si chip with peripheral circuit due to poor insulation ability of thin oxide film, 3) handy or wearable usage with small battery power source, and 4) uses for maximum safety applications, such as healthcare device, which is currently being developed by our group to detect human health markers from a trace of blood [1].

It is well known that pumping pressure of EOF is approximately inversely proportional to square of capillary diamete [2], so decreasing capillary diameter can reduce operating voltage without degrading pumping ability of pressure and velocity.

In this work, low-voltage EOF pumps with very narrow gaps or cascade configuration were fabricated on quartz chips and their pumping abilities were evaluated. As one of the applications of these pumps, a linear stepping actuator was also demonstrated as a pressure source of stand-alone pneumatic-driven microfluidics chips.

2. Experimental

A wide (1mm) width, short (100μm) length, and narrow (0.1~10μm) height gap channel were fabricated by dry-etching process on quartz chip as EOF pump (Figure 1).

J.M. Ramsey and A. van den Berg (eds.), Micro Total Analysis Systems 2001, 230–232.

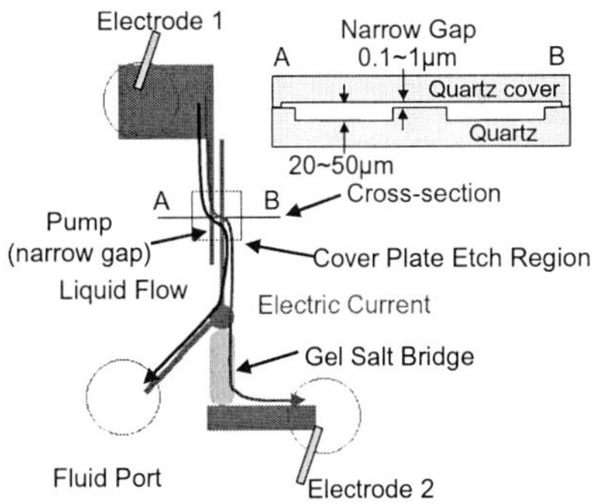

Fig 1. Narrow gap electroosmosis pump fabricated by planer process.

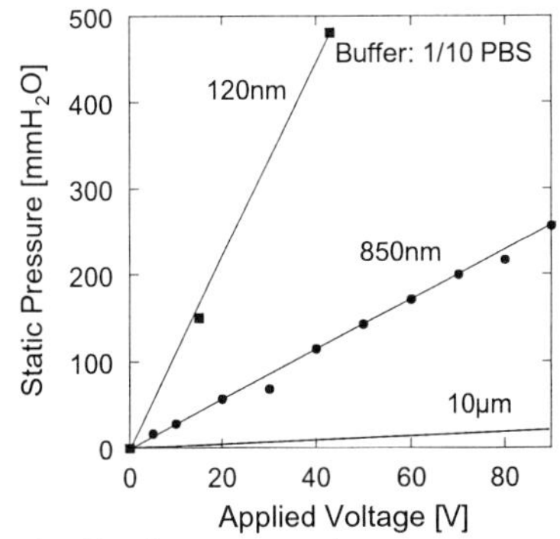

Fig 2. Static pressure by narrow gap EOF pumps at various gap sizes.

For electric contact without pressure leak, photo-polymerizing gel was micro-fabricated as salt bridge in the chip by photolithographic technique. The flow in the channel was monitored by the motion of low zeta-potential polystyrene beads added in operating liquid of 1/10 concentration Dulbecco's phosphate buffer solution (1/10 PBS). Static pressure was measured by balanced counter pressure applied to fluid port in figure 1 when flow in the path to the fluid port was stopped.

3. Results and discussion

Figure 2 shows static pressures by narrow gap EOF pump, which increase with decrease of gap size, and 500 mmH_2O (~5000Pa) was obtained at 40V with 120nm gap. Static pressure of 40 mmH_2O or flow velocity of 200μm/s with 1cm capillary load were obtained at 20V in 850nm gap pump, which is considered to be enough for typical microfluidics applications. The power consumption at this condition was 0.1mW, which is 1/1000 of 10μm gap case for same pressure.

Figure 3 is a photograph of fabricated cascade EOF pump, in which pairs of a narrow gap (5μm) high pressure EOF pump and a wide gap (50μm) low pressure pump with reverse direction were sequentially connected by flow paths with gel electrodes. Each pair has same electric potential at an inlet and an outlet, so it can multiply pressure without accumulation of voltage. No partial inner wall treatment was required. Figure 4 shows static pressures measured by same method as figure 2, together with EOF velocities at no counter pressure, for 6 and 15 stage cascade pumps. The velocities were same in both stage numbers while the pressure for 15 stages was larger than that for 6 stages. These results indicate that the pressure can be successfully multiplied up

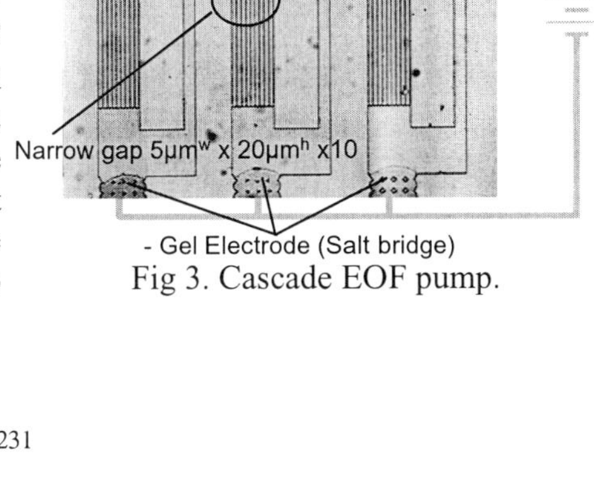

Fig 3. Cascade EOF pump.

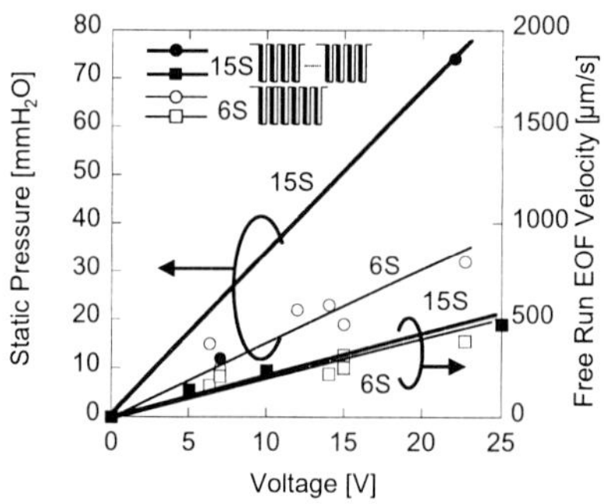

Fig 4. Pressure and velocities for 15 and 6 stage cascade EOF pump.

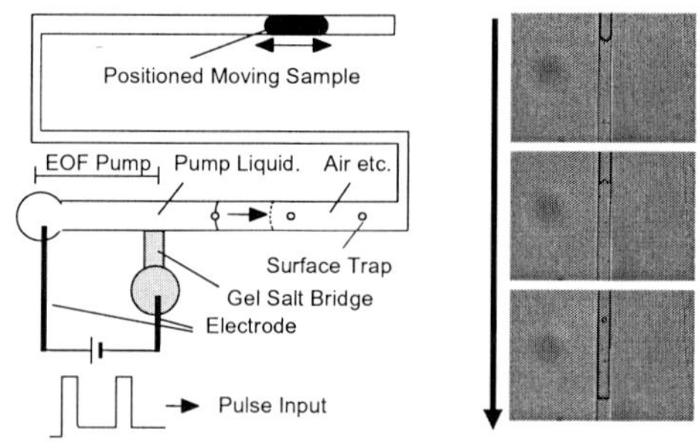

Fig 5. Schematic of linear stepping actuator and its step operations by sequential pumping pressure pulse.

to desired level by simply increasing the number of stages.

These pumps generate pressure so effectively to be used as actuators. Figure 5 illustrates an on-chip air pressure source to moving droplet in remote flow channel. Such pneumatic-driven method is a powerful tool for microfluidics devices [3,4]. However, in the most cases, the pressure sources are located outside chips. Using this work, the source can be integrated on a chip as a stand-alone device. Moreover, we added open-loop positioning mechanism like stepping motors by putting surface traps (pillar) in the channel at moving part of liquid-air surface. The surface was trapped at one of the traps, and successfully moved to the next trap by a pulse input larger than trapping force as shown also in figure 5.

4. Conclusions

Low voltage electroosmosis pumps were developed using narrow gap channel. The combination of narrow gap and wide gap enables the cascade connection of the pumps without accumulating the operation voltage. An integrated linear stepping pneumatic pressure source was demonstrated using these pumps.

References

1. A. Oki, et. al, *Electroosmosis injection of blood serum into biocompatible microcapillary chip fabricated on quartz plate*, Electrophoresis, vol. 22, pp. 341-347, (2001).
2. P. H. Paul, et al., *Electrokinetic Pump Application in Micro-Total Analysis Systems Mechanical Actuation to HPLC*, Proc. Micro Total Analysis Systems 2000, Kluwer Academic Publisher, Dordrecht, The Netherlands, pp. 583-590, (2000).
3. K. Hosokawa, et al., *Formation and Active Mixing of metered nano/picolitter liquid droplets in a microfluidic device*, Proc. Micro Total Analysis Systems 2000, Kluwer Academic Publisher, Dordrecht, The Netherlands, pp. 481-484, (2000).
4. M. Burns, et al., *An Integrated Nanolitter DNA Analysis Device*, Science, vol. 282, pp. 484-487, (1998).

MULTI-MICRO PUMP CHIP TO FLOW GASES AND LIQUIDS

Koji Ikuta, Takao Adachi and Tadahiro Hasegawa
Dept. of Micro Engineering, School of Engineering, Nagoya University
Furo-cho, Chikusa-ku, Nagoya 464-8603, JAPAN
E-MAIL: ikuta@mech.nagoya-u.ac.jp

Abstract

We fabricated the optimized SMA micro pump chip applicable to both liquids and gases. The measured maximum flow and pressure for water were 12 μl/min. and 25 kPa, respectively. And the flow and pressure for air were 11 μl/min. and 5 kPa, respectively. Moreover, the micro pump satisfied insensitive ability to gas bubbles and self-priming. This performance was realized by improving the check valves, the chamber and the SMA actuator design. Our original 3D micro fabrication process named "Hybrid IH process (micro stereo lithography)" to hybridize various micro solid parts with UV-polymer structure was utilized for chip fabrication. Therefore, their packaging without leakage was made successfully.

Keywords: micropump chip, shape memory alloy, Hybrid IH process, Biochemical IC family.

1. Introduction

The authors have been conducting a versatile micro fluidic device named "Biochemical IC Family" for these years [1-3]. As shown in Fig.1, The biochemical IC chip contains different functional devices such as multiple connectors, pumps, one-way valves, reactors, switching valves, chemical concentrators. The total micro chemical system can be constructed by stacking various kinds of chips. Recently, though a large number of studies have been made on miniaturization of chemical components such as channels and reactors in the fields of "micro-TAS" (total analysis system), "Laboratory on the chip" and "DNA chip", these micro devices demand external apparatus such as pumps, valves, sensors for practical use. Accordingly, the total experimental setup with these micro devices becomes larger. On the other hand, the aim of our research is to miniaturize a total micro chemical system, which integrates micro fluidic parts and electronic parts. Our approach is useful for "implantable" micro chemical devices. Therefore, the biochemical IC demand the micro fluidic chip, which can supply the micro fluidics stably.

The micropump chip actuating by the shape memory alloy (SMA) have been already fabricated

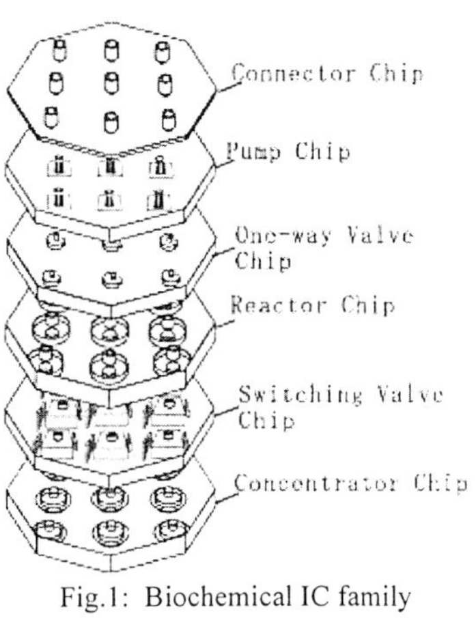

Fig.1: Biochemical IC family

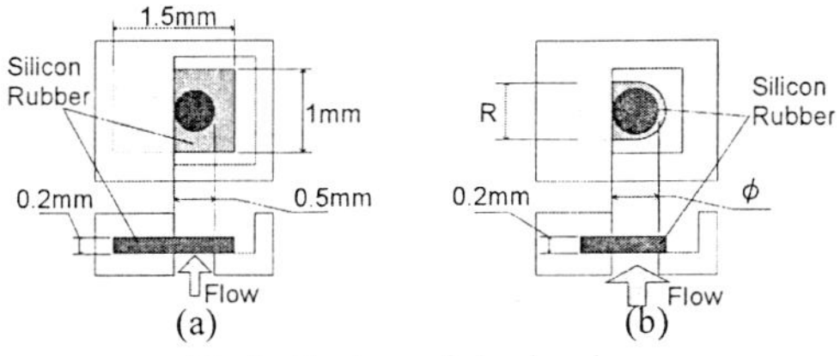

Fig.2: Designs of check valve
(a) Former design (b) Improved design

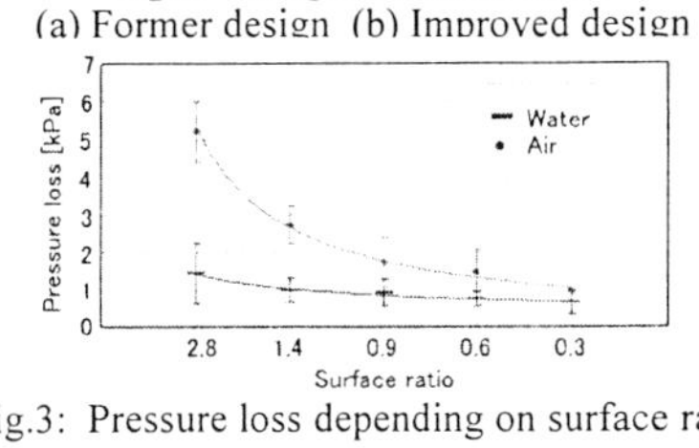

Fig.3: Pressure loss depending on surface ratio

J.M. Ramsey and A. van den Berg (eds.), Micro Total Analysis Systems 2001, 233–235.

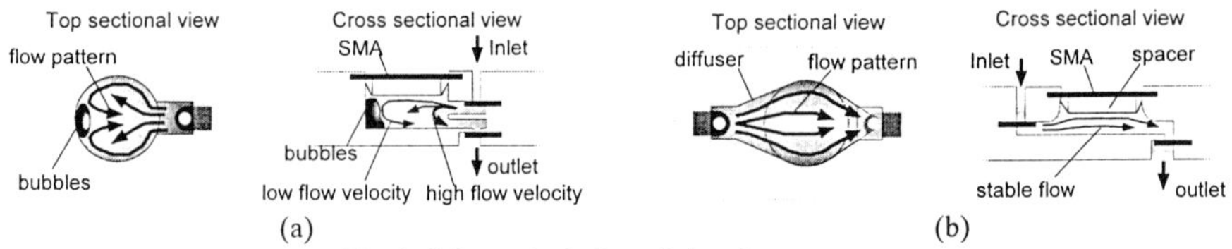

Fig.4: Schematic design of chamber geometry
(a) Former chamber (b) Improved chamber

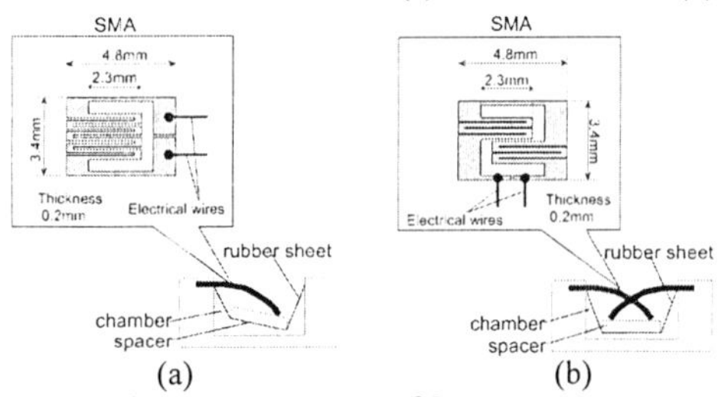

Fig.5: Improvement of SMA actuator
(a) Former design (b) Improved design

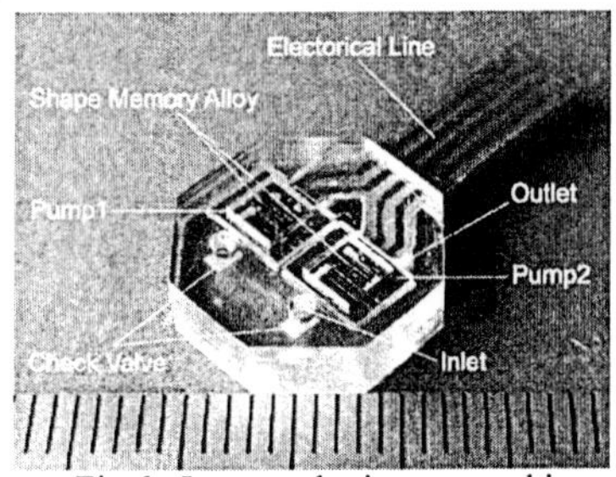

Fig.6: Improved micropump chip

[3]. Each pump has reciprocating pumps, consisting of two check valves, a deformable chamber, a spacer and a SMA actuator. The measured maximum flow and pressure for water were 9.4 ml/min. and 15 kPa, respectively, and its basic performances in the modular micro device system were demonstrated. At the next stage, the micropump should be improved to achieve high performance with stable driving for practical applications [4].

In this paper, we present the optimization of the pump design by improving design to satisfy insensitive ability to gas bubbles and self-priming. Because of the pumping principle, three elements of the micropump were optimized as follows: 1) Pressure loss of the check valves, 2) Flow pattern of the chamber, 3) Compression ratio of the chamber. As a result, these improvements enable us to flow gases as well as liquids.

2. Optimizing design

2.1. Reducing pressure loss of the check valves

Fig.2 shows the check valve designs, which consist of a silicone rubber seat to hybridize with UV-polymer structure. This valve is laid out the cantilever beam. The adhesive force, which is proportional to the contact area of the silicone rubber, affected the pressure loss while opening. Therefore, the ratio between the contact area of the silicone rubber and the cross-sectional area of the channel (surface ratio) was decreased, by means of modifying the silicone rubber design (Fig.2(b)).

The results presented in Fig.3 demonstrate that the pressure loss decreases while surface ratios decrease. Especially, pressure loss decreases remarkably for air, and low pressure loss for the opening function succeeded.

2.2. Optimizing flow pattern inside the chamber

Fig.4(a) shows the former chamber design, which consist of the chamber in a circular shape. Two check valves were laid out at upper and lower sides for miniaturization. This design introduced the problem of trapping gas bubble due to unstable flow patterns, which deteriorated the pump performances. Thus, the chamber in a diffuser shape was modified as shown in Fig.4(b). Two check valves were laid out at both ends of the diffuser chamber. Bubbles can pass through the chamber smoothly due to stable flow pattern.

2.3. Increasing compression ratio of the chamber

The compression ratio was also improved despite the same stroke length of the SMA actuator. The depth of the diffuser chamber got thinner since two check valves were laid out as shown in

Table 1: Experimental result of micropump chip

	Maximum flow	Maximum pressure
Liquid(water)	12.3 μ l/min.	25kPa
Gas(air)	11.0 μ l/min.	5kPa

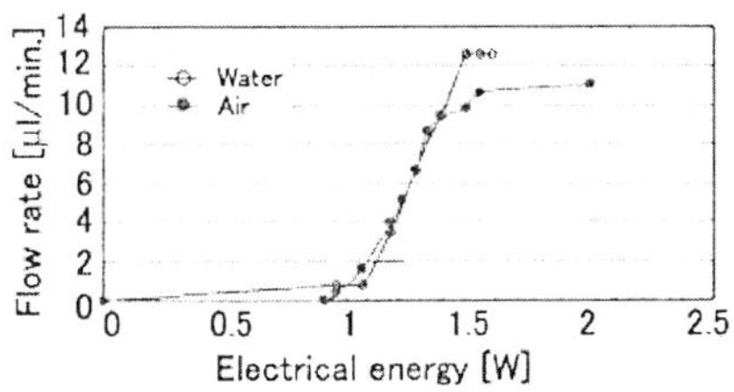

Fig.7: Pump flow rate depending on electrical energy

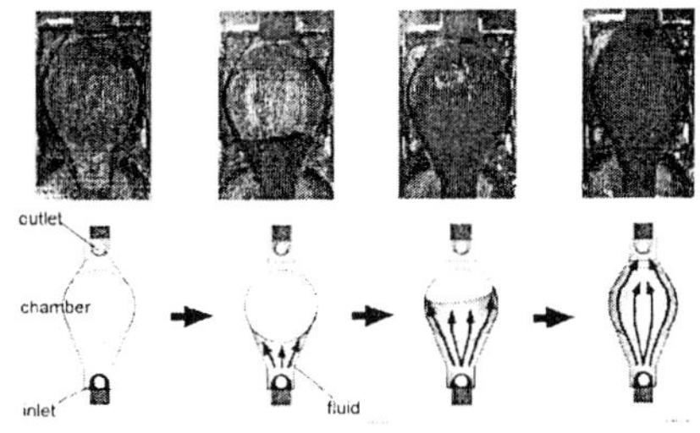

Fig.8: Sequential photographs for sucking liquid
(a) the chamber filled with air
(b)(c) sucking form air to water into the chamber
(d) the chamber filled with water

Fig.4(b). Moreover, SMA cantilevers were laid out on both sides to push down the spacer stably (Fig.5(b)). The former SMA design, as shown in Fig.5(a), cause low volumetric change of the chamber. As these improvements, the compression ratio increased by approximately twice.

3. Fabrication

The our original technique "Hybrid IH process" to make a hybrid structure was utilized for the chip fabrication [2]. This "Hybrid structure" means a composite structure combining UV polymer with other solid parts such as electronic parts. During polymer solidification process, any micro parts such as SMA, silicone rubbers and so on are inserted in turn, and the unsolidified liquid UV polymer can be washed out by rinsing. Since no additional bonding process is needed, chip packaging without any leakage can be made easily. Fig.6 shows the micro pump chip, which contains two improved micro pumps. The shape and size of the chip is an octagon with a diameter of 14 mm.

4. Experimental results

First, flow rate and outlet pressure of the micropump was measured. The measured maximum flow and pressure for water were 12 μl/min. and 25 kPa, respectively (Table.1). The flow for air (11 μl/min.) was almost similar value for water. The pressure for air was 5 kPa. Fig.7 shows the flow rate as a function of the electrical energy. This result suggests that a linear control ability of flow rate for both liquid and gases.

Next, The demonstration of the micro pump, sucking from gas to liquid subsequently with the same pump also succeeded (Fig.8). A pump chamber was filled with air. The pumping gas sucked the water through the pump inlet. The pump then successfully continued the water supply.

5. Conclusions

We fabricated a high performance micro pump chip with optimized design. This chip can flow not only liquids but also gases (12 μl/min., 25 kPa for water. 11 μl/min., 5 kPa for air). Moreover, the micro pump is tolerant of gas bubbles in the liquid chamber and it is possible to prime itself. To achieve this performance, the micro pump improved three elements. The first one is eliminating pressure loss of the check valves. The second one is optimizing flow pattern inside the deformable chamber. And the third one is increasing compression ratio of the deformable chamber. This micropump chip is useful for drug delivery system, poisonous gas sensor and so on.

References

[1] K. Ikuta, Artificial Life, MIT Press. , pp.17-23 (1996).
[2] K. Ikuta, S. Maruo, T. Fujisawa and A. Yamada, MEMS99, pp.376-381 (1999).
[3] K. Ikuta, T. Hasegawa, T. Adachi and S. Maruo, MEMS 2000, pp.739-744 (2000).
[4] R.Linnemann, P.Woias, C.D.Senfft and J.A.Ditterich, MEMS98, pp.532-537 (1998)

ACCURATE MEASUREMENT OF DIELECTROPHORETIC (DEP) MOBILITY OF PARTICLES AND MACROMOLECULES

Carl D. Meinhart and Dazhi Wang
Department of Mechanical & Environmental Engineering
University of California – SB, Santa Barbara, CA 93106
(meinhart@engineering.ucsb.edu)

Abstract

Micron-resolution Particle Image Velocimetry (μ-PIV) is used to measure the motion of nominally 1-μm dia. polystyrene particles suspended in a water/sugar solution in a microdevice. Particle motion is induced by an applied electric field through dielectrophoretic (DEP) forces and through viscous interactions with the fluid. The motion of the fluid is governed by electrothermal forces arising from temperature-induced gradients in the electrical properties of the fluid near the electrodes and through the viscous interaction with the particles. Comparisons between finite element calculations and experimental micro-PIV results confirm quantitatively that electrothermal forces play a major role in the motion of the particles.

Keywords: dielectrophoresis, micro-PIV, microfluidics

1. Introduction

The goal of this research is to extend micron-resolution particle image velocimetry (μ-PIV) techniques to measure spatially and temporally resolved dielectrophoretic-induced motion of polystyrene particles and macromolecules in flowing microchannels [1]. The results from these experiments will determine DEP forces on particles as a function of their motion and spatial position. DEP has been used in a variety of BioMEMS applications, including sorting cancer cells and capturing DNA molecules [2, 3]. Most of the DEP force models simplify the particles to be either spheres or shells. In addition, the viscous drag force is usually modelled as a rigid sphere flowing through a uniform fluid. In many practical applications, such as transporting coiled DNA molecules through microchannels using DEP, these assumptions are not valid. In addition to DEP forces, electrothermal effects can create body forces on the fluid, which can cause the particle to move in a pattern inconsistent with DEP forces [4].

2. Finite element simulation of electric forces

The electric field was calculated in a wedge-shaped geometry by solving $\nabla^2 V = 0$ using Femlab™ software. Fig. 1 shows the voltage potential in the wedge device. The DEP force is calculated by taking the gradient of the magnitude square of the E-field. The force field of F_{DEP} is shown in Fig. 2, for negative DEP. If the particles were only influenced by F_{DEP} and the fluid was stationary, then for steady conditions the DEP force

J.M. Ramsey and A. van den Berg (eds.), Micro Total Analysis Systems 2001, 236–238.

would balance the viscous drag force, and the particle motion would follow the DEP force field shown in Fig. 2. According to Ramos et al. [4], it is also important to account for the electrothermal forces on the fluid. These forces are a result of temperature gradients produce by Joule heating of the fluid. The temperature gradients create gradients in the dielectric and conductivity properties of the fluid, which in the presence of an electric field create a localized body force on the fluid. The fluid motion due to electrothermal effects is estimated following Ramos et al. [4], and is shown in Fig. 3.

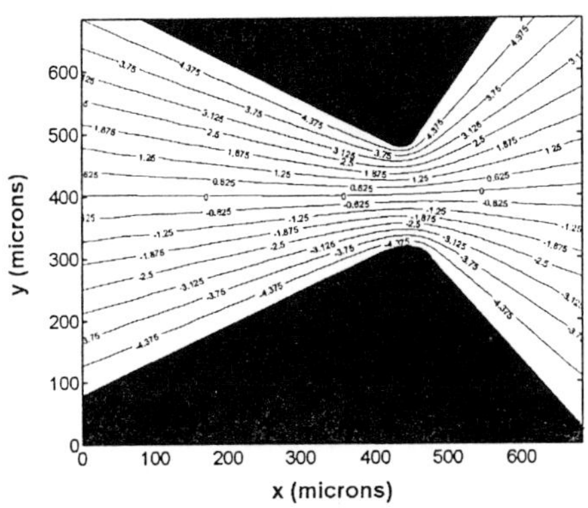

Fig 1. Voltage potential of the DEP wedge device calculated using Femlab™ software.

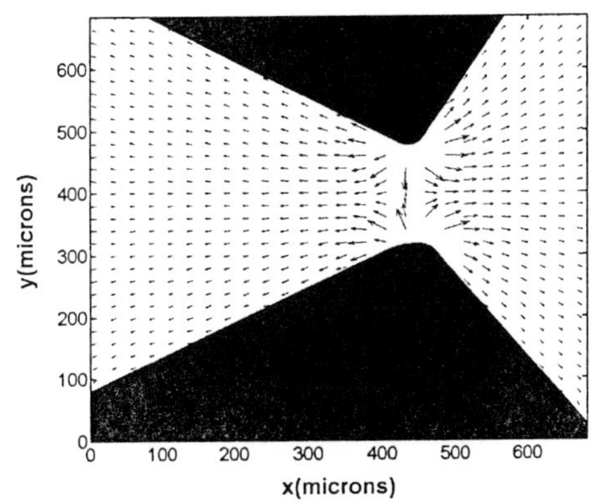

Fig 2. Negative gradient of the magnitude square of the electric field calculated using Femlab™ software. The vectors depict, to within a constant, the force field due to negative dielectrophoresis.

3. Micro-PIV Measurements of particle motion

The trajectories of 1-μm diameter polystyrene particles embedded in a 8.5% sucrose and 0.3% dextrose solution was measured in the wedge device. The device consisted of two brass electrodes sandwiched between two glass wafers. A voltage of 10V RMS at 10 kHz was applied to the electrodes. The Clausius-Mossotti factor is K=-0.5 for the current conditions, producing a negative DEP force. The particle-velocity field is measured quantitatively using μ-PIV and shown in Fig. 4, following Meinhart et al. [1]. The velocity field is characterized by a saddle point at the center of the wedge, with four recirculating vortices. The motion of the particles closely resembles the motion of the fluid due to electrothermal effects. This indicates that electrothermal effects play a significant role in the fluid motion, and the particle motion through viscous coupling.

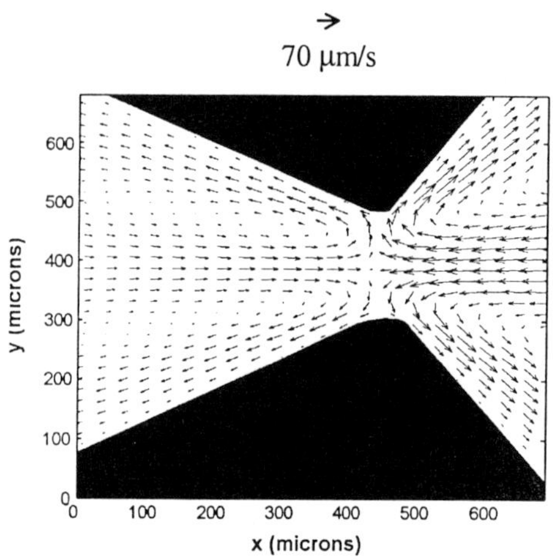

Fig 3. Velocity field of the fluid motion resulting from electrothermal forces (estimated using finite element simulation).

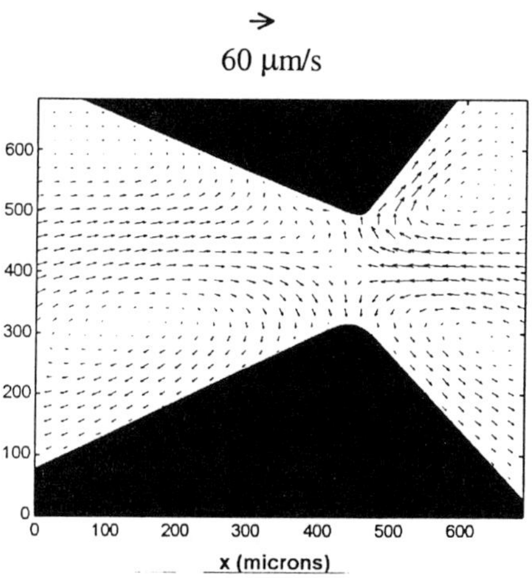

Fig 4. Micron-resolution PIV velocity field of 1-μm dia. polystyrene particles under the influence of dielectrophoresis and viscous coupling to the underlying fluid motion.

5. Conclusions

Micron-resolution Particle Image Velocimetry (μ-PIV) is used to measure accurately trajectories of 1-μm dia. particles under the influence of an alternating electric field. The PIV results agree closely with computer simulations of fluid motion resulting from electrothermal forces near the electrodes. For the experimental parameters considered here, the electrothermal forces appear significant compared to the DEP forces, and therefore should be incorporated in computer simulation models, and in the design of microfluidic devices.

Acknowledgements

This work is supported by DARPA/ARMY DAAD 19-00-1-0400 and DARPA/Air Force F30602-00-2-0609.

References

1. Meinhart, C.D., S.T. Wereley, and J.G. Santiago *PIV Measurements of a Microchannel Flow.* Exp. in Fluids, Vol. 27, pp. 414-419, (1999).
2. Yang, J., Huang, Y., Wang, X-B, Becker, F., and Gascoyn, P., *Cell separation on microfabricated electrodes using dielectrophoretic/gravitational field-flow-fractionation*, Anal. Chem., Vol. 71, pp. 911-918, (1999)
3. Miles, R., P. Belgrader, K. Bettencourt, J. Hamilton, S. Nasarabadi, *Dielectrophoretic manipulation of particles for use in microfluidic devices*, MEMS-Vol. 1, Microelectromechanical Systems (MEMS), Proceedings of the ASME International Mechanical Engineering Congress and Exposition, Nashville, TN, Nov. 14 – 19, (1999).
4. Ramos, A., Morgan, H., Green, N.G., and Castellanos, A., *Ac electrokinetics: a review of forces in microelectrode structures*, J. Phys. D: Appl. Phys. Vol. 31, pp. 2338-2353, (1998).

TEMPERATURE MEASUREMENT AND CONTROL IN MICROFLUIDIC SYSTEMS

David Ross, Michael Gaitan and Laurie E. Locascio
National Institute of Science and Technology
100 Bureau Dr., Gaithersburg, MD 20899, USA
(david.ross@nist.gov)

Abstract

We describe a technique for the measurement of fluid temperatures in microfluidic systems based on temperature-dependent fluorescence. The technique is easy to implement with a standard fluorescence microscope and CCD camera. The efficacy of the method is demonstrated by measuring temperature distributions resulting from Joule heating in a variety of microfluidic circuits that are electrokinetically pumped. Fluid temperatures ranging from room temperature to 70 °C were measured. Simultaneous spatial and temporal resolutions of 1 μm and 33 ms were achieved.

Keywords: Fluorescence, temperature, Joule heating

1. Introduction

Control of fluid temperatures within microfluidic systems can be very important particularly for electrophoretic separations [1,2] and for enzyme activated reactions [3] such as PCR amplification of DNA [4-6]. In spite of its importance, only a few techniques have been developed for the measurement of fluid temperatures in micron-sized geometries. These methods include electrical conductivity [2,7-9], NMR thermometry [10], and Raman spectroscopy [11]. Although these techniques can provide fairly accurate and precise temperature measurements, all three are limited in either their spatial or temporal resolution for microfluidic applications.

Here we describe a simple technique for the measurement of fluid temperatures in microfluidic structures with simultaneously high spatial and temporal resolution. The technique takes advantage of the temperature dependence of the fluorescence intensity of a dilute fluorophore added to the fluid [12]. In the present work, the temperature profiles in several microchannel geometries were measured with a spatial resolution of 1 um and a temporal resolution of 33 ms.

2. Experimental

Microchannels were filled with a Rhodamine B solution and the fluorescence was imaged using a standard fluorescence microscope with an 0.32 numerical aperture objective (defining the spatial resolution of the image), and a CCD video camera with a frame rate of 30 Hz (defining the temporal resolution). Temperature measurements were made by first measuring the fluorescence intensity of a Rhodamine B solution at a

J.M. Ramsey and A. van den Berg (eds.), Micro Total Analysis Systems 2001, 239–241.

known uniform temperature, and then again at an unknown, possibly nonuniform temperature [13]. The ratio of the measured temperatures was used to calculate the temperature based on a previously measured calibration curve (see Ref[13], Fig. 1).

4. Results and discussion

To demonstrate the temperature measurement technique, we have examined temperature gradients resulting from Joule heating in two different microchannel geometries: a 'T' intersection, and a microchannel with a constriction.

Fig. 1 shows the results of a temperature measurement at a 'T' intersection. With a voltage V = -2500 V applied as shown in Fig. 1(a), electroosmosis drives the fluid into the junction from the top and bottom arms and out to the right. Because the current density is twice as high in the single out-flowing channel, the buffer in that channel is hotter. A plot of the fluid temperature down the middle of the out-flowing channel is shown in Fig. 1(b). The variation of the temperature of the buffer after it leaves the junction (the data to the right of the arrow in Fig. 1(b)) can be used to estimate the extent of axial temperature gradients in straight microchannels as relatively cool buffer enters the channel and is warmed by Joule heating. Fitting the data to the form: $T = T_0 - T_1 \cdot \exp(-x/a)$, gives a length scale for temperature variation a = 80 ± 20 μm.

Fig. 2 shows a similar temperature measurement in a microchannel with a constriction. Because the current density is higher in the constriction, more power is dissipated there, and a hot spot is generated. Because of the electroosmotic flow of the

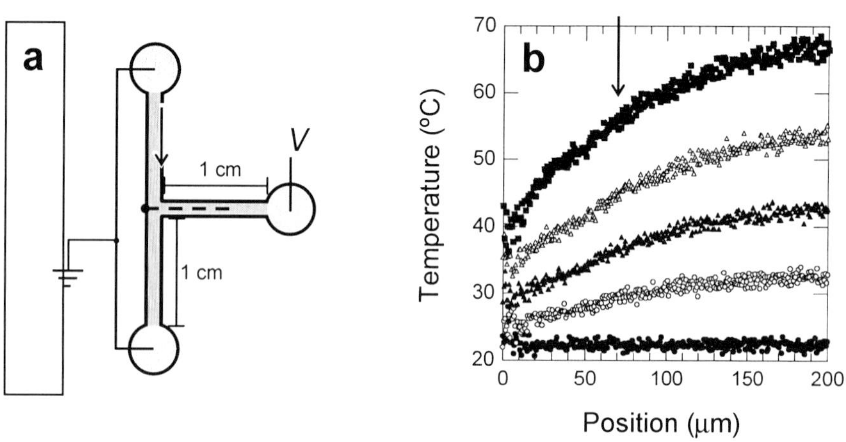

Figure 1. Buffer temperature at 'T' junction of microchannels. (a) Schematic showing arrangement of channels and application of voltage, V = -2500 V. (b) Buffer temperature vs. position along the middle of the channel as indicated by the dashed line in (a). The arrow indicates the position of the right edge of the side channels - see corresponding arrow in (a). Temperature measurements were made 0 s, 0.2 s, 0.4 s, 1.0 s, and 4.0 s after the voltage was switched on.

buffer in the microchannel, the buffer is swept through the temperature ramp shown in Fig. 2(b) at a velocity of about 2 cm/s. Consequently, a device such as this could be used for extremely rapid temperature cycling.

Acknowledgement

The authors wish to acknowledge the financial support of the NRC/NIST postdoctoral Research Program.

References

1. J. O. N. Hinckley, *J. Chromatogr.* **109**, 209 (1975).
2. W. A. Gobie and C. F. Ivory, *J. Chromatogr.* **516**, 191 (1990).
3. Y. Tanaka *et al.*, *J. Chromatogr.* a **894**, 45 (2000).
4. L. C. Waters *et al.*, *Anal. Chem.* **70**, 158 (1998).
5. M. U. Kopp, A. J. De Mello, and A. Manz, *Science* **280**, 1046 (1998).
6. J. Khandurina *et al.*, *Anal. Chem.* **72**, 2995 (2000).
7. D. S. Burgi, K. Salomon, and R. L. Chien, *J. Liquid Chromatogr.* **14**, 847 (1991).
8. J. H. Knox and K. A. Mccormack, *Chromatographia* **38**, 207 (1994).
9. U. Tallarek *et al.*, *Anal. Chem.* **72**, 2292 (2000).
10. M. E. Lacey, A. G. Webb, and J. V. Sweedler, *Anal. Chem.* **72**, 4991 (2000).
11. K. L. Davis, K. L. K. Liu, M. Lanan, and M. D. Morris, *Anal. Chem.* **65**, 293 (1993).
12. M. C. J. Coolen, R. N. Kieft, C. C. M. Rindt, and A. A. Van Steenhoven, *Exp. in Fluids* **27**, 420 (1999).
13. D. Ross, M. Gaitan, and L. E. Locascio, to be published (2001).

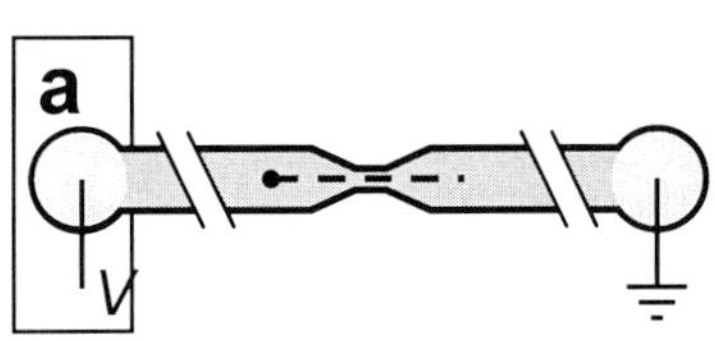

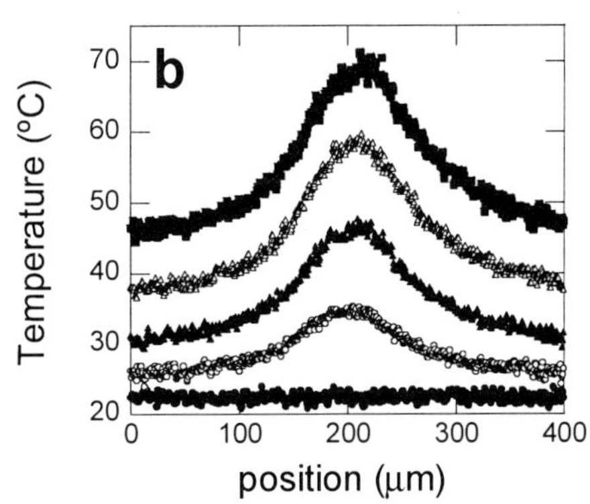

Figure 2. Buffer temperature in microchannel with constriction.
(a) Schematic. The channel used for the measurement contained 14 similar restrictions equally spaced along a length of 1 cm. The applied voltage V = 1600 V.
(b) Buffer temperature vs. position along the middle of the channel as indicated by the dashed line in (a). Temperature measurements were made 0 s, 0.2 s, 0.4 s, 1.0 s, and 4.0 s after the voltage was switched on.

MEMBRANE RECEPTOR PROTEINS AT INTERFACES FOR BIOANALYTICS IN THE MICRO- AND NANOMETER RANGE

Horst Vogel
Department of Chemistry, Swiss Federal Institute of Technology,
CH-1015 Lausanne, Switzerland

Abstract
Cell-based assays are the predominat means of assessing activation of ionotropic receptors and G protein-coupled receptors (GPCRs), but are limited by their inherent complexity. Here we present an approach to immobilize receptors stably and with defined orientation in microarrays on sensor surfaces. Using optical and electrical methods we were able to follow ligand binding, receptor activation and receptor deactivation. Our findings are a novel approach for functional screening of pharmacologically interesting receptor proteins in high density format.

Keywords: Ionotropic receptors, GPCRs, microarrays, functional screening

1. Introduction

The ability of organisms, or individual cells to sense and to react to different external and internal stimuli (light, hormones, odorants, taste, neurotransmitters) is crucial for their survival. After activating cell surface receptors such as ionotropic or G protein-coupled receptors, signals are transduced across the cell membrane and activate a variety of intracellular processes leading to a cellular response. Many of these receptors are targets for the majority of clinically used medicines. Inspite of the importance of these proteins, efficient bioanalytical assays are still missing both for fundamental investigations to understand the molecular basis of signal transduction processes as well as for application in the field of finding new therapeutic compounds. I will report on our recent progress to investigate molecular signal recognition and transduction processes by a combination of surface sensitive physical techiques and molecular biological methods [1, 2].

2. Results and Discussion

Controlled self-assembly of proteins in a functionally active form on sensor surfaces is a prerequisite for the investigation of molecular interactions by surface-sensitive techniques. Receptors, fluorescently labelled by suppressor tRNA technology, are reversibly immobilized on solid supports; ligand binding to and concomitant conformational changes of the receptors are probed by optical techniques such as total internal reflection fluorescence and infrared spectroscopy, fluorescence correlation spectroscopy and surface plasmon microscopy [1-6].

Based on this experience, we have developed a method to detect channel gating of ionotropic receptors in tethered membranes on planar solid surfaces. Electrical fields of several thousand V/m, generated by physiological voltages in the vicinity of small appertures in sub-micrometer thick diaphragms, were sufficient for precise electrophoretical positioning of liposomes and membrane fragments of biological cells across the apertures. Seal resistances greater than 200 Gigaohm, obtained after less than 10 seconds positioning time, allowed

J.M. Ramsey and A. van den Berg (eds.), Micro Total Analysis Systems 2001, 242–243.

voltage-clamp measurements of single ion channels. This so-called 'planar patch-clamp' method is suited for on-chip parallelization of functional assays and the coupling with confocal optical spectroscopies. Standard sample volumes of less than 10 ml reduce diffusion times and the consuption of proteins, ligands and cofactors dramatically. As an alternative to traditional patch clamp measurements, this approach offers novel possibilities for an efficient screening of receptor function [7].

GPCRs constitute an abundant family of membrane receptors of high pharmacological interest. Cell-based assays are the predominant means of assessing GPCR activation, but are limited by their inherent complexity. Functional molecular assays that directly and specifically report G protein activation by receptors could offer substantial advantages. We present an approach to immobilize receptors stably and with defined orientation to substrates. By surface plasmon resonance, we were able to follow ligand binding, G protein activation, and receptor deactivation of a representative GPCR, bovine rhodopsin. Microcontact printing was used to produce micrometer-sized patterns with high contrast in receptor activity. These patterns can be used for local referencing to enhance the sensitivity of chip-based assays. The immobilized receptor was stable both for hours and during several activation cycles. A ligand dose–response curve with the photoactivatable agonist 11-cis-retinal showed a half-maximal signal at 120 nM. Our findings may be useful to develop novel assay formats for GPCRs based on receptor immobilization to solid supports, particularly to sensor surfaces for functional screening [8, 9].

References

1. S. Heyse, T. Stora, E. Schmid, J.H. Lakey, H. Vogel: *Emerging techniques for investigating molecular interactions at lipid membranes.* Biochim. Biophys. Acta 1376, 319-338 (1998).
2. R. Hovius, P. Vallatton, T. Wohland, H. Vogel: *Fluorescence techniques: Shedding light on ligand-receptor interactions.* Trends Pharmacol. Sci. 21, 266-273 (2000).
3. T. Wohland, K. Friedrich, R. Hovius, H. Vogel: *Study of ligand-receptor interactions by fluorescence correlation spectroscopy with different fluorophores.* Biochemistry 38, 8671-8681 (1999).
4. D. Kröger, F. Hucho, H. Vogel: *Ligand binding to nicotinic acetylcholine receptor investigated by surface plasmon resonance.* Anal. Chem. 71, 3157-3165 (1999).
5. T. Stora, J. Lakey, H. Vogel,: *Ion-channel gating in transmembrane receptor proteins: Functional activity in tethered lipid membranes.* Angew. Chem. Int. Ed. 38, 389-391 (1999).
6. S. Terrettaz, W.P. Ulrich, R. Guerrini, A Verdini, H. Vogel: *Immunosensing by a synthetic ligand-gated ion channel.* Angew. Chem. Int. Ed. 40, 1740-1743 (2001).
7. C. Schmidt, M. Mayer, H. Vogel: *A chip-based biosensor for functional analysis of single ion channels.* Angew. Chem. Int. Ed. (2000) 39, 3137-3140.
8. C. Bieri, O.P. Ernst S. Heyse, K.P. Hofmann, H. Vogel: *Chip-based bioassays. Functional micropatterned immobilization of a G protein-coupled receptor and direct detection of G protein activation.* Nature Biotechnol. 17, 1105-1108 (1999), and (News & Views, p 1060).
9. S. Heyse, O.P. Ernst, Z. Dienes, K.P. Hofmann, H. Vogel: *Incorporation of rhodopsin in laterally structured supported membranes.* Biochemistry 37, 507-522 (1998).

THERMOCAPILLARY MIGRATION OF LIQUIDS ON PATTERNED SURFACES: DESIGN CONCEPT FOR MICROFLUIDIC DELIVERY

A. A. Darhuber[1,2], J. M. Davis[1], W. W. Reisner[3] and S. M. Troian[1,3]
[1]Microfluid Dynamics Laboratory, Department of Chemical Engineering
[2]Department of Electrical Engineering
[3]Department of Physics
Princeton University, Princeton, New Jersey 08544-5263

Abstract

We present a novel method of fluidic transport on the open surface of a chemically patterned substrate using thermocapillary actuation. Our experimental and numerical studies provide the desired correlations between the microstream flow rate and tunable parameters like the liquid sample volume, microstream width, and magnitude of the applied thermal gradient.

Keywords: thermocapillary transport, micropatterned surfaces, free surface flow

1. Introduction

Studies have shown that thin liquid films subject to a thermal gradient will flow toward the colder end due to thermocapillary migration [1]. The thermal gradient, dT/dx, along the solid substrate produces a positive thermocapillary stress, dγ/dx, along the air-liquid interface which causes the film to spread. On homogeneous surfaces, the advancing film can undergo lateral break up into a series of regularly spaced fingers with characteristic period λ . Stable spreading and flow control can be attained by confining the flow to hydrophilic stripes whose width $w << \lambda$. Using optical lithography, we have patterned a network of hydrophilic stripes on glass or silicon substrates treated with hydrophobic monolayers. Integrated sub-surface heaters are electronically controlled to maintain the desired temperature distribution which determines the magnitude and direction of the volumetric flow rate. Advantages to this method of fluidic transport include no moving mechanical parts, low operating voltages, no micro machining, no restriction to ionic or highly conductive liquids and accessibility to reaction products for further analysis. Present drawbacks include evaporation, which can be reduced by encapsulation and saturation of the ambient environment, and sensitivity to surface defects within hydrophilic regions, which requires refinement of substrate processing techniques.

2. Theory

Figure 1 represents the geometry of interest. The liquid is confined to the hydrophilic

J.M. Ramsey and A. van den Berg (eds.), Micro Total Analysis Systems 2001, 244–246.

regions (square pad and connecting stripe) by an exterior hydrophobic coating. The stripe contains imbedded heating resistors which in this study subject the overlying liquid to a constant surface shear stress $\tau = d\gamma/dT \cdot dT/dx$.

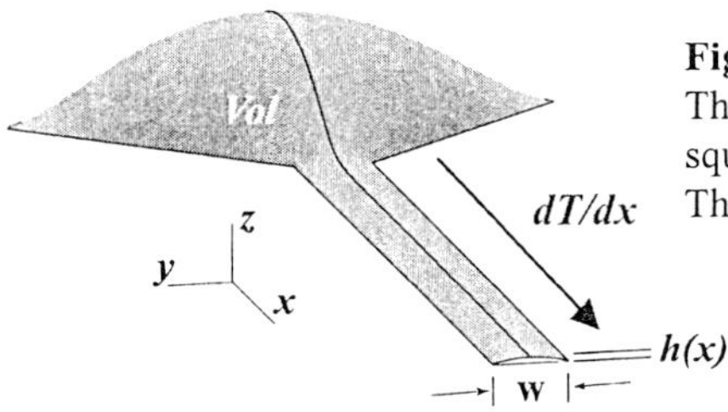

Fig. 1: Schematic diagram of the flow geometry. The liquid sample is extracted from the isothermal square pad onto a differentially heated microstripe. The microstream apex height is designated by $h(x)$.

To optimize our experimental system, we have solved the governing hydrodynamic equation in the lubrication limit for thermocapillary driven flow along a hydrophilic microstripe to obtain the interfacial profile, $h(x,t)$, the volumetric flow rate, $Q(x,t)$, and the front speed, $V(x,t)$. The variation of the liquid surface tension and viscosity, γ and μ, with temperature are included in these computations. The capillary terms $\gamma\nabla^2 h$ represent flow caused by surface curvature variations, which induce slow spontaneous spreading even under isothermal conditions.

$$\frac{\partial h}{\partial t} + \frac{\partial}{\partial x}\left[\frac{\tau h^2}{2\mu} + \frac{h^3}{3\mu}\frac{\partial}{\partial x}\left(\gamma \nabla^2 h\right)\right] + \frac{\partial}{\partial y}\left[\frac{h^3}{3\mu}\frac{\partial}{\partial y}\left(\gamma \nabla^2 h\right)\right] = 0 \qquad (1)$$

3. Experimental Results and Discussion

We have fabricated samples for thermofluidic transport on silicon and glass surfaces using standard micro-electronic techniques. The surfaces are hydrophobized using self-assembled monolayers [3]. To minimize evaporative problems while first exploring the physical and geometrical mechanisms controlling the flow, we have used non-volatile organic liquids like polydimethylsiloxane and tetraethylene glycol which are well confined to the microstripe by the hydrophobic exterior. The small sample volumes and stripe widths used in this study generate rivulets whose thickness is in the tens of microns range. The resultant small Biot number ensures that the temperature at the air-liquid and liquid-solid interface is the same. We have calibrated the linearity of the thermal profile, T(x), in the absence of liquid by direct measurement with fine thermocouples. Measurements of the streaming velocity reveal the functional dependence on the stripe width (see Fig. 2c), the applied temperature gradient and sample volume deposited [4]. The volumetric flow rate always increases with increasing w, dT/dx or sample volume. The experimental measurements compare favorably with numerical solutions of Eq. (1) and can therefore be used to optimize device performance.

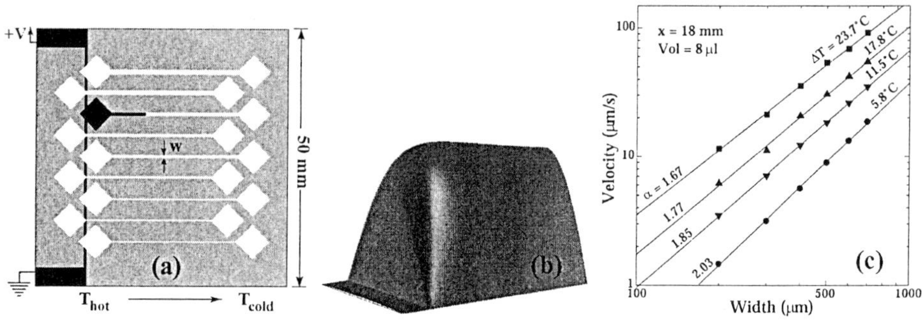

Figure 2 (a) Experimental layout: grey=hydrophobic area, white=hydrophilic area. The black vertical line is the heating resistor which is connected to external contact pads. We have studied the flow for various heater positions. (b) Numerical simulation of liquid rivulet with w=200 μm, dT/dx=14.4°C/cm and inlet height h=7.5μm. Heater was located at stripe entrance x=0. (c) Experimental results for the liquid speed at downstream position x=18 mm as a function of w and ΔT for a total path length Δx=28.4 mm and liquid sample volume 8 μl. The α values denote the fitted power law exponent $V \sim w^{\alpha}$.

4. Conclusion

The thermofluidic concept described above provides a new method for routing liquid microstreams along desired segments of hydrophilic pathways. We have so far achieved a flow speed of 600 μm/s for μ=5 mPa·s, dT/dx=14.4°C/cm and w=800μm, making this a viable technique for microfluidic dosing applications. Applying larger thermal gradients to liquids of lower viscosity and higher dγ/dT will produce higher speeds. In addition to holding or releasing liquid samples dynamically via manipulation of thermal gradients, it is also possible to apply thermal maps to initiate or quench chemical reactions. We are presently studying the use of thermal gradients to enforce micro mixing as well. The application of temperature distributions for routing, reacting and mixing ultra small liquid samples is a unique and attractive feature of our device.

Acknowledgements

This work is supported by the NSF through an *XYZ on a Chip* grant and a Princeton University MRSEC seed grant. JMD holds an NDSEG graduate fellowship.

References

1. D. E. Kataoka and S. M. Troian, *Nature* **402**, 794 (1999).
2. D. E. Kataoka and S. M. Troian, *JCIS* **192**, 350 (1997); **203**, 335 (1998).
3. A. A. Darhuber, S. M. Troian and W. W. Reisner, in press for *Phys. Rev. E.*
4. A. A. Darhuber, J. M. Davis, W. W. Reisner and S. M. Troian, submitted to *Science*.

MICROFLUIDICS FOR TISSUE ENGINEERING MICROVASCULATURE: ENDOTHELIAL CELL CULTURE

K. R. King[1,3,4], H. Terai[2,4], C. C. Wang[1,4], J. P. Vacanti[2,4], J. T. Borenstein[3,4]
[1]Massachusetts Institute of Technology, Cambridge, MA
[2]Dept. of Surgery, Harvard Medical School, Massachusetts General Hospital
[3]The Charles Stark Draper Laboratory, Cambridge, MA
[4]Center for the Integration of Medicine and Innovative Technologies, Boston, MA

Abstract
Development of an integrated blood supply is a critical step towards tissue engineering vital organs. We are attempting to engineer microvasculature using microfluidic networks for the guidance of endothelial cell growth. In this work, we have focused on fabrication of the microfluidic scaffold, *in vitro* seeding, and extended cell culture in the device. Capillary networks were fabricated in biocompatible PDMS, sterilized, coated with cell adhesion molecules, and seeded with cells. Cell-containing devices were then connected to a closed-loop bioreactor for long term culture. We have used the device to demonstrate continuous-flow culture of endothelial cells for up to 4 weeks without occlusion or contamination.

Keywords: cell culture, tissue engineering, PDMS

1. Introduction

Tissue engineering has been proposed as a therapeutic approach to create tissues for patients suffering from loss or failure of organs. One approach involves transplanting cells on biodegradable polymer matrices. This commonly followed paradigm [Fig.1] consists of three components: 1) Cell Acquisition – cells are obtained by tissue biopsy and expanded in culture. 2) Cell Seeding – cells are seeded on polymer scaffolds and cultured *in vitro* to further increase cells number and mechanical strength of the construct. 3) Implantation – the cell/polymer construct is implanted in the body to integrate and serve its organ-specific function while the polymer degrades. [1]

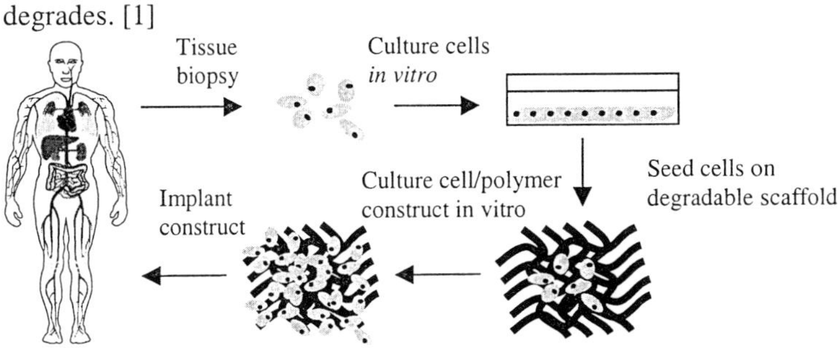

Figure 1. Schematic of an approach to tissue engineering.

J.M. Ramsey and A. van den Berg (eds.), Micro Total Analysis Systems 2001, 247–249.

2. Motivation

A limitation of this approach is related to mass transport. Tissue engineered constructs are currently grown without a blood supply. This leads to hypoxia and nutrient deprivation. Without vasculature, cells in constructs larger than 1-2mm experience significant cell death. Efforts to prevascularize constructs and attempts to promote vessel ingrowth have both met with limited success. [2] We are attempting to engineer microvasculature using biodegradable microfluidic networks for the guidance of endothelial cell growth. [3] Unfortunately, degradable polymer microfluidics have not yet been realized. Therefore, to understand cell culture in confined geometries, we have employed traditional soft lithography microfluidics as a prototype matrix. These cell-containing microfluidics must be capable of supporting long-term culture *in vitro* because *in vitro* expansion of cells prior to implantation can take weeks. The prototype matrix must be designed to supply sufficient oxygen and nutrients while avoiding large shear stresses. Challenges specific to long term microfluidic cell culture include maintenance of sterility and minimization of cell and bubble occlusions as well as enhancement of cell attachment, proliferation, and migration.

3. Experimental

We have developed microfluidic networks that support physiologic flows and pressures by photopatterning SU-8, a high-aspect ratio negative photoresist, onto silicon. This is used as a mold for casting polydimethylsiloxane (PDMS). After removal from the mold, inlets and outlets are cored with blunted syringe needles, and the micropatterned polymer is irreversibly sealed to an unpatterned layer of pyrex or PDMS by oxygen plasma surface treatment. [4] The microfluidic device is autoclave sterilized and perfused with a solution containing cell adhesion molecules (poly-L-lysine, collagen, gelatin, or fibronectin) which are allowed to adsorb for one hour. The fluidic network is then seeded with a $1\text{x}10^6$ – $1\text{x}10^8$ cells/mL cell suspension using a syringe pump at flow rates ranging from 10-100μL/min. The cells are then allowed to attach for 24 hours, after which the device is connected in-line with a sterile bioreactor consisting of a peristaltic pump, oxygenator, bubble trap, and a reservoir of sterile culture medium. [Fig. 2] Finally, the system is run closed-loop in an incubator at standard culture settings.

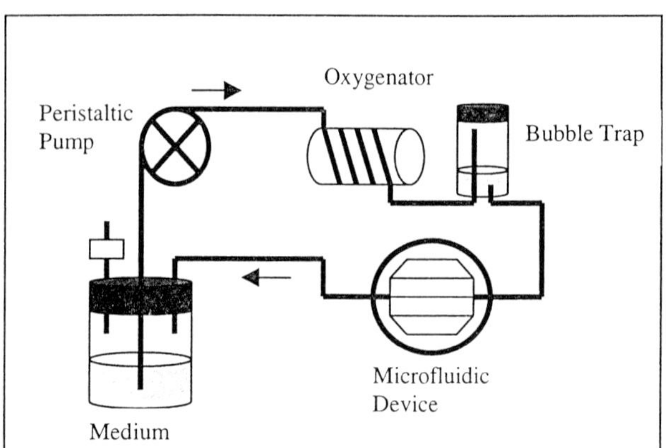

Figure 2. Bioreactor Schematic – Sterile culture medium is pumped peristaltically from a sterile reservoir through an oxygenator consisting of a long length of tubing semipermeable to oxygen. The oxygenator is followed by a small bubble trap, and leads directly to the microfluidic circuit.

4. Results and Discussion

Autoclave sterilization of the microfluidic circuit caused no obvious pattern distortion. Coating the channels with cell adhesion molecules enhanced cell attachment when compared to PBS (phosphate buffered saline) coated control channels. Seeding of cells into channels of widths between 30-200μm was optimized by varying concentrations and flow rates. The continuous-flow bioreactor was used to dynamically culture rat lung endothelial cells at flow rates between 0.01mL/min and 0.1mL/min. Both single channels and complex networks of channels (30-200μm wide and 40μm deep) were successfully seeded and cultured. In 100μm x 40μm single channels, cells have been cultured for more than 4 weeks without contamination or occlusion [Fig. 3].

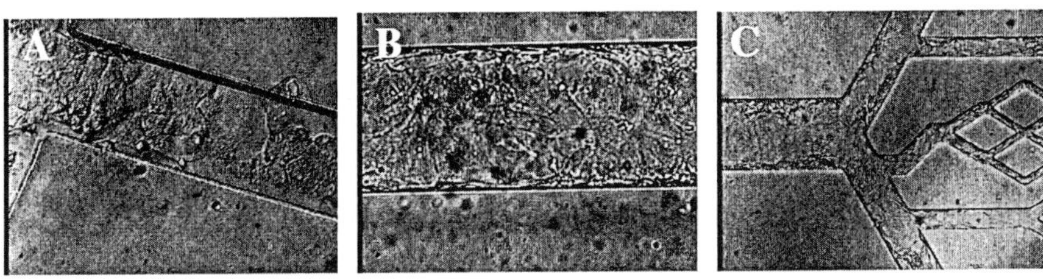

Figure 3. Long term endothelial cell culture in microfluidic networks –Rat lung endothelial cells A) in 100μm x 40μm channels cultured for more than 4 weeks and B,C) in a complex network with channel sizes ranging from 30μm to 200μm wide.

5. Conclusions

While cells have been introduced into microchannels for sorting or analysis purposes, they are not typically cultured long term in microfluidic devices. We have successfully seeded and cultured endothelial cells in microfabricated channels for 4 weeks, demonstrating that cells successfully attach, proliferate, and migrate in closed channels with small geometries. This is a critical step towards the ultimate goal of fabricating microvasculature and enabling the development of bulk organ tissue engineering.

Acknowledgements

This project was sponsored by the Department of the Army, Cooperative Agreement DAMD-99-2-9001. The content of this paper does not necessarily reflect the position or the policy of the government, and no official endorsement should be inferred.

References

1. Langer R., Vacanti, J.P. *Tissue engineering: the design and fabrication of living replacement devices for surgical reconstruction and transplantation*, The Lancet, 354 (1S) pp. 32SI-34SI July (1999).
2. Eiselt, P. et. al. *Development of Technologies Aiding Large-Tissue Engineering,"* *Biotechnol. Prog.* 14, pp. 134-140, (1998).
3. S. Kaihara, et. al. *Silicon micromachining to tissue engineer branched vascular channels for liver fabrication*, Tissue Engineering, 6(2), pp. 105-117, (2000).
4. Duffy D.C. et. al. *Rapid prototyping of microfluidic systems in poly(dimethylsiloxane)*, Anal. Chem. 70, pp. 4974-4984, (1998).

ELECTROPORATION MICROCHIPS FOR GENE TRANSFECTION

Yu-Cheng Lin and Ming-Yuan Huang
1 University Rd., Department of Engineering Science, National Cheng Kung University, Tainan, Taiwan
yuclin@mail.ncku.edu.tw

Abstract

Gene therapy requires effective gene delivery into cells. This work has demonstrated that this novel electroporation microchip can provide *in-vitro* gene transfection with a very small amount of cells and plasmids and requires much lower applied voltage than current commercial equipment.

Keywords: Electroporation, Gene therapy, Gene transfection, MEMS

1. Introduction

Gene therapy is a potential therapeutic modality, which requires effective gene delivery into cells [1]. There are many ways to achieve gene transfer [2]. Electroporation is a technique that introduces foreign materials into cells by applying impulses with an extremely high electric field to create multiple transient pores in the cell membrane through a dielectric breakdown of the cell membrane. Here, a detailed study of a novel electroporation (EP) microchip for *in-vitro* gene transfection was conducted. Electroporation is a technique with which DNA molecules can be delivered into cells in a chamber using high electric field pulses [3].

2. Design and Microfabrication of Electroporation Microchips

The EP microchips were designed in two components, the well-defined cavity for cell accommodation and culturing and the electrodes for providing electroporation electric power. Figure 1 shows the schematic drawings of the EP microchip with 3D and top views. The EP microchips were fabricated with glass sildes (Menzel-Glaser, Germany). A pair of thin film electrodes was fabricated on the glass slide.

3. Plasmids Preparation and Cell Culture

The *pEGFP-N1* plasmids coded with green fluorescent protein (GFP) were extracted and purified from *E.coli* using a Plasmid Isolation Kit. *Huh-7* cells (human hepatocellular carcinoma cells) and *293T* cells (human emmbryonic kidney cells) were used to demonstrate the transfection. For culturing the cells, the microchip was coated with 0.01% polylysine in phosphate-buffered saline (PBS) for 24 hrs. After washing with PBS, the EasiSeal was sealed around the electrode area on the surface of the microchip for cell accommodation.

J.M. Ramsey and A. van den Berg (eds.), Micro Total Analysis Systems 2001, 250–252.

4. Results and discussion

The cell harvest and cell suspension preparation processes for conventional electroporation process using cuvettes are no longer needed when using this EP microchip. Figure 2 shows an image of 24 hour cultured *Huh-7* cells on the microchip surface. A 4 μg GFP plasmid amount was added to the cultured cells. The applied voltage was 10 volts with a pulse duration of 20 ms for one pulse. Figure 3 shows the cell distribution image after electroporation..

The successfully transfected cells would emit a green fluorescence when excited using a 420 - 480 nm light source. Figure 4 shows successfully transfected *Huh-7* cells with green fluorescence found in the area between the electrodes.

The EP microchip was tested using another cell line, *293T*, to show the gene delivery capability as well. Figure 5 shows the images of transfected *293T* cells with green fluorescence. Both *Huh-7* and *293T* were successfully transfected with GFP plasmids using applied voltage as low as 10 volts. Very small amounts of cells and genes would be enough for evaluating the gene delivery and expression. The cell quantity used was 50 μl with the concentration at 10^4 cells/ml and the amount of plasmids was as low as 4 μg.

A detailed study of the novel EP microchip for *in-vitro* gene transfection was conducted. The EP microchip, consisting of the well-defined cell cavity for cell culturing and thin film electrodes to deliver electric power for electroporation, was fabricated on a glass slide using evaporation, photolithography and etching methods. This work has demonstrated that this EP microchip can provide *in-vitro* gene transfection with a very small amount of cells and plasmids and requires much lower applied voltage than current commercial equipment. The EP microchip also has the advantages of simple cell culturing and preparation processes compared to the commercial EP equipment. The large heat dissipation area with respect to the cell concentration eliminates the temperature rising effect in the electroporation process, which can improve the cell survival rate after electroporation. Evaluation of gene transfection can be done in-vitro straight after the electroporation process without further cell harvesting, which provides an easy and better tool for gene transfection tests. This EP microchip will provide many potential applications for gene therapy.

References

1. G.A. Hofmann, and G.A. Evans, "Electronic genetic-physical and biological aspects of cellular electromanipulation," IEEE Eng. Med. & Biol. Mag., 12, pp. 6-25, 1986.
2. T. Nishi, et al., "High-efficiency in vivo gene transfer using intraarterial plasmid DNA injection following in vivo electroporation," Cancer research, 56, pp. 1050-1055, 1996.
3. R.A. Meldrum, M. Bowl, S.B. Ong, and S. Richardson, "Optimisation of electroporation for biochemical experiments in live cells," Biochemical and Biophysical Research Communications, 256, pp. 235-239, 1999.

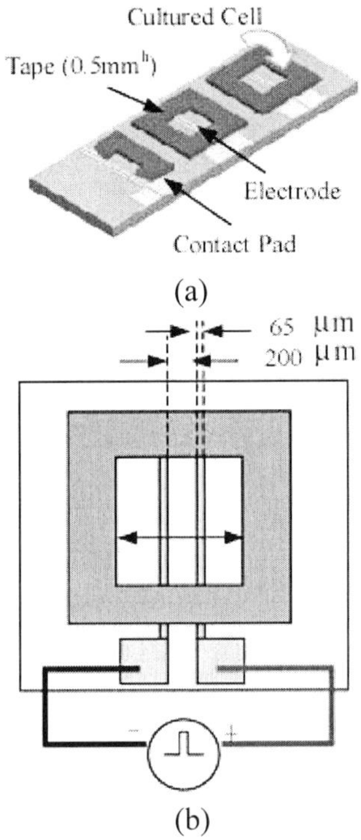

Figure 1: Schematic drawings of an *in-vitro* EP microchip, (a) 3D view and (b) topview. Electrode dimensions are 65 μm wide with 200 μm interval. Cell culture cavity is 10 mm × 10 mm × 0.5 mm.

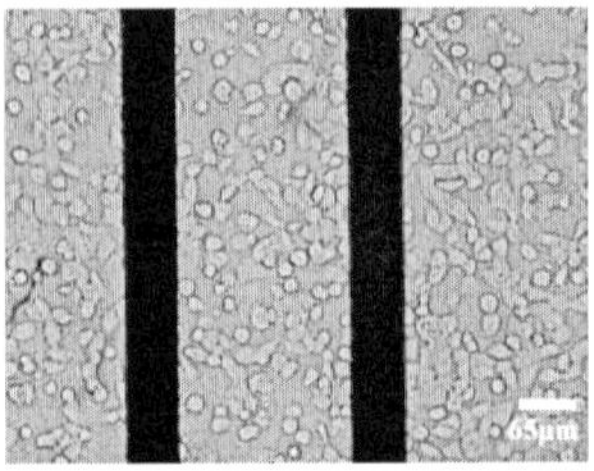

Figure 2: Photo image of 24 hour cultured *Huh-7* cell distribution before electroporation.

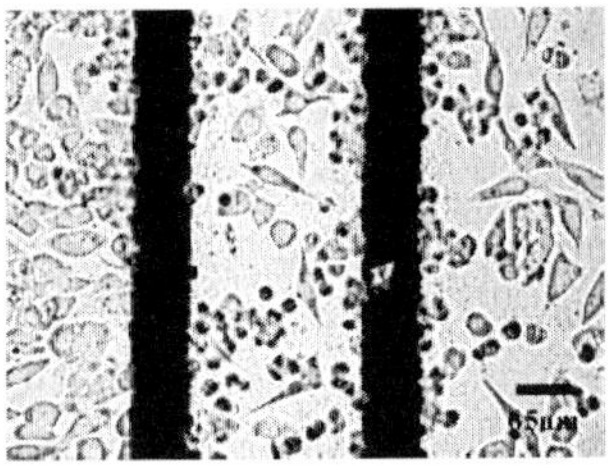

Figure 3: Image of *Huh-7* cell distribution after electroporation using an EP microchip with parallel line electrodes.

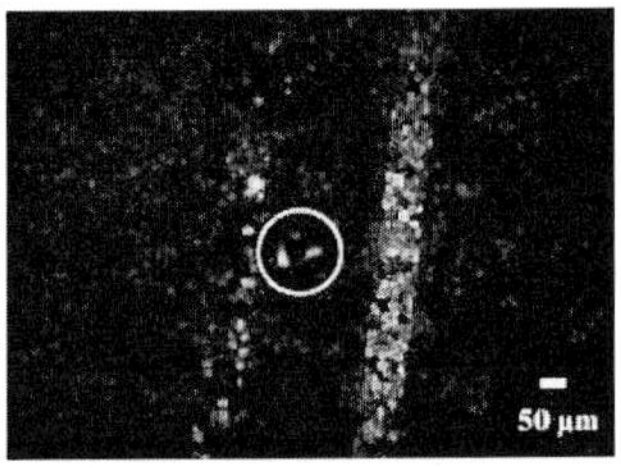

Figure 4: Image of successfully transfected *Huh-7* cells with green fluorescence using an EP microchip. The yellow strips are the gold electrodes.

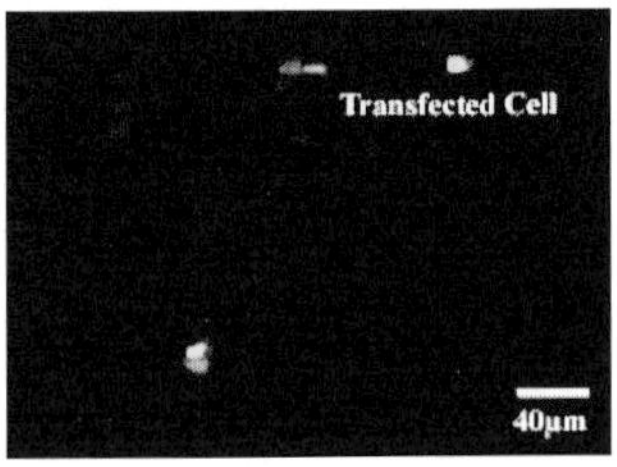

Figure 5: Image of successfully transfected *293T* cells with green fluorescence using an EP microchip.

IMPEDANCE SPECTROSCOPY CELL ANALYSIS IN MICROCHANNELS

Shady Gawad[1], Stefan Metz[1], Laurent Schild[2] & Ph. Renaud[1]
[1]Swiss Federal Institute of Technology, EPFL DMT-IMS, Switzerland.
[2]IPHARM, Université de Lausanne, Switzerland.
shady.gawad@epfl.ch

Abstract

This paper presents a micromachined high throughput flow-cytometer capable of discriminating cells using their electrical parameters as measured by impedance spectroscopy. The system is based on a set of electrodes integrated in a microchannel to measure the differential impedance in two successive segments as the cell passes consecutively through each. This analyzer is intended to drive a sorting actuator realising a subsequent cell separation. Size reduction and integration of functions are essential in achieving precise measurements and high throughput.

Keywords: Cell, impedance spectroscopy, flow-cytometry, electrodes, microfluidic.

1. Introduction

The device is based on a glass-polyimide microfluidic chip with integrated channels and electrodes microfabricated at the length scale of the particles (1-20 μm) to be investigated. A laminar liquid flow transports the suspended particles through the measurement area. Each particle's impedance signal is recorded by a differential pair of microelectrodes using the cell surrounding media as a reference(Fig.1). The integration of microelectrodes in capillary channels allows manipulation and detection [1] at a single particle level without the inherent complexity of optical tools. Therefore the device can be considered for numerous diagnostic and research applications in oncology, haematology or toxicology; where, such this cytometry chip would permit automated and highly controlled measurement of biological cell characteristics at a much lower cost. However, the use of transparent substrate and cover materials allow the combination of electric impedance and optical detection techniques (fluorometry).

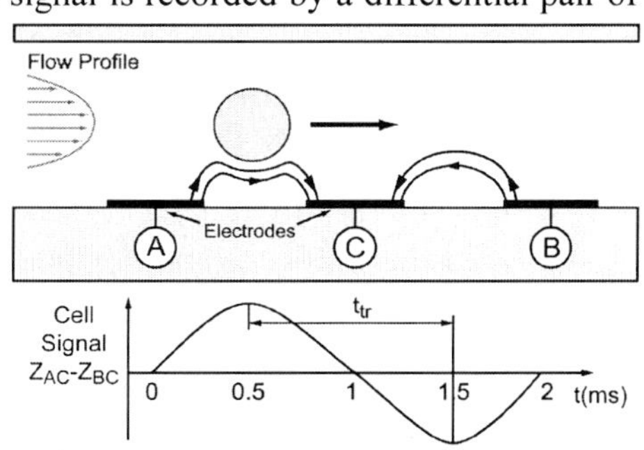

Fig. 1 a) Side view of the microchannel showing a hydrodynamically focussed particle passing over three electrodes (A,B and C). The impedance signal is measured differentially (Z_{AC}-Z_{BC}). **b)** Impedance signal. As the distance between the two measurements areas and time T_{tr} is known, the speed of the particle can be calculated.*

J.M. Ramsey and A. van den Berg (eds.), Micro Total Analysis Systems 2001, 253–255.

2. Theory

The impedance change ΔZ due to a particle passing through the detection area of the microchannel is measured and recorded for several excitation frequencies simultaneously. Placement of the electrodes and concentration of the current lines around the cell is optimized to achieve a high ΔZ/Z ratio. To fully understand the effects of the electrode and channel geometry as well as the dispersion due to the position and size of the measured particle, we have performed 3D FEM electrical simulations of the detection channel [2]. These simulations were useful in determining the best trade-off between electrode design and process technology, and for understanding the particle position and size influence on the detection signal. It showed that the particle position influence could be compensated using at least two excitation frequencies and the measured speed of the particle.

3. Experimental

The chip was microfabricated on a glass substrate using polyimide defined channels and Ti-Pt electrodes (Fig. 2)[3]. The electronic detection circuit produces a continuous excitation signal by summing two or more sine waves (100kHz-15MHz, ~100 mVp-p) at frequencies of specific interest. This method allows for a good signal-to-noise ratio when combined with a synchronous demodulation technique (in-phase and quadrature) [4].

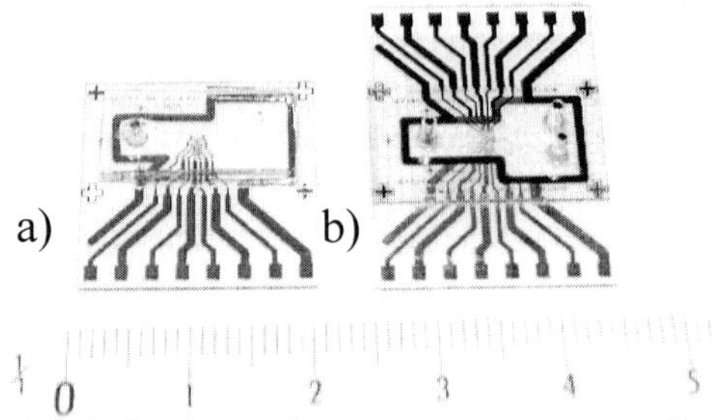

Fig. 2 a) The microfabricated chip with a PDMS cover and molded fluidic connections. **b)** Chip-on-chip configuration designed with two outlets, top and bottom electrodes and an experimental sorting chamber.*

4. Results and discussion

To characterize the device and verify that it could in effect differentiate cell or particle subpopulations in a mixed sample, we used calibrated latex beads of 5 and 8 μm (Serva) and human erythrocytes as well as erythrocyte ghost cells (i.e. cytoplasm replaced by the PBS solution). The chip also provides a measure of each particle speed, which is very useful in determining the particle position in the channel. Plotting of the particle transit time against the measured signal amplitude shows that the average

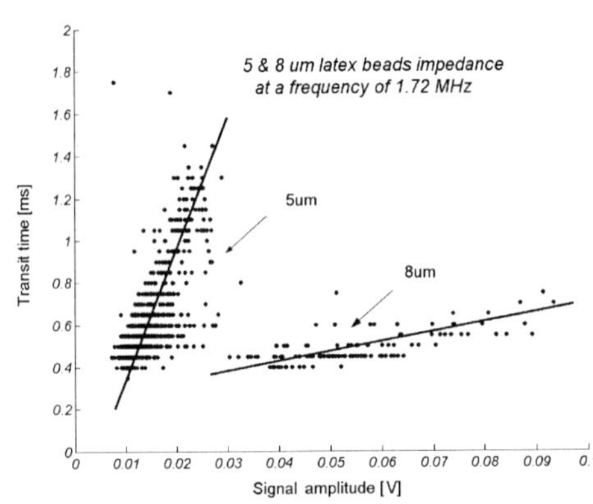

Fig. 3 Transit time vs. measured amplitude signal of 5 and 8 μm latex beads (2000 recorded events). The particles were mixed in a standard PBS solution and passed through the measurement channel.*

velocity of large particles is higher than that of small ones (i.e. they are restricted from the channel wall) and that small position dispersion generates important signal changes for large particles (Fig. 3). Finally, measurements of human erythrocytes and erythrocytes-ghosts cells were performed using pure and mixed samples. They demonstrate that cytoplasm conductance changes could be detected on our chip using only two different excitation frequencies (Fig. 4) and that electrically measured cell sizes are consistent with values obtained optically by confocal microscopy (~6 μm).

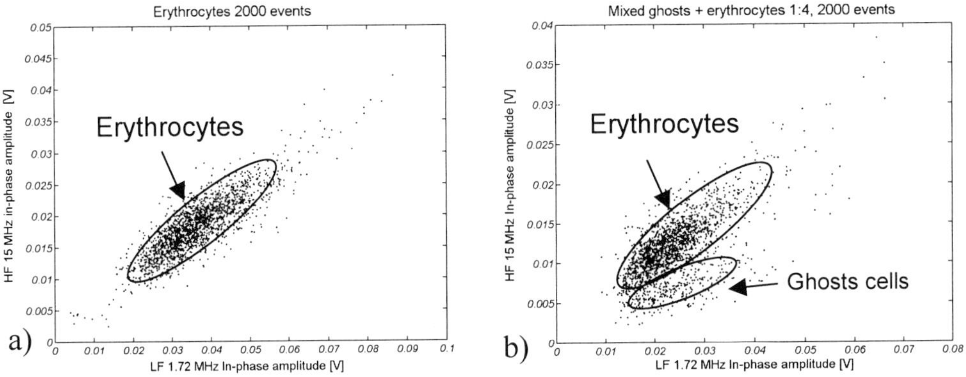

Fig. 4 a) Signal amplitude of 2000 erythrocytes recorded simultaneous for 2 frequencies. **b)** Ghost cells were mixed with erythrocytes in a 1:4 ratio. The ghost cell cytoplasm conductance (similar to external solution) presents a lower HF/LF ratio.*

5. Conclusions

The results show that size and electrical parameters of particles can be measured in a fluidic microchannel and that the device allows single particle tracking for cell sorting applications in the field of hematology, oncology and toxicology.

Acknowledgements

The authors are grateful to the CMI-EPFL cleanroom staff for support in chip processing. This work has been partially supported by LEISTER Technologies.

References

1. Fiedler, S., et al., *Dielectrophoretic sorting of particles and cells in a microsystem.* Analytical-Chemistry, 1998. 70 (9): p. 1909-1915.
2. Gawad, S., et al. *Micromachined Impedance Spectroscopy Flow Cytometer for Cell Analysis and Particle Sizing* to be published Lab-on-a-chip, RSC, September 2001.
3. Gawad, S., et al. Fabrication of a microfluidic cell ananlyzer in a microchannel using impedance spectroscopy. in *IEEE-EMBS Conf. on Microtech. in Med. & Bio.* 2000. Lyon, France, p. 297-301.
4. Hoffman, R.A. and W.B. Britt, Flow-system measurement of cell impedance properties. J. Histochem. and Cytochem., 1978. 27(1): p. 234-240.

AN INTEGRATED SGDBR LASER-BASED BIOSENSOR

Marin Sigurdson[1], Carl Meinhart[1], Kimberley Turner[1]
Larry Coldren[2], Dan Cohen[2], Timothy Stultz[3] & Nelle Slack[3]
[1]Department of Mechanical & Environmental Engineering
[2]Department of Electrical & Computer Engineering
University of California, Santa Barbara, CA 93106
[3]ThauMDx, LLC, Santa Barbara, CA

Abstract

A Tunable Laser Cavity Sensor (TLCS) is being developed for on-chip molecular diagnostics. Two frequency-tunable SGDBR (sampled grating distributed Bragg reflector) lasers – a sensor-laser and a reference-laser – are used to detect minute changes in refractive index resulting from specific antigen-antibody binding on the laser waveguide surface within an adjoining microchannel. Dielectrophoresis is used to increase the concentration of antigens in the binding region of the microchannel.

Keywords: tunable laser, optical sensor, dielectrophoresis

1. Introduction

Current technology in biosensors has yielded sensitive and versatile biochemical sensing systems [1,2]. We concentrate here on a highly sensitive optical detection technology which may be fully contained on a small chip and therefore field deployable. This system directly measures changes in index of refraction due to specific binding in the sensor region and therefore does not require mixing of conjugate antibodies or fluorescent molecules during sample preparation. In addition, the system – laser, electronics, and fluid chip – can be fully integrated and will not require any external optics. By fabricating a one-dimensional array of laser sensors on a single chip, the TLCS is readily scaleable to measuring tens of assays simultaneously.

2. Laser Design and Preliminary Results

Tunable SGDBR lasers (described in [3]) are used for the sensor (Fig. 1). The sensor assembly will be formed by flip-chip bonding an In-P based laser/detector chip to the silicon fluid handling chip such that the fluid flows over the two laser ridges (Fig. 2). The cladding of the sensor-laser waveguide is thinned to allow evanescent wave interaction between the laser cavity and the biofluid in the microchannel. When this is done, the laser frequency is sensitive to a change in fluid refractive index. Such a change can occur as the result of a specific antigen/antibody binding event at the laser-biofluid interface. The sensor-laser and reference-laser are integrated with a wide multimode interference coupler to combine the optical fields and with a small, high-speed photodetector. The combined fields produce a heterodyne signal at the detector equal to the difference in frequency between the two optical fields. The heterodyne signal is in this way sensitive to a binding event.

J.M. Ramsey and A. van den Berg (eds.), Micro Total Analysis Systems 2001, 256–258.

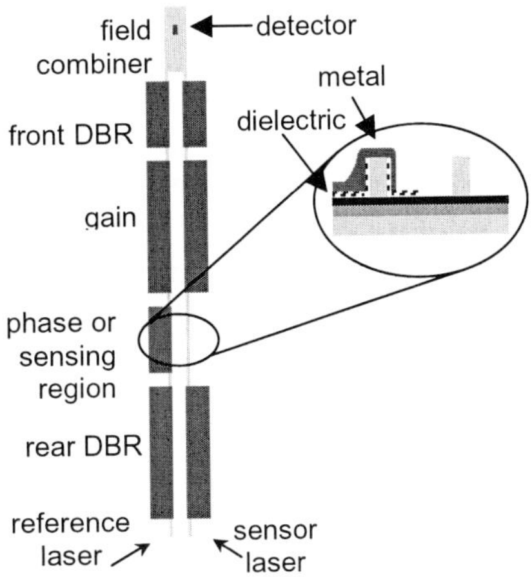

Figure 1. Laser schematic. The optical fields from tandem sampled grating DBR lasers are combined on a photodetector, generating a heterodyne frequency proportional to the refractive index in the evanescent field adjacent to the sensing region.

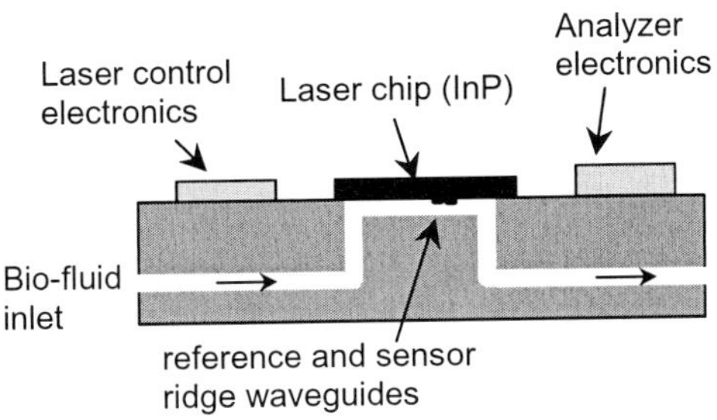

Figure 2. TLCS cross section showing relative location of InP laser chip and Si microfluidics chip. Potential for location of electronics on chip is shown.

The laser sensor was tested in an open reservoir containing one of five different fluids, including air. The heterodyne frequency shifted approximately linearly with the index of refraction of the fluid as shown in Fig. 3. Indeed, the linearity was quite good except in the case of air. This is probably the result of a significant change in the field profile when the surrounding index changed from approximately n = 1.3 to 1.0. We have easily demonstrated here index resolution of 0.01; under ideal conditions, resolution should be close to 10^{-6}.

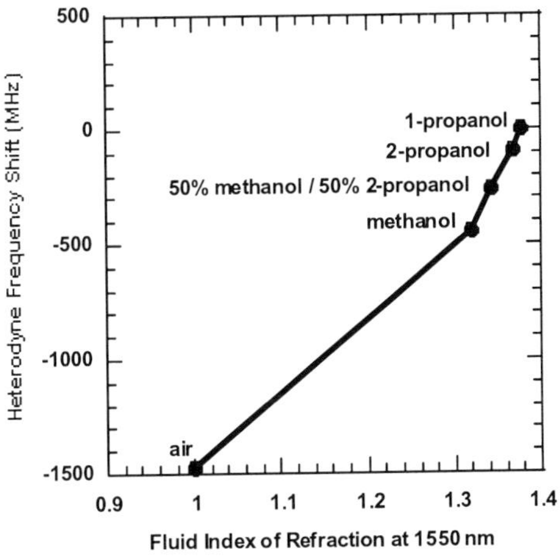

Figure 3. Heterodyne frequency shift as a function of the fluid index of refraction

3. Microfluidics

Dielectrophoresis (DEP) is used to increase antigen concentration near the sensors and therefore enhance binding and device sensitivity. Jones [4] and Miles *et al.* [5] present a background on dielectrophoresis. Finite element simulations were used to predict the DEP forces on the particles (Fig. 4). From these results, we can solve a system of partial differential equations to predict the binding of particles in the interaction region at the microchannel wall (Fig. 5). These results allow us to optimize the design of the DEP electrodes, the microchannel geometry, and the placement of the TLCS's.

In order for the sensor to detect a specific antigen, an appropriate antibody must be immobilized on the sensor surface. The stable, precise immobilization of such an antibody on the InP substrate of the laser chip is the third prong of this research.

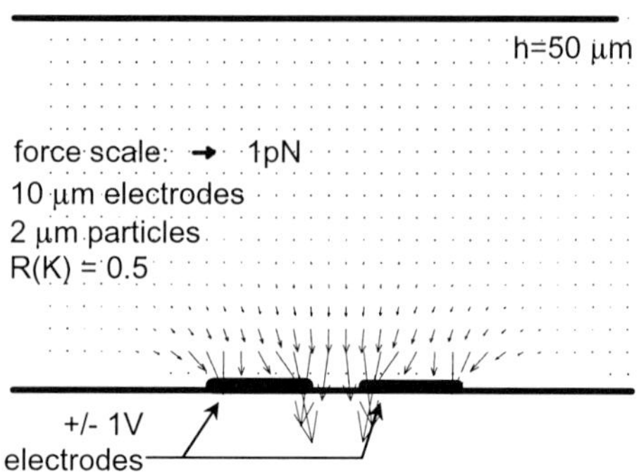

Figure 4. DEP Force vectors acting on 2 micron particles in microchannel simulated in FEMLAB finite element solver. Electrodes are designed to increase particle concentration near the active region of the sensor laser which may be located close to the electrodes or downstream of them.

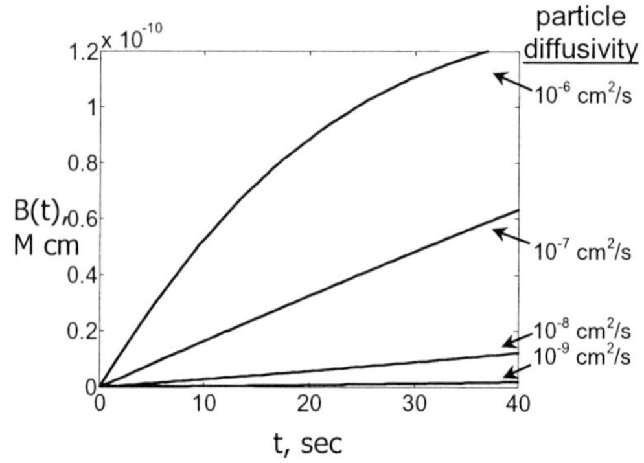

Figure 5. Binding curves for one-step antigen binding to immobilized antibodies on microchannel wall, simulated in Femlab finite element solver. DEP Force is not included in this simulation. $k_a=10^8\ M^{-1}\ s^{-1}$; $k_d = 0.02\ s^{-1}$; $C_T=1$ nM (injected conc.); $R_T=0.167$ nM cm (bound conc.); $V_{MAX}=1$ cm/s; h=50 microns.

Acknowledgement

This research is funded by DARPA grant DAAD 19-00-1-0400.

References

1. Plowman, T. E., W.M. Reichert, C.R. Peters, H.K. Wang, D.A. Christensen & J. N. Herron. 1996. *Biosensors & Bioelectronics*, Vol. 11, pp. 149-160.
2. Myszka, D.G. 1998. Survey of the 1998 optical biosensor literature. J Mol. Recognit. Vol. 12, pp. 390-408.
3. Mason, B., G.A. Fish, S.P. DenBaars, and L.A. Coldren, 1999. Widely Tunable Sampled Grating DBR laser with integrated electroabsorption modulator, *IEEE Photon. Techn. Letts.*, 11, (6), 638-640, June.
4. Jones, T. 1995. *Electromechanics of particles*, Cambridge University Press, New York, NY.
5. Miles, R., P. Belgrader, K. Bettencourt, J. Hamilton, S. Nasarabadi, 1999. Dielectrophoretic manipulation of particles for use in microfluidic devices, *MEMS-Vol. 1, Microelectromechanical Systems (MEMS), Proceedings of the ASME International Mehanical Engineering Congress and Exposition*, Nashville, TN, Nov. 14 – 19.

CONFINEMENT OF FLUORESCENCE EXCITATION FOR SINGLE MOLECULE DETECTION AT HIGH CONCENTRATIONS

S. W. P. Turner, M. Levene, J. Korlach, W. W. Webb and H. G. Craighead
School of Applied and Engineering Physics, Cornell University, Ithaca, NY 14853, USA

Abstract

Nanofabrication techniques are employed to make metal-clad zero-mode optical waveguides to confine the excitation volume in single-molecule fluorescence detection experiments. Fluorescence correlation spectroscopy is used to characterize the effective volume of the system, yielding a smallest observed volume of 20 zeptoliters (20×10^{-21} liters).

Keywords: Single molecule detection, fluorescence correlation spectroscopy, nanofabrication

1. Introduction

Single molecule detection has proven a valuable research tool [1], but has been hindered in practical application by the requirements on purity and analyte concentration brought about by the size of the observation volume. Diffraction limited focal volumes are typically limited to about 0.1 fl, which limits the concentration to approximately 15 nM for an average occupancy of one molecule. Most physiological processes take place at much higher concentrations, so methods for reduction of the excitation volume are important both for single molecule studies of physiological processes, as well as exploitation of biological molecules as actuators and transducers.

2. Zero Mode Waveguides for Confinement

Using electron beam lithography and reactive ion etch techniques, we have fabricated high-aspect ratio holes through evaporated aluminum films on transparent substrates (fig. 1) which, when sufficiently small, act as zero-mode metal clad waveguides across the thickness of the film. Hole sizes varying from 30 nm to 150 nm were fabricated on a single substrate by varying the size of the exposed area and the dose delivered to it. Finite element analysis of Maxwell's equations in these geometries (fig. 2) has confirmed that optical wavelengths cannot be transmitted through small waveguides, and the attenuation length of the radiation is related to the diameter of the guide—approaching the skin depth of the metal for very small holes. Thus, fluorescence is excited in a volume that is confined laterally by the walls of the waveguide, and vertically by the attenuation of the guide. For 50 nm waveguides, the calculations

J.M. Ramsey and A. van den Berg (eds.), Micro Total Analysis Systems 2001, 259–261.

predict an attenuation length shorter than 30 nm. Fluorescence collected from molecules within a waveguide is attenuated over a similar length scale, further confining the effective observation volume.

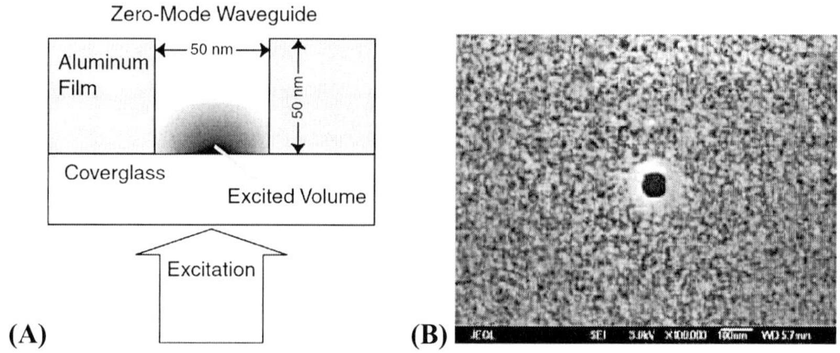

Figure 1. (A) Shows the principle of operation of the zero-mode waveguides. Sufficiently small guides do not transmit either excitation light or emitted fluorescence, resulting in an excitation volume that is confined to the bottom of the hole and an effective observation volume that is further reduced by the rapid attenuation of the emitted light. (B) Shows an electron micrograph of a 75 nm hole in a polycrystalline aluminium film.

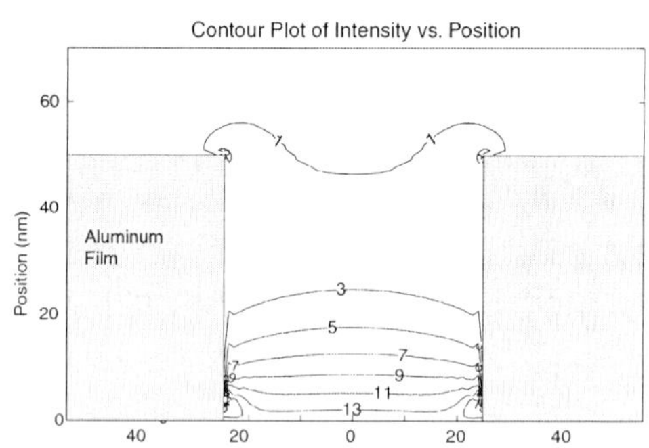

Figure 2. Intensity profile in the center of a zero-order waveguide, obtained by time-domain, finite element analysis of Maxwell's equations for a waveguide with 50 nm diameter and a 50 nm thick aluminium film. The intensity at the exit is arbitrarily set to unity, showing that the intensity of the illumination at the bottom is nearly 15 times brighter than at the exit plane.

3. Experiment

An aqueous solution of fluorescent analogs of nucleotide triphosphates is introduced into the waveguides and excited through the transparent substrate using an upright microscope. Emitted light also cannot traverse the waveguide, so epifluorescence detection is employed with an avalanche photodiode. Fluorescence correlation spectroscopy (FCS) is used to characterize the excitation volume, both through the diffusion time and the average occupancy. The data clearly demonstrate that the excited volumes grow progressively smaller as the waveguide size decreases. By comparing the measured FCS curves (fig. 3) with expected curves (not shown) computed from the Maxwell solutions and a one-dimensional diffusion model, we have confirmed that in the smallest waveguides the excited volume is approximately 20 zeptoliters, which would in principle allow single molecule detection at concentrations up to 100 μM.

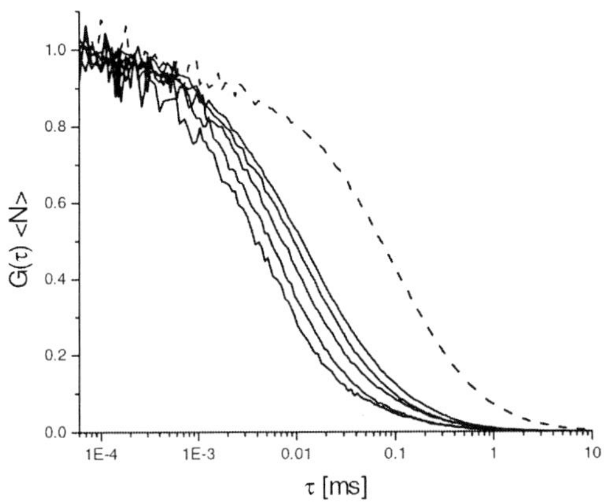

Figure 3. Normalized FCS curves showing the performance of a diffraction limited focal volume in the dashed curve, and the next 5 curves from right to left correspond to guides 80, 70, 60, 50 and 40 nm in diameter

4. Discussion and Conclusions

Zero-mode waveguides are a promising new method of confining the excitation volume for single molecule detection at high concentrations. Electron beam lithography and reactive ion etching have proven capable of fabricating guides as small as 35 nm in diameter. The smallest detection volume observed (for a 40 nm waveguide) was 20 zeptoliters.

Acknowledgements

We would like to thank the Cornell Nanofabrication Facility and its staff, and J. M. Ballantyne for allowing us access to software and computer time to calculate EM solutions in our structures. Major funding is provided by the Department of Energy.

Reference

1. S. Weiss, Fluorescence spectroscopy of single biomolecules, Science 283, pp. 1676-1683 (1999).

DUAL ELECTROCHEMICAL AND OPTICAL DETECTION ON A MICROFABRICATED ELECTROPHORESIS CHIP

Drew P. Manica, Julie A. Lapos, A. Daniel Jones, Andrew G. Ewing
Department of Chemistry, 152 Davey Laboratory
The Pennsylvania State University, University Park, PA 16802, USA

Abstract

The ultimate goal of microchip-based separations is the implementation of all aspects of analysis – sample preparation and introduction, separation, and detection – onto a microfabricated device to improve performance, portability, throughput, cost, and ease of use. Toward the realization of that goal and to broaden the usefulness of microdevices, a novel method for dual electrochemical and fluorescence detection is demonstrated.

Keywords: electrophoresis, electrochemical detection, optical detection, photobleaching, optical gating

1. Introduction

The choice of detection method used for analysis introduces a fundamental limit to the applicability of the device since each detector is selective for a certain chemical characteristic or structural motif. Using orthogonal detection methods vastly increases the amount of chemical information gained from a single separation. Although progressive work has been done to integrate UV absorbance measurements and mass spectrometry together after microchip separations [1], there have been no reports of dual detection using the two simplest and most familiar methods -- electrochemical and fluorescence detection.

Electrochemical detection is highly amenable to the microchip format, providing selectivity for a variety of natively electroactive compounds as well as derivatives [2,3], while fluorescence detection is currently the most common detector employed on microchips. Coupling these two detection methods together would greatly advance microanalysis in the direction of total analytical systems.

Sample introduction for dual detection has been investigated by traditional cross-T injection and also by optical gating. While the channel geometry defines the sample plug in cross-style injections, for optical gating the sample is temporally defined by briefly interrupting a high power photobleaching laser beam upstream from the detection beam [4]. This allows rapid, small-volume serial separations of non-photobleached species from photobleached species. In addition, dual detection can offer a deeper understanding of chemical phenomena such as photobleaching by revealing the different optical and electrochemical properties of the photodegradation products.

J.M. Ramsey and A. van den Berg (eds.), Micro Total Analysis Systems 2001, 262–264.

2. Experimental

To achieve dual detection, a 20 μm-wide thin-film platinum electrode is fabricated directly onto a glass plate and aligned 50 μm from the channel exit on another glass plate prior to thermal bonding. An oxidized silver wire is used as a reference electrode, forming a two-electrode cell. The microchip is positioned on an inverted microscope, where a detection beam is directed near the channel outlet and about 80 μm upstream from the working electrode. A photomultiplier tube collects the fluorescent emission from the separated molecules. A diagram of the dual detection system for optically gated injections is shown in Figure 1. For cross-T injections, a T-shaped chip geometry is used instead of a straight channel, and no probing beam is necessary.

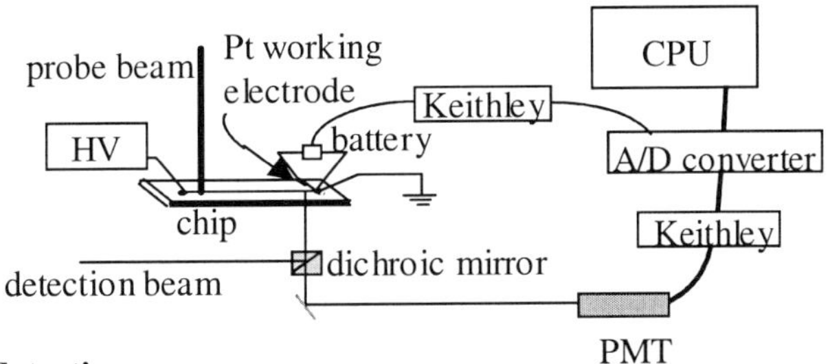

Figure 1. Simplified schematic of the experimental setup for dual detection of optically gated injections

3. Cross-T injection with dual detection

Naphthalene-2,3-dicarboxaldehyde (NDA) and ß-mercaptoethanol (BME) have been chosen as derivatizing agents because reaction of these with primary amines has been reported to produce a fluorescent and electroactive product [5,6]. Preliminary results with cross-T injections of fluorescent and electroactive species demonstrate the ability of the system to perform dual detection.

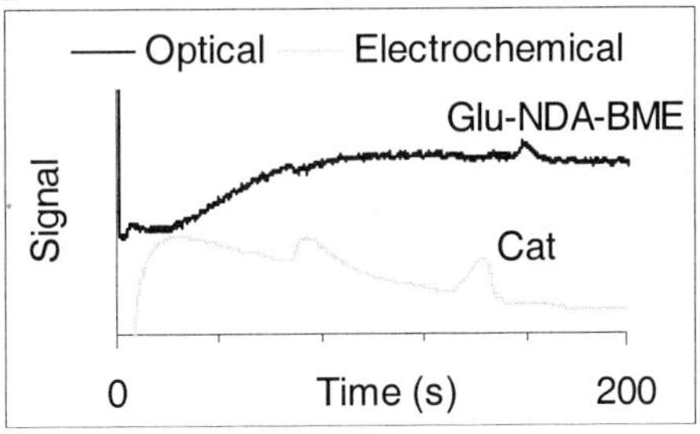

Figure 2. Cross-T injection and dual detection of Glu-NDA-BME and catechol in 10 mM borate buffer, pH 9.3. Separation electric field strength = 110 V/cm. Channel dimensions: 8 μm deep x 31μm wide.

The dual electropherogram in Figure 2 shows detection of 200μM glu-NDA-BME in the optical trace at ~160 s, and 10 mM catechol in the electrochemical trace at ~140 s. The electrochemical peak at ~80 s is the neutral electrochemical response to the methanol used to dissolve NDA. It is noted that, although NDA-BME derivatives have been shown to be electroactive [6], there appears to be no electrochemical detection of glu-NDA-BME. These data, however, represent the initial proof of dual detection for a single separation, and further investigation is underway to improve the separation efficiency and enhance the detection capabilities of the system.

4. Optically gated injection with dual detection

Sample introduction by optical gating produces not only the expected optical signal, but also an electrochemical response. Methods are being developed to more fully understand the dual response phenomenon of photobleaching. It has been proposed that the electrochemical oxidation of NDA derivatives involves the loss of electron from the heterocyclic nitrogen atom [7]. If this is the case, photobleaching may alter the ring structure responsible for fluorescence, but may or may not disrupt the ability to oxidize the nitrogen atom. In this way, dual detection after optical gating may provide a key to understanding the fundamental chemistry of photobleaching.

5. Conclusion

Orthogonal separation methods and channel multiplexing on a chip format have been shown to vastly increase the resolution and throughput of microanalysis, yet little work has been done to multiplex detection methods onto a single chip. Herein, dual electrochemical and optical detection has been demonstrated to widen sensitivity to more diverse numbers of species and may shed light on the different properties of the same analyte under controlled conditions. The effectiveness and versatility of the dual detection system, which can be applied to a cross-T geometry or optical gating for electroactive or fluorescent analytes, enhances an already powerful set of methods that can be used to analyze samples.

References:

1. I.M. Lazar, R.S. Ramsey, S. Sundberg, J.M. Ramsey, *Subattomole-Sensitivity Microchip Nanoelectrospray Source with Time-of-Flight Mass Spectrometry Detection*, Anal. Chem., 71, pp. 3627-3631, (1999).
2. P.F. Gavin, A.G. Ewing, *Continuous Separations with Microfabricated Electrophoresis – Electrochemical Array Detection*, J. Am. Chem. Soc., 118, pp. 8932-8936, (1996).
3. A.T. Woolley, K. Lao, A.N. Glazer, R.A. Matthies, Capillary Electrophoresis Chips *with Integrated Electrochemical Detection*, Anal. Chem., 70, pp. 684-688, (1998).
4. J.A. Lapos, A.G. Ewing, *Injection of Fluorescently Labeled Analytes into Microfabricated Chips Using Optically Gated Electrophoresis*, Anal. Chem., 72, pp. 4598-4602, (2000).
5. B.K. Matuszewski, R.S. Givens, K. Srinizasachar, R.G. Carlson, T. Higuchi, *N Substituted 1-Cyanobenz[f]isoindole: Evaluation of Fluorescence Efficiencies of a New Fluorogenic Label for Primary Amines and Amino Acids*, Anal. Chem., 59, pp. 1102-1105, (1987).
6. M.D. Oates, J.W. Jorgenson, *Determination of Naphthalene-2,3-dicarboxaldehyde Labeled Amino Acids by Open Tubular Liquid Chromatography with Electrochemical Detection,* Anal. Chem., 61, pp. 432-435, (1989).
7. M.A. Nussbaum, J.E. Przedwiecki, D.U. Staerk, S.M. Lunte, C.M. Riley, *Electrochemical Characteristics of Amino Acids and Peptides Derivatized with Naphthalene-2,3-dicarboxaldehyde: pH Effects and Differences in Oxidation Potentials,* Anal. Chem., 64, pp. 1259-1263, (1992).

CELL LYSIS AND PROTEIN EXTRACTION IN A MICROFLUIDIC DEVICE WITH DETECTION BY A FLUOROGENIC ENZYME ASSAY

Eric A. Schilling, Andrew Evan Kamholz and Paul Yager
Department of Bioengineering, Box 352255
University of Washington, Seattle, WA 98195, USA

Abstract

The characterization of a fluorogenic enzyme assay in a simple microfluidic device, the T-sensor, is presented. An integrated microfluidic system for the continuous lysis of bacterial cells and the extraction of β-galactosidase with detection by a fluorogenic enzyme assay is demonstrated.

Keywords: cell lysis, fractionation, fluorogenic enzyme assay, diffusion, T-sensor

1. Introduction

A critical requirement for achieving a micro total analytical system for the analysis of cells and their constituent proteins is to integrate the lysis and fractionation steps on-chip. Such a continuous flow device would be a valuable upstream component of further separations by isoelectric focusing or various chromatographic techniques, as well as for analytical techniques such as immunoassays. Here, a microfluidic system integrating the lysis of bacterial cells and the extraction of a large intracellular enzyme, β-galactosidase, is demonstrated.

2. Enzymatic Reactions in Microchannels

A fluorogenic substrate, resorufin-β-D-galactopyranoside (RBG), was used as an indicator of the presence of β-galactosidase [1]. Cleavage of RBG by β-Gal results in the production of resorufin, a small fluorophore (M.W. 235 Da, λ_{em} = 594 nm). The behavior of this assay was characterized in a microfluidic device simpler than the integrated device discussed below. This simpler device, a two-input T-sensor measuring 1.9 mm across and 10 μm deep [2], introduced β-Gal on one side and RBG on the other (Figure 1). The generation of resorufin was monitored via epifluorescence microscopy at 5000 μm downstream. The resulting data were fit using predictions from a new one-dimensional simulation based on a previously published version of a custom coded model [3]. Values for k_{cat} and K_m used in the model were found to agree quite well with previously published results [4].

J.M. Ramsey and A. van den Berg (eds.), Micro Total Analysis Systems 2001, 265–267.

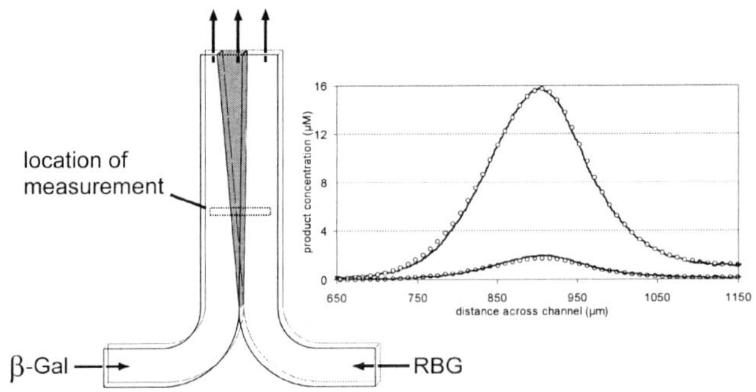

Figure 1. Experimental results of the enzyme assay in a T-Sensor with starting RBG concentrations of 256 µM (high peak) and 25.6 µM (low peak) and starting β-Gal concentrations of 0.46 µM. Fits from numerical simulations are shown with circles.

3. Integrated Lysis and Protein Extraction/Detection Device

A three-inlet, two-outlet microfluidic device was designed for the continuous lysis of bacterial cells and the separation/detection of intracellular proteins (Figure 2). Bacterial cells and lytic agent enter the *lysis channel* in adjacent fluid streams where lytic agent diffuses into the cell stream, lyses the cells and allows intracellular components to diffuse away from the cell stream. Depending on the relative flow rates at each inlet/outlet and on their diffusivity, some of these intracellular components are carried around the bend into the *reaction channel.* Although several different detection techniques could be employed, the above fluorogenic enzyme assay was used to detect β-Gal. The β-Gal (as well as smaller molecular weight components) that diffuses out of the cell stream enters the *reaction channel* where RBG in the adjacent flow stream is cleaved to form resorufin.

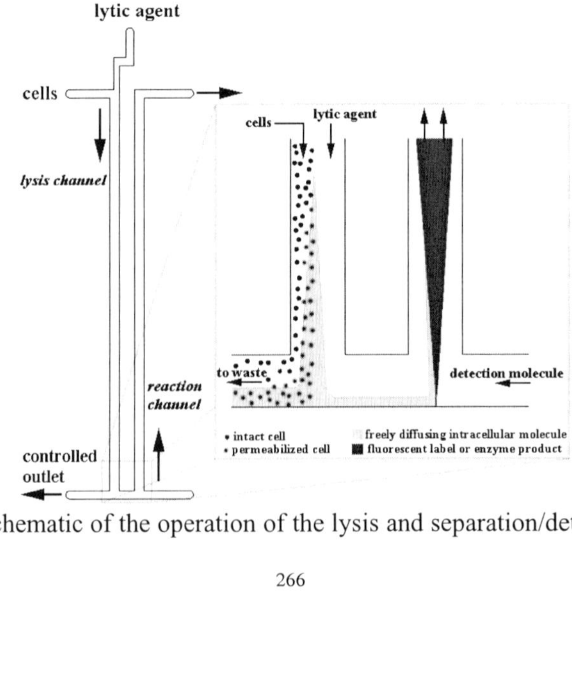

Figure 2. Schematic of the operation of the lysis and separation/detection device.

The integrated device was fabricated by stacking layers of Mylar cut by a PC driven CO_2 laser cutting tool. The main channel width was 1000 μm, the depth was 100 μm and the length was 4 cm. A suspension of *E. coli* cells (~10^8 cells/ml) and Bacterial Protein Extraction Reagent (BPER; Pierce, Rockford, IL) were pumped at 10 nl/s through separate inlets into the *lysis channel* (see Figure 2). These flow rates where chosen based on a CoventorWare (Coventor, Inc., Cary, NC) numerical model used to simulate the diffusion of β-Gal (D = 2.7×10^{-7} cm^2/s) in a similar geometry under pressure driven flow conditions. The flow rate at the *controlled outlet* was 14 nl/s such that the entire cell stream was pumped to waste. The removal of cells was verified by imaging cells under darkfield illumination with background subtraction (Figure 3A) and by fluorescence staining (data not shown). 200 μM RBG was pumped in the third inlet at a flow rate of 10 nl/s. The average residence time in the *lysis channel* was 190 seconds and that in the *reaction channel* was 250 seconds. Figure 3B demonstrates that a strong resorufin fluorescence signal is observed in the *reaction channel*, suggesting the successful extraction of β-Gal.

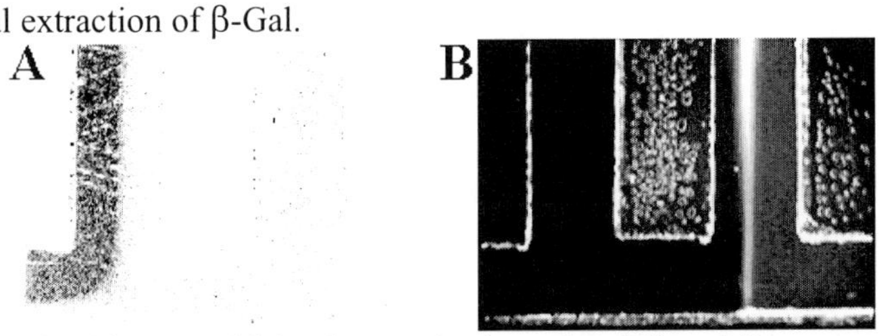

Figure 3. Darkfield image with background subtraction (A) and fluorescence image (B) demonstrating the removal of cells and the enzymatic production of resorufin.

4. Conclusions

This assay demonstrates the feasibility of a continuous flow integrated microfluidic device for the lysis of cells, diffusive separation of certain intracellular components and the detection/quantification of proteins extracted from the cell lysate.

Acknowledgements

This work was supported by the University of Washington Department of Bioengineering, The Washington Technology Center and Micronics, Inc.

References

1. K.D. Wittrup, J.E. Bailey. A single-cell assay of β-galactosidase activity in *Saccharomyces cerevisiae*, Cytometry 9, 1988, p. 394-404.
2. B.H. Weigl, P. Yager. Microfluidic diffusion-based separation and detection. Science 283, 1999, p. 346-347.
3. A.E. Kamholz, B.H. Weigl, B.A. Finlayson, P. Yager. Quantitative analysis of molecular interaction in a microfluidic channel: The T-sensor. Anal. Chem. 71, 1999, p. 5340-5347.
4. A.G. Hadd, D.E. Raymond, J.W. Halliwell, S.C. Jacobson, J.M. Ramsey. Microchip device for performing enzyme assays. Anal. Chem. 69, 1997, p. 3407-3412.

HIGH THROUGHPUT SINGLE CELL PCR ON A SILICON MICROCHAMBER ARRAY

H. Nagai[1,2], Y. Murakami[1], S. Wakida[2], E. Niki[2], E. Tamiya[1]
[1]School of Materials Science, Japan Advanced Institute of Science and Technology (JAIST), 1-1 Asahidai, Tatsunokuchi, Ishikawa 923-1292, Japan
[2]Human Stress Signal Research Center, National Institute of Advanced Industrial Science and Technology (AIST),
1-8-31 Midorigaoka, Ikeda, Osaka 563-8577, Japan
E-mail: hide.nagai@aist.go.jp

Abstract

PCR microchamber array was fabricated to be equipped with 10,000 chambers on a silicon chip. A simultaneous isolation method for thousands of cells using the micro-chamber array was examined and applied to single cell PCR to achieve high throughput analysis

Keyword: Polymerase chain reaction, single cell, microchamber array, cell isolation

1. Introduction

Single cell polymerase chain reaction (PCR), which uses genomes extracted from a single cell as template, is available for a research of viral diseases [1], cancer [2], and immunoglobulin gene rearrangement [3]. However, the isolation methods of single cell are laborious and demand expensive instruments. Therefore, a simple and inexpensive isolation method is important to realize high throughput cell isolation and analysis of cell library, such as peripheral blood lymphocytes. Here, we proposed a novel cell isolation method using a microchamber array. In general, a 1-pL well filled with a 1 pM solution will contain one molecule on average, and the 1 pM is equal to 6×10^8 molecules or materials in a 1 ml solution. Therefore, when the cells are dispensed and inlet into the microchambers through a single drop of defined concentration, the individual cells can be separated in each chamber depending on Poisson Distribution.

In the present research, the simultaneous isolation method for thousands of cells was examined and sequentially applied to single cell PCR to achieve high throughput analysis.

2. Experimental

Cell isolation

A silicon microchamber array chip was fabricated as described in [4]. The volume of the individual microchamber is 85 pL, and the 10,000 chambers are arranged within the area of 20 x 20 mm. The suspension of fluorescence-labeled Jurkat cells (human T-cell lymphoma) was prepared with various concentrations and then dropped onto the microchamber array. The number of microchambers containing the

J.M. Ramsey and A. van den Berg (eds.), Micro Total Analysis Systems 2001, 268–270.

only one cell was counted under a fluorescence microscope, and the optimum number of cells dispensed onto the chip was evaluated to isolate cells efficiently.

Single cell PCR

Jurkat cells suspension was mixed with proteinase K into buffer solution. The cells were isolated in the manner described above. The dispensed solutions were heated up to drying on a heater in order to inactivate proteinase K. A glass slip was covered over the microchambers, and then PCR solution was injected between the chip and the glass slip. TaqMan^{TM} rebosomal RNA control kit was employed as the PCR contents, and Triton X-100 (neutral surfactant) was added into the solution to lyse the cells. The chip was sealed by vanish and clamped into a novel thermal cycle system, which was suitable for the microchamber array chip (Figure 1). Single cell PCR was performed in each microchamber simultaneously, and the amplification of PCR products was estimated using TaqMan^{TM} chemistry that has increased fluorescence with amplification of the target gene.

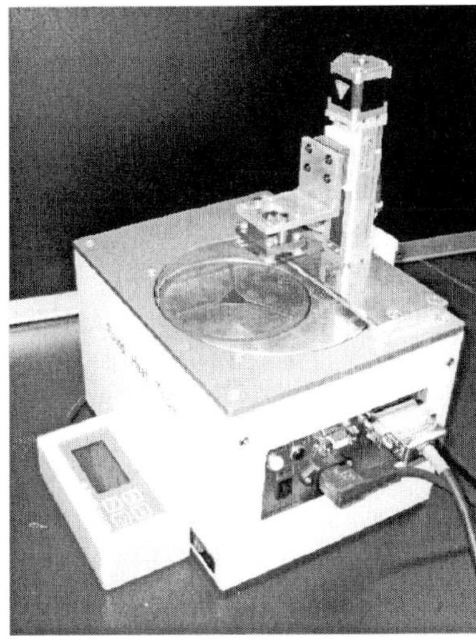
Figure 1. Thermal cycler for the PCR microchamber array.

Results and Discussion

Cell isolation

Table 1 shows the relationship between the total number of dispensed cells and the probability of chamber containing different number of cells. The number of dispensed cells in each chamber obeyed Poisson Distribution. When 1 x 10^4 cells were dropped on the microchamber array, the efficiency of cell isolation was the highest. Consequently, about 1,800 cells can be isolated simultaneously and the ratio of the chamber containing more over two cells was less than 2.5%.

Table 1. The probability of chamber with the isolated cell and more over two cells.

Total dispensed cells	The probability for cell number in chamber /%					Isolation efficiency*/%
	0	1	2	3	4	
5 x 10^4	32.4	34.1	22.4	8.6	1.9	6.8
1 x 10^4	79.3	18.4	2.3	0.3	0	18
5 x 10^3	96.4	3.6	0	0	0	7.2
1 x 10^3	99.3	0.5	0	0	0	5.0

*The ratio of isolated cells in total dispensed cells

Single cell PCR

Single cell PCR was performed after cell isolation with the same condition. The fluorescence intensity increased in 14% of the microchambers as shown in Fig. 2. This result indicates that the PCR amplifications were successful from single cell, because the ratio of the chamber containing more over two cells (2.5%) was lower than 14%. Since the probability of chamber confirmed the PCR amplification were slightly less than the ratio in the above cell isolation experiments, it seems that all isolated cells could not be lysed efficiently and their DNAs could not be used in the PCR.

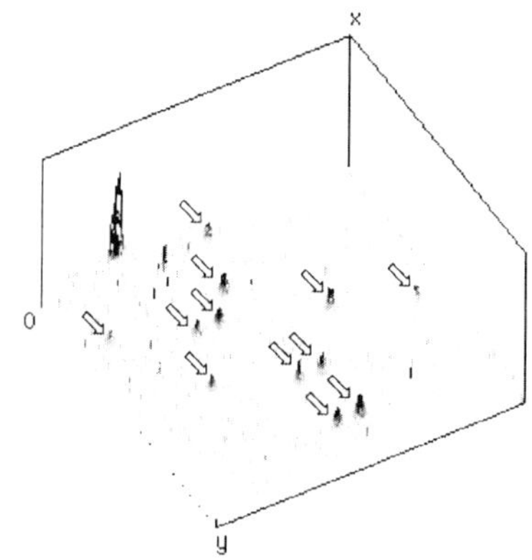

Figure 2. Single cell PCR on the microchamber array chip. Unfilled allows indicate the amplified PCR products.

Conclusions

We demonstrated that the single cell PCR system based on the microchamber array carried out high throughput analysis for more over a thousand cells. The distribution of cells on the microchamber array was adopted to isolate a large number of cells. This single cell PCR methods will realize the analysis of cell library, such as the whole lymphocytes in peripheral blood sample.

Acknowledgements

This work was partially supported by research for the future program of JSPS. Special thanks are given to Toyo Gosei Kogyo Co., Ltd. for kindly providing tetra-methyl ammonium hydroxide solution.

References

1. T. Miyagi, K. Murakami, T. Sawada, H. Taguchi, I. Miyoshi, *A novel single cell PCR assay: detection of human T lymphotropic virus type I DNA in lymphocytes of patients with adult T cell leukemia*, Leukemia, 12, pp.1645-1650, (1998).
2. R. Kuppers, K. Rajewsky, M. Zhao, G. Simons, R. Laumann, R. Fischer, M.L. Hansmann, *Hodgkin disease: Hodgkin and Reed-Sternberg cells picked from histological sections show clonal immunoglobulin gene rearrangements and appear to be derived from B cells at various stages of development*, Proc. Natl. Acad. Sci., 91, pp.10962-10966, (1994).
3. E. ten Boekel, F. Melchers, A.G. Rolink, *Changes in the V(H) gene repertoire of developing precursor B lymphocytes in mouse bone marrow mediated by the pre-B cell receptor*, Immunity, 7, pp.357-368, (1997).
4. H. Nagai, Y. Murakami, Y. Morita, K. Yokoyama, E. Tamiya, *Development of a microchamber array for picoliter PCR*, Anal. Chem., 73, pp.1043-1047, (2001).

ON-CHIP CELL SORTER FOR SINGLE CELL EXPRESSION ANALYSIS

T. Ichiki[1], T. Ujiie[1], T. Hara[1], Y. Horiike[2], and K. Yasuda[3]
[1]Bio-Nano Electronics Research Center, Toyo University, Japan
[2]Department of Materials Science, School of Engineering, The University of Tokyo, Japan
[3]Depertment of Life Sciences, Graduate School of Arts and Sciences, The University of Tokyo, Japan
e-mail address: ichiki@eng.toyo.ac.jp

Abstract

We present an on-chip cell sorter system developed for single-cell analysis. Damage-free selective collection of a single biological cell becomes possible owing to the combined use of the laminar sheath flow and the repulsive dielectrophoretic force in the well-designed microfabricated channels. By using this chip as a core technology, an innovative cell sorting system based on the biologically significant information about the state of each cell becomes attainable in combination with the high-resolution image processing system.

Keywords: single-cell expression analysis, cell sorter, repulsive dielectrophoretic force, image processing

1. Introduction

In the post-genome era, the principal issue is to clarify the relation between genome expression and cell's condition. The single-cell expression analysis, which stands on quite a different viewpoint from the conventional biology, is proposed as a powerful methodology for this purpose [1, 2]. Though the minimum unit of life phenomena is a cell, the statistical measurement of numerous cells has been the common practice in the biology. Since no means for strict control of the origin of each cell, cultivation environment and a cell cycle have been available, averaging has been used to neglect uncertain factors. Establishment of sorting technology of well-defined cells allows us to separate these uncertain factors from experimental results and to have a clear understanding of biological mechanisms. In order to realize this single cell expression analysis, it is necessary to provide a set of technology for preparation, cultivation, manipulation and analysis of a well-defined cell originated from one specific cell. In this paper, we present the development of a cell sorter chip used as the core technology for establishing the cell sorting system that can separate and collect an individual cell according to its biological or biochemical information.

2. Fabrication of Cell Sorter Chips

Figure 1 shows the design of a cell sorter chip. Biological cells are transported from the injection channel toward the central collection channel owing to the laminar sheath flow. Repulsive dielectrophoretic (DEP) force, induced by applying ac voltages to a pair of electrode on both sides of the collection channel, enables selective drive of unnecessary cells to the waste channel. In order to determine proper design and dimensions of the microchannels,

J.M. Ramsey and A. van den Berg (eds.), Micro Total Analysis Systems 2001, 271–273.

fluid flow simulation and calculation of the potential distribution were carried out as typically presented in Figs. 2 and 3. Figure 4 shows the process flow of the chip. Microchannels were fabricated by deep quartz etching technology using high-density fluorine-based plasmas and Cr metal masks [3]. A pair of Pt/Cr electrodes was embedded beside the collection channel using shallow quartz etching and subsequent lift-off process. The processed quartz plate was then bonded with the other one with drilled holes for reservoirs and electrical connection, being dipped in 1% diluted HF, and finally press-bonded under the load of 1.3 MPa at room temperatures.

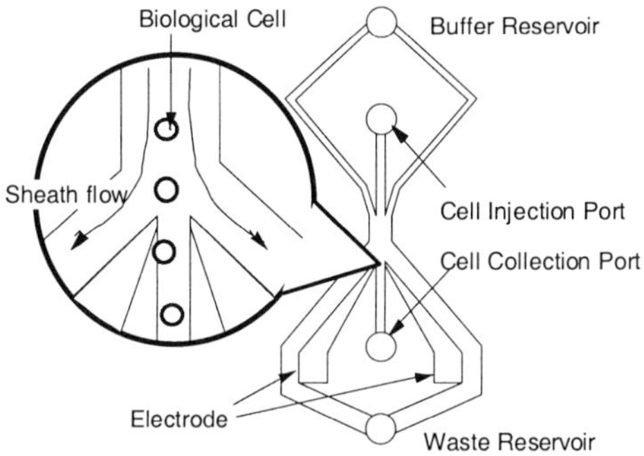

Figure 1: Schematic design of an on-chip cell sorter.

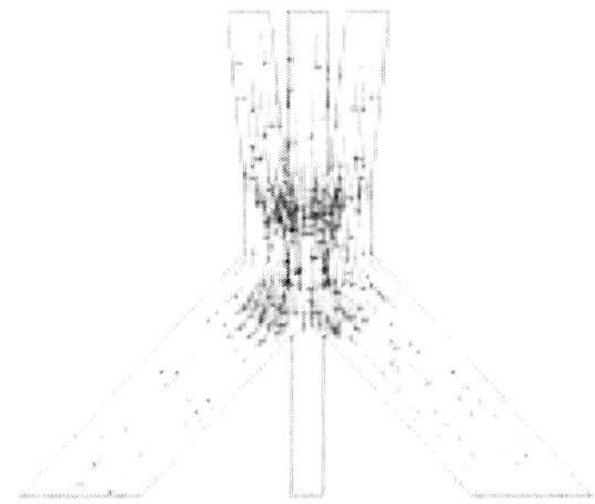

Figure 2: A representative result of fluid flow simulation at the intersection of a cell sorter chip.

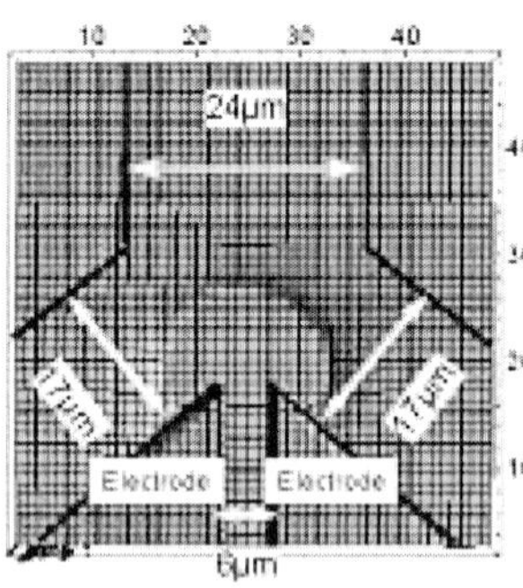

Figure 3: Two dimensional calculation of E^2 was carried out in order to examine the approximate distribution of dielectrophoretic force field around the entrance of the cell collector channel.

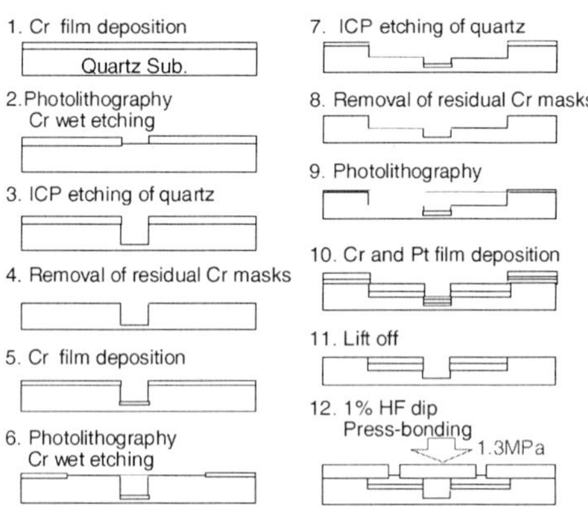

Figure 4: Process sequence of a cell sorter chip.

3. Operation of Cell Sorting System

Performance of the prototype chip was tested using sheep red blood cells suspended in PBS at 1×10^8/ml as a sample. Without applying ac voltage to electrodes, introduced cells flew directly into the collection channel. Figure 5 exhibits the operation conditions for the selective elimination of needless cells, examined with regard to the flow velocity of cells and the applied peak-to-peak voltage. The regional boundary of operation mode is subject to the parabolic correlation since it is determined by the balance between viscous resistance proportional to the streaming velocity and DEP force proportional to the second power of electric field strength. It should be emphasized that this chip allows us to collect necessary cells without the influence of strong electric field. Furthermore, since microchannel has a rectangle cross-section, there is no distortion of an optical image and precise image observation is attainable with an optical microscope (see Fig. 6). Therefore, cell sorting system based on the biologically significant information about the state of cells, which is impossible using the conventional cell sorter, becomes attainable in combination with a precise image processing system.

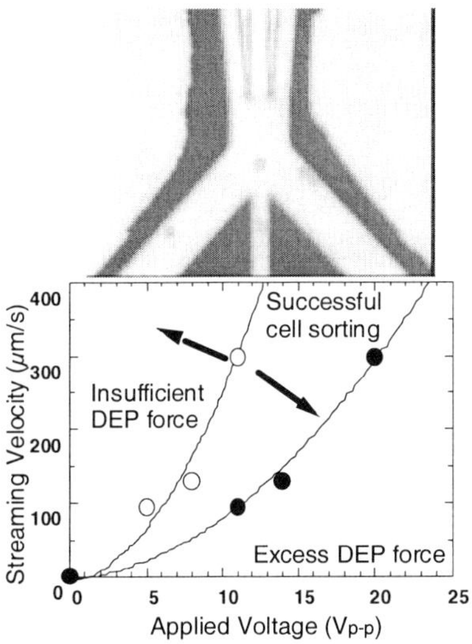

Figure 5: An optical microscopic image of a cell sorter chip separating sheep red blood cells (top). Typical operation conditions of a cell sorter chip for sheep red blood cells (bottom).

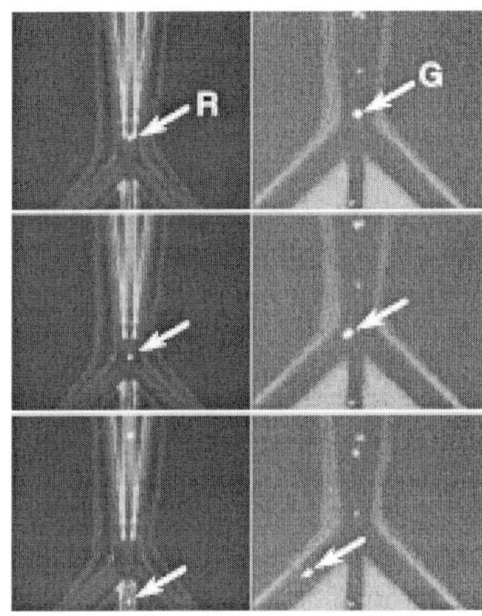

Figure 6: Selective collection of *E. coli* with red fluorescent protein from the mixture of those with red or green fluorescent protains.

4. Conclusions

An on-chip cell sorter device has been developed as an core technology of the innovative cell sorting system for single-cell expression analysis. Future development includes the experimental verification of its usefulness also for delicate samples like nerve cells.

Acknowledgment

This work was supported in part by a Grant-in-Aid for Scientific Research (No. 12838 011) from Japan Society for the Promotion of Science , and High-Tech Research Center organized by the Ministry of Education, Sports and Culture of Japan.

References

1. I. Inoue, Y. Wakamoto, K. Onuma, A. Kashiwagi, K. Kaneko, T. Yomo, and K. Yasuda, Biophys. J. 80, 166a-167a (2001).
2. J. Ziauddin, and D. M. Sabatini, Nature, 411, 107-110 (2001).
3. T. Ujiie, T. Kikuchi, T. Ichiki and Y. Horiike, Jpn. J. Appl. Phys. 39, 3677-3682 (2000).

INTEGRATED LIGHT COLLIMATING SYSTEM FOR EXTENDED OPTICAL-PATH-LENGTH ABSORBANCE DETECTION IN MICROCHIP-BASED CAPILLARY ELECTROPHORESIS

Kyung Won Ro, Bong Chu Shim, Kwanseop Lim, and Jong Hoon Hahn*
Department of Chemistry, Pohang University of Science and Technology,
San 31 Hyoja-Dong, Pohang, 790-784, South Korea
(email: hahn@postech.ac.kr)

Abstract
We have developed an integrated light collimating system with a microlens and a pair of slits for extended optical path-length absorbance detection in a capillary electrophoresis (CE) microchip. The collimating system is made of the same material of the chip, poly(dimethylsiloxane) (PDMS), and integrated into the chip when molding the microchannels of CE. With this system, much enhanced performance in absorbance detection is obtained compared with that of uncollimated system.

Keywords: integrated light collimating optics, absorbance detection, Z-shaped flow cell

1. Light Collimation for Optical-Fiber-Based Absorbance Detection

Insertion of a pair of optical fibers into a capillary electrophoresis (CE) microchip provides a simple means of coupling the fibers to an external light source and a detector [1]. However, the intrinsic divergence of light from an optical fiber degrades the

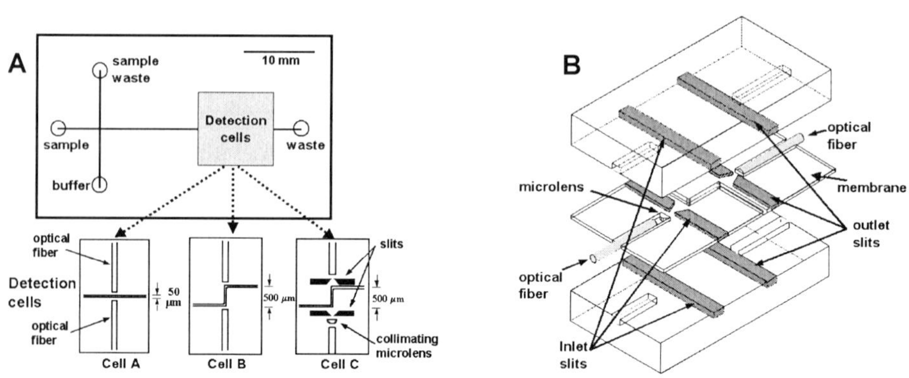

Figure 1. (A) Layout of the absorbance detection microchip fabricated in PDMS. The insets show the three types of detection cells: Cell A, on-channel flow cell; Cell B, uncollimated Z-shaped cell; Cell C, collimated Z-shaped cell with a microlens (r = 50 μm) and slits. (B) Anatomic view of the collimated absorbance detection system.

J.M. Ramsey and A. van den Berg (eds.), Micro Total Analysis Systems 2001, 274–276.

linearity and sensitivity in detection, especially when an extended optical-path-length flow cell is used.

In order to enhance the detectability by extending the optical-path-length without loosing the linearity, we have integrated a light collimating system with a microlens and a pair of slits into a CE microchip equipped with a Z-shaped flow cell.

2. Fabrication of Light Collimating System

By modifying the "membrane sandwich" fabrication method [2], we have fabricated, using three photomasks, a three-dimensionally channeled poly(dimethylsiloxane) (PDMS) microchip, where the centers of the flow channel (depth: 27 μm, width: 50 μm) and two optical fibers of 120 μm O.D. are aligned, a planoconvex PDMS microlens (r = 50 μm) is placed in front of the light-delivering fiber (3 μm core diameter), and two-rectangular apertures (50 × 27 μm) are made behind the microlens and in front of the collecting fiber (50 μm core diameter), respectively (figure 1). The aperture serving as a slit is realized by filling black ink into a specially designed three dimensional microchannel surrounding a rectangular PDMS structure (figure 1B).

3. Results and Discussion

Figure 2 clearly shows that light divergence is very serious along the 500 μm detection cell, resulting in poor linearity in absorbance detection (figure 4). The approximate upper limit of extending the path length without collimation has been known to be only 150 μm [1]. However, in our collimating system, the numerical aperture of the delivering fiber is matched with the focal length of the microlens, and the diverged light is collimated (figure 3). In addition, two rectangular apertures block other stray light.

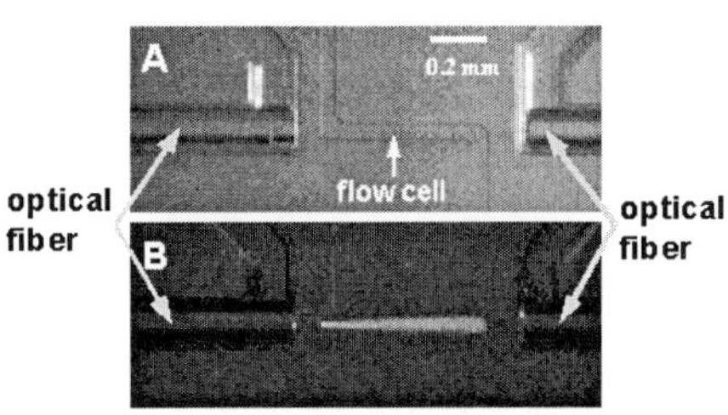

Figure 2. CCD images of uncollimated Z-flow cell system (cell B). (A) Images of optical fibers and flow cell before illuminated. (B) Divergent beam path illuminated by 50 μM fluorescein at 520 nm.

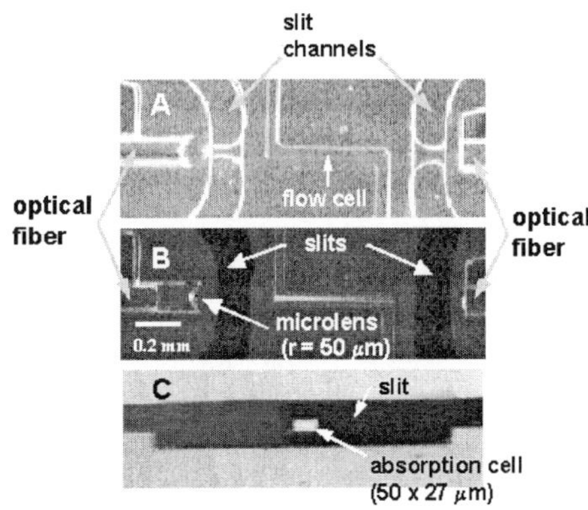

Figure 3. CCD images of collimated Z-flow cell system (cell C). (A) Images of flow cell and slits before filling of black ink. (B) Collimated beam path in a collimated cell and the image of the slits after filling of ink. (C) The cross-sectional image of flow cell through the slits.

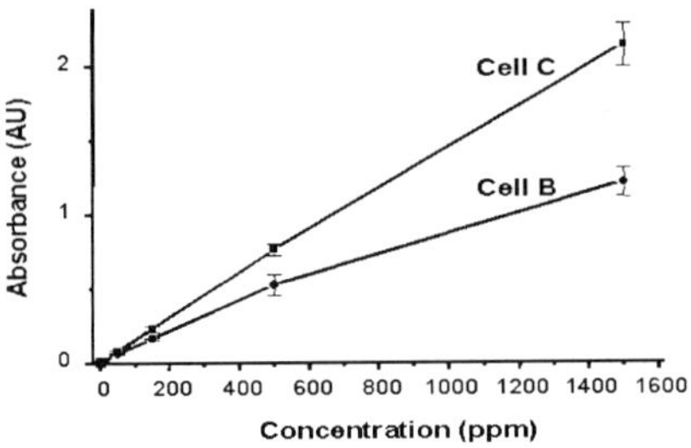

Figure 4. Dependence of the absorbance on the concentration of new coccine in a collimated (Cell C) and an uncollimated (Cell B) detection systems.

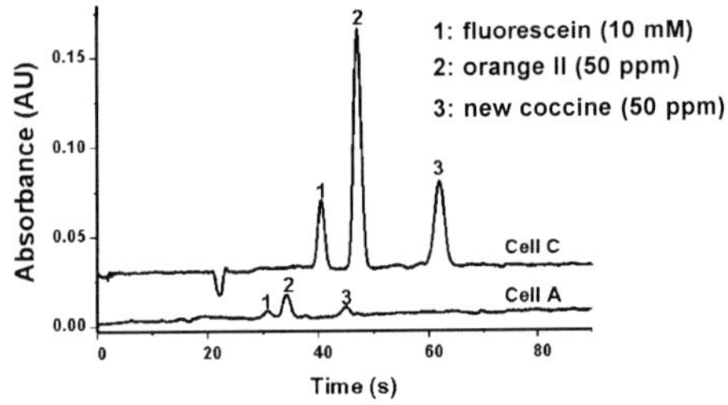

Figure 5. Comparison of detector responses in an on-channel (Cell A) and a collimated Z-shaped flow cells (Cell C). Running buffer: 15 mM borax (pH 10), 25mM SDS. The effective separation channel lengths to the cell A and cell C are 15 mm and 25 mm, respectively.

Figures 4 and 5 illustrate the expected performance of our collimating system: without degrading the linear dynamic range, the detection sensitivity has been increased 10 times, compared with the unextended, 50 μm flow cell. The detection limit and linear dynamic range of new coccine in the collimated cell are 1.3 ppm and 3 orders in concentration, respectively.

4. Conclusions

We have fabricated an integrated light collimating system for the absorbance detection of microchip CE in PDMS. While this work has been demonstrated with a single-mode optical fiber and a laser source, it will certainly work with a multi-mode fiber and other white light sources. In addition, the transition from visible to UV will be readily accomplished, because PDMS is optically transparent down to ~ 230 nm. The ability to use an absorbance detector for microchips should greatly extend the applicability of these devices to various chemical, biological and environmental samples.

References

1. Z. Liang, N. Chiem, G. Ocvirk, T. Tang, K. Fluri, D. J. Harrison, "Microfabrication of a Planar Absorbance and Fluorescence Cell for Integrated Capillary Electrophoresis Devices", *Anal. Chem.*, **68**, (1996), 1040-1046.
2. J. R. Anderson, D. T. Chiu, R. J. Jackman, O. Cherniavskaya, J. C. McDonald, H. Wu, S. H. Whitesides, G. M. Whitesides, "Fabrication of Topologically Complex Three-Dimensional Microfluidic Systems in PDMS by Rapid Prototyping", *Anal. Chem.* **72**, (2000), 3158-3164.

A SILICON MICROCHIP FOR INFRARED TRANSMISSION KINETICS STUDIES OF RAPID HOMOGENEOUS LIQUID REACTIONS

T. M. Floyd[1], M.A. Schmidt[2] and K.F. Jensen[1,3]

[1]Department of Chemical Engineering, [2]Microsystems Technology Laboratory and [3]Department of Materials Science and Engineering
Massachusetts Institute of Technology, Cambridge, MA. U.S.A

Abstract

A silicon microchip for infrared transmission kinetics studies of rapid liquid reactions is demonstrated. The chip consists of a laminar flow mixer, a parallel plate heat exchanger and a probing region for on chip detection. The use of the integrated system is demonstrated with a model reaction, the hydrolysis of methyl formate. The resulting rate constant agrees with the literature value. Moreover, no sample post-processing is required as compared to conventional systems.

Keywords: micromixer, infrared detection, kinetics studies

1. Introduction

Standard infrared (IR) transmission detection for liquid systems is limited by sample size, lack of on-chip mixing, and narrow operating temperatures due to mismatched thermal coefficients of expansion of the fabrication materials. Microfabricated silicon devices have the potential to integrate mixing and sensing with IR spectroscopy. The systems operate over a wide range of temperatures because the entire chip is comprised of silicon, and fusion bonding produces a fixed path length, which facilitates the acquisition of reproducible data. One limitation of the technique is interference fringes observed when bare silicon is used as an IR crystal. This limitation is overcome by engineering an anti-reflective coating for the crystal.

This work reports a fully integrated microchemical system for IR detection (Fig. 1(a)), which consists of a multi-channel mixer (Fig. 1(b)), a parallel plate heat exchanger (Fig. 1(c)) and an IR probing region. The multi-channel mixer allows faster mixing than simple T-contactors, thus providing an opportunity to study faster reactions. The integration of a heat exchanger implies that the reaction mixtures can be cooled. These capabilities, coupled with a heating element in the sample holder, allow kinetics studies to be performed over a wide range of temperatures.

2. Experimental

The microchip consists of two bonded silicon wafers. The channels and probing region of the microchip are defined using photolithography and deep reactive ion etching (DRIE). The channels are ~400 μm deep, whereas the probing region is only 50 μm deep, so a nested etch is required. After the channels and chamber are defined, the patterned wafer and a blank wafer are fusion bonded. Next, the inlet ports are patterned

J.M. Ramsey and A. van den Berg (eds.), Micro Total Analysis Systems 2001, 277–279.

using DRIE, and the microchip is released as a 32 mm diameter cylinder that is compatible with a commercial sample holder (Fig. 2).

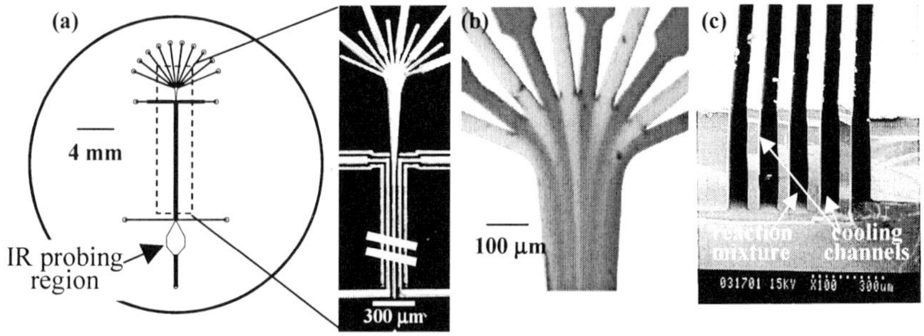

Figure 1: (a) schematic of microchip channels drawn to scale and zoomed top view of contacting region (b) mixing flow visualization (c) scanning electron microscope cross section of parallel plate reactor/heat exchanger

Direct bonding of tubing to the chip or a fluidic manifold with a compressed polymeric gasket provides fluidic access to the microchip. The microchip loads into a sample holder, which contains a heating element and thermocouple for high temperature studies. The chip and holder are placed in the sample compartment of a Nicolet 800 IR spectrometer. Fluids are delivered to the microchip using a syringe pump.

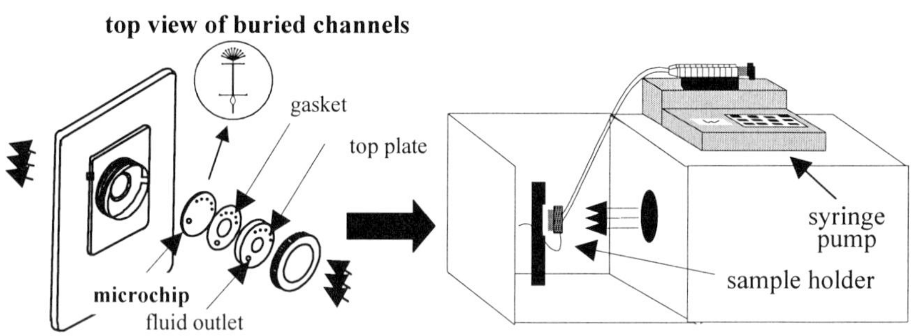

Figure 2: Testing station for the microchip

3. Results and Discussion

Rapid mixing is essential for determining the kinetic parameters of rapid liquid reactions. An acid/base buffer reaction mixing technique showed that the extent-of-mixing is 99% after 25 ms (Fig. 3(a)). This result was confirmed with 3-D computational fluid dynamics (CFD) predictions. CFD contour plots for one of two mixing components with concentrations normalized from 0 to 1 (Figs 3(b)-(d)) reveal the mixing profiles in the system.

The hydrolysis of methyl formate was chosen as a model reaction to demonstrate the system for kinetics studies. The reactants, aqueous methyl formate and NaOH, are delivered to the microchip. When the system has reached steady state, data are collected using a Nicolet 800 IR Spectrometer. (The time to reach steady state is controlled largely by the macroscopic fluid handling.) Light from the spectrometer source passes through the IR probing region. The absorbance of the light is proportional to the concentration of the sample. The residence time for the reaction is manipulated by changing the delivery flow rate of the reactants. A rate constant was extracted from data collected at 294 K. The experimental value of 30.6 L/mol/s compares well with the literature value of 30.1 L/mol/s [1]. In addition, no post processing was required unlike in the original study where samples were titrated.

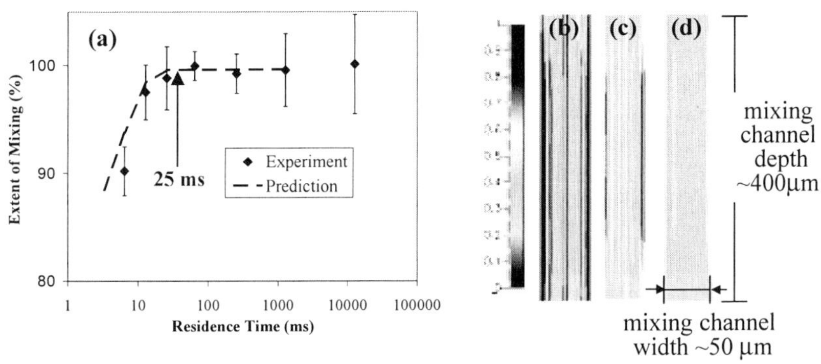

Figure 3: (a) Experimental and CFD predictions for extent of mixing (b)-(d) channel cross sections: normalized concentration contours for one of two components contacted in the laminar flow mixer (b) after focusing, (c) 4 mm down the channel length and (d) 8 mm down the channel length for a flow rate of 1.0 ml/min corresponding to a residence time of ~10 ms

4. Conclusions

A fully integrated silicon microchip for kinetics studies of rapid liquid reactions was fabricated using DRIE and silicon fusion bonding. The mixing time for the device is <25 ms. A rate constant for the hydrolysis of methyl formate was extracted and was in good agreement with the literature value. Due to rapid mixing and good thermal management, the minimum half-life for the reaction was reliably decreased by an order of magnitude while eliminating sample post processing. These results demonstrate the advantage of integrating mixing, reaction, temperature control, and IR spectroscopy on-chip for kinetics studies.

5. Acknowledgements

The DARPA MicroFlumes Program supported this work. TMF thanks the NSF and UNCF•Merck Science Initiative for graduate fellowships.

6. References

1. H. M. Humphreys and L.P. Hammett., "Rate Measurements on Fast Reactions in the Stirred Flow Reactor; the Alkaline Hydrolysis of Methyl and Ethyl Formate", Journal of the American Chemical Society, Vol. 78, 1956, p.521-524.

IN-PLANE UV ABSORBANCE DETECTION IN SILICON-BASED ELECTROPHORESIS DEVICES USING MONOLITHICALLY INTEGRATED OPTICAL WAVEGUIDES

Klaus B. Mogensen[1], Nickolaj J. Petersen[1], Jörg Hübner[2] and Jörg P. Kutter[1]
[1]MIC-Mikroelektronik Centret, DTU, Bldg. 345 East, DK-2800 Lyngby, Denmark.
[2]COM, DTU, Bldg. 345 West, DK-2800 Lyngby, Denmark.

Abstract

A silicon-based capillary electrophoresis device with monolithically integrated UV transparent waveguides and electrically insulated channels is presented. A separation with UV absorbance detection at 254 nm was performed. The applicability of Plasma-enhanced Chemical Vapor Deposited (PECVD) silica as electrical insulation is furthermore investigated.

Keywords: Planar waveguides, capillary electrophoresis, UV absorbance detection.

1. Introduction

Integration of optical waveguides and fluidic components in μ-Total Analysis Systems (μ-TAS) has numerous advantages over using bulk optics in the system; e.g. an inherent stability, a precisely aligned detection core and a higher sensitivity for absorbance detection, because measurement in the plane of the device can be utilized [1-2].

In this paper, we present a silicon-based capillary electrophoresis device with electrically insulated channels and UV transparent optical waveguides. In-plane UV absorbance measurements with a path length of 120 μm across the separation channel have been performed. The waveguides were permanently connected to a light source and a detector by glued-on optical fibers, which made alignment-free operation possible. This is especially advantageous when operating outside the visible range.

Finally, the applicability of plasma-deposited (PECVD) glass as an electrically insulating layer is investigated.

2. Experimental

The device was fabricated on a silicon substrate by means of surface micromachining techniques such as deposition and anisotropic etching of silica glass [2]. A 13 μm thick thermally grown silicon dioxide layer was used in order to achieve electrically insulated channels during electrophoretic separations. This fabrication technology resulted in a CE device that could withstand an operation voltage of 10 kV [2].

J.M. Ramsey and A. van den Berg (eds.), Micro Total Analysis Systems 2001, 280–282.

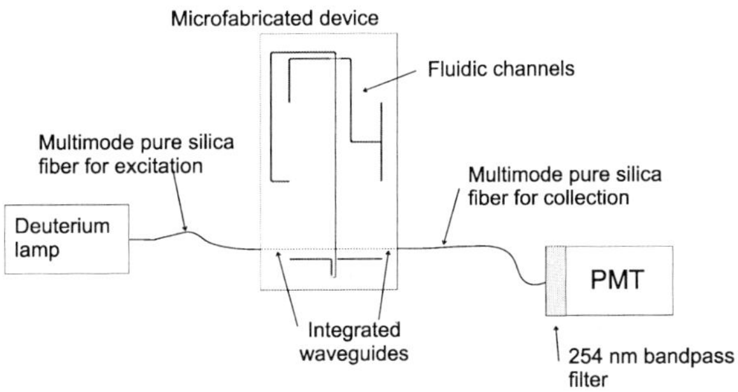

Fig. 1. Schematic view of the instrumental setup. The chip was connected to a deuterium lamp and a PMT with optical fibers, which enabled alignment-free operation.

3. Results and discussion

The chip was tested at 254 nm with 24 μm wide waveguides arranged perpendicular to a 120 μm wide separation channel (Fig. 1). Figure 2 shows a separation of 12.9 mM caffeine, 16.5 mM paracetamol, 9.8 mM ketoprofen and 14.2 mM L-ascorbic acid.

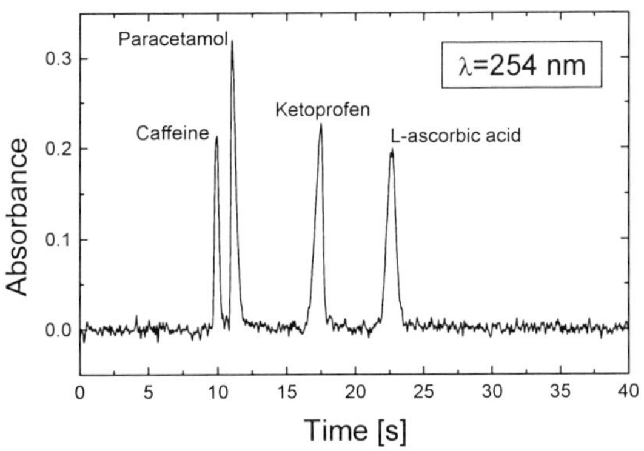

Fig. 2. Electropherogram obtained using absorbance detection at 254 nm with a path length of 120 μm. L_{Sep}=2.89 cm, E = 0.55 kV/cm, injection time: 0.2 s. 25.0 mM borate buffer (pH 9.2).

Collection of stray light during the measurement resulted in a low signal-to-noise ratio. The performance is expected to be improved considerably by optimization of the absorbance cell design in order to suppress detrimental stray light effects in the device.

Electrical insulation towards the silicon substrate was achieved by growing a 13 μm thick thermal silicon dioxide layer. This process yields high quality oxides in terms of defect density, surface roughness, thickness uniformity and electrical breakdown strength and has the further advantage of being a batch process. An inherent disadvantage, beside the long processing time (about 21 days), is that recombination centers are introduced in the silicon crystal during oxidation, which decreases the lifetime of the minority carriers (<5 μs). The process is hence not compatible with monolithic integration of diffusion-limited photodiodes. This shortcoming was the motivation for investigation of the electrically insulating properties of PECVD silica

(Fig. 3). The deposition time of the undoped PECVD glass ranged from 10 to 40 minutes.

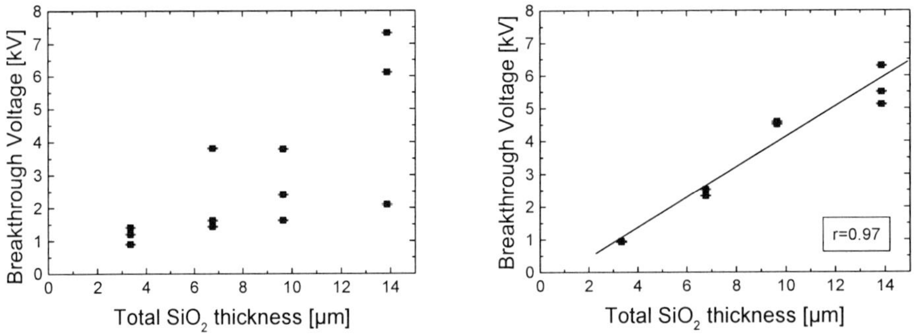

Fig. 3: Left) Breakthrough voltage of as-deposited undoped PECVD glass. **Right)** Breakthrough voltage after annealing at 1100° C in O_2 for 12 hrs.

The results shown in Figure 3 were obtained by gluing reservoirs onto the deposited silicon wafers (p-type, resistivity < 0.025 Ωcm). Electrodes were inserted in two adjacent reservoirs after filling with a 100 mM borate buffer. The insulating layer had to break down in two places (in each reservoir) in order to establish a short circuit. The total thickness given in Fig. 3 is hence twice the thickness of the deposited SiO_2 layer. It is seen that the glass has to be annealed in order to be usable as an insulating layer in CE microdevices. This behavior is expected, because PECVD SiO_2 densifies and stabilizes upon high temperature annealing [4]. The dielectric breakdown strength of the annealed PECVD films was obtained from the fit in Fig. 3 to 4.6 MV/cm.

4. Conclusions

An electrophoretic separation with UV absorbance detection at 254 nm was performed in a silicon chip with integrated optical waveguides. The applicability of annealed PECVD SiO_2 as insulating layers in CE devices was investigated.

References

1. G. Pandraud, T. M. Koster, C. Gui, M. Dijkstra, A. van den Berg, P. V. Lambeck, Sens. Actuators A, Vol. 85, p. 158-162, (2000).
2. K. B. Mogensen, N. J. Petersen, J. Hübner, J. P. Kutter, Electrophoresis. Accepted for publication.
3. K. B. Mogensen, P. Friis, J. Hübner, N. Petersen, A. M. Jorgensen, P. Telleman, J. P. Kutter, Opt. Lett., Vol. 10, p. 716-718, (2001).
4. C. M. M. Denisse, K. Z. Troost, F. H. P. M. Habraken, W. F. van der Weg, M. Hendriks, J. Appl. Phys, Vol. 60, p. 2543-2547, (1986).

DEVELOPMENT OF SERIAL DILUTION CHIP FOR CELLULAR BASED ANALYSIS

Seok Chung[1], Jung-Kyung Kim[2], Junha Park[1], Yongku Lee[1], Hyun-Woo Bang[1], Sung-Jin Park[1], Chanil Chung[3], Dong-Chul Han[1,3] and Jun-Keun Chang[3,4]

[1]School of Mechanical & Aerospace Eng., Seoul Nat' l Univ., San56-1, Shinlim-dong, Kwanak-gu, Seoul, 151-742, Korea

[2]Interdiciplinary Prog. In Medical & Biological Eng. Major, Seoul Nat' l Univ.

[3]Digital Bio Technology Co., Seoul, Korea (http://www.digital-bio.com)

[4]School of Electrical Eng. & Comp. Science, Seoul Nat' l Univ.

(jkchang@amed.snu.ac.kr)

Abstract

Dilution of microfluids is very complex because of the flexibility of microfluids concerned with the channel size, surface characteristics and Reynolds number. With the simulation results of dilution in microchannel, we developed the novel serial dilution chips made of PDMS and glass plate. These chips can be applied to the cell toxicity test, bacteria counting test and Minimal Inhibitory Concentration test of antibiotics and so on.

Keywords: PDMS, Microfluidics, Serial dilution, Cell toxicity test

1. Introduction

The electroosmosis is useful in microfluidics because it make the microflows controllable with voltage and surface characteristics. While hydrodynamic control with flow rate is more difficult because the flow characteristics are changed following the channel size, surface characteristics and Reynolds number. However the syringe controlled flows are more applicable, safe and familier to the researchers in the biological & medical fields.

2. Simulation

In simulation, we fixed the channel dimension, and changed channel depth, flow velocity and surface characteristics. Two dilution methods (StopNGo and LongChannel, DBT^{TM}) were simulated with CFD-ACE software (CFDRC Inc., AL). Figure 1 shows one of the simulated results about the two-liquids dilution.

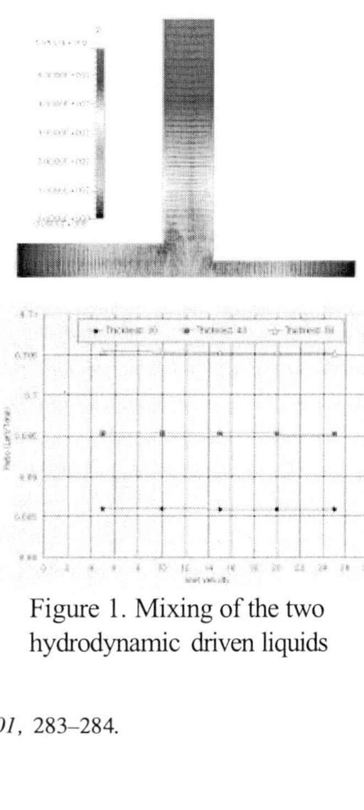

Figure 1. Mixing of the two hydrodynamic driven liquids

J.M. Ramsey and A. van den Berg (eds.), Micro Total Analysis Systems 2001, 283–284.

3. Fabrication & experiments

Figure 2 shows one of the fabricated serial dilution chips using LongChannel dilution method and the multi-layered dilution chip using StopNGO dilution scheme. These chips consist of a connector and a disposable chip with cell culture wells and serial dilution micro channel part. The disposable chip is made of micro-molded PDMS plate bonded onto slide glass. Reliable tubing can be easily aquired with patented tubing protocol.

The diameter of cell culture well is 5 mm, large enough to access in the cell culturing. The number of wells is from 4 to 8, and can be increased following the chip size. The dilution scale is linear or exponential (such as 0.1, $(0.1)^2$, $(0.1)^3$, $(0.1)^4$... and so on).

After autoclaving, HUVAC is cultured in the cell culture well, and we make trypsin and buffer serially diluted in the chip. Figure 3 shows the removed cell proportional to the concentration of trypsin. The flow is generated only with a flow rate controllable syringe pump. The surface characteristics and resistance in the microchannel were precisely considered in the design process and DBT^{TM} microflow control protocols.

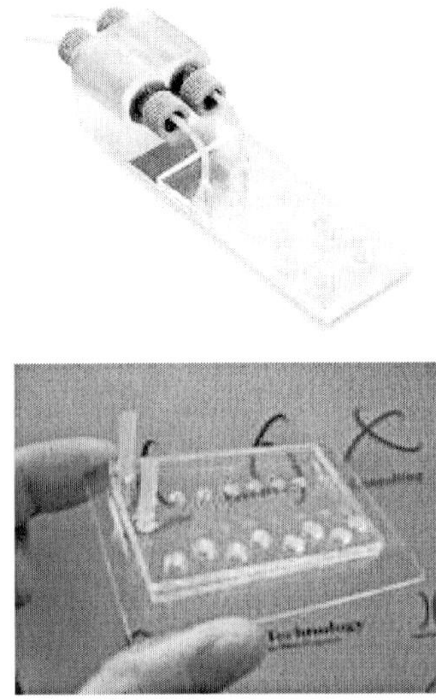

Figure 2 serial dilution chip (BOC-PF-L0204C, above), multi-layered serial dilution chip (BOC-PFM-L0208C, below)

4. Results and discussion

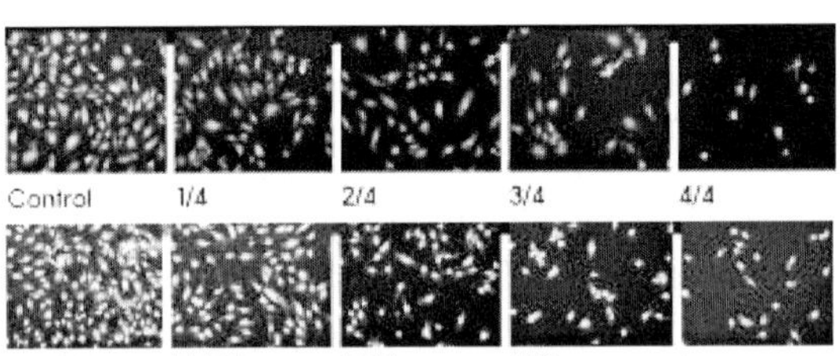

Figure 3 results in the linear scale (using BOC-PF-L0204C, above), results in the 0.1 based exponential scale (using BOC-PF-E0205C, below)

The developed chip is now on commercialisation with tolerance sheets. The experimental data in the medical and biological fileds are being accumulated, and the other application field with functional cells for tissue engineering and micro immuno test are now on study.

References

1. O.T. Guenat et al., Partial electroosmotic pumping in complex capillary systems part 2 : fabrication and application of a micro total anaylsis system suited for continuous volumetric nanotitrations, Sensors & Actuators B, 72, 273-282, (2001)
2. J. Baldyga et al., Interaction between chemical reactions and mixing on various scales, Chemical Eng. Science, Vol. 52, No. 4, 457-466, (1997)

RAPID CELLULAR ASSAYS ON MICROFABRICATED FLUIDIC DEVICES

Christopher T. Culbertson, J. P. Alarie, Maxine A. McClain, Stephen C. Jacobson, and J. Michael Ramsey
Oak Ridge National Laboratory, P.O. Box 2008, Oak Ridge, TN 37831-6142

Abstract
Functional elements are being developed to load cells with fluorescent markers, perform cytometry, lyse cells, and separate the lysate contents on a microfluidic device.

Keywords: cancer, cellular assay, kinase, lysate

1. Introduction

Early diagnosis of cancer is the best prognostic indicator of successful treatment. Cancerous tissue, however, often goes undetected for several years until some clinical symptom manifests itself. To provide earlier diagnosis and to improve our understanding of oncogenesis, high-throughput, sensitive, and inexpensive cellular assays need to be developed to detect the rare mutations in cells that are responsible for tumor formation and eventual metastasis. Microfabricated fluidic devices potentially provide a convenient platform for such assays. While these assays may require multiple chemical processing steps, all of the steps can potentially be integrated into a small inexpensive device. These steps include cytometry for the examination of a large number of cells, microfluidic switches to isolate rare cells of interest, cell lysis, and the separation, detection and identification of the cellular lysate components.

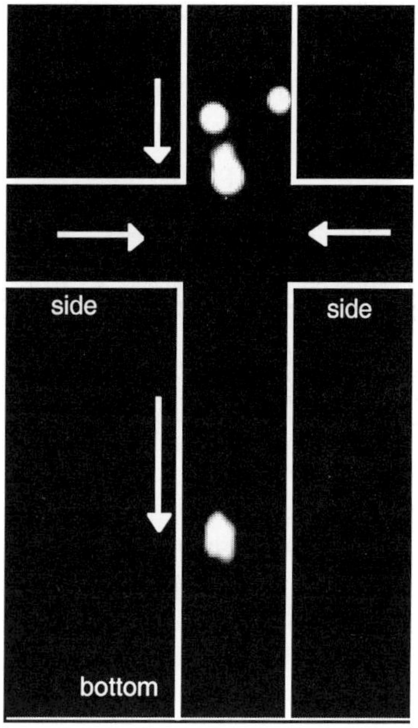

Figure 1. Image of microchip channel containing cells loaded on-chip with calcein AM. The arrows designate the direction of fluid flow.

2. Experimental

A simple cross chip design was used to test several functional elements. The chips were fabricated on soda-lime glass using standard photolithographic, wet chemical etching, and cover plate bonding techniques. The cells were moved using either applied electric fields or hydrodynamically induced flow generated by a syringe pump. An inverted microscope with an epi-illumination attachment and a CCD camera were used to image the cells as they flowed through the chips. A single-point setup was used for the cytometry and for detecting the contents of lysed cells.

J.M. Ramsey and A. van den Berg (eds.), Micro Total Analysis Systems 2001, 285–286.

3. Results and Discussion

Figure 1 is an image of a microchannel with Jurkat cells flowing through it. The Jurkat cells were labeled on-chip using calcein AM. The cells were brought hydrodynamically to a cross intersection where a combination of chemical and electrical lysis was performed. The chemical lysate was added through the side channels and the electric field was applied between the cross and the bottom channel. The cells were quickly lysed (< 66 ms); a necessary requirement for accurate analysis of many intracellular species. Downstream in the analyte channel the released calcein AM was detected. A coincidence detection method was used where both forward laser light scatter and fluorescence were detected (**Figure 2**). In the event of successful lysis little or no signal was seen in the scatter channel.

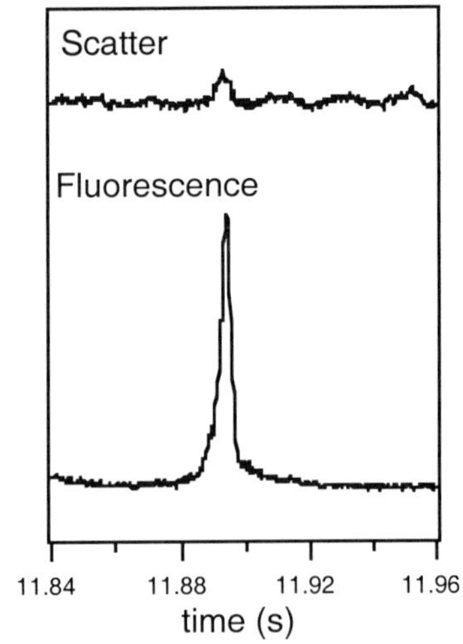

Figure 2. Single-point coincidence detection of calcein AM plume released from cell after lysis. The detection is downstream from the lysing chamber.

We have also improved significantly our ability to perform cytometry on a microfluidic device. **Figure 3** shows the coincident forward laser light scattering and fluorescent detection of *E. coli* labeled on chip with propidium iodide. The cell throughput was 85 Hz. This chip was coated with poly(dimethyl-acrylamide) to prevent cell adhesion. The cells were moved through the chip based on their electrophoretic mobility because the coating essentially eliminated electroosmosis.

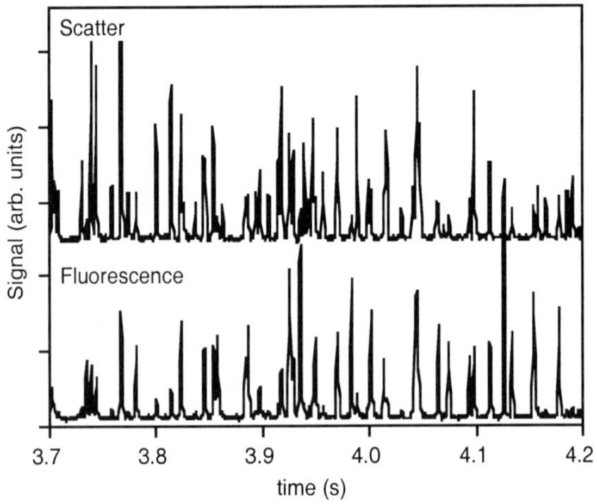

Figure 3. Single-point coincidence detection of *E. coli* labeled with propidium iodide. Each peak represents an individual cell.

Acknowledgements

Oak Ridge National Laboratory is managed and operated by UT-Battelle, LLC under contract DE-AC05-00OR22725 with the U.S. Department of Energy.

A MICRO TOTAL ANALYSIS SYSTEM (μTAS) FOR MONITORING IMMOBILIZED YEAST CELLS USING AN ENZYMATIC CHEMILUMINESCENT DETECTION SYSTEM

Richard Davidsson[1], Martin Bengtsson[2], Björn Johansson[3], Volkmar Passoth[4], Thomas Laurell[2], Jenny Emnéus[1]

[1]Department of Analytical Chemistry, Lund University, Sweden, [2]Department of Electrical Measurements, Lund Institute of Technology, Sweden, [3]Applied Microbiology, Lund Institute of Technology, Sweden, [4]Zentrum für Innere Medizin, Klinik und Poliklinik für Innere Medizin B, AG Molekulare Kardiologie, Germany

Abstract

A miniaturized flow system was constructed for the analysis of cells immobilized on a silicon micro(μ)chip. Cell viability was tested by stimulating the cells with sucrose, followed by monitoring the production of glucose (**GL**) and ethanol (**ET**), using a μ-chip based enzyme catalyzed chemiluminescent (**CL**) reaction.

Keywords: silicon microchip, immobilized cells, flow injection, enzyme, chemiluminescence detection

1. Introduction

The demands on analytical systems of today are not only to identify and determine concentrations of certain compounds, but also to give more complex information, *e.g.* biological activity, toxicity and bio-availability of chemical compounds. Our vision is to develop μ-systems for screening or real-time monitoring of cell events, using biological elements for recognition and detection. The aim of this particular work was to investigate if it was possible to immobilize living cells on a silicon μ-chip and monitor the viability of the cells on-line in a miniaturized flow system.

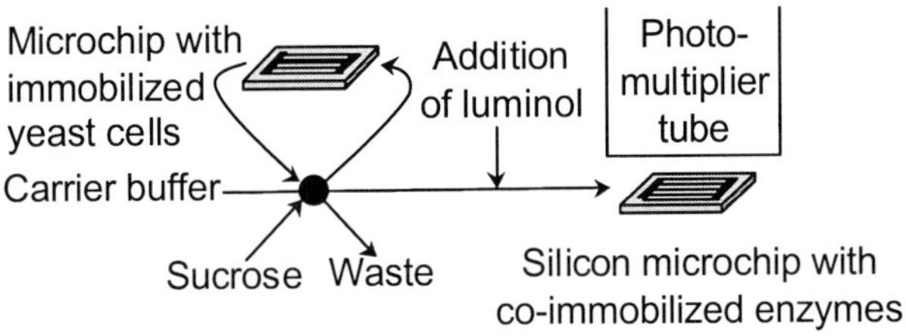

Figure 1. Set-up for the miniaturized flow system used for monitoring of the viability of immobilized yeast cells.

2. Monitoring the viability of immobilized yeast cells

The system is based on yeast (*Saccharomyces cerevisiae*) immobilized on a μ-chip (Fig. 1). When yeast is supplied with sucrose it responds by producing **GL** and **ET**, and this action is used as a viability test. **GL** and **ET** are measured enzymatically using two different μ-chips with co-immobilized enzymes, *i.e.* mutarotase (MUT), glucose oxidase

J.M. Ramsey and A. van den Berg (eds.), Micro Total Analysis Systems 2001, 287–288.

(GOx) and horseradish peroxidase (HRP) to determine **GL**, and alcohol oxidase (AOx) together with HRP to determine **ET**. The detection of **GL** or **ET** is based on that hydrogen peroxide (**HP**) is produced from the action of GOx or AOx. **HP** then participates in the **CL** producing oxidation of luminol catalyzed by HRP, which is monitored directly at the μ-chip surface by a photomultiplier tube.

Experimental

Silicon μ-chips were fabricated by anisotropic wet etching of silicon [1]. Yeast was immobilized according to D'Souza et al. [2] using linear polyethylenimine (PEI). The enzyme immobilization was based on silanization of the μ-chip surface followed by glutaraldehyde activation [3].

Results

A SEM of immobilized yeast cells on a μ-chip is shown in Fig. 2 (left). The production of **GL** (Fig. 2 (right)) and **ET** was successfully monitored on-line, and the results could be correlated to the use of different immobilization technique using SEM.

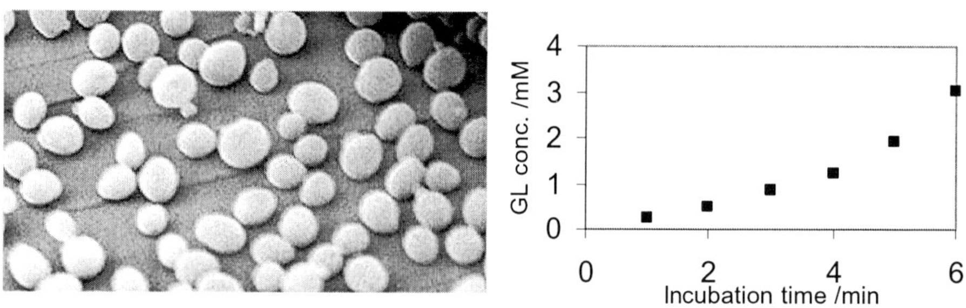

Figure 2. left) SEM of immobilized yeast cells, right) on-line **GL** production.

Conclusions

The results show that we were able to immobilize yeast cells on silicon μ-chips and to monitor their viability on-line. These results are good indications that it should be possible to construct on-line μ-systems with living cells, which may, *e.g.* be genetically modified, obtaining information of biological activity of chemical compounds.

Acknowledgement

Financial support for this work is provided from the Swedish Foundation for Strategic Environmental Research (MISTRA) and the European Commission, INCO-BIOFEED, Contract No. ICA2-CT-2000-10033

References

1. Petersen, K. *Proc. IEEE* **1982**, *70*, 420-457.
2. D'Souza, S. F.; Melo, J. S.; Deshpande, A.; Nadkarni, G. B. *Biotechnol. Lett.* **1986**, *8*, 643-648.
3. Weetall, H. H. *Meth. Enzymol.* **1976**, *44*, 134-148.

IMMUNOMAGNETIC SEPARATION OF RARE CELLS ON CHIP FOR DNA ASSAY SAMPLE PREPARATION

Vasile I. Furdui and D. Jed Harrison
Dept. of Chemistry, University of Alberta, Edmonton, AB, T6G 2G2, CANADA

Abstract

A new method for cell capture and concentration on chip was developed. T cells were successfully separated from human blood. In order to determine the effectiveness of cell capture on chip, DNA was extracted and a T cell specific receptor gene was amplified by PCR. The capture efficiency for cells is about 50%. Our results indicate it is possible to use a magnetic bead bed on chip to capture cells that are present in a 1/10,000 ratio and the material can be recaptured from the chip for further analysis.

Keywords: Immunomagnetic separation, DNA, DNA extraction, PCR

1.Introduction

Clinical diagnostic PCR analysis will require separation of specific cells from blood or other biological fluids and extraction of DNA or RNA. Magnetic bead based separation of cells and magnetic bead extraction of DNA has advantages over other methods: a smaller amount of blood is used, the assay is faster, simpler equipment is required and the capture procedure may be easier to realize on chip. As methods for performing DNA sample preparation, PCR and hybridization on-chip have developed, it has become important to also integrate cell capture and concentration on-chip, to provide fully integrated assay procedures.

2.Results and Discussion

Protein A magnetic particles (1-2 μm) are used, since Protein A has a high affinity for many types of IgG. This approach can then be adapted to many different types of antigenic determinants by varying the bound antibody. After mixing Protein A beads with a solution of anti-human CD3, T cells can be selectively captured because the CD3 receptors are unique to T cells. The glass chip layout in Figure 1 was used to capture T cells present in human blood samples. Anti-CD3 coated magnetic beads were introduced first and trapped with a magnet, then cells were added and sucked downstream with a syringe pump. Figure 2 shows the tracks of beads (black streaks) and the cells captured by the beads (translucent circles). The beads were then collected in a reservoir and removed from the chip for analysis. In order to determine the effectiveness of cell capture on-chip, the cells were lysed, the DNA was captured on DNA Direct Dynabeads, and a T-cell specific receptor (γ) gene [1] was amplified by PCR (300 bp). Capillary gel electrophoresis of the products was performed using a Beckman P/ACE 5010 with 0.4 % hydroxypropyl-methyl-cellulose gel and ethidium bromide stain. Figure 3 shows characterization of the PCR assay, using various quantities of whole blood, where the DNA was isolated off chip on Dynabeads directly from lysed blood samples. The T-cell concentration is roughly 1000 cells /μl in blood

J.M. Ramsey and A. van den Berg (eds.), Micro Total Analysis Systems 2001, 289–290.

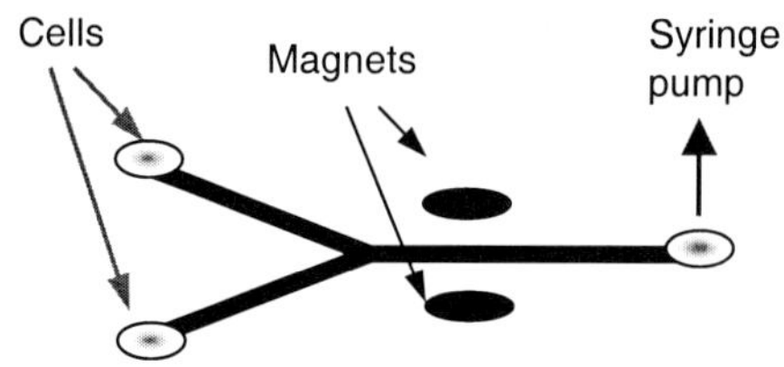

Figure 1: *Y-intersection for immunomagnetic separation of blood cells. Anti-human CD3 magnetic beads are captured by magnets and then cells are introduced from both reservoirs under negative pressure.*

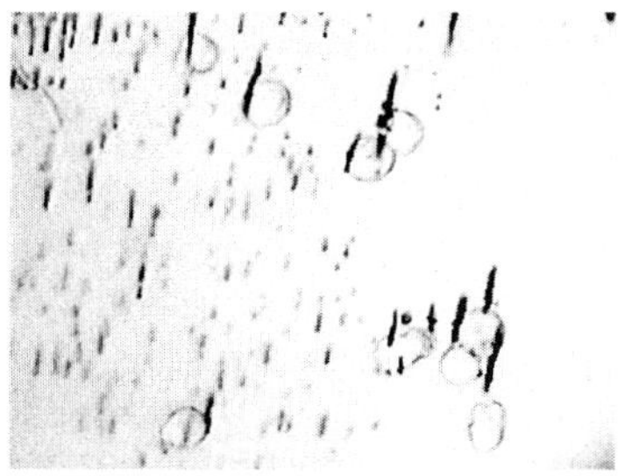

Figure 2: *T cells and Protein A - Anti-human CD3 magnetic beads captured in magnetic field. Beads show up as dark streaks , cells as ovals.*

and the detection limit was about 200 cells. The PCR procedure had not yet been optimized. Figure 4 shows the off-chip PCR analysis of the cells captured on-chip from whole blood using Protein A/anti-CD3 beads. The 2 μL blood sample tested on-chip corresponds to about 2000 T-cells, assayed by PCR. The detection limit for the on-chip capture was 400 cells. This suggests the capture efficiency of cells is about 50 %. Following extensive optimization of the PCR procedure using a T-cell line, detection of as few as 15 T-cells was demonstrated and this procedure will be applied to the chip studies.

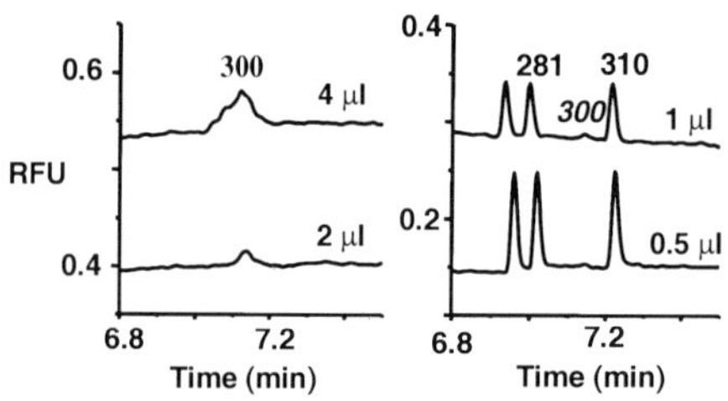

Figure 3: *Capillary gel electrophoresis (CGE) of PCR product generated after extraction of DNA from different volumes of blood. A DNA ladder was added to some samples.*

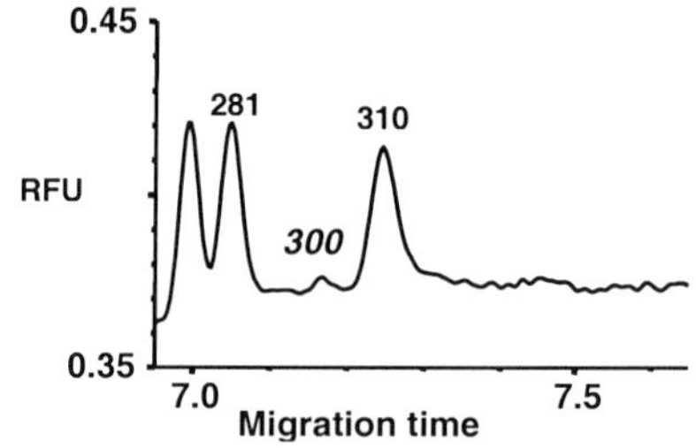

Figure 4: *CGE of PCR product after on-chip cell separation and off-chip DNA extraction from 2 μl of blood. A DNA ladder was added.*

Our results indicate it is possible to use a magnetic bead bed on chip to capture cells that are present in a 1/10,000 ratio, and that the material can be recaptured from the chip for further analysis. The chip can be used with whole blood samples. The results with the optimized PCR procedure suggest that it should be possible to detect a few tens of cells isolated on chip and to extend the isolation to a lower concentration of rare cells. This device represents a first, positive step towards blood sample preparation on-chip for genetic analysis.

References

1. S. Kohler, K.D. Jones, R. A. Warnke, J.L. Zehnder, The American Journal of Dermatopathology, Vol. **22**, No. 4, pp.321-327, (2000).

ULTRA-LOW VOLUME NANOCALORIMETRIC MEASUREMENTS FOR SCREENING CELLULAR METABOLIC ACTIVTY

Erik A. Johannessen[1,2], John M.R. Weaver[2], Peter H. Cobbold[1] and Jon M. Cooper[2].
[1]Dept. Human Anatomy and Cell Biology, University of Liverpool, UK
[2]Dept. Electronic and Electrical Engineering, University of Glasgow, UK

Abstract

A microsystems sensor technology for cellular screening of bioactive compounds is presented which combines extreme sensitivity with low volume, fast response time and improved ease of manufacturing.

Keywords: Ultra-low volume, nanocalorimetry, cellular metabolic activity

1. Introduction

Current strategies for high-throughput cell-based screening (HTS) are focused on the development of highly specific assays[1]. In contrast, the measurement of a non stimulus specific parameter such as temperature, could have widespread applications in the pharmaceutical industry.

2. Theory

Finite element modelling provided numerical solutions of the heat diffusion equation (1) [2], which was employed to recreate the sensor in a computer model to optimise design parameters prior to fabrication using silicon processing methods adapted from

$$\frac{\partial}{\partial x}\left(k_x\frac{\partial T}{\partial x}\right)+\frac{\partial}{\partial y}\left(k_y\frac{\partial T}{\partial y}\right)+\frac{\partial}{\partial z}\left(k_z\frac{\partial T}{\partial z}\right)+S=\rho C_P\frac{\partial T}{\partial t} \qquad (1)$$

the semiconductor industry. The temperature sensor comprised 10 serial connected thin film thermocouples of gold and nickel and had a total volume of 0.7 nL. An increase in temperature was associated with a decrease in the voltage output from the sensor.

3. Experimental

Calibration of the system were achieved by the titration of hydrogen peroxide and catalase within the volume of the sensor, which supported by modelling provided calibration curves for cells of different geometric shapes such as isolated rat cardiomyocytes and primary cultured mice brown adipocytes. These cells were used as model cells in the study and stimulated with 14μM carbonyl cyanide (m)-chlorophenyl hydrazone (CCCP) or 1.4 μM norephinephrine (NE) to increase respiration and heat output.

J.M. Ramsey and A. van den Berg (eds.), Micro Total Analysis Systems 2001, 291–292.

4. Result and discussion

Graphical interpretation of the modelled results identified regions of low and high heat loss, as a function of the spacing of the isotherms (fig.1). The results were used to optimise both thermal insulation of the sensor and localisation of the thermocouples with respect to the temperature gradient. In contrast, after an initial control injection at

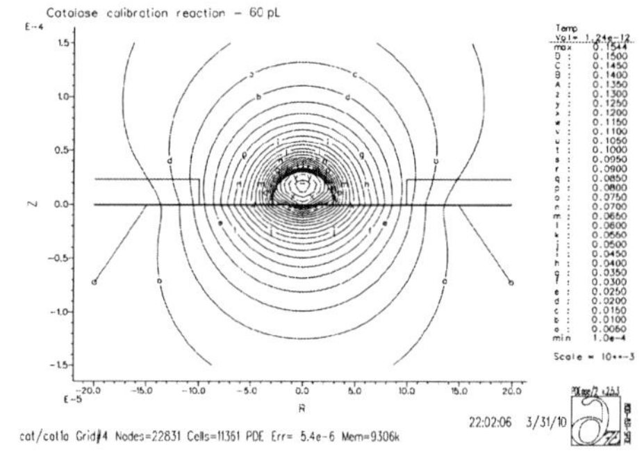

Fig.1 – Cross sectional temperature rise in the sensor from a 10 nW heat source.

500 s, the uncoupled response from 20 cardiomyocytes in the presence of CCCP, was 1.71 nW per cell (n=5). Respiration returned to the basal level which could not be resolved from the control prior to the injection of rotenone at 2400s. The result is comparable to literature values of 1.5 nW/cell [3]. A control experiment with 10 brown adipocytes stimulated with NE gave 1.66 nW/cell (n=5) which was comparable to literature values ranging from 0.2 [4] to 5 nW/cell [5].

5. Conclusion

The temperature measurements offer a generic approach to cell-based assays screening libraries of bioactive compounds where suitable probes are not available or when the cellular actions and signalling pathways are not known.

References

1. Sundberg, S.A. *High-throughput and ultra-high-throughput screening: solution and cell-based approaches.* Current Opinion in Biotechnology **11**, 47-53 (2000).
2. Carslaw, H.S. and J.C. Jaeger. *Conduction of Heat in Solids.* 2 ed, London: Oxford University Press (1959).
3. Loiselle, D.S., *et al. The Effect of Hyperosmolality on the Rate of Heat Production of Quiescent Trabeculae Isolated from the Rat Heart.* J. Gen. Phys. **108**, 497-514 (1996).
4. Tanaka, E., *et al. Regulation of heat production of brown adipocytes via typical and atypical b-adrenoceptors in the rat.* Jpn. J. Physiol. **45**, 1043-1051 (1995).
5. Girardier, L. *Brown fat: An energy dissipating tissue.*, in *Mammalian Thermogenesis*, L. Girardier and M. Stock, Editors. Chapman and Hall: London, 50-98 (1983).

ELECTROLYSIS OF CANCER CELLS USING ISOOSMOTIC BUFFER SOLUTION

Min-Soo Kim[1], Sang-Ho Lee[1], Su-Hyung Cho[2], Chang-Soo Lee[2], Byung Gee Kim[2] and Yong Kweon Kim[1]
[1]School of Electrical Engineering & Computer Science,
[2]School of Chemical Engineering, Seoul National University,
Kwanak P.O. Box 34, Seoul 151-742, Korea

Abstract

A electrolysis of cancer cells is investigated in the isoosmotic buffer solution. Through the regulation of the myo-inositol concentration, the medium can be changed into isoosmotic state or hypoosmotic state.

Keywords: Cancer cells, electrolysis, isoosmotic state, hypoosmotic state

1. Introducion

The cell electrolysis can be used effectively in order to obtain the intracellular materials for celluar analysis such as DNA or protein identification. In this study, nickel micro lysis reactor is performed in the micro lysis reactor, electrolysis is performed in isoosmolar buffer solution and the results are analysed, based on the cell electrical properties [1].

2. Theory

The transmembranevoltage is simply expressed by the equation (1) on assumption that the membrane is absolutely insulator and the membrane charging time is very slow.

$$\Delta\Phi_m = \frac{3}{2} ER\cos\theta \tag{1}$$

E is electric field, R is cell radius and θ is the angle between the field line and the point of interest in the membrane [2]. If the transmembrane voltage is higher that 1V, the cells will be permeable to the outside of the medium. The minimum voltage in the case of cell radius 15μm and electrodes gab 50μm is calculated to be 2.2V.

3. Experimental

The cancer cells (SNU-1) were extracted from the stomach tissue of a 44 years old man and cultured in RPMI 1640 supplemented with 10% heat-inactivated feral bovine serum in humidified incubator with 5% CO_2 in gas phase at 37°C. The microelectrodes (Figure 1(a)) are fabricated by electroplating, high enough to apply the electric fields to cells effectiviely. For the cancer cell lysis, the AC electric fields of $10V_{P\text{-}P}$ were used in the frequency range between 200kHz and 16MHz. The suspension medium is composed of KCl 3.5mM, KH_2PO_4 0.16mM and myo-inositol. The cancer cell electrolysis was

J.M. Ramsey and A. van den Berg (eds.), Micro Total Analysis Systems 2001, 293–294.

performed in the two kinds of mediums with different concentrations of myo-inositol: 100mM (medium-1) and 50mOsmol/kg (medium-2).

4. Results and discussion

After applying voltage of $10V_{P-P}$ 200kHz, the cancer cells in the medium-1 were stretched and lysed between the electrodes tips by positive DEP force (Figure 1(b),(c)). Within 3~4 seconds, the cancer cells were distructed into the fragments. In the frequency region over 1MHz, the cancer cells were not lysed and hold to be eloganted. Secondly, the cancer cells in medium-2 swelled slowly even with no voltage application (Figure 1(d)), and besides, some cancer cells burst out like balloons explosion (Figure 1(e)).

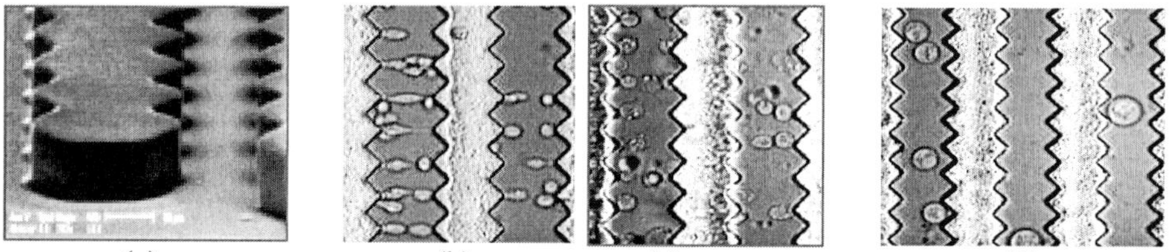

(a) (b) (c) (d)

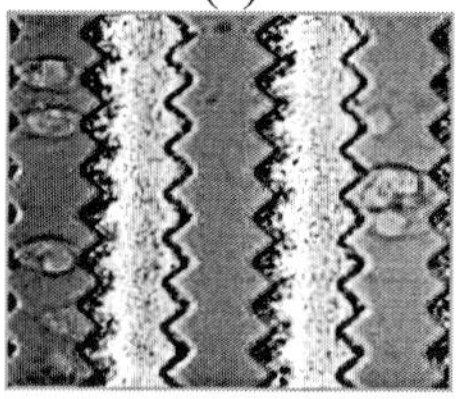

(e)

Figure 1. (a) Nickel microelectrode. Cancer cells electrolysis in isoosmotic state: (b) cell elongation before electrolysis, (c) lysing in isoosmotic state. Cancer cells swells up(d) slowly and some burst up (e) in hypoosmotic state.
Resistance of electrodes in medium : 80~110ohm(0.2~15MHz).
Capatance of electrodes in medium : 80~0.3pF(0.2~15MHz).

5. Conclusion

The cancer cells were electrolysed in the saw typed nickel microelectrodes. The myo-inositol concentration in the suspension medium worked as an important factor on the electrolysis. In hypoosmotic state (low concentration of myo-inositol), the cancer cells are apt to swell and burst up. In case of isoosmotic state (high concentration of myo-inositol), the cells are elongated and lysed easily within 3~4 seconds.

Acknowledgments

This research was supported by the Intelligent Microsystems Program (IMP) that is the project of the "21C Frontier R&D Program"

References

1. U. Zimmermann, U. Friedrich, H. Mussauer, H. Mussauer, P. G essner, K. Hamel, and V. Sukhorukovet, *Electromanipulation of Mammalian Cells: Fundamentals and Application*, IEEE Trans. Plasma Science, Vol. 28, No. 1, (2000).
2. S.W. Lee and Y.C. Tai, *A Micro Cell Lysis Device,* Sensors and Actuators A: Physical, Vol. 73, Issues 1-2, pp 74-79, (1999).

MICROFLUIDIC BIOCHIP INTEGRATED WITH ACOUSTIC WAVE SENSOR FOR SINGLE HEART MUSCLE CELL ANALYSIS

Paul C. H. Li[1], Weijie Wang[1], Ash M. Parameswaran[2]
[1]Department of Chemistry, [2]School of Engineering Science, Simon Fraser University, Burnaby, B.C., Canada, V5A 1S6.

Abstract

We report the fabrication of a microfluidic biochip integrated with an acoustic wave sensor for cell analysis. The sensor serves as a detector for measuring changes in viscoelastic properties (e.g. stiffness) of attached cardiomyocytes (single heart muscle cells) during their contraction and relaxation. The effect of excitation electrode size, use of a hydrogel as a model of contraction, and cell analysis are described.

Keywords: Microfluidic Biochip: Cardiac (Heart) Muscle Cell or Cardiomyocyte: AT-cut quartz: Thickness-Shear Mode (TSM): Impedance Analysis

1. Introduction

A new acoustic method is employed to characterize cardiomyocyte contraction without the need of a force transducer and servo-motor as used in conventional methods [1]. Acoustic measurement of stiffness can be achieved by impedance analysis of a myocyte residing on an AT-cut quartz crystal.

2. Theory

Thickness-shear acoustic wave is allowed to propagate into any solid film (e.g. myocyte) adhered onto the quartz crystal. Any changes in mass and stiffness of the myocyte will alter the resonant frequency of quartz. This sensor, previously called quartz crystal microbalance, has also been used to monitor the growth of biological cells [2].

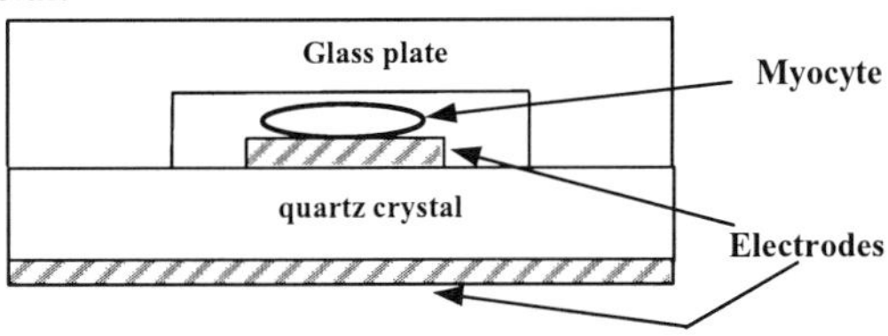

Fig. 1. Cross-section view of a myocyte residing on an acoustic wave sensor integrated in a biochip

3. Experimental

The glass plate was micromachined with a network of channels and chambers [3]. The bottom plate (quartz crystal) was patterned with the electrodes for acoustic wave launching and detection (Fig 1).

The hydrogel, which was synthesized according to a published procedure [4], was stimulated to swell and shrink by pH changes. Rat and rabbit cardiomyocytes, which were freshly isolated, were injected into the channel via a reservoir. Bath solution (containing 2 mM Ca^{2+}) was used to stimulate cell contraction.

A network analyzer (HP4195A) was employed for impedance analysis.

J.M. Ramsey and A. van den Berg (eds.), Micro Total Analysis Systems 2001, 295–296.

4. Results and discussions

First, we found that the original design of electrode size of 100 μm did not always result in sensor oscillations. We then performed subsequent testing on sensor plate with larger excitation electrode.

Second, we selected the hydrogel polymer as a model for the muscle cell. Therefore, the size and contraction rate of the object can easily be altered, which is not possible in the case of the muscle cell. We have found that the series resonant frequency, f_s, of the sensor changed as the pH of the hydrogel was altered (See Fig 2).

The frequency decreased when the cell samples was introduced into the channels . While the cell was made to contract as stimulated by a high Ca^{2+} concentration, it was optically monitored using a microscope. The frequency at minimum impedance, f_s', was found to change periodically during muscle cell contraction. A negative control in which no contracting cardiomyocyte was present showed no recurrent frequency changes.

Figure 2: Changes in f_s of an AT-quartz with hydrogel

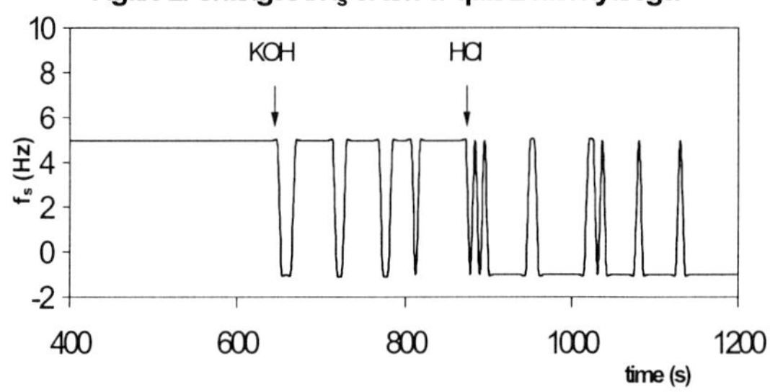

5. Conclusions

When microfluidic channels are used to deliver reagents for interacting with cardiomyocytes, the biochip will ultimately lead to rapid analysis of heart disease drugs by measuring changes in resonant frequency of the sensor onto which single cardiac muscle cells adhered.

Acknowledgements

Financial support from NSERC of Canada is gratefully acknowledged. We also thank Dr. Glen Tibbits and Haruyo Kashihara in the School of Kinesiology at Simon Fraser University for isolation of cardiomyocytes and helpful suggestions. We gratefully acknowledge Mr. Bill Woods for assisting us in the microfabrication process.

References:

1. P.W. Brandt, F. Colomo, N. Piroddi, C. Poggesi and C. Tesi,, *Force regulation by Ca^{2+} in skinned single cardiac myocytes of frog*. Biophys. J. 74, 1994-2004 (1998).
2. Redepenning, T.K. Schlesinger, E.J. Mechalke, D.A. Puleo and R. Bizios, *Osteoblast attachment monitored with a quartz crystal microbalance*, Anal. Chem., 65, 3378-3381, (1993).
3. W. Wang, P. C. H. Li, A. M. Parameswaran, *Micromachined muscle cell analysis chip*, Proc. SPIE, vol 4230, 100-108 (2000).
4. R.H. Liu, Q. Yu and D.J. Beebe, *In-channel processing to create autonomous hydrogel microvalves,* Proc. μTAS 2000 Symposium, 45-48 (2000).

MICROFLUIDIC DEVICES FOR CELL LYSIS AND ISOLATION OF ORGANELLES

Hang Lu[1], Rebecca J. Jackman[1], Suzanne Gaudet[2], Michael Cardone[2], Martin A. Schmidt[3], and Klavs F. Jensen[1]
[1]Department of Chemical Engineering, [2]Department of Biology,
[3]Microsystems Technology Laboratories, MIT, Cambridge, MA 02139, USA

Abstract

The present work targets devices that scale down the cellular analyses to small populations. Electroporation is chosen to lyse the cell membrane, eliminating the use of chemicals, while isoelectric focusing of organelles replaces density centrifugation for separation of organelles. Microfluidic devices with integrated electrodes for lysing and focusing were fabricated on glass substrates with SU-8 and by electroplating. The techniques were explored with human HT-29 cells.

Keywords: cellular analysis, electroporation, mitochondria, isoelectric focusing

1. Introduction

Many protein signal transduction pathways involve subcellular organelles. The currently available equipment and protocols to isolate organelles call for large sample sizes, and are complicated to operate and time-consuming. The aim of this work is to perform such analyses more effectively in miniaturized devices.

2. Methodology

To separate the organelles, the cell membrane has to be broken open first. We chose electroporation because it is cleaner and more controllable than large scale cell lysis (chemical or mechanical). Electroporation lyses cells by breaking down the lipid bi-layer and creating holes in the cell membrane using an applied electric field. By miniaturizing the device, the applied voltage can be reduced significantly while still maintaining the field strength required for dielectric break-down of the membrane.

Isoelectric focusing (IEF) is commonly used for protein separation. Mitochondria, among other organelles, can be focused by this method because the outer membranes also exhibit surface groups (such as proteins) that have fixed isoelectric points. An applied electric field creates and maintains pH gradient in the fluidic channel. Focusing of the particles is achieved by electrophoretic movement of these particles towards their pI value where the particles are no longer charged and therefore not experiencing the electrophoretic forces.

J.M. Ramsey and A. van den Berg (eds.), Micro Total Analysis Systems 2001, 297–298.

3. Experimental

Microfluidic devices with integrated electrodes for lysing and focusing were fabricated using the process shown in Figure 1. Three-dimensional electrodes in the fluid channel are desirable for a more uniform electric field. The fluidic channel was formed on a glass substrate with a photo-definable epoxy (SU-8), and the electrodes were electroplated using thick resist as mold. This process simplifies the formation of fluidic channel and mold for electrodes into one step. A finished device is also shown below. Mitochondrial samples were obtained from cultured human HT-29 cells, and stained.

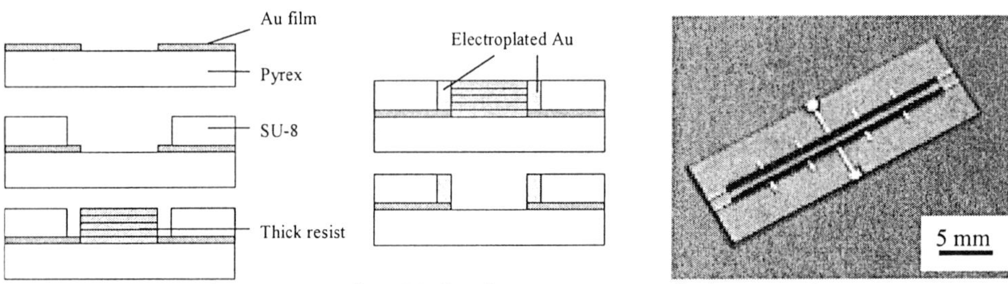

Figure 1. Device fabrication sequence and photograph

4. Results and discussion

For small sample sizes (~1000 cells), we found electroporation is the most controllable method of lysis. Human HT-29 cells were lysed under electric field of a few kV/m under flow conditions. Figure 2(A) shows an image of a cell in bright field before lysis (nuclei and cell membranes are visible), and Figure 2(B) shows a nucleus in the flow channel after applying electric field. This form of lysis is desirable because of the control of lysis that can be achieved by controlling the voltage and the avoidance of contamination by chemical lysing agents. Moreover, the technique can be integrated readily with downstream separation schemes, such as the isoelectric focusing of organelles.

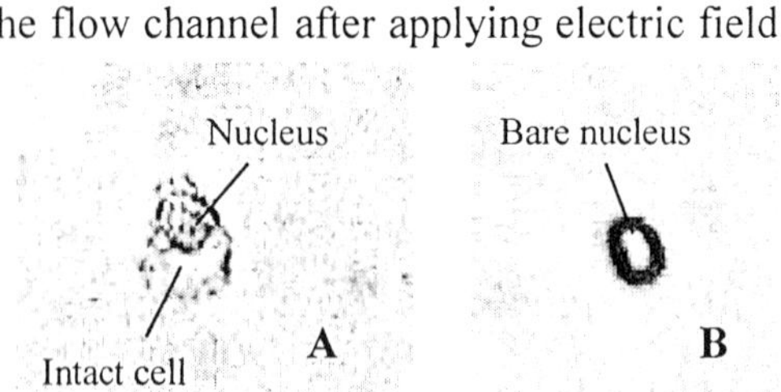

Figure 2. Cell lysis in flow channels by electroporation.

In designing the microfluidic devices for separating mitochondria, the electrokinetic mobilities of mitochondria were determined, so that the time scale of isoelectric focusing matches the residence time of the fluids in the device. The microfluidic device can generate pH gradient in the buffer solution with minimal power input, and the electrophoretic movement of mitochondria in an electric field at different pH has been demonstrated.

Acknowledgement

This work was supported by DARPA as part of the Bio-Info-Micro program at MIT.

ALLERGY SENSOR USING ANIMAL CELLS WITH MICROFLUIDICS

Yasutaka Matsubara, Yuji Murakami, Takeshi Kinpara, Yasutaka Morita, Kenji Yokoyama and Eiichi Tamiya
Japan Advanced Institute of Science and Technology, 1-1 Asahidai, Tatsunokuchi, Ishikawa, 923-1292, Japan
e-mail: yasutaka@jaist.ac.jp

Abstract
In this paper, we report fabrication of allergy sensor on chip using RBL-2H3 mast cells. We fabricate microfluidic system on a PDMS chip. The cells are cultivated on the chip and then release histamine and artificially introduced fluorescent dye into histamine granules with allergen stimulus. This chip can detect fluorescence intensity using microscopy with photomultiplier tube.

Keywords: PDMS, microfluidic, oxygen plasma, Quinacrine, mast cell

1. Introduction

Usually, skin test and RAST methods are used in identification of an allergen. However, these methods are time-consuming and need specific equipments for allergy detection. Hence there is a demand to develop a simple, rapid and safe method for identification of an allergy in the field of medicine. Mast cells secrete histamine by the regulated pathway in response to specific ligands that bind to receptors on the mast cell surface. In this paper, we report fibrication of allergy sensor on chip using RBL-2H3 mast cells. To the best our knowledge, this is first report as allergy sensor using mast cells with microfluidics.

Histamine is difficult to detect electrochemically. Thus, we use fluorescent dye. Quinacrine, an aminoacridine, is a weak base which has been used to identify acidic compartments of various cells microscopically. This dye is effectively introduced into low pH compartments of cells by mass action. When exocytosis occurred with allergen stimulus, the dye is released from cell compartments with histamine. Fluorescent microscopy can detect its fluorescence easily.

2. Experimental

Polydimethylsiloxane (PDMS) chip was made using a thick film of photoresist (SU-8) mold placed on Si wafer (Fig.1). The PDMS chip was rendered to hydrophilic under oxygen plasma and then sterilized with 70% (w/w) ethanol. Mast cells suspended in medium were added into chamber of the chip and then cultured at 37°C and 5% CO_2 in incubator. After 1d cultivation, cells were sensitized with IgE antibody and then PDMS cover lid with injection ports was placed on chip (Fig.1). The fluorescent dye was passed through the flow system of the chip, and allowed for 5min and then washed with PIPES buffer pH 7.4 in order to remove the extra dye from the reaction mixture.

J.M. Ramsey and A. van den Berg (eds.), Micro Total Analysis Systems 2001, 299–300.

Then allergen was injected. The introduced dye was released from the mast cells along with histamine can be detected by fluorescence microscope using down-stream flow system of a chip. The principle of the measurement of detection is shown in Fig.2.

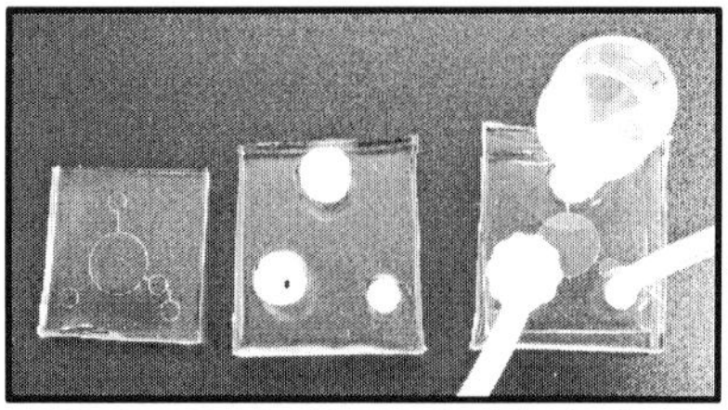

Fig.1. PDMS chip. Left side device is a chip, center is cover lid and right is the flow system of a total chip.

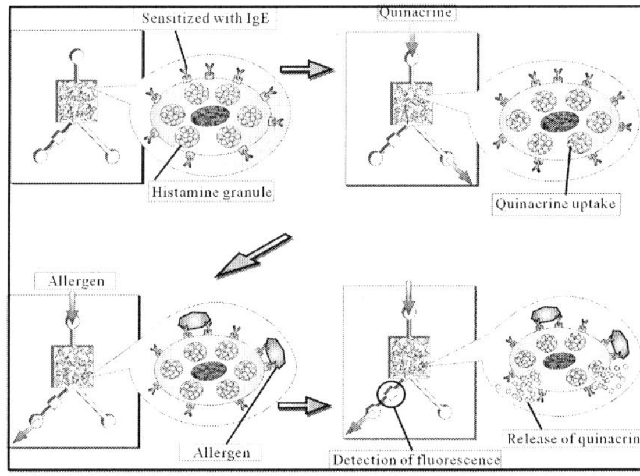

Fig.2. Schematic illustration of flow system

3. Results and discussion

Mast cells could be cultivated for a period of a week on the hydrophilic surface of the PDMS chip, whereas hydrophobic surface could grow for 2 days and then leach out from the surface in 3days (Fig.3). This result clearly suggested that hydrophilic surface allows cells to grow longer periods.

Allergic response of mast cells using quinacrine fluorescent dye was measured on petri dish using above principle (Fig.4). The fabricated PDMS chip (Fig.1) could detect at least 1μM quinacrine dye.

The above two experiments clearly demonstrated that the allergic response of the mast cell can be detected on chip.

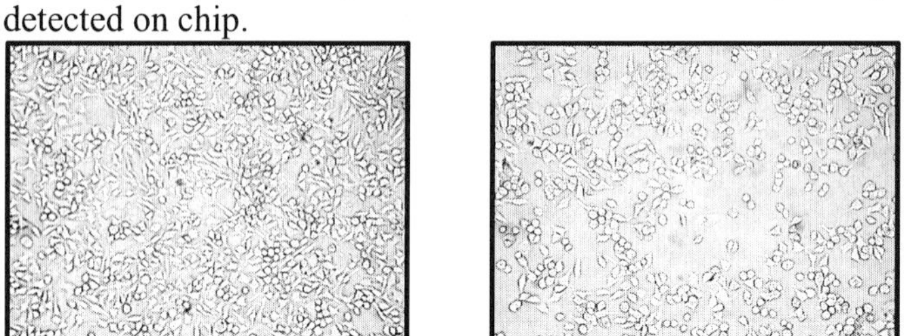

Fig.3: The cells cultivated on the chip.
Left is treated with RIE and right is treated without RIE

Fig.4. Change in quinacrine fluorescence intensity upon addition of allergen.

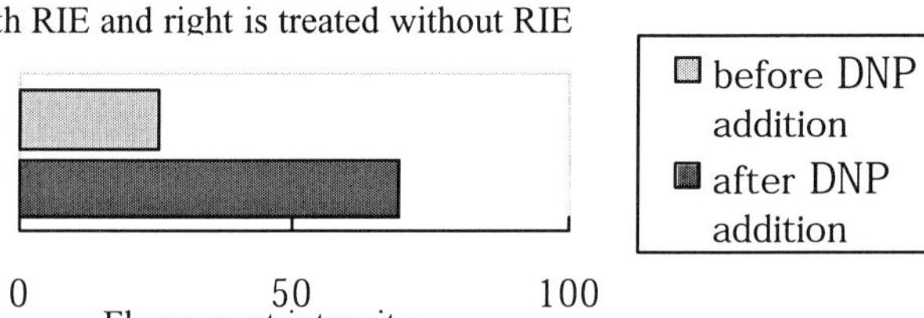

SINGLE CELL LYSIS ON MICROFLUIDIC DEVICES

Maxine A. McClain, Christopher T. Culbertson, Stephen C. Jacobson, and J. Michael Ramsey
Oak Ridge National Laboratory, Oak Ridge, TN 37831-6142 USA

Abstract
Methods to rapidly lyse cells are needed for studying biochemical reaction systems in cells with short half-lives such as protein phosphorylation. Chemical and electrical methods, therefore, were examined to rapidly lyse cells on microfabricated fluidic devices. The most effective cell membrane disruption involved a combination of both chemical and electrical lysing techniques.

Keywords: cell lysis, surfactant, electric field

1. Introduction
Cell lysis is the preliminary step in the analysis of intracellular compounds, such as DNA, RNA or proteins. The lysis process must preserve the integrity of the relevant compounds at their pre-lysis concentrations while releasing them from the cell. Of the various methods to rupture cell membranes in a population of cells, chemical and ultrasonic lysis are the two most common. Electrical lysis is a viable alternative, although it is not commonly used. Of these methods, chemical and electrical lysis are particularly well suited for implementation on a microfluidic platform because their effects can be easily localized.

2. Experimental
The cells used in the experiments were from the Jurkat line of human leukemia cells. The glass chips were fabricated and coated as previously described.[1,2] A syringe pump was used to draw cells into the lysis chamber. The cells were labeled with membrane permeable probes that become fluorescent after the molecule is internalized and hydrolyzed in the cell. The degree of lysis was determined by 2-D imaging and coincident fluorescence and scattering detection.

3. Discussion
Both chemical and electrical lysis were evaluated separately and together. Sodium dodecyl sulfate and several polyoxyethelyene detergents were tested. With SDS cell lysis was accomplished in less than 1 s. Electrical lysis can be performed even more quickly than chemical lysis, which is significant for analyzing transient species. Direct current and alternating current electric fields over a range of voltages were effective in compromising the cell membrane. At low frequencies, DC biased AC fields were effective for lysis and did not produce the same level of Joule heating and bubble formation that the equivalent DC field did. Cell lysis generated with a DC biased AC

J.M. Ramsey and A. van den Berg (eds.), Micro Total Analysis Systems 2001, 301–302.

field and with solubilizing detergent in the side arms is shown in figure 1A. In this figure the cells are brought to a cross intersection hydrodynamically. At the intersection they are exposed to an electric field and lysed within two video frames (<66 ms). In figure 1B, the same region is shown without the electric field applied; no cell lysis is observed. The cell images look elongated in frames 3-7 because the fluid below the cross intersection is traveling at a faster velocity than that above the cross. The speed of membrane disruption correlated with the magnitude of the applied field. The effective range of field strengths was between 150 and 500 V/cm peak-to-peak for square wave 50 Hz AC fields with a 300 to 400 V/cm DC offset. A combination of chemical and electrical lysis provided the best results. The electrical lysis was fast and the chemical lysing agents kept cell membrane debris solubilized.

Acknowledgements
Oak Ridge National Laboratory is managed and operated by UT-Battelle, LLC under contract DE-AC05-00OR22725 with the U.S. Department of Energy

1. Hjerten, S. *J. Chromatogr.* **1985,** *347*, 191-198.
2. Jacobson, S. C.; Hergenröder, R.; Koutny, L. B.; Ramsey, J. M. *Anal. Chem.* **1994,** *66*, 2369-2373.

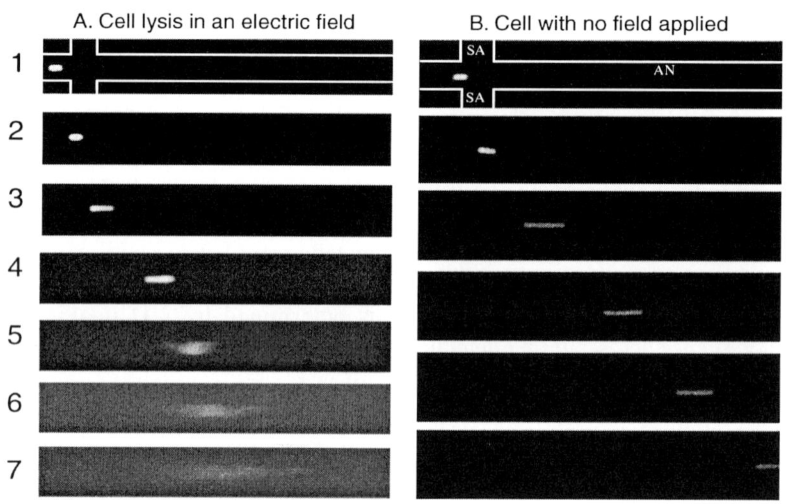

Figure 1. Jurkat cells loaded with calcein are hydrodynamically moved through a cross intersection with (A) and without (B) an electric field applied between the side arms (SA) and the analysis channel (AN). The images of the cells look elongated in the AN channel because of the increase in velocity in this channel.

BIOANALYTICAL DEVICES FOR HIGHLY SELECTIVE MEASUREMENT OF TRANSMITTERS FROM CULTURED CELLS

Osamu Niwa[1], Ryoji Kurita[2], Katsuyoshi Hayashi [1],Tsutomu Horiuchi[1] and Keiichi Torimitsu[3]
NTT Lifestyle & Environmental Technology Labs.[1], NTT Advanced Technology[2], NTT Basic Research Labs.[3], Japan, 3-1 Morinosato, Wakamiya, Astugi, 243-0198 Japan
(E-mail:niwa@aecl.ntt.co.jp)

Abstract

Two types of devices for highly selective measurement of biomolecules were fabricated. One consists of a dual electrochemical flow channel for differential measurement of neurotransmitters. The other is integrated with micro-reactor having a number of micro-pillars that remove interferent effectively in the small volume.

Key words: Bioanalytical device, Cultured cell, Transmitters, Micro enzyme reactor

Introduction

It is particularly interesting to monitor such features of living cells as their action potential or their secretion of various transmitters in real-time. We have developed microanalytical devices for monitoring the transient release of various transmitters from cultured cells [1, 2]. When measuring biological samples, various types of interference, including electrostatic noise and interfering molecules, often pose serious problems as regards monitoring trace level analyte. Here, we report two types of integrated device that achieve high selectivity for the real-time measurement of bio-chemicals.

Differential measurement using dual channel chip

The device shown in Fig. 1 has two small thin layer flow channels. An enzymatic reactor can be integrated upstream of the sensing electrodes if required. Each flow channel has a carbon film based microelectrode. We modified one electrode (A) with bilayer films of osmium polyvinylpyrridine based film containing horseradish peroxidase (Os-gel-HRP) and BSA containing enzymes such as glutamate or histamine oxidase. In contrast, we modified the other electrode (B) with bilayer films of Os-gel-HRP and BSA without other enzymes. Therefore, the differential response (A-B) should be the signal caused by the enzymatic reaction of the analyte and be free from interference. When we used this

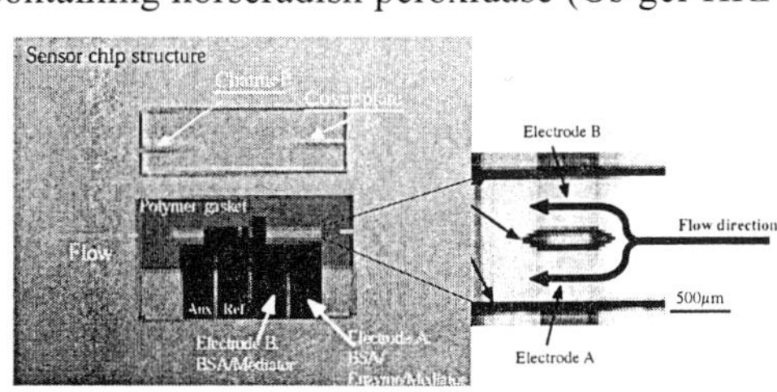

Fig. 1 Dual channel mode bioanalytical device for differential measurement of nerve cells

J.M. Ramsey and A. van den Berg (eds.), Micro Total Analysis Systems 2001, 303–304.

device to measure glutamate released from cultured neurons, the response from electrode (A) contains not only the glutamate signal but also interference, such as capacitance change caused by KCl stimulation and electrostatic noise. In contrast, most of the noise was eliminated from the differential response curve (A-B). We also obtained a similar result when measuring histamine released from 2H3 mast cells. This suggests that very selective cell measurement can be achieved using our integrated bioanalytical devices.

Micro enzymatic reactor integrated chip

Our second device includes a micro-reactor integrated upstream of the detector as shown in Fig. 2. The reactor consists of a number of micro-pillars (Fig. 2). When we modified the micro-pillar arrays with gold film, we were able to use the reactor to pre-oxidize the interferents before the sample reached the detector. The pre-reactor can be used as an enzymatic reactor by modifying the surface with various enzymes. Fig. 3 shows the current response at the glutamate sensing electrode located at the downstream when injecting L-glutamate and L-ascorbic acid. Without the reactor, the glutamate signal almost disappeared due to interference from the ascorbic acid. However, the interference was greatly suppressed by using the micro-pillar based reactor in which L-ascorbate oxidase was immobilized. We also confirmed that the reaction efficiency of the reactor with the micro-pillar structure is much higher than that of one without micro-pillars. This high selectivity is useful not only for in vitro but also in vivo monitoring where the sample usually contains a higher concentration of interferents.

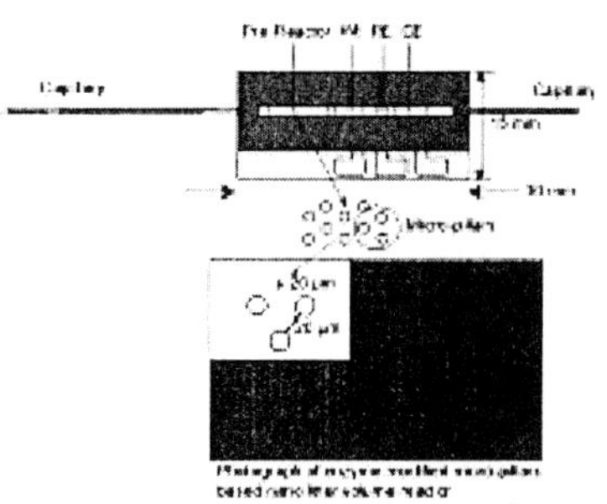

Fig.2 Bioanalytical device integrated with detector and small volume reactor

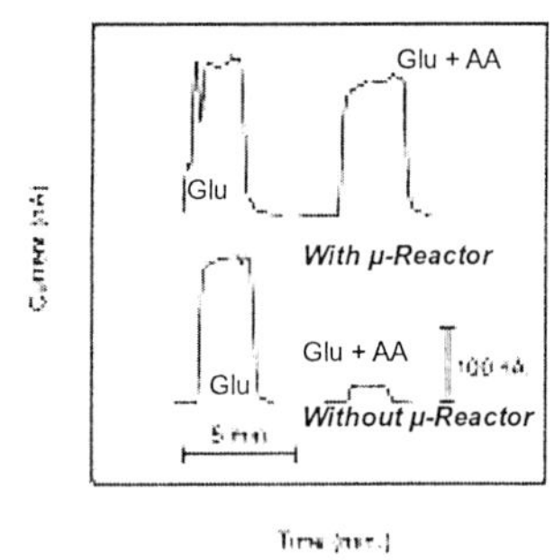

Fig. 3 Improved selectivity for glutamate detection by integrating nano-liter volume enzyme reactor

References

[1] Niwa. O, Kurita. R, Liu. Z, Horiuchi. T and Torimitsu. K, *Subnanoliter Volume Wall-Jet Cells Combined with Interdigitated Microarray Electrode and Enzyme Modified Planar Microelectrode.*, Anal. Chem., 72, **2000**, 949-55.

[2] Niwa. O, Kurita. R, Hayashi. K, Horiuchi. et al., *Continuous measurement of hisutamine from rat basophilic leukemia cells (RBL-2H3) with an on-line sensor using histamine oxidase.*, Sensors & Actuators **B** 67, **2000**, 43-51.

SURFACE MICROMACHINED CELL MANIPULATION DEVICE FOR TRANSFECTION AND SAMPLE PREPARATION

Murat Okandan, Paul Galambos, Sita Mani, Jay Jakubczak
MEMS and Novel Silicon Science and Technology
Sandia National Laboratories
P.O. Box 5800 MS 1080 Albuquerque, NM 87185 mokanda@sandia.gov

Keywords: Cellular manipulation, surface micromachining

Introduction: We have designed and fabricated a micromachined cell manipulation device using Si surface micromachining technology [1]. The device is designed to mechanically disrupt the cellular membrane of a cell, allowing the delivery of large molecules (for example DNA, RNA, proteins or fluorescent molecules) into the cytoplasm without completely lysing the cells. To deliver large molecules into cell, electroporation or chemical permeabilization methods are typically used [2], as well as some recent utilization of micromachined parts [3]. This device is designed to mechanically disrupt the cellular membrane in a controlled and repeatable manner (Figs. 1-3) and operates in a continuous flow mode. The cells flowing through the device are engaged by the manipulator which is operating in a reciprocating motion, powered by high performance comb drives and a compliant displacement multiplier. The mechanical features of the manipulator and the opposing structure are designed to effectively permeabilize the cellular membrane. A third port for addition of chemicals is immediately downstream of the cellular disruption chamber (Fig. 2). Mechanical membrane disruption effectively decouples the membrane disruption process from the chemical addition or electrical manipulation processes, allowing a wider range of chemicals and electrical signals to be used and broadening the field of potential applications for cellular manipulation devices. This device can also be used to lyse cells with a known and repeatable physical mechanism, enabling simplified, highly uniform and consistent cellular sample analysis. To our knowledge this is the first example of a continuous flow, mechanical cellular membrane disrupter to be reported and it is intended to be a capable demonstration tool for the unique integration of mechanical, optical and microfluidic components possible using this fabrication technology. The ultimate "system on a chip" for biomedical applications will require incorporation of microfluidic, mechanical, optical, electrical and chemical subsystem components into an integrated system. The novel device presented here is fabricated utilizing a process that incorporates silicon nitride layers, allowing optical access into the structures as well as the integration of electrical subsystems for electrochemistry, electrokinetics and electrostatics.

Process Technology: Surface micromachining techniques were used to create silicon nitride lined channels integrated with the five polysilicon layers. The silicon nitride insulated channels allow optical access to the structure as well as electrical isolation for integrating electrodes into and around the channels. The absence of these features has been the main limiting factor in the development of Si micromachined microfluidic components [4]. The height of the channels is approximately 6um and lateral dimensions can vary from a few microns to hundreds of microns. High performance comb drives and displacement multipliers were used for mechanical actuation (Fig.2). Fluidic access to the devices is provided by holes etched through the wafer. The sealed chambers created using silicon nitride are released through the Bosch etched holes, eliminating the need for etch release holes and bonding of a cover layer.

Characterization: A dilute solution of red blood cells (RBC) was pumped through the device using a syringe with the manipulator running (Fig.3). Smooth, sub-Hz mechanical actuation of

J.M. Ramsey and A. van den Berg (eds.), Micro Total Analysis Systems 2001, 305–306.

the cell disrupter was observed as well as actuation in the kHz range. The cellular membranes were observed to deform significantly on contact with the disrupter without being lysed. The high performance comb drives are designed to deliver forces on the order of mN, which are then coupled through the displacement multiplier. Current characterization efforts are aimed at demonstrating the incorporation of fluorescent material (fluorescein) into disrupted cells.

Sandia is a multiprogram laboratory operated by Sandia Corporation, a Lockheed Martin Company, for the United States Department of Energy under Contact DE-AC04-94AL95000.

[1] www.mdl.sandia.gov/micromachine

[2] "Microfabricated electroporation chip for single cell membrane permeabilization", Y. Huang, B. Rubinsky, Sensors and Actuators A, v.89, pg.242-249

[3] "An array of hollow microcapillaries for the controlled injection of genetic materials into animal/plant cells", K. Chun, et.al., IEEE MEMS'99, pg.406-411

[4] A.van denBerg, T.S.J. Lammerink and C. Effenhauser in *Microsystem Technology in Chemistry and Life Sciences*, ed. A.Manz, H. Becker, Springer-Verlag, 1999

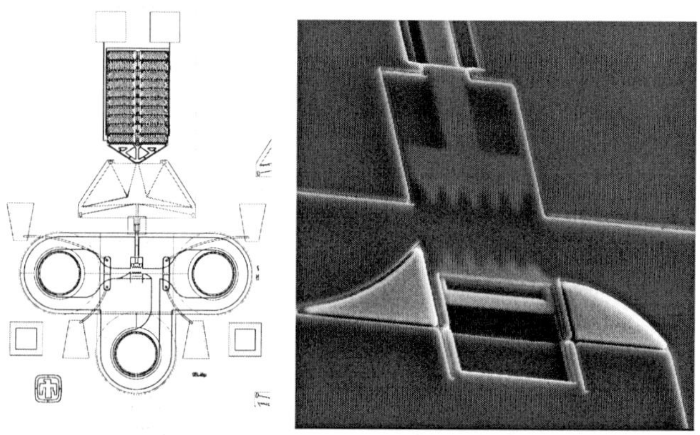

Fig.1 Mask layout of the micromachined cell manipulation device – and SEM close-up of the fabricated mechanical probe which is designed to disrupt the cellular membrane and allow delivery of large molecules into the cellular medium.

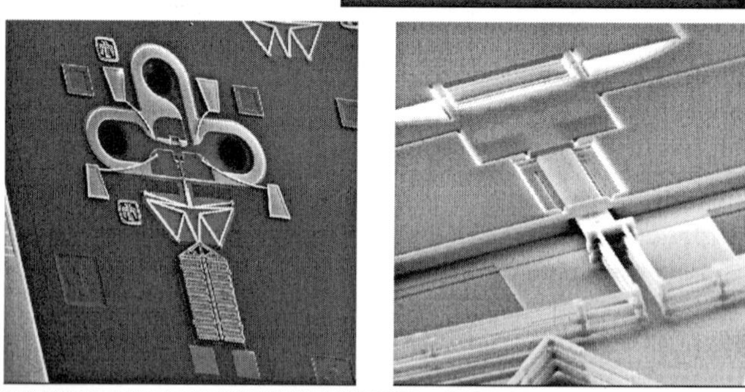

Fig.2 The high performance comb drive and the displacement multiplier that are driving the manipulator are visible in this SEM. The close-up shows the piston that is coupled into the microfluidic chamber, the manipulator is visible inside the silicon nitride channel. The chamber is formed during the process and no bonding steps are necessary to from the cavity.

Fig.3 Red blood cells flowing through the device. The manipulator was run from sub-Hz to kHz range with varying displacements. Current characterization efforts are aimed at delivery of fluorescent material into the manipulated cells. In this picture, material flow is from right to left.

DEVELOPMENT OF AN ON-CHIP CELL SORTING SYSTEM DESIGNED FOR CLONING MAMMALIAN CELLS WITH FUNCTIONAL, HETEROLOGOUSLY EXPRESSED SURFACE RECEPTORS

Anubhav Tripathi, Jeff Wolk, Roy Sandberg, Andrea Chow, Matt Kerby, Ring-Ling Chien and Gary Wada
Caliper Technologies Corp., 605 Fairchild Dr., Mountain View, CA 94043, USA

Keywords: Cell Sorting, cloning, cell-based assays, LabChip

1. Introduction

Cell separation technology has become an important tool in cell biology allowing for the analysis and subsequent cultivation of specific cell subsets. This technique is especially applicable to transfected cells, where only a small proportion of the cells may be expressing the antigen of interest. As an example, efficient selection of high-producing subclones during gene amplification of recombinant Chinese Hamster Ovary can be performed using cell sorting [1].

Several techniques have been investigated to perform efficient and rapid sorting, including fluorescence-activated cell separation (FACS) [2]; magnetic-activated cell separation (MACS) [3]; automated single-cell sorting using dual-beam optical trapping [4]; differential adhesion cell sorting [5] and micro-fabricated fluorescence-activated cell sorter [6]. Each of these techniques has their advantages and disadvantages with regard to cost, efficiency, and labor intensity. We have developed a low cost automated on-chip cell sorter which is efficient and hydrodynamically controlled. It also incorporates selective deposition of individual sorted cells into appropriate wells.

2. Theory

The chip configuration in Figure 1 shows that four wells and a sipper capillary are connected through micro-channels. A pressure controller imposes hydrodynamic pressure at the wells and drives the cell suspension to well #4. Momentum and mass balance equations were used to set the pressures to achieve desired flow rates of cells.

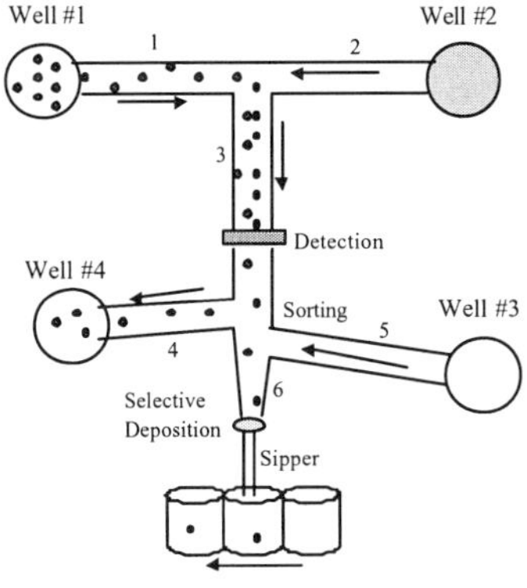

Figure 1-Cell Sorter Chip Schematic

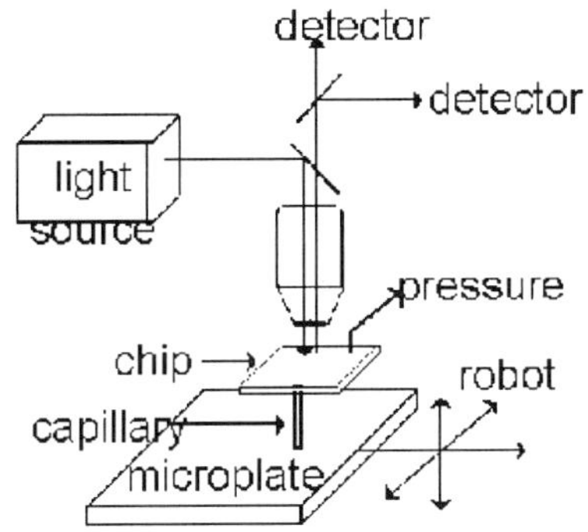

Figure 2- Instrument Schematic

J.M. Ramsey and A. van den Berg (eds.), Micro Total Analysis Systems 2001, 307–308.

A laser-activated fluorescence detector monitors the cells flowing in channel 3. When the signal threshold is exceeded, the detector passes a voltage signal to the pressure controller that diverts the cells to the capillary by momentarily changing the pressures at the wells causing a 600 msec plug of cell buffer to flow toward the capillary.

3. Experimental

Fluorescent 6 um latex beads (Spherotech Inc.) were used to characterize the efficiency of particle sorting. The beads were suspended into Cell Buffer, Hank's balanced salt solution buffered with 20 mM Hepes pH 7.0 containing 13.8% (v/v) Optiprep (Nycomed). THP-1, pre-monocytic cells were labeled with 4 uM calcein-am (Molecular Probes Inc.) fluorescent dye and resuspended in Cell Buffer containing 18% (v/v) Optiprep at 2 x 10^6 cells/ml. For calcium flux experiments, the cells were loaded with fluo-4 and fura red. The cell suspensions (40 ul) were placed into well 1 and cell buffer was loaded into the remaining wells. An argon ion laser was used to excite the cells through an epifluorescence optical detection system, equipped with 520 nm and 680 nm band pass filters for two color detection (Figure 2). A microcomputer processes the signal from the detectors, sends the voltage signal to the pressure controller when a signal exceeds the set threshold, and controls the xyz robot positioning of the microplate for single cell collection.

4. Results and Discussion

Using a mixture of non-fluorescent and fluorescein labeled beads (9:1 ratio) and manual counting of beads in the microplate wells after selection by the on-chip cell sorting, the enrichment of fluorescent particles was 80x into a single well. When the xyz robotic movement of the microplate is used to deposit individual sorted beads into wells, further enrichment was achieved since only one bead was deposited per well. The efficiency of capture of sorted beads into wells was about 90%. When cells were sorted, similar results were obtained. Automated functional cell sorting and cloning should prove to be a useful method for producing receptor transfected cells for cell-based drug screening assays and, perhaps, for expression cloning of novel genes.

5. References

1. N. Borth, M. Zeyda and H. Katinger, Biotechn. & Bioeng., 71(4), 266-273, 2001.
2. W. A. Bonner, H. R. Hulett and R. G. Sweet, Rev. Sci. Instrument., 19, 813-816 (1972).
3. S. Miltenyi, et al., Hematopoietic stem cells: the millhouse manual, Dayon: Alpha Med Press, 201-213 (1994).
4. S. C. Grover, et al., J. of Biomed. Optics, 6(1), 14-22 (2001).
5. M. Steinberg, Science 141, 401 (1963).
6. A. Y. Fu, et al., Nature Biotechnology, 17, 1109-1111 (1999).

GROWTH RATES OF INSECT CELLS IN MICROCHANNELS

G. M. Walker[1], M. S. Ozers[2], D. J. Beebe[1]
[1]Department of Biomedical Engineering, University of Wisconsin-Madison, Madison, Wisconsin 53706, USA
[2]PanVera Corporation, Madison, Wisconsin 53719, USA

Abstract
Cells from *Spodoptera frugiperda* (Sf9) were cultured in multiple microchannels for seven days. Cell growth rates were determined every 24 hours by counting the number of cells in each channel. It was found that cells cultured in microchannels tend to grow at a significantly slower rate than those grown in a conventional 25 mL tissue culture flask.

Keywords: cell culture, microfluidics, cell growth

1. Introduction
While the use of microchannels for cell culture is promising, the area is largely unexplored. Some work has been published in which endothelial cells were cultured in a microchannel to measure their shear stress response [1]. However, most work involving cells in microscale volumes has focused on cell assays for diagnostic purposes [2]. Here we explicitly examine the characteristics of cell growth in microchannels. Studies in which cell viability in a microenvironment is compared to viability in a macroenvironment are needed if one is to ultimately use microchannel environments for cell culture.

2. Experimental
Micromolding was used to create devices out of polydimethylsiloxane (PDMS). Each device contained five microchannels, each channel being 2 cm long, 150 μm high, and varying in width from 200 μm to 1000 μm. Devices were formed by bonding the PDMS mold to a glass slide and placing the assembly into a petri dish partially full of water. Insect cells (Sf9) suspended in TC-100 media were cultured in the devices for a period of seven days in an incubator at 27° C. Each day the total number of cells per channel was counted and compared with the count in a control tissue culture flask. The control cell population was measured using a Neubauer hemacytometer and 0.4% Trypan blue staining. We found that at 48 and 72 hours after seeding, cells in microchannels grew more slowly than the same cells in a 25 mL tissue culture flask (Figure 1). The accepted doubling time for Sf9 cells during the exponential phase is 24 hours. Cells in microchannels fell short of that, growing on average about 50% every 24 hours. Slower growth in microchannels may be indicative of a more *in vivo* like

J.M. Ramsey and A. van den Berg (eds.), Micro Total Analysis Systems 2001, 309–310.

environment. This has been shown to be the case for murine embryos – they exhibit enhanced developmental efficiencies when cultured in microchannels [3].

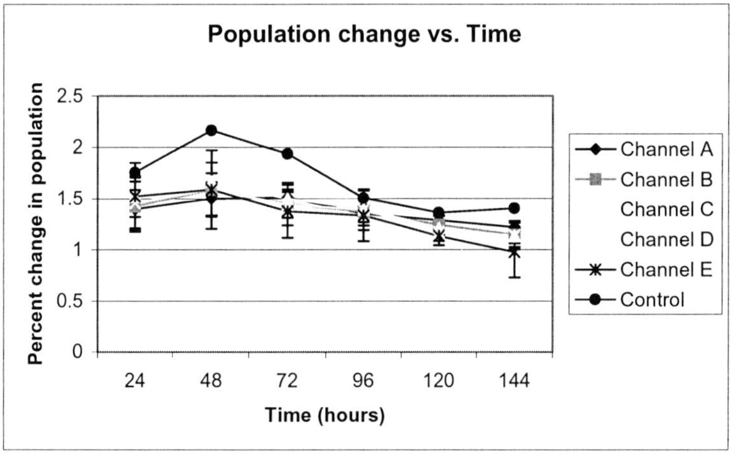

Figure 1. Average growth rates of cells in microchannels of different size. Channel A has a width of 1000 μm, B 800 μm, C 600 μm, D 400 μm, and E 200 μm. The control culture was grown in a 25 mL Falcon tissue culture flask. The control growth rate is significantly different from the microchannels at 48 and 72 hours after seeding.

3. Conclusions

As many techniques in biology begin to be miniaturized via microfluidics [4, 5], cell culture will inevitably follow because it is integral to many other processes (e.g. cell-based assays, recombinant protein production). If microscale culture volumes do affect cellular growth patterns, this will need to be taken into consideration when interpreting the results of any cellular-based research conducted at the microscale.

References

1. Frame, M.D.S. and I.H. Sarelius, *A System for Culture of Endothelial Cells in 20-50 μm Branching Tubes*, Microcirculation, **2**, 4, pp. 377-385, (1995).
2. DeBusschere, B.D., D.A. Borkholder, and G.T.A. Kovacs. *Design of an Integrated Silicon-PDMS Cell Cartridge*, Solid-State Sensor and Actuator Workshop 1998, Hilton Head Island, South Carolina, pp. 358-362, (1998).
3. Raty, S., J.A. Davis, D.J. Beebe, S.L. Rodriguez-Zas, and M.B. Wheeler, *Culture in microchannels enhance in vitro embryonic development of preimplantation mouse embryos*, Theriogenology, **55**, 1, p. 241, (2001).
4. Li, P.C.H. and D.J. Harrison, *Transport, Manipulation, and Reaction of Biological Cells On-Chip Using Electrokinetic Effects*, Analytical Chemistry, **69**, 8, pp. 1564-1568, (1997).
5. Folch, A. and M. Toner, *Microengineering of Cellular Interactions*, Annual Review of Biomedical Engineering, **2**, pp. 227-256, (2000).

CATH.A NEURON CELL ANALYSIS ON A CHIP WITH MICELLAR ELECTROKINETIC CHROMATOGRAPHY

Aaron Wheeler[1], Keisuke Morishima[1], Brian Kirby[2], Andrew Leach[1], and Richard N. Zare[1]
[1]Stanford University Chemistry Department, Stanford, CA 94305
[2]Sandia National Laboratory, Livermore, CA 94551
awheeler@stanford.edu

Abstract

An on-chip electrophoresis analysis of the contents of CATH.a neuron cells is presented. A novel mixed-micelle MEKC separation was required to resolve twelve amino acids and neurotransmitters within five minutes.

Keywords: MEKC, neurotransmitter, amino acid, cell analysis

1. Introduction

Fast, robust methods for analyzing the chemical contents of neurons are necessary to determine the molecular basis of neuron structure and function. The CATH.a cell line is a model for differentiated CNS cells; CATH.a cells express catecholamines and other small amine neurotransmitters, high levels of tyrosine hydroxylase, neurofilaments, and synaptophysin[1]. We have developed methodology to probe the contents of CATH.a cells with separations on microfabricated glass devices.

Since the neurotransmitters of interest have similar electrophoretic mobilities, previous work in this field has employed long separation columns[2] with reduced electroosmotic flow (eof). To effect a fast separation on-chip with a relatively short separation column, Micellar Electrokinetic Chromatography (MEKC) with a unique mixed-micelle buffer was used.

2. Experimental

Separations were performed on glass microfluidic devices prepared using standard microfabrication and lithographic techniques. Analytes (standard solutions and homogenized cell contents) were derivatized with naphthalene-2,3-dicarboxaldehyde (NDA) prior to analysis. A variety of MEKC separation buffers were tried, including varying concentrations of detergents, organic modifiers, and background electrolytes.

Derivatized analytes were injected using pinched injection across a 100 μm double-T injector, and detected with LIF (405 nm Nichia diode laser). The signal was optically and spatially filtered and collected with a PMT, and plotted vs. time on a PC to form electropherograms. Peaks were identified by standard addition.

J.M. Ramsey and A. van den Berg (eds.), Micro Total Analysis Systems 2001, 311–312.

3. Results and discussion

The best separations were performed in borate buffer (100 mM) with a mixed micelle detergent of SDS (80 mM) and the zwitterionic SB-12 (12 mM). Twelve neurotransmitters and amino acids were identified. As previously reported[1], dopamine and norepinephrine were detected, but their precursors (tyrosine and L-Dopa) were not (**Figure 1**). We speculate that future experiments using on-chip derivatization procedures may allow us to detect these precursors.

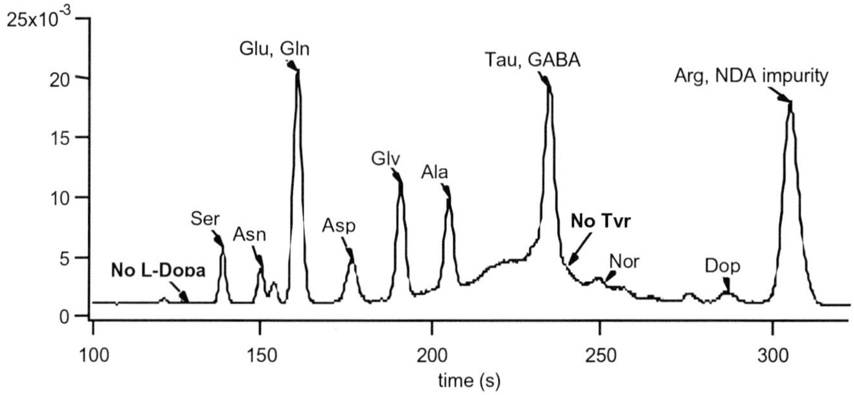

Figure 1: MEKC electropherogram of NDA-derivatized contents of CATH.a cells.

4. Conclusions

We have developed a robust, fast separation method for the analysis of small amine neurotransmitters in CATH.a neuron cells. We anticipate that this technique will be useful for neuroscientists investigating the biochemical basis of neuron structure and function.

Acknowledgements

Support from Beckman Coulter and Sandia National Labs is gratefully acknowledged.

References

1. Suri, C., *et al.*, "Catecholaminergic cell lines from the brain and adrenal glands of tyrosine hydroxylase-SV40 T antigenic mice," *Journal of Neuroscience*, 13, p. 1280-1291, (1993)
2. Gilman, S.D. and A.G. Ewing, "Analysis of Single Cells by Capillary Electrophoresis with On-Column Derivitization and Laser-Induced Fluorescence Detection" *Anal. Chem.*, 67, p. 58-64, (1995)

FUNCTIONAL BIOCHIPS FOR CELL AND MOLECULAR MANIPULATION

Junquan Xu[1], Lei Wu[1], Zhaohuai Yang[1], Weiping Yang[1], Mingxian Huang[1], Jing Cheng[1,2] and Xiao-Bo Wang[1]
[1]AVIVA Biosciences Corporation, San Diego, CA 92121
[2] Department of Biological Sciences and Biotechnology, Tsinghua University, Beijing, 100084, China

Abstract
We report here the application of electromagnetic and dielectrophoresis biochips for processing cell and molecular samples.

Keywords: electromagnetic biochip, dielectrophoresis, integrated biochip system

1. Introduction

Bioanalytical procedures essentially involve processing and detection of cells and molecules in a fluid sample. While substantial advances have been made in miniaturizing these processes on biochip platforms, on-chip sample preparation remains a significant challenge. This paper describes the development of functional biochips for manipulating biological samples.

2. Experimental

Electromagnetic chips comprising 16 electromagnetic units (Fig. 1) were fabricated on silicon substrates using multiple-layer photolithography. Each unit was a microfabricated conductive coil wrapped around a ferromagnetic core and was able to produce magnetic fields for capturing magnetic beads with applied current between 50 and 200 mA. Fluidic chamber was constructed by gluing a molded, plastic housing onto a chip. Teflon tubing attached to the plastic housing served as the input and output for the chamber. For separating T-lymphocytes from blood, buffy coat was diluted twenty fold in PBS buffer and incubated with CD3-labeled magnetic beads (2 micron). For separating white blood cells on the multiple-force chip, human blood was diluted twenty-fold in a sucrose buffer and an AC voltage at 5 MHz and 3 V pk-pk was applied.

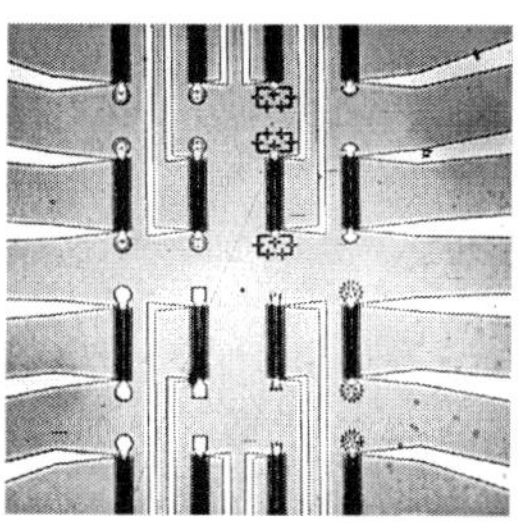

Figure 1. The electromagnetic chip fabricated for cell separation.

Figure 2. T-lymphocytes separated from blood and collected at the end of an electromagnetic element.

J.M. Ramsey and A. van den Berg (eds.), Micro Total Analysis Systems 2001, 313–314.

3. Results and Discussion

For electromagnetic separation of cells, cells within a mixture were pre-incubated with magnetic beads that were labeled with antibodies against target cells. The cell sample was then loaded into the fluidic chamber comprising the electromagnetic chip. The target cells were captured and retained from the flow stream by the energized electromagnetic elements. Using this protocol, we achieved the capture of T-lymphocytes from blood samples with CD3-antibody labeled magnetic beads (Fig. 2). The protocol was also used for purifying mRNA molecules from a cell lysate mixture with oligo $(dT)_n$ labeled magnetic beads (Fig. 4B, 4C).

For an integrated cell and molecular manipulation, a multiple-force chip comprising electromagnetic elements and dielectrophoretic electrodes was used (Figure 3). A protocol was developed for operating such a chip for separating white blood cells (WBCs) by dielectrophoresis (Fig. 4A), lysing the captured WBCs and binding the released mRNA molecules to Oligo $(dT)_{25}$ labeled magnetic beads (Fig. 4B), retaining the magnetic beads on the chip whilst removing other molecules (Fig. 4C) and releasing the beads for off-chip collection. A reverse transcription coupled with PCR amplification for a house-keeping gene G3PDH was performed on the released magnetic beads. Gel electrophoresis was run to detect the PCR product (Fig. 5).

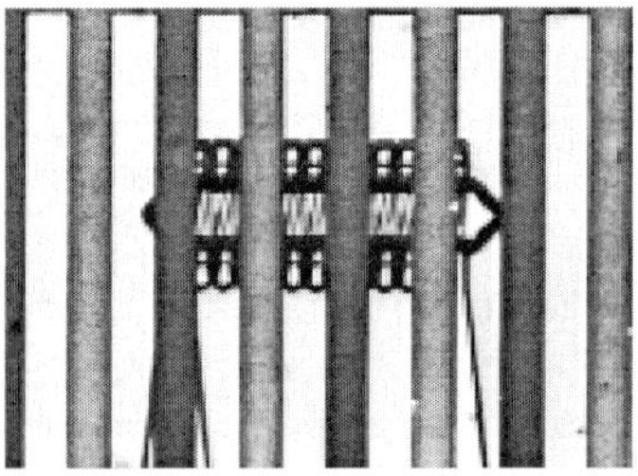

Figure 3. A multiple-force chip with electromagnetic elements underneath and the dielectrophoretic electrodes on the top.

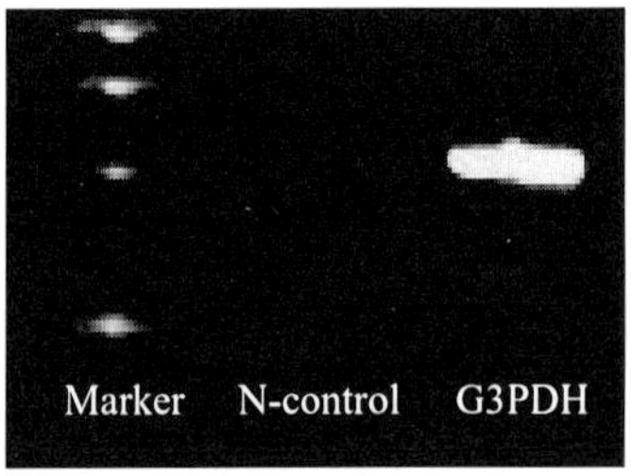

Figure 5. Gel electrophoresis detection of RT-PCR of magnetic beads released from the chamber.

Figure 4. A WBCs were captured on electrodes by dielectrophoresis.

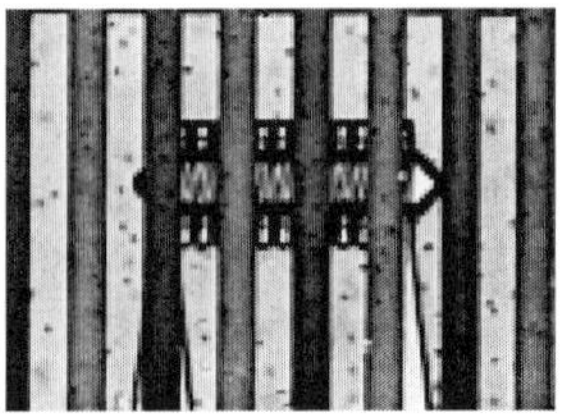

Figure 4. B Lysis buffer and Oligo-dT magnetic beads were introduced.

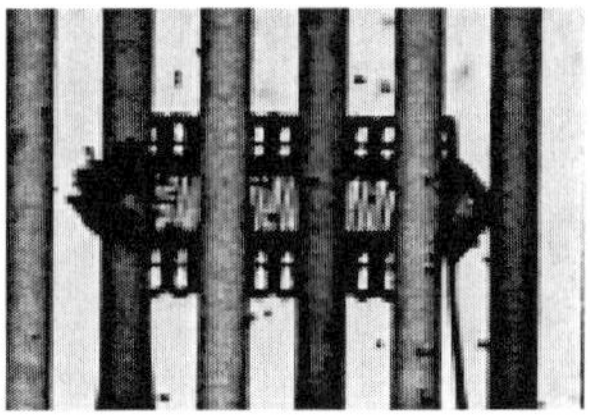

Figure 4. C Magnetic beads with bound mRNA were captured.

4. Conclusions

Functional biochips capable of generating electromagnetic forces were developed. Cell separation and molecular purification were demonstrated by using magnetic beads immobilized with specific capture molecules.

DIFFERENTIAL ANALYSIS OF SINGLE-CELL CULTIVATION USING ON-CHIP MICROCULTURE SYSTEM AND OPTICAL TRAPPING

Y. Wakamoto, I. Inoue, H. Moriguchi, and K. Yasuda
Department of Life Sciences, Graduate School of Arts and Sciences,
University of Tokyo
E-mail: cyasuda@mail.ecc.u-tokyo.ac.jp

Abstract: We developed the system for one cell level differential analysis employing on-chip microculture system and optical trapping. The experimental result showed that system had the ability to reveal the fluctuations in growth rate and division intervals of genetically identical *Escherichia coli* cells even in a homogeneous condition. This indicates the importance of the new approach for one cell level phenomenon.
Key words: differential analysis of single cells, on-chip culture system, optical trapping

1. Introduction

To investigate one-cell expression analysis of cultured cells in particular phase of cell cycle for "drug screening" or to analyze the interaction of cells with their environment for understanding "Division Potential", we need to be able to continuously observe and analyze isolated single cells under the fully controlled condition. We therefore developed a method for continuously observing isolated single cells for generations, that enables cells having the same genome to be compared by an on-chip microculture system and optical tweezers.

2. Experimental

Figure 1(a) is the photograph of the fabricated chamber array made of thick photo-resist SU-8 (MicroChem Corp., MA, U.S.) on 0.2-mm-thick glass slide. The surface of the slide is decorated with biotin and covered with an avidin-decorated cellulose semipermeable membrane (M.W. ~25,000) to separate the chambers from the nutrient medium circulating through the cover chamber. The membrane is fixed on the surface of the glass slide using avidin-biotin attachment, which is strong enough to keep the cells within the microchambers while preventing contamination from the external environment. Buffer in microchambers is continuously exchanged through membrane, maintaining the condition surrounding the cells in microchambers constant. Figure 1(b) shows the procedure of analyzing all cells within 4 generations derived from one ancestral cell, all of which can be assumed to have the identical genetic information. First, a cell group was pre-cultured in chamber 0. Next, a particular cell was transferred into chamber1 (1st generation) using optical tweezers and isolated in chamber 1. When the isolated single cell divided into two daughter cells, each of the daughter cells were transferred into chambers 2a and 2b by optical tweezers as soon as possible and then isolated again. When the isolated daughter cells were divided again into grand-daughter cells, they were transferred into chambers 3a, 3b, 3c, and 3d, and so on. The growth and division of

J.M. Ramsey and A. van den Berg (eds.), Micro Total Analysis Systems 2001, 315–316.

cells were observed continuously with phase-contrast/fluorescent microscopy (Olympus IX-70 inverted microscope with an oil-immersion objective lens, ×100, NA=1.35). Growth rate in length and division intervals of isolated E. coli and its descendant cells were examined and compared.

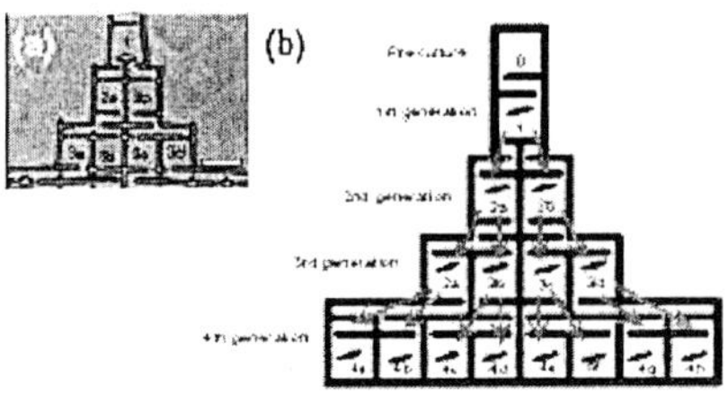

Figure 1
(a) photograph of microchamber array used for observation of direct descendant cells of isolated mother cell. White bar is 50 μm long. (b) over all design of microchamber array

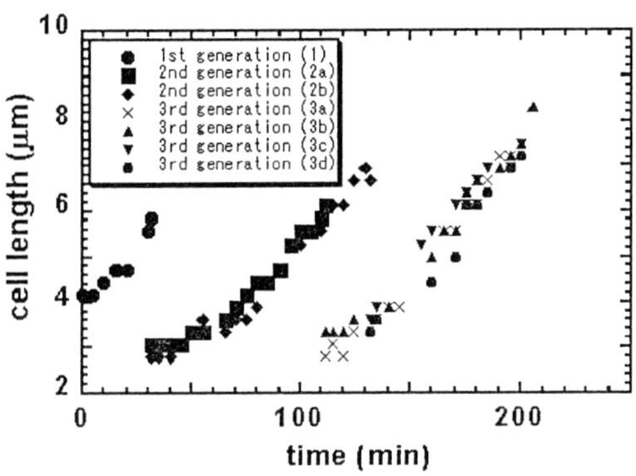

Figure 2. Time course of cell growth of direct descendants of single E.coli under isolation condtions using microchamber array shown in Fig. 1

3. Results and Discussions

Figure 2 shows one of the experimental results of the differential analysis about single cells' growths of generations. This result indicates that growth rate in length and division intervals include large fluctuations even between cells of genetically identical, suggesting both that *E.coli* cells rapidly go out of cell cycle synchrony and that even in homogeneous conditions growth rate and division intervals cannot be constant among cells. This method employing on-chip microculture system and optical tweezers will lead to the new approach to one cell observation, that is differential analysis of single cells.

4. Conclusions

Combining on-chip microculture system and optical tweezers, we developed new method and system for differential analysis of single cells.

Acknowledgements. This study was supported in part by Grants-in-Aids for Science Research from Ministry of Education, Science and Culture of Japan (No.11CE2006(Komaba Complex Systems Life Science Project), and No.12206028).

Reference

1. I. Inoue, Y. Wakamoto, H. Moriguchi, K. Onuma, A. Kashiwagi, K. Kaneko, T. Yomo, K. Yasuda. *Culture system for observation of individual bacteria*, Biophys. J. vol. 80 pp.166a-167a, (2001).

OBSERVATION OF INDIVIDUAL CELL USING ON-CHIP CULTURE SYSTEM

I. Inoue[1], Y. Wakamoto[1], H. Moriguchi[1], K. Okano[2], and K. Yasuda[1]
[1]Department of Life Sciences, Graduate School of Arts and Sciences, University of Tokyo
[2]Central Research Lab., Hitachi Ltd.
E-mail: cyasuda@mail.ecc.u-tokyo.ac.jp

Abstract We developed an on-chip continuous culture system for single-cell analysis. The advantages of the system are: (i) contamination-free continuous cultivation of isolated single cells or groups of cells using semipermiable membrane lid on the microchamber wells with biotin-streptavidin attachment; (ii) 0.2-μm resoltion for phase-contrast/fluorecent real time continuous observation using 5-20 μm deep microchambers on 0.2-mm-thick glass slide; (iii) flexible medium buffer exchange; and (iv) flexible cultivation chamber size. Using the system we examined the chamber size dependency of Escherichia coli's growth, and found no significant difference between 20-μm and 70-μm microchambers.

Keywords: on-chip single-cell culture, avidin-biotin attachment, semi-permiable membrane lid

1. Introduction Recent works on topographic control of cells such as cell orientation, rates of movement, and physical contact of cells were achieved by microfabrication techniques. Though those works showed the potential of microstructures for cell position control, those topological techniques cannot control the swimming cells and cannot prevent contamination from envrionment by themselves. We therefore developed and examined an on-chip culture system for single cell analysis.[1]

2. On-chip Microculture System for Isolated Single Cell Cultivation Figure 1 shows the core of the microculture system. As shown in Fig. 1(a), the microchamber array plate includes an n × n (n=20-50) array of chambers, where each chamber is 20-70 μm in diameter and 5-30 μm deep. The array is etched into an biotin-coated glass slide and covered with a streptavidin-coated cellulose semipermeable membrane (M.W. 25000) separating the chambers from the nutrient medium circulating through the cover chamber. The semipermeable membrane is fixed on the surface of the glass slide by using strepavidin-biotin attachment. The advantage of this system is no need for any heating or any chemical reactin to isolate cells from their environment. An example of the spatial arrangement of the microchamber array is shown in Fig. 1(b), where microchambers are etched at 100-μm intervals in a 0.17-mm-thick glass slide.

3. Chamber-Volume Dependency of Single-cell Growth The chamber-size dependence of Escherichia coli's growth was then examined. This experiment demonstrated another potential applications of this on-chip cluture system, for volume control of the cell-culture chamber. Figure 2 shows the time-course growth of single *E. coli*. The average growth speeds of five sample were 0.06 μm/min both in the small chamber and in the large chamber. The results showed no significant differences in mean values of growth speed and cell division time, even when the mean

J.M. Ramsey and A. van den Berg (eds.), Micro Total Analysis Systems 2001, 317–318.

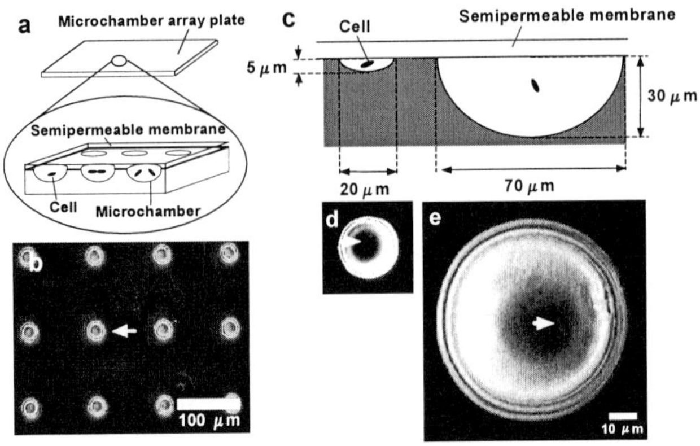

Figure 1. (a) Schematic drawing of the microchamber array plate. (b) Optical micrograph of the microchamber array plate. The arrow indicates the position of one microchamber. (c) Schematic drawing of the two sizes of microchambers. Optocal micrograph of (d) a small chamber and (e) a large chamber. The arrows indicate the positions of *E. coli* in the chambers.

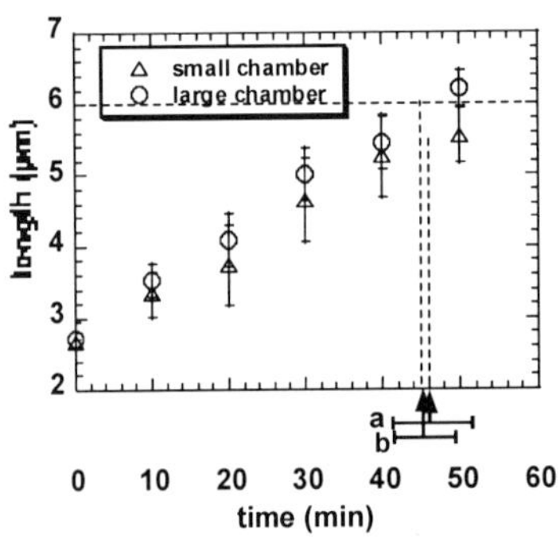

Figure 2. Chamber size dependence of *E. coli* growth. The triangles indicate a small chamber. The circles indicate a large chamber. The plotted points are mean values of cell length at 10-min intervals, with error bars indicating the S.D. of five cells. The arrow and bar labelled "a" indicate the mean time and S.D. of the cell division time for the small chamber; those labelled "b" indicate these values for the large chamber.

free paths were 10 times different.

4. Conclusions The system enabled us to find that single *E. coli* enclosed in a chamber successfully cultivated. Our micro-chamber array system covered with the semi-permeable membrane can simultaneously culture many types of cells. The system is suitable for high-throughput screening of cells because the chamber array has more than 2500 different chambers and the cultivation conditions e.g. pH, antibiotics, and chemical agents are easily changeable. The one-chip culture method described helps to explain the cell growth and cell division mechanisms not only for single cell analysis but also for a group of cells like nerve cells, and might show some unique usage for drug screening studies.

Acknowledgements This study was supported in part by Grants-in-Aids for Science Research from the Ministry of Education, Science and Culture of Japan (No.11CE2006 (Komaba Complex Systems Life Science Project), and No.12206028).

Reference

1. I. Inoue, Y. Wakamoto, H. Moriguchi, K. Onuma, A. Kashiwagi, K. Kaneko, T. Yomo, and K.Yasuda. *Culture system for observation of individual bacteria*, Biophys. J. vol. 80 pp. 166a-167a, (2001).

DETECTION OF NUCLEIC ACIDS USING NOVEL POLYMERS ABLE TO TRANSDUCE HYBRIDIZATION INTO OPTICAL OR ELECTRICAL SIGNAL

M. Boissinot[1], A.-H. Ho[2], M. G. Bergeron[1], G. Corbeil[2] and M. Leclerc[2]

[1] Centre de recherche en infectiologie, CHUL, Sainte-Foy, Quebec, Canada, G1V 4G2.

[2]Département de Chimie, Université Laval, Sainte-Foy, Quebec, Canada, G1K 7P4.

Abstract

We designed novel cationic, water-soluble polythiophene derivatives which can easily transduce oligonucleotide hybridization into a clearly interpretable optical (colorimetric or fluorescent) or electrical signal. These polymeric sensors proved to be specific enough to distinguish between single nucleotide mismatches on 20-mers oligonucleotide probes. This technology has the potential to be integrated into various μTAS applications involving detection of nucleic acids.

Keywords: polymer, biosensor, hybridization

1. Introduction

Sequence-specific methods are needed for the rapid detection of oligonucleotides for genotyping and diagnosis of infections and various genetic diseases. Most current detection methods use a fluorescent or electroactive tag bound to the analyte. An assay that does not require any chemical manipulation of nucleic acids or complex reaction mixtures would be advantageous. Here, we describe new polymer biosensors for nucleic acids hybridization.

2. Experimental

Polymers were obtained by a simple oxidative chemical polymerization [1] of the corresponding monomers (Fig. 1). Hybridizations were carried out using 20 mers oligonucleotides and 149 base pairs amplicons originally designed to differenciate between closely related yeast species *Candida albicans* and *C. dubliniensis* [2].

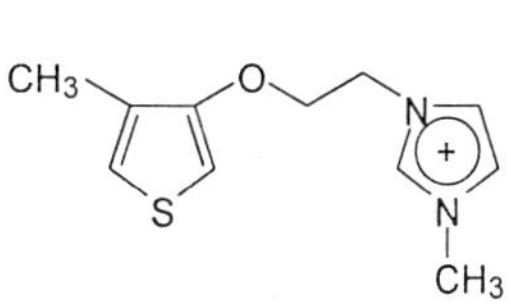

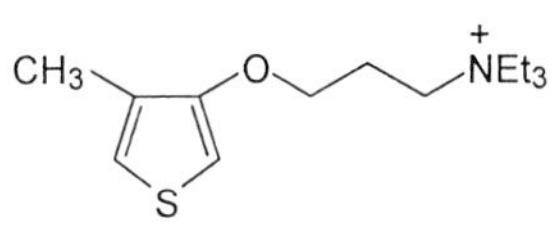

Figure 1. monomer 1 monomer 2

3. Results and Discussion

At 55°C, aqueous solutions (0.1 M NaCl or 10 mM Tris buffer/ 0.1M NaCl) of the cationic polymer **1** is yellow (λ_{max}= 397 nm). Upon addition of a given oligonucleotide (20-mers), the mixture becomes red (λ_{max}= 527 nm) within 5 minutes, due to the formation of a so-called duplex between the polythiophene and the oligonucleotide probe. After 5 minutes of mixing in the presence of 1.0 equivalent of the complementary oligonucleotide, the solution becomes yellow (λ_{max}= 421 nm); presumably due to the creation of a new complex termed triplex, formed by complexation of the polymer with the hybridized nucleic acids. Very similar results have been obtained with polymer **2** and various oligonucleotides. On the basis of previous studies[3], it is believed that these colorimetric effects are made

J.M. Ramsey and A. van den Berg (eds.), Micro Total Analysis Systems 2001, 319–320.

possible due to a different conformational structure of the conjugated polymer in the duplex compared to that observed in the triplex. Circular dichroism measurements revealed a right handed helical structure for the polymer. This is compatible with binding of the polymer to the negatively charged phosphate backbone of the DNA.

These polymers can discriminate between specific and non specific hybridization of nucleic acids which differ by a single nucleotide. Furthermore, the electrochemical properties of these polythiophenes can be exploited for the detection of DNA hybridization events in aqueous solutions. Therefore, as shown in figure 3, the anodic peak current is more important in the case of perfect hybridization, compared to the blank control. In addition, a shift to a higher potential (*ca.* 40-50 mV) is observed when the specific nucleic acid hybridization occurs.

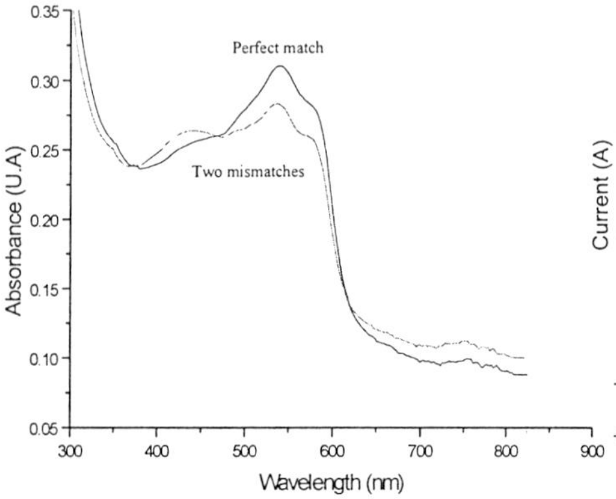

Figure 2. Specific optical detection (using polymer 1) of *Candida albicans* (upper curve) and *Candida dubliniensis* (lower curve) 149 bp amplicons, differing by only 2 nucleotides. Detection was carried out in 45 minutes directly from a PCR product at a concentration normally generated in 100μL PCR volume (ca. 3×10^{12} copies).

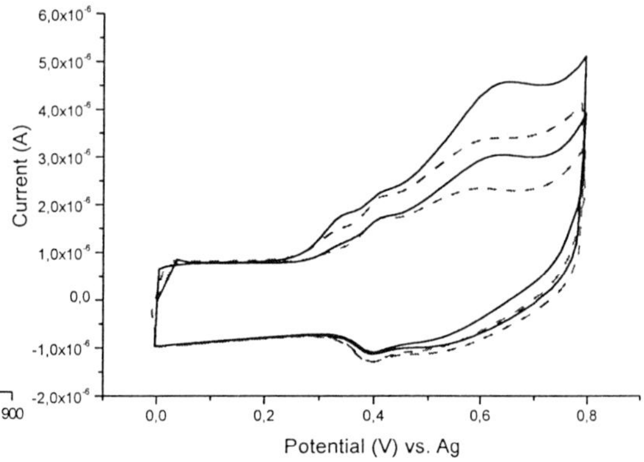

Figure 3. Cyclic voltammogram of polymer 2 on a modified indium tin oxide (ITO) electrode ($S = 50\ mm^2$) in the case of perfect match hybridization (continuous line) and no hybridization (dotted line). Scan rate of 100 mV/sec vs Ag in 0.1 M $NaCl/H_2O$, at room temperature

4. Conclusion

These nucleic acid sensing polymers could act simultaneously as trigger, amplifier and electrical or optical transducers. This rapid, selective, and versatile method does not require any chemical reaction on the probes or the analytes and is based on different electrostatic interactions and conformational structures between cationic poly(3-alkoxy-4-methyl-thiophene)s and single stranded oligonucleotides or double-stranded (hybridized) nucleic acids. These intelligent polymers could be adapted to create numerous novel applications including nucleic acid arrays and microfluidic molecular diagnostics devices.

References

[1] Faïd, K. & Leclerc, M. Functionalized regioregular polythiophenes: towards the development of biochromic sensors. *J. Chem. Soc., Chem. Commun.* 2761-2762 (1996).

[2] Trépanier, H. *et al.* One-Hour Detection of *Candida albicans* and *Candida dubliniensis* in Blood Samples Using the Smart Cycler®. ASM annual meeting. Poster C-389. (2001).

[3] Leclerc, M. Optical and electrochemical transducers based on functionalized conjugated polymers. *Adv. Mater.* **11**, 1491-1498 (1999).

SELF-CONTAINED MICROELECTROCHEMICAL DETECTORS FOR ANALYSIS IN SMALL VOLUMES OF STATIC AND FLOWING FLUIDS

Charles S. Henry,[1] Walter R. Vandaveer, IV,[2] Ibrahim Mubarak,[2] Stephen R. Gray,[2] and Ingrid Fritsch[2]

[1] Dept. of Chemistry, Mississippi State University, Starkville, MS 39762, USA
[2] Dept. of Chemistry and Biochemistry, University of Arkansas, Fayetteville, AR 72701, USA

Abstract

Microfabricated 10 and 53 μm-diameter and 8 μm-deep cavities (geometric volumes of 0.6 pL and 18 pL, respectively) with three individually-addressable electrodes perform electrochemical analysis in small volumes and convective solutions.

Keywords: electrochemistry, detection, convection, microcavity, simulations

1. Introduction

Electrochemical detection is sensitive, operates in turbid samples, and is promising for lab-on-a-chip applications. We use a unique fabrication approach [1,2] to construct complete electrochemical cells in the smallest space reported to-date. These microcavities are ~10 and ~50 μm in diameter and 8 μm deep with a recessed microdisk electrode (RMD) at the bottom, tubular nanoband electrode (TNB) along the wall, and third electrode at the rim of the cavity (Figure 1). We report evaluation of these devices with small volumes and stirred solutions.

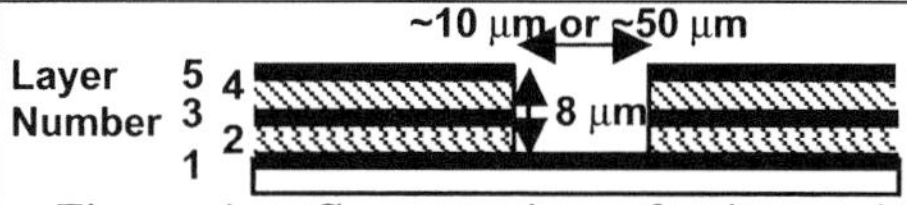

Figure 1: Cross-section of microcavity. Electrode layers 1, 3, 5: gold with chromium adhesion. Insulating layers 2 & 4: ~4 μm polyimide. Substrate: oxidized silicon wafer.

2. Experimental

The cavities were microfabricated [1]. The TNB, coated with Ag/AgI [3], served as a pseudo-reference electrode. The RMD and rim served as working and auxiliary electrodes, respectively. A BAS-100B potentiostat with PA-1 preamplifier ran cyclic voltammetry (CV) and chronoamperometry (CA). Stirring used an open beaker, Corning PC-320 stir plate, and 1/2" magnetic stir bar.

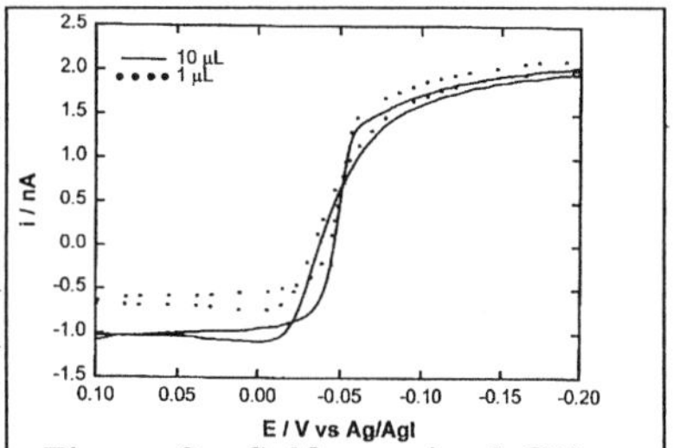

Figure 2. Self-contained CV at 0.1 V/s in 10 μm-diameter cavity for 1 and 10 μL drops of 5 mM $Ru(NH_3)_6^{3+}$ in 0.5 M KNO_3.

3. Results and Discussion

Small volumes. Figure 2 shows self-contained CV responses obtained in 1 μL and 10 μL at a 10 μm-diameter cavity. Responses are similar because both volumes are greater than the electrochemically-perturbed region. Crossover between forward and reverse sweeps is caused by shifting of the potential at the pseudo-reference electrode, which lies within the diffusion layer of the working electrode. Computer simulations verify this interpretation (Figure 3). Future experiments will involve picoliter volumes.

J.M. Ramsey and A. van den Berg (eds.), Micro Total Analysis Systems 2001, 321–322.

Convection studies. Current at the RMD becomes constant (i.e. unaffected by flow) when depth-to-diameter ratio of the cavity increases [2]. This is shown in Figure 4, which compares self-contained CA responses at 53 and 10 μm-diameter cavities in static and stirred solutions. At the 53 μm-diameter cavity, signal at 150 rpm has oscillations and a larger absolute magnitude. The current at the 10 μm-diameter cavity is independent of stirring rate. These results suggest that 10 μm-diameter cavities can be used without recalibrating for variations in flow rates.

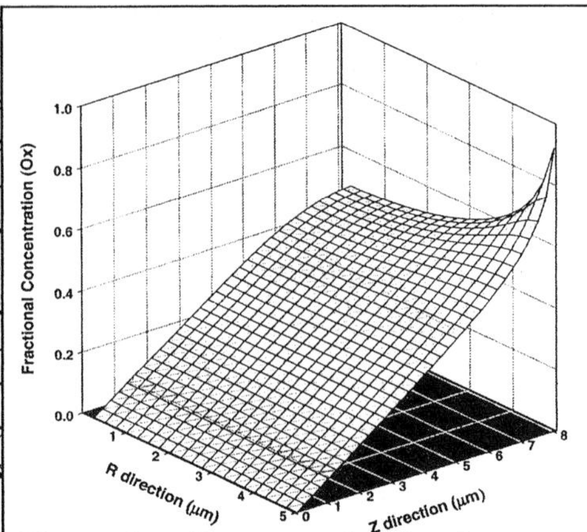

Figure 3. Computer-simulated concentration profile (–0.2 V vs. E°, 0.1 V/s) in a 10 μm-diameter capped cavity. R = radial axis, where center=0 and wall=5 μm. Z = depth, where RMD=0, TNB=4 μm, and rim=8 μm. Starting fractional concentrations: oxidized species=1, reduced=0.

4. Conclusions

Successful small volume analysis and minimal signal fluctuations under convective conditions suggest that self-contained microcavities may be promising candidates for electrochemical detection in micro total analysis systems that involve microfluidics.

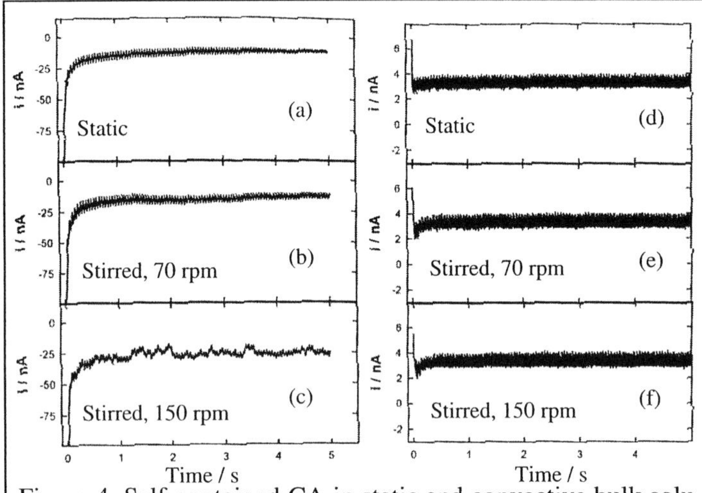

Figure 4. Self-contained CA in static and convective bulk solution. (a-c) 1.0 mM hydroquinone in phosphate buffer (pH 7.0, 0.05 M) at 53 μm-diameter cavity, potential stepped from +0.2 V to +0.8 V. (d-f) 5.7 mM $Ru(NH_3)_6^{3+}$ in 0.5 M KNO_3.at 10 μm-diameter cavity, potential stepped from +0.1 V to -0.5 V.

Acknowledgements

Funding: National Science Foundation (CHE-9308946, EHR-9180762, and CHE-9624114 CA-REER). Microfabrication facilities: Arkansas High Density Electronics Center.

References

1 . C. S. Henry and I. Fritsch, *J. Electrochem. Soc.* **146**, 3367-3373 (1999).
2 . C. S. Henry and I. Fritsch, *Anal. Chem.* **71**, 550-556 (1998).
3 . C. D. T. Bratten, C. D. T., P. H. Cobbold, and J. M. Cooper, *Anal. Chem.* **69**, 253-258 (1997).

OPERATING PARAMETERS FOR AN IN-PLANE, CONTACTLESS CONDUCTIVITY DETECTOR FOR MICROCHIP-BASED SEPARATION METHODS

Jan Lichtenberg, Elisabeth Verpoorte and Nico F. de Rooij
SAMLAB, Institute of Microtechnology, University of Neuchâtel,
Rue Jaquet-Droz 1, CH-2007 Neuchâtel, Switzerland
e-mail: jan.lichtenberg@unine.ch

Abstract

In this paper, we discuss procedures to optimize the operating parameters for a glass-based microchip for capillary electrophoresis (CE) with an integrated, in-plane conductivity detector. Detector layout (electrode size and shielding), electronic parameters for the signal processing unit (operation frequency, lock-in amplifier configuration and filtering), as well as CE conditions (buffer, separation field strength and sample plug formation), are important to achieve good device performance.

Keywords: conductivity detection, capillary electrophoresis

1. Introduction

Contactless impedance detection, described already for classical capillary electrophoresis (CE) [1], is also attracting increasing research interest for chip-based CE systems [2, 3, 4]. Miniaturized in-plane detectors are coupled capacitively to the electrolyte in the channel to measure changes in solution conductivity. Due to the small electrode size, the coupling capacitance is very low, while parasitic capacitances, which reduce sensitivity, are of the same magnitude. As the ratio of the two is frequency-dependent, it is very important to optimize the operating parameters of the complete system.

2. Device Fabrication

The fabrication of the μCE device with integrated conductivity detection (μCE-CD) has already been presented [4]. Briefly, the CE channel network, with its 150-μm, double-T injector and 50-μm-wide, 12-μm-deep channels, is manufactured using conventional wet etching techniques. The detector consists of two 40- to 400-μm-wide, thin-film platinum electrodes, which are positioned close to the channel (Fig. 1). To allow fusion bonding, the electrodes are deposited in recesses formed in the glass.

3. Parameter Optimization

Finding the optimum operating frequency is a tradeoff between limiting the background caused by parasitic capacitances and achieving efficient capacitive coupling between the electrodes and the liquid inside the channel. Furthermore, the performance of the current-to-voltage converter and the lock-in circuit deteriorate with increasing frequency. As Fig. 2 shows, the optimum operating frequency can be determined experimentally by plotting the peak height of a known sample versus the frequency, which showed a maximum response at 58 kHz for the device in Fig 1.

J.M. Ramsey and A. van den Berg (eds.), Micro Total Analysis Systems 2001, 323–324.

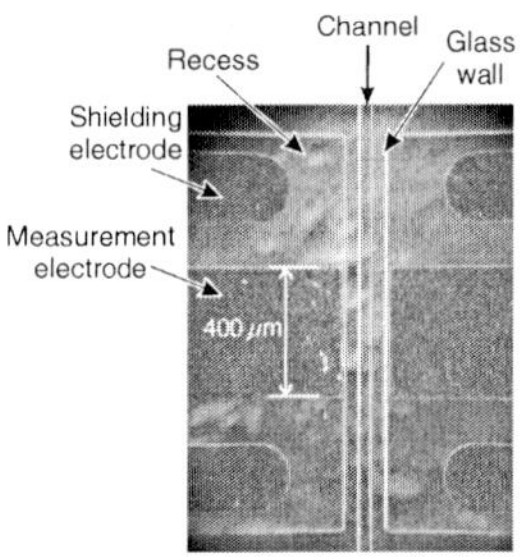

Figure 1: Detector element with 400-μm-wide electrodes. The outer electrodes act as shielding.

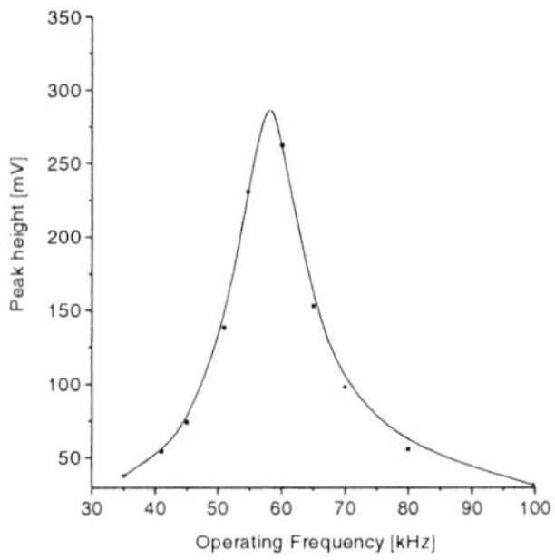

Figure 2: Peak height as a function of the operating frequency for 20 mM MES-His sample in 10 mM MES-His buffer (pH 6.0) at 200 V/cm.

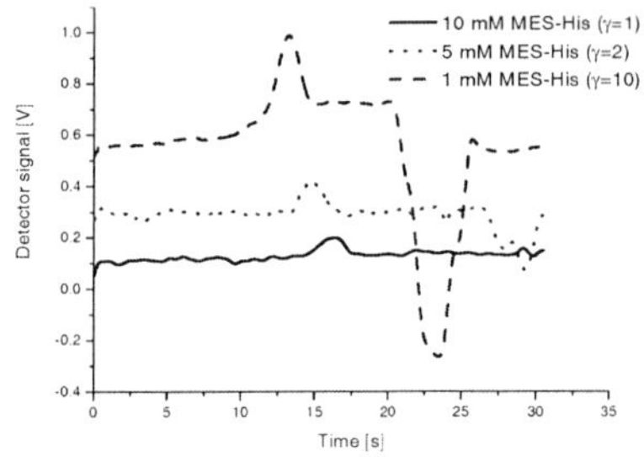

Figure 3: Field-amplification stacking of 100 μM potassium in 10, 5 and 1 mM MES/His buffer. Running buffer: 10 mM MES/His with 50 μM CTAB, 200 V/cm.

4. Results and discussion

CE separations of cationic sample mixtures using the μCE-CD device yielded a limit of detection of around 1 ppm or 35 μM, which is comparable to the 4-electrode device presented in [2]. To improve the sensitivity, field-amplified stacking in a low-conductivity sample matrix was implemented on chip to pre-concentrate the sample prior to separation [5]. Preliminary separations under stacking conditions yielded a signal gain of 3.9 compared to a non-stacking experiment at a relative conductivity γ of 10 (sample matrix 10 times less conductive than running buffer), as shown in Fig. 3.

Acknowledgements

This project was financed by the Swiss Ministry for Education and Science within the EU project "MicroChem" (Brite-Euram BE97-5134). The help of the Samlab technical staff is greatly appreciated.

References

1. A. J. Zemann *et al.*, *Contactless conductivity detection for capillary electrophoresis*, Anal. Chem., **70** (1998) 563–567.
2. F. Laugere *et al.*, *Experimental verification of an improved method for conductivity detection in on-chip capillary electrophoresis systems*, Proc. of Transducers 01, Munich (2001), 1178–1181.
3. B. Graß *et al.*, *A new PMMA-microchip device for isotachophoresis with integrated conductivity detector*, Sensors and Actuators B, **72** (2001) 249–258.
4. J. Lichtenberg *et al.*, *A miniaturized analysis system based on capillary electrophoresis and contactless conductivity detection*, Proc. of Transducers 01, Munich (2001), 408–411.
5. J. Lichtenberg *et al.*, *Sample preconcentration by field amplification stacking for microchip-based capillary electrophoresis*, Electrophoresis, 22 (2001) 258–271.

STUDIES TO IMPROVE THE PERFORMANCE OF ELECTROCHEMICAL DETECTION FOR MICROCHIP ELECTROPHORESIS

R. Scott Martin, Ruri Kikura-Hanajiri,
Nathan A. Lacher, and Susan M. Lunte
Department of Pharmaceutical Chemistry, University of Kansas,
2095 Constant Avenue, Lawrence, KS 66047, USA

Abstract

Strategies to improve the performance of electrochemical (EC) detection for microchip capillary electrophoresis (CE) are described. Contributions to band broadening from the electrochemical detector and analyte adsorption in poly(dimethylsiloxane) (PDMS) -based microchips were investigated.

Keywords: Electrophoresis, electrochemical detection, poly(dimethylsiloxane)

1. Introduction

Electrochemistry as a detection mode for microchip CE is becoming increasingly popular. One of the critical issues in implementing EC detection with microchip CE is decoupling the separation voltage from the detector. This is usually accomplished by configuring the working electrode in an end-channel arrangement, positioned tens of microns from the end of the separation channel. This leads to two sources of band broadening from the detector, one due to the size of the working electrode and one from the end-channel arrangement ($\sigma^2_{\text{EC detector}} = \sigma^2_{\text{electrode size}} + \sigma^2_{\text{end-channel}}$).

In this study, PDMS-based devices were utilized to evaluate the performance of microchip CEEC by investigating the contribution of end-channel detection ($\sigma^2_{\text{end-channel}}$) toward the degree of peak asymmetry and the total amount of band broadening introduced by the EC detector ($\sigma^2_{\text{EC detector}}$). This was accomplished by utilizing a novel, "floating" potentiostat that allows placement of the working electrode directly in the separation channel. Furthermore, the contribution of analyte adsorption to the band broadening experienced with these devices was investigated with the small ion nitrite.

2. Experimental

The PDMS-based microchip CEEC devices utilized carbon fiber working electrodes and were fabricated as previously described [1]. Catechol separations were carried out using a 15 mM boric acid buffer, pH 9.2, at 300 V/cm. Separations involving nitrite were accomplished under reversed polarity (135 V/cm) using a 15 mM boric acid buffer at pH 9, with 7.5 mM DTAB added as an EOF modifier. The effect of the working electrode position on the catechol response was studied using a floating potentiostat (made in-house) operating in a two-electrode format at +1.2 V (vs. Ag/AgCl). Nitrite was detected with a conventional potentiostat (LC-4CE, Bioanalytical Systems) operating in a 3-electrode format at +0.9 V (vs. Ag/AgCl).

J.M. Ramsey and A. van den Berg (eds.), Micro Total Analysis Systems 2001, 325–326.

3. Results and discussion

The floating potentiostat is electrically isolated and transmits data through an optically isolated serial communication system. Because the detection circuitry does not provide a path to ground, a physical decoupling mechanism is not needed. As compared to a typical end-channel alignment, this in-channel design leads to more efficient and symmetrical peaks for catechol, a model analyte (Figure 1). The separation efficiency was increased by a factor of 2.5 and the peak asymmetry factors decreased from 1.56 to 1.07 (perfectly symmetrical peaks equal 1.0). In addition, the floating potentiostat is battery-powered and amenable to portable systems.

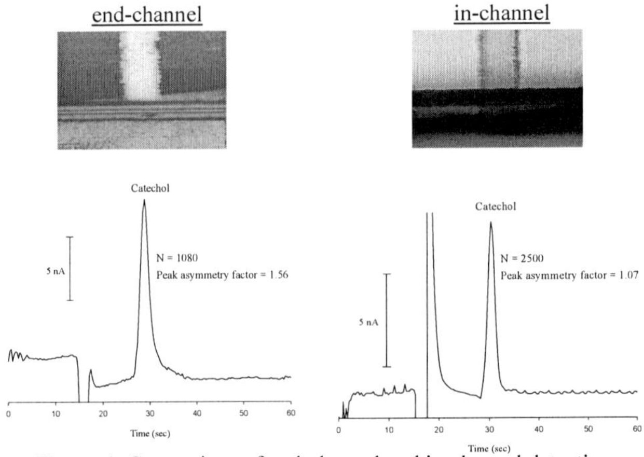

Figure 1. Comparison of end-channel and in-channel detection.

The contribution of analyte adsorption to band broadening in these PDMS-based microchip CEEC devices was also studied with the small anion nitrite. Reversing the EOF with the addition of an EOF modifier permits the detection of the non-adsorbing nitrite, resulting in a peak efficiency of 11,800 plates (Figure 2) and detection limits of 25 nM. Thus, adsorption of analytes into PDMS is a major contributor to band broadening in these microchip CEEC devices.

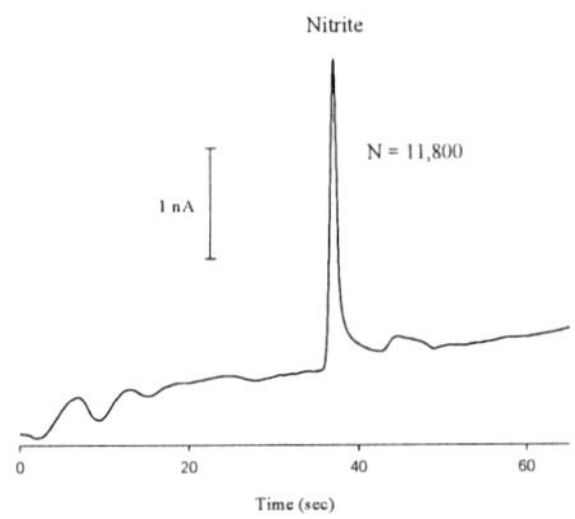

Figure 2. End-channel detection of nitrite.

Acknowledgements

This research was supported by a postdoctoral fellowship (RSM) from the National Institutes of Health and a grant from the National Science Foundation (CHE-9702631). We would also like to thank Ken Ratzlaff (KU-Instrument Design Lab) for developing the floating potentiostat.

References

1. A.J. Gawron, R.S. Martin and S.M. Lunte, Fabrication and evaluation of a carbon-based dual-electrode detector for poly(dimethylsiloxane) electrophoresis chips, *Electrophoresis*, **22**, (2001), 242-248.

CHARACTERIZATION OF SPUTTERED THIN FILM ELECTRODES ON PMMA MICROCHIPS WITH ELECTROCHEMICAL IMPEDANCE SPECTROSCOPY AND CYCLIC VOLTAMMETRY

O. Vogt[1], B. Graß[1], G. Weber[1], R. Hergenröder[1], D. Siepe[2], A. Neyer[2], J.P.Pohl[2]

[1]Institute of Spectrochemistry and Applied Spectroscopy, Dortmund, Germany
[2]University of Dortmund, Germany

Abstract

Sputtered thin-film electrodes (STFEs) of microfabricated PMMA-chips for ITP [1] and CE [2] are characterized by electrochemical impedance spectroscopy (EIS) and cyclic voltammetry (CV). It is shown that the fabrication process strongly influences the electrochemical behavior of STFEs.

Introduction

The active surface of STFE's in microdevices is influenced at least by two processes: the sputter process itself and the bonding of the chip, for which normally some kind of glue and/or elevated temperature and pressure are used. The investigation of the active surface of the electrodes can be done with EIS and CV. Both EIS and CV are non-destructive methods, which can be carried out directly on the bonded microchip.

Experimental

PMMA microchips are fabricated by hot-embossing as described previously [1]. Two designs are investigated: (a) CE-chips (20x95mm, channel 50x50μm) containing Pt/Au-electrodes for conductivity and amperometric detection, (b) ITP-chips (30x70mm, channel 200x200μm) containing electrodes for conductivity detection with different geometrical shape (widths 7μm to 23μm). The electrodes (thickness 200nm) are sputtered onto PMMA after pre-sputtering of a 50nm sub-layer of chromium for better adhesion. CV measurements are carried out in alkaline (0.1M NaOH) and acidic (0.1M H_2SO_4) solutions (–1V to +0.8V vs. Ag/AgCl, scanrate 100mV/s). EIS is done with chip-electrodes in 0.1M KNO_3 (1kHz to 0.1Hz, 0.0V, perturbation 10mV).

Results and discussion

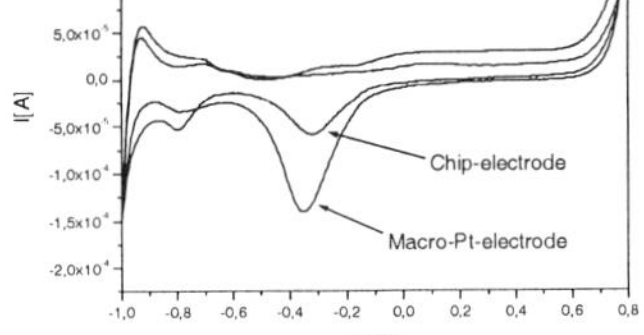

Fig. 1: Typical CV of a macro-Pt-electrode and Pt-chip-electrode (0.1M NaOH).

- Overall shapes of CVs are similar (peak potential, peak widths)
- no peaks originating from the chromium sub-layers are observed
- surface roughness (measured by oxide stripping charge) is higher for chip-electrodes

J.M. Ramsey and A. van den Berg (eds.), Micro Total Analysis Systems 2001, 327–328.

EIS is carried out on a series of 55 ITP-chips to estimate the chip to chip change of electrode roughness after sputtering and bonding.

Fig.2a: EIS data of ITP-chip (Bode-plot).

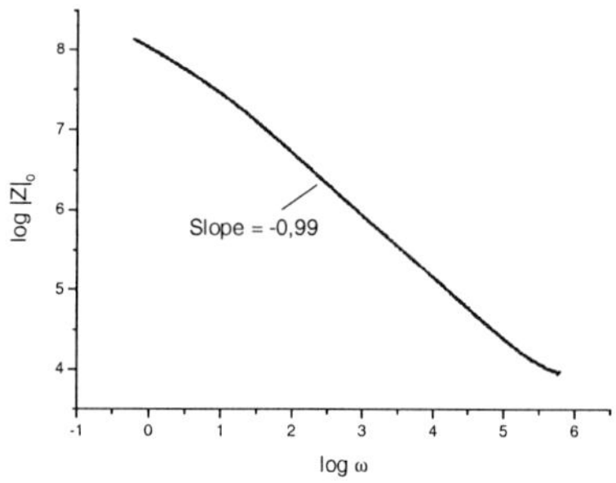

Fig.2b: Distribution of electrode roughness.

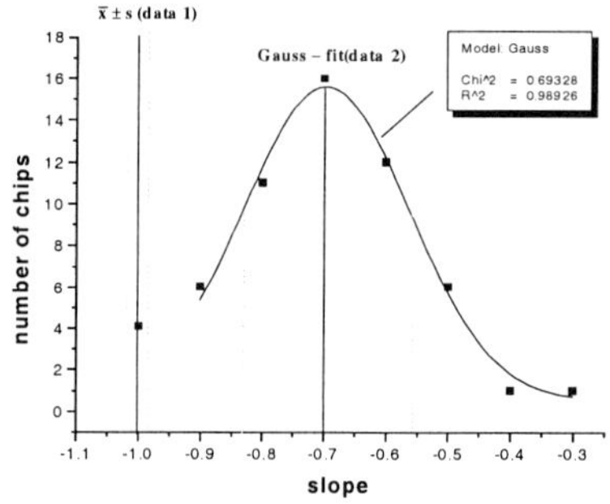

- Surface roughness of sputtered electrodes is estimated from the slope of Bode-plots of each chip (see Fig. 2a, slope is directly depended on surface roughness)
- "Faulty" electrodes, e.g. covered by thin polymer films are detected (Fig 2b, data 1)
- EIS is also well-suited as a tool for quality control of the fabrication process (Fig. 2b, data 2)

The sensitivity of ITP conductivity detection depends on the geometrical shape of the detection electrodes. The sensitivity of four different shapes/arrangements is compared.

Fig.3: Electrode design with 7μm width in a 200μm separation channel.

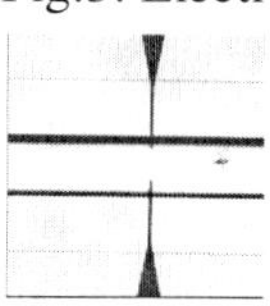

- with small detection electrodes (Fig.3) a gain of the relative step heights in ITP is achieved. These electrodes offer more sensitivity and a better resolution
- parallel electrodes (20μm widths) are better suited for high resistance buffer systems

Conclusions

It is shown that electrochemical methods like EIS and CV are well suited to investigate electrode phenomena in microanalytical devices. The design of the electrodes is an important factor for quality of detection. It is necessary to balance the need of sensitivity and the simplicity of the fabrication process.

References

1. B. Graß, A. Neyer, M. Jöhnck, D. Siepe, F. Eisenbeiß, G. Weber, R. Hergenröder, Sens. Actuators B 72 (2000), 249-258
2. G. Weber, M. Jöhnck, D. Siepe, A. Neyer, R. Hergenröder in: van den Berg A., Olthuis and P. Bergveld (Eds.) *Micro Total Analysis Systems 2000*, Kluwer Academic Publishers, Dordrecht, 383-386

CHEMICAL MICROREACTORS IN COMBINATION WITH MASS SPECTROMETRY

M. Brivio, N. R. Tas, R. H. Fokkens, R. G. P. Sanders, W. Verboom, D. N. Reinhoudt, A. van den Berg.
Laboratory of Supramolecular Chemistry and Technology, MESA$^+$ Research Institute, University of Twente, P.O. Box 217, 7500 AE Enschede, The Netherlands, E-mail: m.brivio@ct.utwente.nl

Keywords: Microreactor, solution phase synthesis, mass spectrometry

Introduction

The application of the "Lab-on-a-Chip" concept to synthetic chemistry [1, 2] is still at its infancy. In this work a borofloat microchip (Fig. 1) was designed and fabricated by means of the powderblasting technique and used as chemical microreactor in which different solution-phase syntheses have been performed successfully. The stream flowing from the outlet of the chip has been transferred directly onto a gold-coated sample plate. The chemical reaction products were analysed using matrix assisted laser desorption ionization [MALDI] – time-of-flight [TOF] mass spectrometry [3]. MALDI-TOF mass spectrometry in combination with post source decay [PSD] experiments shows to be an excellent method in obtaining molecular weights and structural information. The Schiff base reaction, in which primary amines react with aldehydes to give imines, was chosen as a primary study, because the Schiff base reaction is straightforward and products are obtained in high yields. Lab-scale experiments are carried out under acidic conditions (pH = 3-4), however, pH control is not necessary under the "Lab-on-a-Chip" conditions.

Experimental: off-line

The microreactor has been realized by powderblasting technique and direct bonding of the borofloat-borofloat (2.2 mm thick) wafers.

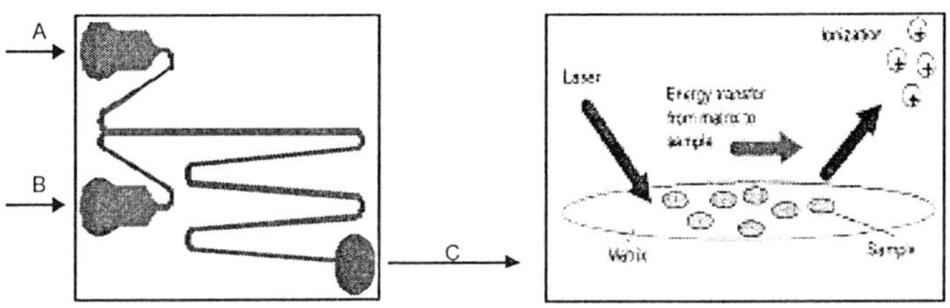

Figure 1. Schematic representation of the chip used in this work and of the MALDI sample plate on which the stream flowing from the chip is analysed.

The reaction channel, fabricated in the top wafer, is about 8 mm long, 200 μm width and 100 μm depth. The flow control was achieved by means of a dual syringe pump with a flow rate of 10 μl/min. The reaction time in the microchannel was approximately 12 sec.

J.M. Ramsey and A. van den Berg (eds.), Micro Total Analysis Systems 2001, 329–330.

Four Schiff base reactions were performed using the microdevice; all of them consisting of separated injection ports (A+B) into the chip device using the dual syringe pump.

a) aniline (A) + benzaldehyde (B): molar ratio = 1:1 (Fig 2)
b) aniline, aminochrysene (A) + anthraldehyde, biphenylaldehyde (B)
c) 4-tert-butylaniline (A) + 4-tert-butylbenzaldehyde (B): molar ratio = 1:1
d) 4-tert-butylaniline (A) + 4-tert-butylbenzaldehyde (B): molar ratio = 1:2

Results and discussion

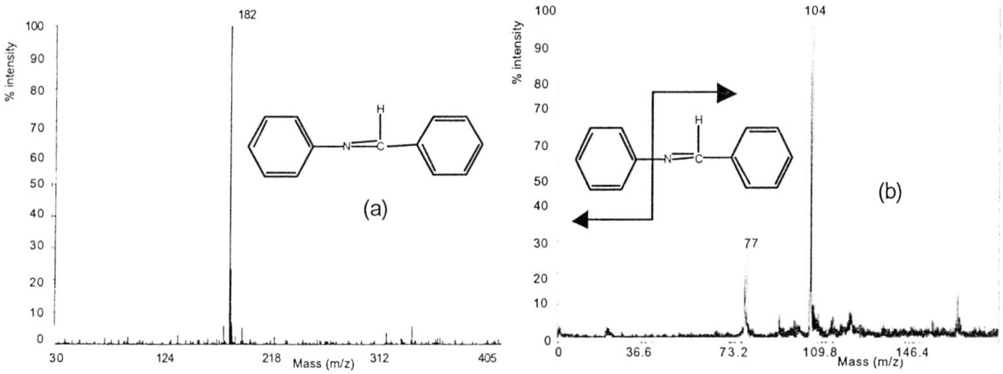

Fig. 2: a) MALDI-TOF mass spectrum of the reaction product from the solution-phase synthesis of Schiff by reacting a solution of 0.2 M aniline and a solution of 0.2 M benzaldehyde, both in methanol; (b) MALDI-TOF PSD mass spectrum of the $[M+H]^+$ = 182 to confirm the structure of the reaction product

The promising results attained in both the microchannels solution phase synthesis and the off-line combination of microreactor and detection units, encouraged the authors to start and successfully carry out additional experiments using on-line coupling involving the same chemistry.

Experimental: on-line

The microreactor for the on-line experiments has been realized by powderblasting technique and anodic bonding of the borofloat-silicon wafers (1.6 mm thick). The reaction channel has the same design as the one used in the off-line measurements. The borofloat bottom wafer has been replaced by a silicon one in order to reduce the thickness of the chip which has been integrated onto the MALDI sample plate (Fig.4).

References

1. S.J. Haswell, R.J. Middleton, B. O'Sullivan, V. Skelton, P Watts, and P. Styring, *Chem. Commun.*, 391-398 (2001).
2. M.C. Mitchell, V. Spikmans and A.J. de Mello, *Analyst*, **126**, 24-27 (2001)
3. R. Zenobi, R. Knochenmuss, *Mass Spectrom. Rev.*, **17**, 337 (1998).

CHIP-MS: A POLYMERIC MICROFLUIDIC DEVICE WITH INTEGRATED MASS-SPECTROMETER INTERFACE

Kevin Killeen, Hongfeng Yin, Sharmila Udiavar, Reid Brennen, Mark Juanitas, Elaine Poon, Dan Sobek, and Tom van de Goor
Agilent Technologies, Inc., Agilent Laboratories
3500 Deer Creek Road, Palo Alto, CA 94304, USA
(kevin_killeen@agilent.com)

Abstract

We report the use of direct write UV laser ablation to create microstructures in polymer film substrates. An integrated chip-MS device is fabricated combining a microfluidic channel with an electrospray tip. Mass spectral data is presented demonstrating protein sample compatibility and excellent spray stability.

Keywords: Laser ablation, microfluidics, integrated devices, electrospray, mass spectrometry

1. Introduction

Several groups have successfully demonstrated a chip-MS interface by a variety of fabrication and assembly methods [1-3]. While these approaches have produced stable electrosprays and quality MS data, they typically do not provide a method for seamless, low dead volume fluidic path from sample reservoir to the electrospray tip.

Direct write UV laser ablation can create complex geometries for specific applications in a rapid and versatile way. In contrast to other fabrication techniques such as chemical etching, direct write laser ablation prototyping can be performed without the need for masks, compressing the cycle times for design iterations.

2. Experimental

Microfluidic channels were ablated in polyimide film using a 355 nm solid state laser. A cover was laminated to the top surface to form a closed channels 10 to 50 μm in width and depth. The assembly was trimmed by laser ablation to create an electrospray tip concentric with the channel with outdide diameters of 30 to 100 μm. The end of the ablated electrospray tip on the polymer microdevice was positioned on axis ~6 mm from the ion collector assembly on an Agilent Technologies 1100 MSD quadrupole mass spectrometer.

3. Results and discussion

Figure 1 is a positive ion mass spectrum for a sample of the protein cytochrome-C introduced into the microfluidic sample inlet. These data were taken with a MS collector bias potential of –2.5 kV. The peaks at M/Z: 765, 816 and 1530 are characteristic of bovine cytochrome-C. Figure 2 shows the short-term signal stability from the MS operated in selected ion monitoring (SIM) mode using a sample of the drug reserpine.

J.M. Ramsey and A. van den Berg (eds.), Micro Total Analysis Systems 2001, 331–332.

The signal level variation observed was than <2% peak to peak, which is comparable to standard capillary nanospray devices and is on par with the best results achieved to date with an integrated microfluidic-electrospray device.

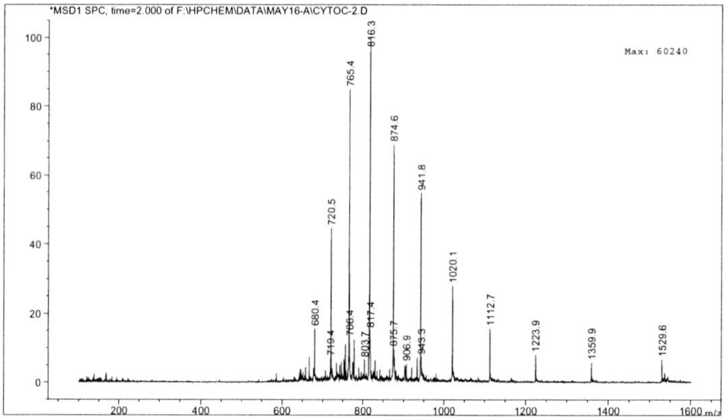

Figure 1. Positive ion mass spectrum of 1 μM cytochrome-C in a 50% methanol/water mixture.

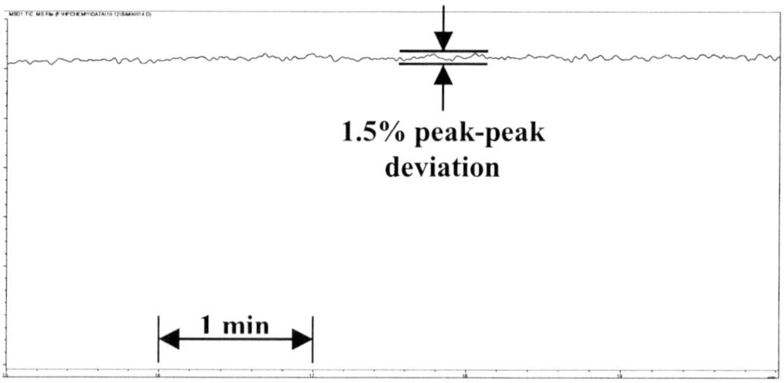

Figure 2. Short term electrospray-mass spectrometer stability test performed using 2ng/ul of reserpine at a flow rate of 200nl/min in 50% methanol/water mixture.

4. Conclusions

Integrated polymer chip-MS devices fabricated by direct write UV laser ablation exhibit excellent spray stability at low flow rates and are compatible with protein samples.

References

1. J. Li, J. F. Kelly, I Chernushevich, D. J. Harrison and P. Thibault, Anal. Chem. 72, pp. 599-609, (2000).
2. L. Licklider, X. Wang, A. Desai, Y. Tai and T. D. Lee, Anal. Chem. 72, pp. 367-375 , (2000).
3. G. A. Schultz, T. N. Corso, S. J. Prosser, and S. Zhang Anal. Chem. 72, pp. 4058-4063, (2000).

MICROFABRICATION OF ELECTROSPRAY IONIZATION EMITTERS IN POLYDIMETHYLSILOXANE MICROFLUIDIC DEVICES USING A SELF-ALIGNING ASSEMBLY SYSTEM

J.-S. Kim and D. R. Knapp
Medical University of South Carolina, Charleston, SC 29425, USA
(knappdr@musc.edu)

Abstract

Electrospray ionization (ESI) emitters for interfacing to mass spectrometry (MS) are produced along the edge of polydimethylsiloxane (PDMS) microfluidic devices by soft lithography. A self-alignment system consisting of posts and matching holes is used to facilitate assembly of the devices.

Keywords: ESI mass spectrometry, soft lithography, self-alignment

1. Introduction

Microfluidic devices using MS detection offer an attractive method for analysis of multicomponent samples. ESI has been the primary MS interfacing method used. There have been a few reports of microfabricated ESI emitters; but most of the reported work has used fused silica tips attached to the devices. To facilitate ESI use, there is need for improved methods of producing ESI emitters as an integral part of the device. We report here further improvements upon our methods for fabrication of ESI emitters as part of a PDMS device [1-3]. We also report a new method for self-alignment of the top and bottom parts of a PDMS microfluidic device during assembly.

2. Experimental

The PDMS microfluidic devices were produced by a soft lithography process [4] using multilayer photoresist (SU-8 100) patterns with layers 50, 200, and 200 μm thick [3]. Reservoirs were cast at the ends of the channels using short pieces of PEEK tubing placed on photoresist "posts" (Fig.1A). Hexagonal holes were cast in the top PDMS part to align with corresponding posts on the bottom part during assembly of the devices. ESI-MS was done on an ABI Mariner ESI Time of Flight MS instrument with the spray potential applied via an electrode inserted into the reservoirs.

3. Results and Discussion

Multilayer photoresist patterns formed on silicon wafers were used as masters for casting PDMS replicas. Lines of photoresist formed microfluidic channels in the underside of the top part of the device which, when bonded to a planar bottom part, formed closed channels (30 μm wide x 50 μm deep) exiting at 500 μm thick points cast along the edge of the device to form emitter tips. At the opposite end of each channel, a

J.M. Ramsey and A. van den Berg (eds.), Micro Total Analysis Systems 2001, 333–334.

reservoir used for sample introduction and placement of a platinum wire electrode was formed by casting around a piece of PEEK tubing placed on a photoresist "post." Figure 1 shows a diagram of the casting of the top PDMS part (Fig. 1A) and the resulting assembled PDMS emitters (Fig.1B).

In addition to an improved design of the ESI emitters, this device also employed a new method for alignment of the top and bottom parts in the device assembly. Using a multilayer photoresist pattern, 450 μm deep hexagonal holes were cast in the top part to match 350 μm high hexagonal posts on the bottom part. The mating surfaces of the two parts, activated by oxygen plasma, were bonded together using a thin layer of methanol to separate the activated surfaces during the alignment. Prior to use of this method, it was very difficult to manually align the channels of a multiemitter device. We have produced 16 channel rectangular devices using this method [4] and are currently extending it to 96 channel circular devices.

The ESI emitters were evaluated for electrospray performance using a solution of angiotensin I and bradykinin on an ESI time of flight mass spectrometer. Work is ongoing to optimize the channel size and point geometry for better ESI performance.

Acknowledgments - Supported by NSF DBI-99006662 and NASA P101007.

References

1. J.-S. Kim, D. R. Knapp, Microfabricated PDMS multichannel emitter for electrospray ionization mass spectrometry, *J. Am. Soc. Mass Spectrom.* 12:463-9 (2001).
2. J.-S. Kim, D. R. Knapp, Microfabrication of polydimethylsiloxane electrospray ionization emitters *J. Chromatogr. A*, in press (2001).
3. J.-S. Kim, D. R. Knapp, Miniaturized multichannel electrospray ionization emitters on poly(dimethylsiloxane) microfluidic devices, *Electrophoresis*, in press (2001).
4. J.R. Anderson, D.T. Chiu, R.J. Jackman, O. Cherniavskaya, J.C. McDonald, H.K.Wu, S.H. Whitesides, G.M. Whitesides. Fabrication of topologically complex three-dimensional microfluidic systems in PDMS by rapid prototyping. *Anal Chem.* 72: 3158-3164 (2000).

Figure 1 - (A) Cross section diagram of the casting process for forming the top PDMS part of an ESI emitter. (B) Diagrammatic representation of an assembled single emitter showing the channel leading from the sample reservoir.

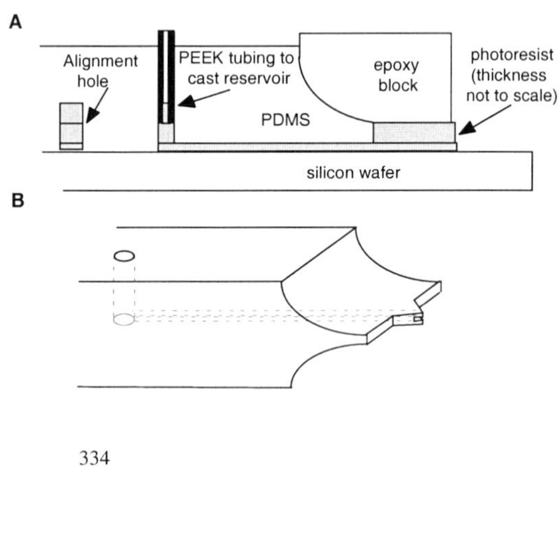

ELECTROSPRAY FROM A PLASTIC CHIP

Malin Svedberg[1], Andreas Pettersson[2], Fredrik Nikolajeff[1] and Karin Markides[2]
[1] The Ångström Laboratory, Department of Materials Science, Uppsala University, P.O. Box 534, 751 21 Uppsala, Sweden
[2] Department of Analytical Chemistry, Uppsala University, P.O. Box 531, 751 21 Uppsala, Sweden

Abstract
We have constructed plastic chips with electrospray tips. By using the chip as an electrospray interface we received a L.O.D. that is lower than 2 fmol/μL for peptides.

Keywords: Electrospray-mass-spectrometry, polymer-microfabrication, plastic-chip, injection-molding

1. Introduction

The interest for plastic chips for analysis purposes in chemical and biochemical applications has increased in recent years, since methods for plastic replication have become available for micro structure technology. Electrospray ionization mass spectrometry (ESI-MS) is an important tool in protein analysis. A few previous studies have reported tips made directly from polymer chips. These chips have bee fabricated by various micromachining methods [1-3]. In this work, we have studied the electrospray from injection-molded chips in poly-methyl-methacrylate (PMMA) and polycarbonate (PC).

2. Experimental

A general sketch of a chip can be seen in Figure 1. Microchannels for the device were wet etched in a silicon wafer using KOH. The channel pattern was then transferred to a nickel mould by electroplating. The structure was further mass-produced into plastic compact discs (CDs) using commercial injection molding equipment. The CDs were cut into smaller pieces and a coverlid in the same material was thermally bonded to the channels, without any glue. Tips were milled from the bonded chip, by manual polishing or by using a high precision milling machine.

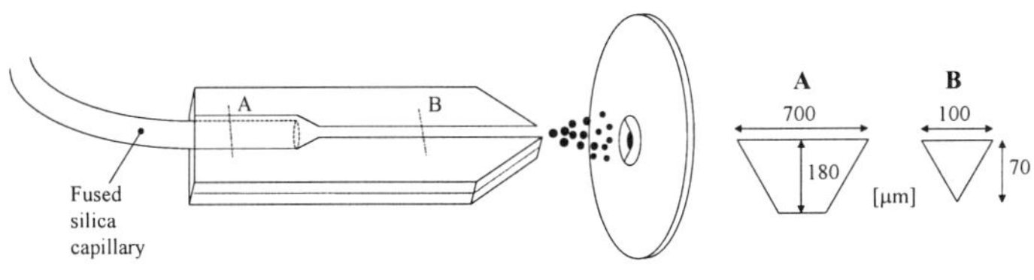

Figure 1. A principal sketch of the final chip.

J.M. Ramsey and A. van den Berg (eds.), Micro Total Analysis Systems 2001, 335–336.

A fused silica capillary (o. d. 150 μm, i. d. 70 μm) was inserted into the chip and sample was injected using a syringe pump. The microchips were tested in an orthogonal time-of-flight mass spectrometer (Jaguar, Leco) using a standard containing 5 peptides in a solvent of water/methanol (50:50 v/v) with 0.1 % acetic acid. Flow rates were between 0.7 and 3 μL/min. An electrospray voltage between 2600 and 2900 V was applied.

3. Results and discussion

Results from a hand-milled tip in PMMA can be seen in figure 2. The mass spectrum is a sum of 1750 spectra. According to Figure 2, there was no problem to measure 2 fmol/μL, which means that the limit of detection is lower than this value.

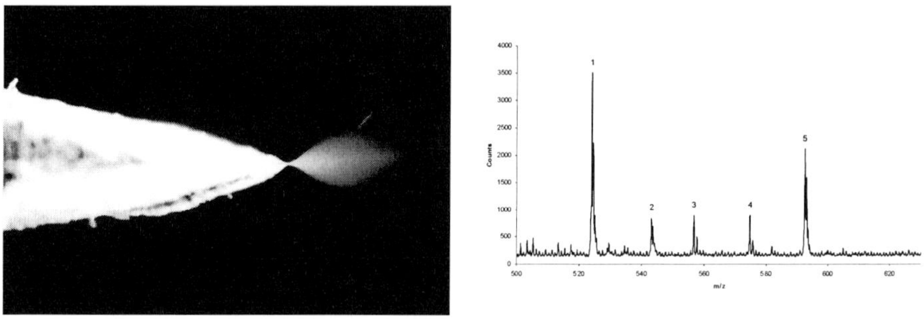

Figure 2. Result from a manual polished PMMA chips. Left: electrospray, right: mass spectrum for 5 peptides at a concentration of 2 fmol/μL

The best sensitivity at the lowest electrospray voltages was received from the chips with the hand-milled tips.

Acknowledgements

The authors want to thank: Carola Strandman at Ericsson AB, Stefan Nilsson at Uppsala university, Lars Lundbladh, Leif Häggbom, Anna-Lisa Tiensuu and Ove Öhman at Åmic AB for help with in the fabrication and evaluation of the plastic chips.

References

1. J. Wen, Y. Lin, F. Xiang, D. W. Matson, H. R. Udseth and R. D. Smith, *Microfabricated isoelectric focusing device for direct electrospray ionization-mass spectrometry*, Electrophoresis, 21(1), pp. 191-197, (1997)
2. C-H. Yuan and J. Shiea, *Sequential electrospray analysis using sharp tip channels fabricated on a plastic chip*, Anal. Chem., 73(6), pp. 1080-1083, (2001)
3. J-S. Kim and D. R. Knapp, *Microfabricated PDMS multichannel emitter for electrospray ionization mass spectrometry*, J. Am. Soc. Mass. Spectrom., 12(4), pp. 463-469, (2001)

MINIATURE ION MOBILITY SPECTROMETER DETECTOR WITH A PULSED IONIZATION SOURCE

Jun Xu, W. B. Whitten, and J. M. Ramsey
Oak Ridge National Laboratory, P. O. Box 2008, Oak Ridge, TN 37831, USA

Abstract

We demonstrate a miniature ion mobility spectrometer (IMS) detector that uses a pulsed corona ionization source. The gas phase detector shows high sensitivity and moderate resoltuion, essential for possible interfacing with other μTAS devices.

Keywords: Ion mobility spectrometry, pulsed ionization, corona discharge, atmospheric pressure chemical ionization

1. Introduction

Miniaturization of ion mobility spectrometry has many advantages for the separation of chemical species at atmospheric pressure and as a detector for other miniature instruments[1]. We have studied a miniature ion mobility spectrometer (IMS) that employs a pulsed corona-discharge ionization source. The drift channel has a diameter of 1.7-2.5 mm and a length of 35-50 mm. The device demonstrated high sensitivity and moderate resolution in detecting ions of H_2O clusters, methyl iodide, SF_6, and DMMP.

2. Experimental

The miniature IMS channel consists of a stack of electrodes separated with insulating spacers and connected by miniature resistors, as shown in Fig. 1. A Ni corona tip is mounted at the end of the drift channel. A high voltage pulse with a width varying from 40 ns to 400 μs is applied to the electrode relative to the drift potential. During the high voltage pulse period, ions are generated and confined to the vicinity of the tip. After the pulse, the ions move in the drift field and separated according their mobilities in the drifting gas. The corona discharge pulse also serves as the ion injection process to start the mobility measurement.

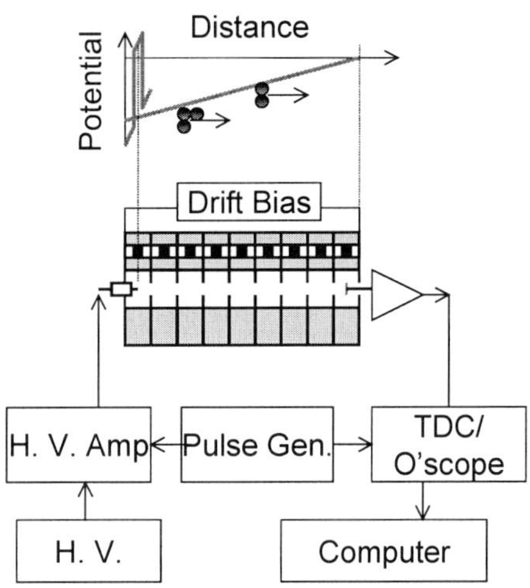

Fig. 1 Experimental setup. Top illustrates the drift potential and the corona pulse.

J.M. Ramsey and A. van den Berg (eds.), Micro Total Analysis Systems 2001, 337–338.

3. Results and discussion

IMS spectra of negative ions generated by pulsed corona discharge ionization of air were measured as a function of drift bias, as shown in Fig. 2.

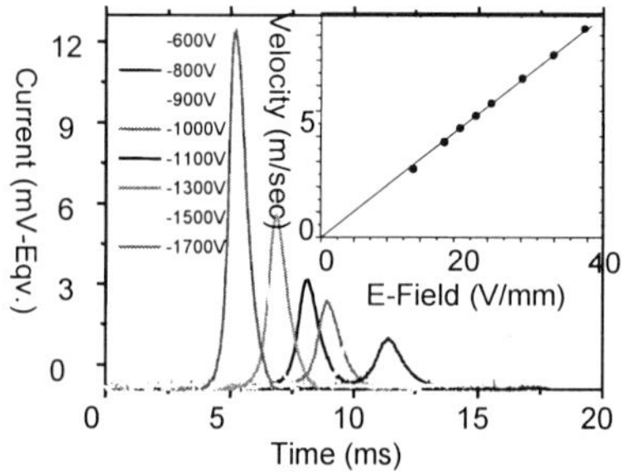

Fig. 2. Mobility spectra of negative ions generated by pulsed corona discharge as a function of drift potentials. The upper corner plots the ion velocity as a function of the drift field.

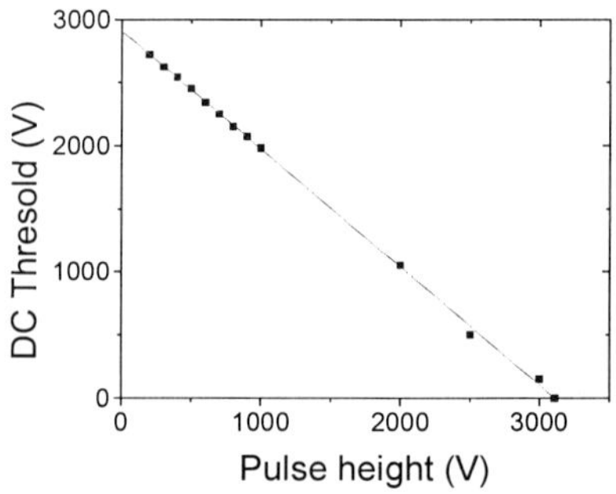

Fig. 3. DC potential threshold for generating IMS ions as a function of pulse height applied to the corona electrode.

A combination of pulsed and DC corona voltage was explored. For a given pulse height, IMS current was measured as a function of dc bias. A threshold for ion production was observed. The DC threshold was found to linearly decrease as the pulse height increased, as shown in Fig. 3. These measurements suggest that the high voltage of the pulse can be reduced by compensating with a DC bias. The pulse height could be reduced to as low as 200 V for a detectable response and as low as 1000 V for a stable response. As pulse height was lowered, the current decreased as the DC voltage was increased to above 2700 V. This decrease is believed to be due to reduction of initial free electrons in the high DC field in the vicinity of the tip.

IMS spectra of both positive and negative ions were measured as a function of pulse width and showed a clear difference between positive ions and negative ions. For positive ions, the ion current increased with pulse width and saturated. For negative ions, the ion current peaked rapidly and then decayed as the width increases. A theoretical model has been developed based on electron ionization and attachment, trapping of electrons and negative ions induced by the pulse bias. The model explains the kinetic data and provides many insights into generation and movement of ions under the pulsed corona discharge conditions used in miniature IMS.

Acknowledgements

This research was sponsored by the US DOE, Office of R&D. ORNL is managed by UT-Battelle, LLC for the U.S. DOE under contract DE-AC05-00OR22725.

Reference

1. Jun Xu, W. B. Whitten, M. J. Ramsey, *Space charge effects on resolution in a miniature ion monbility spectrometer*, Anal. Chem, Vol. 72, pp. 5787-5791, (2000).

VISIBLE AND UV DETECTION THROUGH SQUARE, DEEP PDMS CHANNELS

Laura Ceriotti [1], Jan Lichtenberg[1], Stephanie Clément[2], Philippe Nussbaum[2], Elisabeth Verpoorte[1], René Dändliker[2], Nico F. de Rooij[1]
IMT, [1]SAMLAB, Rue Jaquet-Droz 1, CH-2007 Neuchâtel, Switzerland
IMT, [2]Applied Optics, Rue A.-L. Breguet 2, CH-2000 Neuchâtel, Switzerland
laura.ceriotti@unine.ch

Abstract

In this paper, UV/visible detection performed vertically through 50-μm-deep, square poly(dimethylsiloxane) (PDMS) channels is described. To facilitate optical alignment, a simple detection system with prealigned optical fibers has been developed.

Keywords: Absorbance detection, replica molding, capillary electrophoresis.

1. Introduction

The strong dependence of signal strength on optical pathlength makes absorbance detection in small volumes difficult, since pathlengths, and hence sensitivity, are generally reduced [1]. Detection carried out vertically through the chip, perpendicular to flow, yields insufficient detection limits because commonly used wet-etching techniques restrict channel depths to between 10 and 25 μm or so. Polymer replication techniques using high-aspect-ratio relief patterns dry-etched in silicon offer a route to narrow, deep channels more suited to transverse absorbance detection.

2. Chip and detection system features

PDMS channels are 50 μm wide and 50 μm deep. They are made using a rectangular, metallic mold in which the silicon master is fixed. This allows channel replication and the formation of a thin window of PDMS (about 0.5 mm thick) at the detection point (Fig. 1A). The chip is mounted on a stationary stage between two prealigned optical fibers. These are fixed using ferrules into a bulk piece, which is attached to a

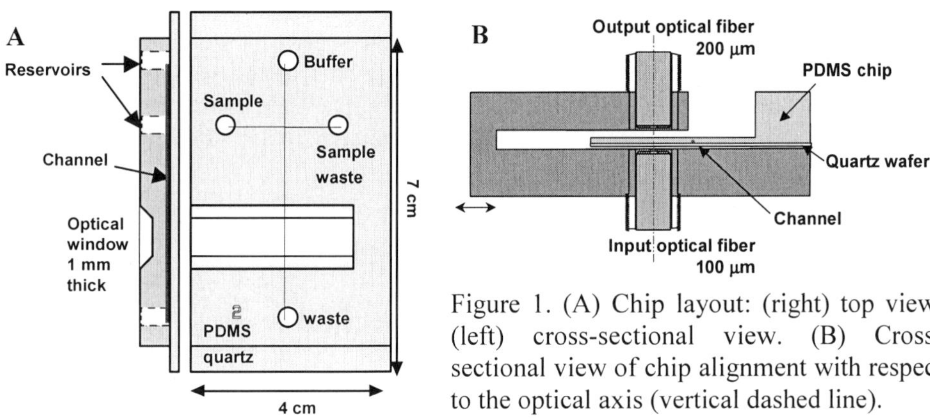

Figure 1. (A) Chip layout: (right) top view, (left) cross-sectional view. (B) Cross-sectional view of chip alignment with respect to the optical axis (vertical dashed line).

J.M. Ramsey and A. van den Berg (eds.), Micro Total Analysis Systems 2001, 339–340.

positioning platform, so that the optical axis can be aligned with the stationary chip (Fig. 1B). Halogen (visible) or deuterium (UV) light sources are used. Transmitted light is transported to a detector with a bandpass filter (visible: 632 nm, UV: 240-400 nm), a photodiode, an amplifier and a low-pass filter.

3. Results and discussion

In initial tests, methylene blue (MB) concentrations down to 5 μM were detected in the visible (Fig. 2). 30 μM MB could still be easily deteminated in the UV, though reduced photodiode response (0.42 A/W in the red versus 0.18 A/W in the UV) limited sensitivity. 0.1 mg/ml (0.5 mM) of the aromatic amino acid, tryptophan, was detected (Fig. 3). The peak is smaller than expected because (a) the bandpass filter is centred at 310 nm, while the $\varepsilon_{max\ abs}$ for tryptophan is 279 nm and (b) the wide spectral bandwidth used probably results in a lower peak height, though it could yield improved S/N ratios [2]. As expected, the signal grows when channel depth increases from 50 to 70 μm.

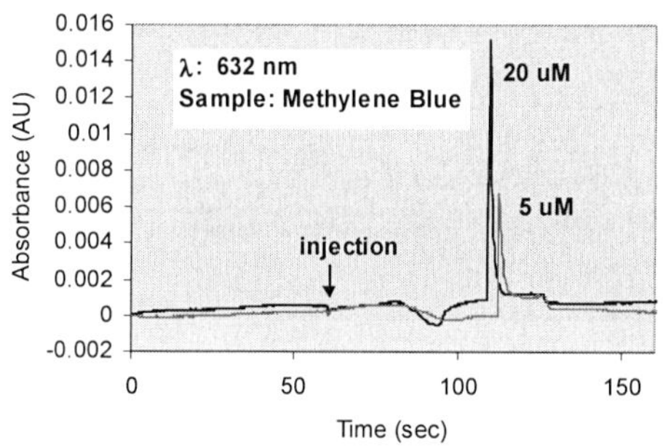

Figure 2. Absorbance peak of MB (ε_{632}= 42960 cm^{-1}/M) . Running buffer: 20mM Tris/HCl, pH 7.6; E_{sep}: 400V/cm; L_{eff}: 2.5cm.

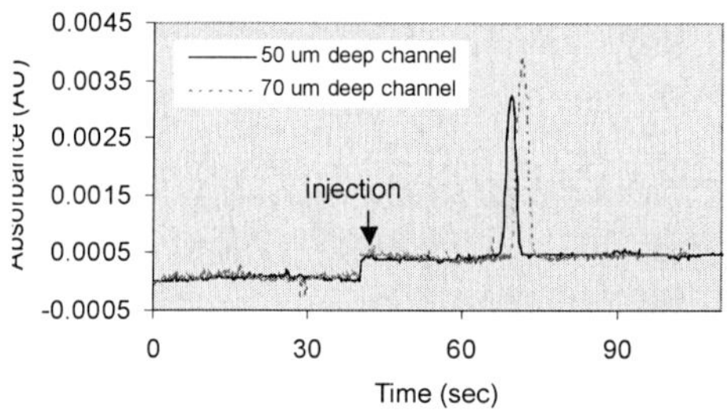

Figure 3. UV absorbance peak of 0.1 mg/ml tryptophan (ε_{279}= 5574 cm^{-1}/M) through 50 and 70 μm deep channels in different chips. Conditions as in Fig. 2.

4. Conclusions

The deep, narrow μchannel profiles possible with replication techniques facilitate the integration of UV/Vis absorbance detection into μdevices. Modification of the optics and electronics, with addition of microlenses and a slit, should lead to performance comparable or better than the system described in [3], which used a modified HPLC detector (no fibers) for transverse absorbance detection in chips.

References

1. Z. Liang, N. Chiem, G. Ocvirk, T. Tang, K. Fluri, D. J. Harrison, *Microfabrication of a planar absorbance and fluorescence cell for integrated capillary electrophoresis devices,* Anal. Chem., 68, 1040-1046, 1996.
2. M. Aiello, R. McLaren, *A sensitive small-volume UV/Vis flow cell and total absorbance detection system for micro-HPLC*, Anal. Chem., 73, 1387-1392, 2001.
3. C. Ericson, J. Holm, T. Ericson, S. Hjertén, *Electroosmosis- and pressure-driven chromatography in chips using continuous beds*, Anal. Chem., 72, 81-87, 2000.

DNA FRAGMENT SIZING AND FLUORESCENCE CORRELATION SPECTROSCOPY IN NANOFABRICATED CHANNELS.

M.E.Foquet[1], J.Korlach[1], W.W.Webb[1] and H.G.Craighead[1]

School of Applied and Engineering Physics, Cornell University, Ithaca, NY 14853, USA

Abstract

A device is fabricated by our sacrificial layer technique [1]. It is used to show how this method provides a way to measure fluorescence from single molecules with a good signal to noise ratio [2]. The size of single DNA molecules is measured by using fluorescence bursts. The channel with submicron dimension allows to probe all the solution that flows through the channel and to have all the molecules in focus on the detector.

Keywords: DNA sizing, Single Molecule Detection, FCS

1. Introduction

The development of a technique to measure DNA fragments faster and with less materials than gel electrophoresis is pursued. The method is also faster than gel electrophoresis. The size of fluorescence burst coming from single DNA molecules is measured to build a histogram of the size distribution in the sample.

2. Experimental

The sacrificial layer technology relies on the use of 3 different materials. A sacrificial layer patterned as the internal space of the microfluidic device, a coating layer that will surround the sacrificial layer and form the walls of the microfluidic device. Finally a chemical etchant is used to dissolve the sacrificial layer, leaving the wall layer intact. A combination of Fused Silica (wall material), Polycrystalline Silicon (sacrificial layer) and TetraMethylAmmonium Hydroxide (TMAH, wet etchant) is used for this application.

An electrosmotic flow is established with electric fields in the 10^5 to 10^6 V/m range. A laser focused on the observation channel provides the excitation light. The profile of the laser beam is made to be constant across the channel.

Figure 1. SEM of a microcapillary.

J.M. Ramsey and A. van den Berg (eds.), Micro Total Analysis Systems 2001, 341–342.

3. Results

A histogram of a measurement is presented in Figure 2. The histogram corresponds to 10,000 burst. This amount to 5 10^{-2} picograms of DNA. The histogram is the result of a 2 minutes acquisition. An array of gaussian peaks can be fitted using a least square method and analysed for fragment size and concentration.

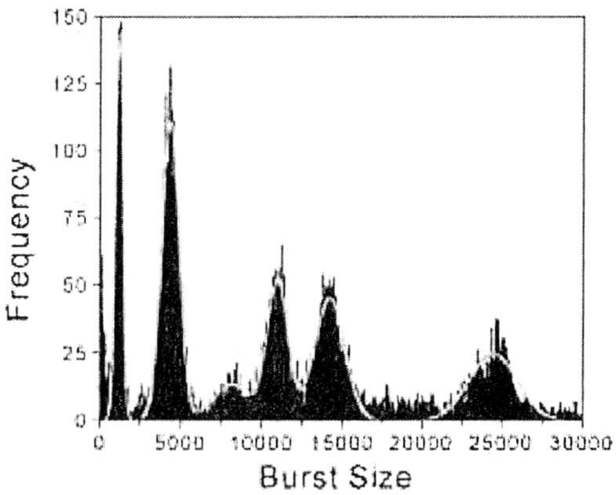

Figure 2: Histogram of the burst size distribution for DNA fragments

4. Conclusions

The use of nanochannels for DNA fragment sizing has been demonstrated. The amount of DNA required to perform an analysis is small. The analysis time is also significantly shorter than conventional gel electrophoresis. Another advantage of DNA fragment sizing is its wide dynamic range. A range of fragment covering several orders of magnitudes in size can be simultaneously analysed.

Acknowledgements

This work was supported by the Nanobiotechnology Center (NBTC), an STC Program of the National Science Foundation under Agreement No. ECS-9876771

References

1. S.W.Turner, A.M.Perez, A.Lopez, H.G.Craighead, Monolithic nanofluid sieving structures for DNA manipulation, J. Vac. Sci. Technol. B, Vol.6, p3835-3837, (1998)
2. M.E.Foquet, S.W.Turner, J.Korlach, W.W.Webb and H.G.Craighead, Nanofabricated device for fluorescence correlation spectroscopy in sub-femtoliter volumes, MicroTAS 20000, Kluwer Academic Publisher, Dordrecht, Netherlands, pp.549-552, (200)

DEVELOPMENT OF A μTAS SCREENING DEVICE FOR DRUG ANALYSIS BY CHEMILUMINESCENCE

Paul A. Greenwood and Gillian M. Greenway
Department of Chemistry, University of Hull, UK

Abstract

A portable μTAS drug screening device has been developed utilizing the tris (2,2'-bipyridine)ruthenium(II) reaction using both chemiluminescence and electrogenerated chemiluminescence. The ruthenium chelate was immobilized onto the electrode using sol-gel technology allowing regeneration of the reagent.

Keywords: chemiluminescence, electrogenerated chemiluminescence, drugs.

1. Introduction

Electrogenerated chemiluminescence (ECL) and chemiluminescence detection are both selective and sensitive to the determination of many drugs using the tris (2,2'-bipyridine)ruthenium(II) reaction[1]. The $Ru(bipy)_3^{2+}$ is oxidized either electrochemically or chemically to $Ru(bipy)_3^{3+}$ before reacting with the tertiary amine groups in the drug to produce light. The aim of this work is to develop a miniaturized, fully portable, battery powered device to monitor drugs in saliva or urine samples. To achieve this aim in a miniaturized system the detection must occur on the chip, it must be possible to move the sample or reagents around the chip in a controlled manner and the reagents must maintain stability. Fiaccabrino *et al.* [2] have previously reported light detection on chip by incorporating a photodiode in the chip design. Electrophoretic separation can also be included in the design of the proposed device. Miniaturized peristaltic pumps can mobilize solutions and reagents can be produced electrochemically or can be immobilized to help stability.

2. Chip Development

The prototype chip design can be seen in Fig. 1. A photodiode was encapsulated in the

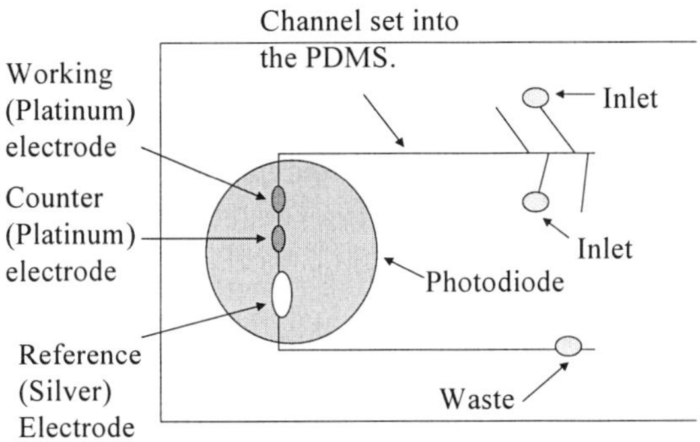

Figure 1: Top view of the 200 μm wide channels and the 3 electrode system passing over the photodiode.

Figure 2: Light scattering occurring within the channels.

J.M. Ramsey and A. van den Berg (eds.), Micro Total Analysis Systems 2001, 343–344.

chip directly over the electrode to capture the light emitted or alternatively light was detected with a fiber optic from a photomultiplier tube. Reagents and samples were moved though the 200 μm wide channels of a polydimethlysiloxane (PDMS) chip using a miniaturized peristaltic pumping system. When designing the system light scattering was observed during the mixing of solutions if chemiluminescence detection was used (see Fig.2), but was minimized when ECL was used to produce light at the working electrode. ECL also enabled accurate positioning of the detection system. The ruthenium chelate can be regenerated on the chip if it is immobilized using sol-gel technology onto the platinum-working electrode. The sol-gel process involves low temperature hydrolysis of an alkoxide precursor (tetramethyl orthosilicate). The silicate matrix was formed by hydrolysis of TMOS, catalyzed by an acid. Nafion was mixed with the matrix and a thin layer was spin coated onto the electrode followed by condensation to yield a polymeric oxo-bridged SiO_2 network. The ruthenium chelate was then introduced to the sol gel matrix by ion exchange.

3. Optimization and Detection

Reaction conditions were optimized using a Yates algorithm. Limits of detection of 3.8×10^{-9} M and 7.7×10^{-8} M were found for atropine and pethidine (Fig.3) with RSDs of

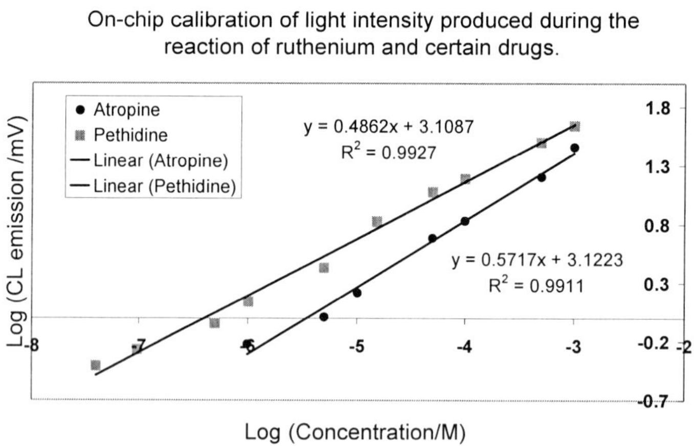

Figure 3: Calibration of Atropine and Pethidine determined on chip using chemiluminescence.

less than 5% when using chemiluminescence detection with a fiber optic and PMT. These results were found to be comparable to those found for codeine by ECL and initial results show a loss of sensitivity of only one order of magnitude when using the photodiode detector.

References

1. G. Greenway, L. Nelstrop, S.Port, *Anal. Chim. Acta.*, **405**, 2000, 43.
2. G.C. Fiaccabrino, N.F. deRooij, M. Koudelka-Hep, *Anal. Chim. Acta.*, **359**, 1998, 263.

INTEGRATION OF MULTIPLE INTERNAL REFLECTION (MIR) INFRARED SPECTROSCOPY WITH SILICON-BASED CHEMICAL MICROREACTORS

R. J. Jackman[1], K. T. Queeney[3], R. Herzig-Marx[3], M. A. Schmidt[2], K. F. Jensen[1]
[1]Dept. of Chemical Engineering and [2]Microsystems Technology Laboratories, Massachusetts Institute of Technology, Cambridge, MA 02139 USA
[3]Department of Chemistry, Smith College, Northampton, MA 01063 USA

Abstract

Anisotropic etching of silicon enables the batch fabrication of crystals suitable for multiple internal reflection (MIR) IR spectroscopy. Integration of these crystals with microfluidics produces devices that permit on-line monitoring of reactions. We use these devices for *in situ* kinetic studies of a model reaction (hydrolysis of ethyl acetate).

Keywords: infrared spectroscopy, integrated optical detection, anisotropic silicon etching, multiple internal reflectance spectroscopy

1. Introduction

A minaturized, integrated system that combines chemical synthesis with on-line detection can reduce the quantity of material required for analysis and increase throughput, as well as enable real-time optimization of reaction conditions and simplify acquisition of kinetic data. We have demonstrated previously the ability to integrate visible and ultraviolet spectroscopy with our microchemical reactors [1]. We present here a simple strategy that allows the integration of multiple internal reflection (MIR; also referred to as Attentuated Total Reflectance (ATR)) IR spectroscopy with our silicon-based microchemical reactors.

Infrared spectroscopy offers chemical sensitivity to a variety of species in the mid-IR fingerprint region. In multiple internal reflection (MIR) spectroscopy (Figure 1) detection capability is not limited by channel dimensions even with depths as small as a few microns, since penetration of the mid-IR light into the channel is ~2.5-15 microns. Moreover, the multiple reflections enhance device sensitivity and offer the potential for probing surface chemistry in these systems [2].

2. Results and Discussion

2.1. Batch Fabrication of Silicon MIR Crystals & Their Integration with Microfluidics

Commercial MIR crystals are produced by individually polishing crystals of the material of interest (which makes them expensive and difficult to integrate). Silicon is, however, IR transparent and can serve as an MIR crystal. Anisotropic etching of silicon produces MIR crystals with beveled edges of 54.7° in a batch process. To produce crystals with suitably polished edges, however, requires a two-step etching process. In

J.M. Ramsey and A. van den Berg (eds.), Micro Total Analysis Systems 2001, 345–346.

the first step, alignment marks are KOH etched to determine the precise orientation of the crystal planes in the wafer, and then a second etch with features oriented to these planes etches the mirror-smooth edges of the crystals.

These crystals become the substrates on which we construct our microreactor channels. Several alternatives exist for interfacing these crystals with fluidic systems (glass or plastic) (Figure 1B).

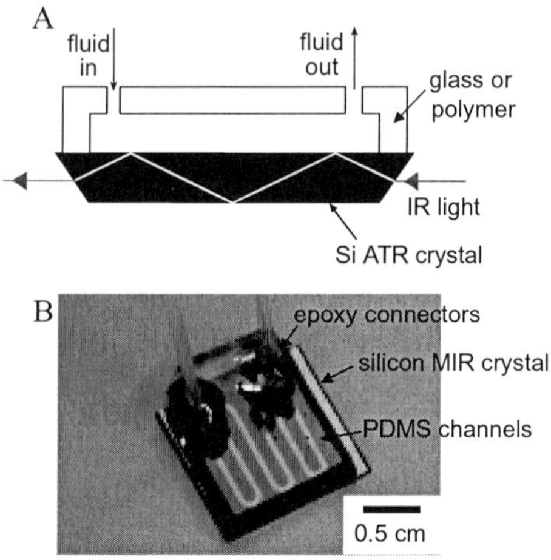

Figure 1. (A) Illustration of MIR Spectroscopy integrated with Microfluidic Device. **(B)** Photograph of Batch Fabricated Integrated Microfluidic Device with MIR crystal

2.2. Characterization of Devices

We are characterizing the performance of these devices by performing model liquid-phase reactions (hydrolysis of ethyl acetate) and surface-enhanced reactions (silane-derivatization of the silicon, with subsequent reactions). Figure 2 illustrates IR spectra for the acid-catalysed hydrolysis of ethyl acetate.

3. Conclusions

The results presented here demonstrate that MIR techniques can be integrated with micro-fluidic systems to create new volume and surface-sensitive techniques for detection and kinetic studies.

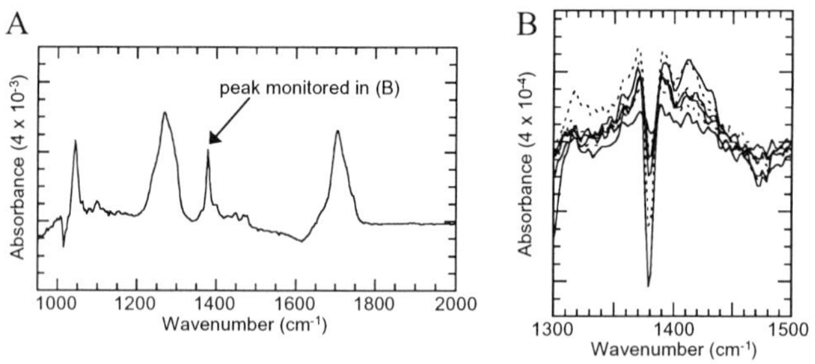

Figure 2. (A) IR spectrum of ethyl acetate solution in Microfluidic Device. **(B)** Monitoring of hydrolysis via depletion of peak at 1379 cm^{-1}.

Acknowledgements

This research was supported in part by DARPA under the Microflumes Project.

References

1. R. J. Jackman, *et al.*, Microfluidic Systems with On-line UV Detection Fabricated in Photodefinable Epoxy, *J. Micromech. Microeng.* **11**, 263 (2001).
2. K. T. Queeney, *et al.*, In-situ FTIR studies of reactions at the silicon/liquid interface, *J. Phys. Chem. B* **105**, 3903 (2001).

A SINGLE POINT EVANESCENT WAVE PROBE FOR ON-CHIP REFRACTIVE INDEX DETECTION

Stephen C. Jakeway and Andrew J. de Mello

AstraZeneca / SmithKline Beecham Centre for Analytical Science, Imperial College of Science, Technology and Medicine, Department of Chemistry, South Kensington, London, SW7 2AY, U.K.

Abstract

A novel method to detect refractive index variations within microchannel environments using an evanescent wave probe is described. Experiments, with and without a fluorescent buffer, demonstrate that 10 mM sucrose plugs can be detected. This method is also applied to the detection of sucrose plugs flowing hydrodynamically through microchannels and for chip-based electrophoretic separations of saccharide mixtures.

Keywords: Refractive index, evanescent wave, microchip

1. Introduction

Although fluorescence spectroscopy has tended to be the detection method of choice for chip-based analysis, the majority of molecules are non-fluorescent. Therefore, several alternative approaches to 'on-chip' detection have been demonstrated. The potential of refractive index (RI) detection (universal, non-destructive) for chip-based analytical systems has been investigated [1,2]. We present a simple method to detect RI changes within microfluidic channels based on the principles of total internal reflection (TIR).

2. Theory

Snell's Law can describe the behavior of electromagnetic radiation at an interface.

$$n_1 \sin\theta_1 = n_2 \sin\theta_2$$

where n denotes the RI and θ is the angle at which the radiation enters or leaves the interface. If n_1 is greater than n_2, then there exists for θ_1 an angle, called the critical angle, where $\sin^{-1}(n_2/n_1)$ is $\pi/2$. If θ_1 exceeds this condition, then all incident radiation is reflected at the interface. However, solving Maxwell's equations for this system shows that an evanescent wave (EW) exists in medium 2 in the vicinity of the boundary and is directly affected by the difference in the refractive indices of the two media. This variation is the basis of the proposed sensing mechanism.

3. Experimental

Details of the optical set-up have been described elsewhere [3]. RI measurements using the EW induced fluorescence system (EWIFS) were either in a Teflon flow cell or microfluidic channel (200 x 80 μm) were filled with the desired sucrose solution. The incidence angle was scanned from 75 to 65 degrees and the fluorescence measured. For

J.M. Ramsey and A. van den Berg (eds.), Micro Total Analysis Systems 2001, 347–348.

repeat injection experiments, the chip was connected to a Rheodyne injection valve with a 5.44 μL sample loop using 152 μm ID PTFE tubing. Solutions were pumped through the chip using a Harvard Apparatus pump at a rate of 10 μL/min.

4. Results and Discussion

Experiments to measure RI variations with and without (Figure 1) a fluorescein solution gave results in close agreement with literature and standard refractometer values. Repeat injections of sucrose with (Figure 2) and without fluorescein were passed through a microfabricated chip. As sucrose destroys the formation of the EW, it is detected due to a large increase in fluorescence from the bulk solution. The same technique was used to detect electrophoretically separated saccharide mixtures. By setting the incidence angle of the laser at a defined angle, hence defining the RI of interest, injections can be made discriminating against lower RI and detecting those of higher RI.

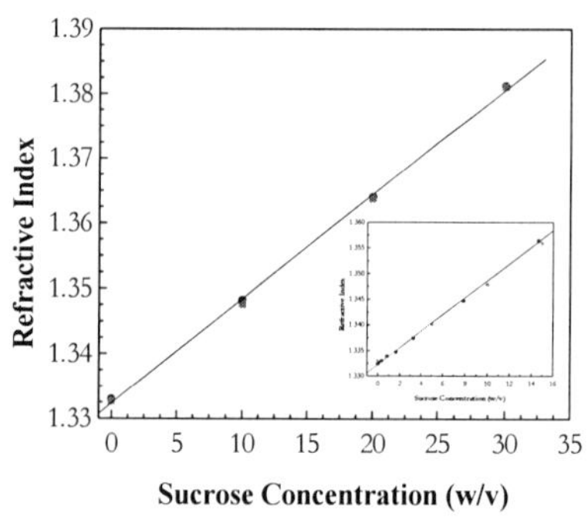

Figure 1
RI measurement of various sucrose conc.: literature (-), refractometer (•), EWIFS (■). The inset shows values from 0 to 500 mM.

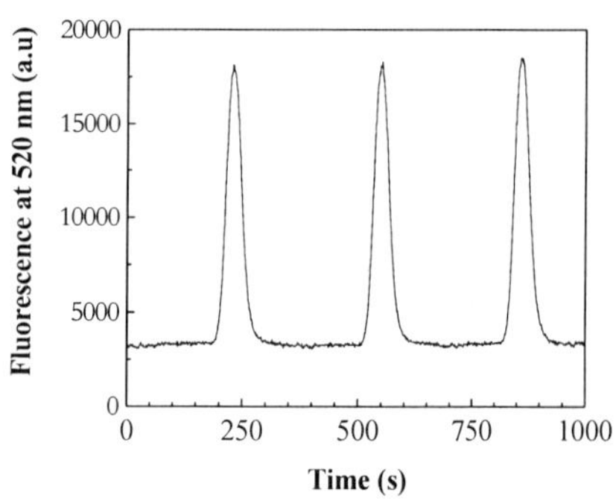

Figure 2
Repeat injections using a Rheodyne valve of 25 mM sucrose in 2 μM fluorescein through a microfluidic channel. The running buffer was 2 μM fluorescein.

5. Conclusions

It has been shown that a single point evanescent probe can be used to detect RI changes on chip. This simple technique provides a first step towards a universal, small-volume detector for chip-based separation systems.

Acknowledgements

The authors would like to thank N. Beard for assistance with the CE chips.

References

1. N. Burggraf et al., *Analyst,* 1998, **123**, 1443-1447.
2. K. Swinney, D. Markov and D. J. Bornhop, *Anal. Chem.,* 2000, **72**, 2690-2695.
3. C. D. Byrne et al, *J. Phys. Chem. B,* 1998, **102**, 10326-10333.

OPTICAL EMISSION DETECTION OF LIQUID ANALYTES USING A MICRO-MACHINED D.C. GLOW-DISCHARGE DEVICE AT ATMOSPHERIC PRESSURE

Gareth Jenkins and Andreas Manz
AstraZeneca / SmithKline Beecham Centre for Analytical Chemistry, Imperial College, London SW7 2AY.

Abstract

Atmospheric pressure D.C. Glow Discharge Atomic Emission Spectroscopy of aqueous samples has been performed on a microchip. The liquid sample itself was used as the cathode for the discharge. The detection of copper in solution has been achieved with the observation of several Cu lines. Stability issues need to be resolved to allow for further optimisation and characterisation of the device.

Keywords: Glow discharge, emission spectroscopy, ELCAD

1. Introduction

DC GD optical emission spectroscopy has been shown as an effective detection method for gas phase micro-systems [1,2]. Integrating such a device with micro-fluidic systems is more problematic due to difficulties in liquid sample introduction on the microscale. A type of glow discharge known as the Electrolyte as Cathode Discharge (ELCAD) [3] in which the liquid sample is used as the cathode, could provide an easy solution to this problem. Cserfalvi et. al. have studied this type of discharge as a means of analysing trace metals in water and have proposed a mechanism involving cathodic sputtering at the liquid-plasma interface [4]. More recently Marcus et. al. have developed a similar method with a flow rate low enough to allow coupling with conventional LC systems and have also proposed a different mechanism in which the volatilisation of the sample is the important process [5]. The work presented here describes an ELCAD type device on a microchip for liquid sample analysis and possible future integration with a μTAS device.

2. Experimental

Figure 1 shows a schematic of the completed structure. The sample inlet is ~700μm wide, and the gas inlet and discharge chamber are ~300μm wide. All channels are 150μm deep. A flowing gas-liquid interface was setup between the liquid sample and carrier gas. The discharge was then created between a Pt wire anode and the liquid sample, which was electrically connected to the HV cathode off-chip. Devices were fabricated in soda-lime glass (Nanofilm Ltd, USA) using a Direct Write Laser Lithography system (HIMT GmbH, Germany). The liquid flow rate was 5μLmin^{-1},

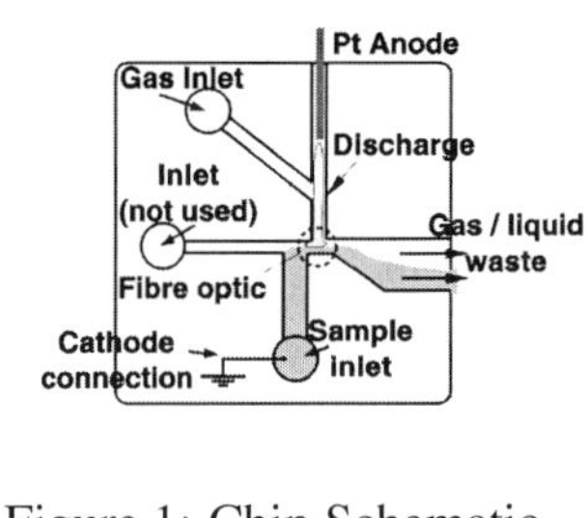

Figure 1: Chip Schematic

J.M. Ramsey and A. van den Berg (eds.), Micro Total Analysis Systems 2001, 349–350.

discharge voltage 1kV, and the current was approximately 1–5mA. An Ocean Optics S2000 miniature spectrometer (Ocean Optics Europe, The Netherlands) was used for analysis.

3. Results and discussion

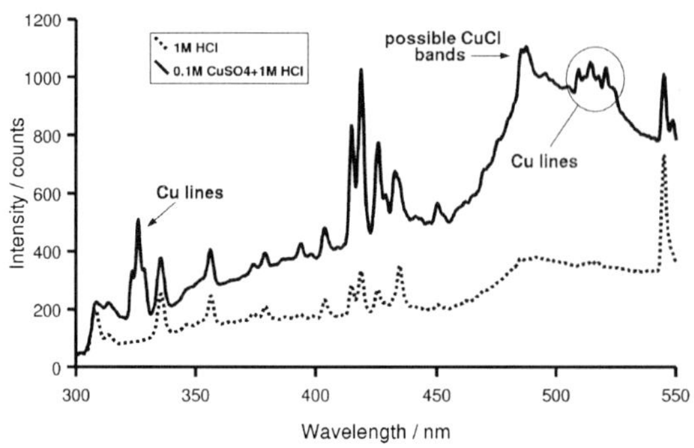

Figure 2: Emission Spectrum of 0.1M $CuSO_4$.

0.1M $CuSO_4$ in 1M HCl was used as the sample and 1M HCl was used as a control. The operation of the discharge was intermittent probably due to the high gas temperature causing vapourisation leading to pressure instabilities. Copper was detected and figure 2 is a spectrum showing Cu lines at 325, 327, 511, 515 and 522nm. Two possible CuCl bands were also seen at 485 and 488nm. Unfortunately, the discharge instabilities have so far prevented the calculation of meaningful detection limits.

4. Conclusions

Continuous GD-AES of copper in solution has been performed on a glass microchip. Although discharge instabilities do need to be eliminated, this detection method has the advantages of being cheap and portable with low power consumption. Further optimisation of the device could eventually yield useful detection limits for a broad range of samples. This type of device could be easily integrated with other micro-fluidic systems as part of a μ-TAS device.

References

1. J.C.T. Eijkel, H. Stoeri, A. Manz, *A molecular emission detector on a chip employing a direct current microplasma*, Anal. Chem., 71, pp. 2600-2606, (1999).

2. J.C.T. Eijkel, H. Stoeri, A. Manz, *A dc microplasma on a chip employed as an optical emission detector for gas chromatography*, Anal. Chem., 72, pp. 2547-2552, (2000).

3. T. Cserfalvi, P. Mezei, Apai, *Emission studies on a glow-discharge in atmospheric air using water as a cathode*, J. Phys. D-Appl. Phys, 26, pp. 2184-2188, (1993).

4. T. Cserfalvi, P. Mezei, *Operating mechanism of the electrolyte cathode atmospheric glow discharge*, Fresenius J. Anal. Chem., 355, pp. 813-819 (1996).

5. Marcus R. K., Davis W.C., *An atmospheric pressure glow discharge optical emission source for the direct sampling of liquid media,* Anal. Chem., 73, pp. 2903-2910, (2001).

Semiconductor Laser Doppler Velocimeter for Micro-Fluidics Applications

Jan Krüger, William Whelan-Curtin, Daniel Hoffmann, Rory Mullane, Alan P. O'Neill and Peter O'Brien

National Microelectronics Research Centre, National University of Ireland, Ireland
jkruger@nmrc.ie

Keywords: velocimeter, micro-fluidics, optical Sensor, integration, non-invasive

Introduction

We present a novel non-invasive technique to measure fluid velocities in micro-channels with a high degree of accuracy and spatial resolution. The experimental arrangement is simple, requiring only a semiconductor laser[1], a photodiode[2] and a signal analyser[3] and requires no additional processing to the micro-fluidic device to facilitate the measurement. The only requirement is that the device enables optical access of the narrow laser beam. The limitation of this technique is that a back-scatter source must be present in the fluid being analysed. However, this makes the technique suitable for fluidic measurements involving cells, beads and applications that have particles suspended in solution. This paper will present our latest experimental data and we will show how the technique can be used to extract useful data to enable effective design of a micro-Cytometer. We will also highlight the integration possibilities and costs benefits of using this technique.

Experimental

It is known that back-scatter in a semiconductor lasers induces fluctuations in its output power and that the fluctuation frequency is directly related to the velocity of the scattering source[4]. The relationship between the fluctuation frequency and the fluid velocity is given by,

$$v = \frac{\lambda f}{2\sin\theta}$$

where v is the fluid velocity, f the frequency fluctuation measured using the spectrum analyser, λ the emission wavelength of the semiconductor laser, and θ is the angle between the laser beam and the fluidic channel. Using a commercially available diode laser we have utilised this effect to measure velocities in micro-fluidic devices. The laser beam is focused on the micro-channel using a low numerical aperture lens and the same lens is used to couple the scattered light from the micro-channel back into the laser cavity – a very low level of back-scattered power is required to induce power fluctuations. A low cost Silicon photodiode is used to measure the fluctuations frequency from the laser back facet and the photodiode is connected to a signal analyser via co-axial cable.

The technique first was used on a simple flow–cell constructed from PMMA. The channel dimensions were 720 µm wide and 1000 µm in height. The laser beam was focused to a tight on the flow-cell - beam width of approximately 20 µm and the power fluctuation frequency measured. The flow-cell position was then translated and the measurement repeated at a number of points across the channel. Figure 1 shows the experimental transverse velocity profile for the flow-cell. It fits well with a parabolic profile predicted using Poiseuille fluidic dynamics. We estimated a Reynolds Number of approximately 20 for this particular flow-cell, indicating the flow-cell was operating in the laminar flow regime during our measurements.

We have also applied this measurement technique to a micro-fluidic device currently being developed for Flow Cytometry applications (Fig.2). A key feature of this device is the hydro-dynamic focusing structure used to squeeze the central sample fluid. This enables single cell flow along the centre of the micro-channel. Our velocity analysis indicates a significant velocity increase

J.M. Ramsey and A. van den Berg (eds.), Micro Total Analysis Systems 2001, 351–352.

at the junction where squeezing occurs, and the experimental profile obtained using the laser velocimeter correlates well with our simulation data using FlumeCAD[5] finite element software (Fig.3-5).

This sensing technology is ideal for integration on to a micro-fluidic platform and is expected to have applications where spatial information on fluid flow and velocity distributions is important.

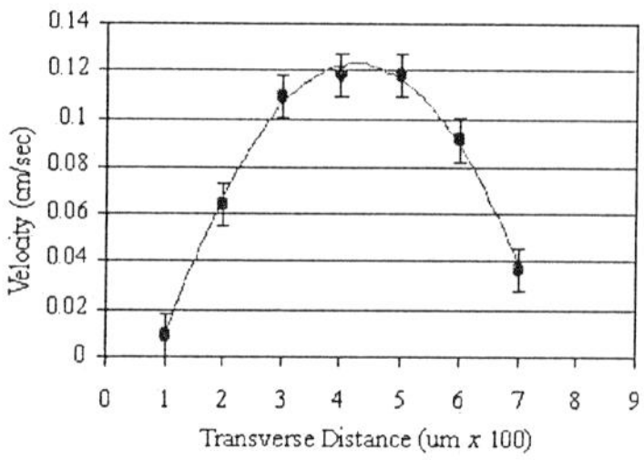

Figure 1 Transverse velocity profile in a 720 μm wide flow-cell. The flow characteristics were laminar with Reynolds Number of approximately 20. The experiment points are fitted to a parabolic profile.

Figure 2 Micro-fluidic device fabricated in SU-8 on a Silicon substrate. There are three input and two output ports. The central sample input is squeezed by buffer fluids from the outer ports. The buffer channels also switch the direction of sample flow between the two output ports.

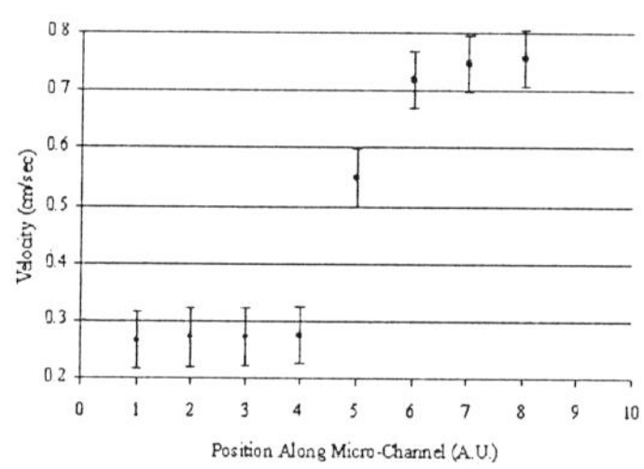

Figure 3 Maximum velocity profile along the micro-channel due to squeezing of the central fluid by the outer buffer fluids. The profile confirms simulation data obtained using FlumeCAD software.

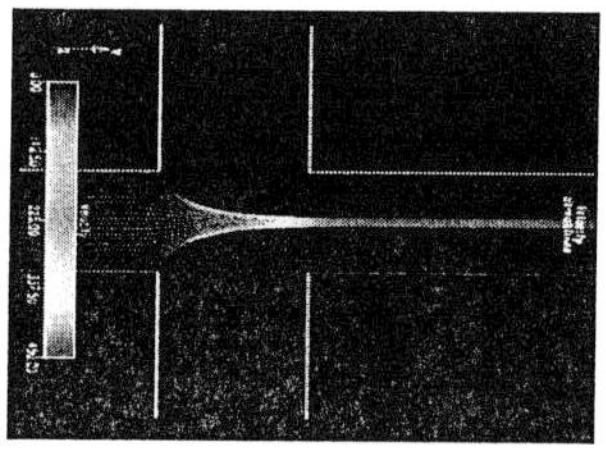

Figure 4 Micro-fluidic simulation of hydro-dynamic focusing using FlumeCAD finite element. The fluid velocity increases across the junction as sample and buffer fluids interact.

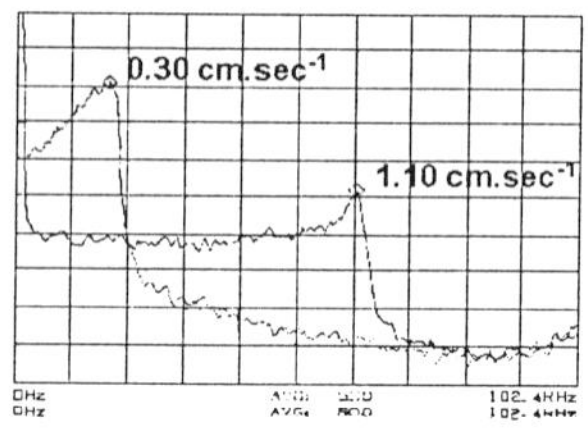

Figure 5 Velocity distributions with and without hydro-dynamic focusing. A peak is observed at the maximum velocities due to the large number of scattering particles passing the laser beam. (***x-axis*** - frequency [kHz] and ***y-axis*** - power fluctuation intensity)

References

[1] Mitsubishi Laser Diode (ML40123N) purchased from Thorlabs Inc.

[2] BPX65 Silicon Photodiode purchased from Radionics Ltd.

[3] HP 35665A Signal Analyser

[4] H. Koelink, M. Slot, F.F.M de Mul, J. Greve, R. Graaf, A.C.M. Dassel and G. Aarnoudse, *Laser Doppler Velocimeter based on the self-mixing effect in a fiber-coupled laser: Theory*, Applied optics **31**,18 (1992)

[5] FlumeCAD Software, Microcosm Technologies

MICROSCALE FIBER-OPTIC SPR SENSORS BASED ON NSOM TECHNOLOGY

Kazuyoshi Kurihara[1], Hiroyuki Ohkawa[2], Yuzuru Iwasaki[3], Tatsuya Tobita[4], Osamu Niwa[3], and Koji Suzuki[1,2]
[1]Kanagawa Academy of Science and Technology,
3-2-1 Sakado, Takatsu-ku, Kawasaki 213-0012, Japan
[2]Keio University, 3-14-1 Hiyoshi, Kohoku-ku, Yokohama 223-8522, Japan
[3]NTT Lifestyle and Environmental Technology Laboratory, NTT Corporation,
3-1, Morinosato Wakamiya, Atsugi, Kanagawa 243-0198, Japan
[4]NTT Advanced Technology Corporation,
3-1, Morinosato Wakamiya, Atsugi, Kanagawa 243-0198, Japan

Abstract

Microscale miniaturization of surface plasmon resonance (SPR) sensors was accomplished using a fabrication technique of fiber-optic probes for near-field scanning optical microscopy (NSOM). We successfully fabricated a micoscale SPR sensor that was sensitive to changes of 10^{-3} refractive index units.

Keywords: SPR, NSOM, Fiber-optic chemical sensor, Miniaturized optode

1. Introduction

SPR is a surface-sensitive analytical method based on the detection of local changes in the refractive index (RI) within an ~200-nm distance of the noble film surface. SPR is very useful for biochemical sensing because the real-time and non-labeling detection of biopolymers is possible. Recent miniaturization of optical chemical sensors [1] has been triggered by the development of a fabrication technique for fiber-optic probes for NSOM [2]. Miniaturization of SPR sensor into the microscale [3] is presented here.

2. Fabrication

The microscale SPR device in Fig.1 is almost the same as the NSOM probe fabricated by the chemical etching method [3] using an optical fiber with a GeO_2-doped core and a

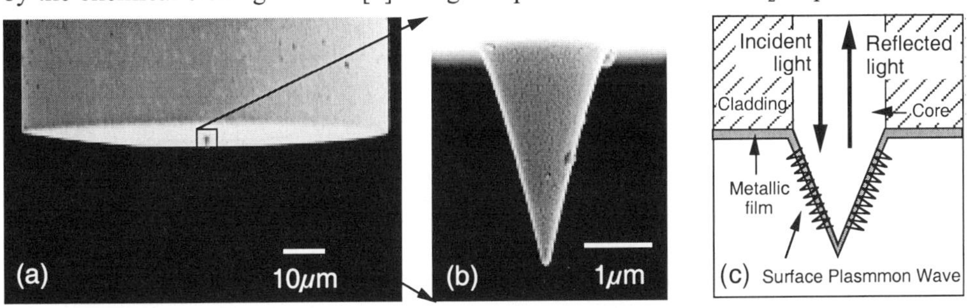

Fig.1 SEM micrographs of a microscale SPR device (a) and the magnified apex region (b). Schematic explanation of the microscale SPR device.

J.M. Ramsey and A. van den Berg (eds.), Micro Total Analysis Systems 2001, 353–354.

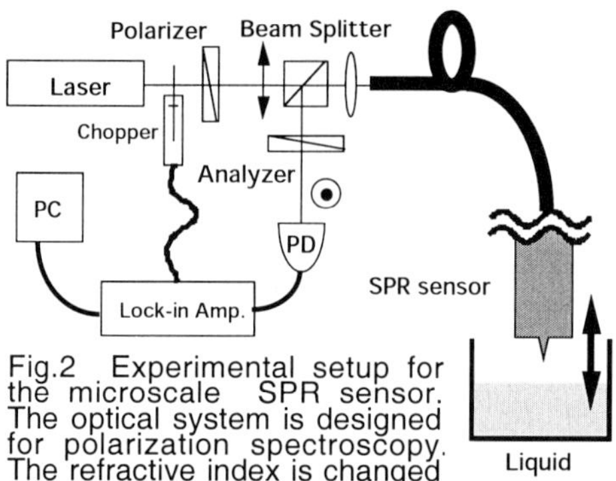

Fig.2 Experimental setup for the microscale SPR sensor. The optical system is designed for polarization spectroscopy. The refractive index is changed based on the immersion liquid.

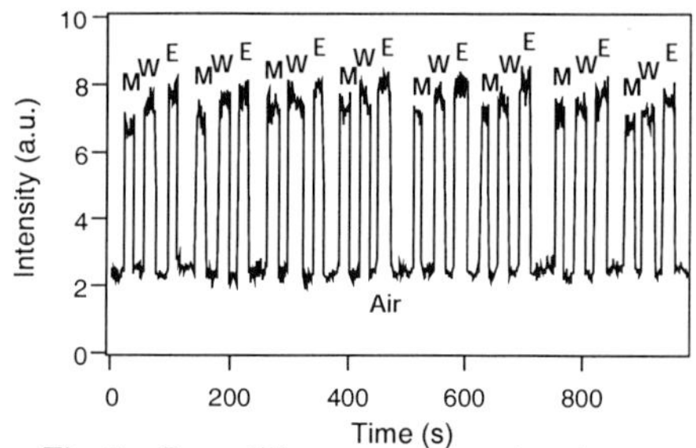

Fig.3 Repetitive response to changes in refractive index of each liquid. Methanol, water and ethanol are labed as M, W and E, respectively. The base line corresponds to refractive index of air.

pure silica cladding. The tapered fiber was obtained by immersion of the fiber's flat end into a buffered hydrogen fluoride (BHF) solution for more than one hour at a room temperature [2]. A metal coating of ~50nm on the surface of the tapered fiber was carefully done using a sputtering method.

3. Experimental and Results

The optical system in Fig.2 was constructed in order to detect changes in the polarization of a linearly polarized laser beam with a wavelength of 633 nm. The beam was coupled to the optical fiber, reflected at the tip of the NSOM probe together with the SPR phenomenon, and finally detected as signals of a polarization perpendicular to that of the incident beam. As shown in Fig. 3, the optical response to changes in the refractive index was observed when the microscale SPR sensor was immersed in three kinds of liquids (methanol, water, and ethanol). We have successfully observed that the microscale SPR device is sensitive to a difference in RI between methanol (n=1.3290) and water (n=1.3330). Its sensitivity is on the order of 10^{-3} RI units.

5. Conclusions

Microscale fiber-optic SPR sensors have been successfully fabricated using NSOM technology. They repetitively respond to changes in the refractive index of liquids.

Acknowledgements

Special thanks go to Professor M. Ohtsu of the Tokyo Institute of Technology, Japan, for teaching the fabrication technique of optical fiber probes in NSOM.

References

1. K.Kurihara, M.Ohtsu, T.Yoshida, T.Abe, H.Hisamoto, and K.Suzuki, *Micrometer-Sized Sodium Ion-Selective Optodes Based on a "Tailed" Neutral Ionophore*, Anal.Chem. **71**, pp. 3558-3566 (1999).
2. M.Ohtsu (Ed.), *Near-Field Nano/Atom Optics and Technology*, Springer-Verlag, Tokyo, (1998).
3. K.Kurihara, H.Iwashige, and K.Suzuki, *Micrometer-Sized Optical Devices for Surface Plasmon Resonance Sensors*, Proc. of Europt(r)ode V, pp.55-56 (2000).

ADVANCES IN ON-CHIP FLUORESCENCE DETECTION FOR DNA ANALYSIS SYSTEMS

Vijay Namasivayam[1, 3], Rongsheng Lin[1, 3], Victor M. Ugaz[1], Sundaresh N. Brahmasandra[1], Brian N. Johnson[1], David T. Burke[2], Carlos H. Mastrangelo[3], Mark A. Burns[1]
[1]Department of Chemical Engineering
[2]Department of Human Genetics
[3]Department of Electrical Engineering and Computer Science
The University of Michigan, Ann Arbor, MI, USA

Abstract
Developing an on-chip fluorescence detection involves construction of ultrasensitive photodetectors that can be integrated onto a fluidic platform. In this work, we report a novel doping and construction of a *P-I-N-N*$^+$ photodiode with an on-chip interference filter and a robust liquid barrier layer. This new design yields a highly sensitive, low-noise photodiode and its applications in several areas of DNA analysis are presented.
Keywords: On-chip fluorescence detection, Ultrasensitive PIN Diodes

Design and Simulations

We have previously demonstrated on-chip DNA detection using P-I-N photodiodes in silicon [1]. The detectors presented in our earlier work could detect double stranded DNA at 100 ng/µl concentrations. In order to perform high-resolution DNA separations and real time PCR, detector sensitivity needed to be significantly improved. We present an improved (n/n$^+$ doped) detector structure (*Fig* 1) with shallow junction depths (0.7µm) that is optimized for increased light absorption in the active region, enhanced photocarrier collection efficiency and reduced dark current noise.

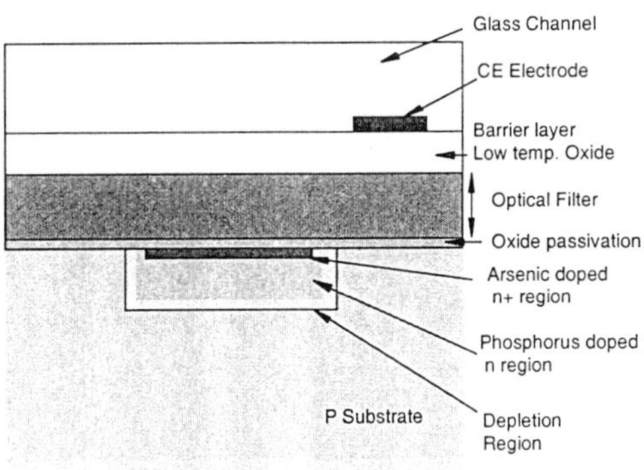

Fig1: Improved Detector Structure

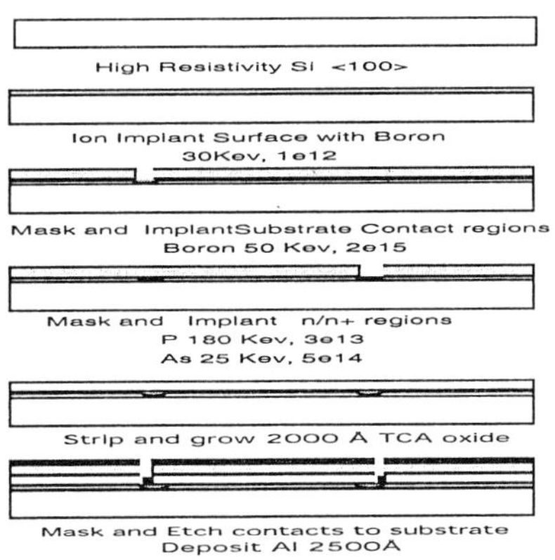

Fig 2: Diode Fabrication

Fabrication and Experimental Characterization

Device fabrication (*Fig 2)* begins with blank Boron implant on p-type silicon wafer. Phosphorus/Arsenic is implanted on the diode (n/n+) region and Boron is implanted on the substrate contact region. A 0.2-µm thick silicon dioxide layer is grown at 900°C and an interference filter (40 stack SiO_2/TiO_2) that blocks wavelengths less than 500nm, is then

J.M. Ramsey and A. van den Berg (eds.), Micro Total Analysis Systems 2001, 355–356.

deposited on top. A 1.5 μm low temperature oxide (LTO) is then deposited as Liquid barrier layer. Quantum efficiency of the diode was not affected by the LTO deposition. Electrodes were deposited on top of the LTO and a channel etched in glass bonded onto the silicon substrate.

Applications

Blue light is used for excitation and a Labview set up is used to measure the voltages generated in the diode. Performance of the photodiodes for DNA detection is typically characterized by doing a calibration. *Fig 3* shows the calibration plot; we can detect DNA concentrations down to *0.9-ng/μl* (2 orders improvement) with a Signal/Noise of 100:1. The linear nature of this plot can be used in quantifying DNA amplification during reactions such as PCR. End point detection of an off-chip PCR is shown in *Fig 4.*

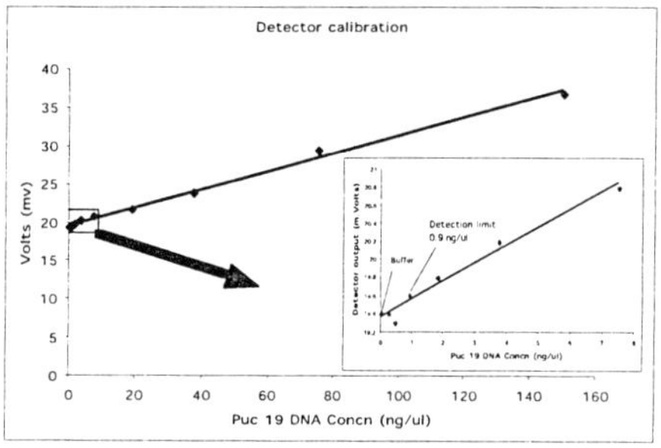

Fig 3: Detector Calibration Plot

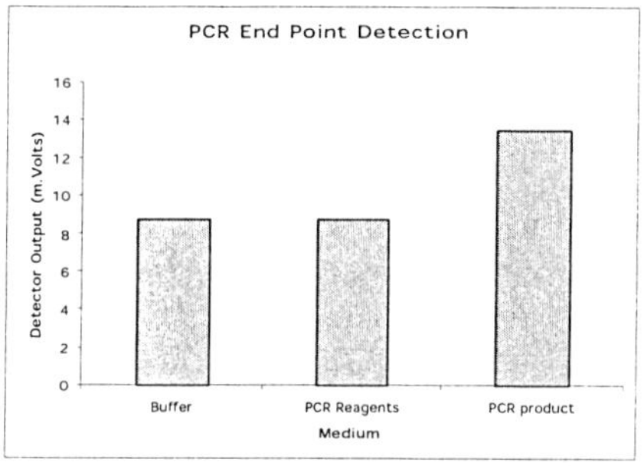

Fig 4: PCR End Point Detection

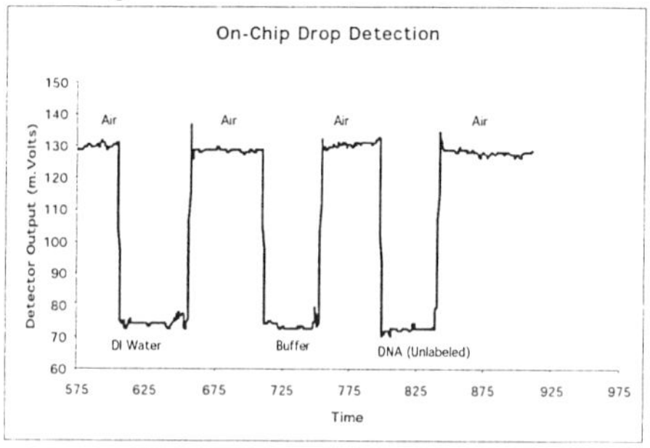

Fig 5: On-Chip Drop Detection

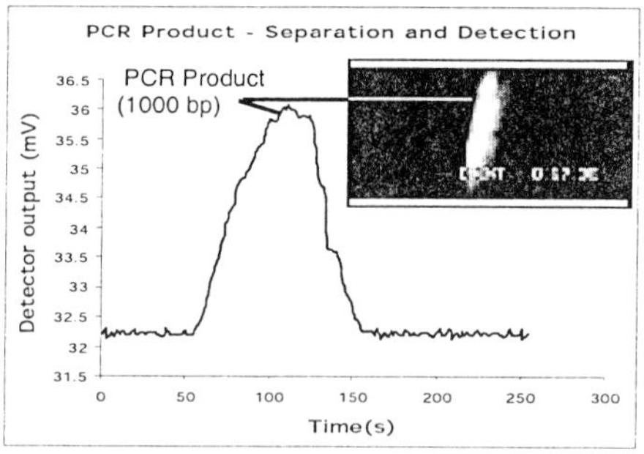

Fig 6: On-Chip Separation and Detection

In microfluidics, these photodiodes can be used in determining the position of a moving liquid drop inside fluidic channels *(Fig 5)*. Detection of migrating DNA bands in a gel can also be accomplished using these photodiodes [2]. *Fig 6* shows the on-chip separation and detection of a PCR product inside a sieving gel (Cross-linked Polyacrylamide). Construction of on-chip diodes is relatively easy, inexpensive and it eliminates the need for external detection schemes.

References

1. M.A Burns et. al, "An integrated nanoliter DNA analysis device" Science v282, 1998, pp.484-487
2. J. R. Webster et. al. " Monolithic capillary electrophoresis device with integrated fluorescence detector", Anal. Chem., 2001 v73. pp 1622-1626

PARTICLE-SHAPE SENSING-ELEMENTS FOR INTEGRATED FLOW CYTOMETER

J.H. Nieuwenhuis[1], S.S. Lee[1], J. Bastemeijer[1], M.J. Vellekoop[2]
[1]Electronic Instrumentation Laboratory – DIMES, Delft University of Technology, Mekelweg 4, 2628 CD Delft, The Netherlands
[2]Vienna University of Technology, Faculty EE&IT, Institute IEMW
J.H.Nieuwenhuis@ITS.TUDelft.nl

Abstract

Two particle-shape sensing-elements are presented that directly measure absolute shape. Measurement results show that without lenses shape is accurately registered when the object is close to the sensor. Finite element simulations show that this condition can be created with a double sheath flow.

Keywords: particle/cell shape, integrated cytometer

1. Introduction

Two different particle-shape sensing-elements have been developed for application in an integrated flow cytometer. These elements are to be placed on the bottom of a transparent micro flow-channel that is illuminated from above (see Fig. 1). When a particle passes over these elements it will partially block the light. One line of this shadow is registered by the sensing-element and by repeatedly reading out the sensor a two-dimensional projection of the particle is obtained.

In a number of applications, particle or cell shape is an important parameter. Sample enrichment is one of these applications; here rare cells are separated from a larger population for further analysis. In contrast to the current light scattering analysis systems, the proposed system yields absolute shape information, which allows for far more accurate determination.

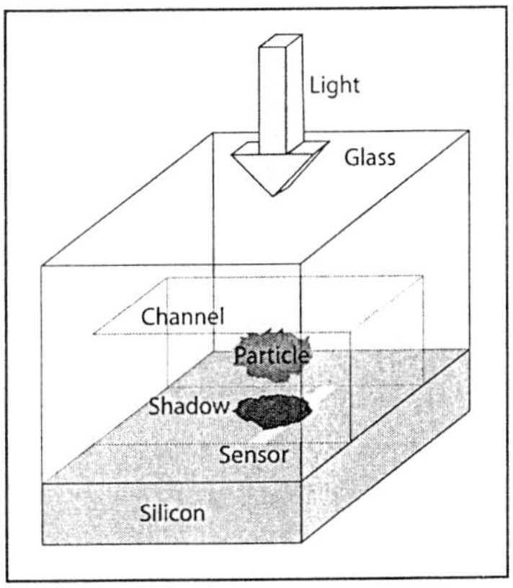

Fig. 1, The sensing element on the bottom of a transparent flow channel that is illuminated from above

2. The Elements

The first sensing-element consists of a miniature double one-dimensional array of photodiodes (see Fig. 2). There is a small offset between the upper and the lower part of the array such that by combining the data of the two parts a virtual one-dimensional array is formed with a pitch of only 2.5 micron.

The second sensing-element consists of an elongated photodiode-structure (see Fig. 4). When a particle passes over this strip photodiode the drop in photocurrent will be proportional to the width of the particle. To be able to uniquely reconstruct the particle shape a cut is placed in the middle as a reference.

3. Comparison

To compare the performance of the two elements images were obtained by manually moving a gold bonding wire (diameter 27 mm) with a small bump (diameter 67 mm) over the two elements.

J.M. Ramsey and A. van den Berg (eds.), Micro Total Analysis Systems 2001, 357–358.

The results are depicted in Fig. 2 and Fig. 4, along with a photo of the bonding wire (Fig. 3). Both sensing-elements deliver an adequate image of the object but there are some differences. The main advantage of the array sensing-element is that it can be used for all kinds of particles, including semi-transparent and non-uniform transparent ones. However, it generates a lot of data that needs to be processed in real-time. The strip sensing-element has the advantage that it only requires two readouts for every image line. A disadvantage is that without any a priori knowledge of the particles it is only suitable for non-transparent particles. A calibration measurement can alleviate this drawback substantially.

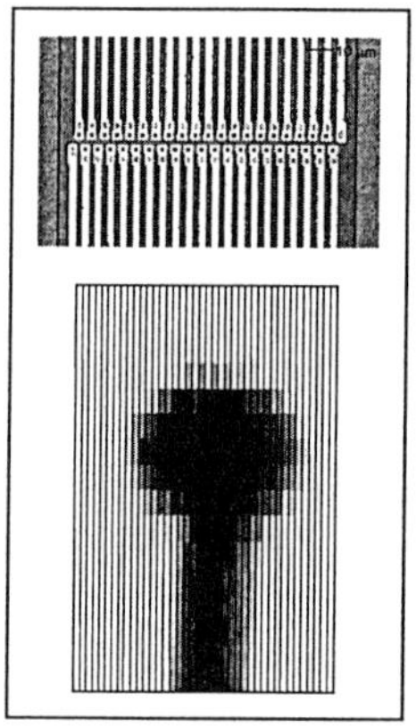

Fig. 2, The array sensing element (top) and the image made with this element (bottom)

Fig.3, A photo of the bonding wire

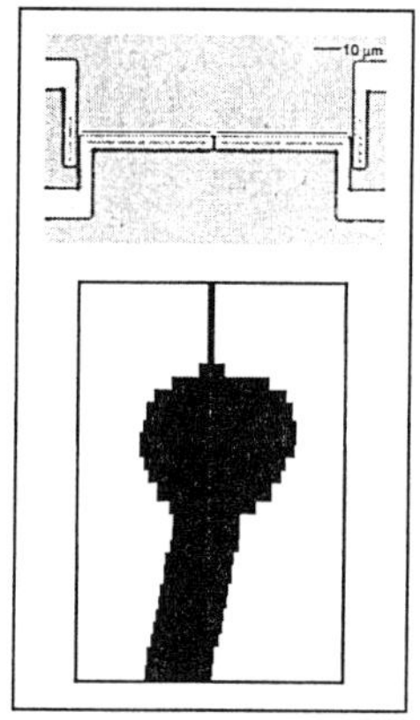

Fig. 4, The strip sensing element (top) and the image made with this element (bottom)

4. Feasibility

The small projection distance is critical for the functioning of the device to avoid serious diffraction distortion of the image. This is realized by the application of a sheath flow [1], a proven technique in flow cytometry. A flow channel can be formed with a simple two-wafer structure. The finite element simulations in Fig. 5 show that the position of the sample inside the channel can be controlled over the complete height of the channel, so the requirement of the small projection distance can be fulfilled.

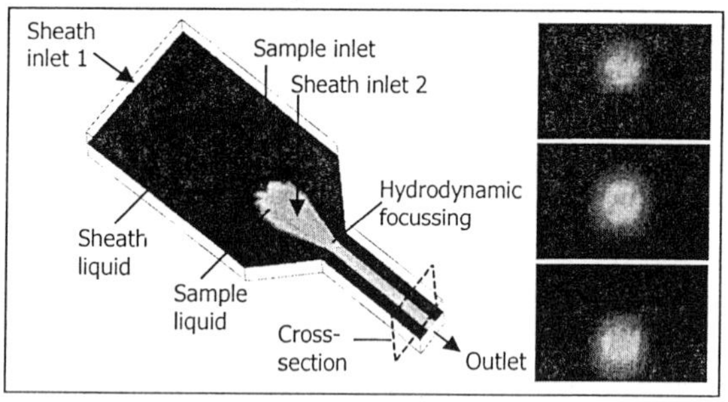

Fig. 5, Finite element simulations showing that the height of the sample can be controlled

5. Conclusions

In this paper two particle-shape sensing-elements are presented for application in an integrated flow-cytometer. Measurements show that both elements work well and the choice for either one depends on the application. This approach is unique in the fact that absolute particle shape information is obtained without any lenses or other optical equipment. To avoid optical diffraction the object should pass close to the sensing-element. Finite element simulations show that this condition can be created in a flow channel with a double sheath flow that can be realized with a two-wafer structure.

References

1. Z. Darzynkiewicz, H.A. Crissman, J.P. Robinson, Methods in Cell Biology, Volume 63 Cytometry, 2001, pp. 26-33

MULTI-CHANNEL OPTICAL DETECTION WITH AN SPR IMAGING SYSTEM AND INTEGRATED WAVEGUIDE BASED SPR DEVICES

Osamu Niwa[1], Tatsuya Tobita[2], Yuzuru Iwasaki[1], Tsutomu Horiuchi[1], HisaoTabei[2], Sabro Imamura[2], Kazuyoshi Kurihara[4] and Koji Suzuki[3, 4]
NTT Lifestyle & Environmental Technology Labs[1], NTT-Advanced Technology[2], 3-1 Morinosato, Wakamiya, Atsugi, Kanagawa 243-0198, Japan, Keio University[3], 3-14-1 Hiyoshi, Kohoku-ku, Yokohama 223-8522, Japan, Kanagawa Academy of Science & Technology[4], 3-2-1 Sakado, Takatsu-ku, Kawasaki 213-0012, Japan
(e-mail: niwa@aecl.ntt.co.jp)

Abstract

Two types of surface plasmon resonance (SPR) measurement system have been developed for multi-channel optical detection for biomolecules. One is an SPR imaging system that images the refractive index change in the array of small micro-wells. The other consists of polymer based integrated waveguide and is small and simple. It can detect the refractive index change in a waveguide by detecting dip shift in the spectrum.

Key word: SPR imaging, Multi-channel measurement, Polymer waveguide

Introduction

Surface plasmon resonance (SPR) sensors have been widely used since SPR provides label-free measurement and relatively high sensitivity for detecting large biological molecules, such as antibodies/antigens and DNA. Recently, we developed an SPR based enzymatic method for detecting low-molecular-weight analytes such as hydrogen peroxide (H_2O_2). H_2O_2 is a product of various types of oxidase. Therefore, we would be able to measure many kinds of molecules simultaneously if we could integrate SPR sensors. Here, we propose two multi-channel SPR systems for bioanalysis.

SPR Imaging System

Fig. 1 is an SPR imaging system. We used an expanded laser beam to illuminate the internal surface of a prism to which we had attached an SPR chip. We monitored the reflected light using a CCD camera. We were

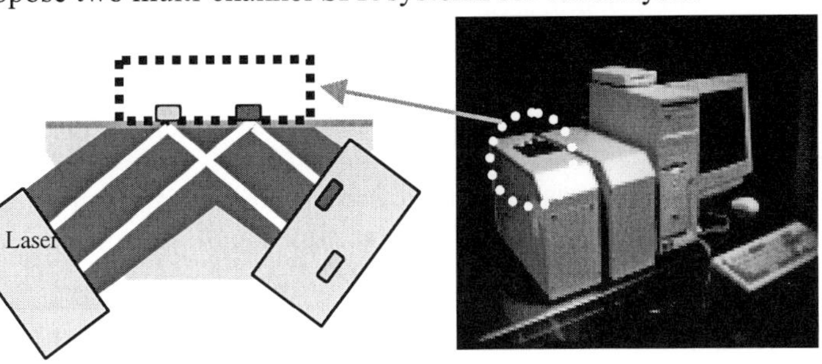

Figure 1. SPR imaging scheme and photograph of system

J.M. Ramsey and A. van den Berg (eds.), Micro Total Analysis Systems 2001, 359–360.

able to detect refractive index changes on the SPR chip surface as reflection intensity by choosing an incident light angle near the SPR dip angle. This system allows us to image the refractive index change on the SPR chip caused by the surface reaction. We fabricated two types of SPR chip. One has arrays of micro-wells on 50-nm-thick gold film. The other also has gold film, but is modified with Os-polyvinylpyrridine wired horseradish peroxidase (Os-gel-HRP) films. HRP reduces H_2O_2 when it is in a reduced state. Fig. 2 shows the images we obtained using these SPR chips. The reflection intensity of the micro-wells differed depending on the samples [Fig.2 (a)].

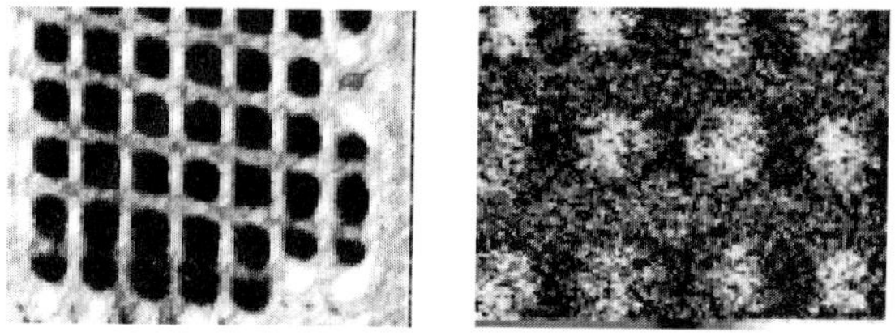

(a) (b)

Figure 2 Image of (a) small volume sample in micro wells and (b) enzymatic reaction.

Fig. 2(b) shows an SPR image of the enzymatic reaction. We electrochemically reduced Os-gel-HRP on the SPR chip and injected H_2O_2. The refractive index of the film changed when the film was oxidized by the enzymatic reaction. Therefore, we were able to image the H_2O_2 concentration distribution by measuring the refractive index change of Os-polymer-HRP film. This system is very useful for simultaneous measurement with a multi enzyme sensor since it allows us to combine various kinds of oxidase.

Optical Waveguide based Multi-channel SPR chip

Fig. 3 shows our other multi-SPR system, which is based on an integrated waveguide. We integrated 4-channel waveguides on one chip and fabricated both the core and cladding of the waveguide from epoxy polymer. We removed part of the cladding and sputter deposited gold film on the core surface for use as an SPR sensing area. We connected one end of the waveguide to a white light source and linked the other to a portable spectrophotometer through optical fibers. We were able to evaluate the analyte concentration by measuring the dip shift in the spectrum. The ease with which many waveguides can be integrated on one substrate makes this device promising for low cost and high throughput analysis.

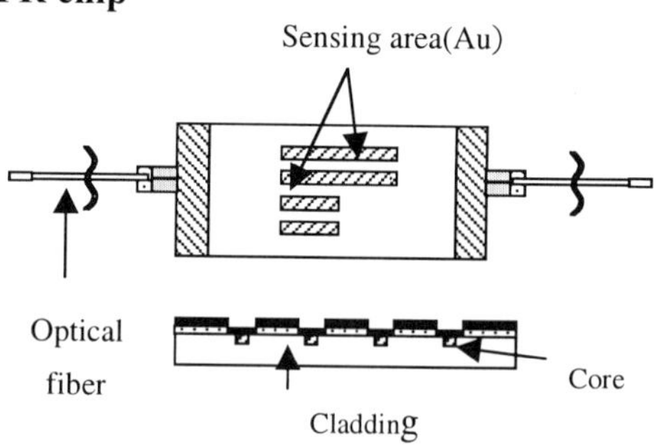

Figure 3 Structure of polymer integratedwaveguide based SPR devices

Conclusion

The SPR systems reported here need low volume samples and inexpensive devices for measurement. Therefore, they will be suitable for high throughput analysis of biomolecules.

NON-CONTACT TEMPERATURE MEASUREMENT INSIDE MICROCHANNEL

M. N. Slyadnev[1], Y. Tanaka[2], M. Tokeshi[1], T. Kitamori[1,2]

[1]Kanagawa Academy of Science and Technology, KSP Bldg, East.307, Takatsu, Kawasaki 213-0012, Japan

[2]University of Tokyo, Dept. of Applied Chemistry, 7-3-1 Hongo, Tokyo 113-8656, Japan

(max_merl@pop12.odn.ne.jp)

Abstract

The paper describes non-contact temperature measurement and 2-D temperature imaging based on temperature dependence of fluorescence. The method is suitable for fast (30 ms) and localized (3 μm) measurements of temperature inside microchannels.

Keywords: Non-contact temperature sensor, temperature control on a chip

1. Introduction

Reliable temperature measurement of reaction media inside microchannel is required for a number of microchip chemical applications. Since heat capacity of the heated volume inside microchannel is extremely small, heat capacity of a measurement device can affect temperature response. In this paper we report a method of temperature monitoring that was implemented on a chip to overcome this difficulties.

2. Experimental

Fluorescence quenching technique was used for non-contact temperature measurement [1] as presented in Fig.1. A quartz glass microchip with microchannel width and depth of 250 and 100 μm was filled with aqueous Rh-3B solution. The sample solution was pumped through the microchannel by a syringe pump at varied flow rates. An Ar^{+}-ion laser (514.5 nm, 200 mW) was used as an excitation source. After proper filtering, fluorescence intensity images were recorded using CCD camera and time-resolved temperature profiles were obtained by lock-in amplifier detection. Heating was induced by photothermal effect with an infrared diode laser (1470 nm, 150 mW) focused inside microchannel. Steady state temperature of the chip was controlled by external Peltier stage and monitored by miniaturized thermocouples inserted in the microchannel.

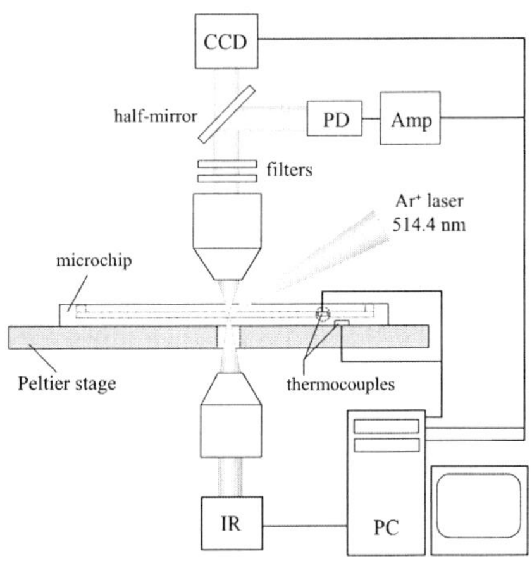

Figure 1. Schematic view of experimental setup for non-contant temperature measurement by fluorescence quenching technique

J.M. Ramsey and A. van den Berg (eds.), Micro Total Analysis Systems 2001, 361–362.

3. Results and Discussion

A non-contact method was designed to monitor fast IR laser-induced thermocycling of solution inside microchannel. Since photothermally heated volume is extremely small (~10 nl), any contact measurement device can influence actual temperature distribution and dynamics. We applied a non-contact measurement technique by fluorescence quenching of rhodamine dye. The intensity of dye fluorescence is temperature-sensitive in a range of temperatures 5–95 °C as shown in Fig.2. Fluorescence intensity distribution images were obtained with CCD camera and converted to the temperature images using the calibration plot.

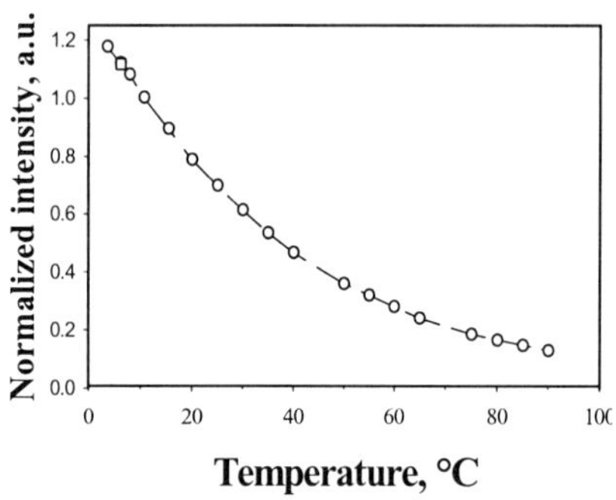

Figure 2. Temperature calibration plot.

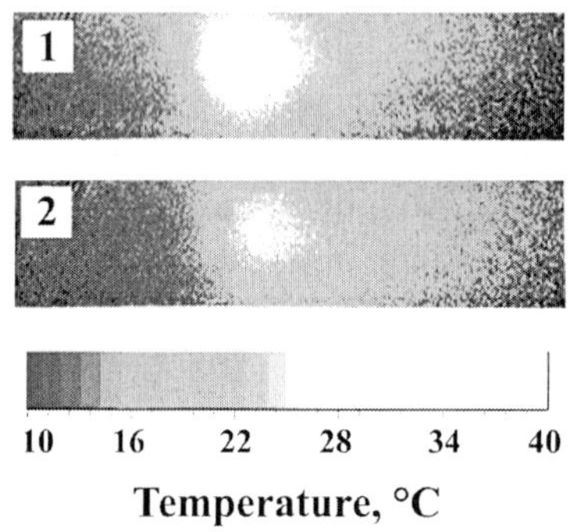

Figure 3. Temperature distribution inside channel. Flow rates: 1 and 10 mm/s.

Results obtained for various solution flow rate are shown in Fig.3 (image dimensions W×H=1000×250 μm). Spatial resolution of the temperature measurements of 3 μm was achieved. A diffusive heat transfer to the channel walls occurs at slow flow rates while convective heat removal dominates at fast flow rates. Fig.4 shows temperature profiles for the microchip with IR laser heating. A time-resolution of 30 ms was demonstrated.

Since measuring technique is non-contact, it does not interfere with heating and cooling and observed profiles closely follow actual temperature rates of the solution. Rapid temperature elevation rates were detected -67 °C/s for heating and 53 °C/s for cooling, respectively. Using this technique the power of IR laser can be adjusted so the solution would reach a desired set-point temperature, as shown in Fig.4.

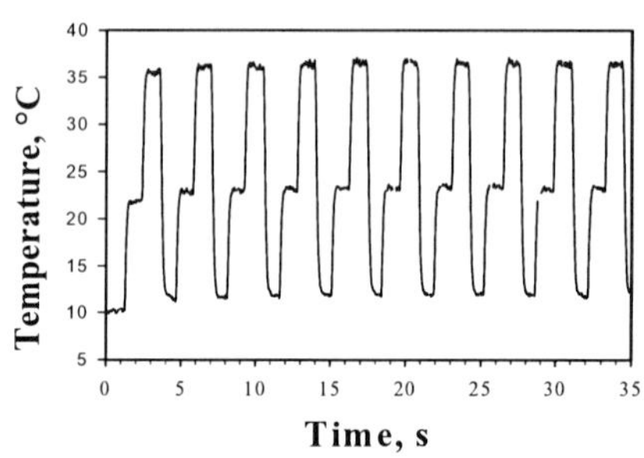

Figure 4. Fast thermocycling by IR laser monitored by fluorescence quenching.

References

1. M.Slyadnev, Y.Tanaka, M.Tokeshi, T.Kitamori, Program of 14th Symposium on Microscale Separations and Analysis, HPCE 2001, P0310M, p.72

A SILICON MICROCHIP FOR BLOOD HEMOGLOBIN MEASUREMENT USING MULTIREFLECTION STRUCTURE

Hidekuni Takao[1], Toshihiko Noda[1], Mitsuaki Ashiki[1], Kazuhiro Miyamura[2], Kazuaki Sawada[1], Makoto Ishida[1]
[1]Dept. of Electrical and Electronic Eng., Toyohashi Univ. of Technology, Japan
[2]Fundamental Technology Research Department, Horiba Ltd., Japan
e-mail: takao@eee.tut.ac.jp

Abstract

In this study, a microchip for blood hemoglobin measurement has been investigated. Multireflection structure is used for sensitivity enhancement in absorbance. Glass/silicon bonded structure was used for the microchip to integrate electron/optical devices on a chip. Calibration curve using red coloring matter with 540 nm source was linear from 0.003 to 0.2 absorbance units.

keywords: Blood hemoglobin, Cyanmethemoglobin, Multireflection, Absorbance Measurement

1. Introduction

Miniaturization of blood inspection system has been demanded for point of care (POC). We have developed silicon microvalves and blood cell counter devices for realization of miniaturized blood inspection system [1], [2]. In this study, a silicon microchip for small size blood hemoglobin measurement system has been developed by using silicon MEMS technology.

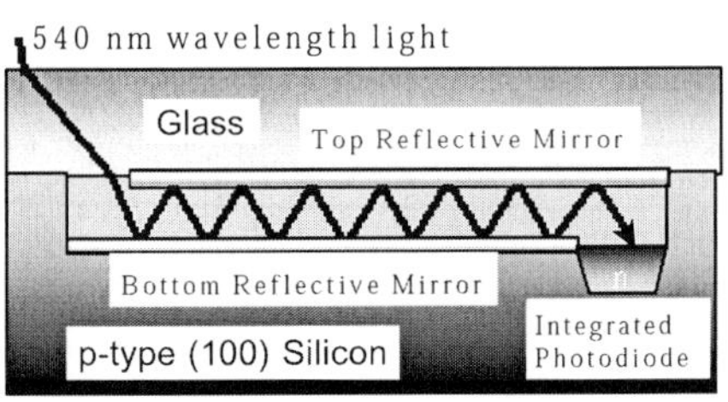

Fig. 1. Absorbance Detection with Multireflection Structure in this device.

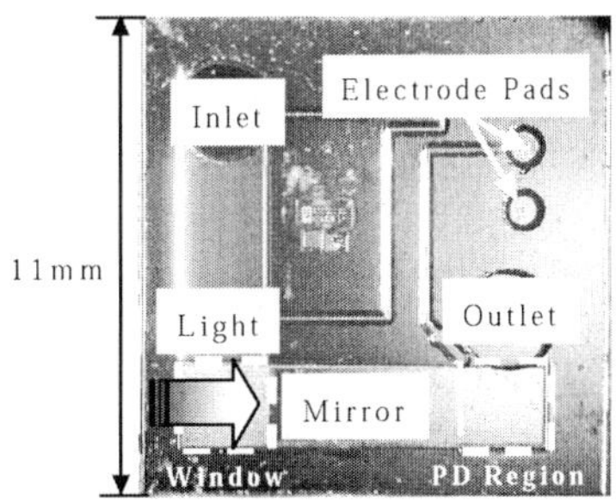

Fig. 2. Photograph of fabricated microchip with bonded cap glass.

2. Theory

Cyanmethemoglobin (HiCN) method is a generalized measurement of blood hemoglobin. So, it is employed in the developed device for general requirement. HiCN is generated by pretreatment of blood sample. HiCN concentration is measured by 540 nm light source because it has a peak of absorbance at 540 nm. In this device, multireflection structure is employed to enhance sensitivity in absorbance (Fig. 1) [3]. Glass is bonded on silicon die including etched flow channel and a photodiode on the bottom. Metal Mirrors are formed on the top and the bottom surface for mutireflection. Effective optical path length can be longer than 5 mm due to multireflection. Fig. 2 shows a photograph of the

J.M. Ramsey and A. van den Berg (eds.), Micro Total Analysis Systems 2001, 363–364.

fabricated microchip. Pretreated blood sample is introduced to the fluid channel with 20 μm-depth. A 540 nm light source is provided from the window.

3. Experimental

Performance for HiCN method was evaluated as shown in Fig. 3. A spectroscope was used as the 540 nm light source. Water solution of red coloring matter was used instead of pretreated blood sample. Wavelength dependence of light absorbance in the solution was similar with that of HiCN. The integrated silicon photodiode was connected to the electrode shown in the figure.

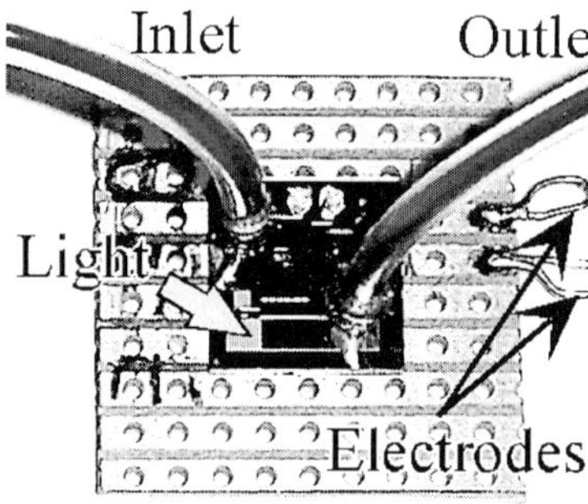

Fig. 3. Measurement of the device with water solution of red coloring matter.

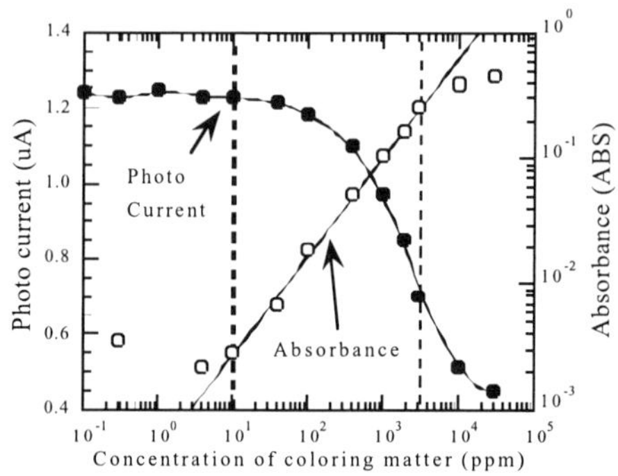

Fig. 4. Calibration curve and photocurrent of photodiode for solution concentration.

4. Results and Discussion

Relationship between photocurrent of photodiode and solution concentration was successfully evaluated (Fig. 4). The obtained calibration curve was linear from 0.003 to 0.2 absorbance units. The lower limit was determined by the dark current of the photodiode. The upper concentration limit was determined by leakage of light source through the cap glass. The dynamic range of the device was 50 dB.

5. Conclusions

Performance of this device will be acceptable for actual measurement of blood hemoglobin. Design and structure of fluid channel, multireflection mirror, and photodiode will be improved in the next version of microchip to obtain more better performance for absorbance measurement.

References

[1] H.Takao, M.Ishida, K.Sawada , "A neumatically-actuated silicon microvalve and its application to functional Fluidic integrated circuits", Proceedings of Transducers'01, 3A3.01, 2001.

[2] D.Satake, H.Takao, M.Ashiki, M.Ishida et al., 'Development of a Durable Sensor for Blood Cell Counter Using MEMS Technology", Proceedings of Transducers'01, 3D1.14P, 2001.

[3] H. S. Moosavi, Y. Jiang, et al., "A multireflection cell for enhanced absorbance detection in microchip-based capillary electrophoresis devices", Electrophoresis, vol.21, pp. 1291-1299, 2000.

TUNABLE THERMAL LENS MICROSCOPE FOR MICROCHIP ANALYSIS

E. Tamaki[1], A. Hibara[1], M. Tokeshi[2] and T. Kitamori[1,2]
[1]Department of Applied Chemistry, School of Engineering, The University of Tokyo, Japan
[2] Integrated Chemistry Project, Kanagawa Academy of Science and Technology, Japan

Abstract

A thermal lens microscope (TLM) system using Xe lamp and monochrometer as exitation light source was developed. By utilizing the heat source localization in microchannel and themal conduction from liquid to glass, thermal lens was formed in a microchannel. By using this system, detection of dye solution and measure of spectrum were demonstrated.

Keywords: thermal lens microscope, integrated chemistry

1. Introduction

Thermal lens microscope (TLM) is one of the photothermal spectroscopic methods that can detect trace amounts of non-fluorescent substances in liquid. The TLM has high sensitivity comparable to fluorescent method and high spatial resolution at sub-μm level [1]. Since a laser beam has been used as excitation light in conventional TLM, the excitation wavelength is limited. In this report, we propose a tunable TLM using a Xe lamp with a monochrometer. When using an incoherent light source as excitation light, it is difficult to localize heat generation, which is easily realized by using Gaussian laser beam. In order to breakthrough the problem, we propose a novel principle utilizing thermal conduction between liquid and microchannel wall to generate thermal lens by incoherent light. Our tunable TLM can measure spectrum of target species in microchannels

2. Design of tunable TLM system

In conventional TLM, the excitation laser is focused with a spot size of approximately 1 μm by the objective lens as shown in Figure 1(a). At the focal point, thermal energy is released according to the laser intensity profile from target substances. The localized heat generation and thermal diffusion generate temperature distribution in/around the focal point. Since refractive index depends on temperature, the temperature distribution acts as a transient optical lens, which is called thermal lens. To detect the thermal lens, another laser beam is coaxially introduced to the thermal lens and divergence change of the probe beam due to the thermal lens is detected

J.M. Ramsey and A. van den Berg (eds.), Micro Total Analysis Systems 2001, 365–366.

In the tunable TLM, a Xe lamp is used as an excitation light source instead of laser. Since the incoherent light cannot be focused tightly, it is difficult to generate high temperature gradient around the beam waist, which are indispensable for thermal lens spectroscopy. Therefore, we have proposed a novel idea to generate thermal lens in microchannel. As shown in Figure 1(b), a microchannel in a microchip provides a small heat source and large thermal bath. That is suitable for high temperature gradient generation. Due to short thermal diffusion length and effective thermal conduction, high temperature gradient can be generated in the microchannel. Since the temperature gradient corresponds to refractive index gradient, quasi-thermal-lens effect can be detected.

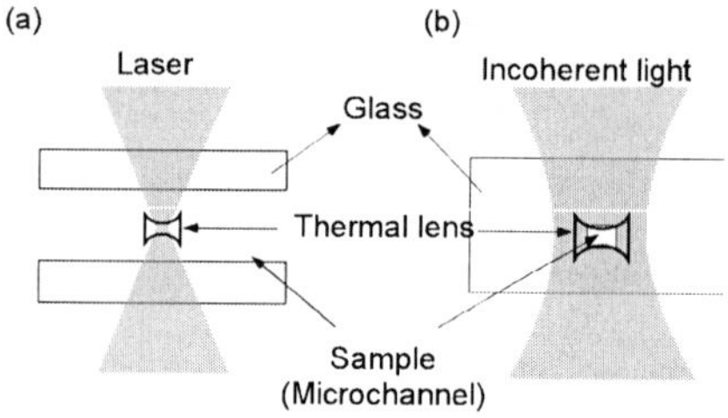

Figure 1 (a) Conventional TLM system. Laser was used as excitation light. The laser was tightly focused and thermal lens was generated around the beam waist. (b) Tunable TLM system. Light was radiated to microchannel and the heat was generated only in the channel. The high temperature gradient was formed by thermal conduction.

3. Experimental

The output light of the Xe lamp is used as the excitation light after passing through a monochrometer and an optical chopper. The wavelengh was scanned from 400 to 550 nm ($\Delta\lambda$=10 nm) at the scan speed of 5nm/min with the monochrometer. He-Ne laser is used as the probe beam. The two beams are aligned coaxially by using a dichroic mirror and focused to a microchannel in a chip through an objective lens. The probe beam intensity is measured by a photodiode after passing a long pass filter and a pinhole. The thermal lens signal is synchronously detected by a lock-in-amplifier.

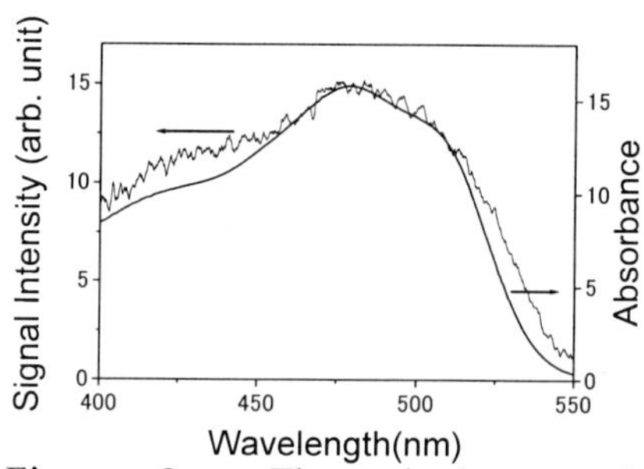

Figure 2 Thermal lens and absorption spectra. The exitation wavelength of Xe lamp for thermal lens was scanned from 400 to 550 nm. The spectrum of thermal lens was compared with absorption spectrum. The sample was SunsetYellow in methanol(10^{-3}M).

4. Results and discussion

Figure 2 shows the thermal lens and absorption spectrum of 10^{-3} M dye solution (SunsetYellow) in methanol. As show in Fig.2, thermal lens spectrum is corresponded to absorption spectrum. This results indicate the system can detect the substances with desirable wavelength. This tunable TLM will be applied to detection of chromatography, electrophoresis and single cell biochemistry.

References

1. K. Uchiyama, A. Hibara, H. Kimura, T Sawada and T. Kitamori *Thermal lens microscope* Jpn. J. Appl. Phys. 39, 5316-5322 (2000).

ULTRAVIOLET ABSORBANCE SPECTROSCOPY IN A 3-DIMENSIONAL MICROFLUIDIC CHIP

Jeffrey Wolk, Michael Spaid, Morten Jensen, Richard MacReynolds, Knute Stevenson, and Ring-Ling Chien
Caliper Technologies Corp., 605 Fairchild Drive, Mountain View, CA 94043.
(e-mail address: jeff.wolk@calipertech.com)

Keywords: absorbance, microfluidic, spectroscopy, ultraviolet

1. Introduction

The strength of absorbance by a compound is linearly dependent on the path length of the detection cell. Because of this, performing high sensitivity absorbance measurements on samples in microfluidic chips, where channel depths are roughly 10-50 μM, is extremely challenging. There have been several attempts to overcome this inherent difficulty, including 1) coupling optical fibers into the chip and directing light along a section of the channel [1], and 2) creating a multi-reflection cell [2]. We have taken another approach, which is to construct a chip containing a three-dimensional fluid path. In our chips, the section of the fluid path running perpendicular to the plane of the chip forms a detection cell with a path length equal to 720 μm, allowing for highly sensitive absorbance measurements.

2. Experimental

The chip we have constructed is shown schematically in Figure 1a. It is assembled from three layers of 0.7-mm thick quartz, which is transparent at the ultraviolet wavelengths where our measurements are performed. The most important feature of the chip is a 100 μm diameter through hole in the middle layer, connecting the fluidic channels on the top and bottom layers. This through hole forms the detection cell used for the absorption measurements. The depth of the detection volume is the thickness of the middle layer plus twice the channel depth, which is $700 + (2 \times 10) = 720$ μm. This implies that the absorption by a sample is a factor of 72 times stronger than the same sample being observed through only a 10 μm deep channel. Importantly, this chip interfaces with the standard optical geometry used for absorption measurements, as shown in Figure 1a. Apertures were placed on the chip to prevent stray light from reaching the detector.

Another important feature of this chip is that it allows for the automated introduction of samples into the detection cell. This chip is an example of the Sipper™ chip technology developed at Caliper Technologies Corp. The bottom layer of the chip has a capillary attached to it that provides a fluidic connection with the channels etched on the chip. A steady vacuum is applied to the waste well of the chip, which draws fluid

J.M. Ramsey and A. van den Berg (eds.), Micro Total Analysis Systems 2001, 367–368.

through the capillary, the detection cell, and finally into the well itself. By using robotics to move various wells of a microtiter plate underneath the capillary of the chip, we are able bring a multitude of samples loaded onto a microtiter plate through the detection volume in a serial fashion.

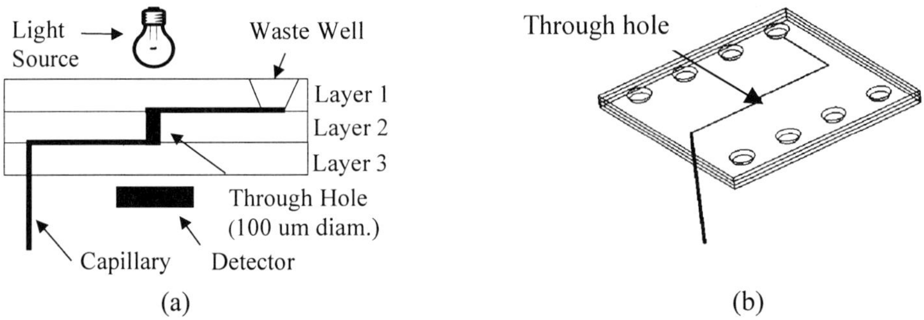

Fig 1a) A schematic representation of the chip and optical set-up. The chip interfaces with the standard optical geometry used for absorption measurements, in which the light is directed through the detection volume from above, and the detector is mounted below the sample. 1b) A view of the actual chip, with all dimensions to scale.

3. Results and discussion

Measurements taken at 254 nm using the absorbance chip just described are shown in Figure 2. The absorbance peaks are due to injections of a 25-mer of DNA ranging in concentration from 20 μM to 0.2 μM. Buffer plugs were injected to achieve separation between the serially injected samples. To confirm that we observed the expected increase in sensitivity, we have also compared the absorption of a 3 μM sample of the 25 mer in the 720 μM path length cell with the absorption of a 3 × (720/10) = 226 μM sample through a 10 μm thick channel. As expected, the two absorptions were equal.

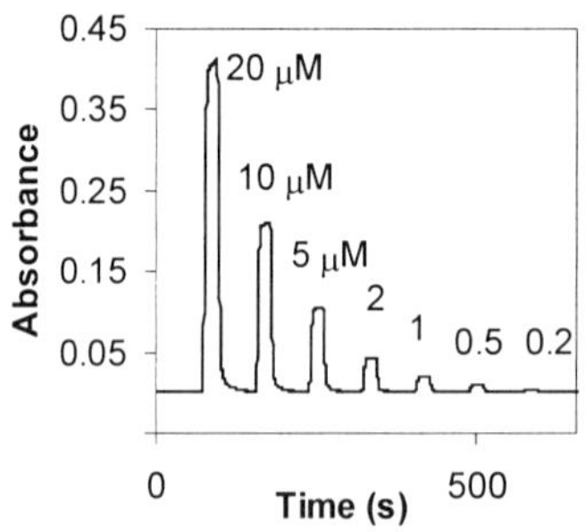

Fig. 2 – Absorbance at 254 nm by a 25-mer of DNA at concentrations ranging over two orders of magnitude.

References
[1] Z. Liang, N. Chiem, G. Ocvirk, T. Tang, K. Fluri, and D. J. Harrison, Anal. Chem., 1996, 68, 1040.
[2] H. Salimi-Moosavi, Y. Jiang, and D. J. Harrison, Electrophoresis, 2000, 21, 1291.

INTEGRATION OF DETECTION SYSTEM ON A CHIP –THERMAL LENS MICROSCOPE USING SELFOC™ MICRO LENS –

J.Yamaguchi[1], A.Hattori[1], M.Tokeshi[2], T.Kitamori[2,3]
[1]Nippon Sheet Glass, Japan
[2]Kanagawa Academy of Science and Technology, Japan,
[3]The University of Tokyo, Japan

Abstract

Miniaturization of sensitive detection system is a key factor for integrated chemical systems. So we constructed a newly designed palmtop-sized thermal lens microscope (PTLM). The PTLM can be applied to the ultra-sensitive detection of non-fluorescent molecules. The PTLM could be realized for the integration of detection by using a very small SELFOC™ micro lens as an objective.

Keywords: thermal lens microscope, photothermal spectroscopy, SELFOC™ micro lens

1. Introduction

The concept of integrated chemical systems implies the miniaturization and integration of all components necessary for chemical analysis. Although the integration of components except detection on microchips was fairly succeeded, the integration of detection has seldom been reported. The integration of electrochemical detection is rather simple than other methods. However, the sensitivity is not very high. Thus, the integration of detection methods with high sensitivity, such as laser induced fluorescence or thermal lens, is desired. Moreover, it is desirable that the detection method of integrated chemical systems has wide applicability.

Considering the mention above, we constructed a newly designed palmtop-sized thermal lens microscope (PTLM).

2. Palmtop-sized Thermal Lens Microscope (PTLM)

The PTLM was realized by using a very small SELFOC™ micro lens as an objective. The outline is shown in figure 1.

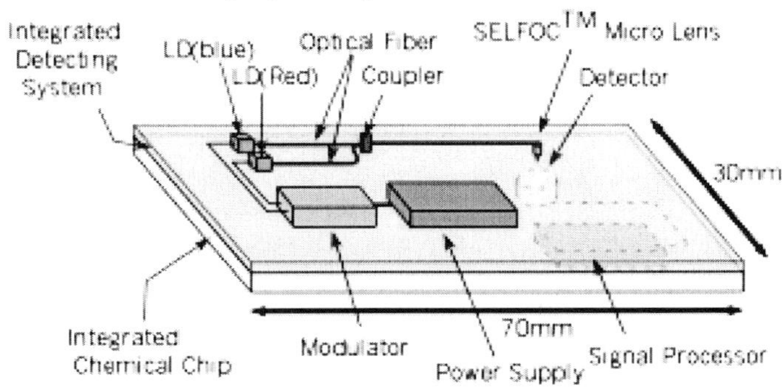

Figure 1. Outline of the PTLM

J.M. Ramsey and A. van den Berg (eds.), Micro Total Analysis Systems 2001, 369–370.

The SELFOC™ micro lens has a radial refractive index gradient and adequate chromatic aberration. These characteristics of the SELFOC™ micro lens can be controlled arbitrary by changing materials and size of lens. Since the end face of the SELFOC™ micro lens is flat, the optical arrangement on a microchip is very easy (fig.2). The cylindrical SELFOC™ micro lenses (diameter: 1 mm, length: 2.56 mm) was bonded to the end of the single-mode optical fiber.

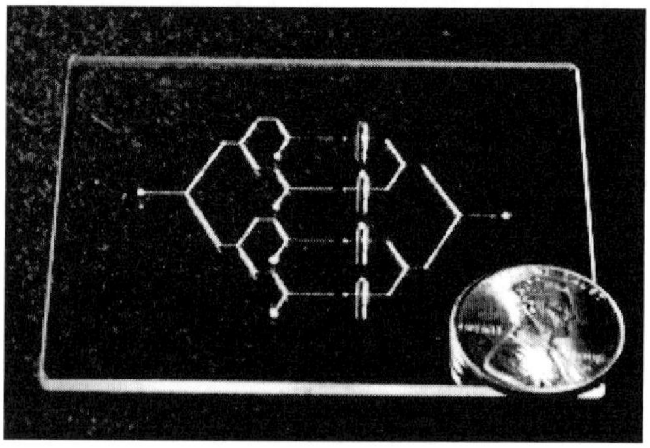

Figure 2. SELFOC™ micro lenses on the microchip

3.Results and discussion

For TLM measurement, it is necessary to shift the focal points of excitation and probe beams [1]. Thus, the TLM has a complicated optical unit for adjustment of focal points [2]. If the SELFOC™ micro lens is used as an objective, the complicated optical unit is omitted because it has chromatic aberration.

The excitation (532nm) and probe (633nm) beams were made to be coaxial using a dichroic mirror and introduced into the single-mode optical fiber (unused coupler). The excitation beam was mechanically chopped (1kHz). Using dye (sunset yellow) aqueous solutions, the performance evaluation of the PTLM was carried out. Figure 3 shows a calibration curve of sunset yellow. The calibration curve shows good linearity in the range of 1.0×10^{-6}-1.0×10^{-4} M. The detection volume is equal to the focal volume of the excitation beam and it is calculated to be 5 fL ($f=10^{-15}$) [1]. The PTLM shown here could be realized for the integration of detection with high sensitivity.

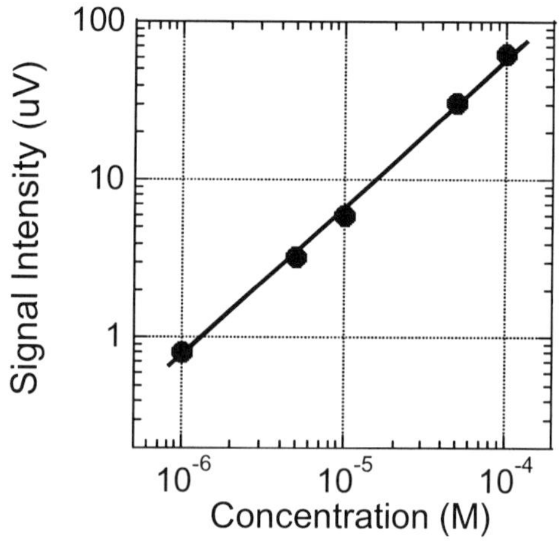

Figure 3. Calibration curve of sunset yellow.

4.References

1. K. Uchiyama, A. Hibara, H. Kimura, T. Sawada and T. Kitamori, *Thermal Lens Microscope*, Jpn. J. Appl. Phys., 39, pp. 5316-5322 (2000).
2. M.Tokeshi, M.Uchida, A.Hibara, T.Sawada and T.Kitamori, *Determination of Subyoctomole Amounts of Nonfluorescent Molecules Using a Thermal Lens Microscope: Subsingle-Molecule Determination*, Anal. Chem., 73, pp. 2112-2116 (2001).

SPATIAL CONCENTRATION PROFILING WITHIN CHANNEL NETWORKS IN TRANSPARENT MICROREACTORS USING A DIGITAL IMAGING METHOD

Xunli Zhang, Xizeng Feng, Paul D.I. Fletcher and Stephen J. Haswell
Department of Chemistry, The University of Hull, Hull HU6 7RX, UK
(X.Zhang@hull.ac.uk)

Abstract

A method for *in situ* concentration profiling within a microchannel network in optically transparent microreactors has been developed. This technique provides a convenient and simultaneous method for monitoring the spatial and temporal evolution of chemical reactions in microreactors. In addition it also enables the study of microfluids in microreactor and μ-TAS devices to be carried out.

Keywords: Spatial concentration profiling, microreactors, digital imaging

1. Introduction

The method involves capturing digitised microscope images of the channel networks in which optically absorbing molecules are present or being formed and analysing the images pixel by pixel to yield a spatially resolved array of absorbance values [1]. The array is then converted to concentration data based on the Beer-Lambert law, to provide spatial concentration profiles. In this example a Rhodamine B dye undergoing diffusion in a T shape reactor is used to illustrate the methodology.

2. Experimental

An Axiovert 100S inverted microscope coupling with a digital CCD camera was employed to obtain digital micrographic images of the channel in which Rhodamine B aqueous solution and/or water is being pumped under electroosmotic flow.

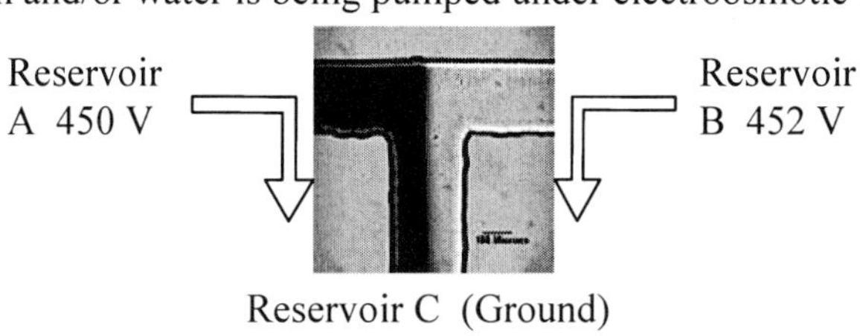

Figure 1. Micrograph of the T-shaped channel network when Rhodamine B aqueous solution flowing from reservoir A and water from reservoir B, showing clearly the diffusive interface.

The image is transferred into Microsoft EXCEL as numerical transmitted light intensities for subsequent conversion into absorbance values and hence pixel-by-pixel

J.M. Ramsey and A. van den Berg (eds.), Micro Total Analysis Systems 2001, 371–372.

concentration values. The glass microreactor was fabricated according to published procedures [2]. It consists three reservoirs where A was filled with Rhodamine B aqueous solution (2.0 mM, 50 ml), B and C with water (50ml each), see Figure 1. Platinum electrodes (0.26 mm diameter) and were placed in each reservoir and connected to a Paragon 3B power supply (Kingfield Electronics) for electro kinetic pumping. The voltage and time variables were controlled by a LabVIEW program.

3. Results and discussion

A spatial concentration profiling of a 860 × 860 μm^2 area of the T-shaped channel network is shown in Figure 2, which corresponds to the optical image obtained from the microscope (Figure 1). Concentration profiles in a channel cross section along either X or Y axis can be obtained at specific positions. From Figure 2 the diffusive interface can be seen to be increasing along the Y axis as expected. The approach is now being used as an important tool to study molecular interaction as a function of both dispersive and electrophoretic mixing for a range of plug and continuous flow regimes. A model reaction system based on PADA complex with nickel is currently under investigations.

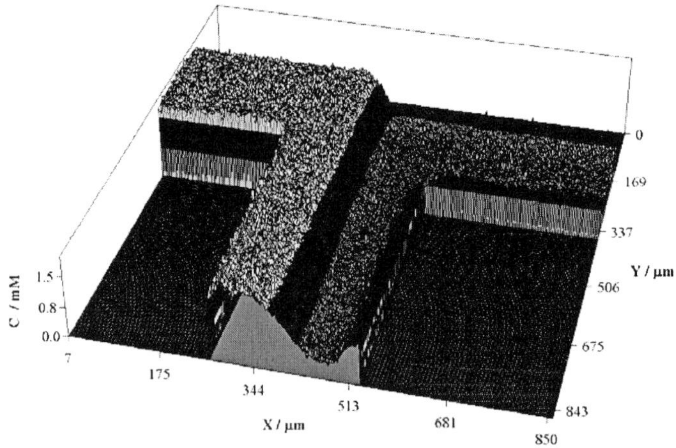

Figure 2. Spatial concentration profiling of an 860 × 860 μm^2 area of the T-shaped channel network.

4. Conclusions

This technique described provides a convenient and simultaneous method for monitoring the spatial and temporal evolution of chemical reactions in microreactors.

Reference

1. I. Broadwell, P.D.I. Fletcher, S.J. Haswell, T. McCreedy and X.L. Zhang, *Lab-on-a-Chip*, (2001) In press.
2. T. McCreedy, *Anal. Chim. Acta.*, **427** (2001) 39–43.

HIGH THROUGHPUT CONTINUOUS SAMPLE INTRODUCTION INTERFACING FOR MICROFLUIDIC CHIP-BASED CAPILLARY ELECTROPHORESIS SYSTEMS

Qun Fang, Guang-Ming Xu and Zhao-Lun Fang
Institute of Microanalytical Systems, Chemistry Department, Zhejiang University, Hangzhou 310028 , China

Abstract

World-to-chip interfacing on a microchip was developed using a novel flow-through sampling reservoir. Efficient sample change was achieved in the continuous CE separation and determination of FITC-labeled amino acids with LIF detection at maximum throughputs of 80/h with <1.5% carryover, and precision of 1.5% rsd for arginine (n=11).

Keywords: Microfluidic chip, world-to-chip interface, capillary electrophoresis

1. Introduction

Microfluidic chips for performing CE separations have made great progress in recent years [1]. Despite the strong impetus in their development, "world-to-chip" interfacing still requires further development if such devices are to be used routinely under "real-life" conditions. In this work, non-interrupted continuous introduction of hundreds of samples was achieved on a microfluidic CE chip using a novel flow-through sampling reservoir.

2. Experimental

The design of the 1.7×30×60 mm crossed-channel glass chip with the flow-through sample reservoir is shown in Fig. 1. The reservoir (3 in Fig. 1) was formed by fixing a 2-mm i.d. tubular plastic section on the plate surrounding the access hole with a bottom inlet. A filter paper collar was wrapped tightly around the plastic tubing to guide the overflowing sample (FITC-amino acid solution) into a semicircular waste trough during sample introduction, maintaining constant fluid level in the reservoir. Electrokinetic sampling from the reservoir and CE separations were performed in the pinched mode simultaneously with the sample introduction. LIF detection was performed at 488-nm.

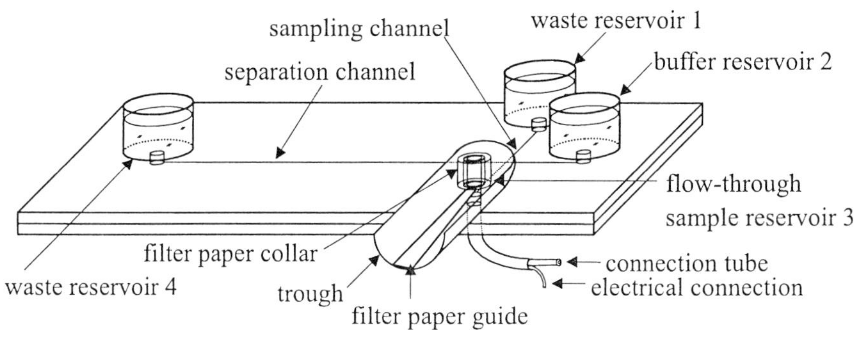

Fig. 1 Structure of flow through sampling reservoir for microfluidic chip

J.M. Ramsey and A. van den Berg (eds.), Micro Total Analysis Systems 2001, 373–374.

3. Results and discussion

In order to achieve efficient sample changeover with low carryover, high throughput and separation efficiency, in addition to controlling the liquid level in the sampling reservoir (see **Experimental**), the dead volume of the flow-through reservoir was minimized while the diameter of the other reservoirs were made sufficiently large. This allowed the introduction of large number of samples without changing their liquid levels significantly. Equalizing the liquid levels in the reservoirs is important for suppression of Poiseuille flows in the separation channel.

The performance of the system was demonstrated in the separation and determination of FITC-arginine, glycine, phenylalanine and glutamic acid by introducing a train of different samples without electrical interruption. Employing a separation channel of 4 cm (2 cm effective length), 1.4 kV separation voltage, throughputs of 40-80/h were achieved with <1.5% carryover, and precision of 1.5% for arginine (n=11). Studies on the reproducibility of the system for extended periods during the continuous monitoring of the derivatizing process of amino acids by FITC for 160 analytical cycles over a period of 4 h (Fig. 2) yielded a precision of 4.8% rsd (n=160) for fluorescein (internal standard). Optimum separation efficiencies in plate numbers were in the range 1×10^4 - 2×10^4.

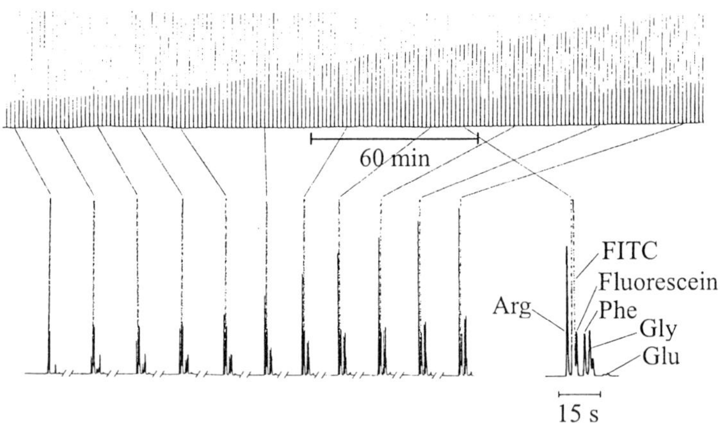

Fig. 2. On-line monitoring of FITC-amino acid derivating reaction

4. Conclusions

The proposed interface proved to be an efficient means for achieving sample change on microfabricated chip systems, while neither disturbing the chip position nor the electrical connections. This system should prove to be particularly useful for the on-line monitoring of chemical processes involving multiple constituents.

Acknowledgement

This work was supported by Natural Science Foundations of China.

Reference

1. A. van den Berg; W. Olthuis,; P. Bergveld, (Eds.), *Micro Total Analysis Systems 2000,* Kluver Academic Publishers, Dordrecht, The Netherlands **(2000).**

TOWARDS EVAPORATION-DRIVEN HPLC ON A CHIP: AN ALTERNATIVE TRANSPORT PROCESS FOR MICRO ANALYSIS SYSTEMS

N. Goedecke and A. Manz
AstraZeneca / SmithKline Beecham Centre for Analytical Chemistry, Imperial College, London SW7 2AY

Keywords: High Performance Liquid Chromatography (HPLC), open-tubular liquid chromatography (OTLC), evaporation-driven liquid transport

Abstract
Within this project we developed an alternative transport process to drive HPLC on a chip. The combination of liquid evaporation and capillary forces at the channel termini provides a continuous liquid transport along the micro channel.

1. Introduction
Several approaches to integrate HPLC on a chip have been undertaken in recent years [1-2]. A general problem is the connection of the chip to the external pressure-generating pump at high pressure without leakage or material break down. With the presented approach the pressure is generated inside the channel itself by evaporating the liquid at the channel termini. Applying this principle to an open channel it is possible to achieve a permanent liquid transport.

2. Experimental
Even though the transport principle and the experimental set-up are very basic (see Figures 1 and 2) evaporation-driven processes have not yet been introduced to OTLC. According to the principle shown in Figure 1 we designed, produced and tested glass/glass chips. These chips, self filled by capillary forces, started immediately to evaporate liquid at the channel termini. As the liquid level lowers in the channel termini infinitesimally while evaporating, they refill by capillary forces. A micro fan (processor cooler) supplied the channel termini with "fresh" air from the set-up environment. The electrodes were connected to a high voltage power supply for sample injection.

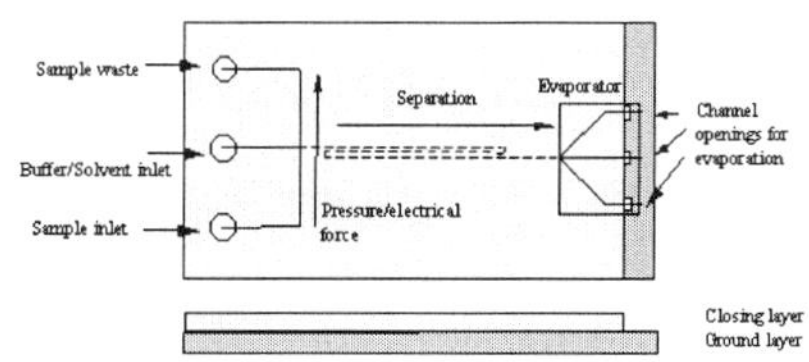

Fig.1: Experimental principle

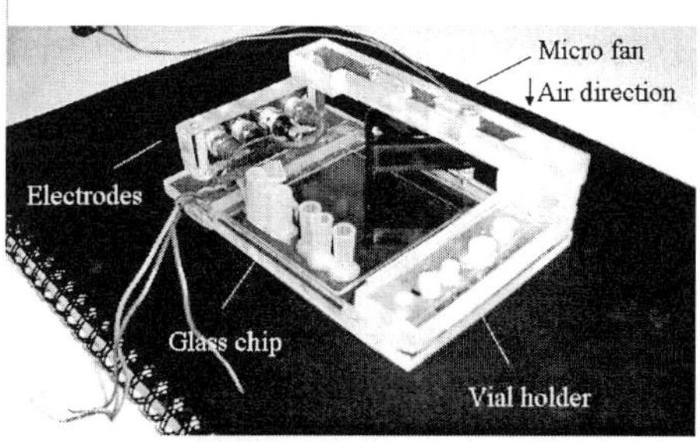

Fig. 2: Chip holder with glass chip

J.M. Ramsey and A. van den Berg (eds.), Micro Total Analysis Systems 2001, 375–376.

3. Results and Discussion

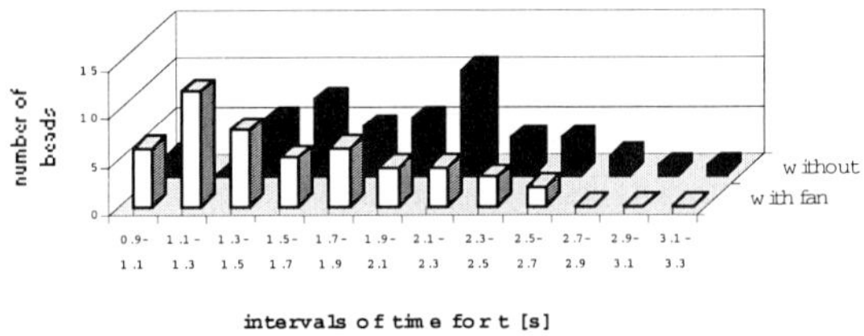

Fig.3: Difference in velocity between continuous air stream over the vapourization area, the corresponding travel distance for the beads was 650µm

With the first generation chip we demonstrated that the liquid moves within the channel without any additional means apart from capillary forces and evaporation. To image the movement latex beads were introduced into the system. The video data obtained from the process showed that the beads travelled along the channel towards the evaporation zone. Figure 3 shows the travel times of randomly chosen particles with the fan switched both on and off, sorted into fractions after time intervals of 0.2s. The travel times demonstrated that the particles moved with different velocities within the channel and with a significant difference in the velocity distribution between both cases. If an air stream is fanned onto the evaporation area, then the liquid flow becomes continuous. The average passing-by time for each particle decreased by 0.42s, equivalent to an increase in speed of 20%. Accordingly all further experiments were undertaken with a permanent air stream onto the evaporation zone.

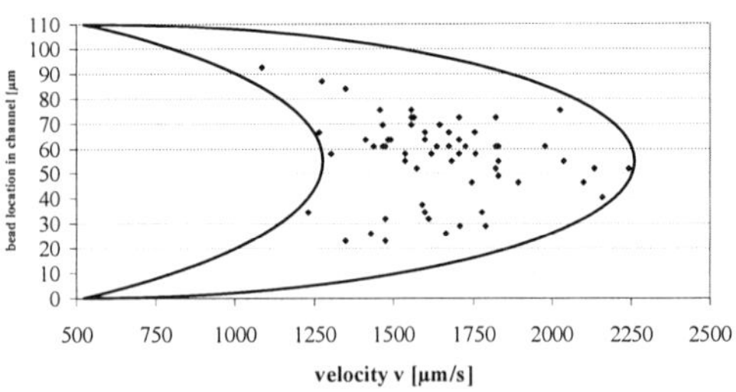

Fig.4: Parabolic flow profile of latex beads in a continuous evaporation driven liquid flow

Another video document was analysed frame-by-frame to measure the velocity and location of each bead within the channel. Figure 4 shows the results of this attempt. The dots represent a group of beads within the liquid profile. It can be found that the further the beads are from the centre the slower they moved, which is proportional to the friction/shear force appearing in the channel.

4. Conclusions

Evaporation driven liquid transport in combination with capillary forces seems to be a promising new alternative towards HPLC on a chip. To gain a continuous liquid flow it is necessary to fan fresh air onto the evaporation area.

5. Referenzes

1. Ocvirk, G., et al., *High-performance liquid-chromatography partially integrated onto a silicon chip.* Anal. Methods and Instrumentation, 1995. **2**(2): p. 74-82
2. Paul, P.H., et al. *Electrokinetic pump application in micro-total analysis systems; mechanical actuation to HPLC.* Proc. Micro Total Analysis Systems. 2000. Kluwer Academic Publishers, Dortrecht, The Netherlands, pp.583-590

SIMPLE AND EFFECTIVE METHOD FOR HOLD AND CONNET OF CHEMICAL IC CHIP-SET

Tadahiro Hasegawa and Koji Ikuta
Dept. of Micro Engineering, School of Engineering, Nagoya University
Furo-cho, Chikusa-ku, Nagoya 464-8603, JAPAN
E-MAIL : ikuta@mech.nagoya-u.ac.jp

Abstract

We developed a simple and effective coupling method for interconnecting modular micro fluidic devices such as "Biochemical IC Family" proposed by Ikuta [1]. Thin silicone rubber films formed from a polymer mold are utilized as reusable couplers. This simple coupler provides a good seal under high pressure (>400 kPa). Moreover, this connection can minimize a dead volume between the biochemical IC chips containing multiple micro fluidic devices. This coupling method is useful not only for the biochemical IC chips but also other micro fluidic devices.

Keyword: Micro coupler, Thin silicone rubber film, Holder unit, Biochemical IC chip

1. Introduction

For integrated fluidic devices, the important issue is still unsolved. Some fluidic couplers have addressed to interconnect external fluidic reservoirs to a micro fluidic device, by inserting capillary tubes [2, 3]. As a connection between micro fluidic devices, they have the disadvantages of dead volume, not accurate positions between multiple channels, and complicated fabrication. On the other hand, we have demonstrated that a luminous enzyme of the firefly was synthesized by combining various biochemical IC chips, using the proposed coupling method (Fig.1)[4]. In this paper, the coupling method will be presented, whose validity will be demonstrated by performing leakage tests.

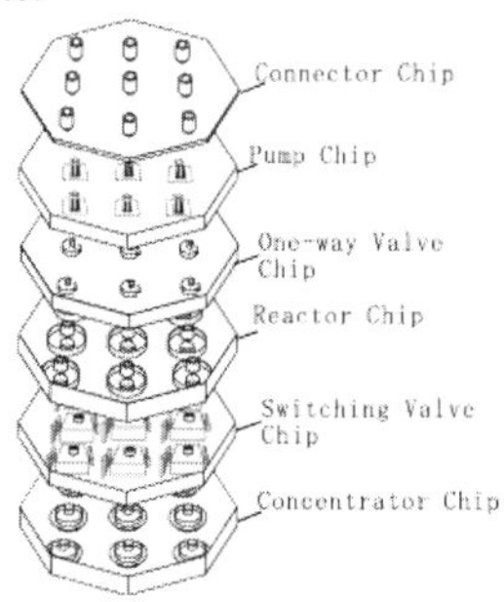

Fig.1 Concept of Biochemical IC Chip-set family

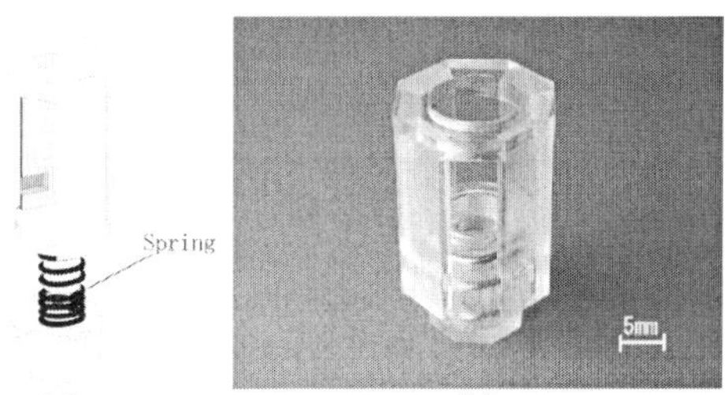

(a) (b)

Fig.2 Folder Unit for Biochemical IC
(a) Basic design (b) Prototype unit

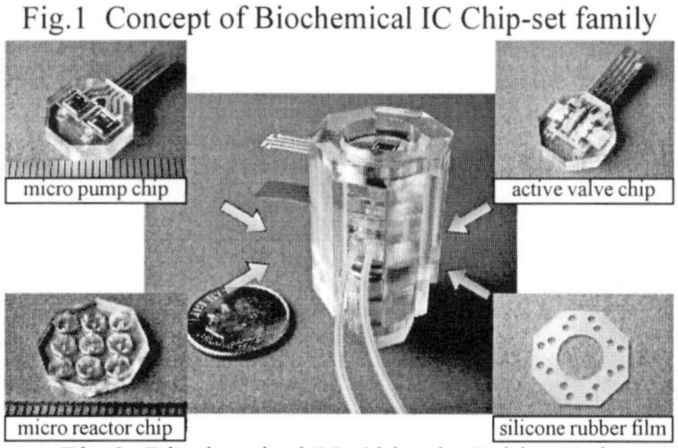

Fig.3 Biochemical IC Chips in Folder Unit

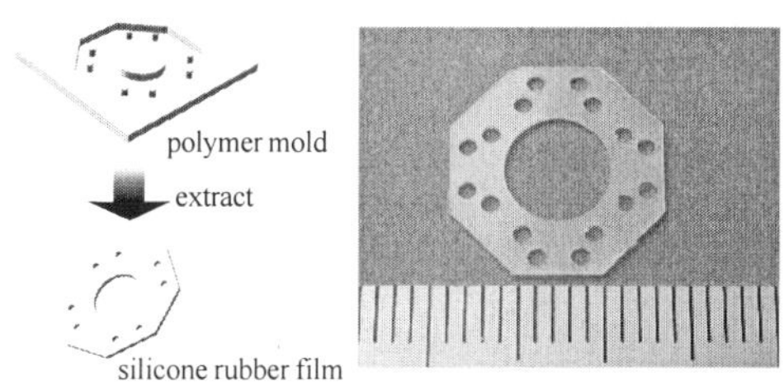

(a) (b)

Fig.4 Thin silicone rubber film formed from a poymer mold
(a) fabrication process (b) prototype

J.M. Ramsey and A. van den Berg (eds.), Micro Total Analysis Systems 2001, 377–378.

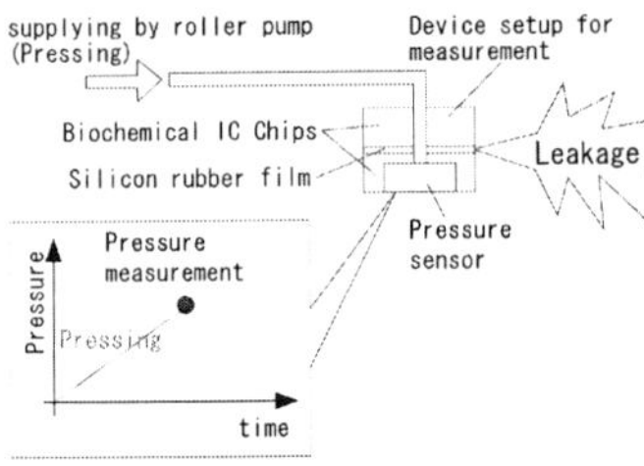

Fig.5 Experimental setup for leakage tests

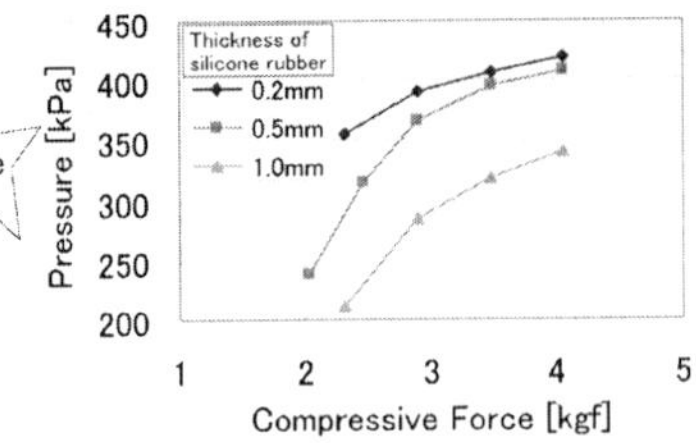

Fig.6 Proof pressure depending on compressive force by changing the spring force

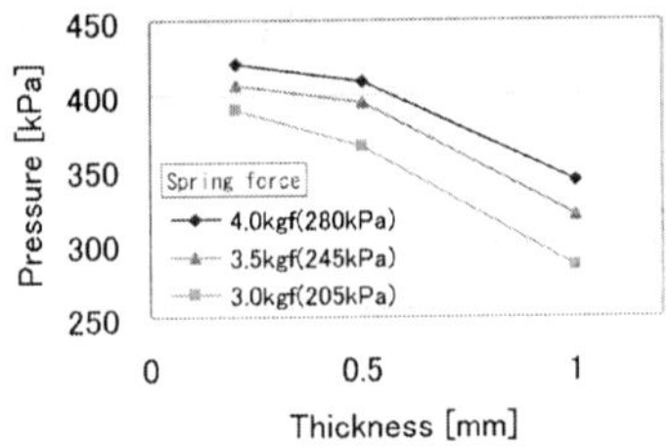

Fig.7 Leakage test for different thicknesses of silicone rubber films

2. Coupling Method

The proposed coupling method utilizes thin silicone rubber films and the holder unit. Fig.2 shows the holder unit to stack up the biochemical IC chips. Thin silicone rubber films are sandwiched between the chips. All of the chips are held together by the spring force, to prevent fluid leakage (Fig.3). Thin silicone rubber films, formed from a polymer mold, using microstereolithography, is fabricated easily in various shape (Fig.4). Moreover, multiple channels between the IC chips are adjustable precisely, since thin silicone rubber films and the IC chips are of the same shape (14x14 mm).

3. Characterization

Fig.5 shows the experimental setup for leakage tests. The connector chip, thin silicone rubber film and the chip with a pressure sensor are stacked up into the holder unit and held by spring force. Thin silicone rubber films with different thicknesses (200, 500, 1000 μm) are made. This proof pressure was measured continuously until leakage, while the device setup was pressurized by the roller pump. As shown in Fig.6, the maximum proof pressure increased, too while the spring force increased. The silicone films of 200 μm thickness can withstand water pressures up to 420 kPa without leakage. This proof pressure is sufficient for various biochemical experiments. The result presented in Fig.7 suggests, that maximum proof pressures increase while the thickness of silicone rubber films decrease. Thus, the dead volume can minimize between IC chips under high pressures. Moreover, the device setup under pressure at 250 kPa remained sealed for over 12 hrs.

4. Conclusions

The simple and effective coupling method for the Biochemical IC family has been developed and high hermetic seal has been verified by the experiment. The above mentioned proof pressures is the highest of the pumping pressures (25kPa) by the micro pump chip and the pressure for concentration (100kPa) by the micro concentration chip.

Basic features of this coupling method are summarized as follows;

1) A good seal under high pressures is achieved between micro fluidic devices.
2) A dead volume between micro fluidic devices can minimize by using thin silicone rubber films.
3) Multiple channels between micro fluidic devices are matched up easily.
4) Total micro chemical system is easy to reconstruct by combing each IC chips.
5) Thin silicone rubber films in various shape are fabricated easily with a polymer mold.

The above results demonstrate that the silicone rubber film coupling method is applicable to various micro fluidic systems.

References

[1] K.Ikuta, K.Hirowatari, T.Ogata, MEMS'94, pp.1-6
[2] N. J. Mourlas, D. Jaeggi, N. I. Maluf, and G. T. A. Kovacs, Transducers'99, pp.1988-1989
[3] Tze-Jung Yao, Sangwook Lee, Weileun Fang, Yu-Chong Tai, MEMS'00, pp.624-627
[4] K.Ikuta, T.Hasegawa, T.Adachi, and S.Maruo, MEMS'00, pp.739-744

MICRONEEDLE ARRAY FOR TRANSDERMAL BIO-FLUID SAMPLING AND DRUG DELIVERY

E. Mukerjee[1], S. D. Collins[1], R. L. Smith[1] and R. Isseroff, M.D. [2]
[1] MicroInstruments and Systems Laboratory (MISL)
[2]Dermatology, School of Medicine
University of CA, Davis, CA 95616, USA
(smith@ece.ucdavis.edu)

Abstract

This paper presents a microfabricated array of silicon microneedles, integrated with fluidic microchannels, designed for the extraction of extracellular fluid constituents from just beneath the skin. Biofluid sample extraction is obtained by means of capillary and extracellular fluid osmotic forces. Skin piercing by the needles has been validated by *in vivo* confocal micoscopy, electrical impedance measurements, and biopotential recording experiments. Sample delivery and collection from/to the microchannels is demonstrated *in vitro*.

Keywords: Microneedles, delivery, sampling, transdermal

1. Introduction

A number of microneedle constructs have been recently demonstrated, mostly for drug delivery applications. The silicon microneedle array presented here differs by size and shape, fabrication method, and intended application. It is fabricated in a manner that facilitates attachment to the skin, allows for a large degree of freedom in dimensions, density and placement of the needles, enables delivery of sample to/from microchambers or microchannels (where analysis or other functions can be carried out) and facilitates the integration of microelectronics for signal processing or control functions.

2. Design and Fabrication

The microneedles extend 100-200 μm normal to the plane of the silicon chip surface. Attachment to the skin is made by simply pressing the chip to the skin surface and affixing in place with a Bandaid® or strap. Samples are delivered to microchannels on the opposite side of the chip through a center bore hole, 6-20 μm in diameter. The microchannels are sealed with a glass cover, enabling visual inspection and optical interrogation. Sensors, e.g. microelectrode arrays, are integrated within the channels.

Each geometrical feature of the needle (shank length, hole size, hole position, sharpness, needle spacing) is governed by a separate, independent processing step, allowing for a high degree of structural variability to meet different application

J.M. Ramsey and A. van den Berg (eds.), Micro Total Analysis Systems 2001, 379–380.

requirements (e.g. blood sample retrieval versus extracellular fluid). Fabrication involves double sided, micromachining of silicon, using a combination of dry (DRIE) and wet (HNA) etching methods. Microchannels and bore hole are formed using DRIE. Then, glass is bonded to the silicon to seal the holes and channels. Next, rectangular pillars are etched on the reverse side, aligning the bore hole slightly off-center of the pillar. Finally, the pillars are isotropically etched to sharpen them to needle tips.

3. Experimental Results and Discussion

Figure 1(a) is an SEM photo of a 40 X 40 array of microneedles, ≈ 200 μm in height. Needle bore hole is visible in top view SEM of needle tip (Figure 1(b)). Figure 3 presents a series of images taken of a puncture wound made by a microneedle at increasing depths beneath the skin surface. Insertion of the needle tips into a fluid results in immediate drawing of sample into the microchannels by capillary action. Likewise, injection of fluids, e.g. drugs, through the needle tip is achievable, by diffusion, or by pumping. Experiments with transdermal extraction of extracellular fluid are underway.

(a) array of 200μm tall needles

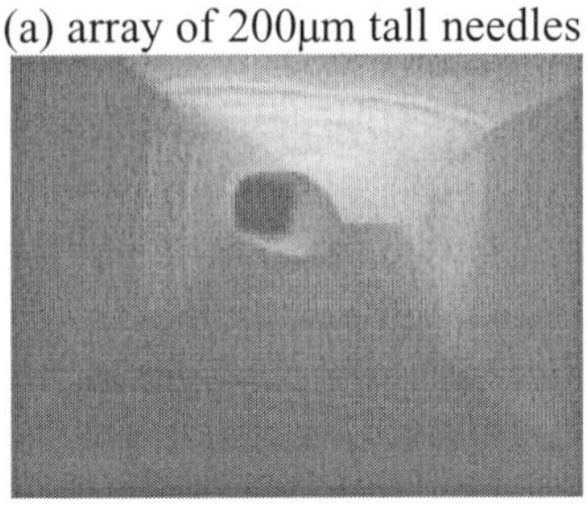

(b) top view of microneedle

Figure 1. Micrographs of microneedles.

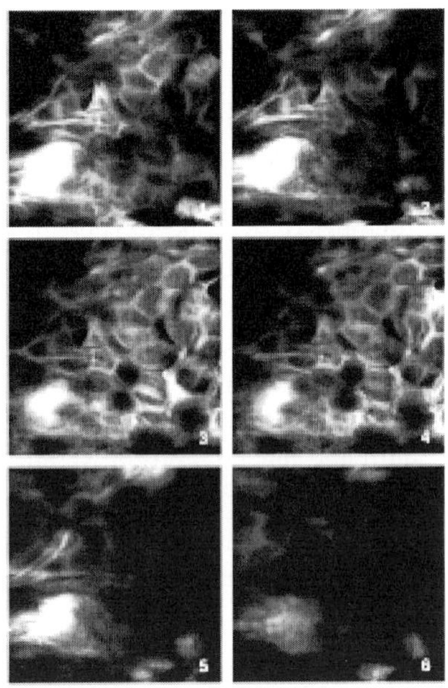

Figure 2. Microneedle puncture wound, imaged at increasing depths in the skin.

Acknowledgements

This work is funded by DARPA Contract N66001-00-C-8091. The *in vivo* images of needle pucture wounds in skin were collected using the Optiscan Stratum, courtesy of Dr Amanda Reese, Optiscan Pty Ltd.

FLUOROCARBON COATED MICROMACHINED GAS SAMPLING DEVICE

B.H.Timmer[1], J.G. Bomer[1], K.M. van Delft[1], R.P. Otjes[2], W. Olthuis[1], P. Bergveld[1], A. van den Berg[1]

[1]MESA[+] Research institute, University of Twente, Enschede, The Netherlands
[2]Energy Research Center of the Netherlands, ECN, Petten, The Netherlands

Abstract

A thin micromachined alternative for commercially available hydrophobic microporous membranes used in diffusion gas samplers has been realized. The thin membrane is made of Si_3N_4 and is etched porous. Underneath the membrane a channel is created. The membrane is rendered hydrophobic with a fluorocarbon coating. Bubble point and contact angle measurements have been performed on the membrane.

Keywords: FC-coated Si_3N_4 membrane, gas sampler, membrane covered channels

1. Introduction

The past few decades have shown an increasing interest in environmental issues such as acidification and climate change. This has resulted in a demand for detectors capable of measuring very low gas concentrations. At ECN a system is developed for the measurement of ambient ammonia concentrations with a detection limit of 6 ng/m^3 [1]. Miniaturization of this device and integration of its key components on a chip may lead to several advantages such as decrease in reagent and waste volumina and response time. A method to miniaturize the diffusion gas sampler is presented.

The gas sampler consists of two channels. A gas channel and a liquid channel, separated by a water repellent gas permeable microporous membrane like commercially available PTFE membrane. For fast response times the membrane thickness becomes the limiting factor when using very shallow channels.

2. Experimental

A micromachined alternative for the membrane is developed using a hydrophobic coated Si_3N_4 layer. A 1-µm thick LPCVD Si_3N_4 layer is deposited on a silicon substrate. Standard photolithography is used to define the pores. The nitride layer is etched anisotropically using dry RIE etching with a CHF_3/O_2 plasma. Subsequently the silicon underneath the nitride layer is etched isotropically using SF_6/N_2 gas. The etched channel has a depth of about 15 µm. A 150 µm deep gas channel and through holes are made in Pyrex glass using powderblasting [2]. The two wafers are anodically bonded together. The process scheme is given in figure 1.

The designed pores are 2*2 and 5*5 µm^2. A 3M FC722 fluorocarbon coating is applied by dip-coating. The resulting effective pore-size of the 5 µm pores in the membrane is measured using the bubble point pressure (ΔP [Pa]) equation (1) with r_p the pore radius (m), γ the liquid air/surface tension (N/m) and Θ the contact angle [3].

$$r_p = \frac{2\gamma}{\Delta P} \cdot cos\Theta \qquad (1)$$

J.M. Ramsey and A. van den Berg (eds.), Micro Total Analysis Systems 2001, 381–382.

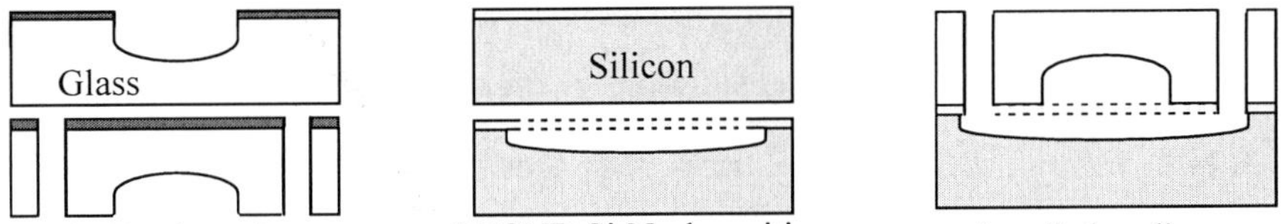

- Powderblasting of gas channels and through holes
- LPCVD Si_3N_4 deposition
- Etching of membrane and 15 µm deep channel underneath
- Anodic bonding

Figure 1. Process scheme

3. Results and Discussion

After rendering the membrane hydrophobic, the resulting effective pore-size is measured using the bubble point method. The bubble point pressure increases with subsequent application of hydrophobic coating layers, while gas can still easily pass through the membrane (figure 2). Two possible effects responsible for this increase are a decrease in effective pore-size and a higher membrane wetting pressure caused by the hydrophobicity of the coating. SEM pictures of the holes show no significant narrowing.

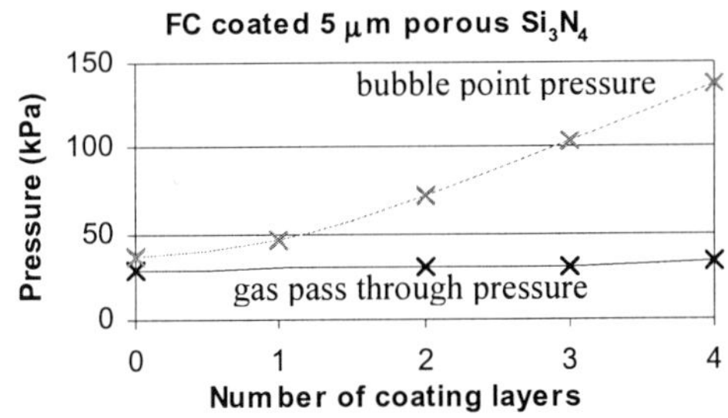

Figure 2. Bubble point measurements

Contact angle measurements have been performed to determine the hydrophobicity of the membrane before and after applying the coating (figure 3), and indicate a change from 56±5° to 109 ±3° after dipcoating. Gas detection response measurements are being performed using CO_2 plugs through the gas channel and Demi water flowing through the sample channel. The ion concentration change in the sample fluid is detected conductively.

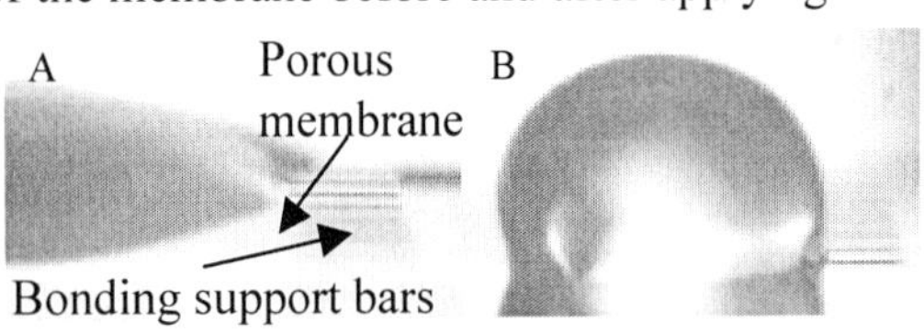

Figure 3. Contact angle before (A) and after (B) applying hydrophobic coating

4. Conclusions

A micromachined channel covered with a 1-µm thin hydrophobic coated membrane is fabricated. The water repellency can be adjusted by applying multiple layers of FC coating. The small membrane thickness makes it very suitable for fast gas sampling.

References

[1] J.W. Erisman, R. Otjes, A. Hensen, P. Jongejan, P. v.d. Bulk, A. Khlystov, H. Möls and S. Slanina, "*Instrument development and application in studies and monitoring of ambient ammonia*", Atmospheric Environment 35, pp 1913-1922, 2001

[2] H. Wensink, J.W.Berenschot, H.V. Jansen and M.C. Elwenspoek, "*High resolution powder blast micromachining*", Proceedings of the IEEE Micro Electro Mechanical Systems (MEMS), pp 769-774, 2000

[3] M. Mulder, "*Basic principles of membrane Technology*" Kluwer Academic Publishers, Dordrecht, The Netherlands, Second edition, Chapter IV.3, pp162-18, 1996

WELL-PLATE FORMATS AND MICROFLUIDICS - APPLICATIONS OF LAMINAR FLUID DIFFUSION INTERFACES TO HTP SCREENING

Bernhard H. Weigl*, Christopher J. Morris, Natasa Kesler, Pat Saltsman, Ron Bardell
Micronics, Inc., 8463 154th Avenue, N.E., Redmond, WA, 98052
*Corresponding author (e-mail: bernhardw@micronics.net)

Abstract
Microfluidic disposables are presented that are compatible with standard well plate readers and robotic filling systems. The disposables perform extractions and sample cleanup procedures using the diffusion-based separation principle.

Keywords: passive microfluidics, diffusion-based detection and separation, T-Sensor, H-Filter

1. Introduction

Laminar Fluid Diffusion Interfaces (LFDIs) are generated when two or more streams flow in parallel in a single microfluidic structure[1-5]. LFDIs have been applied to, among other examples, DNA desalting, the extraction of small proteins from whole blood samples, and the detection of various constituents in whole blood. Other LFDI-based microfluidic chips have been designed to allow the uniform and controlled exposure of cells to lysing agents, thus allowing the differentiation of cells by their sensitivity to specific agents in an on-chip cytometer coupled directly to the lysing structure.

Here, we present self-contained lab-on-a-chip arrays that are compatible with common well plate formats. These devices comprise both the well arrays (currently 96 or 384), as well as an underlying microfluidic circuitry that allows separations, complex chemical reactions, cell lysing, and other functions to be performed directly on the well plate. These plates are operated in a passive, hydrostatic-pressure mode and are compatible with standard well-plate robotic systems and readers.

2. Theory

For the design of a laminar diffusion interface card such as an H-Filter™, a one-dimensional diffusion model was used to model the concentration of a species as it diffuses into a receiver fluid. This helped yield an optimum diffusion time. Using an electrical-equivalent model (or "macro-model"), a microfluidic system was designed so that two gravity-driven fluids would flow on either side of their resulting laminar fluid diffusion interface for the right amount of time. The design was constrained so that well centers were in an array compatible with the standard 96-well plate format.

J.M. Ramsey and A. van den Berg (eds.), Micro Total Analysis Systems 2001, 383–384.

3. Results and discussion

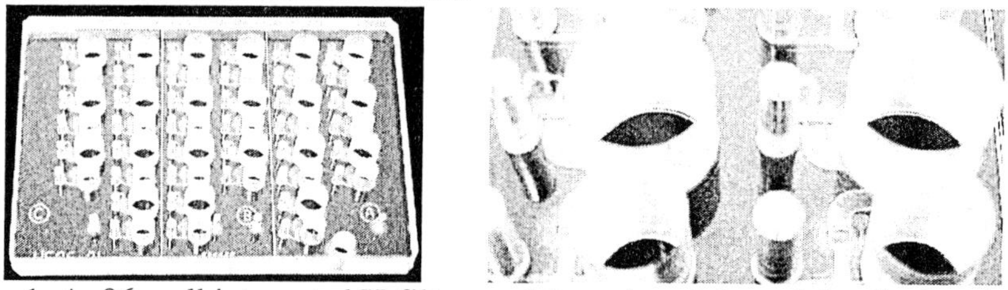

Figure 1. A 96-well integrated H-filter card (top view, left, and detail, showing the microfluidic elements underneath the wells, right)

The 96-well-compatible card shown in Fig. 1 was fabricated using Micronics' unique rapid prototyping process which yields low-cost, plastic, disposable microfluidic chips. Each group of four wells form an H-filter as shown in Fig. 2. Figure 2(b) shows a sample tank filled with a solution containing fluorescein and blue dextran (MW 2,000,000), and a receiver tank filled with a clear buffer. During the separation process, the two lower tanks are filling with a "product" of blue dextran and lower concentration of fluorescein, and a lighter-colored, extracted fluorescein "waste". This process can be used to either clean up a sample containing a lower molecular weight contaminant, or extract a lower molecular weight component from a sample also containing larger contaminants.

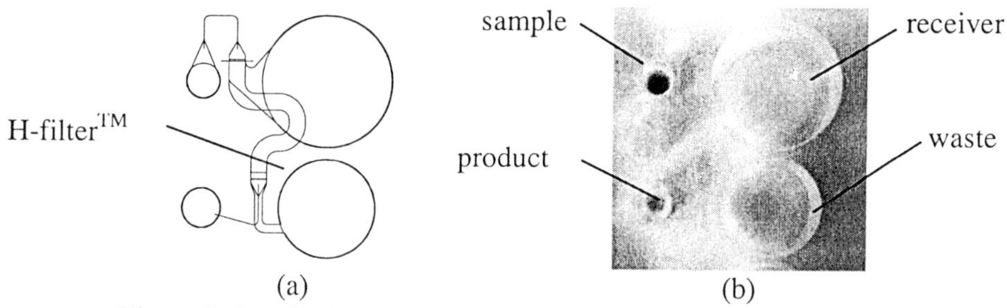

Figure 2. Detailed view of an H-filter on the card shown in Fig. 1.

References

1. B H Weigl and P. Yager, *Microfluidic diffusion-based separation and detection,* Science, Vol 283, pp 346, 1998.
2. Kricka, L.J., *Miniaturization of analytical systems.* Clin Chem 44, 2008-14, (1998).
3. Yager, P. et al., *Applying microfluidic chemical analytical systems to imperfect samples,* µTAS '98, D. J. Harrison and A. van den Berg, eds., Kluwer Academic Publishers, Dordrecht, 207-212 1998.
4. Kamholz, A. E. et al., *Quantitative Analysis of Molecular Interaction in a Microfluidic Channel: the T-Sensor,* Analyt. Chem., Vol. 71, part 23, 1999.
5. A. Hatch, A. Kamholz, K. Hawkins, M. Munson, E. Schilling, B. H. Weigl. P. *Yager, A rapid diffusion immunoassay in a T-Sensor, Nature Biotechnology, Vol 19, pp 461, 2001.*

ROOM TEMPERATURE LOW-COST UV-CURED ADHESIVE BONDING FOR MICROFLUIDIC BIOCHIPS

Ramgopal Divakar[1], David Butler[2], and Ian Papautsky[1,2]
Depts. of [1]Electrical and Computer Engineering and [2]Biomedical Engineering
University of Cincinnati, PO Box 210030, Cincinnati, OH 45221

Abstract

In this paper, we present a simple method for bonding of micromachined glass biochips. Our approach relies on medical-grade UV-cured adhesives to enable low-cost, room temperature, biocompatible, and optically transparent bonding. This method can be extended to other substrate types, such as silicon, ceramic, or plastic. Using this method, we fabricate biochips to dispense biochemical stimuli to tissue engineered patellar tendon implants in culture to improve their biomechanical properties.

Keywords: Adhesive bonding, microfluidic glass biochips, tissue engineering

1. Introduction

During the past decade, numerous biochemical analysis techniques have been demonstrated in microscale biochips [1]. Such biochips are often fabricated by etching trenches in a glass or plastic substrate and capping it with another substrate. Glass and many plastics offer excellent features, including biocompatibility and transparency for optical detection or microscopy. However, glass biochips can be expensive due to the need for highly flat, polished, high quality glass wafers. Although plastic biochips are less expensive, startup costs associated with injection molding or hot embossing can be significant. Fusion and adhesive bonding methods are typically used to form biochips [2,3]. Fusion bonding typically requires high temperatures for glass and extremely clean and smooth surfaces [2]. Adhesive bonding with polyimide or fluoropolymers (e.g., CYTOP) is not able to withstand high pressure and thus is not suitable for some microfluidic applications [3]. In this work, we present an approach that provides a low-cost alternative to the current fabrication methods that uses glass microscope cover slips as substrates and medical grade UV-cured adhesive for room temperature bonding.

2. Fabrication

Initially, glass cover slips (Fisher Scientific) are RCA cleaned and annealed at 590 °C. A layer of Cr/Au is used for masking in subsequent etching of trenches with 10:7:33 mixture of $HF/HNO_3/H_2O$. Input and output ports are drilled with diamond plated bits. Bonding is accomplished by spin coating the capping cover slip with 150 cP adhesive (Loctite). The cover slips are then aligned and the adhesive is cured by a 3-min exposure using an aligner. Residual adhesive is removed form the channel lumen with a 7:3 mixture of H_2SO_4/H_2O_2. The device is packaged with commercial HPLC connectors using 5,000 cP UV-cured adhesive. For our application of precisely

J.M. Ramsey and A. van den Berg (eds.), Micro Total Analysis Systems 2001, 385–386.

dispensing biochemical stimuli to tendon implants in culture, the bottom biochip surface is coated with a hydrophobic fluoropolymer to enhance droplet formation.

(a) (b)

Figure 1. (a) Packaged biochip (channel network is filled with dye). (b) SEM of a channel cross-section (250 μm in upper width and 90 μm in depth).

3. Results and Discussion

Biochips with channels of different dimensions (250-500 μm wide, 50-100 μm deep) have been fabricated (Figure 1). Although H_2SO_4/H_2O_2 solution dissolves adhesive on contact [4], it does not harm adhesive sandwiched between the cover slips and thus does not affect channel geometry. The bonded glass substrates can be separated by heating the biochip to 500 °C. Following an RCA clean, the substrates can be bonded again. Biochip packaging can be removed with acetone without affecting the device. The biochips can then be re-packaged with new fittings. Biochips fabricated with this simple process are very reliable and are produced with nearly 100% yield.

Figure 2. Silicone mold for implant incubation. Each mold has 4 wells (2 shown). The top well contains an implant between support posts.

In our application, tissue engineered tendon implants are grown in silicone molds (Figure 2) as described in [5]. Glass biochips are placed on top of individual wells of a silicone mold inside an incubator while an automated syringe pump is used as a pressure source to precisely deliver sub-microliter volumes of biochemical stimuli. Prior to use with implants, biochips are sterilized by autoclaving under standard conditions.

4. Conclusions

A simple bonding method for fabricating micromachined glass biochips was described. Our approach provides a low-cost alternative to the current biochip fabrication methods. The main advantages are: (a) low cost using glass cover slips, (b) rapid room temperature bonding, (c) biocompatible, and (d) optically transparent. The described bonding approach can be extended to other substrate types, such as silicon, ceramic, or plastic, if one of the substrates is UV transparent.

References

1. A. van den Berg, et al., *Micro Total Analysis Systems 2000*, Kluwer, 2000.
2. P. Simpson, et al., *Biomedical Microdevices*, vol. 1, pp. 7-25, 1998.
3. A. Han et al., *MEMS '00*, Miyazaki, Japan, 2000.
4. Loctite Corp., *Adhesives for Medical Device Assembly*, Rocky Hill, CT, 2000.
5. D. L. Butler and H. A. Awad, *Clinical Orthopedics*, vol. 367, pp. S324-S332, 1999.

A MICRO STENCILING PROCESS FOR WAFER SCALE METALLIZATION OF PLASTIC SUBSTRATES

Swomitra K. Mohanty, Mason Graff, Kumar Ravula, Arum Han and A. Bruno Frazier
School of Electrical and Computer Engineering, Georgia Institute of Technology
Atlanta, GA 30332-0250

Abstract

In this work, the fabrication of micro stencils for metal patterning on non-conventional substrates was demonstrated. Stencil feature sizes ranging from 10 μm to 270 μm with aspect ratios in the range of less than 1:1 to 15:1 were fabricated using ICP etching of silicon. The stenciling process was demonstrated for the deposition of Ti / Al, Ti / Cu, Ti / Au onto silicon, glass, polyethylene, polycarbonate, polystyrene, polyimide, PMMA, and other materials interesting for micro fluidic system development. The results demonstrated some dependency of the deposition rate on the stencil feature size and aspect ratio (stencil feature height : width). The characteristics of the deposited metal features are described. Adhesion studies were performed using the Scotch Tape test.

Keywords: Micro-stencil, microfluidic systems, MEMS packaging

1. Introduction

In recent years, the use of plastics as materials of construction for MEMS systems has increased dramatically. This increase has been driven primarily by the field of micro fluidics, particularly for biological and chemical analysis systems. Most of these systems would benefit from the ability to integrate electrical functionality into the plastic packaging containing the microfluidic elements. Since plastics are non-conventional substrate materials for common microelectronics processing technologies such as photolithography, a method is needed for defining fine line metallization onto substrates formed from non-conventional microelectronic materials. Stenciling has become a valuable tool in the fabrication and patterning of electronics in small parts. The most common type of stencil is the stainless steel stencil commonly used by the electronics packaging industry. These types of stencils are made using chemical etching or laser cutting. While these methods are well established and useful, they are limited to a minimum feature size of approximately 50 μm with aspect ratios (height:width) of 0.75. Micro scale shadow masks have been demonstrated for patterning applications with the first type being constructed by drilling a 100 μm hole in a silicon wafer [1]. More recently work has been done on micro stenciling to achieve features smaller than 50 μm. Researchers have demonstrated the use of chip-scale silicon stenciling for metal patterning onto integrated circuits [2].

2. Methods

The micro-stencil was fabricated using a 3-inch silicon wafer, 150 μm thick. The wafer was patterned with the desired stencil design using SU-8 negative photoresist. After patterning, the wafer was etched using the ICP Bosch etching process. SU-8 was found to be resilient in the Bosch process, thus allowing the original feature size of the design to be retained. It also served as a structural reinforcement for the stencil. The stenciling process was performed using electron-beam evaporation and sputtering. The 'stencil wafer' was clamped (SU-8 side down) over the substrate and patterned. After deposition, the stenciled feature was scanned to characterize the deposition profiles for arc and line stencil features. A comparison was made for each of the metals deposited looking at the aspect ratio versus normalized thickness. Titanium, Al, Cu, and Au were stenciled on to polyethylene, polystyrene, polycarbonate, PMMA, and polyimide. The titanium was used as an adhesion layer for each of the three

J.M. Ramsey and A. van den Berg (eds.), Micro Total Analysis Systems 2001, 387–388.

metals. An oxygen plasma ash was used to prepare the surface for patterning. The Scotch Tape test was used to check the adhesion of the metals to the various plastics.

3. Results

Using ICP etching we were able to fabricate a stencil mask with aspect ratios ranging from 0.55 to 15. Results of the surface scans demonstrated that 1:1 aspect ratio features allowed for slightly thicker depositions (~10%) than the 15:1 aspect ratio features. Figure 1a demonstrates the use of micro stenciling to obtain 100 μm and 250 μm arcs using e-beam deposition of Ti / Au. In Figure 1b, the micro stenciling process is used to pattern 10 μm and 25 μm lines using e-beam deposition of Ti / Au. The deposition profile did change with increasing stencil feature aspect ratio. Larger features with aspect ratios of less than 5:1 had a relatively flat profile, while the 10 μm features with aspect ratios of 15:1 showed a distinct roundness. Figure 3 shows a profile comparison between 120 μm and 10 μm features. Electron beam evaporation was found to be superior to sputtering for the micro stenciling process resulting in increased edge acuity and overall pattern definition without the requirement for complete contact between the stencil and the substrate. The micro stencils have been repeatedly used for over 50 runs and can be cleaned using the standard metal etchants used by the microelectronics industry.

Adhesion of the metals to the plastic substrates showed that polyethylene, polycarbonate, polystyrene, and polyimide had excellent adhesion to all metals deposited. Each sample withstood the Scotch Tape test. PMMA did not perform as well. Adhesion on PMMA was substantial and difficult to rub off, however some parts did lose adhesion with the application of scotch tape.

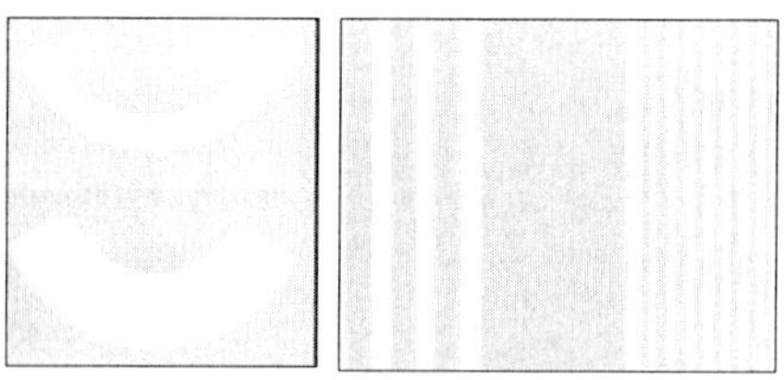

(a) (b)

Figure 1. Image of (a) micro stenciled arcs and (b) lines.

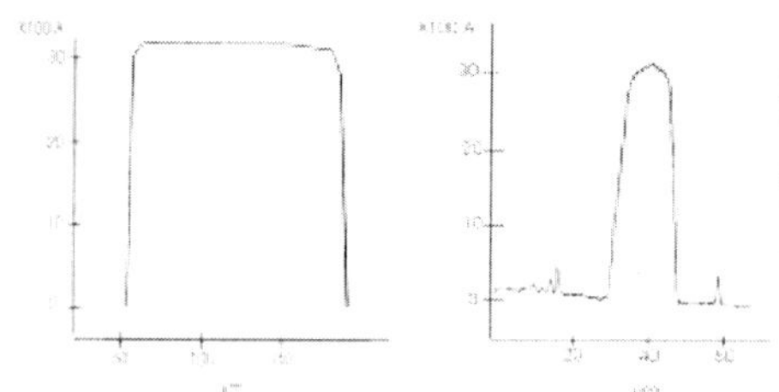

Figure 2. Deposition profiles of 120 micron (left) and 10 micron (right) stenciled features.

4. Conclusion

We have demonstrated high aspect ratio micro stenciling using a micromachined silicon stencil. Fabrication and patterning with the stencil is simple, making this an attractive method for integrating electronic features onto plastic parts. It could also serve as an alternative method for metal patterning in more traditional applications in the electronics packaging industry for feature sizes of less than 10 μm.

References

[1] Dhler G H, Hasnain G, and Miller J N, Appl Phys. Lett. pp. 704-6, 49 (1986).
[2] A. Tixier, Y. Mita, J. P. Gouy and H. Fujita, Journal of Micromech. and Microeng., pp. 157-162, 10 (2000).

SELF- ASSEMBLED MICROLENSE ARRAYS ON MICROFLUIDIC SYSTEMS BY SELECTIVE PLASMA SURFACE MODIFICATION

Jeonggi Seo and Luke P. Lee
Berkeley Sensor & Actuator Center, Department of Bioengineering,
University of California at Berkeley, CA 94720, USA
(seojohn@uclink.berkeley.edu)

Abstract

Self-assembled microlens arrays on microfluidic system are fabricated by selective plasma surface modification of patterned photoresist. Spin-coated UV-curable polymer forms microlens only on patterned photoresist due to the different surface energy between the substrate and the surface of photoresist that has been modified by the control of plasma gas.

Keywords: Plasma surface modification, self-assembled microlenses, wetting

1. Introduction

Integration of micro-optical components is essential to the development of an advanced micro-Total Analysis System (μTAS). Microlenses were fabricated by the reflow method of photoresist, or etching transfer method on substrates by ion beam or reactive ion etching. Recently, several authors have demonstrated the fabrication of microlenses using hydrophobic effect [1-2]. However, their methods for patterning hydrophobic domains have drawbacks since they require extra process steps including Teflon® thin film depositions, etching, and annealing. In this paper, we present a new simple method of fabricating microlenses by selective plasma surface modifications on patterned photoresist, which is compatible with IC fabrication process without using multiple processing steps of Teflon® thin film.

2. Microlens by selective plasma surface modification

Plasma surface modification can be used to fabricate microlenses using dewetting/wetting effects. Polymer materials of microlenses attach only to hydrophilic surface areas while repelling from hydrophobic surface areas. Thus, it is important to create highly selective method of surface modification for wetting and dewetting effects on a substrate. The templates for microlenses are prepared by developing the photoresist layer on a glass substrate as shown in Fig. 1. The templates are aligned to the microfluidic channels that are on the other side of the substrate. The consecutive O_2 and CF_4 modification change the undeveloped photoresist layer to hydrophobic surfaces and the developed areas to hydrophilic surfaces. After the surface modification, an UV-curable polymer is coated uniformly on the substrate. The different surface energy

J.M. Ramsey and A. van den Berg (eds.), Micro Total Analysis Systems 2001, 389–390.

between the developed areas and undeveloped areas forms self-assembled microlens arrays.

3. Experimental results

We characterized the contact angles of photoresist layers, Si wafers, and glass wafers before and after hard baking of the photoresist layer and after the modification with three different plasma gases: O_2, SF_6, and CF_4 as shown in Fig. 2. Figure 2 shows the effects of two plasma combinations, which cause the selective surface modification. After the O_2 modification, the original contact angles on both of the photoresist layer and the Si substrate decrease. However, after the CF_4 and SF_6 modification succeeding the O_2 modification, the contact angle on the photoresist layer increases to about 120°. Thus, the difference between the contact angles on the photoresist layer and the Si substrate can be around 110°. UV light is applied to cure the UV-polymer after the formation of microlens as shown in Fig. 3.

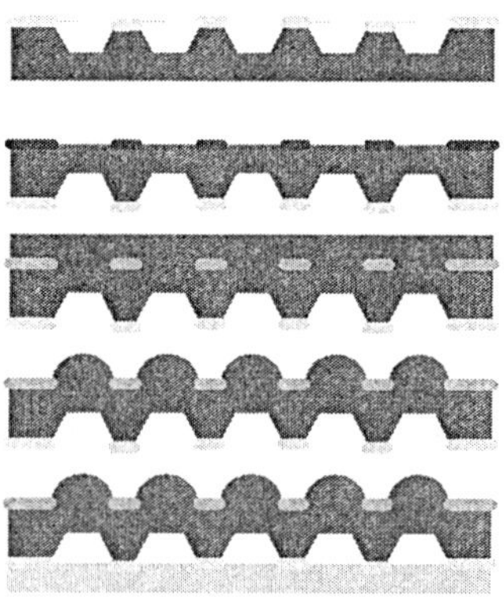

Figure 1. Fabrication steps of self-assembled lens array on microfluidics

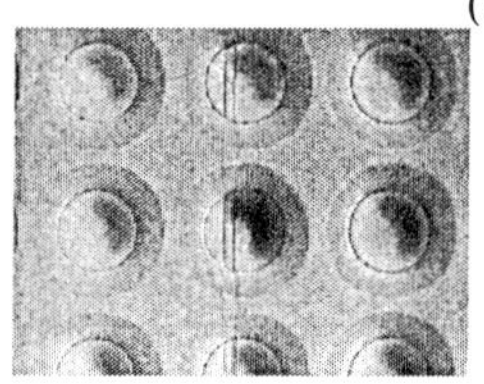

Figure 2. Self-assembled microlens array

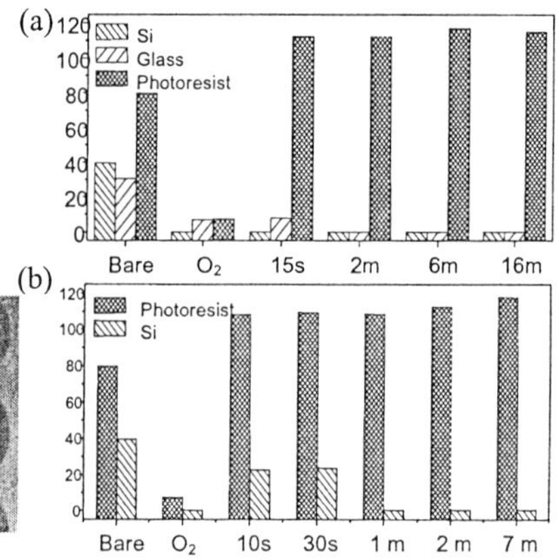

Figure 3. Contact angle measurements after plasma surface modification: (a) O_2 and CF_4 and (b) O_2 and SF_6

4. Conclusions

Self-assembled microlens arrays on a microfluidic system are fabricated by the selective plasma surface modification of patterned photoresist. Selectivity of dewetting/wetting properties is controlled by plasma surface modification. This method is compatible with IC fabrication process without using Teflon® thin film and ideal for advanced integrated optical microfluidic systems.

References

1. D.M. Hartmann, O. Kibar, and S.C. Esener, *Characterization of a polymer microlens fabricated by use of the hydrophobic effect*, Opt. Lett., 25(13), pp. 975-977, (2000).
2. L. P. Lee, *Key elements of biomedical polymer optoelectromechanical systems*, Ph.D. dissertation, Univ. of California at Berkeley, (2000).

DUAL EXPOSURE GLASS LAYER SUSPENDED STRUCTURES (DEGLASS): A NOVEL FABRICATION PROCESS FOR GLASS MICRO-FLUIDIC NANOSTRUCTURES ON PLANAR SUBSTRATES

D. M. Tanenbaum[†], Y. Chen[2], and H.G. Craighead[2]
[1]Department of Physics, Pomona College, Claremont, CA 91711
[2] Applied and Engineering Physics, Cornell University, Ithaca, NY 14853
(email: dtanenbaum@pomona.edu)

Keywords: DEGLaSS, hydrogen silsesquioxane (HSQ), electron beam lithography

1. Introduction

Microfluidic systems that can be easily integrated with microelectronics have the potential to generate a variety of new sensor and separation technologies. Typical fabrication techniques require a combination of lithography, dry etching, multi-layer thin film deposition, thermal or anodic bonding, and specialized materials with high etching selectivities. We propose a new approach to fabrication of microfluidic systems.

2. Theory

The DEGLaSS process uses a single layer of spin on glass (SOG) material with two negative tone exposures at two different exposure energies to define the channel ceilings and channel walls respectively. A low energy exposure has a very short penetration depth, similar to a top surface imaging process, and defines the channel ceiling. A second exposure is performed with a higher energy that penetrates through the entire glass layer thickness, resulting in channel walls that support the ceiling. There is no processing between the exposures, and the order of the exposures does not matter. Both patterns are developed simultaneously in TMAH which removes all the unexposed material, including the material inside open ended channels. Samples can be transferred through an ethanol bath to a critical point CO_2 dryer to prevent ceiling collapse due to surface tension in the drying of large open cavities. After development the three-dimensional cross-linked SOG structures can be densified converting them to an amorphous glass structure. Any two dimensional pattern can be used for both the support structure and the ceiling structure, as long as the ceiling structure does not contain isolated unsupported regions which will either wash away or fall flat to the substrate. Thus ceiling structures can be patterned with irrigation holes to enhance development, or supported underneath by pillar arrays to enhance structural rigidity or to create artificial sieves.

3. Experimental Results

J.M. Ramsey and A. van den Berg (eds.), Micro Total Analysis Systems 2001, 391–392.

We have demonstrated the DEGLaSS process using low energy electron beam exposures of Hydrogen Silsesquioxane (HSQ). HSQ was chosen for its high resolution and sensitivity in negative-tone electron beam lithography.[1] Variable energy electron beam lithography allows control of the electron penetration depth over three orders of magnitude, from 10 nm to 10 μm with a single exposure tool. We have demonstrated the process in HSQ films with a thickness between 200 nm and 1.2 μm. We have used 1-2 keV electrons to define our channel ceilings, and 3-10 keV electrons to define our channel walls. The 1 (2) keV exposures produce suspended ceilings 40 (110) nm-thick. Channels have been demonstrated with lengths up to 150 μm without irrigation holes with reasonable development times. The channel ceiling and walls can be fabricated with a resolution near 100 nm depending upon electron energies channel heights. The HSQ structures can be converted to SiO_2 using thermal annealing in an oxygen ambient or an exposure to an oxygen plasma. The channels are optically transparent, and demonstrations of dye molecules flowing in the channels will be presented.

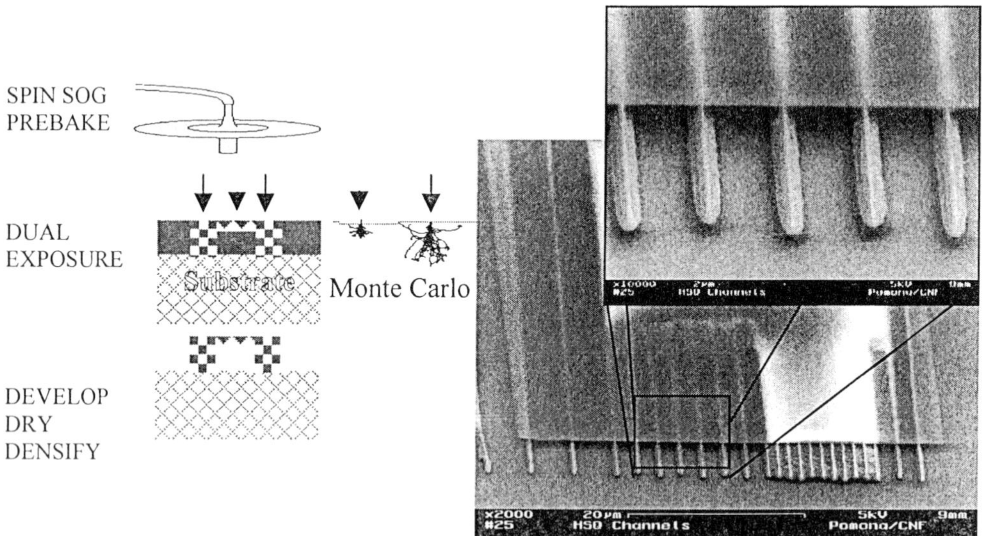

Figure 1. DEGLaSS Process Flow and Fabricated Channels.

4. Conclusions

The DEGLaSS process can be easily expanded to include more exposure energies allowing more intricate internal structures to be fabricated inside channels, such as those needed for separation technologies, flow sensors, valves, etc. The DEGLaSS process eliminates the need to do reactive ion etching or anodic bonding and can be performed on any planar substrate, even above existing integrated circuits, since the ceiling and walls are fabricated from the same SOG in an additive process.

NEW DEVELOPMENTS IN BULK MICROMACHINING BY POWDER BLASTING

Henk Wensink, Miko C. Elwenspoek
MESA+ Research Institute, University of Twente, P.O.Box 217, 7500 AE Enschede, The Netherlands, H.Wensink@tn.utwente.nl

Abstract

Powder blasting is a fast, inexpensive and accurate directional etch technique for brittle materials like glass, silicon and ceramics. We have developed several improvements for a successful implementation of powder blasting as a micromachining technique.

Keywords: Powder Blasting, Glass Micromachining, Roughness.

1. General Capabilities of Powder Blasting

Powder Blasting is a fast process, the etch rate in silicon and Pyrex is about 25 μm/min using one nozzle in our set-up. The relatively large powder particles (5-30μm), which are responsible for the high etch rates, restrict the minimum feature size to about 25μm. For small (<100μm) structures, aspect ratios of 1.5-2.5 can be realized (depending on the mask material). Cleaning the sample and e.g. direct bonding after blasting is not a problem. Several successful μfluidic devices have already been fabricated using powder blasting [1, 2, 3].

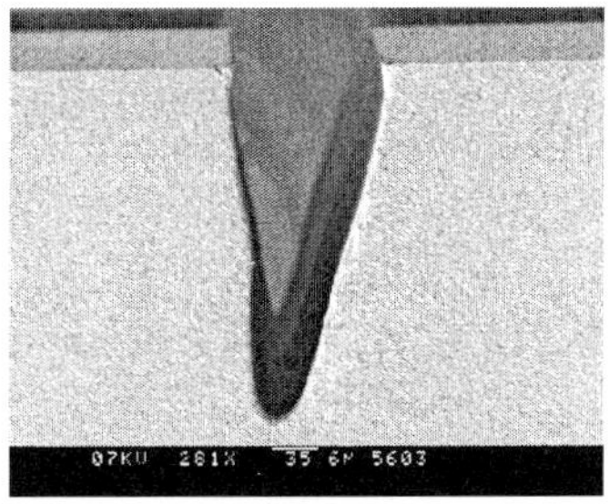

Figure 1. High aspect ratio channel in Pyrex.

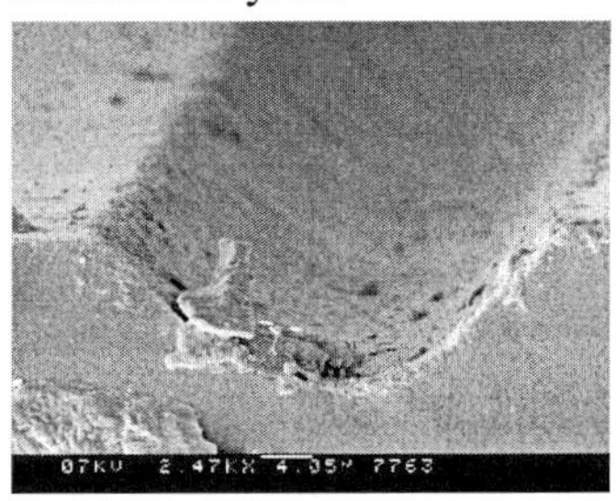

Figure 2. A 30μm channel in Pyrex.

2. Mask Materials

The accuracy of powder blasting also depends on the type of mask that is used. The powder blasting resist foil Ordyl BF410 can be used for standard applications. We developed electroplated copper as a new mask material [4], which allows us to the increase aspect ratio (Figure 1). It is also suitable for feature sizes of less than 50 μm (Figure 2).

3. Blast Lag

The cross-section in Figure 3 shows the general rounded v-shape of the powder blasted channels. This results in what we call the blast lag: wide channels become deeper compared to smaller channels. By using smaller particles the effective channel width become wider since they can impact more closely to the mask border. The sidewalls become steeper and the blast lag is decreased [5].

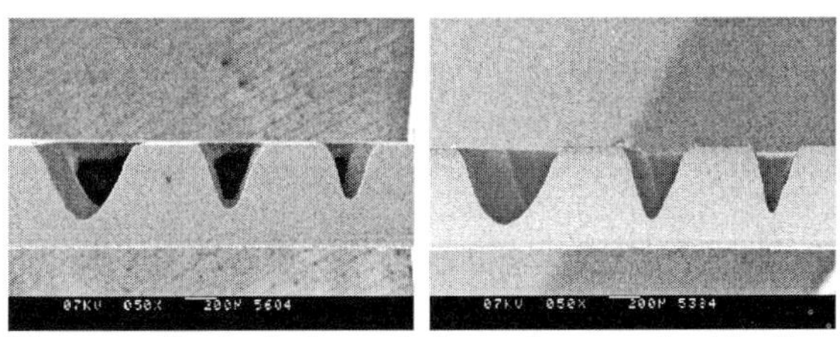

Figure 3. Typical channel shape for powder blasting, with decreased blast lag in right picture.

J.M. Ramsey and A. van den Berg (eds.), Micro Total Analysis Systems 2001, 393–394.

4. Roughness

Powder blasted surfaces are rough due to the nature of the etch process. Rough micro channels have shown to increase μfluidic dispersion in certain cases [6], and also to increase the electro-osmotic flow. It is therefore important to be able to tune the surface roughness. Changing the kinetic energy (through the particle size) of the powder particles can already influence this roughness (Table 1). The roughness has also been successfully manipulated by an HF-etch and a high temperature anneal step after powder blasting [7].

Table 1 Changing roughness

Treatment	Roughness [μm]
9 μm powder	1.2 ± 0.2
+ HF etch	2.2 ± 0.1
+750°C anneal	0.58 ± 0.1
29 μm powder	2.5 ± 0.6
+ HF etch	3.9 ± 0.5
+750°C anneal	1.8 ± 0.4

5. Conclusions

Powder blasting has become a mature micromachining technique for glass and silicon. It fits very well between the common micromachining techniques due to its lithographic masking abilities, process similarities and compatibility. Aspect ratio and roughness of powder blasted channels can be optimized by process settings. The capabilities of powder blasting allow new ways of micromachining materials that otherwise are difficult to etch.

Acknowledgements

This work is part of the research programme of the 'Stichting voor Fundamenteel Onderzoek der Materie (FOM)', which is financially supported by the 'Nederlandse Organisatie voor Wetenschappelijk Onderzoek (NWO)'.

References

1. T. T. Veenstra, J. W. Berenschot, J. G. E. Gardeniers, R. G. P. Sanders, M. C. Elwenspoek, and A. van den Berg, *Use of Selective Anodic Bonding to Create Micropump Chambers with Virtually No Dead Volume*, Journal of The Electrochemical Society, 148, (2001) pp. G68-G72.
2. S. Schlautmann, H. Wensink, R.B.M. Schasfoort, M.C. Elwenspoek, A.van den Berg, *Powder-blasting technology as a powerful tool for micro-fabrication of CE-chips with integrated conductivity sensors*, Accepted for J. Micromech. Microeng.
3. R.E. Oosterbroek, M. Goedbloed, A. van den Berg, *Optimization and realization of electro-osmotically driven microsynthesis systems*, published at this conference.
4 H.Wensink, H.V.Jansen, J.W.Berenschot, M.C. Elwenspoek, *Mask materials for powder blasting*, J. Micromech. Microeng. 10 (2000) pp.175-180
5 H.Wensink, J.W.Berenschot, H.V.Jansen, M.C.Elwenspoek, *High Resolution Powder Blast Micromachining* Proc. 13th Int. workshop on Micro Electro Mechanical Systems (MEMS2000), Miyazaki, Japan, january 23-27, pp.769-774
6 M.T. Blom, E. F. Hasselbrink, H. Wensink, A. van den Berg, *Solute Dispersion By Electroosmotic Flow In Nonuniform Microfluidic Channels*, published at this conference.
7 H.Wensink, S.Schlautmann, M.H. Goedbloed, M.C.Elwenspoek, *Fine Tuning the Surface Roughness of Powder Blasted Glass Surfaces*, Proc. Sensor Technology Conference 2001, Enschede, The Netherlands, may 14-15, pp.101-106

TEMPERATURE-ADDRESSABLE MICROELECTRODE ARRAY

Haesik Yang, Chang-Auck Choi, Chi-Hoon Jun, Youn Tae Kim
Contributor, Microsytem Team, ETRI, 161 Kajong-Dong, Yusong-Gu, Taejon 305-350, Korea (hsyang@etri.re.kr)

Abstract

A new temperature-addressable microelectrode array with low power consumption has been developed using surface micromachining. It consists of 96 temperature-controllable microelectrodes, and each microelectrode contains a vacuum-sealed cavity, an n^+-doped poly-Si microheater, and an electrically isolated electrode. The vacuum-sealed cavity was employed for thermal isolation and low power consumption and was fabricated by using the DECTOR (DEep Cavity using Trench Oxidation and Release) process. The performance of the microelectrode was evaluated electrochemically. The microelectrode in electrolytes showed better heating/cooling characteristics.

Keywords: Microelectrode array, vacuum-sealed cavity, microheater, surface micromachining

1. Introduction

Biological reactions and interactions depend highly on temperature. To control them, many temperature-controllable microsystems have been developed. In this study, a new temperature-addressable microelectrode array with low power consumption has been developed using surface micromachining.

2. Experimental

A Si_3N_4 and a TEOS oxide films were deposited on a 5″ silicon wafer by LPCVD, followed by a dry-etching process to define trench patterns. A Si trench array with 5 μm depth was formed and doped with $POCl_3$. After removal of TEOS oxide, the Si trench array was oxidized thermally (Fig. 1a). Si_3N_4 stripping was followed by LTO deposition and patterning. A thick polysilicon film was deposited, and then etch hole was formed after post-annealing. The sacrificial oxides which was filled in the cavity region was removed by HF wet etching through the etch window. Thick poly-Si supporter was made thinner by dry etching. LTO deposition under 0.24 Torr made the cavity sealed in a vacuum state. For an n^+ poly-Si microheater formation on the vacuum-sealed cavity, poly-Si deposition was followed by $POCl_3$ doping and patterning. Another LTO film was deposited to isolate microheater electrically from electrode. After the formation of interconnection lines, electrode area, and pads with Pt, a passivation film was deposited. Finally, the electrode area was uncovered by dry etching. Figure 1b shows a side view of a temperature-controllable microelectrode. The diameters of electrode and vacuum-sealed cavity are 100 and 200 μm, respectively.

J.M. Ramsey and A. van den Berg (eds.), Micro Total Analysis Systems 2001, 395–396.

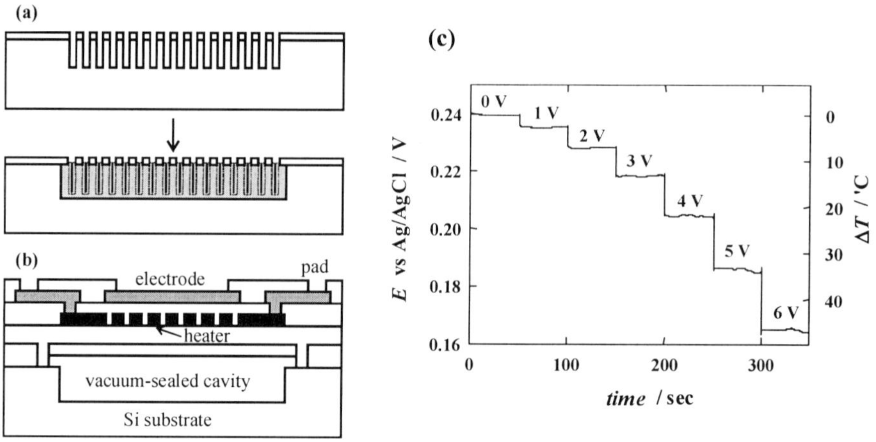

Figure 1. (a) Thermal oxidation of Si trench array. (b) Side view of a microfabricated temperature-controllable microelectrode. (c) Open circuit potential change of a temperature-controllable microelectrode in 5 mM $K_3Fe(CN)_6$ + 5 mM $K_4Fe(CN)_6$ / 0.5 M KNO_3 (R_{heater} = 1.6 kΩ).

4. Results and discussion

An increment of electrode surface temperature causes thermally induced heat-convection and redox potential shift. To measure the temperature of electrode surface indirectly, open circuit potential of a temperature-controllable microelectrode was traced in a solution of ferricyanide/ferrocyanide redox couple (Fig. 1c). As the applied voltage of microheater increases, open circuit potential decreases. A decrement of 1.6 mV corresponds to an increment of 1 °C of electrode temperature. At 6 V, the temperature increased by ca. 47 °C and the current measured at microheater is 3.6 mA. The power consumption was about 23 mW, and this value is very low compared to the case of no vacuum-sealed cavity where temperature change is less than 20 °C for even 140 mW. The open circuit potential is stabilized within several seconds after every voltage step. Moreover, the open circuit potential returns to the original potential and is stabilized in less than 10 seconds after reverse voltage step to any voltage, manifesting the satisfactory heating/cooling performance of the resulting microelectrode array.

References

1. C.-H. Jun, C.-A. Choi, W.-I. Jang, Y. T. Kim, *Silicon surface micromachining of a deep vacuum cavity structure and its application to a microflow sensor*, Proc. of the SPIE on micromachining and microfabrication process technology VI, vol. 4174, pp.164-171, (2000).

A DISPOSABLE MICROPUMP CONCEPT FOR SMALL AND CONSTANT FLOW RATES

Carlo S. Effenhauser, Herbert Harttig, and Peter Krämer
Roche Diagnostics GmbH, Roche Diabetes Care, New Technologies DR-N
Sandhoferstr. 116, D-68305 Mannheim, Germany
(carlo.effenhauser@roche.com)

Abstract

An inexpensive and simple pumping principle is described that is capable of delivering both small and constant flow rates (10-1000 nl/min) over a longer period of time (days to weeks).

Keywords: Micropump, continuous monitoring

1. Introduction

Continuous monitoring of glucose levels represents a very promising means to reduce the risk of late complications and improve the quality of life of diabetic patients. Both implantable sensors (1) and microdialysis based systems (2) are current avenues of research. In microdialysis based systems, due to small catheter cross section and small flow rates employed (high glucose recovery), typical volume flow rates of 0.1-1 μl/min have to be delivered over several days to a few weeks. Major pump requirements in such a system are small and constant flow rates, small size and low weight, low energy consumption, simple and reliable construction and operation, very inexpensive manufacturing process enabling disposability, compatibility with planar system architectures, and pump sterilization. Even though an impressive arsenal of μ-pump concepts has been described in the literature (3), many of them have been optimized to high flow rates and cannot be employed as a disposable due to fabrication costs.

2. Pump Concept and Results

For that reason, we have developed a novel pump concept that is based on controlled evaporation of a liquid through a membrane into a gas space containing a sorption agent (see Figure 1, left). As long as the sorption agent keeps the vapor pressure in the gas phase below saturation, fluid evaporated from the membrane will be replaced by capillary forces inducing flow through the catheter. The dimensions of the Plexiglas pump body of about 3x2x2 cm were chosen to allow convenient experimentation. In a feasibility study, a total volume of 300 μl of Ringer's solution has been continuously pumped over a period of 6 days (constant average flow rate: 35 nl/min, see Figure 1, right). The maximum transportable fluid volume is limited by sorption capacity and amount of the desiccant. The decrease of the flow rate after 140 h exhibited in Figure 1 reflects the saturation of the sorption agent in a quantitative way.

3. Summary and Conclusions

J.M. Ramsey and A. van den Berg (eds.), Micro Total Analysis Systems 2001, 397–398.

Main features of the pump are very low fabrication costs, high reliability (no moving parts), and suitability for integration in planar system architectures. Furthermore, if applied in a patient monitoring system attached to the skin surface, no external energy source is required, as the energy will be taken from the patient at a fairly constant temperature. Continuous flow can be achieved in contrast to many other pump mechanisms leading to discontinuous, pulse-type flow. The current construction of the pump, though not yet "truly" microtechnology-based, exhibits a significant potential for further miniaturisation and integration into complex microfluidic systems.

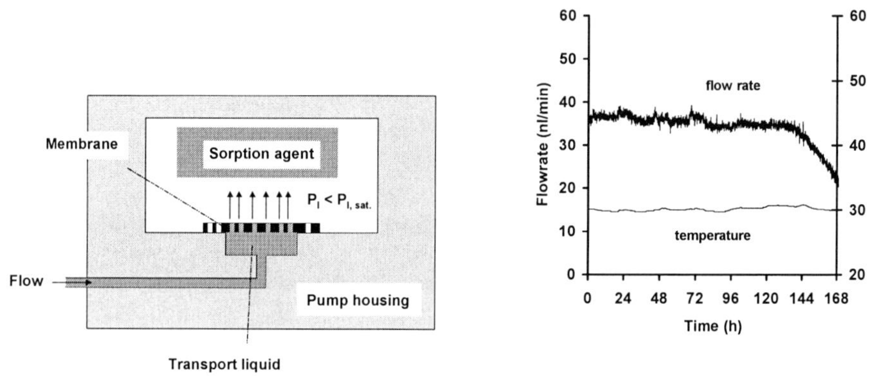

Figure 1: Left: Principle set-up of a membrane evaporation based pump. The hydrophilic membrane was fixed over an opening in the pump body by adhesive tape containing a punched hole (1 mm). The chamber above the membrane was filled with a desiccant (2 g molecular sieve MS 518, Grace Davison, sorption capacity 17 % w/w). Right: Flow rate of the pump as a function of time (raw data). The flow rate was monitored by a gravimetric technique. A BGS 65 polysulfone membrane (USF Filtration and Separations Group) was chosen. Also shown is the temperature in the vicinity of the pump.

References

1. 1. A. Heller, Annual Review of Biomedical Engineering 1 (1999) 153-175
2. H.L. Lutgers et al., The Netherlands Journal of Medicine 57 (2000) 7-12
3. S. Shoji, Topics in Current Chemistry 194 (1998) 163-188

INDIRECT ELECTRO-OSMOTIC PUMPING FOR DIRECT SAMPLING FROM BIOREACTORS

Rosanne M. Guijt[1], Jan Lichtenberg[2], Erik Baltussen[1], Elisabeth Verpoorte[2], Nico F. de Rooij[2] and Gijs W.K. van Dedem[1]

[1]Kluyver Institute for Biotechnology, Delft University of Technology, Julianalaan 67, 2628 BC Delft, the Netherlands
[2] SAMLAB, Institute of Microtechnology, University of Neuchâtel, Rue Jaquet-Droz 1, 2007 Neuchâtel, Switzerland

Abstract
In this contribution, electro-osmotically induced, pressure-driven flow along a micromachined channel is presented, using two adjacent channels at either end. Imperfect bonding in the narrow glass wall between the main channel and each side channel allows electrical contact. A voltage applied between the side channels results in an electric field in the main channel, thereby enabling capillary electrophoresis (CE) and generating an electro-osmotic flow (EOF). Electrical displacement of liquid in one section of the channel induces hydraulic fluid flow in the other, field-free sections of the channel, a phenomenon we call indirect electro-osmotic pumping (IEOP).

Keywords: Indirect electro-osmotic pumping, field-free sampling

1. Introduction
Microfabricated analytical devices hold great promise for bio(chemical) analysis. In the development of a μTAS, manipulation of sample and reagents within a micro-capillary network still represents a considerable challenge. Besides fluidic transport, electro-osmotic flow (EOF) can also be used as a driving force for induced electro-osmotic pumping (IEOP) [1,2]. In this contribution, a microdevice using adjacent channels for IEOP, and for CE without electrodes in the sample or buffer reservoirs, is presented.

2. Theory
Bonding defects in narrow glass walls between parallel channels allow the formation of a porous membrane in order to establish electrical contact. Application of an electrical potential over the two side channels, both placed parallel to a main channel, generates an electric field in this main channel. The EOF generates a pressure difference that is compensated by flow in the field-free sections of the main channel. Flow in the field-free zone thus results from IEOP, while flow in the electric field between both side channels is based on both electrophoretic migration and EOF.

3. Experimental
Application of a large voltage drop over the thin glass wall causes electrical breakdown of this insulating layer, and establishment of electrical contact as a result. All channels are filled with 1M NaCl, and the current is limited to control the break-through process.

J.M. Ramsey and A. van den Berg (eds.), Micro Total Analysis Systems 2001, 399–400.

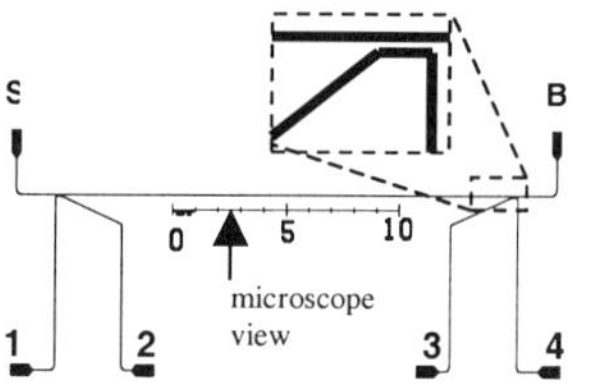

Fig. 1 Layout of the single channel device, dimensions 15x30 mm. The glass wall between the adjacent channels is 9 µm. S-B is 27 mm.

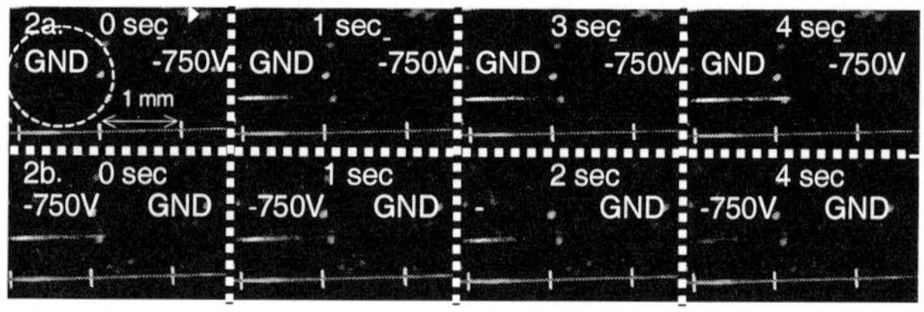

Fig. 2 Reservoir S at the left, reservoir B at the right, laser spot at the circle in the left image of figure 2a., microscope position as given in Fig.1. **2a.** F is drawn from S through the field-free channel using IEOP. **b.** Inverted polarity, F migrates back towards S.

For IEOP experiments, both side channels were filled with 1 M NaCl, the main channel S-B and B with 32 mM carbonate buffer pH 9, and S with fluorescein in buffer (F) (Fig.1). Voltages of up to –800 V were applied to reservoir 4 (1 was grounded), and the migration speed of F in the field-free section was measured. In another experiment, F was pumped back and forth; this was visualized using a microscope and video camera. A laser beam was focused into the channel and a potential was applied (Fig. 2).

4. Results and discussion

The migration speed of fluorescein in the field-free zone is linearly related to the field applied (0-400 V/cm) between reservoirs 1 and 4 (R^2=0.98). Leakage of F into the side channels is not observed. EOF as driving force for the pumping is confirmed by both the increased pumping speed at higher applied fields, and the inverted transport direction with inverted polarity. Images of the migration of F through the main channel are given in Figure 2. Preliminary observations indicate that injection and electrophoretic separation of a sample is possible in devices with cross injection using IEOP.

5. Conclusion

Electrical contact between adjacent channels is caused by bonding defects, and can be extended by application of a break-through voltage. This allows the generation of an electric field in a glass channel by indirect electrical contact over the leaky junction. The EOF in the electric field also induces a pressure-driven liquid flow in the field-free sections of the channel by means of IEOP.

Acknowledgements

RG acknowledges the Dutch Foundation for Science and Technology (STW) for financial support in the framework of the BIOMAS project.

1. C.T. Culbertson, R.S Ramsey and J.M. Ramsey, *Electro-Osmotically Induced Hydraulic Pumping on Microchips: Differential Ion Transport*, Anal.Chem. 2000, 72, 2285-2291
2. T.E. McKnight, C.T. Culbertson, S.C. Jacobson and J.M. Ramsey, *Electro-Osmotiocally Induced Hydraulic Pumping with Integrated Electrodes on Microfluidic Devices*, Anal.Chem. 2001 A.S.A.P. July 9th 2001, A-E

SQUEEZING MICROPUMPS FOR ELASTOMER MICROCHIPS; PICOLITER-HANDLING CAPABILITY

Kwanseop Lim[1], Suhyeon Kim[2], Kihoon Na[1], Je-kyun Park[2], Jong Hoon Hahn[1]

[1]Department of Chemistry, Pohang University of Science and Technology, Pohang, South Korea
[2]LG Electronic Institute of Technology, Seoul, South Korea
(hahn@postech.ac.kr)

Abstract
We have developed a squeezing micropump for elastomeric micropumps, which is featured by capabilities of bi-directional pumping, highly reproducible and ultra-small-volume handling down to picoliters.

Keyword : squeezing micropump, elastomer microchip, picoliter-handling

Introduction

Electroosmosis and on-chip pneumatic micropumps are generally used for delivering liquids through microchannels. Their pumping conditions, especially pumping rate, are, however, sensitive to the physical properties of a pumping medium, and individual calibrations are often required for different pumping media[1]. We have developed squeezing micropumps for microfluidic systems made of elastomeric polymers. This is not only insensitive to the physical properties of a pumping medium, but also free of external connections. Moreover, it is featured by capabilities of bi-directional pumping highly reproducible, and ultra small volume handling down to picoliters.

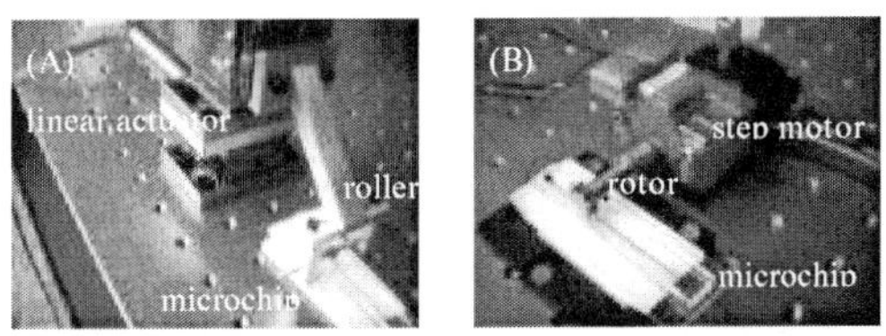

Figure 1. Photographs of squeezing micropumps; (A) roller type and (B) peristaltic type.

Squeezing Micropump I (Roller Type)

The first type of squeezing device we have developed is a roller actuated by a motorized x-translator. Its pumping mechanism is illustrated in figure 2. The squeezing force is adjusted by a manual z-translator. A PDMS chip (the upper plate thickness = 1 mm) with a microchannel (50 μm width x 30

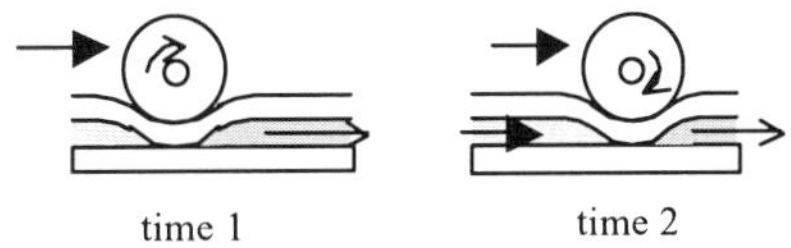

Figure 2. Pumping mechanism of a roller type squeezing micropump.

J.M. Ramsey and A. van den Berg (eds.), Micro Total Analysis Systems 2001, 401–402.

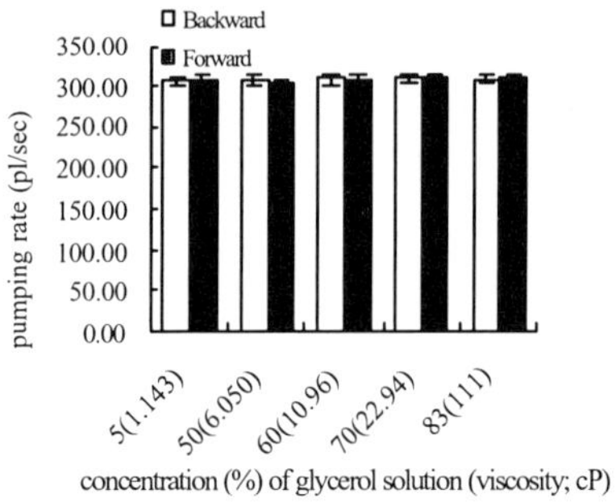

Figure 3. Pumping rates for various viscous solutions.

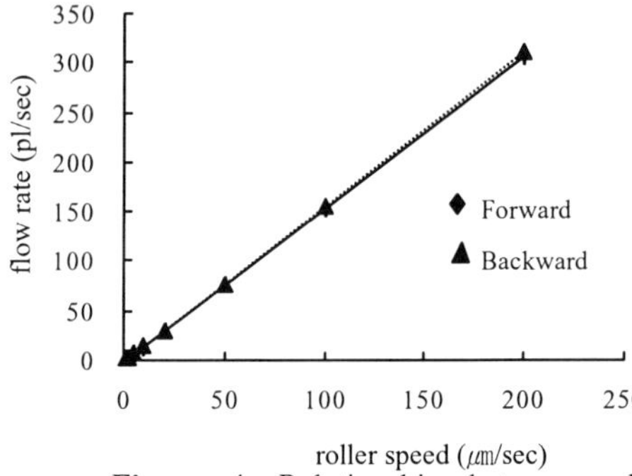

Figure 4. Relationship between roller speed and pumping rate.

μm depth) is used to evaluate the pump. PDMS is very elastic and only mild squeezing is necessary for sure pumping. Figures 3 and 4 show the excellent performance of our squeezing pump: the pumping rate is independent of the viscosity of the pumping medium, and the pumping is highly reproducible and truly bi-directional. The linearity range in the relationship between roller speed and pumping rate is limited only by the speed range of the linear actuator used.(figure 4) The minimum volume delivered by the pump can be as small as 100 picoliters.

Squeezing Micropump II (Peristaltic Type)

By replacing the roller in the squeezing micropump I with a gear-shaped rotor, we have realized a micro-peristaltic pump. The rotor is directly coupled with a stepping motor. This peristaltic type micropump nearly equal the roller type one in performance. However, it has a much wider range of pumping rate and volume capacity of pumping media, but in contrast to the roller type, it cannot avoid the intrinsic pulsation in flow.

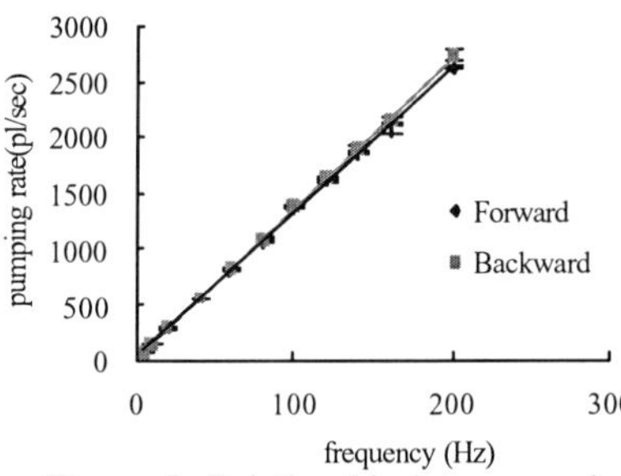

Figure 5. Relationship between pulse frequency and pumping rate in the peristaltic type squeezing micropump.

Reference

1. H. Andersson, W. van der Wijngaart, P. Nilsson, P. Ennksson, G. Stemme, "A valveless diffuser micropump for microfluidic analytical sysyems", Proceedings of the u-TAS 2000 symposiumEnschede, Netherlands, 14-18 May 2000.

FABRICATION OF MICRO FLOW CELL FOR MULTI-ANALYTE SANDWICH IMMUNOASSAY USING TEFLON MICRO CHECK VALVES

N. Honda[1], S. Shoji[1], M. Isomura[2], Y. Ashihara[2], K. Akahori[1], H. Sato[1]
[1] Department of Electronics, Information and Communication Engineering
Waseda University, 3-4-1 Ohkubo, Shinjuku, 169-8555 Tokyo Japan
[2]Frontier Research Division, Fujirebio Inc.
51,Komiya-cho, Hachioji-shi, 192-0031, Tokyo Japan
E-mail : no-honda@fujirebio.co.jp / shojis@mn.waseda.ac.jp

Keywords: Multi-analyte Sandwich Immunoassay, Clinical Diagnostics, Micro Flow Cell, Check Valve, Teflon

Abstract

A micro flow cell for sandwich immunoassay using Teflon (Asahi Glass Co.: CYTOP^{TM}) check valves for sample and reagent inlets is designed and fabricated. The check valves are used not only to prevent reverse flow of injected solutions but also to realize sequential injection of the sample and reagent by simple applied pressure control. The valve structure is designed to control the width and thickness of the sample. We examined a fluorescent sandwich immunoassay for human AFP (α-fetoprotein) as a preliminary experiment and fast immunoassay within 30 minutes compared to the conventional one was realized in the micro flow cell.

1. Design of Micro Flow Cell

A micro flow cell for the multi-analyte bioaffinity assay system was designed. It has two Teflon micro check valves. Some different antibodies for clinical diagnostic were immobilized on center surface of the channel as bioaffinity probes, which realizes sequential multi-analyte assay. The schematic functions of the micro flow cell are illustrated in Figure 1. (A) Sample is injected through Inlet 3 and antigen in the sample is captured with the arrayed antibodies. (B) Reagent of antibodies labeled fluorescent dye is introduced from Inlet 2 and is reacted with the captured antigen. Before the optical detection, the remaining reagent is washed out with the carrier solution from Inlet1.

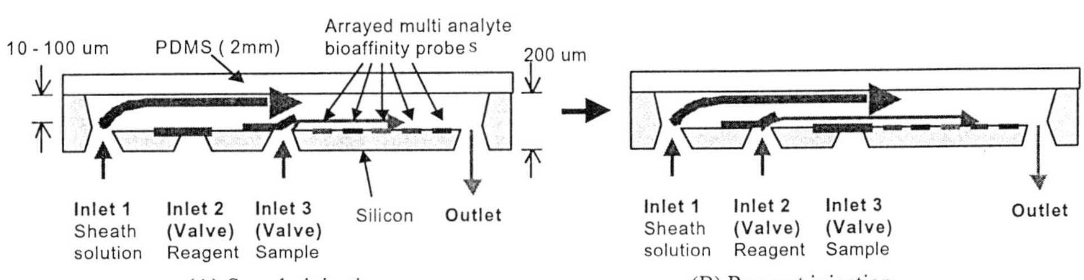

Figure 1 Schematic of microflow cell for multi-analyte for sandwich immunoassay using check valves.

J.M. Ramsey and A. van den Berg (eds.), Micro Total Analysis Systems 2001, 403–404.

2. Teflon Micro Valves

The Teflon check valve is fabricated using positive photoresist as a sacrificial layer to form the movable part. The magnified top and cross-sectional views of the valve structure and expected flow distribution of sample solution in the flow cell are shown in Figure 2. Both sides of the movable part of the Teflon film are fixed to the substrate. The liquid is introduced only the center of the channel. The thickness of the injected liquid flow is also kept thin by controlling the valance of the carrier flow rate and injected liquid flow rate. Since the immobilized antibodies are located at the center of the channel, efficient reaction with small sample volume is expected. Figure 3 shows the forward and reverse flow characteristics of the check valve (from –5kPa to 12.5kPa and from 12.5kPa to –5kPa).

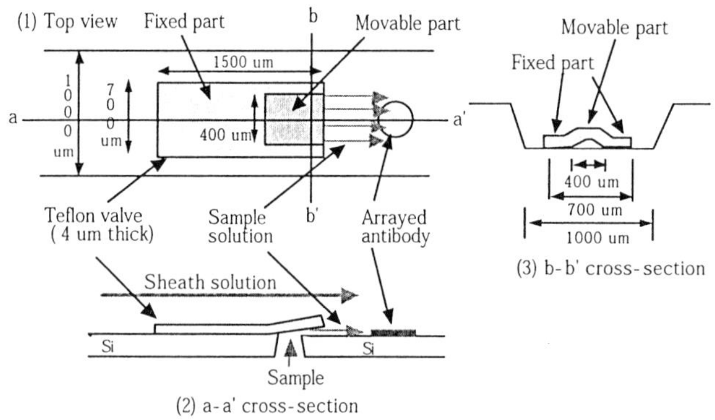

Figure 2 Structure of the Teflon check valve and expected flow distribution in the flow cell

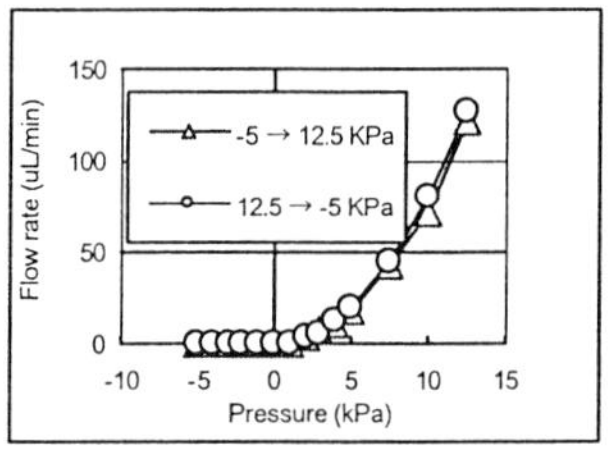

Figure 3 Forward and reverse flow characteristics of the check valve

3. Fluorescent Sandwich Immunoassay

The fluorescent sandwich immunoassay for AFP was chosen as the preliminary experiment. Anti-AFP monoclonal antibody was immobilized chemically on the SiO_2 chip coated with poly(Lys,Phe). For the first step, 5μL mixture of AFP and anti-AFP antibody labeled with R-phycoerythrin was injected to the flow cell. After washing procedure, fluorescence intensity on the chip surface was measured. Immunoreaction was reached plateau within 30 minutes. The reaction time is shorter than that of conventional ELISA method (Typically 180 minutes). A calibration curve for AFP was shown in Figure 4. The sensitivity of AFP was lower than that of expected. This is due to poor S/N ratio of the used fluorescent detector system. We are currently improving the optical detection method and also developing multi-analyte bioaffinity assay adding IL-6 and TSH measurements.

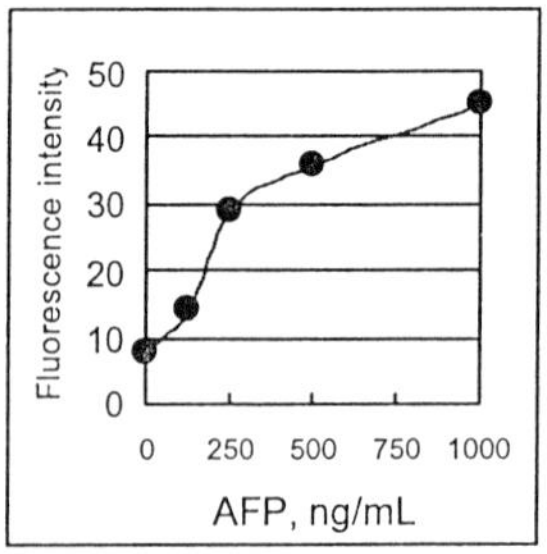

Figure 4
Calibration curve for AFP Immunoassay using prototype of micro flow cell

Acknowledgement

A part of this work is supported by the Japanese Ministry of Education Science and Culture under a Grant-in-Aid for the COE of Waseda University (2001-2005) and under a Grant-in-Aid for Scientific Basic Research (B) No.12450167.

TOTAL ANALYSIS SYSTEMS ON A CARTRIDGE

Antoni Ivorra, Mar Folgueral and Jordi Aguiló
Centre Nacional de Microelectrònica, 08193 Bellaterra, Barcelona, Spain
(antoni.ivorra@cnm.es)

Abstract

A new technique for the implementation of low-cost miniaturized fluidic systems is presented. Active valves and pumps are integrated into a disposable cartridge that is controlled by a hand-held unit by means of magnetic fields induced by electromagnets.

Keywords: passive cartridges, hand-held device, magnetic actuation, elastomer

1. Introduction

As it has been pointed out by Weigl et al. [1], there has been a failure to obtain true hand-held and low-cost μTAS. Apparently, there exists an unsolvable trade-off between portability and low-cost; when the fluidic active elements (valves and pumps) are integrated into the miniaturized parts, the fabrication steps and materials become too much expensive for the market constraints, while, on the other hand, when external fluidic elements are used, the system size is significantly increased resulting in desk-top sized analysis instruments.

Here, a novel strategy for the implementation of miniaturized fluidic systems that overcomes the previous limitations is presented [2]. The key idea of this strategy is to divide the active and the passive parts in a valve or pump. Conventional electromagnets (active parts) are located into a hand-held device (fig. 1) and ferromagnetic elements and elastomer membranes (passive parts) are integrated on a low-cost disposable plastic cartridge.

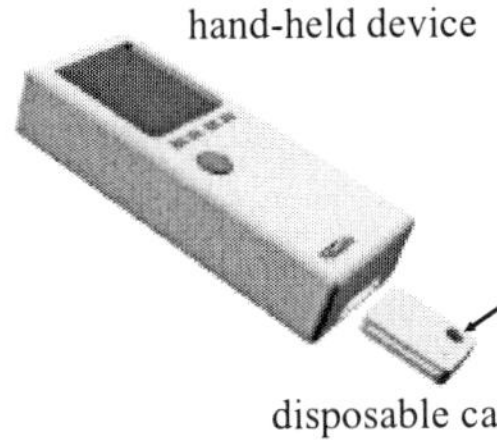

Figure 1. Hand-held analysis system based on passive cartridges

2. Valves, pumps and cartridges

The actuation principle for an active valve can be understood from fig. 2: the elastomer sheet seals the input channel until a strong enough magnetic field is caused by the external electromagnet. When this occurs, the ferromagnetic element pulls up the

J.M. Ramsey and A. van den Berg (eds.), Micro Total Analysis Systems 2001, 405–406.

membrane and creates a vacuum chamber that can be used as a fluidic path (opened valve) or as suction element for pumping purposes.
Bi-directional pumps have been implemented by placing a suction element between two active valves (flow rates beyond 100μl/s).

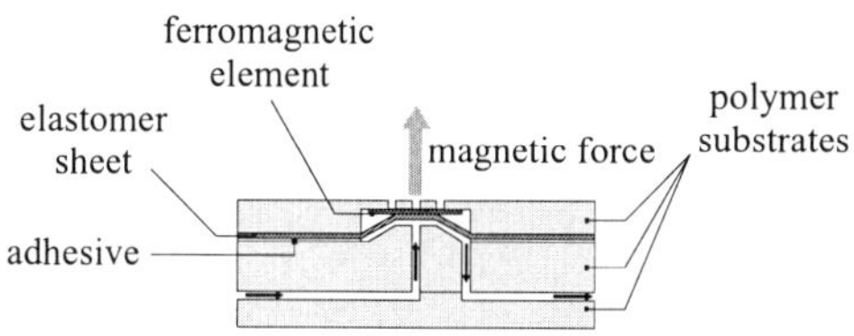

Figure 2. Active valve cross section.

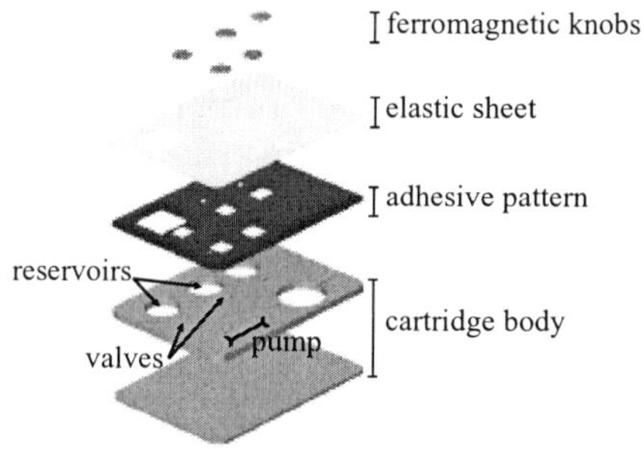

Figure 3. Cartridge parts.

Figure 3 shows how the cartridges are implemented. Channels and reagent reservoirs are created by the assembly of rigid polymer pieces made by injection moulding, hot embossing, laser ablation or any other process. Then, an elastomer membrane (thickness ~ 100μm) is stuck on it and ferromagnetic knobs (diameter ~ 7mm, thickness ~ 0.5 mm) are glued where the active elements are located.

3. Prototypes

Cartridge prototypes have been made by conventional mechanical machining on policarbonate (PC) and polyester (PS) substrates (50×80×5 mm). By means of computer controlled electromagnet actuation, multiple fluidic operations have been performed. For example, it has been possible to transfer quantified amounts of liquid from a reservoir to another one.

Therefore, it has been demonstrated that hand-held systems built with this new strategy will be able to perform several lab standard procedures (to mix, to dissolve, to measure liquid volumes...) on a low-cost plastic cartridge.

This technique is being used for the development of a hand-held chlorine analyzer based on colorimetry quantification.

References

1. B.H. Weigl, R. Bardell, T. Schulte and C. Williams, *Passive Microfluidics Ultra-Low-Cost Plastic Disposable Lab-on-a-Chip*, Proceedings of the uTAS Symposium, Enschende, 14-18 May 2000, pp.299-302, (2000).
2. TASOC Web site: **www.cnm.es/gab/tasoc.html**

IN-LINE MICRO BALL VALVE THROUGH POLYMER TUBING

Kwang W. Oh, Rong Rong, and Chong H. Ahn
MicroSystems and BioMEMS Lab
Center for Microelectronic Sensors and MEMS
Department of Electrical & Computer Engineering and Computer Science
University of Cincinnati, OH 45221-0030, USA

Abstract

A magnetically driven in-line micro ball valve through polymer tubing has been realized in this work. The structure is symmetric so that it can be used bi-directionally and it is very simple making it easily connectable with other microfluidic systems.

Keywords: Ball valve, microvalve, polymer tubing

1. Introduction

This paper presents an in-line micro ball valve using polymer tubing, as shown in Figure 1. Recently, a ball microvalve using electromagnetic actuation was reported [1]. However, tight sealing was difficult to achieve using the silicon valve seat. In this work, to realize a tighter and mechanically more stable valve seat, polymeric material was used for that valve seat. This enhances the sealing of the outlet orifice and results in lower leakage.

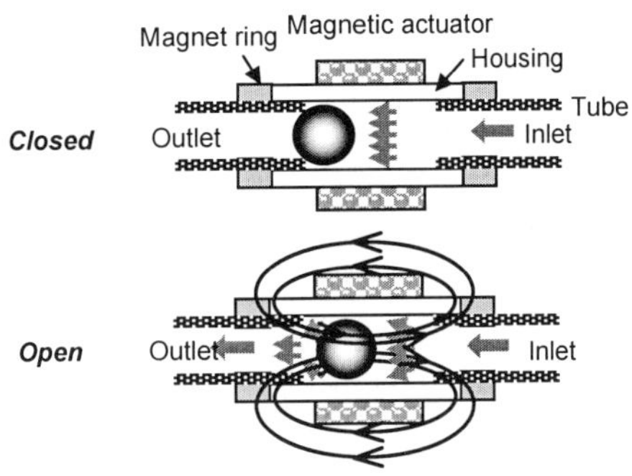

Figure 1. Schematic diagram of an in-line micro ball valve through polymer tubing.

2. Design

Figure 2 shows an analytically calculated magnetic force on the spherical metal ball exerted by the magnetic actuator along the x-direction. In addition, the force of attraction on the ball is plotted as a function of displacement by the MagNet® simulation. The ball has to be placed slightly inside the end of the magnetic actuator, where the maximum magnetic force is achievable. Figure 3 shows computational fluid dynamics simulation results at static position where the ball is 100 μm and 300 μm distant from the outlet. The simulation results

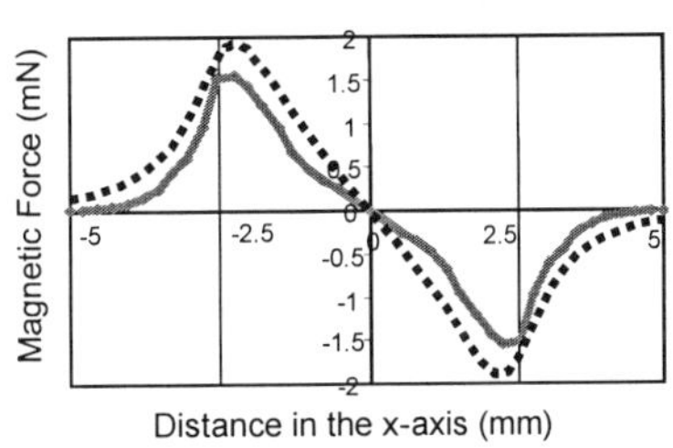

Figure 2. Analytically calculated result (dashed line) and the MagNet® simulation result (solid line). a = 0.8 mm, l = 5 mm, I = 1 A, N = 500, the diameter of wire = 0.2 mm, the diameter of nickel ball = 760 μm.

J.M. Ramsey and A. van den Berg (eds.), Micro Total Analysis Systems 2001, 407–408.

show that the flow rate increases with increasing gap of the metal balls from the outlet.

3. Experimental Results

Based on these simulation results, an in-line micro ball valve was realized, which consisted of biomedical grade silicone tubes (625 μm ID and 1190 μm OD), a nickel ball with a diameter of 760 μm, and a Teflon tube for housing. Figure 4 shows the experimental result of the leakage flow rate as a function of the input pressure with DI water. Although this device is not appropriate for the application in low ranges of input pressure, by adding an electromagnet ring or a permanent magnet ring around the outlet position, much less leakage flow rate may be achievable. Figure 5 shows flow measurement result at an input pressure of 0.3 psi. If the actuation current is larger, the ball can travel further upstream, which makes a larger gap between the outlet orifice and the ball, and fluids flow at higher flow rates. So, the flow rate was controllable by the actuation current.

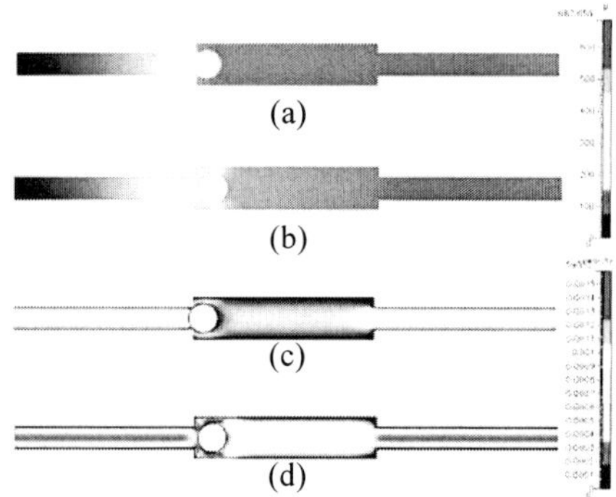

Figure 3. CFD-ACE® simulation results: the pressure drop profiles when the balls (a) 100 μm and (b) 300 μm distance from the outlet, respectively; The fluidic velocity profiles when the ball is (c) 100 μm and (d) 300 μm distance from the outlet, respectively.

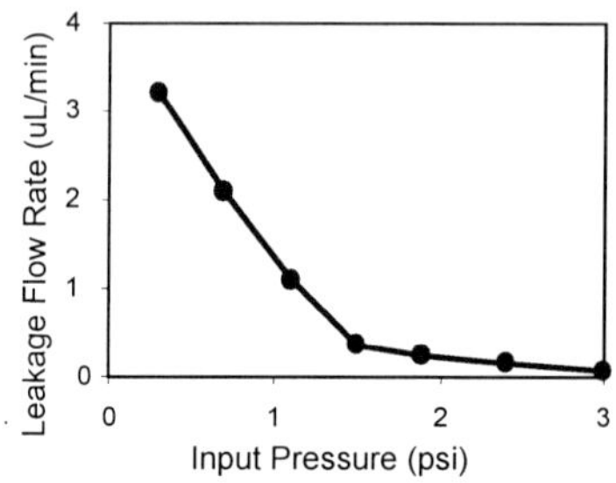

Figure 4. Leakage flow rate vs. the input pressure for DI water.

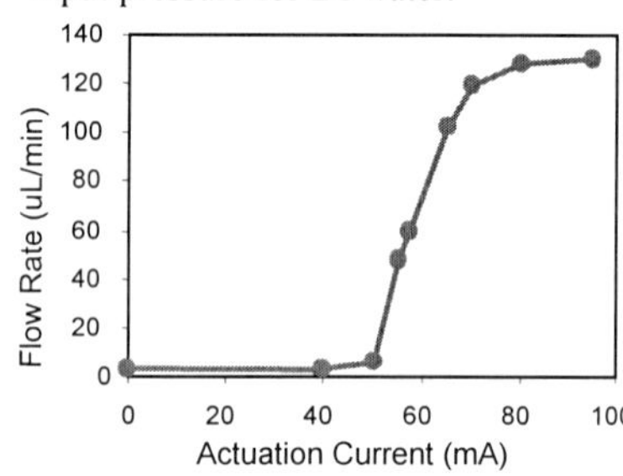

Figure 5. Flow rate vs. the input pressure for DI water at 0.3 psi input pressure.

4. Conclusions

An in-line magnetically driven normally-closed ball microvalve using polymer tubing has been realized in this work. The structure is symmetric so that it can be used bi-directionally and it is very simple making it easily connectable with other microfluidic systems. Thus, the new micro ball valve developed in this work will have tremendous applications in various bio/chemical microfluidic systems.

Acknowledgements

This work was partially supported by a DARPA grant under AF F30602-97-0102 from the *MicroFlumes/BioFlips* program, DoD, USA.

References

1. O. Krusemark, A. Feustel, and J. Muller, *μTAS '98*, pp. 399-402, 1998

Thermolysis Reaction Actuating Pumps,TRAP

Kuo-Yao Weng
Microfluid Technology Department / ERSO
Industrial Technology Research Institute , Hsinchu , Taiwan
E-mail: ALLAN_WENG@ITRI.ORG.TW

Abstract
Pumping forces provided by the thermolysis reaction of ammonium bicarbonate are implemented to drive 2.53 micro liters of the deionized water section. Pumping speeds of up to 3 mm/s in a 300 x 300 μm cross-section channel were achieved. The higher decomposition temperatures are, the faster the liquid section movements are.

Keywords: Thermolysis reaction, Micro pumps, Pumps

1. Introduction
The majority of most microfabricated pumps are mechanically based, contain moving parts, and are designed for moving continuous streams of liquids [1]. A few pumping schemes have been based on discrete droplets such as thermocapillary pumping [2] , thermopneumatic pumping [3] and electrowetting [4]. Here, the mechanism of thermolysis reaction actuator pumping are presented. Pumping speeds of up to 3 mm/s in a 300 x 300 μm cross-section channel were achieved.

2. Theory
Ammonium bicarbonate powder was used to produce carbon dioxide gas for the chemical leaven. The thermolysis reaction is given by:

$$(NH4)HCO3 \text{ [Heating] } \text{---}\rightarrow NH3(gas) + CO2(gas)+H2O$$

Heating the powder above 60 deg. C produces carbon dioxide gas, ammonia gas and water. By sealing the powder in one end of a microchannel, this gas production can be used to pump liquid. Similar thermolysis reactions can also be applied to pumping applications[5].

3. Experimental
Figure 1 shows a schematic of the structure of the thermolysis pump. The microchannels were fabricated using a hot embossing machine. Channels were first etched in silicon by deep silicon reactive ion etching (STS) with an oxide mask to produce 300x 300μm channels. Next gold was sputtered to form a seed layer and nickel was electroplated. The bulk silicon was removed in KOH to complete the embossing mold. The PMMA microchannels were bonded to a blank PMMA sheet by direct fusion bonding. 7mg of ammonium bicarbonate were placed into the reservoir, which was mechanically drilled and sealed with tape.

J.M. Ramsey and A. van den Berg (eds.), Micro Total Analysis Systems 2001, 409–410.

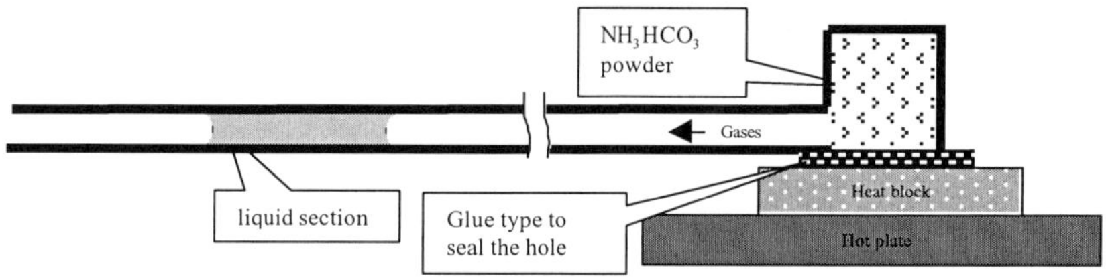

Fig. 1 The cross section of operating the micro fluidic

4. Results and discussions

A hot plate was used to locally heat the reservoir area. Three hot plate temperatures, 80,100,120 deg. C, are used to observe the displacements during the pumping period. In each case, the thermolysis pumping was compared with the pumping effect of an air filled reservoir due to thermal gas expansion. The relationship between the liquid section displacement and heating time is showed in the fig. 2 Clearly, the pumping forces are dominated by the gas produced from the thermolysis reaction.

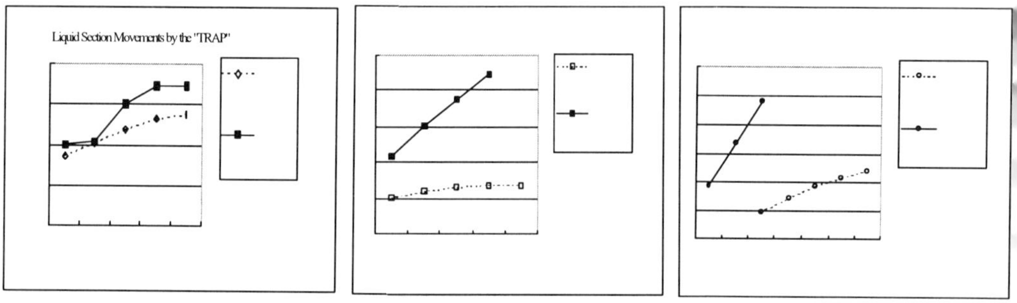

Fig. 2 Liquid section position historical track; Left: the hot plate setting 80 deg. C, Middle : the hot plate setting 100 deg. C, the hot plate setting 120 deg. C

Condensed water droplets are observed inside the channel surface of the cool end by using the ammonium bicarbonate, which was decomposed the liquid water. Some organic chemicals like AZDN, NTA, are adopted to solve the issues further.

5. Conclusions

The mechanism of thermolysis reaction actuator pumping is successfully demonstrated in the micro channels. Pumping speeds which are dependent on decomposition temperature, up to 3 mm/s in a 300 x 300 μm cross-section channel were achieved.

References

1. G. T. A. Kovacs , "*Micromachined Transducers Sourcebook*", McGraw-Hill 1998.
2. T. S. Sammarco, M. A. Burns , J. Micromech. Microeng. pp. 42-55 , No.10 (2000)
3. Handique, K., D. T. Burke, C. H. Mastrangelo, and M. A. Burns, Proceedings of the Solid State Sensors and Actuators Workshop 98, Hilton Head, 346-349 (1998).
4. M.G. Pollack, etc., Applied Physics Letters, 77 (11), 2000.
5. The extended data showed in the patent (pending).

A PLASTIC MICRO INJECTION MOLDING TECHNIQUE USING REPLACEABLE MOLD-DISKS FOR DISPOSABLE MICROFLUIDIC SYSTEMS AND BIOCHIPS

Jin-Woo Choi, Sanghyo Kim, Ramachandran Trichur, Hyoung J. Cho, Aniruddha Puntambekar, Robert L. Cole, Jeffrey R. Simkins, Suresh Murugesan, Kabseog Kim*, Jeong-Bong (JB) Lee*, Gregory Beaucage, Joseph H. Nevin, and Chong H. Ahn

MicroSystems and BioMEMS Lab
University of Cincinnati, Cincinnati, OH 45221-0030, USA
*University of Texas at Dallas, Richardson, TX 75083-0688, USA

Abstract

This paper presents an innovative plastic micro injection molding technique using replaceable disk-mold for applications to microfluidic systems and biochips. Precisely patterned and microfabricated wafer-type mold inserts can be easily loaded into the injection molding machine. Processing time for one chip was as fast as 10 seconds while hot embossing techniques require at least several minutes. Less than a few μm of precise patterns were also achieved. A promising new material has also been introduced and characterized using the developed injection molding technique as well as demonstrated for a disposable biochip.

Keywords: Injection molding, replaceable mold-disk, plastic biochip

1. Introduction

In past few years, fabrication of plastic microfluidic devices has become an important issue in BioMEMS and lab-on-a-chip applications for low cost and high throughput manufacturing. Hot embossing techniques have been widely adapted for plastic microfabrication [1-2]. Injection molding is also an attractive technique for plastic microdevices because it has shorter processing time than hot embossing has. This paper presents an innovative plastic micro injection molding technique using replaceable disk-mold for applications to microfluidic systems and biochips. Precisely patterned wafer-type mold inserts can be easily loaded into the injection molding machine as described in Figure 1.

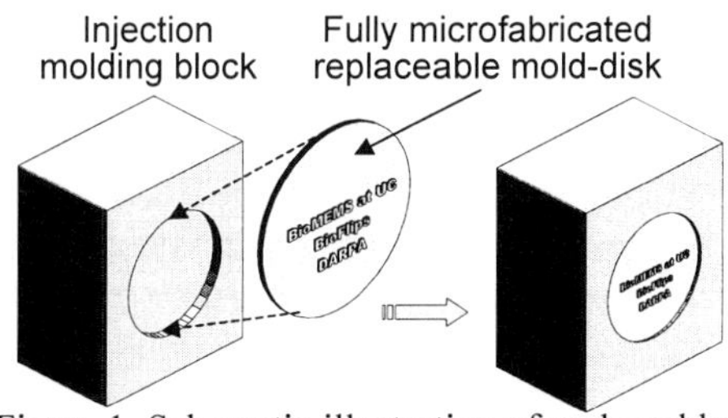

Figure 1. Schematic illustration of replaceable mold-disk technique.

2. Injection Mold Design

Replaceable mold-disks have been fully microfabricated on 3" nickel wafers using photolithography and an electroplating technique. The custom-made molding blocks and replaceable disk-molds enable to replace only metallic mold insert for any new structure instead of replacing the whole molding block. When we need to replace

J.M. Ramsey and A. van den Berg (eds.), Micro Total Analysis Systems 2001, 411–412.

the metallic mold insert, we can easily do by ejecting the existing metallic mold insert and inserting the new metallic mold inserts as shown in Figure 2.

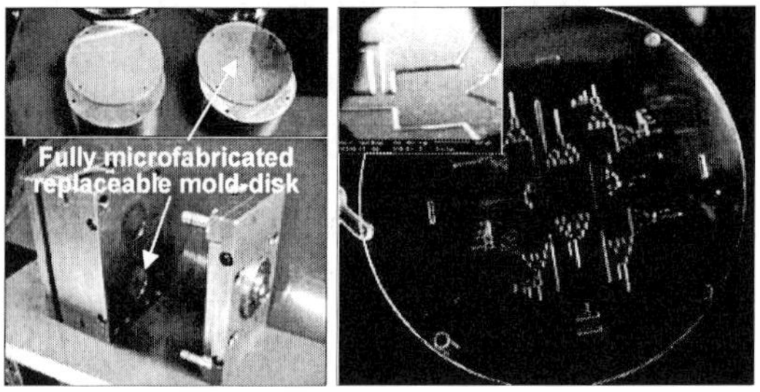

(a) (b)

Figure 2. Photograph of: (a) the developed injection molding blocks for replaceable mold-disk technique and (b) the fabricated disposable biochip made out of olefin copolymers.

3. Experiments

Fast and precise microfabrication has been successfully achieved with polycarbonate, PMMA, and olefin copolymers. Plastic injection conditions have been optimized in terms of injecting temperature, flow rate of melted plastic material, molding pressure, and dual control of molding block temperature to achieve precise microstructures. Characteristics of olefin copolymers have been examined after injection. The material revealed hydrophobicity that is essential for structurally programmable microfluidic systems (sPROMs) [3]. The material also shows an excellent optical property which is strongly required for fluorescein-based biochemical analyses and bio-optical applications (Figure 3).

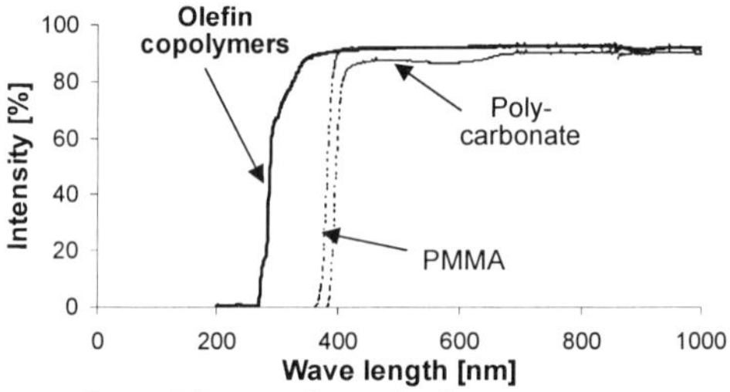

Figure 3. Measured optical transparency and absorption characteristics for different wave lengths after injection. UV range is transparent for olefin copolymers.

4. Conclusions

An innovative plastic injection molding using replaceable mold-disk technique has been developed and characterized for disposable microfluidic systems and biochips. Precise patterns and fast injection have been successfully achieved using the developed injection molding technique. The new technique promises low cost and mass productive biochips and bio-analysis systems.

Acknowledgements

This research was fully supported by a DARPA grant under contract AF F30602-00-1-0569 from the *BioFlips* program, DoD, USA.

References

[1] H. Becker and U. Heim, "Hot Embossing as a Method for the Fabrication of Polymer High Aspect Ratio Structures," *Sensors and Actuators A*, Vol. 83, 2000, pp. 130-135.

[2] M. Heckele, W. Bacher, and K. D. Muller, "Hot Embossing – The Molding Technique for Plastic Microstructures," *Microsystem Technologies*, Vol. 4, 1998, pp. 122-124.

[3] C. H. Ahn, A. Puntambekar, S. M. Lee, H. J. Cho and C.-C. Hong, "Structurally Programmable Microfluidic Systems," *Proc. of Micro Total Analysis Systems (μ-TAS) 2000*, Enschede, The Netherlands, May 14-18, pp. 205-209, 2000.

INJECTION MOLDING HIGH-ASPECT-RATIO MICROSTRUCTURES FOR LAB-ON-A-CHIP APPLICATIONS

Donggang Yao and Byung Kim
Department of Mechanical and Industrial Engineering
University of Massachusetts, Amherst MA 01003

Abstract

Potential applications of plastics materials in lab-on-a-chip applications are numerous. The low-cost mass-production of plastic microstructures remains a challenge, however. This paper provides experimental results of replicating high-aspect-ratio micro sampling wells using a novel injection molding process called rapid thermal response molding, in which the mold surface temperature can be rapidly changed. The effect of the heating temperature on the replicability was studied. A sharp increase in the replicated depth was observed when the heating temperature approached the melting temperature. Complete replication of the 400-μm deep and 40-μm thick microfeature was achieved with a 2-s heating duration and a total cycle time of 15 seconds.

Keyword: Micro molding, microstructure, lab-on-a-chip

1. Introduction

One of the hurdles limiting the current technical capability in injection molding microstructures is that the molten polymer in a tiny cavity instantaneously freezes upon contacting with the relatively cold cavity wall. The problem gets worse when microstructures with high aspect ratios are to be molded, which are frequently encountered in lab-on-a-chip applications. A rapid thermal response (RTR) molding process has been developed to improve replication and reduce cycle time. In RTR molding, the mold surface can be rapidly heated to the melt temperature in several seconds and then rapidly cooled. This enables complete replication of microstructures while short cycle time is maintained.

2. Experimental

The geometry of a portion of the microstructure to be molded is shown in Figure 1. The actual microfeature consists of 576 micro wells, 225 μm x 225 μm, and 400 μm deep. The rib separating the adjacent wells is 40 μm thick. A set of mold inserts designed for rapid heating and cooling were made first, of which one has the to-be-copied microfeature. The thermal responses of the cavity surface with different heating time are shown in Figure 2. The material used in the molding experiment is HDPE from Allied Chemical. The melting point and recommended ejection temperature for this material are 125 °C and 95 °C respectively. With all other parameters fixed, the heating time is the only parameter studied in the experiments.

J.M. Ramsey and A. van den Berg (eds.), Micro Total Analysis Systems 2001, 413–414.

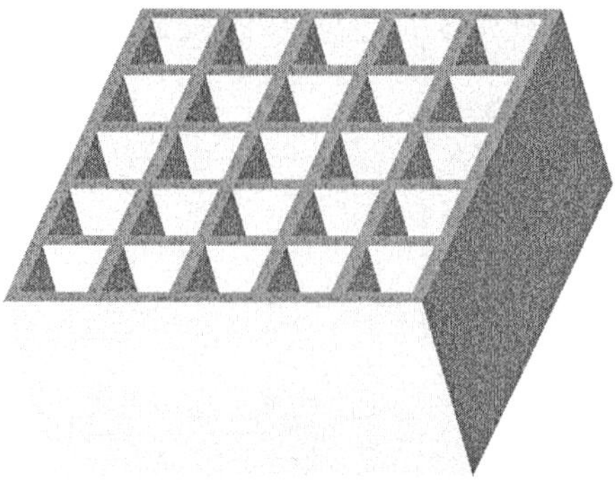

Figure 1: Microstructures

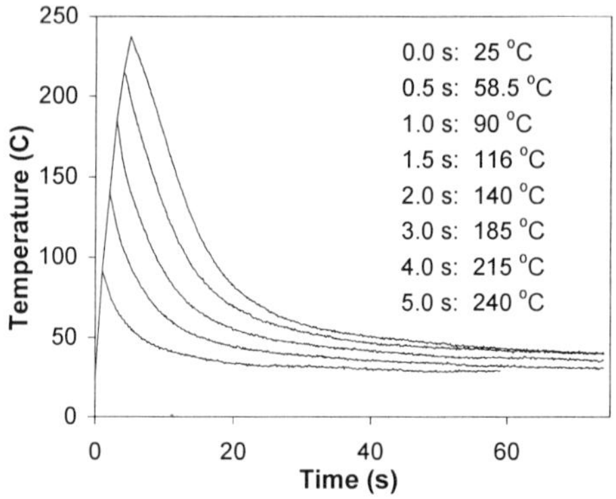

Figure 2: Mold surface temperature

3. Results and Discussion

The replicability of the microstructure is plotted as a function of the heating temperature, as shown in Figure 3. The experimental results agree with the simulation very well, and both data indicate that a sharp transient occurs very close to the melting point. The micrograph of molded micro wells is shown in Figure 4. The height of the microstructure was completely replicated at heating temperature of 140 °C and above.

Figure 3: Micrograph of molded wells

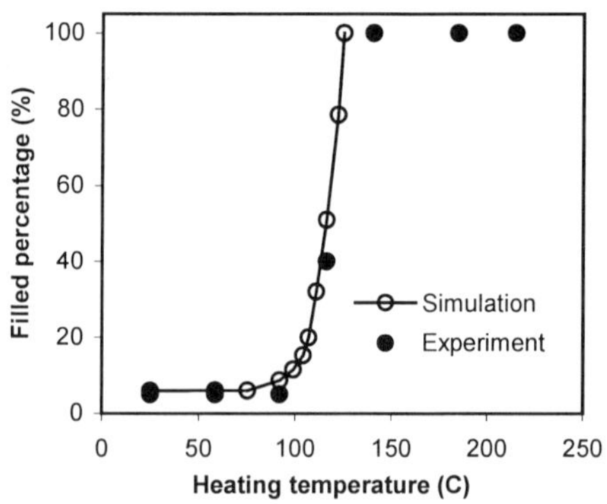

Figure 4: Replicability versus heating temperature

4. Conclusions

Replication of high-aspect-ratio microstructures was investigated using the rapid thermal response molding technology. When the heating temperature is higher than the melting point, complete replication can be achieved. Rapid thermal response molding is an ideal process for replicating high-aspect-ratio microfeatures, because the mold temperature can be rapidly raised to above the melting temperature. The high-aspect-ratio micro sampling wells were successfully replicated within a very short cycle time.

DNA SEPARATIONS IN OXIDE-PASSIVATED THIN-WALLED PLASTIC MICROCHANNELS WITH UV POLYMERIZED ACRYLAMIDES

Jun Zheng[1], C. H. Mastrangelo[1], M. A. Burns[2] and D. T. Burke[3]
[1]Center for Integrated Microsystems, Department of Electrical Engineering and Computer Science, [2]Department of Chemical Engineering, [3]Department of Human Genetics, University of Michigan, Ann Arbor, Michigan 48109, USA

Abstract

The formation of cross-linked sieve matrices such as acrylamide gel in plastic based capillary electrophoresis devices is difficult because of the high permeability of oxygen through thin plastic capillary walls. In this paper we fabricated parylene thin wall channels which have been coated with an Al_2O_3 layer as an oxygen barrier. Initial results indicate much improved gel quality and DNA separation results than otherwise obtainable without the barrier coating.
Keywords: Parylene Channel, Acrylamide

1. Introduction

Recently the development of high-speed, high throughput DNA sequencing techniques has attracted much attention. Compared with conventional slab-gel electrophoresis devices, capillary electrophoresis(CE) chips offer faster separation and higher resolution [1]. The majority of these devices are fabricated using bulk micromachining of glass substrates. More complex DNA analysis systems, however demand a more versatile fabrication platform [2-4]. As a low cost, easy to fabricate and bio-compatible plastic material, parylene thin-walled microfluidic components show great potential, but they have limitations associated with the high permeability rates of gases through the thin walls such as O_2. Cross-linked acrylamide gel, which is widely used in slab-gel electrophoresis today, offers the longest read length in DNA analysis [1]. However acrylamide or similar gels cannot polymerize in the presence of O_2. Due to the porous characteristics of parylene films, UV gels cannot polymerize in parylene channel.

2. Materials and Devices

In order to overcome this problem, this paper explores using UV gel in parylene channels with an aluminum oxide layer overcoat. The alumina layer serves as an O_2 barrier. The simplified process for the passivated parylene channel process is a low temperature process (<120°C), on a silicon substrate using a sacrificial layer of photoresist. This process flow includes 3 masks. The final parylene channel is about 20ìm high, 400ìm wide and 1.3cm in length plus two reservoir area on both ends. Considering both permeability and stress between parylene and alumina films, we optimized the alumina layer at a thickness of 300nm which is deposited by E-beam evaporator. Fig.1 shows the cross section of this device.

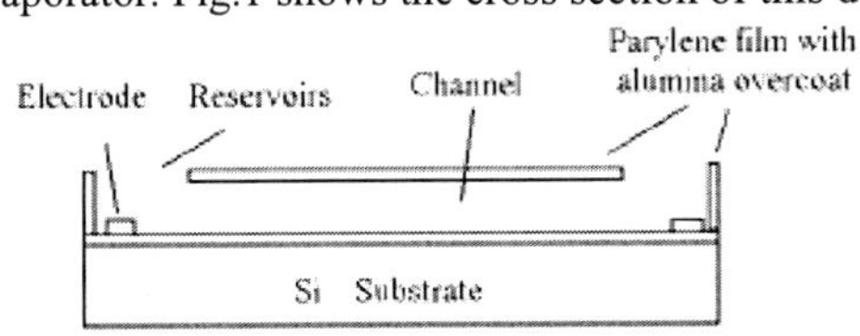

Figure 1. Cross-section of parylene channel

J.M. Ramsey and A. van den Berg (eds.), Micro Total Analysis Systems 2001, 415--416.

In order to achieve a uniform alumina overcoat inside the channel, especially around the sidewall, we rounded the sidewall by pre-deposition baking (photoresist). The cross-section of this channel is shows in Fig.2.

(a) (b)

Figure 2. SEM of cross-section of Parylene Channel: (a) Channel Profile (b) Detailed Layer Structure

3. Separation and Discussion

Fig.3 shows the DNA separation results of on an alumina passivated parylene channel using polymerized acrylamide as the sieve matrix. The DNA sample used here is Bio-Rad 20 bp ladder. The DNA dye is YOYO, made by Molecular Probes. The theoretical plates is about 161645. UV gel is polymerized following manufactures procedures. The separation voltage is 30V/cm. The total separation was completed in 45min to go through 1.3cm distance

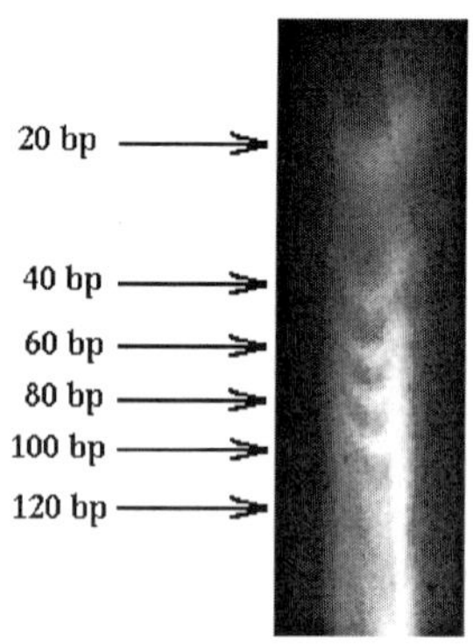

Figure 3. 20 bp ladder separation in UV gel

References:

1. Liguo Song, Dehai liang, etc. Electrophoresis 2001, Vol22., 729-736
2. Haihong, Arthur W.Miller, etc. Anal. Chem. March 1, 2000, 1045-1052
3. Carlen ET, Proc. IEEE Thirteenth Annual Inter. Conference. on MEMS, p551-5
4. J. R. Webster, C.H.Mastrangelo, etc; Analytical Chemistry; 2001; 73(7); 1622-1626.

THE MINIATURIZATION OF NMR PROBES HYPHENATED TO CAPILLARY SCALE SEPARATIONS

Timothy L. Peck[1] and Jonathan V. Sweedler[2]
[1]Magnetic Resonance Microsensors Corporation, 101 Tomaras Avenue, Savoy, IL 61874, USA (tpeck@microNMR.com), [2]School of Chemical Sciences, University of Illinois, 600 S. Mathews Avenue, Box 63-5, Urbana, IL 61801 (jsweedle@uiuc.edu).

Abstract
Microanalytical chemical and biochemical systems have greatly impacted the way samples are evaluated, new drugs are discovered, and processes are chemically monitored (e.g. for pollutants and environmental contamination). An important component of such systems is the detector. This paper addresses the miniaturization of nuclear magnetic resonance (NMR) detection and the hyphenation of NMR to a variety of capillary-scale analytical separation techniques, including liquid chromatography, electrophoresis, and isotachophoresis. The ability of these hyphenated microanalytical systems to efficiently analyze smaller quantities of sample with higher throughput, higher performance, and at a lower cost with less reagent waste is viewed with great excitement by pharmaceutical, biotechnology, and analytical products companies searching for tools to enable competitive advantage in their respective industries.

Keywords: Nuclear Magnetic Resonance, Microcoil, LC-NMR, CE-NMR

Introduction

Nuclear magnetic resonance (NMR) spectroscopy is one of the most powerful detection techniques available for determining the three dimensional structure of complex chemical and biochemical molecules. However, while one of the most widely used analytical detection techniques, NMR remains one of the most expensive and least sensitive. To address this limitation, a wide array of technologies are being applied that include improved high temperature superconducting materials capable of operation at extremely high magnetic field (> 20 T), miniaturized radiofrequency receiver coils with high fill factor that closely conform to the sample for maximum energy transfer, susceptibility-matched materials for enhanced magnetic field homogeneity, optimized pulse sequences and data acquisition algorithms, and sample management techniques that enable concentration of a maximum quantity of analyte in the NMR detection volume. Academic and industrial research scientists at the University of Illinois and Magnetic Resonance Microsensors Corporation have focused on the development of miniaturized, capillary-based NMR detection probes and the hyphenation of capillary-based separation techniques for sample management [1]. Not only do these systems offer greater mass sensitivity and higher throughput, but also afford greatly enhanced chromatographic performance and fluidic integrity.

Theory

A fundamental limitation of NMR is an inherently poor detection sensitivity when compared with other spectroscopic techniques. The signal-to-noise ratio (S/N) per unit volume in an NMR experiment varies with the sensitivity of the receiver coil (B_1/i), the strength of the square of the static magnetic field (B_0), and inversely with experimental noise (Eq. 1) [2]. Solenoidal RF coils provide excellent sensitivity, and are readily

J.M. Ramsey and A. van den Berg (eds.), Micro Total Analysis Systems 2001, 417–419.

fabricated at capillary (millimeter and smaller) sizes. While the resistance of a solenoidal coil is relatively constant with decreasing size, sensitivity increases approximately inversely to coil diameter in a skin-effect dominated regime [3]. Consequently, as shown in Figure 1, the S/N per unit volume increases inversely with coil size in a significant portion of the capillary regime.

$$S/N \propto \frac{B_0^2 (B_1/i) \upsilon_s}{V_{noise}} \tag{1}$$

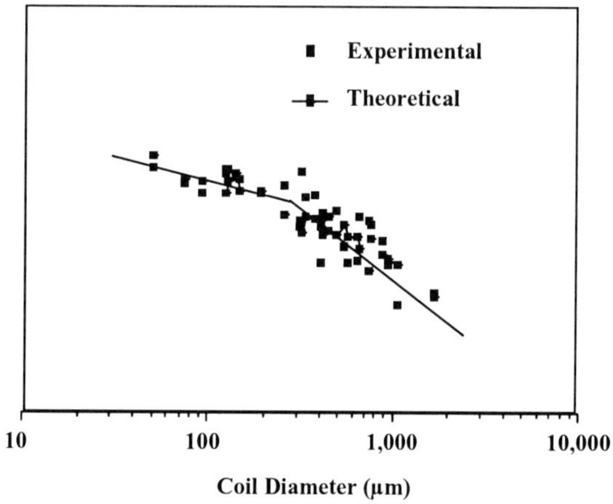

Figure 1. The S/N per unit volume for a solenoidal NMR receiver coil operated in the skin effect regime varies approximately inversely with coil diameter [3].

Results and Discussion

While ***microcoils*** (NMR receiver coils with diameters of ≤1 mm and wrapped directly onto the capillary that contains the sample) have been employed for twenty years, their application to capillary-scale analytical separations and to high resolution NMR spectroscopy did not occur until the mid-1990s [1,3-11]. Using technology developed at the University of Illinois (UIUC), a family of microcoil probes are now commercially available by MRM Corporation for on-line NMR detection in capillary liquid chromatography. In ongoing academic research at UIUC, microcoils serve as on-line detectors for capillary zone electrophoresis, capillary electrochromatography, and capillary isotachophoresis. While the immediate demand for LC-NMR outweighs the other hyphenated combinations, the long-term potential of these complementary techniques, early preliminary results, and the potential for combination of multiple, on-line chromatographic techniques in a single analytical system holds great promise.

While NMR offers a wealth of chemical information, NMR normally requires fairly pure compounds to assay. Hence, the hyphenation of capillary LC separations with on-line NMR detection is a powerful tool to analyze complex mixtures, drug metabolites, and impurities. Capillary-based liquid chromatography and microcoil NMR have compatible flow rates and sample volume requirements. The combination of the Waters CapLC™ and the MRM CapNMR™ flowprobe provides excellent separation capability in addition to UV-Vis and NMR detection for mass-limited samples. A typical CapLC on-column flow rate is 5 μL/min or less, the autosampler-injected analyte volume is 0.1 μL or more, and accurate flow is achieved through capillary of 50 μm i.d.. The NMR flow cell has a total volume of 5 μL, and the microcoil observe volume is 1 μL. A typical injected sample amount for CapLC-μNMR analysis of a small molecular weight compound is a few μg (nmol) or less [9]. Capillary LC-NMR enables steep solvent gradients of water/organic mobile phases as high as 30% per min. Examples will be shown where sample components are applied to the CapLC column, separated and removed using a steep solvent gradient, then individually stopped in the NMR flow cell to acquire a spectrum after a relatively brief equilibration time.

The other complementary capillary separation technique described involves capillary electrophoresis (CE). We describe our progress in interfacing CE with nanoliter volume NMR flowprobes. By electrophoretically concentrating dilute microliter-volume samples into higher concentration nanoliter-volume bands, capillary isotachophoresis (cITP) extends the range of feasibility for microcoil NMR experiments [7]. In addition to concentration factors of up to two orders of magnitude, cITP offers the capability to separate charged components in mixtures so that purified analytes are efficiently presented to the NMR detector. A wealth of diagnostic information such as temperature, pH and background electrolyte can easily be measured using NMR microcoils [8]. Lastly, using multiple NMR detection coils within a single probe offers enhancements to peak identification and sample throughput [10,11].

Conclusions

The application of microcoils for on-line NMR detection in analytical microseparations provides an ideal size and functionality match which enables one of the most information-rich detection techniques to be optimized in sensitivity through miniaturization of the NMR receiver coil and through coupling to sample management techniques that provide sample purification and preconcentration. The eventual integration of a multiplicity of chromatographic techniques, a multiplicity of NMR detection coils, and further miniaturization to chip-based systems in high field magnets promises further significant sensitivity gain in this emerging field.

References

[1] M.E. Lacey, R. Subramanian, D.L. Olson, A.G. Webb, J.V. Sweedler, Chemical Reviews **99**, 3133-3152 (1999).

[2] D.I. Hoult and R.E. Richards, J. Magn. Reson. **24**, 71 (1976).

[3] T.L. Peck, R.L. Magin, and P.C. Lauterbur, J. Magn. Reson. B, **108**, 114 (1995).

[4] D. Olson, T. Peck, A. Webb, R. Magin, and J. Sweedler, Science **270**, 1967 (1995).

[5] N. Wu, T.L. Peck, A.G. Webb, R.L. Magin, and J.V. Sweedler, J.Am.Chem.Soc. **116**, 7929 (1994).

[6] N. Wu, A.G. Webb, T.L. Peck, R.L. Magin, and J.V. Sweedler, Anal. Chem. **67**, 3101 (1995).

[7] R.A. Kautz, M.E. Lacey, A.M. Wolters, F. Foret, A.G. Webb, B.L. Karger, J.V. Sweedler, J. Amer. Chem. Soc. **123**, 3159-60 (2001).

[8] M. Lacey, A.Webb, J. Sweedler, Analytical Chemistry **72**, 4991-4998 (2000)..

[9] M.E. Lacey, Z.J. Tan, A.G. Webb, J.V. Sweedler, J. Chromatogr. A, in press.

[10] G. Fisher, C. Pettuci, E. MacNamara, and D. Raftery, J. Magn. Reson. **138**, 160 (1999).

[11] Y. Li, A. Wolters, P. Malawey, J. Sweedler, and A. Webb, Anal. Chem. **71**, 4815 (1999).

PLASMON-WAVEGUIDE RESONANCE SPECTROSCOPY: A NEW TECHNIQUE FOR STUDYING MOLECULES IMMOBILIZED ON SOLID SURFACES

Gordon Tollin and Zdzislaw Salamon
Department of Biochemistry and Molecular Biophysics, University of Arizona, Tucson, AZ 85721, USA

Abstract

Plasmon-waveguide resonance (PWR) spectroscopy allows kinetic, thermodynamic and structural characterization of a wide variety of systems in which molecules are immobilized on solid surfaces. The technique involves resonant excitation by light of electronic oscillations (plasmons) in a thin metal film (e.g. silver) coated with a dielectric layer (e.g. silica). This generates a surface-localized electromagnetic field, which can probe the optical properties of materials immobilized at the film surface. Resonances can be excited by both parallel and perpendicular polarized light, thereby allowing the anisotropic properties of ordered molecular arrays to be characterized. The application of this methodology to ligand binding by the human brain delta-opioid receptor incorporated into a single lipid bilayer on the resonator surface will be described.

Keywords: biomembranes, receptors, lipid bilayers, ordered molecular arrays, molecular orientation.

Introduction

Plasmons (oscillations of conduction electrons) are produced by illuminating a metal film deposited on the surface of a prism at the critical angle for total internal reflection with polarized light. This generates a surface-localized evanescent electromagnetic field that can interact with molecules deposited on the outer surface of the film. Resonance spectra, corresponding to the angular dependence of reflected light intensity, are determined by the refractive index (n), absorption coefficient (k) and thickness (t) of the film. Maxwell's equations provide an analytical relationship between the spectral parameters (angular position, width and depth) and these properties, allowing their evaluation by curve-fitting procedures. A variant of this methodology, called plasmon-waveguide resonance (PWR) spectroscopy, can excite resonances with both parallel and perpendicular polarized light, allowing characterization of anisotropic properties of ordered molecular assemblies such as biomembranes. In the PWR method, the metal film is coated with a dielectric (e.g. silica), which acts as an optical amplifier, a shield for the metal layer, and a surface that allows various types of molecular immobilization strategies. The exceptionally narrow line widths of PWR provide a greatly enhanced sensitivity and spectral resolution. PWR can follow both binding and structural events in real time, providing information

J.M. Ramsey and A. van den Berg (eds.), Micro Total Analysis Systems 2001, 420–422.

on alterations in film thickness, in molecular orientation, and in the orientation of chromophores within the immobilized molecules.

Experimental

The human brain delta-opioid receptor (hDOR) mediates pain responses via ligand binding. The cloned and expressed receptor was purified and incorporated into a pre-existing lipid bilayer (phosphatidylcholine/phosphatidylglycerol; 3:1 mole ratio) by adding aliquots of a detergent solution into the aqueous compartment of a PWR cell, thereby diluting the detergent to below its cmc and allowing mass transfer from the micelle to the bilayer. Subsequent addition of ligand solutions into the aqueous volume of the cell resulted in ligand binding to the receptor, which was monitored by PWR spectroscopy [1]. Three types of ligands were used: agonists (which stimulate G-protein activation on the membrane interior), antagonists (which block agonist activity) and inverse agonists (which prevent endogenous G-protein activation that occurs in the absence of agonist binding).

Results and discussion

The incorporation of the receptor into the bilayer resulted in changes in the position, width and depth of the PWR spectrum. Theoretical fitting allowed values for the two refractive indices (n_p and n_s) and the membrane thickness (t) to be obtained as a function of added receptor. By extrapolating the hyperbolic incorporation curves to infinite concentration, the molecular dimensions of the receptor could be calculated (area per receptor = 1200 ± 100 Å^2; length = 68 Å), in good agreement with values obtained by other methods. Addition of aliquots of solutions of agonists, antagonists and inverse agonists to the PWR cell caused further spectral changes that were characteristic of the ligand type. Ligand binding constants were consistent with values obtained by other methods. Fitting these spectral data yielded the following conclusions.

1- Agonist binding results in an increase in refractive index anisotropy ascribed to ordering of the transmembrane helices of the receptor molecule and of the lipid acyl chains in the membrane interior. An increase in membrane thickness also occurs due to an elongation of the receptor molecule, as well as a mass density increase due to an influx of lipid molecules caused by this elongation process.

2- Antagonist binding results in neither receptor elongation nor mass density increase, but does cause an even larger increase in refractive index anisotropy, suggesting that binding of this type of ligand produces structural changes that are limited to the transmembrane region of the receptor molecule.

3- Inverse agonist binding leads to smaller increases in receptor length, mass density and refractive index anisotropy than those produced by agonists.

Conclusions

The results demonstrate that binding these three ligand types produces different but characteristic structural changes. Only agonist and inverse agonist binding results in structural changes that are perpendicular to the membrane plane, consistent with the ability of these ligands to transduce information across the membrane, thus providing new insights into receptor function.

Acknowledgements

Work supported by the NSF (MCB-9904753) and the U.S. Public Health Service (DA-06284 and DA-05787).

References

1. Z. Salamon, S. Cowell, E. Varga, H.I. Yamamura, V.J. Hruby and G. Tollin, *Biophys. J.* **79**, 2463-2474 (2000).

SELF-ASSEMBLED AND SELF-SORTED CHEMICALLY ACTIVE BEADS ON UNSTRUCTURED AND STRUCTURED SUBSTRATES

Helene Andersson[1], Christina Jönsson[2], Christina Moberg[2], Göran Stemme[1]

[1]Department of Signals, Sensors and Systems, [2]Department of Chemistry, Organic Chemistry, Royal Institute of Technology, S-10044 Stockholm, Sweden

Abstract

A technique for generating a general screening platform consisting of dots of immobilized beads on silicon has been developed via self-sorting and self-assembly of different kinds of beads. The dots are defined by a teflon-like film, which due to its hydrophobic characteristics also prevents cross-contamination of liquid from different dots. To enable functionalization of individual dots with different target molecules simultaneously a new way of microcontact printing has been explored where different target solutions are printed in parallel using one stamp. In order to show that this platform can be designed for both biochemical assays and organic chemistry, streptavidin-, amino- and hydroxy-functionalized beads have been self-sorted and self-assembled both on separate and common platforms. Beads of different sizes and material have successfully been immobilized in line patterns as narrow as 5 μm. The same chemistry has also been applied on silicon channels in order to enable selective bead trapping in microfluidic devices without the use of physical barriers.

Keywords: microcontact printing, self-assembly, self-sorting, beads, microfluidics

1. Introduction

Reagents attached to surfaces have a wide variety of chemical applications, such as platforms for high throughput screening and combinatorial chemistry. These platforms allow for example minimized use of samples and reagents, facile separations, easy handling, and parallel analysis of multiple samples. Therefore, developing techniques for immobilization of different putative ligands, ranging from biomolecules to catalysts, at well-defined locations on a surface (unstructured or structured) is important.

In this study, a technique has been developed for generating a dot array of self-assembled and self-sorted polymer beads for analytical and biochemical screening [1]. Figure 1 shows a conceptual drawing of the different steps involved in generating the dot array. The dots are separated by a lithographically defined hydrophobic teflon-like film preventing cross-contamination between the dots (figure 1 (a)). The dots are then chemically modified by parallel microcontact printing (μCP), which is a novel technique developed in this study, enabling different chemical modification of different dots on the array simultaneously, see figure 1 (b). The stamp is designed to modify each row of dots with the same compound. Beads with different functional groups are then self-arranged onto the matching dots, see figure 1 (c). In order to show that this platform can be designed for both biochemical assays and organic chemistry, streptavidin-, amino- and hydroxy-functionalized beads have been self-sorted and self-assembled both

J.M. Ramsey and A. van den Berg (eds.), Micro Total Analysis Systems 2001, 423–425.

on separate and common platforms. The self-sorting and self-arrangement are based on surface chemistry only, which has not previously been reported.

The same chemistry has also been applied on silicon channels in order to enable selective bead trapping in microfluidic devices without the use of physical barriers [2]. First, the surface chemistry of the internal walls of the channels is modified by microcontact printing [3]. The chip is then submerged in a bead solution and beads self-assemble based on surface chemistry and immobilize on the internal walls of the channels.

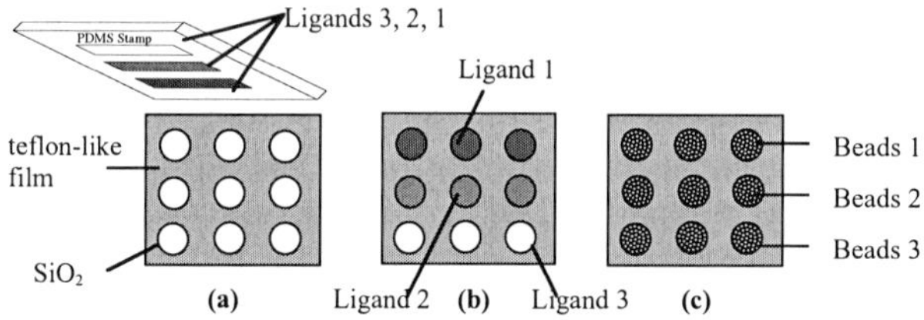

Fig. 1. A conceptual drawing showing the different steps involved in generating the dot array.

2. Experimental

Preparation of substrates: Standard photolithography and lift-off methods were used to pattern the silicon wafers with octafluorocyclobutane, which was deposited in a standard inductively coupled plasma machine. *Microcontact printing:* PDMS stamps were inked with biotin-labeled bovine serum albumin (BSA) 1mg/ml in 0.01 M PBS solution for 30 min, rinsed and dried. The stamp was applied under its own weight onto the substrate for 1 s-15 min. The substrate was then sonicated and washed in PBS and water. 4-(Chlorosulfonyl)benzoic acid in ethanol (1 mg/0.1 ml) was applied on a PDMS-stamp for about 5-15 min. Then the stamp was applied to the substrate for 15 min. The substrate was sonicated for 5 min in ethanol, washed with methanol and dichloromethane and dried. The 4-(chlorosulfonyl)benzoic acid was transformed into anhydride [3]. *Parallel microcontact printing:* The PDMS stamp design used for parallel μCP, consists of three patches separated by 500 μm wide channels and 50 μm deep. The stamp was inked simultaneously with biotin-labeled BSA (patch 1 and 3) and 4-(chlorosulfonyl)benzoic acid (patch 2) for 30 min. The channels separate the different samples both during loading and printing. The 4-(chlorosulfonyl)benzoic acid was transformed into anhydride [3]. The substrate was then washed in dichloromethane. *Immobilization of beads:* Three different types of beads were used: magnetic polystyrene beads coated with streptavidin or amino groups, 2.8 μm in diameter (Dynal Biotech, NO), and polystyrene beads functionalized with hydroxy groups, 0.825 μm in diameter (Bangs Laboratories, IN, US). A drop of an aqueous slurry of beads was added to the substrate and the chip was incubated for 1 h. The chip was washed in water and in some cases (especially when using hydroxy-functionalized beads and Teflon-coated substrates) aqueous detergent solution was necessary to remove unbound beads.

3. Results and Discussion

Streptavidin-, amino- and hydroxy-functionalized beads have successfully been immobilized in monolayers with a surface coverage of 13 beads/100 μm^2, 13 beads/100 μm^2, and 130 beads/100 μm^2, respectively. The beads are firmly attached to the substrate and liquid can be added and withdrawn without problems. The immobilized beads can, however, be removed from the substrate by sonication in water for 30 seconds. Beads have also successfully been immobilized to quartz and polyethylene substrates using the same methodology. Streptavidin-coated beads have also been immobilized into a high resolution pattern by using a structured stamp with features as small as 5 µm. In figure 2 it is shown that beads are successfully arranged into an almost perfect match of the printed pattern.

(a) 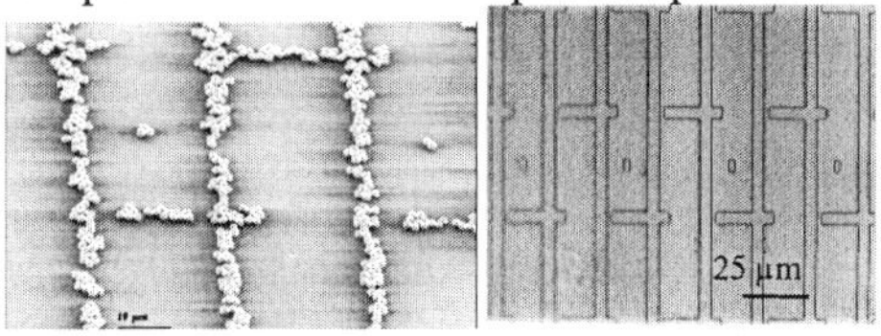 (b)

Fig. 2. Streptavidin-coated beads immobilized in a high-resolution pattern. In (b) a photo of the stamp is shown.

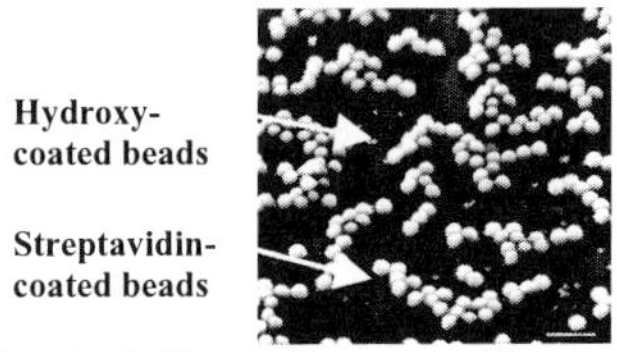

Fig. 3. Self-sorted streptavidin-coated beads; only few hydroxy-functionalized beads were immobilized to this (false) location.

Figure 3 shows the results of a self-sorting experiment, streptavidin-coated beads self-sorted and immobilized themselves onto a biotin-functionalized surface, only very few hydroxy-functionalized beads were immobilized to this (false) location.

Figure 4 shows 100 µm wide and 50 µm deep channels separately immobilized with (a) magnetic streptavidin-coated beads, (b) hydroxy-functionalized beads, and (c) amino-functionalized beads. The surface coverage on the bottom of the channels is about the same as obtained for unstructured substrates. The coverage of beads on the channel walls is good except for the upper part (5-10 µm) of the channel wall where fewer beads are present. The high-resolution pattern of lines (Fig. 2 (b)) of self-assembled streptavidin beads has also been generated on the bottom of a 500 µm wide and 50 µm deep channel. The channels were sealed with a PDMS film, which was applied like an adhesive tape. Flow tests were performed in sealed channels with the different immobilized beads to confirm that the immobilized beads could withstand the forces generated by water flowing in the channels.

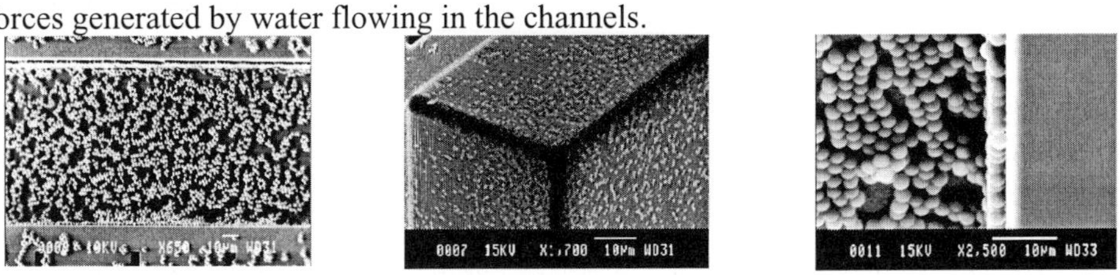

Fig. 4 Immobilized (a) streptavidin-coated magnetic beads, (b) hydroxy-functionalized beads and (c) amino-functionalized beads in etched channels.

4. Conclusions

The combination of the different techniques presented in this study enables controlled individual bead immobilization on unstructured and structured substrates based on different surface chemistries.

References

[1] H. Andersson, C. Jönsson, C. Moberg, G. Stemme, *Talanta,* In Press, [2] H. Andersson, C. Jönsson, C. Moberg, G. Stemme, *Electrophoresis*, In Press, [3] H. Andersson, C. Jönsson, C. Moberg, G. Stemme, *Sensors and Actuators B.*, 2001, 79, 78-84.

DEVELOPMENT OF A NEW, CHIP-BASED METHOD FOR LIPOPROTEIN ANALYSIS BY CAPILLARY ELECTROPHORESIS

B. H. Weiller,[1,3] T. Shibata,[1] L. Ceriotti,[1] M. A. Roberts,[2] D. Rein,[2] J. B. German,[2] J. Lichtenberg,[1] E. Verpoorte[1] and N. F. de Rooij[1]
[1]Institute of Microtechnology, CH-2007 Neuchâtel, Switzerland
[2]Nestlé Research Center, CH-1000 Lausanne, Switzerland
[3]The Aerospace Corporation, PO Box 92957, Los Angeles, CA 90009

Abstract

This paper describes the development of a new, chip-based assay for lipoproteins by capillary electrophoresis. The separation of low-density (LDL) and high-density (HDL) lipoproteins by capillary zone electrophoresis (CZE) was first demonstrated in fused-silica capillaries with both UV absorption and laser-induced fluorescence (LIF) detection. This separation was accomplished using tricine buffer with methylglucamine. Similar results were obtained in glass microdevices although with much shorter analysis times (< 40 sec) and higher resolution. To our knowledge, these results show the first separation and detection of lipoproteins in a microfluidic device. Atomic Force Microscopy (AFM) was used to characterize the surface roughness of the microfluidic channels directly.

Keywords: Lipoproteins, blood, HDL, LDL, capillary zone electrophoresis, CZE

1. Introduction

One of the most important clinical measurements made today is of low-density (LDL) and high-density (HDL) lipoproteins in blood for cardiac risk assessment.[1] LDL and HDL are complex, nanometer-sized particles that are carriers of cholesterol in the bloodstream. Lipoproteins consist of cholesterol, proteins, triglycerols, and a phospholipid monolayer.[2] There is interest in developing a microfluidic device for this measurement in order to reduce the measurement time, cost and sample volume required and ultimately, to perform sample preparation on chip.[3,4]

Capillary zone electrophoresis (CZE) has proven to be a valuable technique for protein analysis. Microfluidic devices for CZE have been very successful for a wide range of bioanalytical measurements. However, the implementation of this powerful analytical technique has been limited by surface adsorption of lipoproteins and the actual separation of LDL from HDL using CZE has not been previously reported in capillaries or chips. This paper describes the development of a new method for the separation and detection of HDL and LDL particles from native serum samples by CZE in fused-silica capillaries and glass microdevices without fixed coatings.

2. Experimental

LDL and HDL fractions were prepared by ultracentrifugation and dialysis of fresh blood samples. The fractions were characterized by size exclusion chromatography. Capillary

J.M. Ramsey and A. van den Berg (eds.), Micro Total Analysis Systems 2001, 426–428.

electrophoresis was carried out in a Beckman P/ACE system with UV or LIF detection. CE chips were fabricated in-house from Pyrex with standard processes [5] or purchased (Micralyne). Both chips have very similar designs with double-tee injectors of 200 or 100 μm, respectively, but are made by different processes. Chip-based separation and detection was accomplished using high voltage power supplies (FuG), relay network (Gunther), Ar^+ laser (Ion Technologies) and PMT (Hamamatsu). Lipoprotein fractions were stained with NBD ceramide (Molecular Probes) [6], and separations carried out using tricine buffer (40 mM, pH = 9.0) with added methylglucamine (40 mM) as a dynamic coating. AFM measurements (Nanoscope) were taken at the bottom of channels etched in unbonded witness plates.

2. Results and Discussion.

The separation of HDL and LDL by CZE was first demonstrated in fused-silica capillaries. With UV detection, LDL eluted as a relatively sharp peak (~10 min.) and HDL eluted as a broad peak at slightly later time. Laser-induced fluorescence was also used to detect lipoproteins stained with NBD-ceramide with the same buffer system to give very similar results (Figure 1). Peak assignments have been confirmed by varying the ratio of HDL to LDL. This data demonstrates the feasibility of HDL and LDL measurement by a simple CZE method in capillaries without fixed coatings. Similar results have been obtained using unfractionated serum samples, suggesting the method can be used for direct analysis of serum.

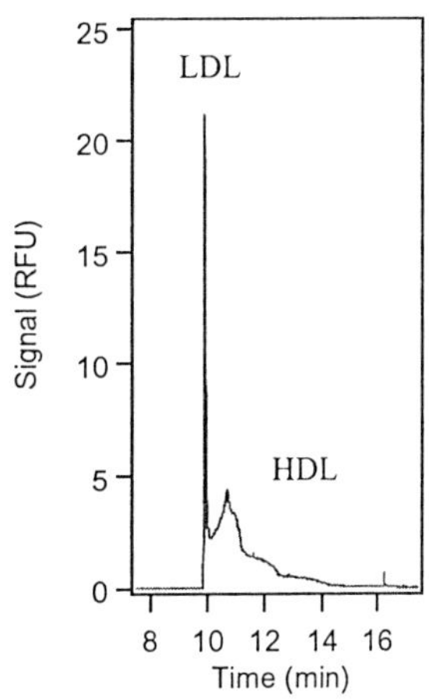

Figure 1. CE of Lipoprotein Fractions in Fused Silica Columns. Separation conditions: L_{eff} = 50 cm, E = 333 V/cm, plug length = 6 mm, column diameter = 75 μm.

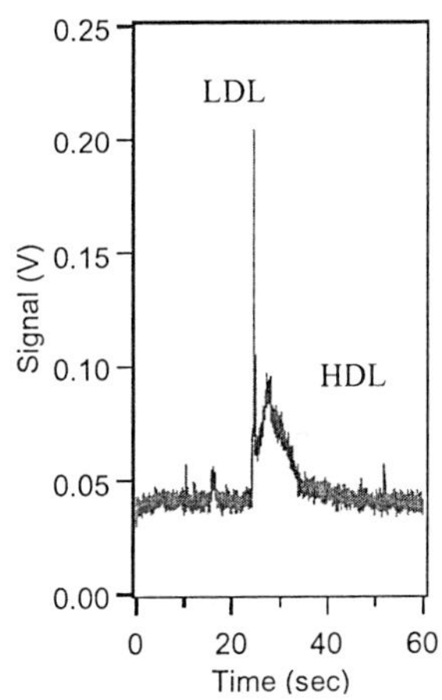

Figure 2. CE of Lipoprotein Fractions in Glass Chips. Separation conditions: L_{eff} = 4.2 cm, E = 714 V/cm, plug length = 200 μm, channel dimensions = 20 μm deep by 50 μm wide.

Using the procedure described above, it is possible to separate and detect HDL and LDL samples in glass microdevices with LIF detection (Figure 2). CE data from the microdevices is very similar to the capillary columns, except that the retention times are greatly reduced to ~ 30 sec. To our knowledge, this represents the first separation and detection of lipoproteins in microfluidic devices.

Atomic Force Microscopy (AFM) was used to characterize the surface roughness of different microfluidic channels directly (Figure 3). The two types of chips used have surface roughnesses of $R_q = 13 \pm 2$ nm and $R_q = 2.1 \pm 0.2$ nm, respectively. The commercial chips are smoother as a result of the different fabrication process used. Published data for fused silica capillaries gives $R_q = 0.42 \pm 0.06$ nm [7]. Work in progress is aimed at determining if the surface roughness of the microdevices has an effect on the separation and sample recovery.

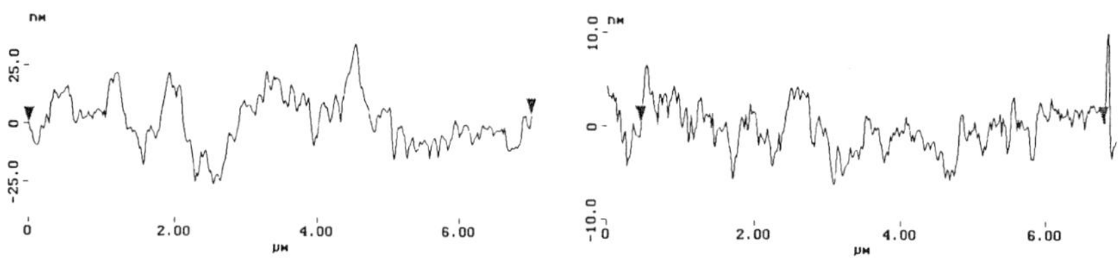

Figure 3. Surface Roughness Measurements of Microfluidic Channels by AFM. The left trace was taken from a glass channel etched in-house, with a full scale of 75 nm. The right trace was taken from the commercial glass channel, with a full scale of 20 nm.

4. Conclusions

These results show that HDL and LDL can be measured by CZE in capillary columns and in glass chips by a simple, new method that does not require the use of fixed coatings. The surface roughness of etched microchannels was measured directly by AFM. Future work will examine the effect of surface roughness on chip performance.

Acknowledgements

We gratefully acknowledge financial support from the Swiss Commission for Technology and Innovation (CTI), Project No. 3969.1.

References

1. Buchwald, H.; Boen, J.R.; Nguyen, P.A.; Williams, S.E.; Matts, J.P. *Atherosclerosis* **2001**;*154*, 221-7.
2. Lehninger, A. L.; Nelson, D. L.; Cox, M. M. *Principles of Biochemistry*, Second ed.; Worth Publishers: New York, 1993.
3. Colyer, C. L.; Tang, T.; Chiem, N.; Harrison, D. J. *Electrophoresis* **1997,** *18*, 1733-41.
4. Roberts, M.; Geiger, W.; German, J. B. Am. J. Clin. Nutrition 2000, 71, 434-7.
5. Lichtenberg, J.; Verpoorte, E.; de Rooij, N.F. *Electrophoresis* **2001**, *22*, 258 – 271.
6. Schmitz, G.; Mollers, C.; Richter, V. *Electrophoresis* **1997,** *18*, 1807-13.
7. Kaupp, S.; Wätzig, H. *Electrophoresis* **1999,** *20*, 2566-7

MICROFLUIDIC NETWORKS FOR PATTERNING BIOMOLECULES AND PERFORMING BIOASSAYS

D. Juncker [1,2], A. Bernard[1], I. Caelen[3], H. Schmid[1], A. Papra[4], B. Michel[1], N. de Rooij[2], E. Delamarche[1]

[1]IBM Research, Zurich Research Laboratory, 8803, Rüschlikon, Switzerland
[2]University of Neuchâtel, Institut de Microtechnique, CH-2000 Neuchâtel
[3]Centre Suisse d'Électronique et de Microtechnique, 2000 Neuchâtel, Switzerland
[4]Risø National Laboratory, 4000 Roskilde, Denmark

Abstract

Microfluidic networks (μFNs) are passive, self-filling devices which can be used for depositing lines of biomolecules on surfaces at high resolution. We have explored the use of μFNs replica molded in poly(dimethylsiloxane) (PDMS) or micromachined in Si. The density of biomolecules patterned on a substrate can be controlled through the channel geometry and surface chemistry, and through the volume and concentration of injected biomolecules. For instance, hundreds of micrometer long density gradients of adsorbed biomolecules could be formed. Further, we used μFNs to perform bioassays in an array-format with a spot size of 20 x 20 μm^2. We present advanced μFNs and a sealing strategy relying on free surface energy. Hard Si μFNs can thus be sealed onto a hard substrate by simply pressing the two surfaces against each other.

Keywords: Microfluidic networks, patterning of proteins, bioassay, wetting.

Introduction

Biomolecules patterned on surfaces can be used in small-scale bioassays to detect a specific binding partner in a sample solution, as it is done with DNA microarrays. Here, we present microfluidic networks (μFNs), a soft lithography tool for patterning biomolecules onto a variety of substrates [1]. μFNs are arrays of open channels which are closed upon sealing to a substrate. The filling ports, connected to the microchannels,

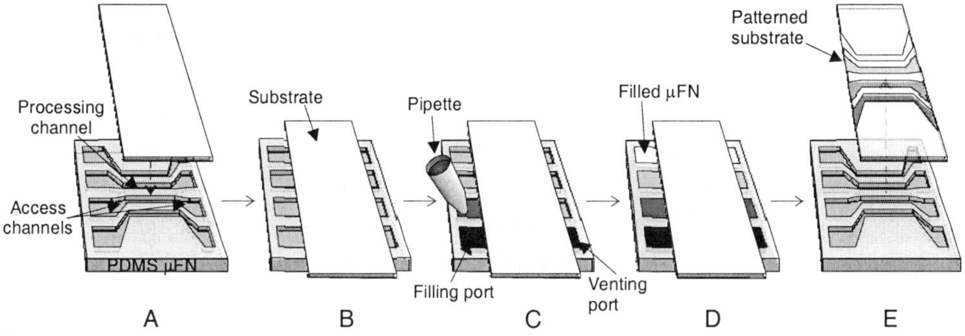

Figure 1: Procedure to pattern a substrate using a single-level μFN. (A) A μFN featuring open conduits is partially closed with a flat substrate. (B) The μFN is sealed onto the substrate. (C) The filling ports are individually addressed with different solutions that fill the channels up to the venting ports by capillary pressure. After completion of the reactions (D), the patterned substrate is separated from the μFN (E).

J.M. Ramsey and A. van den Berg (eds.), Micro Total Analysis Systems 2001, 429–431.

are individually loaded using a pipette with solutions containing distinct molecules. The inner walls of the microchannels are hydrophilic and thus a capillary pressure is generated at the liquid-air interface; this pressure draws the liquid into the microchannels, Fig.1. The small dimensions of the conduits require only minute amounts of reagents and entail quick adsorption of the biomolecules onto the substrate. These immobilized molecules thereby replicate accurately the channel pattern to a resolution of better than 1 μm.

Controlling the Areal Density of Adsorbed Proteins

In a bioassay, the readout signal should be proportional to the concentration of the analyte. When such an assay is performed at the solid-liquid interface, a precise control of the areal density of the marker molecule immobilized on the solid surface is a prerequisite. In a microchannel, the surface/volume ratio can get very large and may lead to depletion of both the molecules in the solution and those adsorbed to the surface. We overcome this problem by two means: first, by selectively grafting poly(ethylene) glycol to the three channel walls of the μFN, making them protein repellent [2]. Second, by increasing the volume of liquid flowing through the channel by means of flow-promoting structures. Alternatively, the depletion of proteins in solution may be used to create a few hundred micrometers long density gradients of immobilized proteins, Fig.2. These gradients can in turn serve to qualify the efficiency of a binding reaction as a function of the areal density of an immobilized binding partner [3].

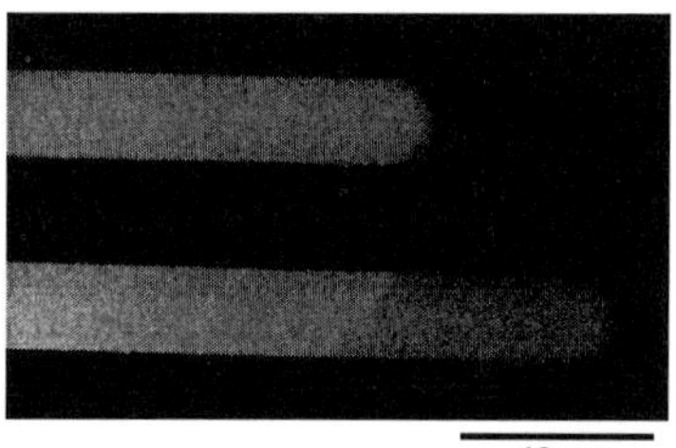

Figure 2: Short-scale density gradients of immobilized rhodamine-tagged IgGs patterned using 1.5 μm deep Si μFNs. The curvature of the gradient front reflects the poiseuille flow profile.

Micromosaic Immunoassays

μFNs are also suited for conducting bioassays with an array format along a procedure which we call micromosaic immunoassay (μMIA) [4]. This procedure allows us to generate a high areal information density by simultaneously testing n samples against m markers, Fig.3. First, the markers are immobilized as straight lines onto a surface using a μFN. Next, the uncovered areas of this surface are passivated. Finally, a μFN rotated by 90 degres is sealed onto the substrate, and the samples are flown through the microchannels across the marker molecules.

Design and Fabrication of Advanced μFNs

Conventionally, μFNs have been made in PDMS, but they can also be fabricated from silicon [5]. Si μFNs can support high-aspect-ratio structures, allow a simpler tailoring of the chemical surface properties, and can be cleaned more easily. We fabricated such μFNs with vias, which can be serviced from the back-side, are more compact and reliable. They can be used on PDMS substrates, which seal the conduits or, alternatively, on hard substrates. In the latter case, the wetting properties of the raised surfaces need to be tailored such that a free surface energy based seal is created. It is the case if both surfaces are hydrophobic. For this purpose, a flat stamp inked with perfluorotrichlorosilanes is printed onto the μFN, rendering the contacted areas highly

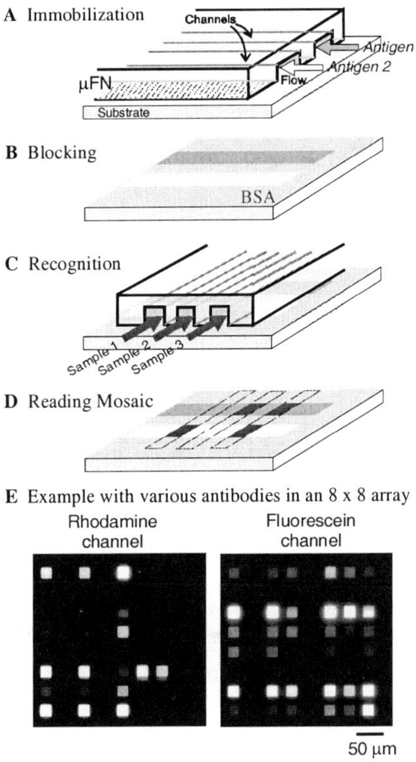

Figure 3: Procedure for making and reading a μMIA (A-D) with an example (E). The two micrographs reveal the bound antibodies tagged with green fluorescein and red rhodamine fluorophores, respectively.

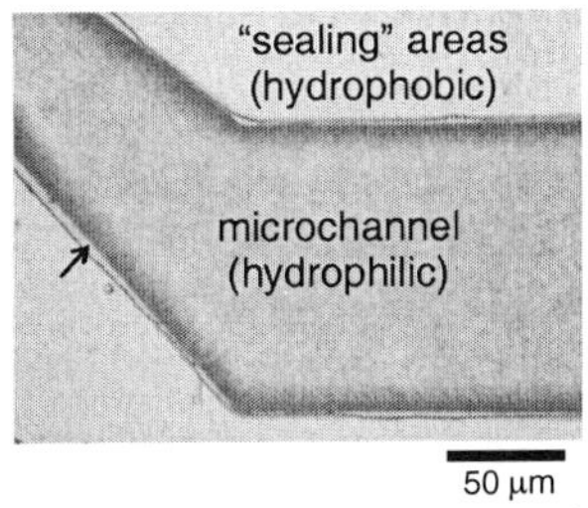

Figure 4: Free surface energy based confinement of a liquid. The liquid spontaneously fills the microchannel, but only minimally invades the "sealing" areas (arrow), in spite of a gap remaining between the two opposing surfaces.

hydrophobic. Such a Si μFN can now be sealed to hard, hydrophobic substrates, *e.g.* polystyrene, Fig.4.

Conclusions

The recent advances in the fabrication and the novel mosaic assay concepts are empowering the μFN as potential device for performing diagnostics. This passive μFN can do without mechanical moving parts, electrical circuitry and complex control devices, yet retains a high simplicity of usage; this combination represent its key asset in view of quickly performing multiple, miniaturized bioassays.

References

1. E. Delamarche *et al.*, Patterned Delivery of Immunoglobulins to Surfaces Using Microfluidic Networks, *Science* **276**, 1997, pp. 779-781.
2. Caelen *et al.*, Formation of Gradients of Proteins on Surfaces with Microfluidic Networks, *Langmuir* **16**, pp. 9125-9130, 2001.
3. A. Papra *et al.*, Microfluidic Networks made of Polydimethysiloxane, Si, and Au coated with Polyethylene Glycol for Patterning Proteins onto Surfaces *Langmuir* **17**, pp. 4090-4095, 2001.
4. A. Bernard *et al.*, Micromosaic Immunoassays, *Anal. Chem.* **73**, 2001, pp. 8-12.
5. D. Juncker *et al.*, Soft and Rigid Two-Level Microfluidic Networks for Patterning Surfaces, *J. Micromech. & Microeng.*, in press, 2001.

HIGHLY SENSITIVE SILICON MICROCHIP BASED FLOW IMMUNO BIOSENSOR USING IMMOBILIZED AFFINITY PROTEINS AND CHEMILUMINESCENCE DETECTION

Jenny Emnéus[1], Julia Yakovleva[2], Richard Davidsson[1], Anna Lobanova[2], Sergey Eremin[2], Thomas Laurell[3], Martin Bengtsson[3]

[1]Department of Analytical Chemistry, Lund University, Box 124, S-221 00, Lund, Sweden, e-mail: Jenny.Emneus@analykem.lu.se

[2]Division of Chemical Enzymology, Department of Chemistry, M.V. Lomonosov Moscow State University, Moscow 119899 Russia, e-mail: Julia.Yakovleva@newmail.ru

[3]Department of Electrical Measurement, Lund Institute of Technology, Lund University, PO Box 118, S-221 00, Lund, Sweden, e-mail: Thomas.Laurell@elmat.lth.se

Abstract

Highly sensitive micro flow immuno biosensors (μFIBs) for atrazine are presented, performed as competitive binding assays in two different assay formats on silicon μ-chips. Different polymer matrices, conjugation chemistries, carrier and regeneration buffers are explored to obtain the best immobilization and assay conditions for antibodies (**Ab**) and protein G (**p G**) in terms of assay stability, sensitivity, and reusability.

Keywords: silicon microchip, flow injection, immuno biosensor, surface modification, chemiluminescence detection

1. Introduction

High surface area immobilized enzyme μ-reactors in silicon has been pursued since 1994 [1] and the surface could vastly be enhanced by anodization of the silicon μ-structure in hydrofluoric acid, providing a high surface area layer on the silicon μ-reactor [2]. The use of μ-structures in other μ-analytical applications where surface area is a fundamental need is a natural development and it seems promising, therefore to expand this concept to, *e.g.* the field of immuno biosensors (IB). The proper choice of the **Ab** immobilization method to retain activity and stability is one of the most important steps in the development of IB. Several challenges are associated with immobilizing **Ab** on silicon μ-surfaces, *e.g.* the ability of silica to exhibit strong non-specific protein adsorption via ionic interactions, Van-der-Waals forces, polar-polar interactions is a well known phenomenon [3].

A multi-purpose μ-flow injection system, useful for monitoring emerging luminescence directly from the surface of a μ-chip, containing various biological

J.M. Ramsey and A. van den Berg (eds.), Micro Total Analysis Systems 2001, 432–434.

components, is discussed. The focus here will be on μFIBs and different surface modification strategies for immobilization of **Ab** and **pG**.

2. Method

In the first μFIB format the analyte (**Ag**) and a tracer (**Ag***, analyte labeled with horseradish peroxidase (HRP)) is injected into the μFIB in Fig. 1, competing for immobilized **Ab** binding sites on the μ-chip. The second μFIB format is a generic system (usable for many **AbAg** systems) where **Ag**, **Ag*** is competing for **Ab** off-line and then injected in the μFIB, containing a **pG** μ-chip, capable of trapping **Ab**, **AbAg**, and **AbAg***. In both formats the **AbAg*** is detected by injection of HRP substrate (luminol, H_2O_2 and *p*-iodophenol (PIP)), generating a chemiluminescent (CL) signal monitored by a photomultiplier tube (PMT, Hamamatsu, Japan). The μFIBs are then regenerated before new assay cycles can be performed.

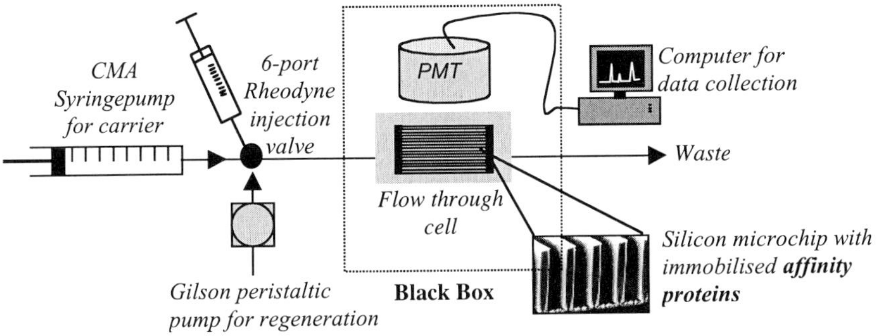

Figure 1. The μFIB set-up containing a μ-chip with immobilized **Ab** or **pG**.

3. Experimental

Polyclonal anti-atrazine **Ab** (R. Abuknesha) and recombinant **pG** (Biomedicals INC, Ohio, USA) were immobilized on silicon μ-chip surfaces, fabricated by anisotropic wet etching [4]. Different immobilization protocols and polymer combinations were investigated, using covalent or non-covalent attached 3-aminopropyltriethoxysilane (**APTES**), 3-glycidoxypropyltrimethoxysilane (**GOPS**), linear polyethyleneimine (**LPEI** MW 750,000), branched polyethyleneimine (**BPEI** MW 25,000), polyvinyl alcohol (**PVA**) or dextran (**DEX**), followed by adsorption or covalent attachment (via glutaraldehyde (**GA**)) of the affinity proteins.

4. Results and discussion

The obtained LODs for μFIBs with covalently attached **APTES-GA-Ab**, physically adsorbed **LPEI-GA-Ab** and covalently attached **GOPS-BPEI-GA-Ab** were 45, 4 and 0.8 ng L^{-1}, thus giving the most sensitive assay for **GOPS-BPEI-GA-Ab**. The **pG** based μFIBs were in general less sensitive with LODs for **LPEI-GA-pG**, **APTES-DEX-pG**, and **APTES-GA-PVA-pG** of 22, 6, and 39 ngL^{-1}, respectively. However, all assay can be considered sensitive with LODs below the European Commission directive to measure atrazine in drinking water at 0.1 μg L^{-1}.

The long-term stability (LST) is an important feature and could be substantially increased, by using Tris/HCl carrier instead of inorganic buffer such as phosphate or carbonate. In general the best overall performance was exhibited for **LPEI-GA**-based chips in terms of both LOD and LST, where no loss of antibody activity was detected over a one-month period. Figure 2 shows calibration graphs obtained for μFIBs with **LPEI-GA** modified **Ab** and **pG** based μ-chips.

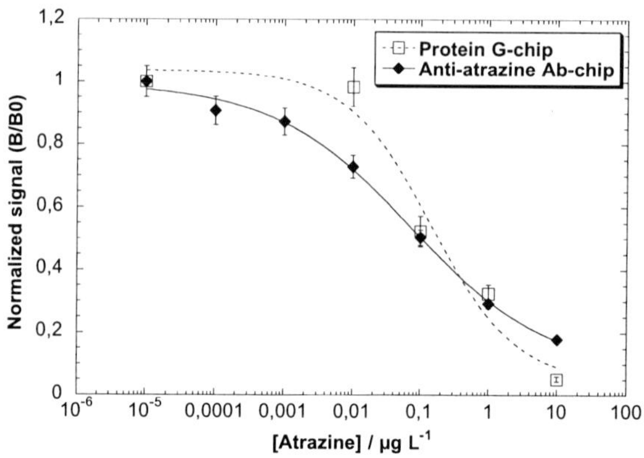

Figure 2. μFIB assays for LPEI-GA protein G and antibody modified μ-chips

5. Conclusions

Numerous immobilization chemistries have been reported for **Ab** immobilization however most efforts are devoted to obtaining good calibration curves, while finding the best immobilization procedure gains less attention. Our work was focused on the optimization of **Ab** immobilization protocols, resulting in the best stability, reusability, and sensitivity of the μFIBs. The best immobilization procedure, offering a simple protocol, allowing good regeneration, reusability, and sensitivity was obtained with physically adsorbed LPEI, followed by covalent attachment of **Ab** (**LPEI-GA-Ab** surface).

Acknowledgements

Financial support from the Swedish Foundation for Strategic Environmental Research (MISTRA) and the EC contract (IC15CT98-0910) as well as **Ab** obtained from Dr R. Abuknesha (King's College, London, UK), are gratefully acknowledged.

References

1. T. Laurell and L. Rosengren , *Sensors and Actuators B*, 1994, *19*, 323-326.
2. T. Laurell, J. Drott, L. Rosengren and K. Lindström, *Sens. Actuators (B)*, 1996, *31*, 161-166.
3. Andrade J.D., Hlady V. *Ann. N.Y. Acad. Sci*. 1987, *516*, 158-163.
4. Petersen, K. Proc. IEEE 1982, 70, 420-457

HETEROGENEOUS COMPETITIVE IMMUNOASSAY FOR HUMAN IgG IN DISPOSABLE BIO-PASSIVATED PDMS MICROFLUIDIC CHANNELS USING AN ON-CHIP INTERNAL FLUORESCENT STANDARD

Vincent Linder[1,2,3], Elisabeth Verpoorte[1], Wolfgang Thormann[2], Nico F. de Rooij[1], Hans Sigrist[3]

[1]Institute of Microtechnology, Univ. of Neuchâtel, 2007 Neuchâtel, Switzerland
[2] Dept. of Clinical Pharmacology, Univ. of Bern, 3010 Bern, Switzerland
[3] CSEM SA, 2007 Neuchâtel, Switzerland

Abstract

A multi-layer surface treatment of a PDMS/glass microfluidic chip was developed to suppress adsorption to the channel walls. The outermost layer of the coating was patterned with antibodies that were used to carry out a heterogeneous immunoassay. Human IgG was analyzed using an electrokinetically driven competitive immunoassay with Cy5-labeled human IgG as tracer. An internal standard was implemented on-chip to meet the high level of confidence in assay results required by clinical diagnostics. This work demonstrates that it is possible to control key aspects of a chip-based immunoassay, including the fluidic properties, antibody activity and sample delivery.

Keywords: PDMS, heterogeneous immunoassay, IgG, coating, internal standard

1. Introduction

Poly(dimethylsiloxane) (PDMS) enables rapid and accurate replication of polymer-based microfluidic networks. However, massive adsorption of apolar molecules to PDMS channel walls is often observed. An adsorption-suppressing treatment for the PDMS surface has been implemented, based on the deposition of a three-layer sandwich coating consisting of biotinylated IgG, neutravidin and biotinylated dextran. The first IgG layer is physisorbed to the PDMS, whereas the second and third layers are fixed using highly specific biotin-neutravidin binding. By replacing biotinylated dextran with any biotinylated reagent, the modified surface can be readily patterned with biochemical probes such as antibodies. The combination of probe immobilization chemistry with low non-specific binding (NSB) enables affinity-binding assays within channel networks [1,2]. Here, an example of an electrokinetically (EK) driven, heterogeneous immunoassay (IA) for human immunoglobulin G (hIgG) in disposable microfluidic chips is described.

2. Experimental

Channels laid out in simple cross geometries were prepared by sealing a replicated PDMS microfluidic network on a microscope slide. The passivation coating was deposited by sequential suction pumping of solutions containing biotinylated goat anti-mouse IgG, neutravidin and biotinylated dextran throughout the channel network. The last solution can be substituted with one containing biotinylated antibodies to impart

J.M. Ramsey and A. van den Berg (eds.), Micro Total Analysis Systems 2001, 435–437.

biospecificity to the surface and obtain a platform for immunoreaction, which we call TERASYS (TERnary Affinity SYStem) [3].

In Figure 1a, biotinylated antibodies are immobilized at the center of the intersection. The immobilization follows the laminar flow pattern of antibody and dextran solutions established when these solutions are drawn from their designated reservoirs using vacuum. IA was carried out by EK pumping of a sample in a TERASYS device, followed by removal of unreacted analyte (Figure 1b,c). Quantitative on-chip fluorescence detection was recorded using a 428™ scanner (Affymetrix), as shown in Figure 1d.

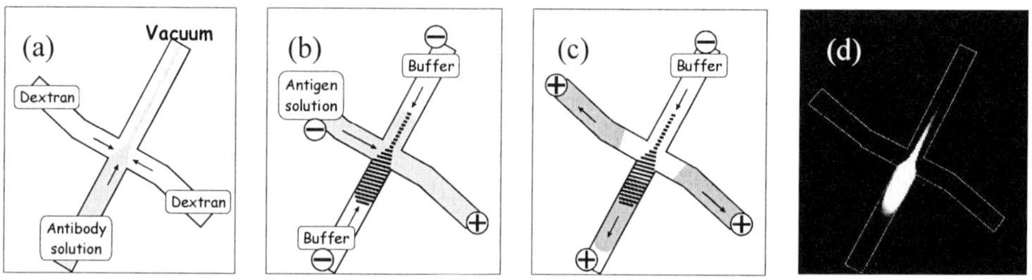

Figure 1: IA at pH 7.4, arrows indicate IgG transport direction. (a) Flow patterning of antibodies in TERASYS chips. (b) Antigen incubation; immunoreaction takes place where antigen is in contact with immobilized antibodies (hatched area). (c) Removal of unbound antigen. (d) Fluorescence image used for quantification.

3. Results and discussion

TERASYS chips were prepared for hIgG determination by competitive IA using Cy5-labeled hIgG (Cy5-hIgG). An internal standard was implemented in the TERASYS chips, by immobilizing biotinylated goat anti-mouse IgG and biotinylated goat anti-human IgG in equal amounts in the sensing area. Samples containing 665nM Cy3-mouse IgG (Cy3-mIgG) and 665nM mixtures of Cy5-hIgG/hIgG (different ratio) were then reacted on the chips. Recording fluorescence using 532 and 635 nm laser lines provided individual measurements for Cy3 and Cy5. An abnormal or missing Cy3 signal showed that TERASYS was not working. Minor chip-to-chip variations were compensated for by using the Cy5/Cy3 signal ratio. The reference signal obtained for 665nM Cy3-mIgG in physiological buffer remained unchanged chip-to-chip and sample-to-sample (RSD 13%, n=12).

Competition in the hIgG IA took place, but had a strong dependence on the running buffer pH. EK transport of sample was very different in physiological buffer (pH 7.4) than under alkaline conditions. Unlabeled and labeled IgGs were observed to migrate very slowly in a direction opposite to electroosmotic flow (EOF) at pH 7.4. However, at pH 9.3 the observed migration was with the EOF. Typical net mobilities of Cy3-mouse IgG were in the order of $6.5 \cdot 10^{-5}$ $cm^2 s^{-1} V^{-1}$ towards the cathode at pH 9.3 and towards the anode at pH 7.4.

Fluorescent labeling of IgG introduced additional negative charges to the IgG molecules. Therefore, when migrating against EOF at pH 7.4, Cy5-hIgG reached the intersection first and could react initially with immobilized antibodies without competition. Competitive IA could therefore take place only using the remaining free binding sites, resulting in poor assay sensitivity. However, at pH 9.3 unlabeled IgG migrated with EOF, so that initial hIgG binding took place under non-competitive conditions, yielding enhanced assay sensitivity, as shown in Figure 2.

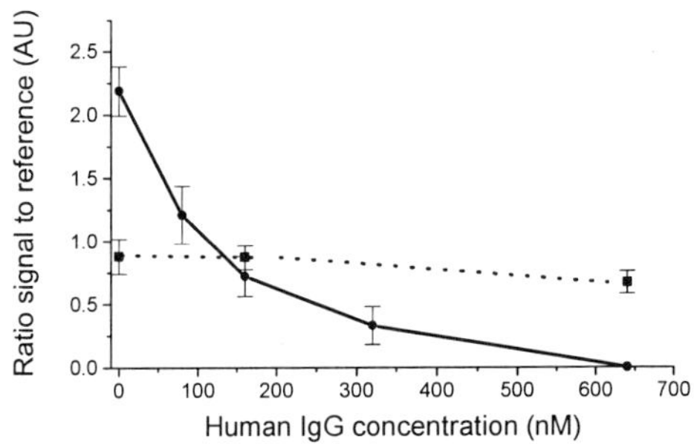

Figure 2: Calibration curve for human IgG competitive immunoassay. Sensitivity was enhanced when changing pH from 7.4 (dotted line) to 9.3 (solid line).

The IA in alkaline conditions enabled detection of hIgG below 80 nM (12 µg/mL), with 14% RSD (a value calculated by averaging RSD from 4 sets of data, $n_{tot} = 17$). The incubation time (not optimized) was set to 7 min to insure saturation of all binding sites on the surface. The heterogeneous format of the assay may enable the analysis of more dilute samples by allowing longer incubation times.

4. Conclusion

Heterogeneous IA for hIgG has been demonstrated on disposable chips. The use of a Cy3-mIgG internal standard brings essential information about chip quality and therefore levels off chip-to-chip variation, which gives more confidence in the assay results. The direction of electrokinetic transport of sample was inverted from physiological to alkaline media, so that choosing the right pH enhanced assay sensitivity. This example showed how differences in mobility between partners of a competitive immunoreaction could be exploited in a microfluidic device to reach an hIgG limit of detection below 80 nM.

References

1. A. Dodge, K. Fluri, E. Verpoorte, N.F. de Rooij, *Electrokinetically Driven Microfluidic Chip with Surface Modified Chambers for Heterogeneous Immunoassays*, Anal. Chem., in press, (2001).
2. T. Yang, S. Jung, H. Mao, P. Cremer, *Fabrication of Phospholipid Bilayer-Coated Microchannels for On-Chip Immunoassays*, Anal. Chem., pp. 165-169, 73, (2001).
3. V. Linder, E. Verpoorte, W. Thormann, N.F. de Rooij, H. Sigrist, *Surface Bio-Passivation of Replicated PDMS Microfluidic Channels and Application to Heterogeneous Immunoreaction with On-chip Fluorescence Detection*, Anal. Chem., in press, (2001).

A MICROFABRICATED PROBE WITH INTEGRATED COILS AND CHANNELS FOR ON-CHIP NMR SPECTROSCOPY

C. Massin[1], A. Daridon[2], F. Vincent[1], G. Boero[1], P.-A. Besse[1], E. Verpoorte[2], N.F. de Rooij[2], and R.S. Popovic[1]
[1]Institute of Microsystems, EPFL, Lausanne, Switzerland
[2]SAMLAB, Institute of Microtechnology, University of Neuchâtel, Switzerland
charles.massin@epfl.ch

Abstract

We present the development of a monolithic NMR probe fabricated on a glass substrate, combining an integrated high-quality factor, multi-turn planar coil with a microfluidic network. Concentrations of a few percent ethanol in water can be detected by ^1H NMR in a 30-nL volume with only 3 scans.

Keywords: NMR spectroscopy, planar microcoils, microfluidics, detection method

1. Introduction

Nuclear Magnetic Resonance (NMR) spectroscopy is a powerful analytical technique that provides extensive functional and structural information about compounds. While optical and electrochemical methods of detection are commonly used in Micro Total Analysis Systems (μTAS), little work has been presented using NMR as a detection method for microfluidic systems. Trumbull et al. have investigated coupling chip-based capillary electrophoresis with NMR spectroscopy [1]. They showed that the integration of NMR into a μTAS is feasible from a fabrication standpoint, but satisfactory NMR spectra could only be obtained from samples of high concentrations. Dechow et al. [2] fabricated planar coils on both glass and GaAs substrates with an etched cavity on the backside and acquired an NMR spectrum of a 60-nL drop of silicone oil placed in the open sample chamber. While these initial studies demonstrated the feasibility of combining NMR with microfluidics, we present a new probe design which results in improved detection sensitivity.

2. Theory

The main limitation in micro-NMR spectroscopy is a poor signal-to-noise ratio (*SNR*). The signal is maximized when the size of the coil best matches the size of the sample, while the noise is dominated by the thermal noise from the series resistance of the coil. To analyze small samples, microcoils may be used advantageously. For NMR detection in the nL sample volumes typically encountered in μTAS, multi-turn microcoils with low series resistance should be fabricated in close proximity to microfluidic channels to maximize *SNR*.

J.M. Ramsey and A. van den Berg (eds.), Micro Total Analysis Systems 2001, 438–440.

3. Experimental

Two 525-μm-thick glass wafers with etched inner faces are aligned and fusion-bonded to form the microfluidic substrate (Figure 1 a & b), as presented elsewhere [3]. To optimize the coupling between the coil and the NMR sample, the top surface is thinned by polishing, leaving a 65 μm glass membrane over the 150-μm-deep channels. The NMR detection coils are then fabricated on top of this membrane, using a process [4] based on SU-8 photoepoxy and copper electroplating (Figure 1 c). Figure 2 shows a picture of a 5-turn, 500 μm inner diameter coil on top of the microfluidic substrate.

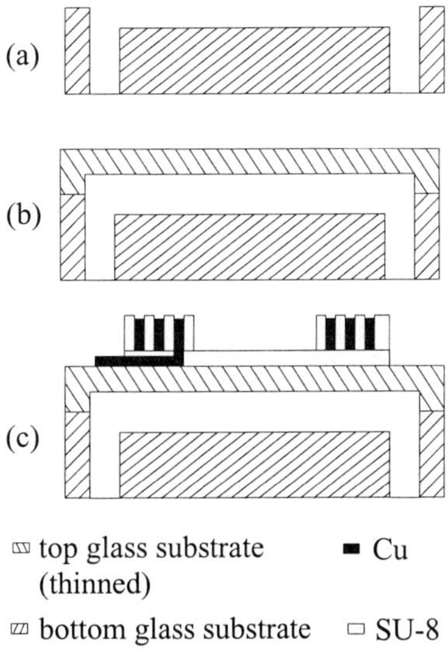

Figure 1: Simplified fabrication process of planar microcoils on a glass substrate with integrated channels.

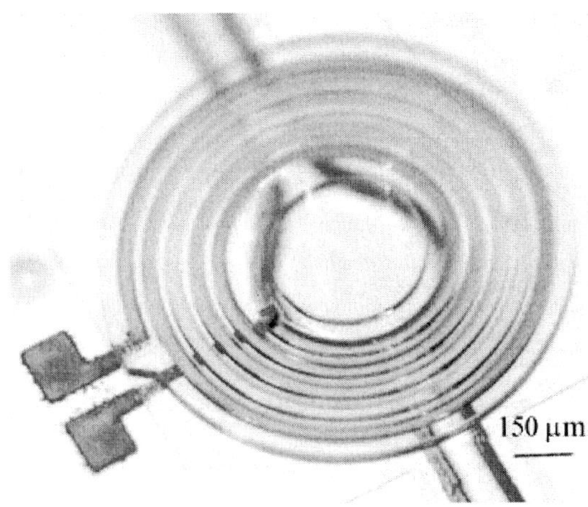

Figure 2: Photo of a 5-turn, 500 μm inner diameter, electroplated coil on a glass substrate containing an etched flow cell (30 nL) with in- and outlet channels.

4. Results and discussion

For the 3-turn, 500-μm-inner-diameter coil with 6 μm trace thickness used in this experiment, the measured inductance is 8.8 nH, series resistance, 1.0 Ω, and parasitic capacitance, 90 fF, giving a Q factor of 17 at 300 MHz. The channel geometry below the coil results in a sample volume of about 30 nL. ^{1}H-NMR spectra are acquired in a 7 T superconducting magnet with 90°, 10 μs, 440 mW pulses. The *SNR* for water is 117 per scan with a linewidth of 30 Hz. This represents a factor of 5 increase of the *SNR* compared to previously published work [1]. As a result of the high *SNR*, volume concentrations of a few percent ethanol in water can be detected with only 3 scans (Figure 3). On the other hand, the achieved spectral resolution does not yet allow the resolution of spin coupling, and will need to be improved in the future.

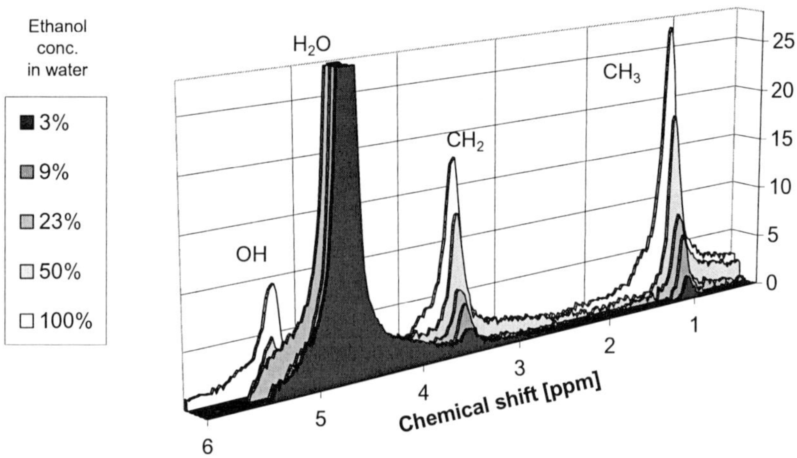

Figure 3: ^{1}H NMR spectra of ethanol at different concentrations in water. Spectra acquired with a 3-turn, 500-μm-diameter coil over a 30 nL sample chamber.

5. Conclusions

A monolithic NMR probe combining a high-Q factor multi-turn planar coil with microfluidic channels etched in a glass substrate has been fabricated and tested in an NMR experiment. Further improvements may allow mM detection range in volumes of a few tens of nanoliters.

Acknowledgements

The devices were fabricated in the SAMLAB (IMT, Univ. of Neuchâtel) cleanroom, and at the Center of Microtechnology (CMI) at EPFL. We thank the technical staff of both centers, as well as G. Vaucher and Y. Pilloud for helpful support. This work was supported by an EPFL Presidential Fund and the Swiss National Science Foundation (Proj. Nr. 2100-61549.00).

References

1. J.D. Trumbull, I.K. Glasgow, D.J. Beebe, R.L. Magin, *Integrating microfabricated fluidic systems and NMR spectroscopy*, IEEE Transactions on Biomedical Engineering, vol.47, no.1, pp.3-7, (2000).
2. J. Dechow, A. Forchel, T. Lanz, A. Haase, *Fabrication of NMR-microsensors for nanoliter sample volumes*, Microelectronic Engineering, vol.53, pp. 517-19, (2000).
3. J. Lichtenberg, E. Verpoorte, N.F. de Rooij, *Sample Preconcentration by Field Amplification Stacking for Microchip-based Capillary Electrophoresis*, Electrophoresis, vol. 22, pp. 258-71, (2001).
4. C. Massin, G. Boero, P. Eichenberger, P.-A. Besse, R.S. Popovic, *High-Q factor RF planar microcoils on glass substrates for NMR spectroscopy*, Proceedings of Transducers'01, Springer Verlag, vol.1, pp. 784-787, (2001).

SPR IMAGING MEASUREMENTS OF DNA MICROARRAYS IN PDMS MICROFLUIDIC CHANNELS ON GOLD THIN FILMS

Hye Jin Lee, Terry Goodrich and Robert M. Corn
University of Wisconsin-Madison, Department of Chemistry, 1101 University Avenue, Madison, WI 53706, USA

Abstract

The application of microfluidic channels created in polydimethylsiloxane (PDMS) to surface plasmon resonance (SPR) imaging experiments is described. The microfluidic channels are used to fabricate "1-D DNA line arrays" and "2-D DNA hybridization arrays" for the detection of DNA and RNA hybridization adsorption onto chemically modified gold surfaces. A total sample volume of 500 μL is used in the 1-D line array experiments, and a total sample volume of only 1 μL is required in the 2-D DNA array experiments.

Key words: **SPR imaging, DNA microarrays, PDMS microfluidic channels.**

1. Introduction

SPR imaging is emerging as a powerful tool for the label-free detection of the affinity binding of biological molecules onto chemically modified gold surfaces. We have recently described a fabrication methodology for creating DNA microarrays on gold surfaces [1]. In these experiments, the hybridization adsorption of single stranded RNA (ssRNA) target molecules onto an array of surface bound ssDNA probe molecules was observed. Target molecules could be detected down to a concentration of 10 nanomolar with a total sample volume of approximately 500 μL. For some biological applications, such as gene expression and the detection of microbial species, it would be preferable to use a significantly smaller sample volume. One way to achieve this is to employ microfluidic networks that express the concept of microscale total analysis systems (μ-TAS) [2]. In this paper, we demonstrate (i) a general method for fabricating 1-D ssDNA line arrays using parallel microfluidic channels on chemically modified gold surfaces, and (ii) a microliter detection volume methodology that creates 2-D DNA hybridization arrays by employing the 1-D ssDNA line arrays in conjunction with a second perpendicular set of microfluidic channels for sample solution delivery. The microfluidic channels were created in PDMS polymer using a well-established, simple, rapid and low cost fabrication methodology [3].

2. Fabrication of 1-D ssDNA line arrays using microfluidics

The fabrication methodology for creating ssDNA microarrays is as follows:

J.M. Ramsey and A. van den Berg (eds.), Micro Total Analysis Systems 2001, 441–443.

(1) PDMS microchannels were attached to a gold thin film that was modified with a self-assembled mononlayer of 11-mercaptoundecylamine (MUAM). (2) A surface pattern was created by flowing a heterobifunctional linker, sulfosuccinimidyl 4-(N-maleimidomethyl) cyclohexane-1-carboxylate (SSMCC), through the microchannels. SSMCC reacted with the amine-terminated self-assembled monolayer to produce a maleimide-terminated monolayer that can react with thiol-labeled biomolecules. (3) A variety of 5' thiol-modified ssDNA probes were each flowed into one of the PDMS microchannels and left to react with the maleimide-terminated gold surface for four hours. This created a "1-D" line array of different ssDNA probe molecules on the surface. (4) The PDMS was removed from the gold thin film and the MUAM background was exposed to a solution of PEG-NHS (*N*-hydroxysuccinimidyl ester of methoxypoly(ethyleneglycol)propionic acid), which prevents non-specific adsorption.

Figure 1 shows SPR difference images of the hybridization adsorption of complementary ssDNA from solution onto 1-D DNA line arrays composed of 3 different 18-base ssDNA probes. In these measurements, the total sample volume was 500 μL. When the array was exposed to a target ssDNA solution, only the complementary array elements were visible in the SPR difference image due to hybridization adsorption. Exposure of the surface to a mixture of all three targets resulted in a SPR difference image that contained all of the array elements. The nearly equivalent SPR signal level of each ssDNA line demonstrates the regularity of the microfluidic surface modification chemistry.

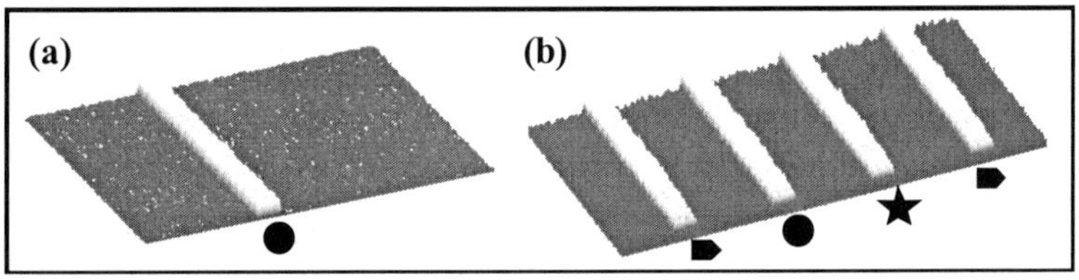

Figure 1. SPR difference images showing hybridization of target DNA onto an array of surface bound probe DNA. Hybridization onto the probe DNA array is indicated by a change in the percent reflectivity.

3. Fabrication of 2-D DNA hybridization arrays using microfluidics

The fabrication of 2-D DNA hybridization arrays was achieved by using the 1-D ssDNA line arrays created on the gold surface in conjunction with a second set of PDMS microchannels that were attached onto the surface perpendicular to the ssDNA lines. When complementary ssDNA was flowed into the microchannels, a 2-D microarray of double-stranded DNA (dsDNA) intersections were created. These 2-D microarrays can be used to detect hybridization adsorption onto the surface either with fluorescence microscopy or SPR imaging. Figure 2 shows SPR imaging measurements for the detection of perfectly complementary DNA hybridizing onto three different ssDNA probe spots. The amount of

DNA in this sample was about 20 femtomoles in a total sample volume of 1 μL. The line profile in figure 2 (b) shows that the specific hybridization adsorption of the complementary DNA target onto each probe was achieved without non-specific adsorption onto the background gold surface in the microchannel. As in the previous measurements, nearly equivalent SPR signal levels were obtained for each ssDNA hybridization adsorption onto all target molecules. The detection limit of this 2-D DNA array was about 5 femtomoles, which should be sufficient for a number of biological applications, including mRNA detection for gene expression and rRNA detection for complex microbial samples.

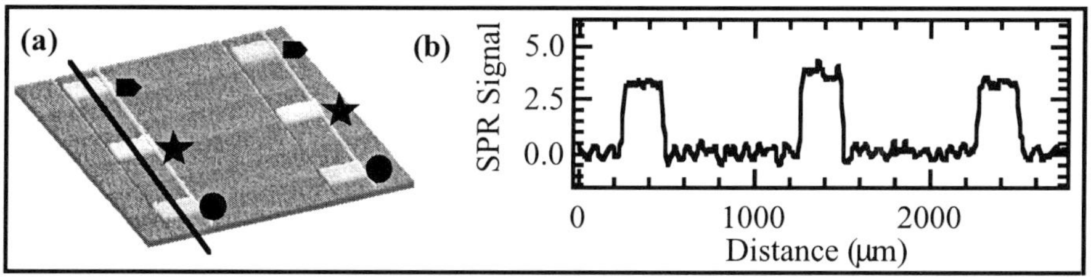

Figure 2. (a) SPR difference images and (b) line profiles showing hybridization of perfectly complementary DNA onto a 2-D microarray of surface bound probe DNA. Each array element is 300 μm by 300 μm. Each shape marker refers to a different ssDNA probe.

4. Conclusions

The use of microchannels described in this paper represents a new methodology for the creation of 1-D and 2-D DNA microarrays on gold surfaces for SPR imaging measurements of DNA-DNA and DNA-RNA interactions. The 1-D line arrays allow us to create surface patterns as narrow as 50 μm as well as to place up to 50 different probes on a single DNA array. The 2-D DNA microarrays can be used for microliter volume detection of binding events from biological samples with SPR imaging.

Acknowledgements

This research is funded by the National Institute of Health (Grant R01-GM59622-02) and the National Science Foundation (Grant CHE-9820742).

References

1. B. P. Nelson, T. E. Grimsrud, M. R. Liles, R. M. Goodman and R. M. Corn, *Anal. Chem*, 73, 1-7, (2001).
2. A. Manz, E. Verpoorte, D. E. Raymond, C. S. Effenhauser, N. Burggraf, and H. M. Widmer, "μ-TAS: Miniaturized Total Chemical Analysis Systems" at μ-TAS Workshop, MESA Monograph (Eds), Enschede (NL), *1*, (1994).
3. D. C. Duffy, J. C. McDonald, O. J. A. Schueller, G. M. Whitesides, *Anal. Chem.* 70, 4974 -4984, (1998).

MAGNETIC METHOD FOR DNA DETECTION ON AN ARRAYED SOLID STATE DEVICE

C.R. Tamanaha[1], R.J. Colton[3], M.M. Miller[3], M.A. Piani[2], J.C. Rife[3], P.E. Sheehan[3], L.J. Whitman[3]
[1]Geo-Centers, Inc., Fort Washington, MD 20749
[2]NOVA Research, Inc., Alexandria, VA 22308
[3]Naval Research Laboratory, Washington, DC 20375
Cy.Tamanaha@nrl.navy.mil

Abstract
A unique solid-state giant magnetoresistive (GMR) biosensor is the centerpiece of an integrated sample processing system for the detection of target DNA molecules through hybridization. In addition to the electromagnetic instrumentation, surface chemistries and assays have been developed, as well as a cartridge-based microfluidics network in which pumps and valves are operated by an independent actuation mechanism.

Keywords: Biosensor, DNA, fluidics, GMR, magnetic labeling

1. Introduction

Current DNA array technologies often rely primarily on optical detection methods such as chemiluminescence, fluorescence, and colorimetric assays [1]. We have developed a fundamentally different technique to quantitatively detect and identify biological molecules. Our technique is based on the labeling of DNA molecules with magnetic microbeads and their subsequent detection with giant magnetoresistance (GMR) magnetic field sensors [2,3]. A complete analytical system called the Bead ARray Counter (BARC), which includes electronic instrumentation and fluidics, is being built around this technology. The fluidics component is essential to the efficiency at which the analytical instrument operates and the samples are analyzed. In addition, a successful early detection system suitable for field use requires components that are rugged and easy to use. It seems clear that in the interests of cost, construction, and maintenance, many microfluidics devices should be based on polymeric materials utilizing a minimal number of inexpensive parts [4]. We also feel that these same objectives can be further met by a synergistic design that couples disposable microfluidic components (e.g. pumps and valves) with reusable actuator devices. Our recent progress toward completion of the BARC system is described here.

2. Instrumentation

Over the last few years, BARC has witnessed an evolution in GMR sensor chip design. The most recent chip, BARC III, incorporates an array of 64 GMR sensors that show promise for improved detection efficiency and parallel analysis of multi-analyte DNA samples (Figure 1). Miniature fluidic devices and actuators have also been developed to automate the assay procedures. Microchannels, valve-less pumps, and

J.M. Ramsey and A. van den Berg (eds.), Micro Total Analysis Systems 2001, 444–446.

pinch-valves have been cast into a polydimethylsiloxane (PDMS) membrane and packaged into a palm-sized cartridge along with the BARC sensor chip (Figure 2). Operation of the pumps and valves occurs when the cartridge is inserted into the BARC instrument and is mechanically coupled to an independent actuator block.

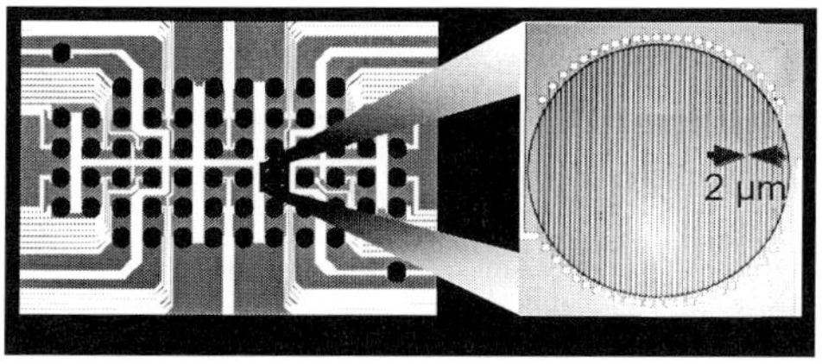

Figure 1. Micrograph showing the 64-sensor array on the BARC III chip. Each 200 μm-diameter sensor spot includes a single serpentine 2 μm-wide GMR sensor strip. A 0.2 μm-thick layer of Si_3N_4 protects the chip from the aqueous assay environment.

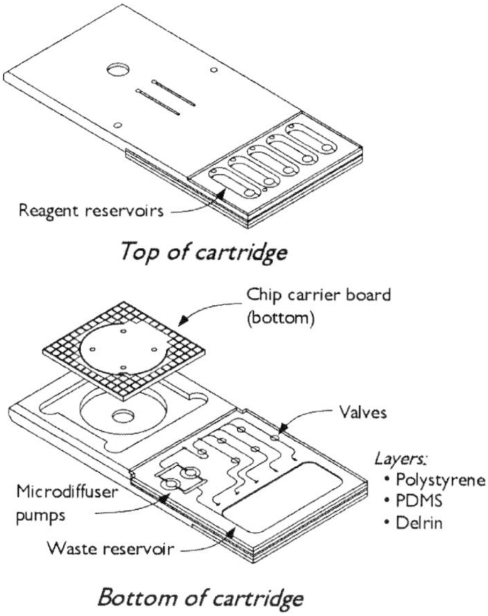

Figure 2. Schematic drawing of a 5.1 cm x 10.2 cm cartridge used to hold a PCB mounted BARC chip and a PDMS membrane containing a network of 125 μm-deep channels, valves, pumps and reservoirs. The cartridge interfaces with an independent pump and valve actuator block for automated dispensing of fluids.

3. Experimental assays

Distinct single-stranded DNA capture probes are immobilized on each sensor through thiol-gold chemistry. Sample target DNA, complementary to the immobilized probes, is allowed to hybridize on the chip followed by labeling with magnetic microbeads. The microbeads are then subjected to a magnetic field and detected by the GMR sensors.

Refinement of our hybridization and labeling methods has improved the overall sensitivity to 10 fM for a 15-minute hybridization time with 30 bp oligonucleotides (Figure 3). Further efforts to improve the sensitivity are currently underway; for example, we have been experimenting with peptide nucleic acid (PNA) capture and detection probes. Initial results with PNA indicate that an increase in sensitivity of one to two orders of magnitude can be expected.

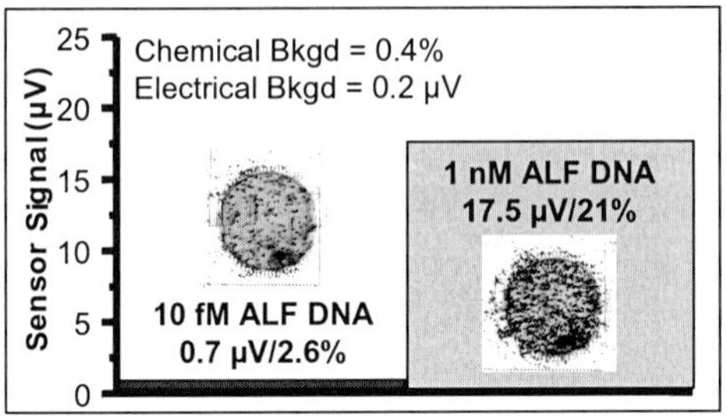

Figure 3. BARC III single-sensor assay results showing Anthrax lethal factor DNA detection for 10 fM and 1 nM concentrations. The complete assay required about 40 minutes. Both the sensor signal and optical micrographs of the sensor are shown. The percent of the sensor covered by beads in each case is also indicated.

4. Conclusions

Through the synergism of various macro and micro components, a complete and effective non-optical system for the analysis of DNA samples has been demonstrated. The use of state-of-the-art solid-state GMR sensors, non-fouling surface and assay chemistries, and a specially designed fluidics system has all contributed to the high detection sensitivity achieved by the BARC instrument.

Acknowledgements

This work was supported in part by Defense Advanced Research Projects Agency.

References

1. G. Ramsay, "DNA Chips: State-of-the-Art," *Nature Biotech.* **16**, 40-44 (1998).
2. D.R. Baselt, G.U Lee, M. Natesan, S.W. Metzger, P.E. Sheehan, and R.J. Colton, "A Biosensor Based on Magnetoresistance Technology," *Biosens. Bioelectron.* **13**, 731-739 (1998).
3. R.L. Edelstein, C.R. Tamanaha, P.E. Sheehan, M.M. Miller, D.R. Baselt, L.J. Whitman, and R.J. Colton, "The BARC Biosensor Applied to the Detection of Biological Warfare Agents," *Biosens. Bioelectron.* **14**, 805-813 (2000).
4. G.M. Whitesides, A.D. Stroock, "Flexible Methods for Microfluidics," *Physics Today* **54** (2001). 14 June 2001 <http://www.physicstoday.org/pt/vol-54/iss-6/p42.html>.

A MICROCANTILEVER-BASED DETECTION PRINCIPLE FOR μTAS

Jacob Thaysen and Anja Boisen
Mikroelektronik Centret, Technical University of Denmark, Ørsteds Plads, DK-2800 Kgs. Lyngby

Abstract

We present a new cantilever-based bio-chemical sensor integrated into a microliquid handling system. The cantilevers have integrated piezoresistive readout, which enables simple measurements on even non-transparent liquids. The chip design consists of 10 cantilevers inserted in a channel. The chip has been used for in-situ online monitoring of the etching of a metal film and the immobilization of molecules with sub-monolayer resolution.

Keywords: Cantilever-based sensor, piezoresistive readout, micro liquid handling system

Introduction

Detection of molecular interactions on a surface by a mechanical sensing principle has been investigated by several groups in the recent years [1,2]. When molecules are immobilized on the surface of a microcantilever, a bending of the microcantilever can be observed due to a surface stress change. Normally, cantilevers for atomic force microscopy are placed in liquid cells with milliliter volumes and the bending of the cantilevers are measured by optical techniques. By optical detection, it is difficult to perform simultaneous measurements on several cantilevers. Moreover, the milliliter sized liquid cells allow temperature gradients and turbulence in the liquid, which causes noise in the system.

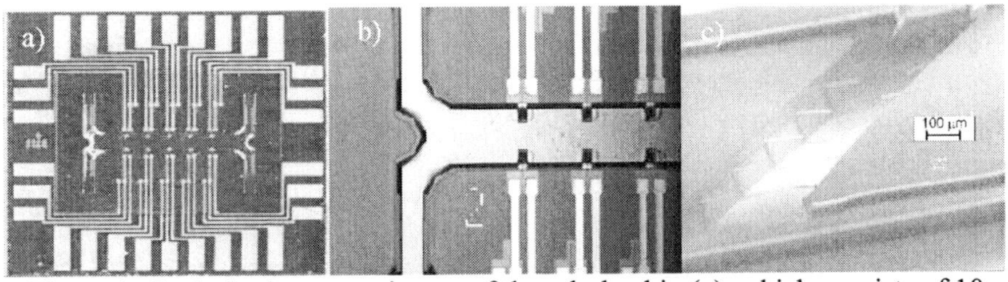

Figure 1: Optical microscope image of the whole chip (a), which consists of 10 cantilevers and 4 in/outlets and a close-up of the channel (b), where six cantilevers are visible. A SEM micrograph shows a side view of the channel (c).

Integrated system

We present a new cantilever-based bio-chemical sensor integrated in a microliquid handling system. The cantilevers have integrated piezoresistive readout

J.M. Ramsey and A. van den Berg (eds.), Micro Total Analysis Systems 2001, 447–449.

which, compared to optical readout, enables simple measurements on even nontransparent liquids, such as blood. The sensor consists of 10 cantilevers in a channel. The channel is 115 μm high, 400 μm wide and 3 mm long, yielding a 0.14 μl volume (figure 1). Moreover, the channel has four in/outlets such that two liquids, for example a test and a reference liquid can flow side by side in the channel. By functionalizing the 10 cantilevers differently, it is possible to detect several different biochemical reactions in a sample simultaneously. The chip fabrication is based on conventional microfabrication technology and is described in [3].

Experimental

In the experiments performed with the chip, half of the cantilevers have been used as references. Inserting a piezoresistor from a measurement cantilever and a piezoresistor from a reference cantilever into a Wheatstone bridge, the reference cantilever is used as a common-mode rejection of the noise from the surroundings.

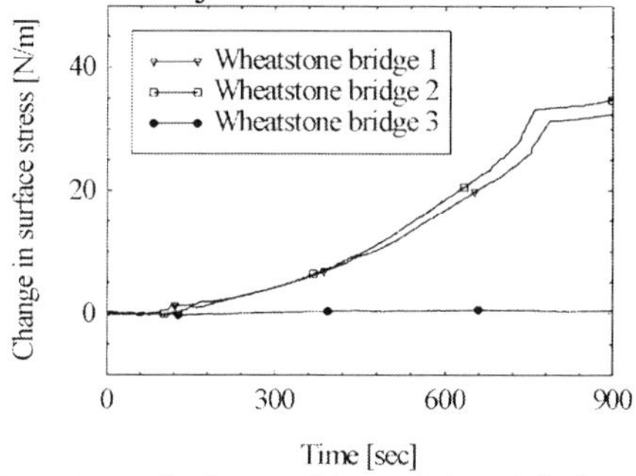

Figure 2: Change in surface stress during etching of the nickel on the measurement cantilevers.

The chip has been used for the detection of the build-in stress in nickel films on the measurement cantilevers. The nickel a thermal induced tensile stress level of σ_{Ni}= 845 MPa. In figure 2, the change in surface stress as a function of time is shown during the etching of the 30 nm thick nickel layer on the measurement cantilevers. Wheatstone bridge 3 shows the signal when two reference cantilevers are inserted into a Wheatstone bridge. The measured tensile stress is found to σ_{meas}= 1000 MPa, which is in good agreement with the calculated value.

Graph (a) in figure 3 shows the change in surface stress as a function of time during the immobilization of cysteine molecules onto gold coated measurement cantilevers. The measurements are obtained simultaneously from two different cantilevers in the channel. Surface stress changes have also been monitored during thiol modified DNA-oligo immobilization on gold coated measurement cantilevers. The surface stress change is following a Langmuir type adsorption model. The experiment is shown on graph (b) in figure 3. The measured rate constants are in good agreement with the literature [4]. The result of these experiments confirm that the cantilever-based sensing principle can be able to detect sub-monolayer formations on the sensor surface. Other groups have reported on the detection of DNA hybridization using cantilever-based sensing [5,6]. These results show that our sensor has the potential as a fast point-of-care device.

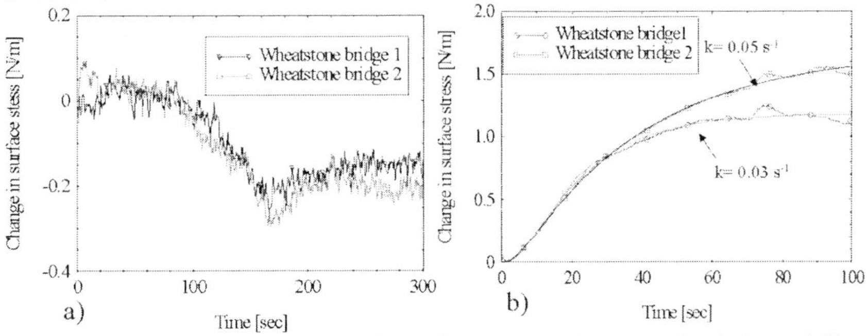

Figure 3: Graph (a) shows the change in surface stress when cysteine is immobilized on the gold coated surface. Graph (b) shows the change in surface stress changes as a function of time when a sample of thiol modified DNA-oligos are flowing in the channel.

Conclusions

Cantilever-based sensors have been integrated into a microliquid handling system. Since the cantilevers have integrated readout, no optical system is required. This eases the readout from large arrays of cantilevers and makes it possible to measure on even non-transparent liquid. The experiments performed with the chip include the monitoring of change in surface stress during etching of a nickel film, the immobilization of cysteine molecules and DNA-oligos. The high sensitivity of the cantilever-based sensor makes it possible to obtain sub-monolayer resolution.

Reference:

1. S.J. Shea, T.A. Brunt, A.R. Ramadan, T. Raymetn, and M.E. Welland, *Atomic force microscopy stress sensors for studies in liquids,* J. Vac. Sci. Technol. B 14 (2) 1996 pp. 1383-1385
2. R. Raiteri, W. Knoll, P. Skádal G. Nelles, and H. Butt. A novel transducer for direct sensing of bioaffinity interactions based on microfabricated cantilevers, J. Phys, Chem, 1995, 99, pp. 15728-15732
3. J. Thaysen, R. Marie and A. Boisen, Cantilever-based bio-chemical sensor integrated in a microliquid handling system, Proceedings MEMS 2001, pp. 401-404
4. M. Yang, H.C.M. Yau and H.L. Chan, "Adsorption kinetics and ligand-binding properties of thiol-modified double stranded DNA on gold surface, *Langmuir*, 14, 1998, pp. 6121-6129
5. J. Fritz, M.K. Baller, H.P. Lang, H.Rothuizen, P. Vettiger, E.Meyer, H. Güntherodt, Ch. Gerber, J.K.Gimzewski, *Translating biomolecular recognition into nanomechn ics* Science, Vol. 288 2000 April 14, pp. 316-318
6. K. Hansen , H. Ji, G. Wu, R. Datar, R. Cote, A. Majumdar and T. Thundat, *Cantilever-based optical deflection assay for discrimination of DNA single nucleotide mismatches,* Accepted in Analytical Chemistry 2001

HYBRID NANOSTRUCTURED MICROCANTILEVERS FOR ENHANCED CHEMIMECHANICAL TRANSDUCTION AND SURFACE ENHANCED RAMAN SPECTROCOPY

Michael J. Sepaniak[1], Nickolay V. Lavrik[1], Chris A. Tipple[1], and Panos G. Datskos[2]

1. Department of Chemistry, University of Tennessee, Knoxville, TN 37996
2. Engineering Tech. Div., Oak Ridge National Laboratory, Oak Ridge, TN 37831

Abstract

Hybrid microelectromechanical systems (MEMS) based on microcantilevers (*MC*s) that exhibit very large chemi-mechanical responses are described. The principle method for enhancing response involves introducing metallic nanostructure to one side of the *MC*. This serves to increase surface area and create new modes of analyte-induced surface stress. Methods of fabricating these cantilevers are described. Bending of these *MC*s upon exposure to both gas phase and liquid phase analytes have been shown to be as much as two orders of magnitude greater than for conventional smooth surface cantilevers. The utility of these systems for sensing and actuation are discussed and demonstrated. The nanostructured cantilevers also exhibit potential as substrates for surface enhanced Raman scattering (SERS). The capability of these devices to function as hybrid MEMS, combining large mechanical responses to analytes with an ability to obtain distinctive SERS spectra, will be shown. **Keywords: Microcantilever sensors, chemi-mechanical actuation, surface enhanced Raman scattering**

1. Introduction

Microfabricated cantilevers like those traditionally used in scanning probe microscopy have become popular MEMS devices for chemical and biochemical sensing. With these *MC* transducers, changes in surface stress caused by surface-analyte interactions are converted into mechanical response. The absolute sensitivities of *MC*s to minute amounts of adsorbate are very high and picogram quantities of chemicals can often cause detectable responses. Deflection of the cantilevers requires differential interactions of analyte with opposite sides of the cantilever. Thus, modification of one side of the *MC* with receptors or other chemical recognition phases is often performed. Our efforts have included modifying cantilever surfaces with GC phases [1], sol-gels [2], cyclodextrins receptors [3], and bioaffinity reagents [4]. The tip deflection, z_{max}, of a smooth-surface cantilever resulting from differential stress at its opposing sides, $\Delta\sigma$, is given by the Stoney equation

$$z_{max} = \frac{3l^2(1-v)}{Et^2}\ddot{A}\sigma$$

where v and E are, respectively, the Poision's ratio and Young's modulus for the cantilever and l and t are its length and thickness, respectively [5].

Despite the aforementioned virtues of *MC* devices as sensors, they often suffer from only modest concentration sensitivity, limited dynamic range, and poor

J.M. Ramsey and A. van den Berg (eds.), Micro Total Analysis Systems 2001, 450–452.

measurement selectivity. The present study shows that a dramatic increase in chemi-mechanical response to analytes results from asymmetric nanostructuring of *MC*s. These nanostructureed surfaces are compatible with the incorporation of self assemble monolayers or thicker layers of selective receptor phases and also offer the possibility of adding SERS capability to the MEMS device.

2. Experimental

Commercially available silicon or silicon nitride *MC*s (typical l = 300 μm, t = 500 nm) were modified by one of two processes (involving assembling gold nanobeads or dealloying Au:Ag composites) as shown in Figure 1 and described elsewhere [3,4]. These cantilevers were mounted in a flow cell and exposed to various analytes. Deflection of the cantilever tip is monitored by displacement of a reflected laser beam on a position sensitive detector (Figure 2).

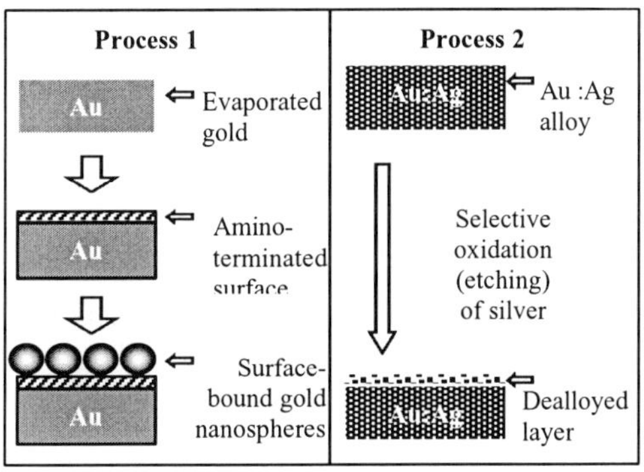

Figure 1. Depiction of nanostructuring processes.

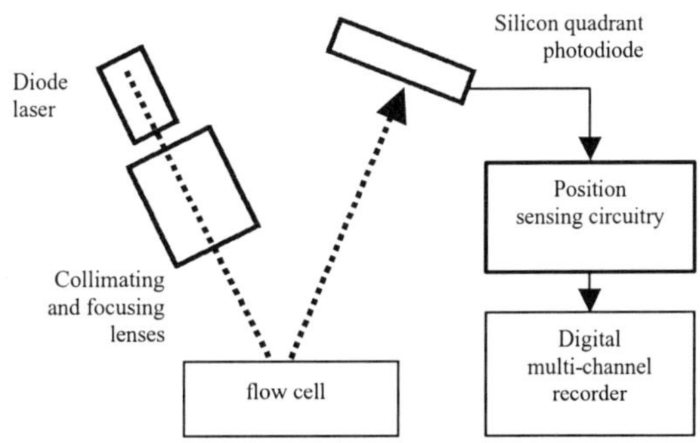

Figure 2. Block diagram of flow cell and optical arrangement for detecting *MC* deflections.

3. Results and Discussion

An effect of nanostructuring the surface of one side of a *MC* is to increase the surface area for binding of analyte and/or receptor phases. In addition to increasing both mass response and dynamic range, this creates new mechanisms for surface stress involving steric effects that magnify $\Delta\sigma$ and, hence, further increase z_{max}. For example, a designed pH-sensitive *MC* exhibited responsivity of up to 1 micron per pH unit (Figure 3) and represents a promising platform for enzyme-based sensors. Large responses to protein based bioaffinity interactions on these *MC* surfaces have also been observed (Figure 4) [4]. These responses are far greater than those observed using conventional receptor-modified *MC*s.

Acquisition of surface enhanced Raman scattering (SERS) signals generated on the cantilever surfaces can provide additional analytical possibilities for studies of specific and nonspecific surface binding events and in highly selective sensing applications. The structures that have yielded large chemi-mechanical *MC* responses have also been shown to be SERS active (Figure 5). Overcoating the dealloyed surface with Ag, a more SERS active metal than Au under the conditions used for the figure, resulted in improvements in observed SERS spectra.

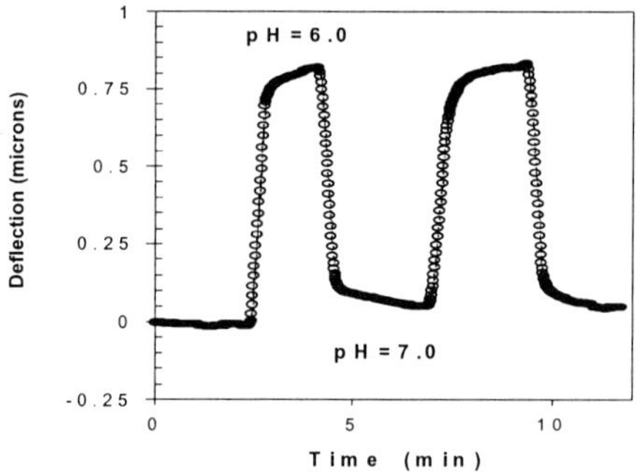

Figure 3. Deflections of a silicon nitride *MC* coated with nanostructured gold surface in response to stepwise changes of pH.

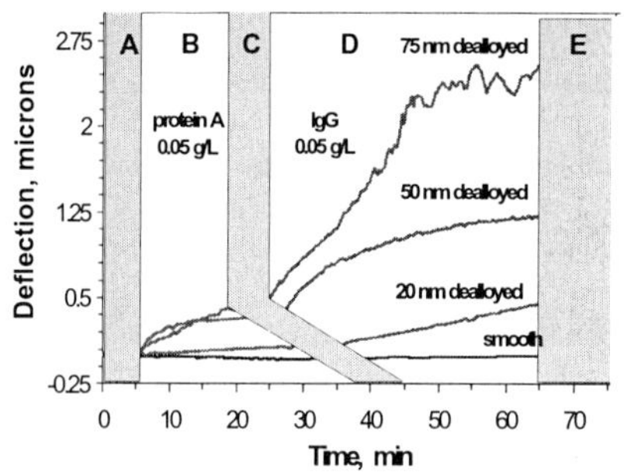

Figure 4. Responses of *MC*s to Protein A then IgG. Regions A, C, and E correspond to PBS buffer. Smooth *MC* shows no response while the nanostructured *MC* yields μm-level deflections.

Taking advantage of the micron-scale chemi-mechanical responses of the nanostructured *MC*s, our recent studies focus on development of a *microfluidic chemi-transistor* device that would produce substantial changes in chemical microenvironments in response to minute chemical stimuli. Different optical read-out schemes for the integral acquisition of analyte SERS signatures, monitoring of chemi-mechanical transduction, and actuated control of flow are under investigation. It is anticipated that chemically-actuated, microfluidic systems will constitute a platform for future bioanalytical/medical devices, including systems powered solely by chemi-mechanical effects.

Figure 5. SERS spectra obtained on the nanostructured surfaces used in *MC* work; polyaniline adsorbed on a silver island film (1), 50 nm dealloyed gold (2), and gold nanobeads assembly (3) – Laser 5 mW at 647 nm.

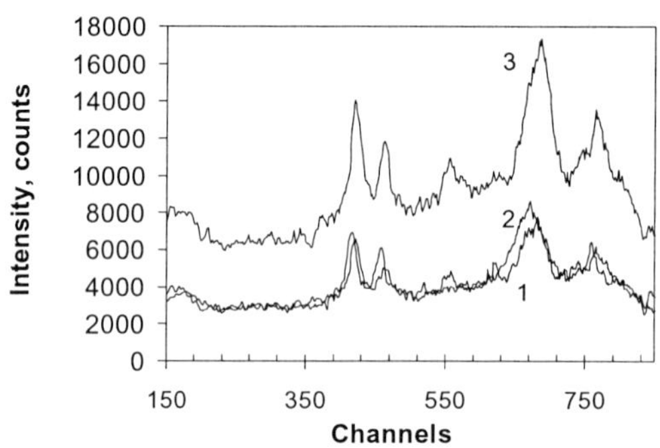

References

1. T. Betts, C.A. Tipple, P.G. Datskos, and M.J. Sepaniak, *Anal. Chim. Acta*, **442**, 89-95 (2000).
2. B. Fagan, C.A. Tipple, M.J. Sepaniak, and P.G. Datskos, *Talanta*, **53**, 599-606 (2000).
3. N.V. Lavrik, C.A. Tipple, P.G. Datskos, and M.J. Sepaniak, *Chem. Phys. Lett.* **336**, 371-376 (2001)
4. N.V. Lavrik, C.A. Tipple, P.G. Datskos, and M.J. Sepaniak, *Biomedical Microdevices*, **3**, 33-42 (2001).
5. F.J. Von Preissig, *J. Appl. Phys.*, **66**, 4262-4269 (1989).

ALTERNATIVE APPROACHES TO MICROFLUIDIC SYSTEMS DESIGN, CONSTRUCTION AND OPERATION

D. J. Beebe, G. Mensing, J. Moorthy, C. M. Khoury and T. M. Pearce
Department of Biomedical Engineering, University of Wisconsin-Madison

Abstract

An alternative approach (μFluidic Tectonics) to microfluidic device construction that utilizes liquid phase photopolymerization and responsive materials is presented. Ultra rapid (3 min) microchannel fabrication is demonstrated. Filtering, flow control, and mixing components are demonstrated.

Keywords: channel network, rapid prototyping, microfluidic, photopolymerization

1. Introduction

Many approaches to the construction of microfluidic systems have appeared in the last few years including glass and silicon etching and bonding, laser machining, micromolding and others. None have emerged as a clear "winner." Each has advantages and disadvantages for a given application. The cost of manufacturing for most approaches is still relatively expensive when compared to a more standard manufacturing method such as injection molding. Injection molding is of limited use for microfluidic applications due to limited resolution (although this is continually improving) and functionality (typically only passive, inflexible materials are used). Here we present an alternative approach to the design, construction and operation of microfluidic systems that we call μfluidic tectonics (μFT) that compares to injection molding in cost, but allows for a wide variety of functionality.

2. Theory & Methods

μFT is based on liquid phase photopolymerization. By eliminating the standard lithography steps (spin, bake, expose, bake, develop) the speed of construction can be significantly enhanced. μFT devices can be constructed in as little as 3 minutes with no expensive equipment required. The basic process flow to construct microchannels using μFT is as follows: (1) fill cartridge, (2) expose, (3) evacuate and flush.

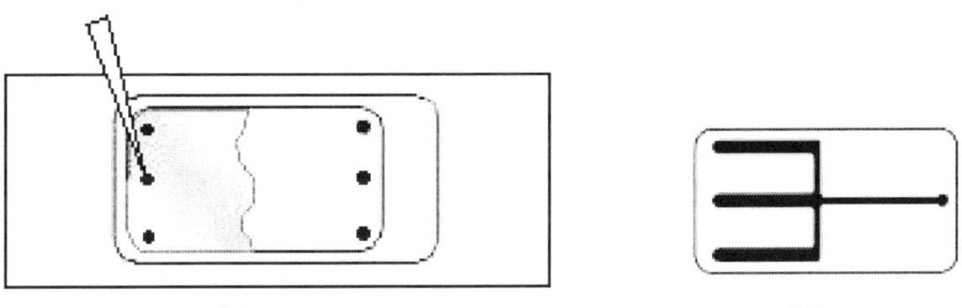

(a) (b)

Figure 1 (a) The cartridge is filled with "construction material" (IBA, TeGDMA, DMPA) via capillary action. The mask (b) is then placed on top of the cartridge and the cartridge/mask sandwich is exposed to UV (~365 nm) for ~30 seconds. A syringe is used to evacuate the unpolymerized material realizing the channel network shown in Figure 2 (a).

J.M. Ramsey and A. van den Berg (eds.), Micro Total Analysis Systems 2001, 453–455.

3. Results and Discussion

The methods described above combined with previously published methods [1,2,3] can be used to create microfluidic devices including passive and multi-layered channel networks, passive and active components, and pressure sensitive assays.

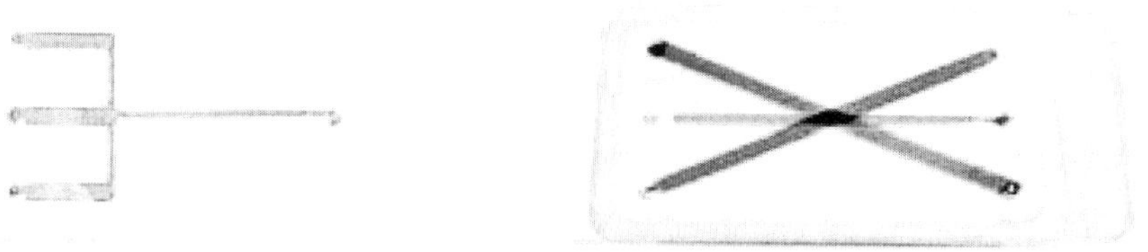

(a) (b)

Figure 2 (a) Single layer channel network (channel height = 250 μm). (b) Three layer channel network. All channels are filled with dyed water for visualization. Although not shown, the process allows interconnections between any layers.

While traditional and "transparency" masks can be used, the fabrication process can be further expedited by simply drawing the mask onto the cartridge using a UV resistant ink eliminating the need for computer generated masks.

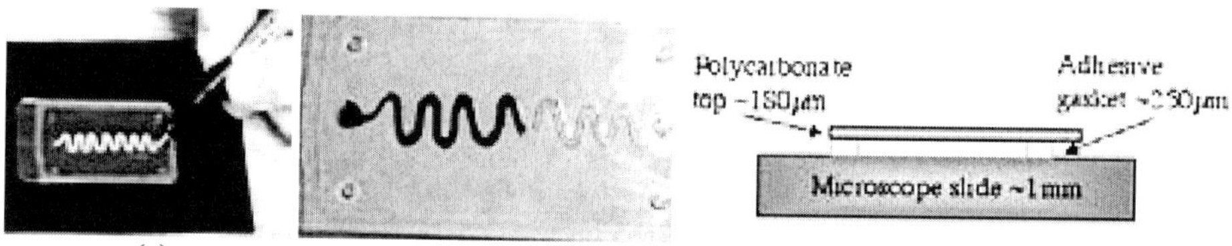

(a) (b) (c)

Figure 3 (a) Writing the mask directly on the cartridge using a pen with UV resistant ink. (b) Capillary filling of finished channel after exposure and removal of mask via wiping with a paper tissue. (c) Side dimensional view of the cartridge used in Figures 1-3.

In order to realize complete systems with the μFT platform, the construction processes must support the creation of various components. As shown in Figures 4, 5, and 6 functional valves, filters and mixers have been realized.

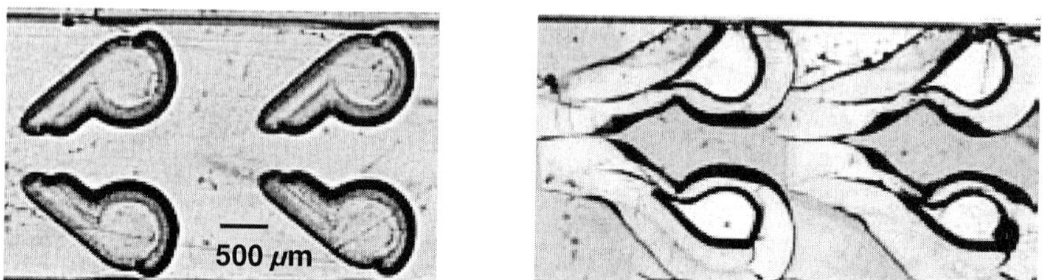

(a) (b)

Figure 4 (a) Redundant bistrip valves constructed of both responsive (Acrylic Acid-based) and non-responsive (HEMA-based) photopolymerizable materials. (b) Valves after activation – in this state the valves operate as a fluid diode allowing flow in one direction.

Figure 5 – (a) Exit of filter. Only 1-3 μm beads pass through. (b) Entrance to filter. Larger 30 μm beads accumulate at the filter. (c) SEM of internal filter structure. (a,b) Fluorescent images using WIG filter (excitation 520 – 550 nm and emission beyond 580 nm).

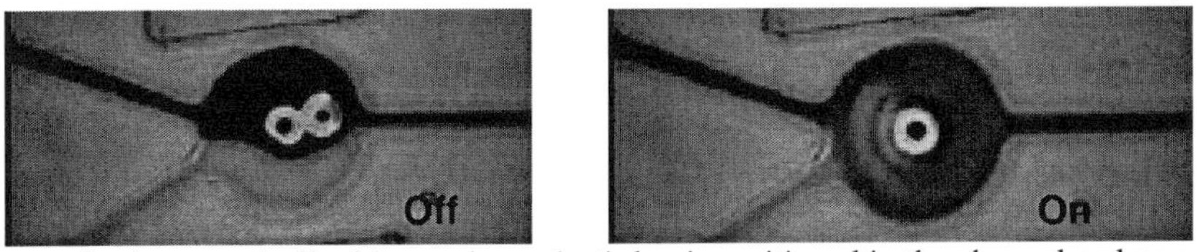

Figure 6 – A magnetically driven mixer. A stir bar is positioned in the channel and an axle (pin) is polymerized to hold the bar in position. A magnetic stirrer is used to drive the mixer.

A component is constructed sequentially (e.g. for the valve - first the channels are formed, then the non-responsive part of the valve and finally the responsive part of the valve). Each step involves flowing in the prepolymer mixture, exposure, and flushing (less then 3 min/step).

4. Conclusion

μFluidic Tectonics provides a single construction platform that allows for the creation of a wide variety of microfluidic devices. The design flexibility combined with the ease of fabrication and low cost (similar to injection molding) enhances the microfluidic toolbox. We have also used μFluidic Tectonics to demonstrate systems capable of regulating the pH of a fluid stream autonomously [4] and to produce organic microfluidic readouts [5]. All channel and component construction uses the same basic process, thus integration of various components into systems is straightforward.

Acknowledgements

We wish to thank Qing Yu and Mary Kraft of Prof. Moore's group at UIUC for many helpful discussions. Supported by DARPA-MTO (#F30602-00-1-0570).

References

1. Liu, R. H., Q. Yu and D. J. Beebe, "Fabrication and characterization of hydrogel-based microvalves,", *Journal of Microelectromechanical Systems,* in press, 2001.
2. Zhao, B., J. S. Moore and D. J. Beebe, "Suface-directed liquid flow inside microchannels," *Science,* Vol. 291, pp. 1023-1026, 2001.
3. Beebe, D. J., J. Moore, Q. Liu, R. H. Liu, M. L. Kraft, B-H. Jo and C. Devadoss, "Microfluidic Tectonics: A comprehensive construction platform for microfluidic systems," *Proceedings of the National Academy of Science*, Vol. 97, No. 25, pp. 13488-13493, 2000.
4. Eddington, D. T., R. H. Liu, J. S. Moore and D. J. Beebe, "A self-regulating microfluidic system," μTAS 2001, October 21-25, Monterey, CA.
5. Moorthy, J. and D. J. Beebe, "Autonomous Colorimetric Readout for Microfluidic Devices," μTAS 2001, October 21-25, Monterey, CA.

MODELING ELECTROKINETIC TRANSPORT FOR THE DESIGN AND OPTIMIZATION OF MICROCHANNEL SYSTEMS

S. K. Griffiths and R. H. Nilson
Chemical and Materials Process Modeling Department
Sandia National Laboratories, Livermore, California 94551-0969

Abstract

Sample band spreading in microchannel systems is examined at the component level using analytical and numerical methods. Solutions describing growth of the band variance in straight channel segments, conventional turns and low-dispersion turns are reviewed. The use of these results in optimum channel design is discussed.

Keywords: electrokinetic, electroosmotic, flow, transport, dispersion, turns

1. Introduction

Direct numerical simulation of electrokinetic transport in microchannel systems is a challenging task because the solutions must resolve length scales ranging from the Debye layer thickness of perhaps 10 nm to column lengths of 10 cm or more. Given this seven-decade span of internal and physical lengths, even steady-state solutions to the full governing equations are generally unobtainable.

A useful alternative to full system simulation is the analysis of components such as turns, tees and straight channel segments. Through a good understanding of species transport at these component scales, system-level performance can be assessed and optimized. Here we review several such analyses and provide new design criteria for conventional turns used in folded separation columns.

2. Straight Channel Segments

Electroomosmotic fluid velocities are quite uniform across a channel; all variations in the axial speed are confined to small regions comparable in thickness to the electric Debye layer and adjacent to the channel walls. Electroosmosis thus enables long-range species transport with relatively little band spreading due to hydrodynamic dispersion.

To quantify this spreading, we seek a late-time solution in a coordinate system that moves with the sample band. This transformation reduces the governing equations to a set of ODEs. These are solved to yield the coefficient of dispersion relating the band variance to the elapsed time, species diffusivity and Peclet number. For a rectangular channel and negligible zeta potential, the results are well approximated by [1]

$$\left(\frac{\sigma}{a}\right)^2 = \frac{2Dt}{a^2}(1+\alpha \mathrm{Pe}^2) = \frac{2L^*}{\mathrm{Pe}}(1+\alpha \mathrm{Pe}^2) \quad \text{where} \quad \frac{1}{\alpha} \approx 210 + \frac{5}{\lambda^{*3/2}} + \frac{12}{\lambda^{*2}} \tag{1}$$

$\mathrm{Pe} = Ua/D$ is the Peclet number, a is the channel width, $\lambda^* = 2\lambda/a$ is the normalized Debye layer thickness and $L^* = L/a = Ut/a$ is the normalized length of travel. The

J.M. Ramsey and A. van den Berg (eds.), Micro Total Analysis Systems 2001, 456–458.

term $2L^*/\mathrm{Pe}$ is the diffusive contribution to spreading; the remaining term represents spreading due to dispersion. Note that the plate height is related to the variance through $H=\sigma^2/L$. Similar results have been derived for non-negligible zeta potentials [2].

3. Conventional Turns

Conventional turns are a well-known source of dispersive band spreading in microchannel systems [3]. Illustrated in Fig. 1, this spreading can impair performance in separation processes and, as a result, turns are often avoided in microchannel system design. However, even high-performance separation columns can be folded if the influence of the turns is well understood and appropriate designs are employed.

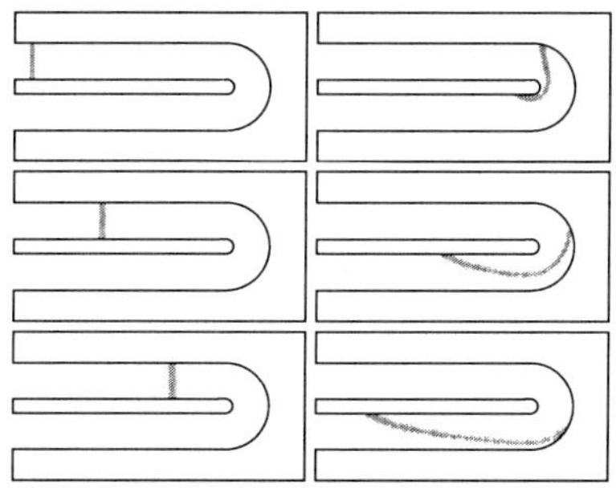

Figure 1. Simulated electroosmotic transport of a species band through a small-radius 180-dgree turn. Nonuniform electric field and unequal path lengths severely skew the band. Species transport is described by a Monte Carlo method using tracer particles. Such numerical results confirm the analytical solution given in Eq. 2.

Banding spreading in turns can be quantified by examining the behaviors of the induced spatial variance in the extremes of low and high Peclet numbers [4]. The normalized turn-induced variance, obtained using analytical methods alone, is

$$\left(\frac{\sigma}{a}\right)^2 = \frac{2r^*\theta}{\mathrm{Pe}} + \frac{\theta^2\,\mathrm{Pe}}{15r^*\theta + 3\mathrm{Pe}} \tag{2}$$

where $r^* = r/a$ is the normalized turn radius and θ is the included turn angle. The first term in this result is the diffusive contribution to the variance; the second is the contribution due to dispersion.

4. Folded Columns

Conventional turns in folded serpentine separation columns do not significantly degrade system performance provided that the dispersive turn-induced spreading is small compared with the diffusive spreading in the turns and adjacent straight channel segments. From Eq. 2 and the diffusive contribution from Eq. 1, this condition can be expressed in terms of an acceptable minimum turn radius

$$r^* \geq \frac{\mathrm{Pe} + 5L^*}{10\theta}\left(\sqrt{1 + \frac{20\mathrm{Pe}(5\theta^2\mathrm{Pe} - 3L^*)}{3(\mathrm{Pe} + 5L^*)^2}} - 1\right) \tag{3}$$

where $L^* = L / a$ is the normalized length of a straight channel segment associated with each turn. Turn-induced spreading accounts for less than 10% of the total increased band variance when this condition is satisfied. Based on a 40 μm channel width, a Peclet number of 400, a straight segment length of 30 mm and a minimum turn radius of 4.9 mm, a 200 mm column can be folded onto a region of 13 by 42 mm using an optimized configuration. The resulting column yields a plate height of just 0.22 μm.

5. Low-Dispersion Turns

Although large-radius turns offer satisfactory performance in many applications, the area required for channel layout may sometimes be excessive. In this case, the use of special turns having low-dispersion geometries may be beneficial [5,6].

Such geometries have been discerned using a minimization algorithm along with an inverted method treating spatial coordinates, instead of the electric potential, as the dependent variables [6]. The resulting low-dispersion turns, illustrated in Fig. 2, yield an increased band variance that is two to three orders of magnitude below that of equivalent conventional turns. Using low-dispersion 180-degree turns, a 100 μm channel up to 900 mm long can be folded onto a region covering only 10 by 10 mm; this compact column is capable of plate heights down to 0.04 μm.

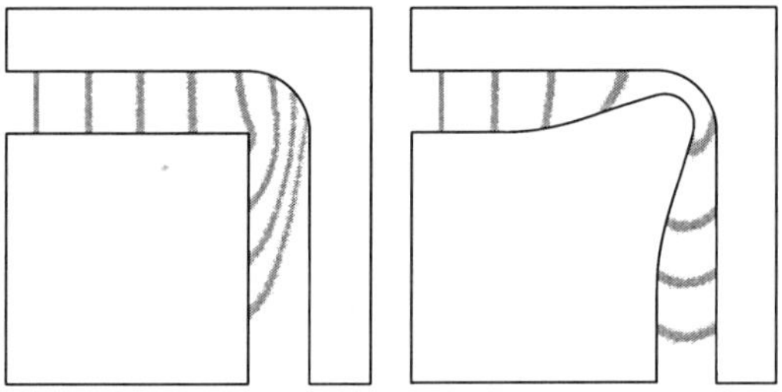

Figure 2. Simulated species transport through conventional and low-dispersion 90-degree turns. Low-dispersion turns and tees are useful in folding columns and in the preparation, splitting, mixing and routine transport of sample bands.

6. Summary

Analytical solutions describing sample band spreading in straight channel segments and conventional turns are reviewed. These provide fundamental understanding, enabling optimum design, as well as quantitative results for use in system-level numerical simulations. A criterion for the acceptable radius of conventional turns is presented; low-dispersion turns for compact column configurations are also discussed.

References

1. S. K. Griffiths and R. H. Nilson, *Anal. Chem.* **71**, 5522-5529 (1999).
2. S. K. Griffiths and R. H. Nilson, *Anal. Chem.* **72**, 4767-4777 (2000).
3. C. T. Culbertson, S. C. Jacobson, J. M. Ramsey, *Anal. Chem.* **70**, 3781-3789 (1998).
4. S. K. Griffiths and R. H. Nilson, *Anal. Chem.* **72**, 5473-5482 (2000).
5. B. M. Paegel, L. D. Hutt, P. C. Simpson, R. A. Mathies, *Anal. Chem.* **72**, 3030-3037 (2000).
6. S. K. Griffiths and R. H. Nilson, *Anal. Chem.* **73**, 272-278 (2001).

FABRICATION OF MODULAR MICROSYSTEMS FOR ANALYZING *K-RAS* MUTATIONS USING LDR

Steven A. Soper,[1] Michael C. Murphy,[2] Robin L. McCarley,[1] Dimitris Nikitopoulos,[2] Xuezhu Liu,[1] Bikas Vaidya,[1] James Barrow,[1] Yannick Bejat,[2] Sean M. Ford[3] and Jost Goettert[4]

[1]Department of Chemistry, Louisiana State University, Baton Rouge, LA 70803

[2]Department of Mechanical Engineering, Louisiana State University, Baton Rouge, LA 70803

[3]MEZZO Systems, INC., LBTC, South Stadium Drive, Baton Rouge, LA 70803

[4]Center for Advanced Microstructures and Devices, Louisiana State University, Baton Rouge, LA 70803

Abstract

We are developing microsystems for genetic-based assays, which allow assembling devices with discrete functions to carry out complex assays. Specifically, the microfluidic devices are designed to perform ligase detection reactions (LDR) for identifying mutations in *K-ras* genes associated with colorectal cancer. The functions of the devices consist of PCR, LDR, and interrogation of LDR products using either micro-electrophoresis or DNA microarrays. The microfluidic devices are fabricated using *LIGA* to manufacture molding dies (metal electroforms) consisting of high aspect ratio microstructures (>10:1). Our devices are made from polymers, such as PMMA or PC. In this presentation, we will discuss the fabrication and operational characteristics of PCR/LDR thermal cycling modules as well as micro-hybridization units. The thermal cycling devices consist of isothermal zones with fluid packets shuttled through a spiral channel of ~1.2 m in total length (total number of thermal cycles = 20).

Keywords: Polymer micromachining; PCR; Genetic Diagnostics; LDR.

1. Introduction

A powerful molecular-based analysis tool for the detection of mutational markers is PCR/LDR or LCR. LCR consists of using a thermophilic (*Tth*) ligase in a DNA amplification technique known as the ligase chain reaction (LCR), or when coupled with a primary PCR reaction, a one-step ligase detection reaction (LDR). These techniques can be used for DNA diagnostics and, in particular, screening for rare point mutations in selected genes [1]. The high discriminatory power of LCR results from the *Tth* ligase sensitivity to mismatches at the ligation juncture, with error rates < 1 in 10^3.

We are developing miniaturized microfluidic platforms for performing LCR/LDR-type assays of *K-ras* oncogenes. The microfluidic devices are fabricated in polymers, such as poly(methylmethacrylate) (PMMA) or polycarbonate (PC) using hot-embossing. The molding dies are metal electroforms that are machined using *LIGA* processing, which can fabricate high aspect ratio microstructures (HARMs). In addition, the design approach is modular, with each module containing a discrete function. Assembly of the modules allows the integrated unit to perform PCR/LDR

J.M. Ramsey and A. van den Berg (eds.), Micro Total Analysis Systems 2001, 459–461.

assays for cancer detection in an automated fashion. In this presentation, we will discuss the fabrication of thermal cycling units appropriate for PCR or LDR. These units are continuous flow devices containing isothermal zones [2] and fabricated from PC. We will also discuss chemical modifications of PC and PMMA surfaces to support electrodynamic pumping and/or immobilization of biologics.

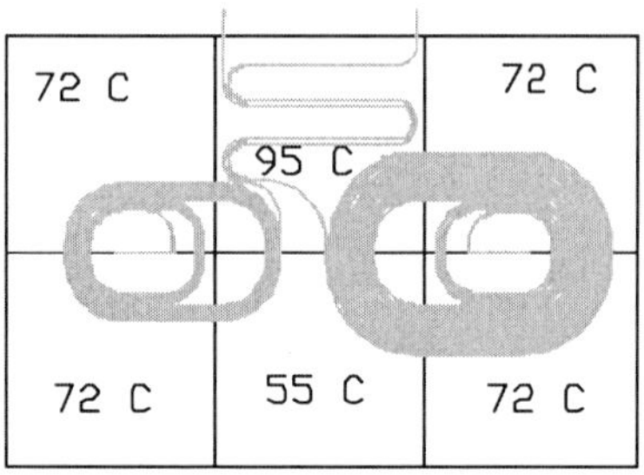

Figure 1. Topography of the continuous flow thermal cycling unit. There are two separate thermal cyclers with different channel spacings. The channel dimensions are 50 x 150 μm with the total channel length ~1.2 m. The channel spacings are 50 μm and 250 μm.

2. Results

The basic architecture of the thermal cycling unit is shown in Figure 1. It consists of a two units with spiral channels ~1.2 m in length moving through three isothermal zones. Assuming a linear flow velocity of 0.2 cm/s, the travel time of a sample packet through each thermal cycle would be 35.7 s and 48 s for the smaller and larger spirals, respectively. The molding dies fabricated for this device consist of raised Ni-microstructures electroplated onto a stainless steel plating base. The process for making the molding dies has been outlined previously [3] and basically involves *LIGA* processing. SEMs of the Ni-electroform and molded part can be seen in Figure 2. Hot-embossing was selected to fabricate the final parts from the Ni-electroforms as opposed to injection molding since high purity polymers can be directly embossed without the addition of additives and the maximum temperature to which the polymer is subjected is lower compared to injection molding.

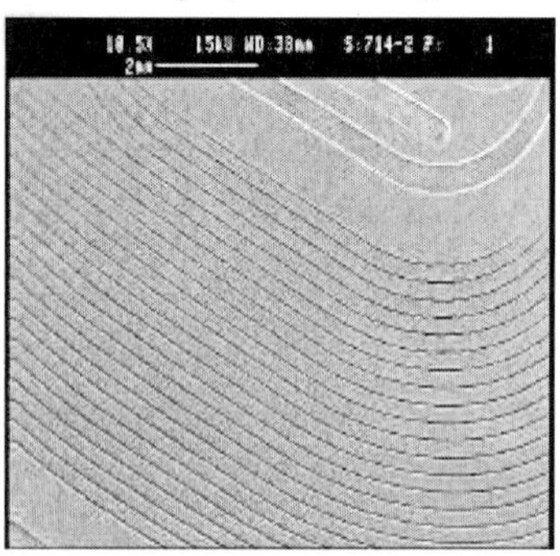

Figure 2. Scanning electron micrographs of embossed PC chip (A) and the Ni-electroform molding die (B).

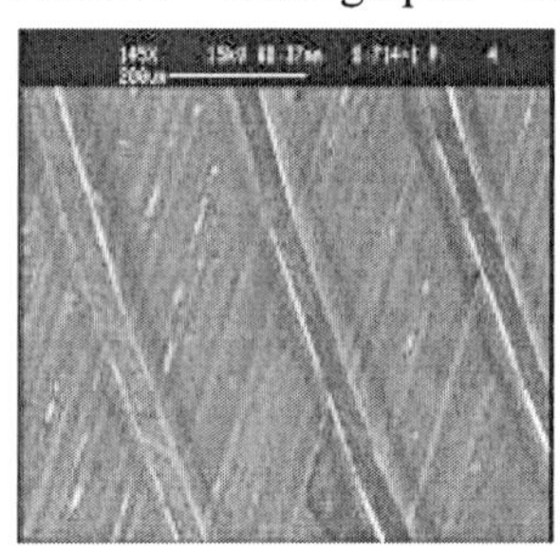

(A) **(B)**

In order to support flow through this polymer microchip, it is impractical to consider micropumps and pressure driven flow due to the large pressure drop that is generated by the extended length channel. We therefore investigated electrodynamic pumping, where the electroosmostic flow (EOF) generated by the polymer would shuttle liquids through this channel. Unfortunately, many polymers show very low EOFs that cannot support pumping for all ions in one direction (anode to cathode). Therefore, we developed modification chemistries to alter the electroosmostic flow [4]. In Figure 3 is shown EOF profiles for native and modified PMMA as a function of pH. As can be seen, modifying the chemical composition of the PMMA surface had a profound impact on the magnitude and direction of the EOF. In addition to EOF control, these surface modifications could be used to immobilize biologics, such as DNAs for hybridization-based assays. Shown in Figure 4 is a line image of a DNA microarray fabricated on a PMMA microchip.

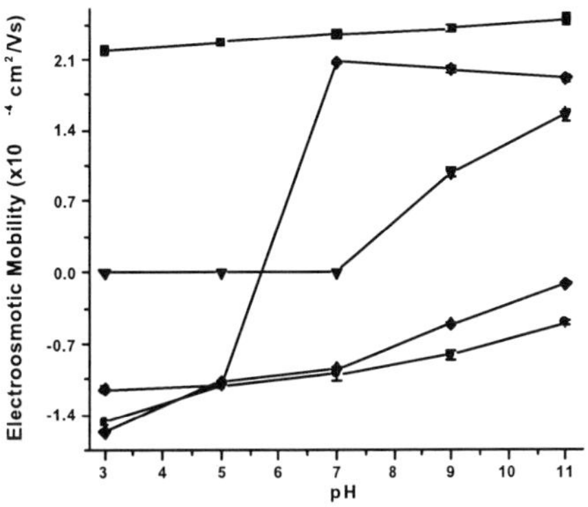

Figure 3. EOF profiles as a function of pH for native PMMA (■); C18 modified PMMA (◆); PC (▼); NH_2 modified PMMA (▲) and COOH modified PMMA (●). The EOFs were measured using the current monitoring method.

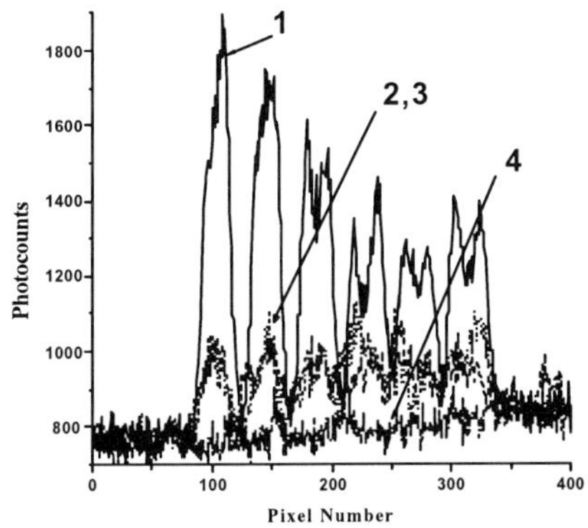

Figure 4. Fluorescence line image of 6 spots from a hybridization array fabricated on a PMMA microchip. The four traces are, (1) perfect matched duplex, (2) 1-base mismatch, (3) 3-base mismatch and (4) totally mismatched duplex.

3. Conclusions

We have effectively demonstrated the ability to hot-emboss HARMs in polymers using metal molding dies fabricated by *LIGA*. The dies were used to make modular microsystems appropriate for performing genetic-based assays.

Acknowledgements

The authors would like to think the National Institutes of Health (HG01499 and CA84625) for financial support of this work

References

1. F. Barany, *PCR Methods Appl.* **1**: 5-16 (1991).
2. M.U. Kopp, A.J. deMello, A. Manz, *Science* **280**: 1046-1048 (1998).
3. S. Qi, S. Ford, X. Liu, J. Barrow, K. Kelly, A. McCandless, K. Lian, S.A. Soper, *Anal. Chem.* (2001, submitted for publication).
4. A.C. Henry, T.J. Tumari, M. Galloway, Y.Y. Davidson, C.S. McWhorter, S.A. Soper, R.L. McCarley, *Anal. Chem.* **72**:5331-5337 (2000).

MICROFABRICATED CAPILLARY ARRAY ELECTROPHORESIS: HIGH-THROUGHPUT DNA SEQUENCING AND POLYMORPHISM ANALYSIS

Brian M. Paegel,[1] Charles A. Emrich,[1] Robert G. Blazej,[1] Christopher J. Elkin,[2] James R. Scherer[1] and Richard A. Mathies[1]
[1]Department of Chemistry, University of California, Berkeley, CA 94720 USA
[2]Department of Energy Joint Genome Institute, Walnut Creek, CA 94598 USA

Keywords: DNA Sequencing, polymorphism detection, capillary array electrophoresis

Abstract

Microfabricated capillary array electrophoresis (μCAE) circuitry has been developed and optimized allowing high-throughput DNA sequencing and polymorphism analysis in a compact, circular, wafer-scale device. The 96-lane processor, which incorporates hyper-turns and fluidically balanced injectors, produces ~41,000 bases of M13 extension products to an accuracy >99% (phred 20) in only 24 minutes. Polymorphism ratio sequencing (PRS) with the same device enables rapid scanning of the entire human mitochondrial genome in a single run.

1. Introduction

The completion of the rough draft of the human genome has highlighted the need for more advanced genomic analysis technologies. The expanded throughput and sample handling capabilities of μCAE devices designed and optimized for DNA sequencing will provide the technologies needed for further genomic and comparative sequencing efforts. Our group pioneered the development of μCAE devices for rapid DNA sizing and sequencing [1]. The current device represents the latest convergence of technologies developed to perform DNA sequencing in a compact, 96-lane processor. The versatility of the device is demonstrated by the analysis of DNA variations using the new PRS method.

2. Experimental

The μCAE sequencing chip shown in Figure 1A contains 48 doublet structures composed of two separation channels that share common cathode and waste access ports. The injector (Fig. 1B) channel widths have been optimized for equal gel filling of the sample, waste, and cathode arms. The separation channel length has been extended to 16-cm by adopting a serpentine layout. The turn regions (Fig. 1C) utilize our hyper-turn geometry that minimizes dispersion introduced by the turns [2].

Separation matrix is loaded and removed using a high-pressure loading device [3]. An elastomer buffer reservoir is affixed to the top of the chip to address all cathode and waste wells with common buffer compartments. An electrode ring addresses all sample access ports and reservoirs with the appropriate voltages. Sample is injected at 500 V/cm to the waste reservoir for 60 sec while floating the cathode and anode. The sample is separated at 200 V/cm toward the anode with a 200 V/cm back-biasing field applied at the sample and waste.

J.M. Ramsey and A. van den Berg (eds.), Micro Total Analysis Systems 2001, 462–464.

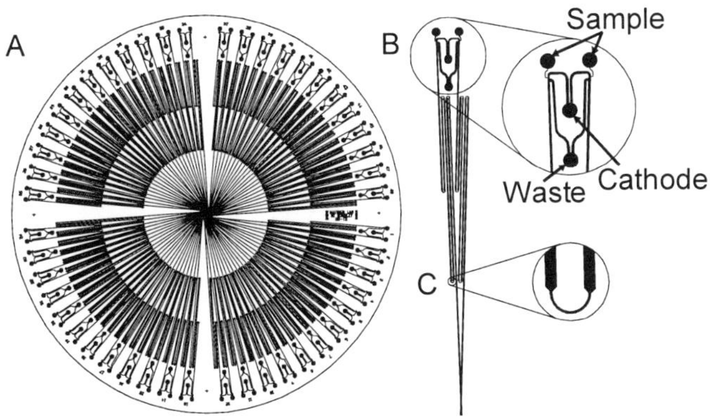

Figure 1. A) Layout of the 150-mm dia. μCAE DNA sequencing device. B) Expanded view of an injector. The sample arm, waste arm, and separation channel widths are 85, 200, and 300 μm, respectively. C) Expanded view of a hyper-turn. The channel tapers to a final width of 65 μm over a distance of 150 μm. The radius of curvature is 250 μm.

3. Results and Discussion

The microdevice produced an average of 430 bases to an accuracy of >99% (phred 20) per lane in 24 minutes for a standard M13mp18 single-stranded DNA sequencing sample with a single lane failure [4]. A 5-min excerpt of the 96-lane output corresponding to base positions ~300-325 is presented in Figure 2A.
In addition to analyzing standard sequencing samples, the μCAE device was used to scan the human mitochondrial genome using PRS, a novel relative sequencing method [5]. By labeling bases from reference and sample DNA with different dyes, the PRS method enables facile detection of polymorphisms even in low quality sequence regions. Nearly complete coverage of the entire genome was completed with a single device in 28 minutes. Figure 2B shows a portion of the PRS output from a single lane.

4. Conclusions

The μCAE platform is the next paradigm in high-throughput DNA sequencing and diagnostics. Genomic and relative sequencing demonstrate the capability of the 96-lane device to produce a variety of data in a fraction of the time required by current commercial systems.

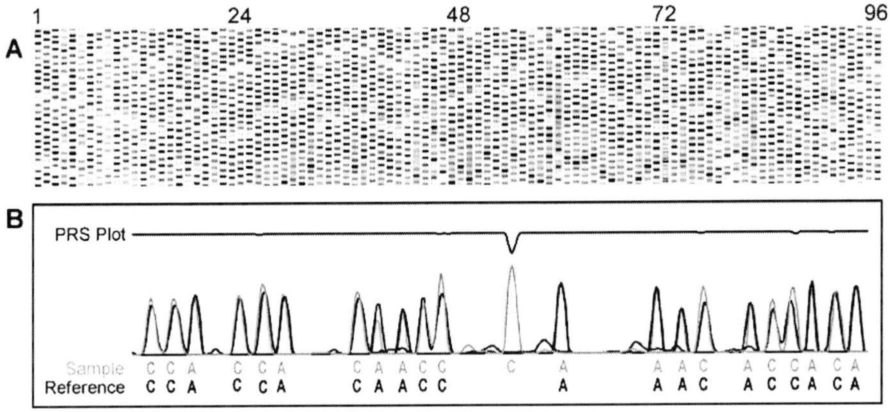

Figure 2. BaseFinder processed output from the μCAE device. A) 96-lane data showing base positions ~300-325 of the M13mp18 standard sequencing sample. Lane numbers are shown at top. 4.5% linear polyacrylamide, 7M urea, 1X TTE, 50° C.
B) Four color PRS output from human mitochondrial DNA correctly identifies a C to T transition at position 6,045. Window shows trace positions 174 to 205. Sample and reference DNA was PCR-amplified from the genomic DNA of two CEPH individuals. The PRS plot displays the difference between the sample and reference signals. 4.7% linear polyacrylamide, 7M urea, 1X TTE, 50° C.

Acknowledgments

This research was funded by the National Institutes of Health and the Department of Energy. Support from Amersham-Pharmacia Biotech is also gratefully acknowledged. B. M. P. was supported by a National Institutes of Health NRSA Trainee fellowship from the Berkeley Program in Genomics (NIH T32 HG00047).

References

(1) A. T. Woolley, R. A. Mathies, "Ultra-High-Speed DNA Sequencing Using Capillary Electrophoresis Chips," *Analytical Chemistry* 1995, **67**, 3676-3680.

(2) B. M. Paegel, L. D. Hutt, P. C. Simpson, R. A. Mathies, "Turn geometry for minimizing band broadening in microfabricated capillary electrophoresis channels," *Analytical Chemistry* 2000, **72**, 3030-3037.

(3) J. R. Scherer, B. M. Paegel, C. A. Emrich, G. J. Wedemayer, I. Medintz, R. A. Mathies, "High-Pressure Matrix Loading Device for Electrophoresis Microchips," *Biotechniques* 2001, (in press).

(4) B. M. Paegel, C. A. Emrich, G. J. Wedemayer, J. R. Scherer, R. A. Mathies, "High-Throughput DNA Sequencing in a 96-Lane Microfabricated Electrophoresis Device," 2001, (in preparation).

(5) R. G. Blazej, B. M. Paegel, C. A. Emrich, I. L. Medintz, J. R. Scherer, R. A. Mathies, "Identification of DNA Polymorphisms Using a Ratio Sequencing Approach," 2001, (in preparation).

INTEGRATED MICROFLUIDIC DNA ARRAY BIOCHIPS

Robin H. Liu^, Ralf Lenigk^, Kenneth R. Luehrsen*, Huinan Yu*, Haixu Chen^, Dale Ganser^, Justin Bonanno^, Piotr Grodzinski^

^Motorola Labs, *Motorola Life Sciences, Motorola Inc., Tempe, AZ, USA

Abstract

Two microfluidic DNA array devices that integrate massively parallel microfluidic channels with conventional DNA microarray technology have been developed. These devices allow DNA hybridization assays to be performed in a miniaturized and highly parallel fashion. The microfluidic arrays are a new DNA diagnostics platform that has several desirable features such as single-use and cost-effective, highly parallel (up to 52 protocols in parallel), and processed rapidly. Additionally, only nanoliter-volumes (~100 nL) of DNA sample/reagents are consumed per array.

Keywords: Microfluidic DNA array, microarray, hybridization, plastic biochips

1. Introduction

Conventional high-density DNA microarray chips for genetic analysis [1] do not take advantage of the miniaturized system environment and often suffer from small number of arrays (up to four arrays) per chip, large volume of sample/reagent consumption (~ 60 μL per array), lengthy hybridization (~ 18 hr), and high cost. While ultra-high-density arrays are powerful in expression analysis, low or medium density but highly parallel arrays are useful in many other applications such as clinical diagnostics and pharmocogenomic applications. The ability to perform massively parallel assays with nanoliter quantities of sample/reagent per assay would provide substantial cost savings, and hence is highly desirable. To match the overall throughput of microarray sequencing and further provide cost reductions in genetic analysis process, we developed *single-use* microfluidic channel arrays, capable of performing hybridization of nucleic acids with improved speed and reduced sample/reagent consumption, and analyzing hundreds of samples on one single chip. Specifically, we focused on a PDMS/glass biochannel arrays and a plastic array biochip, both of which can be easily integrated with the future sample preparation module on a chip.

2. PDMS Biochannel Arrays

The PDMS biochannel array chips are composed of two key components: 1) a disposable, microfabricated PDMS device that consists of tens or hundreds of open microchannels (~200 μm wide, ~50 μm deep) as short as a few centimeters; 2) a microarray glass-based chip (Motorola CodeLink^{TM}) containing a number of DNA oligonucleotide probes that allow specific or non-specific capture of DNA via hybridization (Fig.1). The PDMS channels were fabricated using a micromolding technique and then bonded on a microarray glass slide using a selective plasma surface

J.M. Ramsey and A. van den Berg (eds.), Micro Total Analysis Systems 2001, 465–467.

treatment. With channel volumes in the range of nanoliters (~ 100 nL), the overall volume for a reaction and the reagent cost are reduced by a factor of 100.

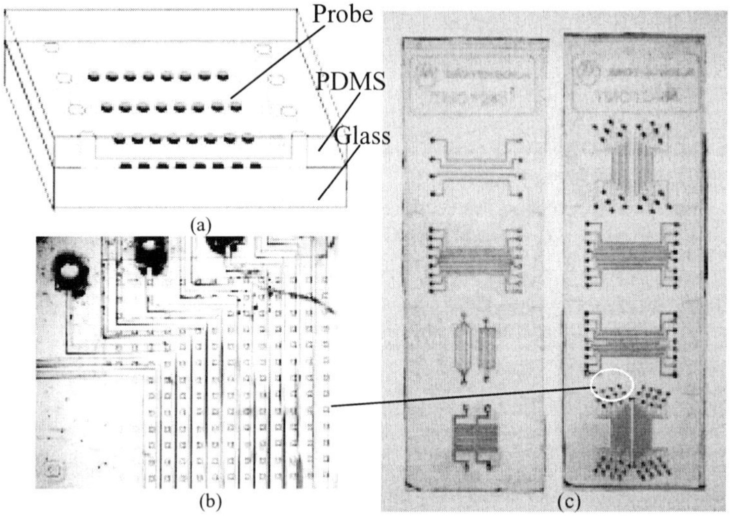

Fig. 1 (a) Schematic of a highly parallel biochannel array chip consisting of a PDMS microchannel layer and a microarray glass chip. (b) A close view of the high-density PDMS channel array running across gel pad based DNA array. The microchannels are 250 μm wide and 50 μm deep. (c) Micrograph showing two biochannel array chips, each has various channel array designs. The chip on the right-hand side consists of 52 channels in parallel.

Single base extension (SBE) enzymatic reaction was performed in the biochannels with single nucleotide polymorphism (SNP) DNA arrays. PCR was used to amplify potential SNP targets of human genomic DNA before the hybridization reaction. 100nL of DNA hybridization reaction mixture contains 10ng human genomic DNA targets. Hybridization reactions were completed within minutes (~ 3 min) during a thermal denaturation process (i.e., 2 min at 85°C, 1 min at 60°C). Fig. 2 shows some of scanned fluorescent images of biochannel arrays, where high intensity of the spots indicates positive signals ("hits"). Note that in Fig. 2(a) a thermal bubble micropump was used to induce oscillation flow inside the channel (i.e., to move the DNA sample plug back and forth) during the hybridization reaction, which in turn improves the mixing and hybridization. 95% accuracy was achieved for SNP assays in the biochannels.

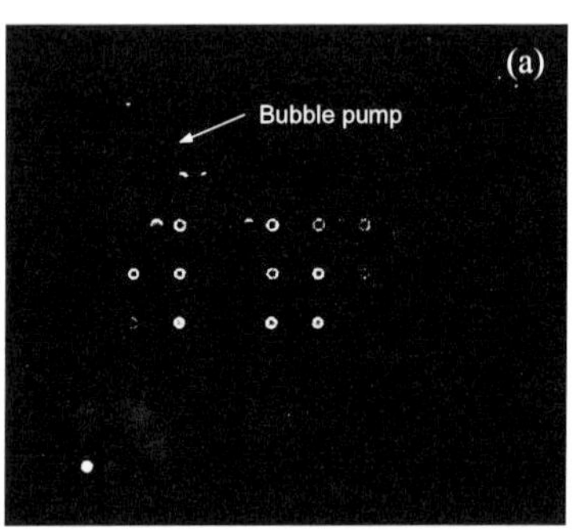

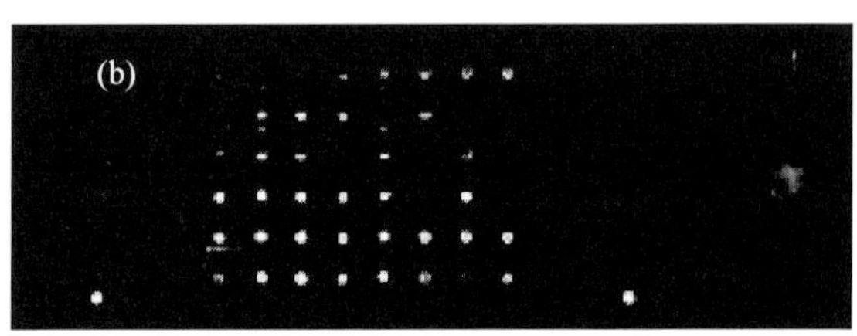

Fig. 2 (a) Fluorescent image of single base extension (SBE) single nucleotide polymorphism (SNP) DNA hybridization assay in a PDMS biochannel coupled with a thermal bubble micropump. (b) Fluorescent image of SBE assays in parallel PDMS biochannel arrays.

3. Plastic Array Biochips

Another approach to form microfluidic DNA array is to dispense oligo probes directly into plastic microchannels that were fabricated by plastic micromolding, followed by sealing with a cover layer using a double-side adhesive tape (Fig. 3a). A novel plastic molding technique was developed to construct 3D micro-dome structures (20 μm high) inside the channel (40 μm deep), as shown in Fig. 4, in order to increase the surface of the DNA probes exposed to target solution and induce complex flowfield inside the channel to enhance hybridization. Prior to deposition the channels were coated with a thin gel film allowing the covalent attachment of oligonucleotides. A microarrayer with 130 μm diameter stealth pins was used to print probes specific for three pathogenic bacteria surrogate strains (E. coli, S. salivarius, P. aeroginosa). Synthetic target labeled with a fluorophor was loaded into the channels, and then the device was enclosed in a humidity-controlled chamber at 60 °C. After 20 minutes the chip was removed, washed and scanned with a laser scanner (Fig. 3b). Compared to the conventional method using glass slides, the assay time was significantly reduced.

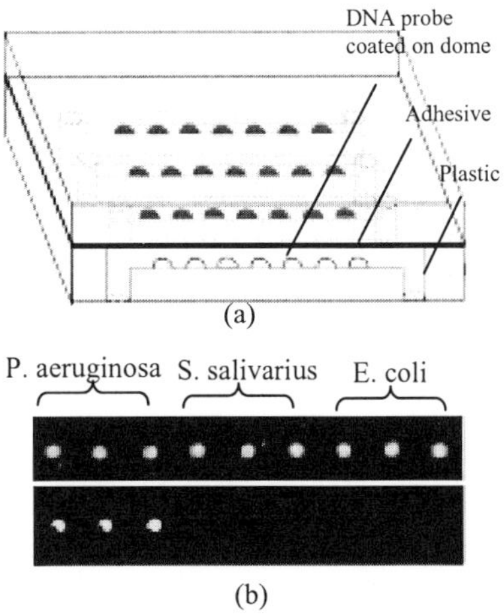

Fig. 3 (a) Schematic of the plastic array biochip; (b) Fluorescent image of bacteria detection assay showing hybridization of a mixture of three synthetic targets to the probes in parallel plastic channel arrays.

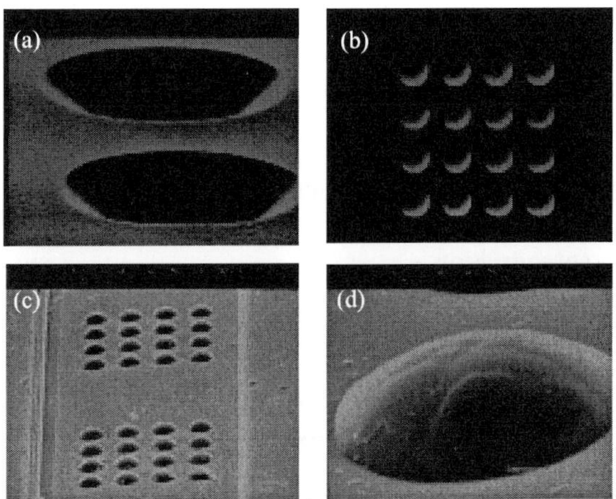

Fig. 4 (a) (b) Si stamper fabricated by selectively control of isotropic wet etching; (c) (d) Plastic microchannel (40 μm deep) with dome structures (20 μm high) molded using the Si stamper shown in (a) and (b).

4. Conclusions

The marriage between microfluidics and DNA microarrays results in a number of advantages, including increased hybridization efficiency, substantially reduced sample/reagent consumption (~ nL), and the ability to perform hybridizations in a highly parallel manner. The use of PDMS and plastic microfabrications allow producing integrated microfluidic DNA array chips for high throughput, low cost applications.

Reference:

1. D. J. Graves, "Powerful Tools for Genetic Analysis Come of Age", *Tibtech*, vol. 17, March, 1999, pp 127-133.

Air-Liquid Two-Phase Microfluidic System for Low-Cost, Low-Volume, and Low-Power Micro Flow Cytometer

D. Huh[1], Y-C. Tung[2], J. B. Grotberg[1], S. Skerlos[2], K. Kurabayashi[2, 3], S. Takayama[1, 3]
[1]Department of Biomedical Engineering, [2]Department of Mechanical Engineering, and [3]Macromolecular Science and Engineering, University of Michigan, Ann Arbor, MI 48109-2125, USA (e-mail: takayama@umich.edu)

Abstracts

This paper reports on the design and initial experimental investigation of a disposable air-liquid two-phase microfluidic system for use in microscale flow cytometry (μFC). Air sheath flow is advantageous in developing miniaturized and integrated flow cytometers because it reduces volume and power requirements. Air-liquid two-phase stratified flow was achieved by a combination of suction (vacuum pump) and injection (syringe pump) in a thin polymeric microfluidic channel. A photo-multiplier tube (PMT)-based detection system mounted on an inverted epi-fluorescence microscope was used to demonstrate counting of the number of fluorescent particles passing through the observation channel.

Keywords: Micro flow cytometer, air sheath flow, disposable microfluidic channel

1. Introduction

Flow cytometry is the measurement of physicochemical characteristics of microscopic biological particles using a coupled fluidic-fluorescence measurement system [2]. The process of existing flow cytometers depends heavily on the single-file flow of bioparticles achieved by hydrodynamic focusing [1]. Up to date, many efforts have been directed towards miniaturizing and improving the design of flow cytometers that use aqueous sheath flow [3], [4]. This paper describes a novel polymeric microfluidic system that uses an air sheath, instead of aqueous solutions, to aerodynamically focus sample streams in the observation channel.

2. Characteristics of two-phase microfluidic system

Air-liquid two-phase microfluidic system of μFC has three axisymmetric flow compartments that generate a stable core stream of suspending particles inside the observation channel (Figure 1). Compared to conventional cytometers that require up to 1000 mL of sheath liquid for 1 mL of sample, the air sheath fluidic system needs no liquid other than the sample fluid and would allow drastic reduction of overall size and weight of the cytometer. Furthermore, power requirements for pumping sheath fluid can be potentially lowered through the use of a pressurized air canister. The fluidic module of μFC is fabricated out of poly(dimethylsiloxane) (PDMS) using soft lithography [5], which permits rapid prototyping of disposable, low-cost, non-toxic, and optically transparent microfluidic channels.

J.M. Ramsey and A. van den Berg (eds.), Micro Total Analysis Systems 2001, 468–470.

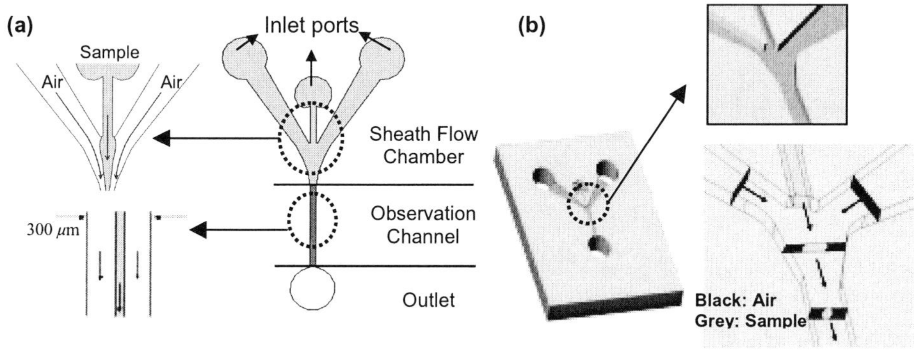

Figure 1. (a) Flow compartments of μFC. In the sheath flow chamber, sample fluid is injected/enveloped by air sheath flow leading to tight aerodynamic focusing. The focused sample core flow passes along the observation channel (total length of microfluidic channel = 14 mm). (b) Three dimensional rendering of microfluidic module of μFC for aerodynamic focusing of sample stream (channel thickness = 100 μm).

3. Experimental

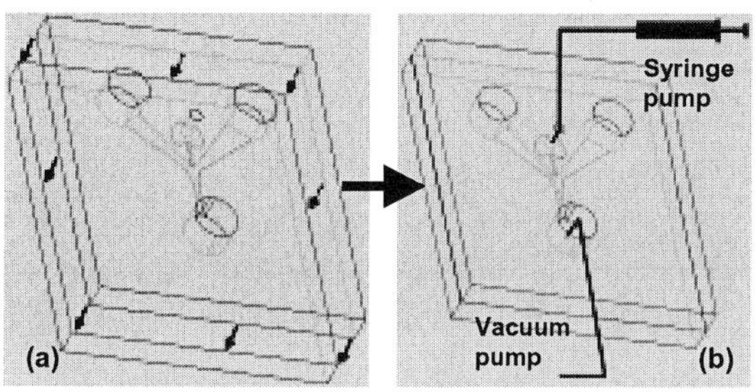

Figure 2. (a) Channel PDMS slab (upper) is sealed to blank PDMS layer (lower). (b) Sample inlet port and outlet are connected to syringe pump and vacuum pump, respectively by using syringe needle and silicon tubing.

A PDMS slab with channel features molded into one face was attached to a blank PDMS layer to produce a sealed fluidic system (Figure 2-(a)). Sample liquid was injected into the middle inlet port by a syringe pump at three different flow rates (20 mL/h, 10 mL/h, and 6 mL/h), while the air sheath flow was driven by suction of a vacuum pump (vacuum pressure: 75 mmHg) connected to the outlet (Figure 2-(b)). Images of the air-liquid flow were captured by a CCD camera. Optical detection of red fluorescent micro beads suspension (size range: 15 μm ~ 19 μm, Spherotech Inc., IL) at a concentration of 0.0036% weight/volume was demonstrated using an epi-fluorescent microscope-based setup. Fluorescent emissions from the beads in the sample core flow of the observation channel were detected at 5 MHz by a photo-multiplier tube (PMT) (P30CWAD5-01, Electron Tubes Inc., NJ), and sampled at 50 kHz by a data acquisition system (PCI-6111 DAQ card, National Instruments, TX) connected to a PC. The data were processed off-line to yield spectrometric measurements of the number of beads.

4. Results and discussion

Surface chemistry of microfluidic channel walls affects fluid flow [6]. As such it was found that the relative affinity of the PDMS channel wall for water significantly influenced the aerodynamic focusing. In hydrophilic channels prepared by plasma oxidization of the PDMS, spreading due to wetting of liquid occurred at the entrance of the observation channel and led to the atomization of the liquid column (Figure 3-(a)).

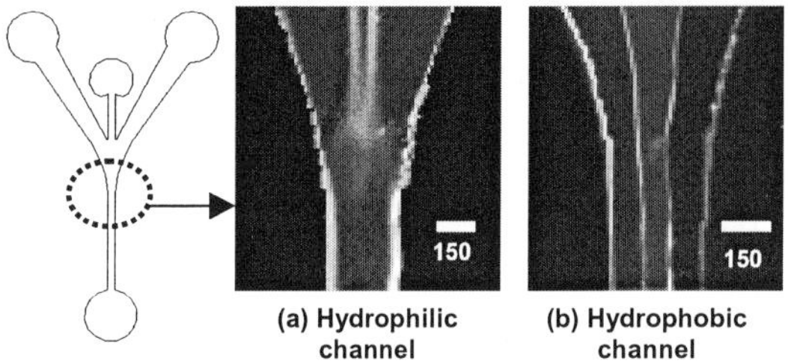

Figure 3. Flow behavior of aqueous liquid column in (a) hydrophilic channel and (b) hydrophobic channel at the entrance of observation channel. The unit of size bar is μm.

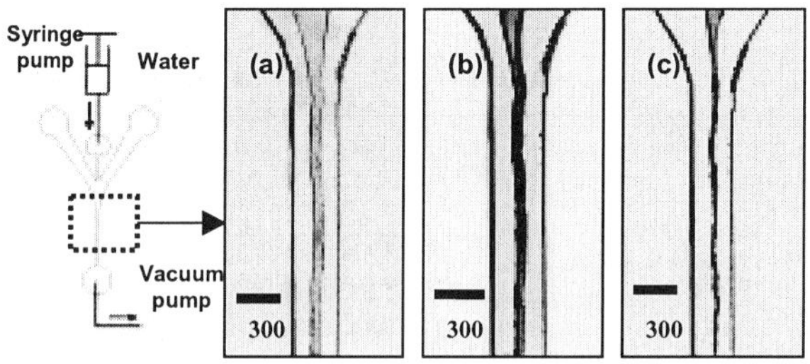

Figure 4. Stably focused aqueous sample core flow along the microfluidic channel of μFC. (a) flow rate = 20 mL/h, (b) flow rate = 10 mL/h, (c) flow rate = 6 mL/h. The unit of the size bar is μm.

In contrast, untreated hydrophobic PDMS channel walls repelled the aqueous liquid and assisted focusing of the sample core stream (Figure 3-(b)). Moreover, this sample stream was maintained steadily along the length of the observation channel. The width of the liquid column decreased as the flow rate of the syringe pump was lowered (Figure 4). In this range, fluorescent particle counting was readily achievable. The average of nine counting measurements performed over 16 minutes yielded an average of 87 beads per second at a sample flow rate of 20 mL/h, which agrees well with the expected number of beads based on data from the manufacturer (55 beads/sec ~ 113 beads/sec).

5. Conclusions

An air-liquid two-phase microfluidic system for developing a microscale flow cytometer was constructed and its operational feasibility was demonstrated. Specifically, the low-cost, low-volume, and low-power requirements of this fluidic system will pave the way for the development of hand-held cytometers for point-of-care clinical diagnosis and *in situ* environmental monitoring of the aqueous ecosystem.

Acknowledgments We thank the Leukemia and Lymphoma Society, Horace H. Rackham School of Graduate Studies, NSF (CTS-0092716), and NIH for funding.

References

1. H. M. Shapiro, Practical Flow Cytometry, 3rd Edition, New York: Wiley-Liss, (1995).
2. H. B. Steen & T. Lindmo, Flow Cytometry - High-Resolution Instrument for Everyone, *Science* **204**, pp. 403-404, (1979).
3. D. Sobek, A. M. Young, M. L. Gray, & S. D. Senturia, A microfabricated flow chamber for optical measurements in fluids, *Proceeding of the IEEE* **2**, pp. 219-224, (1993).
4. R. Miyake, H. Ohki, I. Yamazaki, & T. Takagi, Investigation of sheath flow chambers for flow cytometers (Micro machined flow chamber with low pressure loss), *Transactions of the Japan Society of Mechanical Engineering* **40**, pp. 106 – 113, (1997).
5. D. C. Duffy, J. C. McDonald, O. J. Schueller, & G. M. Whitesides, Rapid prototyping of microfluidic systems in poly(dimethylsiloxane), *Analytical Chemistry* **70**, pp. 4974-4984, (1998).
6. B. Zhao, J. S. Moore, & D. J. Beebe, Surface-directed liquid flow inside microchannels. *Science* **291**, pp. 1023-1026, (2001).

A PARTICLES AND BIOMOLECULES SORTING MICRO FLOW SYSTEM USING THERMAL GELATION OF METHYL CELLULOSE SOLUTION

K. Tashiro[1], S. Ikeda[1], T. Sekiguchi[1], S. Shoji[1], H. Makazu[2], T. Funatsu[2], and S. Tsukita[3]
[1]Department of Electronics, Information and Communication Engineering,
[2]Department of Physics,
Waseda University, 3-4-1 Ohkubo, Shinjuku, 169-8555 Tokyo Japan
[3]Department of Cell Biology, Faculty of Medicine, Kyoto University
E-mail : kou.tashiro@nifty.com / shojis@mn.waseda.ac.jp

Abstract

A novel micro flow system for particles and biomolecules sorting using thermal gelation of methyl cellulose (MC) solution was proposed and evaluated. The fabricated system is to sort out the target biomolecules from the samples in a simple Y-shape microchannel. The cross-linked MC (1wt%, 1000cP) changes its state from sol to gel by elevating the temperature up to 55℃. This phenomenon was applied to plug the channel by the IR laser irradiation. The switching time was within 1 second both open-to-close and close-to-open. By using the MC as a carrier flow, biomolecules sorting system was realized without any mechanical valve structures.

Key Words: Particles and biomolecules sorting, Micro flow system, Thermal gelation, Methyl cellulose

1. Introduction

Particles and cell sorting micro fluidic systems have been developed using laminar behavior in microfabricated flow channels. These devices have two-dimensional structures which have two carrier inlets and one sample inlet located in the middle [1]. The sample flow was controlled by the flow ratio of the right and left carriers. A multi outlet flow switch was also demonstrated with the similar principle [2]. We proposed a novel micro flow switch for particles and biomolecules sorting system using thermal sol-gel transformation of MC solution. A first prototype of the one inlet and two outlets flow switch was fabricated and tested. [3] The switching behaviors were observed in the preliminary experiments. However, the switching time was about 5 seconds both open-to-close and close-to-open. We have been refined the channel structure and have been studied the optimum concentration of MC solution.

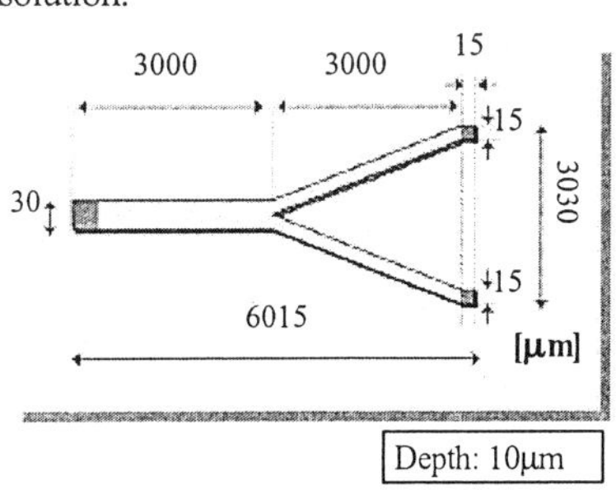

Fig.1 Dimensions of the Y-shape micro channel

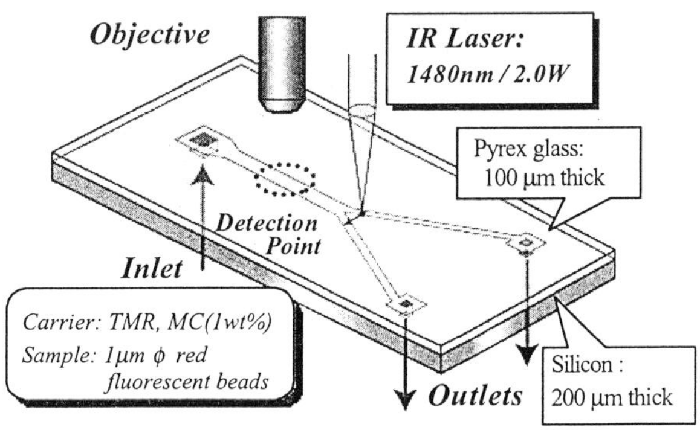

Fig.2 Schematic of the micro flow system

J.M. Ramsey and A. van den Berg (eds.), Micro Total Analysis Systems 2001, 471–473.

2. Principle and structure of micro flow system

Y-shape micro channels were fabricated by anodic bonding of Pyrex glass and silicon substrates. Its dimensions are shown in Fig.1. The schematic of the micro flow system is illustrated in Fig.2. It consists of a micro channel, a optical detector and a movable focused IR laser. The MC carrier flow including the sample is introduced from the inlet. The detection point was located at 200 μm upstream from the branch. Fig.3 shows the principle of the flow control. By irradiating the focused IR laser at the right outlet near the branch, the MC solution is gelled and blocks the flow. Then the sample flows to the left outlet. (Fig.3 (b), (c))

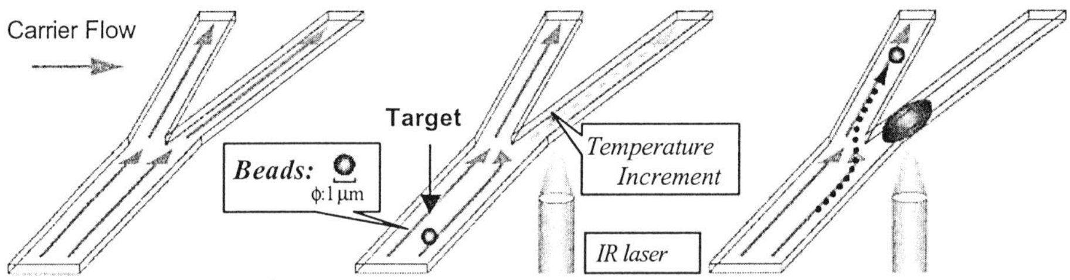

(a) Initial state (b) IR irradiation (c) MC thermal gelation

Fig.3 Principle of the particles and biomolecules sorting system using MC thermal gelation

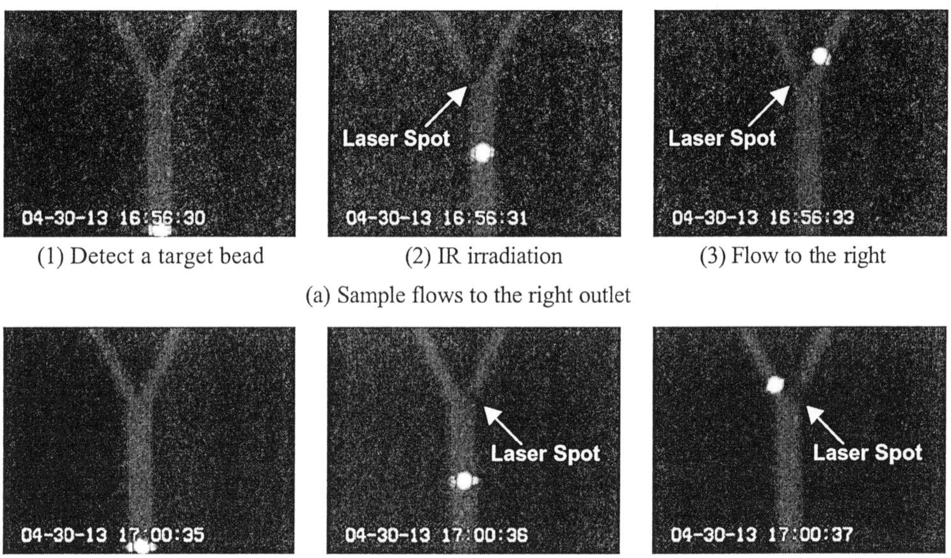

(1) Detect a target bead (2) IR irradiation (3) Flow to the right

(a) Sample flows to the right outlet

(1) Detect a target bead (2) IR irradiation (3) Flow to the left

(b) Sample flows to the left outlet

Fig.4 Measured tomograms of a bead flow inside the Y-shape channel

3. Experimental Results and Discussion

The optimized concentration and viscosity of MC solution is about 1wt% and 1000Cp. In the case of 1wt% MC solution, the critical temperature is about 55 ℃. To evaluate the system, beads of red fluorescent 1 μmφ microshpere were used as the samples. Tetramethylrhodamine-5-maleimide: TMR was mixed with MC solution to observe the carrier flow. Fig.4 shows continued video fluorescent images of the sorting behavior of a single bead, (a) sort to the right and (b) sort to the left. The MC block is observed as the black spot because the TMR loses the fluorescence by heating. The results indicated that sorting was completed less than 3 sec after detecting the sample. The necessary IR laser power to form the block in the fabricated micro channel is about 2 W.

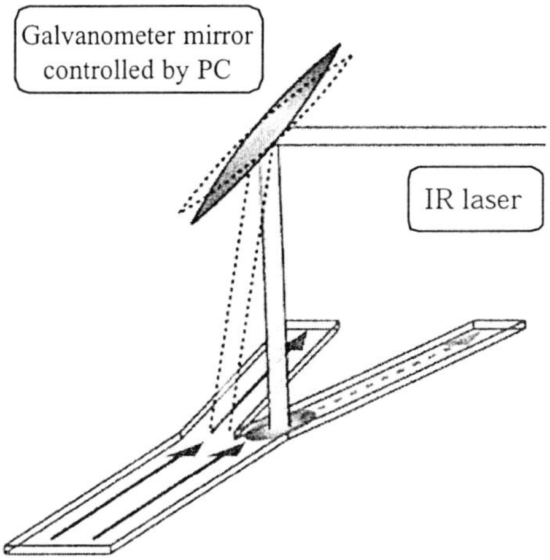

Fig.5 The image of using galvanometer mirror with a single IR laser source

The MC solution is slightly shrunk at the gel state during the sol-gel transformation. Scanning the IR laser is much useful to control the gelled MC area for the plug and to prevent the leakage. The galvanometer mirror controlled by PC as illustrated in Fig.4 is used for this purpose as well as the switching with a single IR laser source.

4. Conclusion

A novel particles and biomolecules sorting micro flow system using thermal gelation of methyl cellulose (MC) solution was developed. The switching time was within 1 second both open-to-close and close-to-open. The biomolecules sorting system was successfully realized without any mechanical valve structures by using the MC as a carrier flow. We are currently trying to apply this system to DNA, protein and biological cell sorting.

Acknowledgement

A part of this work is supported by the Japanese Ministry of Education Science and Culture under a Grant-in-Aid for the COE of Waseda University (2001~2005) and under a Grant-in-Aid for Scientific Basic Research (B) No.12450167.

Reference

[1] P.Telleman, U.D.Larsen, J.Philip, G.Blankenstein, A.Wolff, "Cell Sorting in Microfluidic Systems", Proc. Micro Total Analysis Systems'98, Banff, pp39-44 (1998)

[2] U.D.Larsen, G.Blankenstein, J.Branebjerg, "Flow Switch for Analyte Injection and Cell/particle Sorting", Proc. Micro Total Analysis Systems'96, Basel, pp113-115 (1996)

[3] K.Tashiro, S.Ikeda, T.Sekiguchi, H.Sato, S.Shoji, H.Makazu, K.Watanabe, T.Funatsu, S.Tsukita, "Micro Flow Switches Using Thermal Gelation of Methyl Cellulose for Biomolecles Handling", Transducers'01, pp932-935 (2001)

FLOW PATTERN CONTROL BY FLOW BALANCING IN MICROCHANNEL FOR HIGH SPEED & HIGH PURITY SEPARATION OF MICROBE IN MICROCHIP

Fumihito Arai*[1,2], Akihiko Ichikawa*[1], Toshio Fukuda*[1], Koji Horio*[3], and Kouichi Itoigawa*[4]

*1 Nagoya University, *2 PREST, JST,
*3 Moritex Corporation, *4 Tokai Rika Co., LTD.

Abstract

We developed a new system for random separation of a single microorganism, such as a living cell and a microbe, in the micro fluidic device under the microscope using laser manipulation and the microchannel flow. The main flow in the microchannel splits at the exit of the sample chamber. One flow goes to the extraction port with high speed to take out the target. Another flow goes to the drain. The target was transported by the laser manipulation from the sample chamber to put on the main flow. Since the force balance of the laser-trapped object is important for stable manipulation of the target, here we propose the stagnation point control around the sample chamber exit. Pressure balance was adjusted and the stagnation point was controlled. Some preliminary experiments are conducted to show the effectiveness. The target was trapped by the laser, transported to put on the main flow around the split area through the stagnation, and was taken out from the extraction port. Separation time was reduced less than 10 seconds.

Keywords: Separation, Microbe, Yeast cell, Laser tweezers, Flow pattern control

1. Introduction

Recently, there has been great interest in the high throughput screening of microorganisms, for example, for finding the novel microbes and for investigating the unknown character of the microbes. It is important to manipulate and separate randomly suspended microbes with high speed and high purity. We have developed a selective separation system of the microbe suspended in the liquid. We integrated the laser tele-manipulator for local position control of the target, and the dielectrophoresis for exclusion of the other objects around the target by shaping the potential field in the microchannel. Then the target was isolated and was transported out by the high speed microchannel flow. These different methods are cooperatively used in the microchannel for high speed & high purity separation[1],[2]. In this method, the microchannel design and pressure balancing as well as transportation trajectory of the laser manipulator are all important.

J.M. Ramsey and A. van den Berg (eds.), Micro Total Analysis Systems 2001, 474–476.

2. New Microchannel Design to Reduce Separation Time

The previous method is based on the flow speed distribution in the microchannel, and utilized the stagnation point at the blanch [1],[2]. Here we aim to reduce the separation time much more by improving the flow pattern of the separation cell. We propose a new separation cell, which has microchannel pattern as shown in Fig.1. The main flow comes close to the exit of the sample chamber and splits. We visualized the flow pattern around it as shown in Fig. 2. There is the stagnation near the split area. The flow speed is reduced at the stagnation and we can achieve stable laser manipulation.

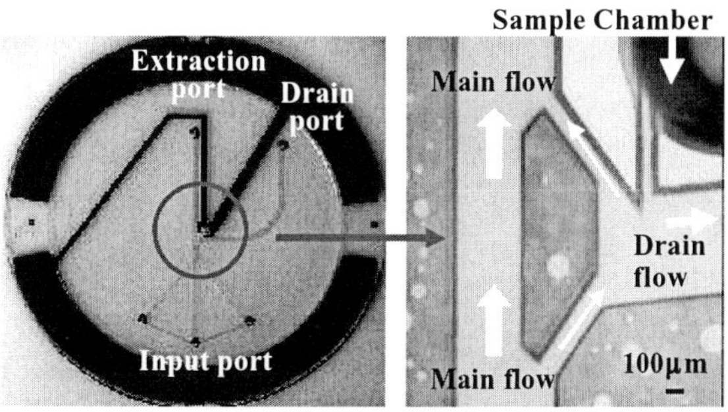

(a) Whole photo (ϕ 40mm)   (b) Central part

Fig. 1 New separation cell to reduce separation time

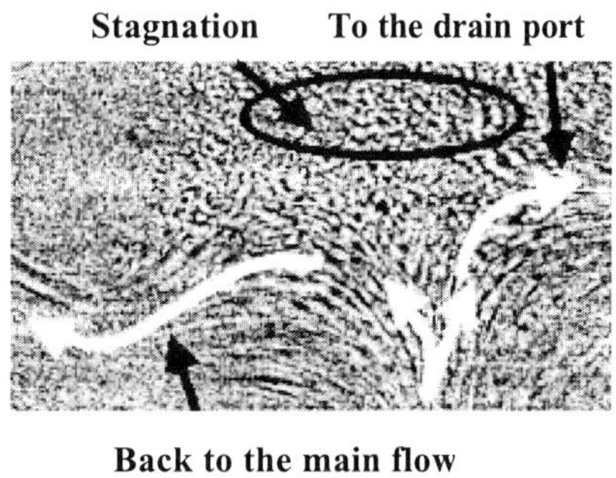

Fig. 2 Streamline at the split area

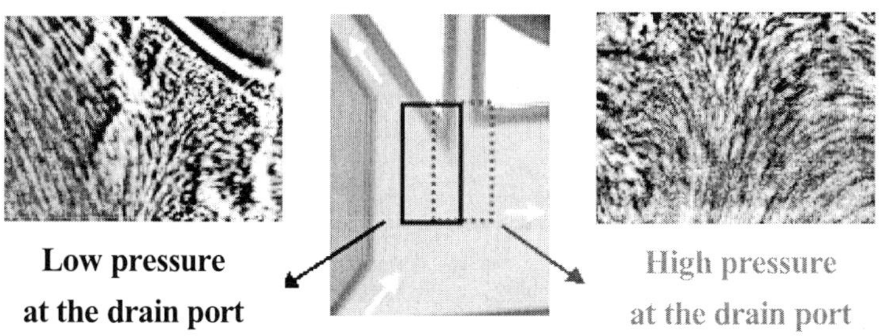

Fig. 3 Stagnation point control around the exit of the sample chamber

Table.1 Pressure difference between the input port and drain port

Flow Position	Drain Status	Differential Pressure [kPa]
	Open	4.41
Flow1	Compressed	4.14
Flow2	Compressed	4.02
Flow3	Compressed	3.84
Flow4	Compressed	3.31

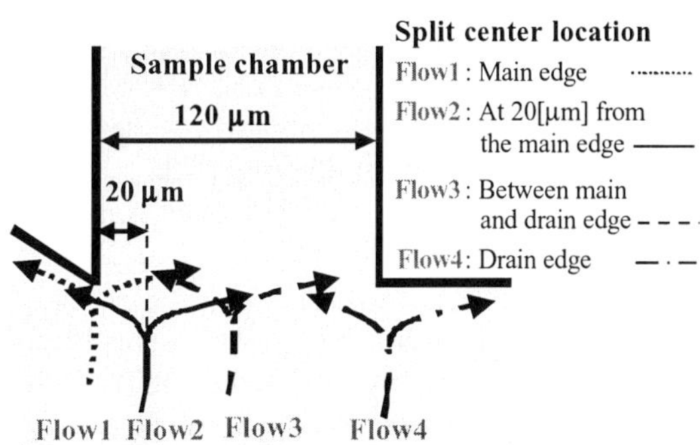

Fig. 4 Flow pattern change around the exit

We discovered that it is possible to control the position of the stagnation by balancing the pressure at the input port and the drain port of the cell as shown in Fig. 3. In Fig.3, stagnation is formed at the left side of the exit of the sample chamber, when the pressure of the drain port is reduced. We measured the pressure difference between the input port and drain port. The results are shown in Table 1. Depends on the pressure difference, we can observe the different flow pattern around the exit of the sample chamber as illustrated in Fig. 4.

Under an appropriate flow environment like Flow 2 in Fig. 4, we can transport the laser trapped object along the edge line of the microchannel wall stably as illustrated in Fig. 5. Figure 6 shows the streamline around the split area. It was possible to control the flow pattern by adjusting the pressure difference between the input port and drain port. Based on the proposed method, we transported the Yeast (Size: ϕ 5μm) to the main flow within 10 seconds with high purity. We succeeded in high speed & high purity separation by the new microchannel and flow pattern control.

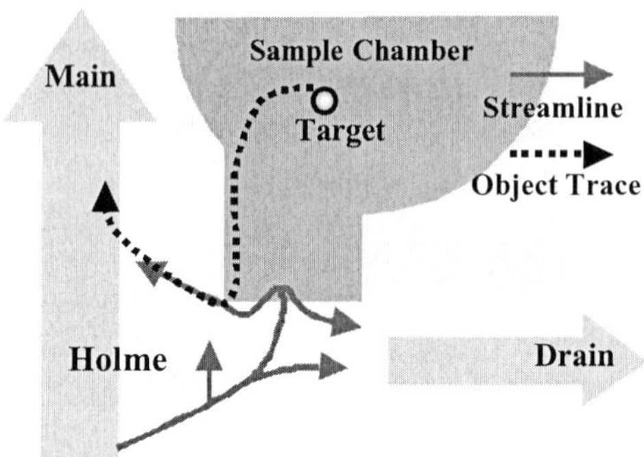

Fig. 5 Transportation trajectory of the target

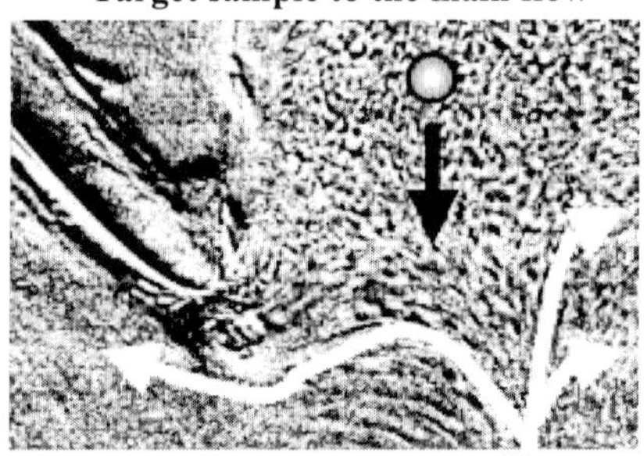

Fig.6 Streamline around split area

3. Conclusions

An arbitrary selected single Yeast was extracted with high speed and high purity, even though there are a large number of Yeasts in solution. Exploration of the microbes will be encouraged in this century for effective utilization of earth resource. We confirmed the effectiveness of the proposed method based on the strategy of controlling the flow pattern and stagnation point in the microchannel. Our method is extremely useful in the pure cultivation of the microbe and will be a promising method for biologists in screening useful microbes.

References

1 F. Arai, et al., Separation of Target Microbe in Microchip by Flow Balancing and Visually Guided Laser Tele-manipulator, Micro Total Analysis Systems 2000, Kluwer Academic Publishers, 2000, p.29-32.

2 Arai, F., Ichikawa, A., Ogawa, M., Fukuda, T., Horio, K., and Itoigawa, High Speed Separation System of Randomly Suspended Single Living Cells by Laser Trap and Dielectrophoresis, Electrophoresis 2001, 22 No. 2, pp.283-288 (2001)

HIGH-DENSITY REAGENT STORAGE ARRAYS FOR HIGH-THROUGHPUT SCREENING

Sherri Biondi, Jeffrey Wolk, Gloria Cheng, Ravi Vijayendran, Tex Horning
Caliper Technologies Corporation, 605 Fairchild Drive, Mountain View, CA 94043-2234

Keywords: LibraryCard™, High-throughput screening, nanoliter dispensing

1. Introduction

Two of the challenges in ultra high throughput drug screening are the expense of compound usage and the logistics of compound plate handling. Typically, tens of microliters of compound are required for each screen, compounds must be used within hours of coming out of storage and thousands of plates must be handled for each screen.

We have been developing the LibraryCard™ reagent array as a method for storing and accessing library test compounds. This method minimizes many of the problems in traditional screening by storing each compound as a sub-picomole dried spot. Approximately 25,000 compounds can be stored on a substrate the size of a 96-well plate. To run screening assays, the stored compounds are redissolved in nanoliter volumes of fluid and aspirated directly into a Caliper LabChip® device. Once on the device, the dissolved compound is mixed with the appropriate reagents for the screening assay and analyzed.

2. Experimental

LibraryCard™ reagent arrays are made using commercially available DNA microarraying techniques. The library compounds are spotted onto silanated glass substrates using an array of straight pins. The compound solutions are typically 100uM in DMSO with a small percentage of excipients added to improve spotting consistency and redissolution. Once spotted, slides are dried and stored in a dark, low humidity, oxygen-free environment.

We have built an instrument which can use LibraryCard™ arrays for high-throughput screening with a Caliper LabChip® device (Figure 1; For more information about LabChip® devices see reference [1]). To use this instrument, the array is placed in an x,y,z-controllable fixture beneath a LabChip® device and aligned to the sipper capillary. During a screen, the robot moves the array so that the sipper capillary alternates between an aqueous buffer well and a position 30μm directly above each array spot. Each compound spot is redissolved in a seven nanoliter drop of buffer which hangs by surface tension from the sipper capillary. Once redissolved, the compound is aspirated into the LabChip® device where it is mixed with the other components of the screening assay.

J.M. Ramsey and A. van den Berg (eds.), Micro Total Analysis Systems 2001, 477–479.

Figure 1. A LibraryCard™ array is shown beneath a LabChip® device. Fluid from the glass capillary is used to redissolve and aspirate the reagent spots on the array.

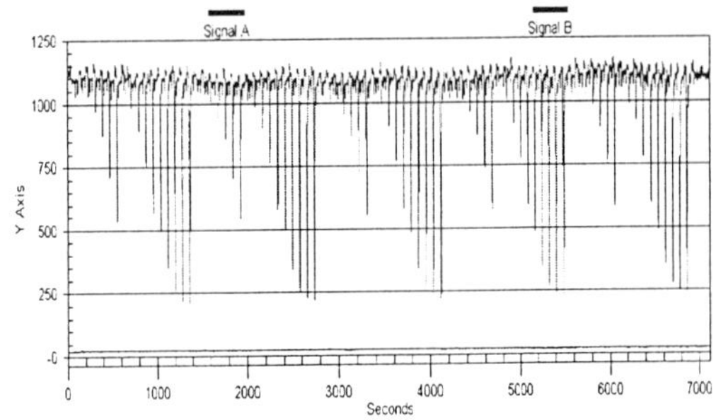

Figure 2. Dose-response data for a fluorogenic assay using LibraryCard™ arrays as an inhibitor reagent source.

3. Results and discussion

Both phosphatase and kinase assays have been performed with this platform and the results have compared well to plate-based screening systems. For the fluorogenic phosphatase assay, the enzyme and substrate are introduced and mixed on the LabChip® device while the potential inhibitors are stored on the LibraryCard™ array. In the absence of an inhibitor, the enzyme cleaves the phosphate group from the substrate which results in a fluorescent signal. When an inhibitor is redissolved from the array and aspirated into the LabChip® device, the phosphate group on the substrate remains uncleaved and the result is a signature dip in fluorescence. The data in Figure 2 illustrates a dose-response for three phosphatase inhibitors where each inhibitor was spotted on the substrate at five different concentrations repeatedly. The reproducibility of this system (accumulated between both spotting and sipping) can be seen in this data where the coefficient of variation for these inhibitors is on average less then 4%. The data in Figure 3 is from a phosphatase screen of a collaborator's compounds (only a subset of the data is shown). We were able to identify each of the inhibitors in this library. The inhibitor potency compared well to that measured using a plate-based screening method where the variation between the two methods was on average less then 25%. The log P's for each compound were also determined and varied between 1 and 5.5 indicating that we were well able to redissolve compounds with a range of hydrophobicity

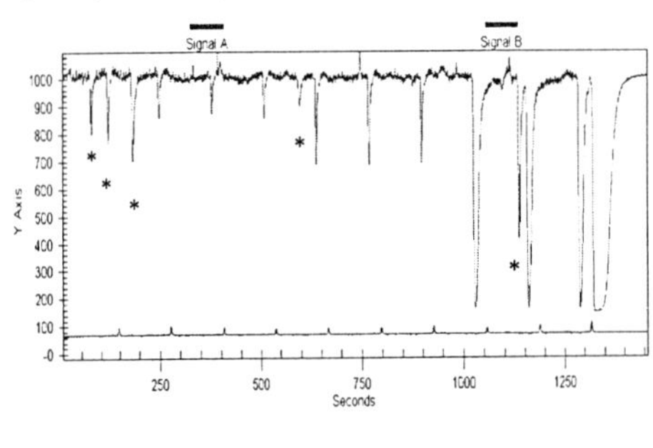

Figure 3. Phosphatase screening data using LibraryCard ™ arrays with a collaborator's compounds. The inhibitors marked with a '*' are the hits found in this library. The remaining inhibition signatures are spiked control compounds for 15%, 40% and 100% inhibition.

Kinase assays have also been performed with this platform. The kinase assay is more complex than the fluorogenic phosphatase assay. Here, the enzymatic reaction is detected by the change in electrophoretic mobility of a fluorescent substrate. As with the fluorogenic assay, the enzyme and substrate are introduced and mixed on-chip and the inhibitors are redissolved and aspirated from the LibraryCard™ array. After incubation,

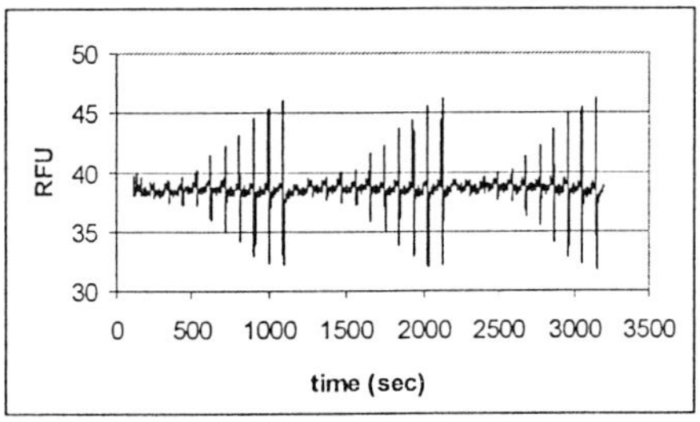

Figure 4. Dose-response of a kinase inhibitor redissolved and aspirated from a LibraryCard™ array.

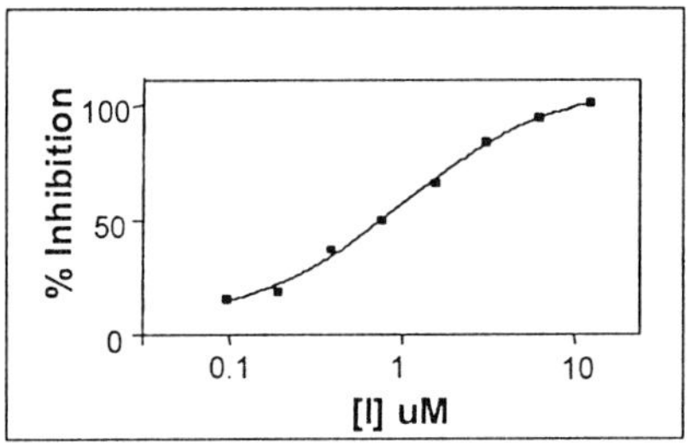

Figure 5. Inhibition is plotted as a function of on-chip inhibitor concentration for the kinase dose-response data.

the mixture passes into a separation channel where an electric field is applied. This causes the substrate and product to migrate at different speeds. In the absence of an inhibitor, the fluorescent signal remains constant. In the presence of an inhibitor, however, a peak and a dip appear in the signal due to a change in the relative amounts of substrate and product in the separation channel. The magnitude of the peak and dip can be used to determine the potency of the inhibitor. In Figure 4, the raw dose-response data is shown for a kinase inhibitor sipped from a LibraryCard™ array. The percent inhibition is plotted as a function of spotted inhibitor concentration in Figure 5. The IC_{50} was calculated from this data (1 μM) and compared well to the IC_{50} calculated from a comparable well-based assay (2.5 μM).

4. Conclusions

We have described a method for manufacturing and using LibraryCard™ reagent arrays in high throughput screening. Both phosphatase and kinase assays have been demonstrated using this platform. The data quality is reproducible and compares well to traditional screening methods. Implementation of a dried reagent array as the compound sourcing medium could deliver dramatic savings in compound usage and plate handling in high-throughput screening operations.

References

1. A Chow et al., *High-Throughput Screening on Microchips*, Proc. Micro Total Analysis Systems 2000, Kluwer Academic Publisher, Dordrecht, The Netherlands, pp. 489-492, (2000).

HIGH THROUGHPUT PHARMACEUTICAL FORMULATION EVALUATION AND ANALYSIS USING CAPILLARY ELECTROCHROMATOGRAPHY ON A MICROFLUIDIC CHIP

Michael Finot[1], Abebaw B. Jemere[1], Richard D. Oleschuk[1], Lori Takahashi[2], D. Jed Harrison[1]

[1] Dept. of Chemistry, University of Alberta, Edmonton, Alberta, CANADA, T6G-2G2

[2] Affymax Research Institute, Santa Clara, California, USA, 95051

Abstract

A 1 mm long chromatographic column was fabricated on a quartz substrate, packed with 1.5 μm diameter ODS1 silica beads, and used for reversed phase capillary electrochromatography (CEC) for evaluating the solubility of pharmaceutical compounds in drug formulations. Given its high efficiency, short analysis times, and small sample requirements, on-chip CEC presents a promising solution to the challenges of high throughput drug screening.

Keywords: drug screening, capillary electrochromatography, solubility studies

1. Introduction

The advent of combinatorial chemistry and high throughput screening has resulted in a huge increase in the number of lead compounds discovered in pharmaceutical research. Quantitative analysis steps down-stream of the initial screening process have become one of the new bottlenecks in the progress of a lead. For example, the need for increased sample throughput in formulation assessment for bioavailability has become apparent. Thus, one stage in pharmaceutical development is the determination of drug solubility, and the effects of various formulations on the solubility of lead compounds. Many derivatives of each lead compound must be tested with several different formulations during this process in order to identify viable compounds. Currently this is accomplished using HPLC methods, but ultra-high throughput screening will exceed the capacity of existing methods in the near future. Microfluidic devices, which may be multiplexed, offer a solution to high throughput formulation demands, and provide a convenient method of replacing conventional HPLC in solubility studies.

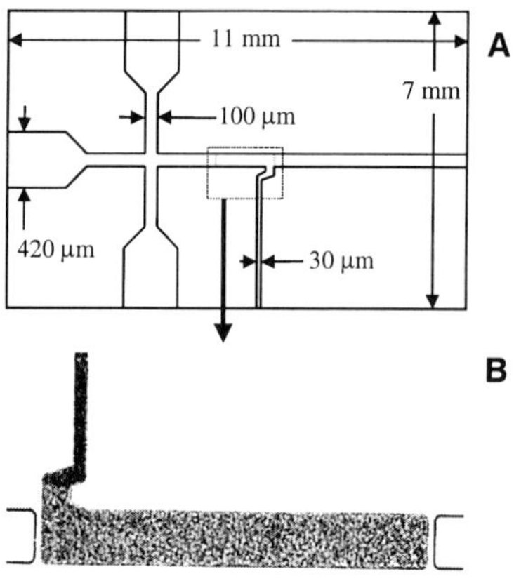

Figure 1 *1mm chromatographic column with cross injector.*

J.M. Ramsey and A. van den Berg (eds.), Micro Total Analysis Systems 2001, 480–482.

In our previous work we have successfully demonstrated the performance of solid phase extraction (SPE) and CEC experiments on 0.2 mm packed columns. [1] The column used in this study was 1 mm long and also features a cross injector. Figure 1A displays a schematic diagram of the device employed in these studies. As observed in Figure 1B, the column displays good uniform packing with no voids. We are able to trap the beads within this region of the channel, as the diameter of the ODS1 beads is greater than the depth of the channel over top of the weirs.

Coumarin compounds were used as a model to test the CEC chip's ability to evaluate solubility in several common drug formulation matrices. The formulations tested represent common commercial compositions, and the pharmaceutical compounds resemble real drugs, so the assays are representative of true sample analysis

2. Experimental

Calibration curve standards were prepared by serial dilution of a 10 mM stock solution of the coumarin compounds in acetonitrile (ACN), or in specific filtered formulations, with the mobile phase buffer. Specific formulations consisted of various mixtures of Tween 80, propylene glycol, ethanol, and saline solution. Diluted samples were then injected into the chromatographic column using voltage controlled plug shaping. Once eluted from the column, samples were detected using laser induced fluorescence with UV excitation.

3. Results and discussion

Our initial efforts were focused on the evaluation of the long column, cross injector device design. Once successful operation of the device was demonstrated testing in drug screening applications was investigated. Figure 2 shows an example of the chromatograms obtained during a typical calibration curve experiment for a duplex run. The compounds, C440 and C460, exhibit corrected retention times of 1.02 s and 1.42 s. Variation in the retention times was typically less than 2% over the course of a calibration run. The resolution of the two peaks is 0.77 and is sufficient for individual peak analysis. Plate numbers of 110 and 78 (110,000/m, 78,000/m and 108/s, 55/s) were calculated for C440 and C460 respectively. Figure 3 displays a typical calibration curve for C440 obtained from these chromatograms. The calibration curves exhibit excellent linearity with R^2 values greater that 0.9900 for both peak area and peak height analysis. The results show that the device responds as expected to incremental changes in concentration. Solubility of the test compounds in each formulation was then determined from the calibration curves.

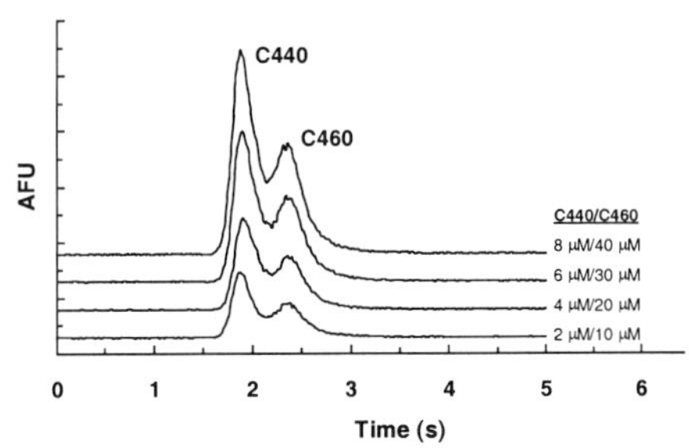

Figure 2. *Typical set of chromatograms obtained for a duplex calibration run. 65% ACN, 35% borate buffer pH = 8.2, separation voltage –1000 V.*

The results obtained from solubility studies of C440 are presented in Table 1. The formulations (F1-F3) display different solubility for each compound with F1 displaying the highest solubility and F3 the lowest solubility in each case. Migration times varied by less than 3.5% between the different formulations. Hence, we conclude that no significant effects on the migration times of the compounds are observed as a result of the various formulation compositions. In most cases, the data show excellent agreement between replicate duplex experiments, demonstrating the reproducibility of the technique. Also, good agreement between individual and duplex experiments shows that it is possible to combine analysis of both compounds into a single experiment, therefore increasing sample throughput.

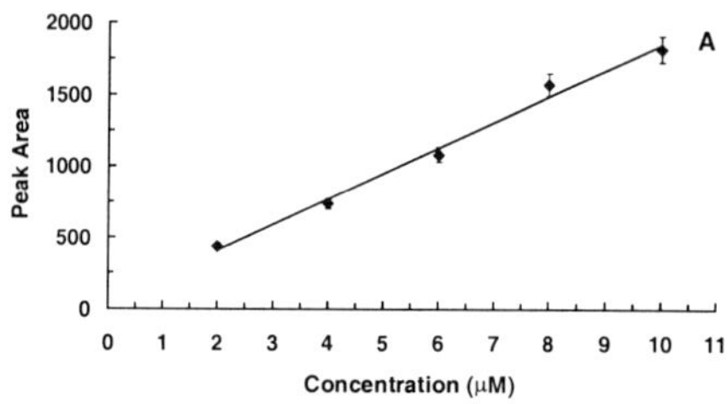

Figure 3. *Typical calibration curve for C440. n = 3, error bars represent 5%.*

Table 1.

Trial	Formulation 1		Formulation 2		Formulation 3	
	Area (μM)	Height (μM)	Area (μM)	Height (μM)	Area (μM)	Height (μM)
C440 Duplex	5.4	5.7	3.6	3.9	1.3	1.4
C440 Duplex	5.6	6.4	4.6	5.1	1.4	1.2
C440 Single	7.1	7.4	4.0	3.9	1.3	1.1

4. Conclusions

The results we have presented successfully demonstrate the feasibility of performing reverse phase CEC using the packed column device. We are able to perform CEC of a mixture of pharmaceutical compounds, and use the data to generate linear calibration curves for quantitative and semi-quantitative analysis of these compounds. As a result, we have successfully employed on-chip CEC in the determination of the solubility of pharmaceutical compounds in various drug formulations.

5. Acknowledgements

The authors would like to thank the Natural Sciences and Engineering Research Council of Canada and the Affymax Research Institute for funding this work, P. Meyers and Phase Separations UK for donating the ODS1 stationary phase material, and J. Taylor for help with the LIF detection system and helpful discussions.

6. References

1. Oleschuk, R.D.; Schultz-Lockyear, L.L.; Ning, Y; Harrison, D.J. *Anal. Chem.* **2000**, *72*, 585-590

MULTIPLEXED MICROFLUIDIC ANALYSES FOR PROTEOMICS USING LABCARD DEVICES

Hossein Salimi-Moosavi, Jerry Wallweber, Hasan Tahir, Herb Hooper, Hrair Kirakossian, Sharat Singh
ACLARA BioSciences Inc., Mountain View, CA, 94043
(HMoosavi@aclara.com)

Abstract

Using Proprietary eTag^{TM} technologies developed at ACLARA BioSciences, we have developed an immunoassay for proteins employing both competitive and sandwich assay formats. Each eTag reporter released per binding event represents one protein. The amount of released eTag reporter is directly proportional to the concentration of the target protein. In a multiplexed protein assay, many different eTag reporters representing different proteins are released and separated using conventional capillary electrophoresis or LabCard^{TM} devices. The advantages of this eTag technology over common proteomic techniques are highlighted.

Keywords: Proteomics, eTag Reagent, Cytokines, Microfluidic, CE

1. Introduction

Modern genetic analysis and drug discovery depend increasingly on rapid, accurate, and cost-efficient analysis of samples with large number of data-points per sample. Hence, there is a growing need for fast and reliable methods for genomic and proteomic analysis [1-2]. ACLARA's proprietary eTag reagents, coupled with electrophoresis, is a versatile and high performance technology for multiplexed quantitation of gene expression, genotyping, and functional proteomics. eTag reagents are a set of electrophoretic mobility tags, or "e-tags" that can code specific oligonucleotide probes or proteins in multiplexed sets. These reagents can release an electrophoretically tagged reporter having a unique and designed electrophoretic mobility among a set of e-tag reporters, as a result of specific enzymatic activity or protein-protein interaction in an assay. Released e-tag reporters are subsequently separated by capillary electrophoresis. In addition, microfluidic systems have been proven to be effective tools for rapid separation and analysis of biological compounds [3-5]. Therefore, combination of the eTag technology with ACLARA microfluidic LabCard devices provides a powerful and unique tool for modern proteomic and genomic analysis.

2. Experimental

Conjugation of electrophoretically tagged moieties to antibodies: Match pairs of antibodies for different cytokines were purchased from different commercial sources

J.M. Ramsey and A. van den Berg (eds.), Micro Total Analysis Systems 2001, 483–485.

(Pierce, R&D Systems). The first antibody of each match pairs was labeled with molecular scissors while the second antibody of that particular match pair was labeled with a specific e-tag moiety.

Immunoassay: The Immunoassays were performed in a sandwiched-based format. In this assay, the antibodies labeled with molecular scissors and antigens (cytokines) were added to the assay buffer, 10% PBS and 40 mg/ml BSA (pH=7.2), and incubated for 30 min at room temperature. Then, the second antibodies for each match pairs labeled with e-tag reagents were added to the assay mixture and incubated for another 30 min at room temperature. Finally, the introduction of molecular scissors released the e-tag reporter upon each binding event.

Separation of e-tag reporter: The released e-tag reporters were separated by conventional capillary electrophoresis using ABI3100 and on the ACLARA plastic LabCards. Figure 1 depicts the layout of the LabCards which were used for the separation of e-tag reporters.

3. Results and discussion

The eTag technologies provide a very unique and powerful tool for the analysis of a wide range of biological compounds. Here, each e-tag reporter released in a particular biological event represents one analyte and/or one particular event. Therefore, the separation and analysis of that particular e-tag reporter provides a very important analytical and biological information about the sample. Since the e-tag reporters are small molecules, they can be easily separated and analyzed in a highly multiplexed fashion (such as 50-100 plexs) in a single analysis. In addition, two or more biological events can be monitored in the same sample. For example, one can look at the gene expression and protein expression at the same time for the same cells.

We have developed an assay for the analysis of cytokines which are very important proteins for drug screening and diagnostics. Each cytokine is represented by an e-tag reporter. We have demonstrated a multiplexed cytokine assay. Figures 2 and 3 show the separation of the e-tag reporters in conventional capillary electrophoresis and ACLARA LabCard devices respectively. In this approach, proteins are analyzed by separation of small molecules rather than cumbersome separation of proteins. In addition, we get signal amplification by multi-labeling of each antibody with a given e-tag moiety and therefore get multiple release of e-tag reporters. The sensitivity for this cytokines assay varies, and is in the range of 0.2 - 5 pM. The linear dynamic range for these proteins vary and are in the range of 1-5000 pM. Cytokines (Angiogenin and TGF-β1) were measured in human serum samples. The results compared favorably with ELISA (as a conventional technique). In addition, no significant serum effect was observed at 10% or lower serum concentration.

4. Conclusion

The eTag technology provides simple, rapid, sensitive, and high-throughput multiplexed quantitation of both protein abundance and gene expression. The reconfigurability, multiplicity, and rapid separation nature of the eTag reagents and LabCard will enable diverse applications in proteomics and genomics.

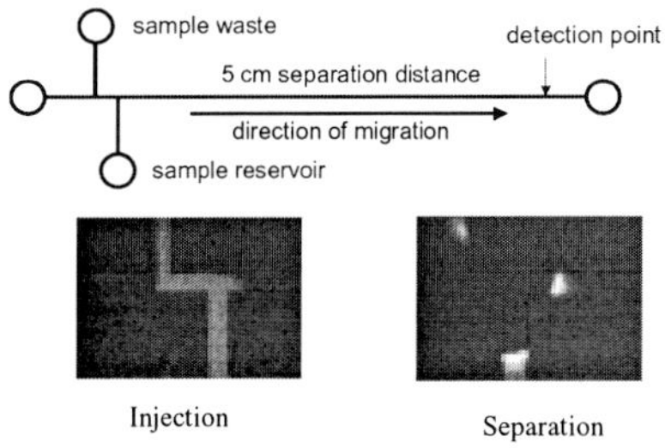

Figure 1. Layout of the Microfluidic Device used for the separation of the released eTag reporter.

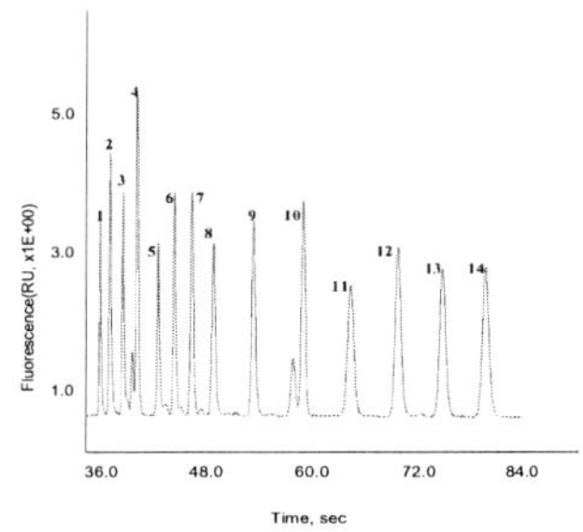

Figure 2. Separation of the 14 eTag reporters using microfluidic LabCard.

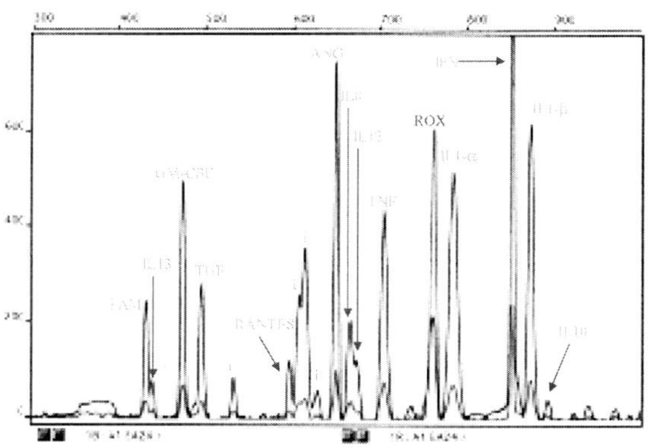

Figure 3. Separation of eTag reporters in a 12-Plex cytokine assay using conventional capillary electrophoresis.

Acknowledgments

We thank Jing Wei and Steve Williams for separation of eTag reporters on LabCard devices. Partial support for this work was provided by DARPA BioFlips Program.

References

1. G. MacBeath and S. L. Schreiber, *Science*, 289, 1760 (2000).
2. S. Steiner and F. A. Witzmann, *Electrophoresis*, 21, 2099 (2000).
3. D.J. Harrison, K. Fluri, K. Seiler, Z. Fan, C. S. Effenhauser, A. Manz, *Science,* 261, 895 (1993).
4. C. S. Effenhauser, A. Manz, H. M. Widmer, *Anal. Chem.,* 65, 2637 (1993).
5. N. Chiem and D. J. Harrison, *Clin. Chem.*, 44, 591 (1998).

REGULATION OF pH IN A MICROFLUIDIC STREAM

D. T. Eddington[1], R. H. Liu[2], J. S. Moore[2], and D. J. Beebe[1]
[1]Department of Biomedical Engineering, University of Wisconsin–Madison, Madison, WI 53706, USA
[2]Beckman Institute, University of Illinois at Urbana-Champaign, Urbana, IL 61801, USA

Abstract

Using stacked micromolding methods and liquid phase polymerization techniques, we have constructed an organic self-regulating microfluidic system. We demonstrate the system's ability to autonomously regulate an input stream (pH 2) to pH 7 under a range of flow conditions. Volume expansion and contraction of a pH sensitive hydrogel govern the flux of compensating buffer (pH=12) delivered to the output channel at a point upstream of the hydrogel. System performance is enhanced by utilizing micro scale phenomena – vertical laminar fluid stacking.

Keywords: microfluidic, pH, regulation, hydrogel, PDMS

1. Introduction

In contrast to control systems engineered by man, *in vivo* control is achieved through organic mechanisms. Here, we report an organic feedback scheme for microfluidic systems with the ability to regulate pH. This system has applications in many fields including protein sequencing, DNA analysis, and drug delivery. Feedback control can be achieved while maintaining system elegance through the use of hybrid system designs combining a layered micromolded channel network with *in situ* construction of stimuli responsive components. By using a stimuli responsive hydrogel post as the single active element in the system, we eliminate the need for external power and electronics. The hydrogel acts both as a pH sensor and an actuator, which changes volume at different pH values. The volume change deforms a thin PDMS membrane to partially occlude an orifice and regulate the feedback stream of basic buffer solution as shown in Figure 1.

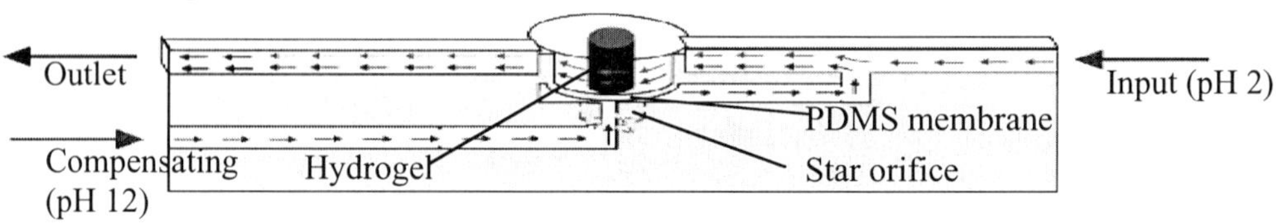

Figure 1. A general schematic of the pH regulation system is shown above.

2. Experimental

Our initial valve design consisted of a small circular orifice that displayed an oscillatory response under constant input stream flow. These observations suggested a functioning feedback loop that lacked stability. Several key parameters dictate system stability including chamber height, membrane thickness, fluid path lengths, laminar flow orientation and orifice morphology. Chamber height (optimal < 80 μm) and

J.M. Ramsey and A. van den Berg (eds.), Micro Total Analysis Systems 2001, 486–488.

membrane thickness (optimal < 40 μm) modifications improved the time response of the system, but the circular orifice produced an on/off control of the feedback stream. When the hydrogel expands, it completely seals the small orifice, stopping the compensating flow. Lack of feedback shrinks the hydrogel, continuing the oscillatory cycle.

Incorporation of a star shaped orifice provided a throttle for the compensating stream eliminating the oscillatory response of the system; Figure 2a compares the circular and star shaped orifices. During system operation, the diameter of the hydrogel post ranges from 300 to 600 μm with the tip-to-tip distance of the star measuring 500 μm. The orifice is sealed when the hydrogel is fully expanded. The star orifice produces a flow gradient of compensating buffer dependent on the diameter of the hydrogel occluding the orifice. The area under the 30 μm PDMS membrane that seals the star is directly correlated to the size of the hydrogel deforming the membrane. For a high input stream flow, the hydrogel shrinks, causing the membrane to seal off a smaller area of the star opening, allowing more compensating buffer solution to pass through. The incorporation of the star orifice, the optimization of other geometrical parameters, and the use of laminar fluid stacking described below, result in the stable regulation of the output stream over a range of flow rates.

3. Results and Discussion

The compensating buffer is injected into the input stream from the bottom. Because the system operates at a low Reynolds number ($R_e \sim 100$), laminar flow causes the two streams to stack vertically with the input stream on top and the compensating buffer on the bottom as shown in Figure 2b.

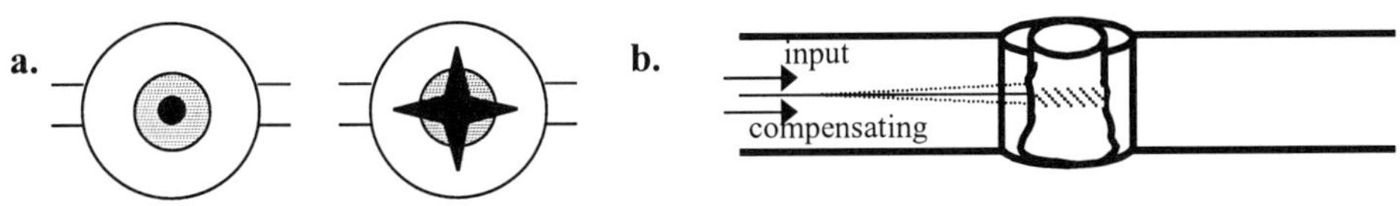

Figure 2. a. Schematic of top views of the actuation chamber with the hydrogel (gray) occluding the orifice (black). The star shaped orifice allows the throttling of compensating buffer, while the circular orifice only allows the compensating buffer to be "on" or "off". **b.** Schematic of the vertically stacked input and compensation streams as they flow by the hydrogel. The slice of the hydrogel that experiences and volume change is indicated by the hash marks.

Vertical alignment of the streams is crucial for the system to operate correctly. The stacking causes the hydrogel post to form into a slightly conical shape as shown in. The top of the post is continually exposed to the input stream and the bottom of the post to compensating buffer. Note that if the compensating and input streams mix rapidly, the entire hydrogel would expand and contract uniformly to deform the membrane. However, this device uses laminar flow to create a small slice in the post that changes volume, effectively minimizing the active region of the hydrogel actuator. Previous experiments correlate a smaller hydrogel volume with improved time response. Diffusion creates the small slice experiencing the pH gradient at the interface of the two streams. This reduces the "size" of the hydrogel post by a factor of 10 and also improves the time response by a factor of 10. Without vertical stacking, the system requires fifty minutes to reach steady state and with the vertical stacking the system reaches steady state in just a few minutes. As shown in Figure 3, the pH is regulated by the hydrogel sensor/actuator at pH 7 for pH 2 buffer flow rates between 25-100 μL/min. The system works within these flow conditions for this particular parameter space. By modifying the flow lengths, orifice morphology, or chamber dimensions the system can be altered to regulate to other values of pH.

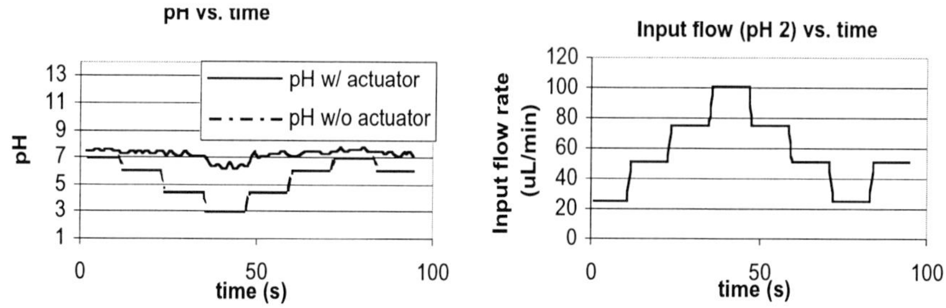

Figure 3. Graphs showing **a,** pH of the output stream vs. time for two separate devices, one with a hydrogel sensor/actuator and one without. **b.** Flow rate of input vs. time (which is the same for the two devices described in figure 3a).

4. Conclusion

The self-regulation system described above can be altered to regulate a variety of variables, by modifying the chemistry of the hydrogel. Other hydrogels have been developed that change with exposure to temperature, light, and biologicals [2-4]. Due to the ease of fabrication and lack of external power requirements, the concepts described here can be readily incorporated into existing microsystems. The organic approach to achieving control of complex function at the micro scale combined with the development of biological responsive materials [2-3] will allow for the creation of biomimetic systems that capture nature's ability to be sensitive and specific while maintaining the robustness and design flexibility of synthetic systems.

Acknowledgements

This work was supported under grants from DARPA-MTO (# F30602-00-1-0570 & F33615-98-1-2853) (Program manager: Dr. Mike Krihak)

References

1. D. Beebe, Jeffrey S. Moore, Joseph M. Bauer, Qing Yu, Robin H. Liu, Chelladurai Devadoss, Byung-Ho Jo., *Functional hydrogel structures for autonomous flow control inside microfluidic channels,* Nature, **404**, pp. 588-590, (2000).
2. T. Miyata, N. Asami, T. Uragami, *A reversible antigen-responsive hydrogel,* Nature, **399**, pp. 766-769, (1999).
3. K. Kataoka, H. Miyazaki, M. Bunya, T. Okano, Y. Sakurai, *Totally Synthetic Polymer Gels Responding to External Glucose Concentration: Their Preparation and Application to On- Off Regulation of Insulin Release,* Journal of the American Chemical Society, **120**, pp. 12694-12695, (1998).
4. Holtz, J. and S. Asher, *Polymerized Colloidal crystal hydrogel films as intelligent chemical sensing material,.* Nature, **389**, pp. 829-832, (1997)

A GLASS/PZT ULTRASONIC MICROFLUIDICS PLATFORM

Chung-Hoon Lee, Yuqian Dong, Amit Lal
Department of Electrical and Computer Engineering
University of Wisconsin - Madison

Abstract

In this paper we present an acoustic pumping technology that can effectively compete with electrokinetic pumping in glass substrates with respect to operating voltage, power, and bubble insensitivity. The technology is based on acoustic streaming along a microchannel embedded in glass acoustic waveguides (horns or beams) and acoustically driven by bonded piezoelectric PZT (Lead-Zirconate-Titanate Oxide) plates (Figure 1), resulting in predominantly longitudinal motion along the device, as opposed to bimorph motion investigated by other piezoelectric microfluidic devices. The pumping relies on acoustic field gradients inside the glass channel near the wall, which are generated by the waveguide shape and/or placement of the piezoelectric ceramics. The design of the horn shape allows a well-defined acoustic gradient, which in turn leads to a predictable pumping force.

Keywords: Acoustic streaming, microfluidic, piezoelectric, pump

1. Introduction

Electrokinetically driven flows are widely used [1] but require relatively high voltages. Magnetically [2] and thermally driven flows can require high currents. Several groups [3,4] have demonstrated ultrasonic pumps, but required bulk micromachining and uncommon thin piezoelectric films. Furthermore, although work on glass-PZT devices has been done, all have focused attention on high-frequency waveguide modes in the channel, not the bulk mode presented here. Our approach allows very low frequency operation (< 50 kHz) resulting in low power consumption and simpler electronics than other approaches requiring frequencies of tens and hundreds of megahertz.

2. Fabrication

The channels are wet-etched in boro-silicate glass slides (1- 20 microns deep and 100 microns wide) using a photoresist mask. The horn shape was obtained by etching the pattern into glass and then thermally and mechanically shocking along the break lines. This is a novel innovation over our previous technique of anisotropically etching silicon horns [6]. The process is amenable to mass production and has a high yield of ~90%. The two horns are fusion bonded and piezoelectric ceramic was attached using thin film solder. The fabrication devices has a $\lambda/2$ resonance at 44 kHz (Figure 2).

3. Experimental Results

The glass/PZT technology presented here advances ultrasonic liquid manipulation by demonstrating the following: (1) Pumping without electrical contacts to the liquid: A typical flow-rate versus drive voltage curve is shown in Figure 2 following a quadratic dependence on drive voltage as expected. Velocities as high as 1 cm/s are observed at higher drive voltages of 50 Vpp at 44 kHz. When the structure is driven at its longitudinal mode resonance, the shear viscous coupling in the channel walls results in a time-averaged streaming, which is a weak

J.M. Ramsey and A. van den Berg (eds.), Micro Total Analysis Systems 2001, 489–491.

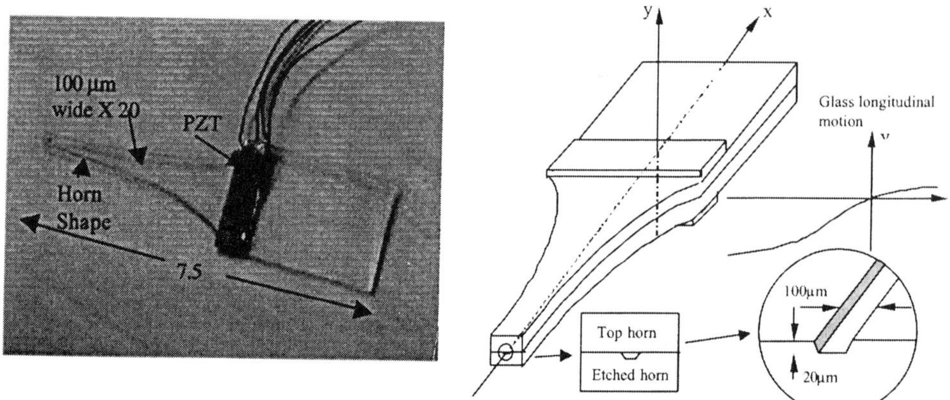

Figure 1. Left shows one of many fabricated devic es with four wires soldered to the four electrodes corresponding to two PZT plates. Right shows the schematic showing glass slides shaped into horn shapes with etched channels and bonded PZT ceramics. Brass grommets were attached on the back to connect silicone tubes.

function of viscosity. This is fundamentally different than our previous result utilizing acoustic streaming at the orifice [6]. (2) Liquids are handled in glass which is low cost, has well-understood optical transmittance properties, and known surface chemistry. (3) The glass structure is driven using inexpensive commercially available piezoelectric PZT ceramics that is amenable to optimization using an analytical theory developed earlier [5]. (4) Low voltage (< 10 Vpp) and power drive (10 milliWatts) is obtained using CMOS drivers. (5) Liquid atomization for mass-specs (Figure 3): At slightly higher intensities than necessary for pumping, surface capillary waves are excited on the liquid meniscus at the high amplitude tip, which when increased to high amplitude result in atomization due to Rayleigh surface-tension instability. (6) Bubble insensitive pumping (Figure 4): As the shear viscous motion of the wall occurs at the wall, bubbles formed in the channel are easily pumped by the force at the liquid around the bubble wetting the wall.

Acknowledgements

We would like to acknowledge the staff of the Wisconsin Center for Advanced Microelectronics (WCAM). We also thank DARPA-MTO for partial support of this work under contract #F30602-00-2-0572.

References

[1] Bousse, L., "Electrokinetic Microfluidic Systems," Proceedings of SPIE, Sept. 20-21, 1999, vol. 3877, pp. 2-8.

[2] Zhang, W., Ahn, C. H., " A Bi-Directional Magnetic Micropump on a Silicon Wafer," Technical Digest of the Solid State Sensors and Actuators Wokshop, Hilton Head Isla nd, 1996, pp. 94-97.

[3] Zhu, X., et al., "Micromachined Acoustic-Wave Liquid Ejector," Technical Digest of the Solid State Sensors and Actuators Wokshop, Hilton Head Island, 1996, pp. 280-283.

[4] Moroney, R.M., White, R. M., Howe, R. T., "Microtransport induced by ultrasonic Lamb waves," Applied Physics Letters, vol.59, no.7, 1991, pp. 774-6.

[5] Lal, A., White, R. M., "Optimization of the silicon/PZT longitudinal mode resonant transducer," Microelectromechanical Systems, Proceedings of the ASME World Congress 1997, Dallas, pp.61-67.

[6] Lal, A., White, R. M., "Ultrasonically driven silicon atomizer and pump," Solid State Sensor and Actuator Workshop June 3-7, 1996, Hilton Head Island, USA.

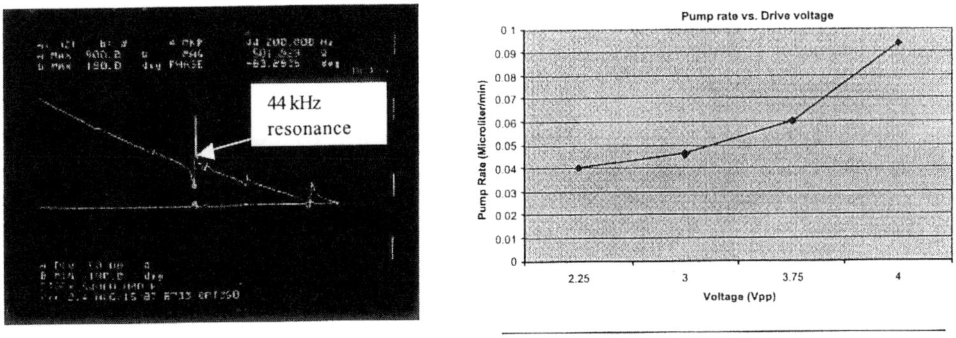

Figure 2. Left shows the HP4194 impedance measurement of the 44 kHz resonance. Low frequency resonance allows simple CMOS drivers to be used. Right shows the typical volume pump rate vs. drive voltage.

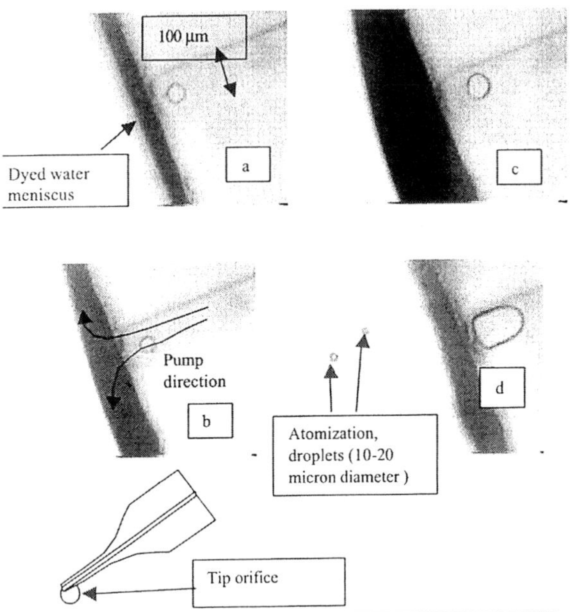

Figure 3. (a) pumping pushes yellow-dyed water to the tip orifice. (b,c) Water volume increases indicating pumping of 0.1 µL/min. (d) Last frame shows atomized drops after the drive amplitude was increased to induce capillary-wave atomization. Atomization rate was greater than 0.1 µL/min.

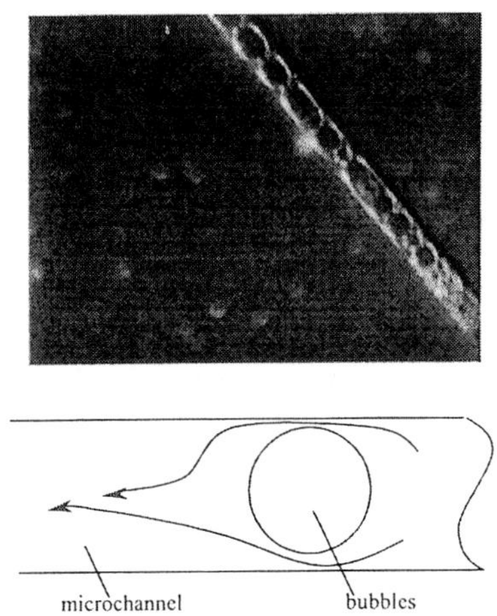

Figure 4. High power drive causes cavitation inside channel resulting in large number of bubbles, which are also pumped at velocities of 1-2 mm/s, with water flowing around them at higher velocities due to wall viscous layer pumping indicating bubble insensitivity to device operation.

FERROFLUIDS FOR DISPOSABLE MICROFLUIDIC SYSTEMS

R. Pérez-Castillejos, J. Esteve, M.C. Acero and J.A. Plaza
Centre Nacional de Microelectrònica (IMB-CNM), Campus UAB, Bellaterra
E-08193 Barcelona, Spain
(Raquel.Perez@cnm.es)

Abstract

During the last decade increasing interest has been put in microfluidics and μTAS because of their improved efficiency in terms of response time, sample volumes, throughput, cost and automation. Despite the big effort put in the development of the microactuators for fluidic handling, some difficulties derived from the miniaturisation remain unsolved. As an alternative to mechanical, pneumatic and electric driving of liquids, a ferrofluidic system for microfluidics was presented recently [1]. The novelty of the concept lies in the use of ferrofluids (stable paramagnetic liquids) as the driving agent of the microfluidic system. In this paper, test and characterization of passive mixers based on this concept (MicroFerroFluidicSystem, μFFS) are presented.

Keywords: Ferrofluid, micromixer, microfluidics

1. Introduction

Ferrofluids can be considered a homogeneous superparamagnetic liquid so that strong forces can be generated when ferrofluids are used as a magnetic actuator [2], to form a microvalve [3, 4] or a micropipette [5]. Our proposal takes profit of using a ferrofluidic drop as a dynamic plunger inside of a complex passive fluidic network [1]: as a liquid, no problems of leakage, wear or friction arise; as a stable material, no sedimentation is observed, and as magnetic, it can be controlled through permanent magnets reducing the power consumption to every switching step. Moreover, it is possible to have a ferrofluid to fit almost each application.

2. MicroFerroFluidicSystem, μFFS

A MicroFerroFluidicSystem consists in a passive microfluidic network and an external magnetic actuator to manipulate the ferrofluid. Ferrofluid movement (controlled by an external magnetic field) creates a pressure drop that induces the displacement of the reagents into the common channel. Displacement sequence will obviously depend on fluidic design.

As an advantage of being liquids, ferrofluids can be introduced into the fluidic network as another reagent so that the fabrication of the microfluidic system simplifies dramatically as the actuator (the ferrofluidic plunger) is introduced into the device after fabrication and in the same way all other reagents are also introduced. Although the results presented here were obtained

J.M. Ramsey and A. van den Berg (eds.), Micro Total Analysis Systems 2001, 492–494.

from devices fabricated by silicon-based technologies, it is under process the fabrication of the passive microfluidic network by cheaper technologies as molding, and thus allowing cheap, disposable devices for single-use applications.

3. Micromixers

Many different functional elements for microfluifics can be designed by applying the μFFS concept and as an example, in this paper results on the test of passive mixers are shown. In order to test the behavior of passive mixers based on ferrofluidic actuation two main designs were fabricated and tested. In figure 1 both the designs are shown: first one (a) includes columns that divide the main stream of the flux, and second one (b) is an arrow-shaped channel. Circular holes are used as reagent reservoirs as well as inlets and outlets. Devices were fabricated by a 30μm DRIE step of a silicon wafer to define the channels, and holes were drilled in a 0.5mm-thick glass wafer before anodically bonding both wafers.

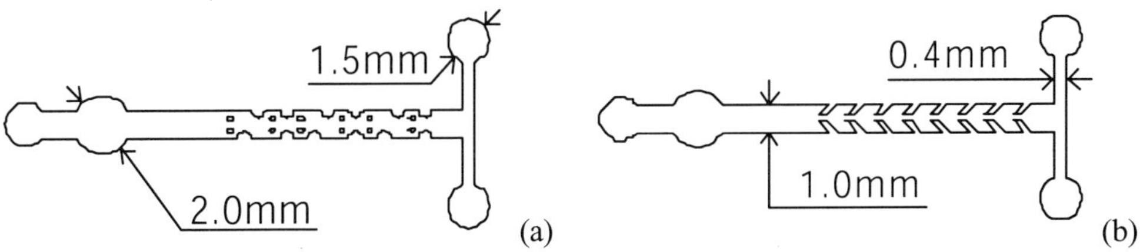

Figure 1. Schematic of the passive mixers designed, fabricated and tested: (a) 200μm-diameter columns and (b) sharp restrictions divide the main stream through the channel.

4. Results and discussion

In the experiments, a hidrocarbon-based ferrofluid (M_S=350G, η=5cp) by Liquids Research Ltd., UK [6] has been used together with differently colored water-based inks. Magnetic actuation on the ferrofluidic plunger was generated by miniaturized permanent magnets: diameter of 3mm and height of 1.5mm; the remanent flux was 12,000G.

From the results it is observed that the ferrofluidic plunger is perfectly controlled by the external magnetic field applied, i.e. the permanent magnet, as well as the ferrofluid displacement along the channel defines the shift of the reagents, avoiding any problem of priming or capillarity. Once the ferrofluid is introduced into the channels network and fixed by an external magnetic field, capillarity results to be fully prevented (Fig. 2). In both designed mixers operation it can be observed how laminar flow develops at the beginning of the inks loading cycle and at the end of the channel, inks start to mixing. More than one single cycle can be easily performed by just shifting the ferrofluidic plunger in the opposite direction it followed before which improves the mixing operation in microchannels.

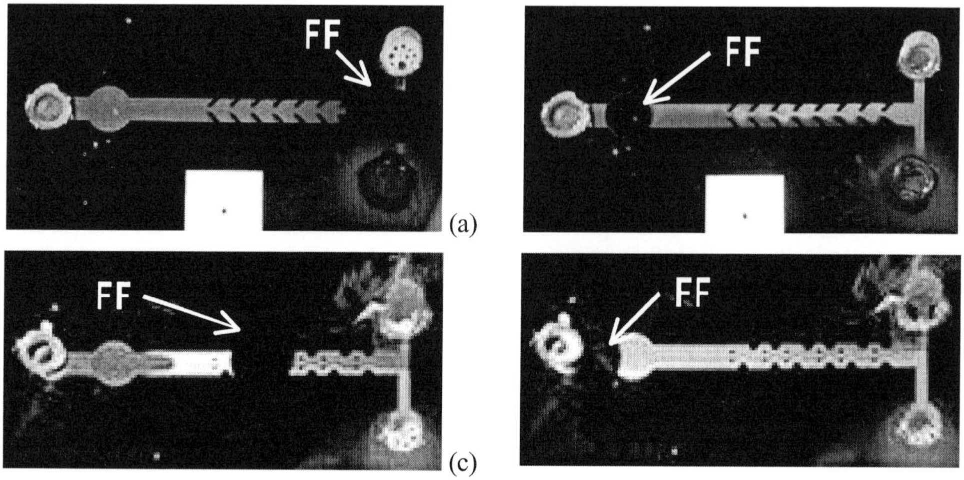

Figure 2. Samples of the movie showing the operation of the arrow-shaped mixer (a, b), and the operation of the mixer with columns (c, d).

5. Conclusions

The MicroFerroFluidicSystem concept has been proven to be a useful method to control the liquids behavior in microchannels, avoiding both capillary and priming problems. As an example of the μFFS possibilities when combined with smart passive microfluidic devices, results obtained from the operation of two different mixers have been presented showing good skills for mixing liquids. Optimization of the mixers design parameters is under development.

Acknowledgements

This work was funded by the Brite-Euram III program (MICROMAG project).

References

1. R. Pérez-Castillejos et al., *Smart passive microfluidic systems based on ferrofluids for μTAS applications*, Proc. Transducers 2001 Eurosensors XV, Springer-Verlag Berlin Heidelberg, Germany, pp-954-957, (2001).
2. R. Pérez-Castillejos et al., *The use of ferrofluids in micromechanics*, Sensors and Actuators A 84, Elsevier, The Netherlands, pp. 176-180, (2000).
3. A. Menz et al., *Fluidic components based on ferrofluids*, Proc. 1st Int. IEEE-EMBS Conf. Microtechnologies in Medicine and Biology, Lyon, France, (2000).
4. H.Hartshorne et al., *On-Chip ferrofluidic microvalves and micropumps*, Proc. IFBT, 2000.
5. R. Pérez-Castillejos et al., *MICROMAG: basic research on the use of magnetic fluids in microsystems*, Proceedings of Nanotech'99, Montreux, November, 1999.
6. Liquids Research Ltd., *www.liquidsresearch.com* [Accessed 10 July 2001]

PERFORMANCE OF A CLOSED-CYCLE CAPILLARY POLYMERASE CHAIN REACTION MACHINE

J. Chiou[1], P. Matsudaira[1], A. Sonin[2], and D. J. Ehrlich[1]
[1]Whitehead Institute, 9 Cambridge Center, Cambridge, Massachusetts 02142
[2]Department of Mechanical Engineering, Massachusetts Institute of Technology, Cambridge, Massachusetts 02139

Abstract
A novel capillary polymerase chain reaction machine was characterized in almost 200 successful 23 minute, 30-cycle PCR runs, with up to 78% amplification efficiency of a 500 base pair target.

Keywords: Polymerase Chain Reaction, PCR

1. Introduction

Polymerase chain reaction (PCR) is an exponential process that can up to double the number of copies of product DNA each cycle. MEMS versions of PCR machines, both of static-fluid and unidirectional-flow-through design, have dramatically improved the time efficiency of PCR. A further improvement in PCR yield and time efficiency might be possible with a miniature bi-directional flow device. This device would offer ultra-low thermal mass cycling combined with ability to highly customize the temperature history of the sample.

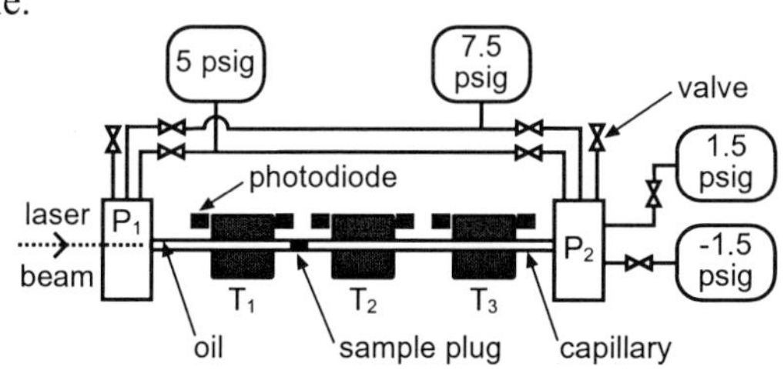

Figure 1: the device.

2. Machine Description

We have constructed a novel thermocycling machine based on a microcapillary equipped with bidirectional pressure-driven flow and in-situ optical position sensors (Fig. 1). A computer-controlled pressure differential moves a 1-μL sample of reaction mixture between heat zones T_1, T_2 and T_3 in a 1 mm I.D., oil-filled capillary. A laser beam waveguides down the capillary to the drop/oil interface. The resulting scattered light is sensed by the photodiodes to provide position feedback. Since the temperature transition time is that required to move and heat/cool the drop alone, the theoretical speed of thermal cycling is very high. Dwell times and accelerations can be adjusted independently. A further benefit is that vast amounts of data can be acquired relevant to all classes of flow-through PCR devices.

J.M. Ramsey and A. van den Berg (eds.), Micro Total Analysis Systems 2001, 495–496.

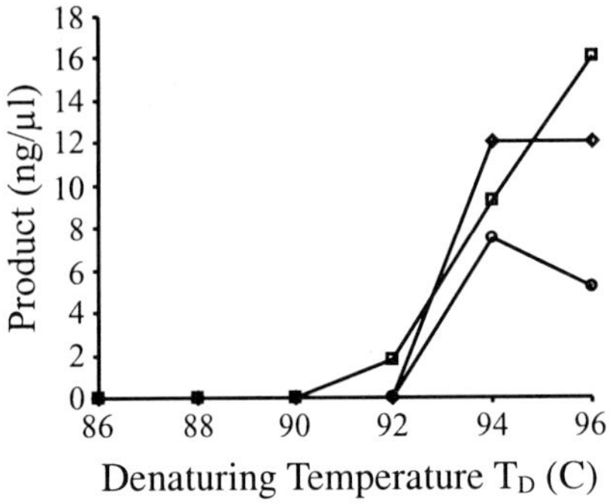

Figure 2: denaturing temperature optimization. Yield increases up to a plateau at 94 C, as expected.

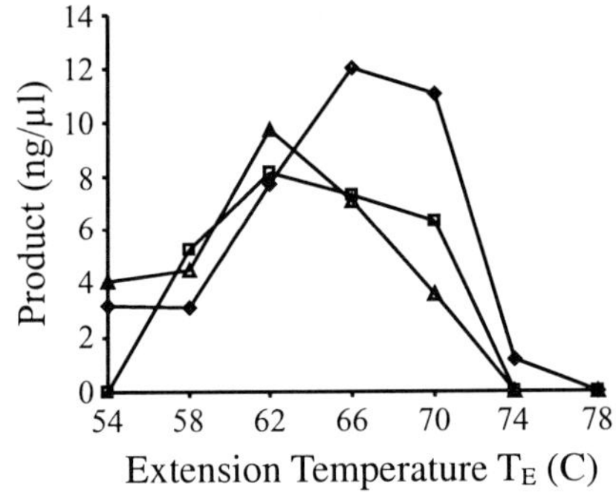

Figure 3: Optimal extension temperature is slightly lower than expected for Taq Polymerase (70-80 C) due to large amount of the enzyme used.

3. Experimental

We have previously described preliminary 10-cycle performance of the device [1]. Here, we have extended the study to full 30-cycle PCR of a 500-bp product (Figs. 2, 3). Almost 200 successful 23 minute, 30-cycle PCR runs were performed under various conditions, with up to 78% amplification efficiency.

4. Scaling Model

Based on the success of these studies we have developed a multi-parameter model that predicts the results of scaling of our device to small size and greater integration (Fig. 4). We predict significant advantages for the scaled device relative to other MEMS approaches in terms of speed and reprogrammability for versatile thermal cycling applications.

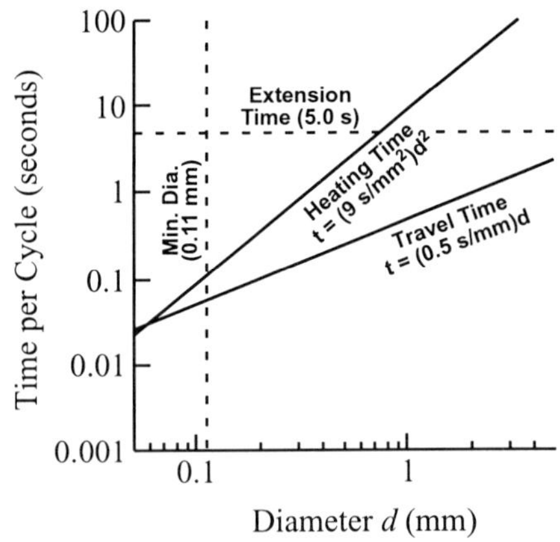

Figure 4: dependency of theoretical time limits in our device on capillary inner diameter d for PCR with a 500 bp product.

Acknowledgements

This work was supported by the National Institutes of Health (grant HG01389) and the Air Force Office of Scientific Research (grant F49620-98-0235).

References

1. J. Chiou, P. Matsudaira, A. Sonin, and D. Ehrlich. A Closed-Cycle Capillary Polymerase Chain Reaction Machine. *Analytical Chemistry,* 73(9): 2018—2021, 2001.

A MICROFABRICATED BIO-MAGNETIC SEPARATOR BASED ON CONTINUOUS HYDRODYNAMIC PARALLEL FLOW

Nikolas Chronis, Wilbur Lam, and Luke Lee
Berkeley Sensor and Actuator Center, Department of Bioengineering
University of California, Berkeley, CA 94720, USA

Abstract

A new micromachined, disposable biomagnetic separator concept is proposed and developed. This design utilizes an H-shape fluidic configuration and exploits the concept of continuous parallel flow in an external magnetic field. With this cost-effective hybrid integrated design, any desired cellular population can be rapidly and continuously isolated from its suspension.

Keywords: bio-magnetic separation, hydrodynamic parallel flow

1. Introduction

Immunomagnetic separation has seen a resurgence of interest in recent years. Recently, several micromachined bio-magnetic separators have been reported utilizing fluidic channels and electromagnets [1]. However, the monolithic fabrication of those devices involves the complex and expensive integration of an electromagnet within each chip. The current design enables hybrid integration, and therefore satisfies the criteria for cost-effectiveness, disposability and design flexibility.

The proposed separator is based on the concept of hydrodynamic parallel flow in a magnetic field. Two buffer solutions, one containing bio-magnetic beads and one without, are directed into an H-shape fluidic network, forming a two-stream parallel flow. An electromagnet on the side of the channel containing the buffer solution without any beads causes migration of the beads into that stream (fig.1&2).

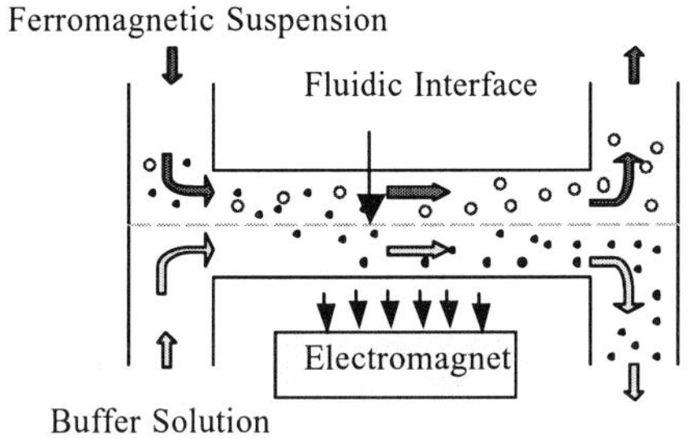

Figure 1: Hydrodynamic parallel flow separation (magnetic beads shown in black)

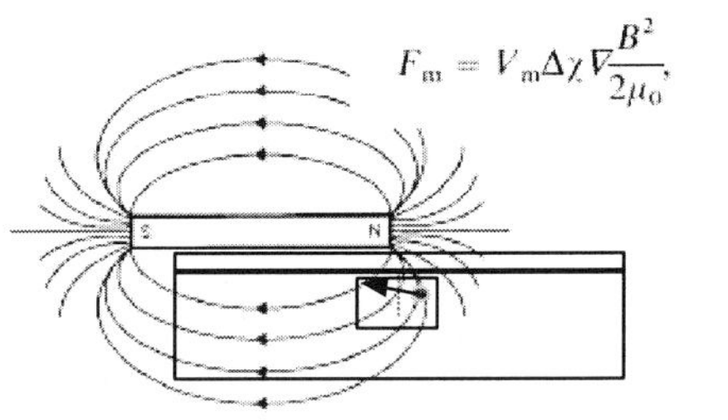

Figure 2: Developed magnetic force on a magnetic bead

2. Theoretical background

Modeling the flow behavior of the ferromagnetic suspension is a crucial step towards high separation yields. For bw solid volume fraction, interparticle interactions can be neglected and Newton's equations of motion for a single bead can be applied:

J.M. Ramsey and A. van den Berg (eds.), Micro Total Analysis Systems 2001, 497–498.

$$\mathbf{F}_D + \mathbf{F}_M = m_p \mathbf{a}_p$$

$$6\pi\mu R_p\left(\mathbf{v}_f - \mathbf{v}_p\right) + \frac{V_m \Delta\chi}{2\mu_o}\nabla(\mathbf{B}\cdot\mathbf{B}) = \frac{4}{3}\pi R_p^3 \rho_p \frac{d\mathbf{v}_p}{dt}$$

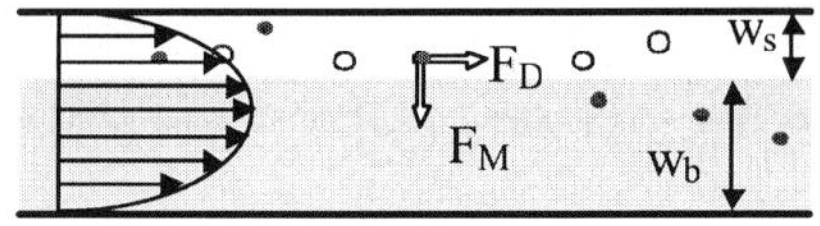

Figure 3: Particle trajectory under the action of F_M and F_D

Where F_M, F_D are the magnetic and hydrodynamic drag forces and v_f, v_p, $\Delta\chi$, μ_o, B are the velocity of the fluid, the velocity of the particle, the magnetic susceptibility, the permeability of the vacuum and the magnetic flux density correspondingly. For a pressure-driven fully developed flow (fig.3) and ignoring inertial forces, the trajectory of a single bead can be extracted for the case of a constant magnetic force perpendicular to the flow. The percentage of beads that reaches the fluidic interface –*separation yield*- can therefore be calculated as a function of the ratio of the widths of the two streams w_b/w_s and parameterized by the magnetic flux density (low, medium, high) (fig.4).

3. Fabrication and experimental setup

The fluidic chip was micromachined with standard photolithographic techniques. An SU-8 mold was used to transfer the pattern to a silicone elastomer (PDMS). A glass slide with four drilled holes allowing fluidic access was then bonded to the PDMS layer. The width ratio of the two streams was adjusted by modulating the hydrostatic pressure in each inlet. An external electromagnet was finally attached on the top of the chip.

4. Results and discussion

Precise control of the design parameters can result in high separation yields (fig.4). For large values of the w_b/w_s ratio the theoretical model predicts that separation up to 100% is feasible. Partial separation had also been verified experimentally (fig.5).

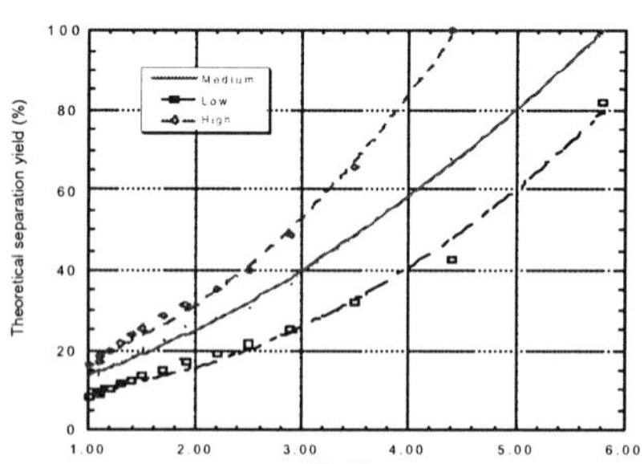

Figure 4: Theoretical separation yield

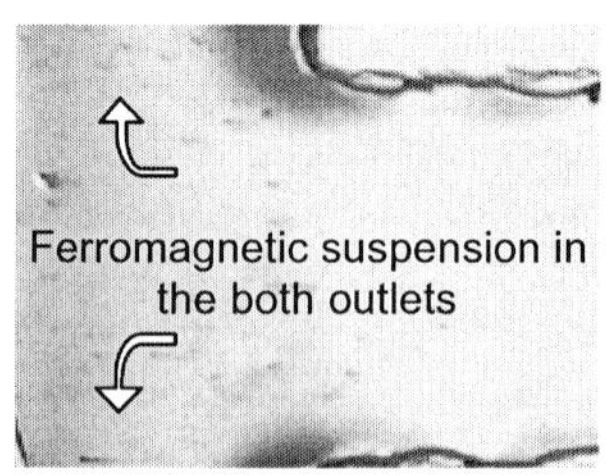

Figure 5: Experimental partial separation (Bio-Mag anti-CD45 beads)

5. Conclusions

A microfluidic, biomagnetic separator based on hydrodynamic parallel flow has been successfully developed and theoretically evaluated. This design can be applied for sorting rapidly and continuously any biological particles, like human blood cells.

References

1. T.M. Liakopoulos, J-W. Choi, C.H. Ahn, "A biomagnetic bead separator on glass chips using semi-encapsulated spiral electromagnets," Transducers '97, Chicago, USA, p.485-8

A MICROFLUIDIC CHIP FOR MULTI-STEP, HETEROGENEOUS IMMUNOLOGICAL REACTIONS

Arash Dodge, Elisabeth Verpoorte, Nico F. de Rooij
Institute of Microtechnology, University of Neuchâtel, Rue Jaquet-Droz 1, CH-2007 Neuchâtel, Switzerland, arash.dodge@unine.ch

Abstract

In previous work, we demonstrated a simple heterogeneous immunoreaction in a microfluidic platform, based on the interaction between physisorbed Protein A (PA) and rabbit Immunoglobulin G (rIgG) [1]. In this contribution, a more complex device based on a sub-nL reaction chamber is presented for multi-step immunological reactions.

Keywords: Electrokinetically driven flow, heterogeneous immunometric assay

1. Introduction

Heterogeneous immunoassays are still a mainstay in bioanalytics, because of their inherent capability to concentrate sample from very dilute solutions onto an immobilized layer of probe molecules in a reaction vial or well. These assays can be performed in different configurations, but all require multiple solution transfers. Integration of these procedures into microfluidic devices yield complex, multichannel networks. Figure 1 shows the chip layout of our device, used to perform a two-step heterogeneous assay.

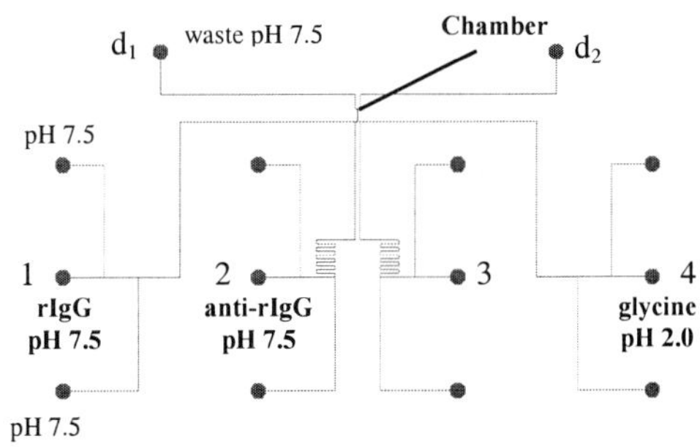

Figure 1: *Chip layout with different solutions and reagents. Chamber is 1 mm long, 18 μm deep and 50 μm wide, for a total volume of 760 pL. Outer dimensions of the device: 2.7 x 4.8 cm (drawing to scale). All unlabeled reservoirs are filled with Tris/HCl buffer (pH 7.5).*

2. Experimental

The immunoassay used to test the chip (Figure 2) is based on the affinity between rIgG and goat anti-rIgG. In this case, a base layer of PA serves to tether the probe antibodies to the surface. Immobilization using PA prevents rIgG denaturation, and positions the rIgG more conveniently for further binding [2]. Prior to analysis, the reaction chamber is manually coated by drawing a silanizing agent from reservoirs 2 and 3 to d_2 by application of vacuum. PA is then physisorbed to the modified surface along the same path. The chamber size is finally defined by drawing 1M NaOH from reservoirs 2 to 3 and from d_1 to d_2, limiting it to the 1 mm long vertical segment. For the analysis, rIgG is first pumped electrokinetically through the chamber until all PA sites are occupied. Then, Cy5-

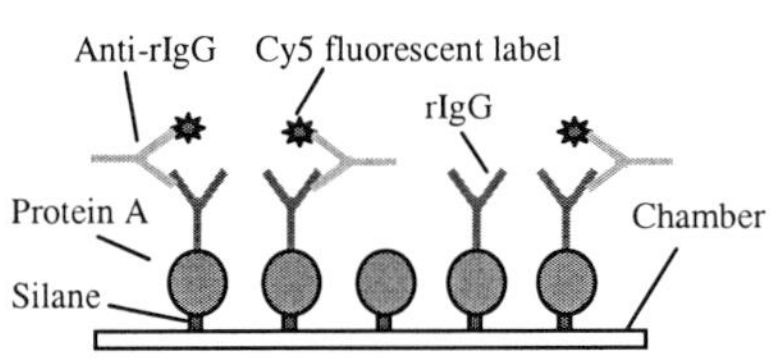

Figure 2: *Binding schematic*

J.M. Ramsey and A. van den Berg (eds.), Micro Total Analysis Systems 2001, 499–500.

labeled goat anti-rIgG is incubated in a continuous flow mode with the rIgG layer. Finally, a chaotropic agent, 25 mM glycine/HCl buffer (pH 2.0), breaks the rIgG/PA bonds and elution of labeled sample is detected. A cleaning sequence is performed to remove excess reagent between each step.

3. Results and discussion

Figure 3 shows fluorescence microscope images of electrokinetic solution flows into the chamber during reagent addition or wash steps. A fluorescent solution of 104 μM tetramethyl rhodamine dextran was used to reveal flow patterns.

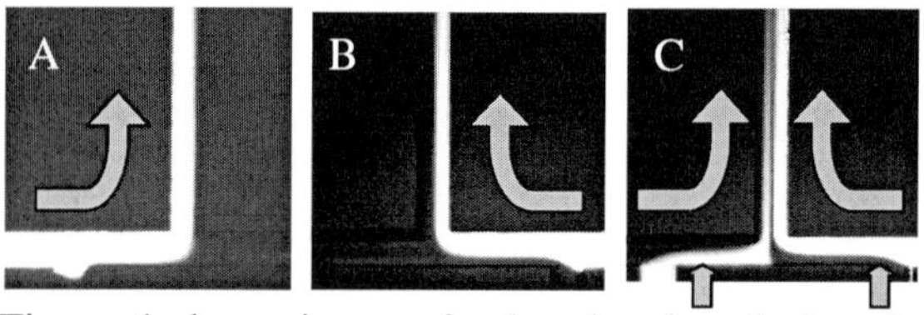

Figure 3: *Flow control in the chamber. Arrows indicate flow. a) rIgG flow b) glycine elution c) four flows corresponding to washing sequence.*

Figure 4 shows images in the chamber during the last two sequences of a full cycle. After the incubation step, followed by the cleaning sequence, there is a strong signal at the chamber surface, indicating bound sample. Introduction of glycine buffer results in zero signal, because PA/rIgG bonds were broken and bound sample was eluted from the chamber, as observed previously [1]. Figure 5 shows the elution step recorded at the top of the chamber.

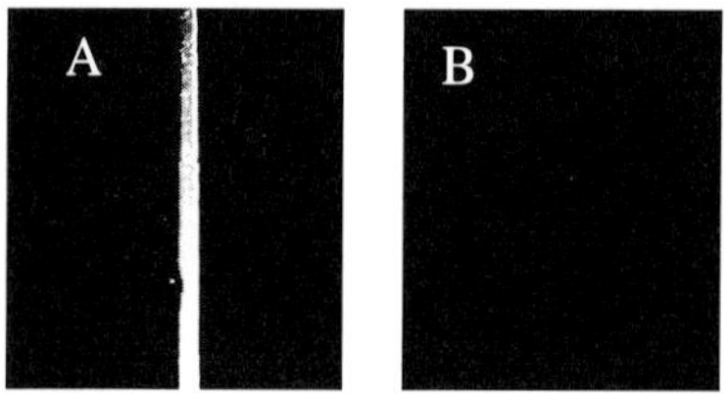

Figure 4: *Chamber images a) after incubation of 1 μM Cy5 labeled anti-rIgG sample and cleaning sequence b) after elution.*

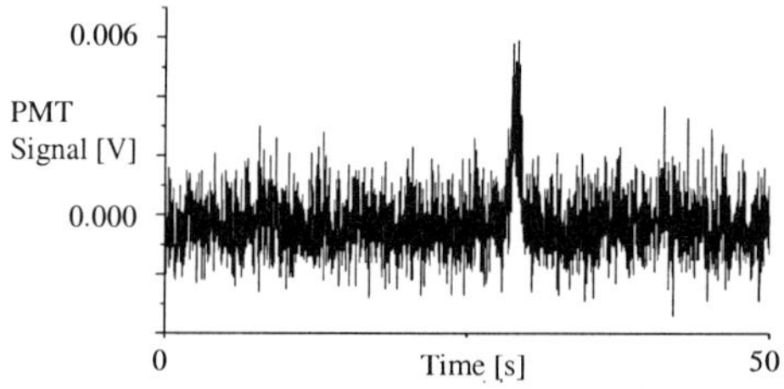

Figure 5: *Peak obtained for the elution of 1 μM Cy5-labeled anti-rIgG flowing for 300 seconds over a layer of rIgG bound to PA.*

4. Conclusions

A controllable microfluidic platform capable of performing multi-step heterogeneous assays has been established. Preliminary results indicate that this device could be used for quantitative analysis.

Acknowledgements

The authors wish to thank Vincent Linder for helpful discussions. A. Dodge acknowledges financial support from the MedTech program, of the Commission for Technology and Innovation (Switzerland), project No.572.

References

1. A. Dodge, et al., *Electrokinetically driven microfluidic chips with surface modified chambers for heterogeneous immunoassays*, accepted for Anal. Chem, (2001).
2. T. Dubrovsky, et al., *Immunological activity of IgG Langmuir films oriented by protein A sublayer*, Sensors and Actuators, B23, pp. 1-7, (1995).

MICROCHIP-BASED SELECTIVE PRECONCENTRATION USING PROTEIN A IMMUNOAFFINITY CHROMATOGRAPHY

Abebaw B. Jemere, Richard Oleschuk, Justine Taylor and D. Jed Harrison
Dept. of Chemistry, University of Alberta, Edmonton, AB T6G 2G2, Canada

Abstract:

A microchip structure packed with protein A-coated microspheres (see Figure 1) was used for selective sample clean up and preconcentration. Theophylline was used as a model analyte to demonstrate feasibility of the on-chip heterogeneous immunoassay.

Keywords: Immunoassay, protein A immunoreactors, heterogeneous immunoassay, beads

1. Introduction

Reported concentration limits of detection for microchip-based CE homogeneous immunoassay (CEIA) are usually at the nM level, which is 2 to 3 orders of magnitude less sensitive than required for most clinical applications. One way to decrease the limit of detection in CEIA is to increase the number of sample molecules available in the detection volume. Recently, we have reported [1] on a microchip design that utilizes micro weirs to reversibly trap reagent-laden beads within a microchip while still allowing solution to flow through the packed beads. Using a C-18 bead packed bed, a detection limit of 70 fM was obtained for a fluorescent dye BODIPY, Figure 2. However, C-18 is not specific. In this work, we describe the potential of the same chip design for selective sample enrichment involving heterogeneous immunoassay. When an antibody-antigen complex is applied to the column, the protein A bed specifically binds the Fc portion of the antibody. The concentrated complex is then released from the column by changing the mobile phase conditions.

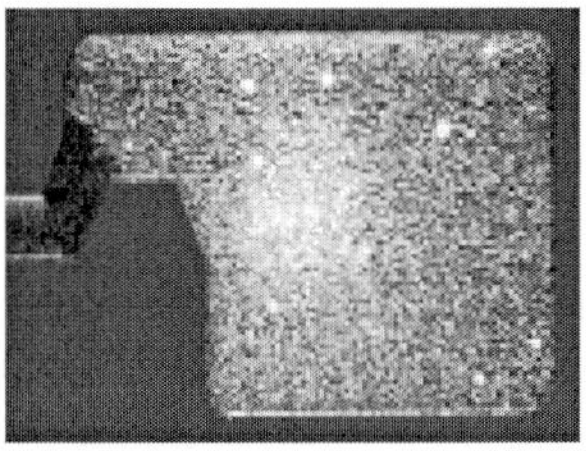

Figure 1. Photograph of packed bed of 1.4 μm diameter protein A-coated bead. Chamber volume is 330 pL.

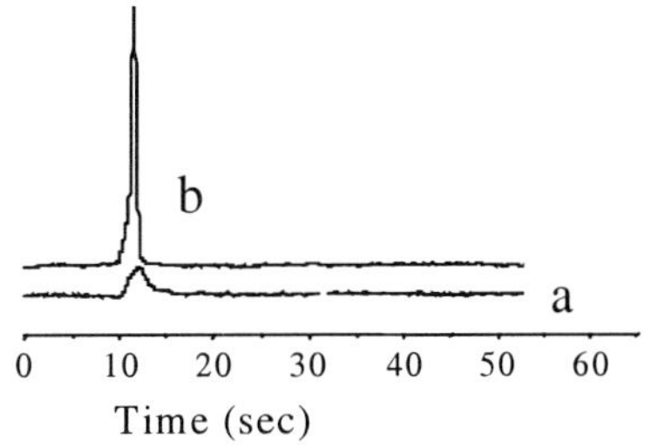

Figure 2. Elution traces of (a) 1 pM and (b) 10 pM BODIPY concentrated for 3 min. (a) S/N = 45; (b) S/N = 434. Samples were preconcentrated for 3 min from an aqueous solution.

2. Experimental

The microchip structure, shown in Figure 1, was made on a glass substrate using a standard photolithography technique. It has two weirs that are etched down 1 μm from the rest of the chip, as described elsewhere [1]. In the presence of a suitable surfactant

J.M. Ramsey and A. van den Berg (eds.), Micro Total Analysis Systems 2001, 501–502.

(0.05% Tween-20), the protein A-coated beads move freely and were directed into the bed using electrokinetic pumping, pressure pumping or magnetic fields. Once the surfactant is removed, the beads agglomerate, causing them to be secured in place.

3. Results and Discussion

Figure 3 shows three traces obtained during the three steps involved in the direct preconcentration assay for theophylline. The small peak at around 20 s {trace (b)}, resulting from the buffer wash, may have been from the unbound complex or unreacted tracer. Although the acidic elution condition used (citric acid buffer, pH 3.5) decreases the fluorescence signal of FITC and the electroosmotic flow velocity during the elution step, successively preconcentrated sample resulted in reproducible signals (inset in Figure 3).

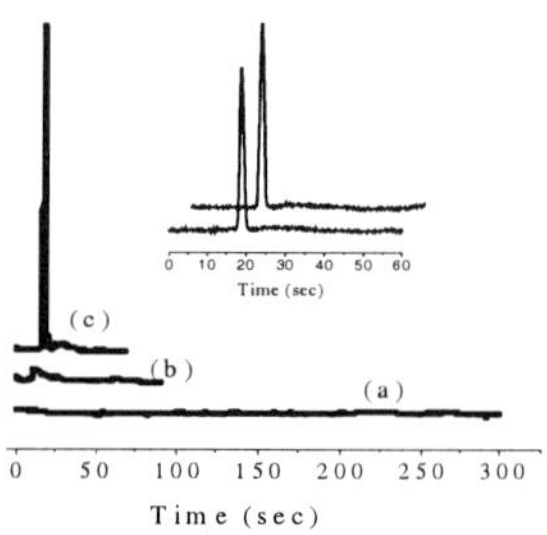

Figure 3. (a) Loading a premixed solution of 200 μl Th-FITC and anti-theophylline in binding buffer (Tris buffer pH 8.8); (b) Binding buffer flush; (c) Eluted complex peak. Inset traces have been offset slightly.

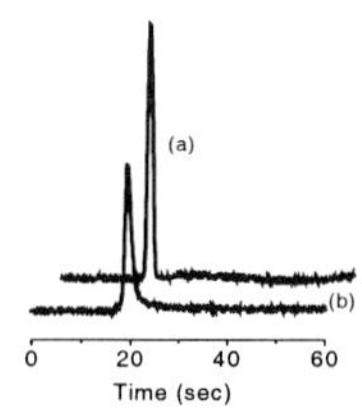

Figure 4. For (a) 200 μl FITC-Th added and (b) 100 μl FITC-Th added. Traces have been offset slightly.

As expected, a decrease in the concentration of tracer (theophylline-FITC) incubated with a fixed concentration of antibody resulted in a proportional decrease in the signal of the eluted peak (Figure 4).

4. Conclusions

The present data shows that an integrated bead based immunoaffinity chromatography can be used for selective sample preconcentration. Efforts are ongoing to determine the lowest possible detection limit achievable after sample preconcentration using a pH insensitive label (Alexa Fluor^{TM}) labeled antigen. This approach should provide a means to determine minute quantities of clinically important analytes.

References

1. R. Oleschuk, A.B. Jemere, L.L. Schultz-Lockyear, F. Fajuyigbe and D.J. Harrison, *Utilization of Bead Based Reagents in Microfluidic Systems,* Proc. Micro Total Analysis Systems 2000, Kluwer Academic Publisher, Dordrecht, The Netherlands, pp. 11 – 14, (2000).

AFFINITY-BASED BIOANALYSIS USING FREELY MOVING BEADS AS MATRICES FOR HETEROGENEOUS ASSAYS

G.-L. Lettieri, E. Verpoorte, N.F. de Rooij
Sensors, Actuators, & Microsytems Laboratory, Institute of Microtechnology,
University of Neuchâtel, Rue Jaquet-Droz 1, CH-2007 Neuchâtel, Switzerland
Gian-Luca.Lettieri@unine.ch

Abstract

We present here a new way of handling beads for affinity-based bioanalysis without being limited by micromachined barriers, possible problems with solution delivery, or removal of beads. The ultimate goal is to perform an affinity-based analysis with bead-type reagents captured in well-defined vortices formed at diffuser elements.

Keywords: vortices, affinity-based analysis, preconcentration

Previous experiments [1] showed that well-defined vortices, formed at diffuser elements (channel expansions) by applying electro-osmotic flow (EOF) counter to pressure-driven flow, could be exploited to capture and preconcentrate particles (**Fig. 1,** inset) without using micro-machined barriers [2]. In this work, we demonstrate that bead clusters can be formed and held in place in a particular flow pattern while being sequentially perfused by different solutions. A preliminary binding experiment using 2-μm, streptavidin-coated beads (SC beads) and biotin-4-fluorescein (B-4-F) was carried out in a microdevice. The streptavidin-biotin interaction was chosen for its large affinity constant and application in immuno-reactions. SC beads were caught at a diffuser simply through formation of moving clusters (**Fig. 1**). Figure 1 shows a schematic of the biotin-streptavidin binding experiment. Channels 1 and 4 are filled with sodium tetraborate (10 mM, pH 9). Reservoir 2 is filled with B-4-F (10 μM in sodium

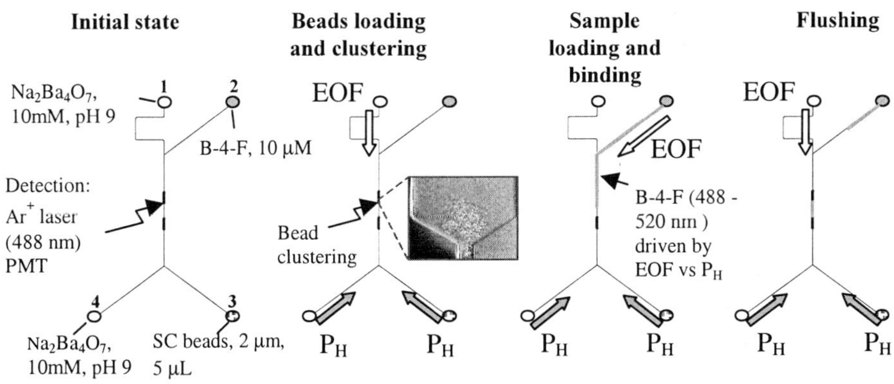

Fig. 1 *Schematic of the experiment. Between reservoir 2 and 4, the narrow channel has two expansions and two contractions.*

J.M. Ramsey and A. van den Berg (eds.), Micro Total Analysis Systems 2001, 503–504.

tetraborate buffer). SC beads (10^9 beads/mL) are introduced from reservoir 3, using a flow induced by application of hydrostatic pressure (column of buffer, 7-10 mbar) to this reservoir. Once the beads reach the second diffuser, they are stopped by applying an electric field (~200 V/cm) between reservoir 1 and 3, giving rise to an opposing EOF. While the SC beads are clustering at the expansion, the potential is switched to reservoirs 2 and 3 (U_{23}= 1.5 kV), and EOF is used to load the B-4-F up to the clustering point. The beads are then perfused by B-4-F to accomplish the binding. After 200 seconds, the potential is switched again to reservoirs 1 and 3 in order to flush the excess B-4-F away, while the newly bound B-4-F/SC beads are still clustering. Figure 2 shows the fluorescence signal recorded by the photomultiplier tube, located at the diffuser, as a function of time. The difference in signal intensities in Fig. 2a) after removal of B-4-F, indicates that the binding took place. Simultaneous video imaging (not shown) of the beads confirms this. Interestingly, PMT signals also indicate that B-4-F is preconcentrated at the diffuser (Fig. 2b)). Once the analysis is finished, the beads can be washed away by releasing the potential. Efforts are underway to determine the response curves for different concentrations of B-4-F and different reaction times.

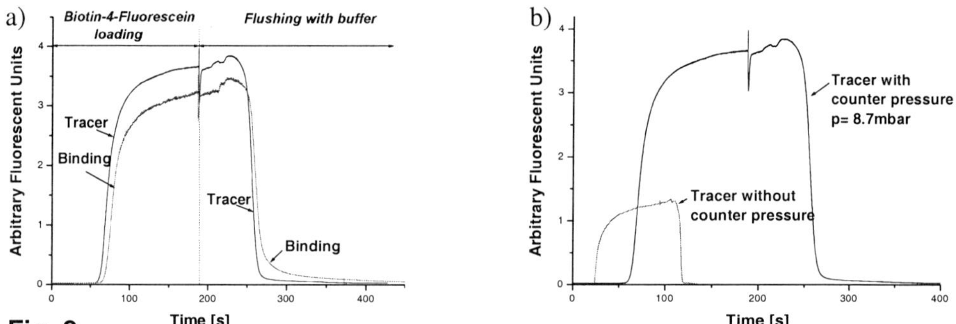

Fig. 2

a) *Fluorescence as a function of time with binding and without beads (B-4-F=Tracer). The initially high intensity corresponds to the B-4-F. Removal of excess tracer with buffer is accompanied by a sudden decrease in intensity.*

b) *Reference curves (blank) of B-4-F with and without counter pressure. Comparison of this signal indicates that application of a counter EOF can lead to preconcentration of reagent at a channel expansion, as well as bead cluster formation.*

Acknowledgements

This work was supported by the Centre Suisse d'Electronique et de Microtechnique (CSEM), Neuchâtel, Switzerland.

References

[1] G.-L. Lettieri et al., *Consequences of Opposing Electrokinetically and Pressure-Induced Flows in Microchannels of Varying Geometries*, Proc. µTas 2000, p. 351-354.

[2] R. Oleschuk et al., *Trapping of Bead-Based Reagents within Microfluidic Systems: On-Chip Solid-Phase Extraction and Electrochromatography*, Anal. Chem. 2000, 72, pp. 585-590.

ELECTROPHORESIS VELOCITY MEASUREMENT OF LYMPHOCYTES UNDER SUPPRESSION OF PH CHANGE OF PBS IN MICROCAPILLARY

Akio Oki, Sakuichiro Adachi, Yuzuru Takamura, Hiroyuki Onoda, Yoshitaka Ito[1] and Yasuhiro Horiike

Dept. of Materials Engineering, The University of Tokyo, 7-3-1, Hongo, Bunkyo-ku, Tokyo, Japan

* R&D Center, Shindengen Kogyo Corp., Hannou, Saitama, Japan

Abstract

pH changes in PBS(Phosphate Buffer Solution) in a microcapillary including reservoir and waste were measured using embedded ISFETs (ion sensitive field effect transistor). The pH change was measured in the capillary suppressed by inserting salt bridges near the reservoir, thus allowing to measure electrophoresis velocity of lymphocytes in the capillary

Keyword: Lymphocytes, Microcapillary chip, ISFET, pH

1. Introduction

Electrophoresis velocity distribution of B and T cells in a lymphocyte was investigated utilizing their different charges in a microcapillary chip. When voltages were added through Pt electrodes between a reservoir and a waste of the microcapillary filled with PBS, however, it was found that lymphocytes in the reservoir became inactive within about two minutes after measurements due to notable increase of pH in the reservoir. These phenomena resulted from electrolysis of PBS in the reservoir with small volume of 2μl, because the reservoir volume was small to save limited amount of the lymphocyte. In this paper, suppression of the pH change in the capillary and following measurement of electrophoresis velocity of T and B cells are reported.

2. Chip fabrication

To fabricate a chip for measurement of pH changes in the capillary, first a reversal pattern of a 30μm x 30μm cross section was plasma-etched employing a $C_3F_8/70\%CF_4$ mixture in a quartz plate. This pattern was molded on a PET plate. A cap PET plate in which ISFETs and a reference electrode were embedded was bonded on the capillary- patterned PET plate. Figure.1 shows photographs of two chips measuring pH variations in the reservoir and the waste, (a), and in the capillary, (b).

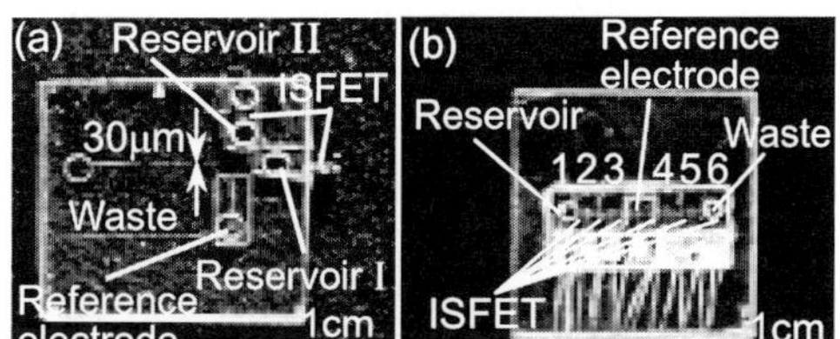

Fig. 1 Two kinds of measurement chips of pH variations in the reservoir (a), and in the capillary ,(b).

3. Results and discussion

Figure 2 shows measurement time dependence of pH variation in the reservoir, where pH measurements in the waste were done by exchanging polarity of voltages. Added voltages were controlled to maintain a constant current of 7μA. II and I mean

J.M. Ramsey and A. van den Berg (eds.), Micro Total Analysis Systems 2001, 505–506.

cases with and without insertion of a salt bridge in the capillary near the reservoir, respectively. In the case of I, acid and alkaline concentrations increased rapidly on anode and cathode sides with time, while in the case of II, changes in pH were suppressed more than 150 sec. due to stop of diffusion of acid or alkaline solutions into the capillary by the salt bridge. Indeed, six ISFETs in the Fig. 1(b) chip without use of the salt bridge demonstrated time dependent variations of pH in the capillary.

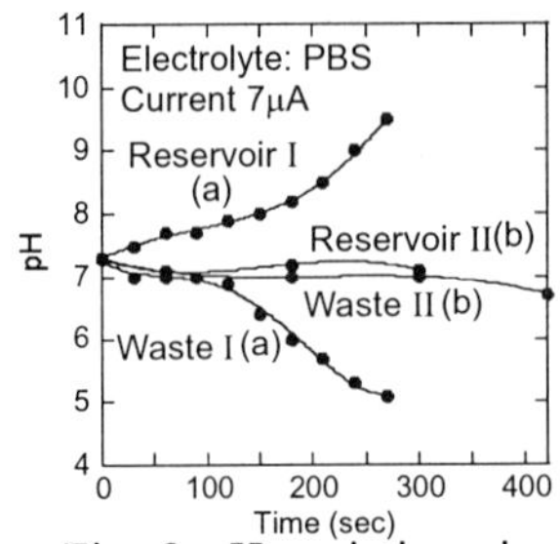

Fig. 2 pH variations in the reservoir I, (a) and the reservoir II, (b).

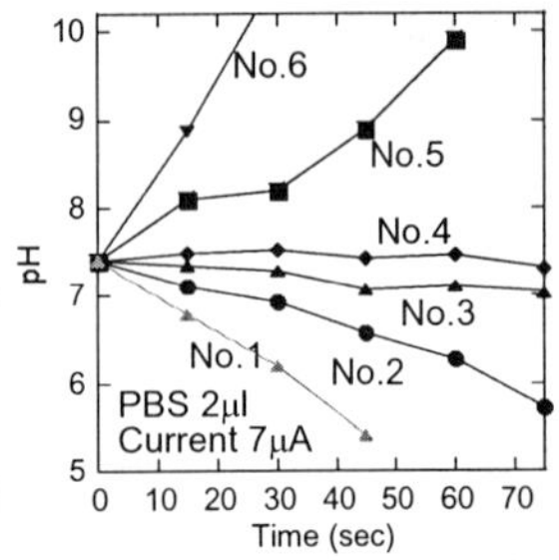

Fig. 3 pH variations in six positions of the capillary. Each point was obtained by 15 sec. measurement and 15 sec. pause.

Finally, a quartz microcapillary chip whose inner wall was coated by the polymer micelle [1] to reduce zeta potential and adsorption of lymphocytes was fabricated for measurement of electrophoresis velocity distribution of B and T cells. As shown in Fig. 4, a red laser light was irradiated to a channel through a slit and electrophoresis velocities were measured from the slit width divided by the flowing time of one lymphocyte. Lymphocytes were injected from the (b) reservoir separated by the salt bridge from voltage added reservoir. The electrophoresis mobility distribution of B and T cells shown in Fig. 5 was almost same as a result measured by a site flow meter. Based on this result, lymphocytes collected on the salt bridge were injected in to a capillary, being separated to B and T celles successfully as shown in Fig. 6.

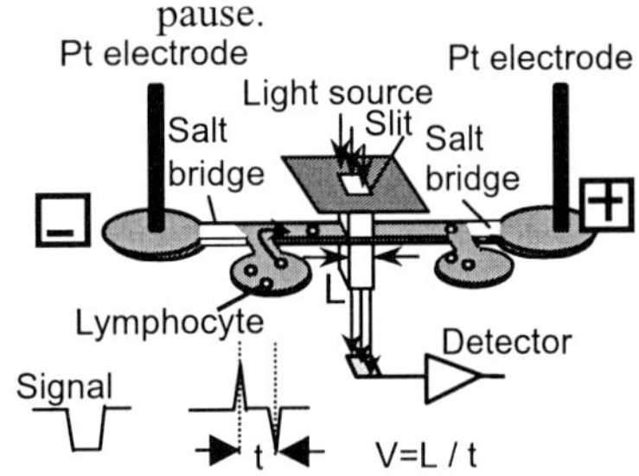

Fig. 4 Illustration of measurement system for lymphocyte electrophoresis velocities V.

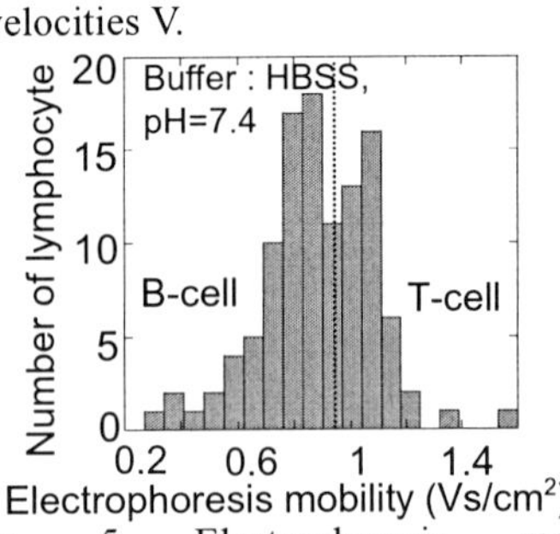

Fig. 5 Electrophoresis mobility distribution of B and T cells where the inner wall of the quartz made capillary.

4. Conclusions

Insertion of salt bridges between the reservoir/waste and the channel suppressed pH changes in the capillary, thereby enabling us to separate B and T cells in lymphocytes based on their electrophoresis velocity distribution measurement in the capillary.

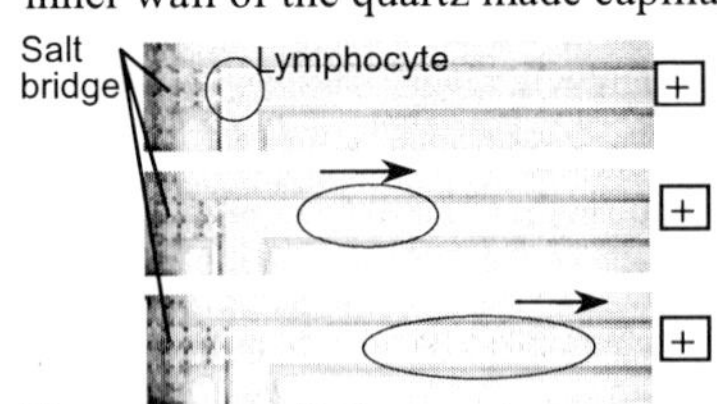

Fig. 6 Pulse injection of lymphocytes into the polymer micelle coated capillary

References

[1] K. Kataoka, et al.; Langumuir **15** No.16, 5212-5218 (1999)

A SIMPLE AND HIGHLY SENSITIVE IMMUNOASSAY ON A CHIP WITH MULTIPLEXING CAPABILITIES

Pieter K. Roos and Cameron D. Skinner
Concordia University, Dept. of Chemistry & Biochemistry
1455 Boulevard de Maisonneuve W. Montreal, Canada, H3G 1M8

Abstract

We have developed a simple and rapid immunoassay that uses antibody-coated fluorescent microspheres for high sensitivity detection of antigens. Detection limits in the low pg/ml (10^{-14}M) are obtainable with this method. This assay has been demonstrated in a flow cytometer, a capillary sheath flow cuvette and a microfluidic chip format. Multiple antigens can be simultaneously quantitated within a single sample by judicious choice of bead dye concentration and colour combination.

Keywords: Immunoassay, Fluorescent microspheres, Cytometry

1. Introduction

Many clinical samples require a large number of screening tests for a variety of analytes. Immunoassays provide highly specific analyte detection and identification in these complex matricies but lack sensitivity when scaled to smaller volumes. Enzyme-linked immunosorbent assays (ELISA) use an enzyme linked to the detection antibody to produce a large number of easily detectable molecules for each antigen molecule. We are using this same strategy, except that the detection antibody is attached to a highly fluorescent microsphere. These microspheres (100 to 500 nm) carry the equivalent of a few hundred to several thousand fluorescent molecules.

2. Theory

The principle of the assay is: larger (1 micron) capture spheres are coated with antibodies specific to one epitope on the antigen. The smaller detection spheres are coated with an antibody to a second (different) epitope. Coating is accomplished through well-established secondary antibody binding schemes using commercially available ELISA antibody pairs. When the two sets of beads are mixed they will form sandwiches via the antigen linkage as illustrated below.

The complex can be detected by flowing the solution past a multiwavelength fluorescence detector (cytometers, microchips, capillaries etc.). Only when a signal is observed on both spectral channels is an antigen counted. Unbound (excess) beads do not generate sufficient signals in both channels to be counted. This means that extensive washing steps are unnecessary, unlike ELISA. Different colour, and intensity, combinations can be used to produce a multiplexed assay.

Interleukin-6 (IL-6) is produced by the body in response to infection and the presence of some cancers and so provide a biologically useful model for the assay. However, the assay is applicable to any antigen that has an ELISA antibody pair available. In principle polyclonal antibodies may also be used, however aggregation and lower sensitivity may be problematic.

J.M. Ramsey and A. van den Berg (eds.), Micro Total Analysis Systems 2001, 507–508.

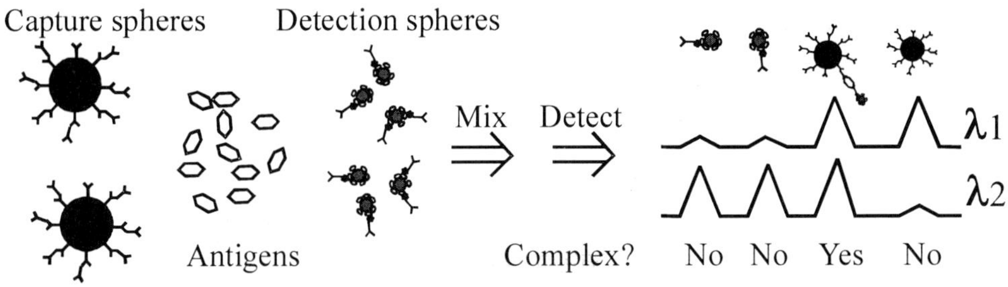

3. Experimental

Goat-anti-mouse coated capture spheres were incubated (all incubations = 30 min., not optimised) with monoclonal antibodies against IL-6. After incubation, IL-6 was added and allowed to bind to the capture spheres. Detection spheres, with monoclonal antibody bound via biotin-streptavidin interaction, were incubated with the sample. No washing steps were necessary with the bead-based assay.

4. Results and discussion

The assay was performed in microtitre plate, flow cytometer, sheath flow cuvette and microfluidic chip formats and compared to ELISA. These methods yielded detection limits (D.L.) of ≈ 10 pg/ml, compared with 1 pg/ml for ELISA. The detection limits are largely determined by the non-specific binding and kinetics of the beads and assay rather than the instrumentation. In the microfluidic chip format, the complexes were electrokinetically focussed and detected as recently described by Schrum *et al.*[1]

The current microfluidic optical system uses a confocal excitation/detection system that leads to uneven excitation spot illumination. This increases the variability of the fluorescence signals and so limits the multiplexing capability. We are addressing this issue with a novel optical configuration that will reduce the SD of the signals.

5. Conclusions

The immunoassay functions very well with existing commercial flow cytometer technology as well as sheath flow cells and microfluidic devices. Commercially available antibodies were used with simple, well-established protocols that can be reproduced easily. Detection limits similar to the ELISA are obtained and can be surpassed with appropriate optical modifications to existing instrumentation. To fully exploit the multiplex capabilities of the assay simultaneous spectral and intensity measurements are required and can be most easily met with a multichannel detection spectrometer.

Acknowledgements

Concordia Faculty Research and Development Program (FRDP), Fonds pour la Formation de Chercheurs at l'Aide à la Recherche (FCAR) and Microlyne for the kind donation of the microfluidic chips.

References

1. D.P. Schrum, C.T. Culbertson, S.C. Jacobson and J.M. Ramsey, *Microchip Flow Cytometry Using Electrokinetic Focusing*, Anal. Chem., pp. 4173-4177, (1999).

PROTEIN ANALYSES WITH ELECTROCHEMICAL AND NANOELECTROSPRAY DETECTION ON DISPOSABLE PLASTIC MICROCHIPS

J.S. Rossier[1,2], F. Reymond[2], I. Arnaux [1], V. Gobry[1], Z. Wu[1], T. Rohner[1], X. Bai[1] and H.H. Girault[1]
1) Laboratoire d'Electrochimie, EPFL, 1015 Lausanne, Switzerland
2) DiagnoSwiss S.A., Rte de l'Ile-au-Bois 2, 1870 Monthey, Switzerland ;
E-mail : info@diagnoswiss.com

Keywords: polymer μ–chips, plasma etching, biosensor,diagnostics, proteomics

1. Introduction

The development of μ–-total analysis systems for proteomics faces features that are specific to the diversity in protein properties such as hydrophobicity and heterogeneity. Particular problems such as non specific adsorption on the capillary walls (glass or plastics) or the use of surfactants or solvent mixtures render the use of electrokinetic pumping and free flow electrophoresis more problematic than in DNA analyses. Therefore, we developed concepts where disposable chips are fabricated with various properties and where the chips are functionalised for affinity or chromatography purposes with electrochemical or mass spectrometry detection. Finally, a novel sub-nL injection method without electrokinetic means has been developed in order to enable the integration of injector, separation channel and detector on a single μ–chip.

2. Enzyme Linked Immunosorbent Assays with Electrochemical Detection

Micro-chips are advantageously adapted for the development of immunoassays. Antibodies are attached to the chip surface, and immunosorbent assays (ELISA) are performed with very short incubation time (<5 min) due to the small diffusion distance between the bulk solution and the surface (<20 μm) [1]. The use of electrochemical detection is also advantageous because it is proportional to the concentration of redox molecules present in the channel even in these small volumes (<60 nL) [2]. Electrochemistry is also sensitive enough since it enables the detection of 10 pM of antigen with a dynamic range of three orders of magnitude.

3. Coupling Plastic Micro-chips to Mass Spectrometry for Protein Analyses

Laser photoablation and plasma etching have also been used to fabricate hydrophilic μ–channels with an open end surrounded by an hydrophobic surface (See Figure 1). Thanks to this principle, the chips can be used as disposable nano-electrospray for the analysis of proteins by mass spectrometry (See Figure 2).

4. Development of a Pressure Pinched Injection without Electrical Means

A novel injection method has been developed in order generate sub nL plugs in μ–channel by means of a multiport valve. This injection pattern has the advantage to be

J.M. Ramsey and A. van den Berg (eds.), Micro Total Analysis Systems 2001, 509–510.

independent of the composition of the sample and/or of the wall properties of the channel. Its development enables the integration of injector, separation column and detector on the same chips for chromatographic purposes such as e.g. ion chromatography for. On-chip fluorescence and conductivity detection of injected plugs have been used for the demonstration of the concept.

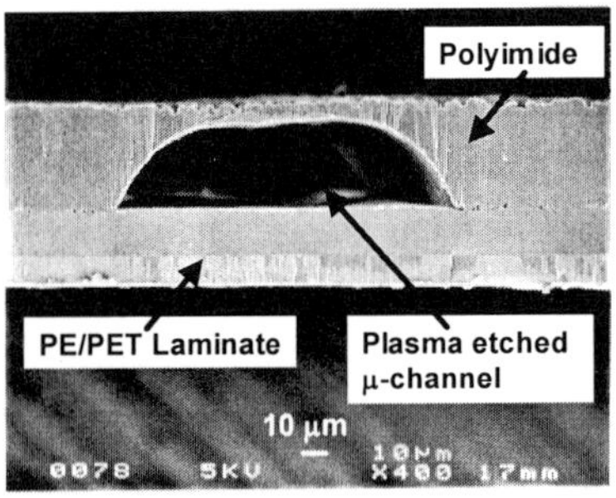

Figure 1: SEM view of a cross section of plasma etched polyimide μ-channel (120 x 35 μm^2) used for immunoassays and nano-electrospray ionisation of proteins.

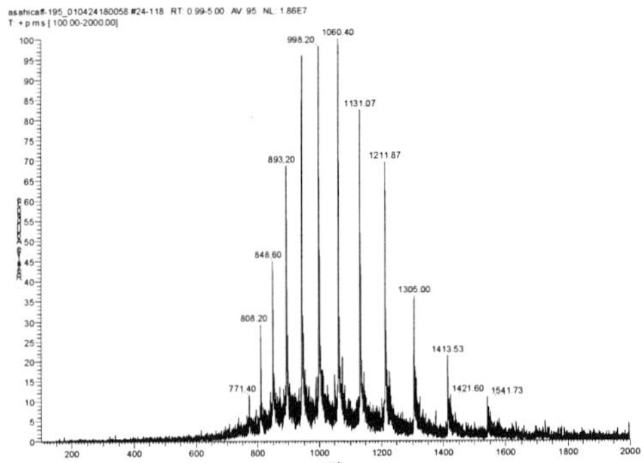

Figure 2: MS analysis of a 4 μM myoglobine sample sprayed with the chips presented in Figure 2 (spray voltage: 1.4 kV; spray solution: 50 % methanol, 49 % water, 1 % actetic acid; acquisition time: 3 min)

5. Experimental

Disposable μ–chips for protein analysis have been fabricated by UV-laser photoablation of polymer [3]. Mass production is then performed by plasma etching, a method developed for the fabrication of printed circuit boards [4]. Micro-channels with integrated electrodes can be fabricated with different surface properties (surface charge, roughness, hydrophobicity, cristallinity) [5], and μ–fluidic devices have been developed for the electrophoretic separation and detection of electroactive species [6], coupled to mass spectrometry as well as for the μ–patterning of affinity reagents [7].

References

1. J.S. Rossier, G. Gokulrangan, H. Girault, S. Svojanovsky, G.S. Wilson, *Langmuir* 16, 8489 (2000).
2. J.S. Rossier, M. Roberts, R. Ferrigno, H. Girault, *Anal. Chem.* 71, 4294 (1999).
3. M. Roberts, J.S. Rossier, P. Bercier, H. Girault, *Anal Chem* 69, 2035 (1997).
4. See: www.dyconex.com.
5. J.S. Rossier, P. Bercier, A. Schwarz, S. Loridant, H. Girault, *Langmuir* 15, 5173 (1999).
6. J.S. Rossier, R. Ferrigno, H. Girault, *J. Electroanal. Chem.* 492, 15 (2000).
7. A. Schwarz, J.S. Rossier, E. Roulet, N. Mermod, M.A. Roberts, H. Girault, *Langmuir* 14, 5526 (1998).

INTEGRATED IMMUNOASSAY SYSTEM USING MULTICHANNEL MICROCHIP FOR SIMULTANEOUS DETERMINATION

Kiichi Sato[1,2], Maho Yamanaka[1], Hiroko Takahashi[2], Kenji Uchiyama[2], Manabu Tokeshi[2], Hajime Katou[3], Hiroko Kimura[4], Takehiko Kitamori[1,2]

[1] Department of Applied Chemistry, School of Engineering, The University of Tokyo, Hongo, Bunkyo, Tokyo 113-8656, Japan

[2] Integrated Chemistry Project, Kanagawa Academy of Science & Technology, Sakado, Takatsu, Kawasaki, Kanagawa 213-0012, Japan

[3] Mechanical Engineering Research Laboratory, Hitachi Ltd, Kandatsu, Ibaraki 300-0013, Japan

[4] Department of Forensic Medicine, School of Medicine, Juntendo University, Hongo, Bunkyo, Tokyo 113-8421, Japan

Abstract

Immunoassay system was integrated on a multichannel microchip. The chip, which has branching channel patterns with weirs, was fabricated by single step wet etching. To prevent contamination of up-flowed samples, pneumatic valves made of silicone rubber were set on a chip. By using the system, 4 assays were successfully performed at a time. We concluded that the system should be practicable.

Keywords: Clinical diagnosis, bead-bed immunoassay, multichannel microchip

1. Introduction

Immunoassay is known as one of the most important analytical methods and widely used in clinical diagnoses and biochemical studies. The conventional immunoassay, however, requires a long assay time, and involves troublesome liquid-handling procedures. We developed a bead-bed immunoassay system on a microchip (Fig. 1) and applied to a clinical diagnosis. This integration reduced the time necessary for the reaction to ~ 1% [1], and the system was successfully applied to a practical serodiagnosis of colon cancer with short assay time and high sensitivity [2]. Here, integrated immunoassay system using multichannel microchip was developed for simultaneous determination of multiple samples.

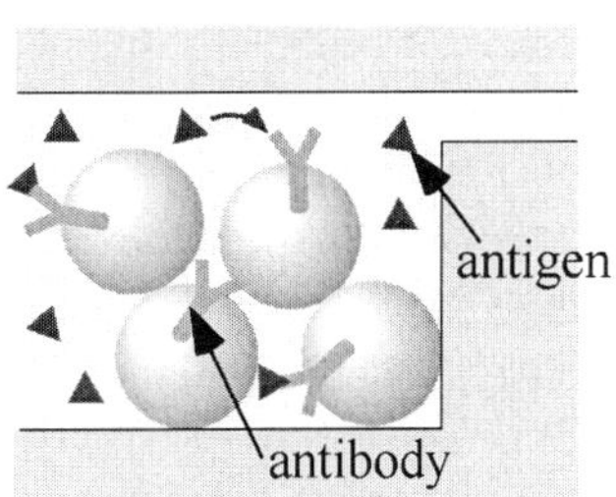

Fig. 1. Cross section of immunoassay chip

2. Microchip fabrication

To dam up the beads, weir or cage structure is necessary. We developed a simple single step wet etching technique to make weirs and channels at a time on a glass microchip, instead of multistep etching and/or dry etching. The photo mask used has

J.M. Ramsey and A. van den Berg (eds.), Micro Total Analysis Systems 2001, 511–512.

branching channel patterns (10 μm in width) with 160 μm gaps. The patterned Pyrex glass substrate was etched with hydrofluoric acid solution, and gapped channel was connected by etching. By changing the gap width and etching time, the microchip with optimal weirs could be obtained.

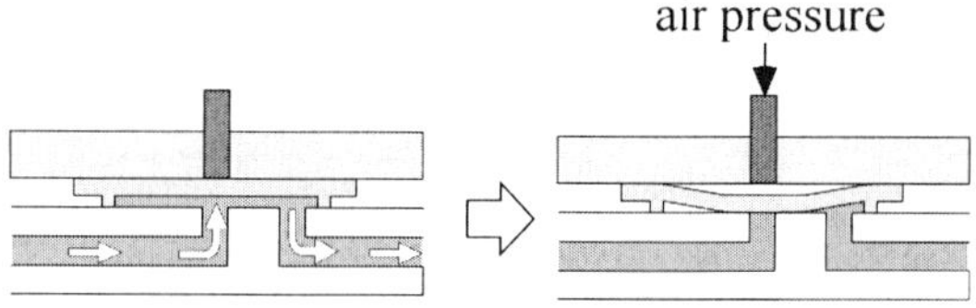

Cross section of the valve

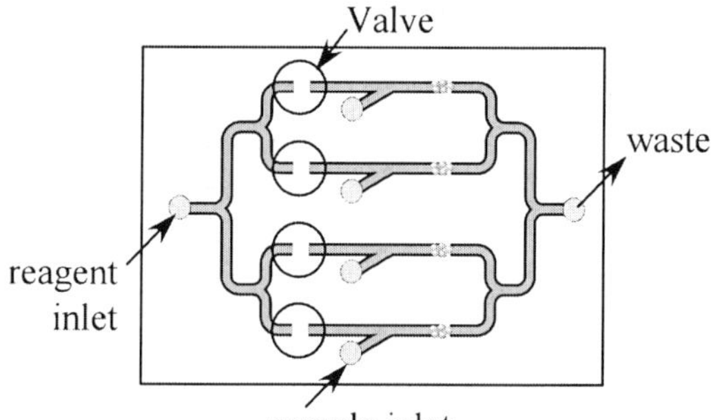

Fig. 2. Valves on a chip

3. Valves on a chip

To prevent contamination of up-flowed samples, air-pressure driven valves made of silicone rubber were set on a chip (Fig. 2). The valve was mounted on the upstream channel from sample inlet, and it closed during sample injection to introduce each sample precisely. Moreover, flow of each channel could be controlled individually, because the valves were controllable separately.

4. Simultaneous determination using multichannel immunoassay microchip

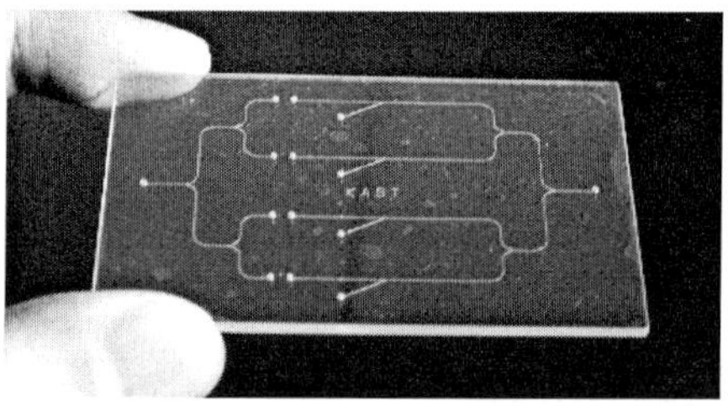

Fig. 3. Four-channel microchip

Polystyrene beads, precoated with anti-interferon gamma (IFNγ) antibody, were introduced into a 4-channel microchip (Fig. 3) from sample inlet. Closing the valves, 4 samples containing IFNγ were introduced from each sample inlet. After 10 min reaction, the sample inlets were closed and the valves were opened, and then phosphate buffer, biotinyl anti-IFNγ antibody, the buffer, streptoavidin-colloidal gold conjugate, and the buffer were injected from reagent inlet successively. After reactions surfaces of the resulting beads were monitored with thermal lens microscope.

The 4-channel system with the valves could analyze 4 samples at a time successfully. The RSD value for each channel was less than 10%, which was the same level as that of single channel chip. Interchannel bias was also small (differences of the averages were less than 10%). We concluded that the system should be practicable.

References

1. K. Sato, M. Tokeshi, H. Kimura, T. Ooi, M. Nakao and T. Kitamori, *Anal. Chem.* **72**, 1144-1147 (2000).
2. K. Sato, M. Tokeshi, H. Kimura and T. Kitamori, *Anal. Chem.* **73**, 1213-1218 (2001).

LOW DENSITY LIPOPROTEIN (LDL) ANALYSIS IN A CHIP-BASED SYSTEM

Takayuki Shibata[1,*], Laura Ceriotti[1], Bruce H. Weiller[2], Matthew A. Roberts[3], Elisabeth Verpoorte[1] and Nico F. de Rooij[1]
[1]IMT, Rue Jaquet-Droz 1, P.O. Box 3, CH-2007, Neuchâtel, Switzerland
[2]The Aerospace Corporation, P.O. Box 92957, Los Angeles, CA 90009
[3]Nestlé Research Center, Lausanne, Switzerland

* *Present address:* Department of Mechanical Engineering, Faculty of Engineering, Ibaraki University, 4-12-1 Nakanarusawa, Hitachi, Ibaraki 316-8511, JAPAN

Abstract

This paper demonstrates low density lipoprotein (LDL) analysis in a chip-based system without coatings, and is the continuation of work by Weiller et al. (in this Proceedings). It is shown that peak efficiency can be improved remarkably by adding a small amount of sodium dodecyl sulfate (SDS) to the sample. The method proposed here could be adopted for rapid analysis of different LDL forms.

Keywords: Low density lipoprotein (LDL), capillary electrophoresis (CE), SDS

1. Introduction

Oxidation of low density lipoprotein (LDL) is an important indicator of premature artherosclerosis. Since changes in electrophoretic mobility are a reliable measure of LDL modification, capillary electrophoresis (CE) can be used for LDL analysis as reported previously [1]. LDLs are large particles and have apolar surface regions; therefore, coatings have been adopted in conventional systems to prevent particle adsorption and consequent loss of detector signal. This paper proposes a new, fast and efficient chip-based method for LDL analysis which does not require permanent or dynamic coatings.

2. Experimental

The channel layout of the Pyrex chip used in this work has a 200-μm-long double-T intersection and 50-μm-wide, 20-μm-deep channels. Human LDL fractions were prepared from fresh serum and stained with the lipophilic dye, NBD-ceramide [2]. LDLs were detected at 4.2 cm from the injector by laser-induced fluorescence (LIF) using a 40 mM Tricine (pH 9.1) running buffer and an electric field strength, E, of 760 V/cm.

3. Results and discussion

As shown in Fig. 1, a peak efficiency of 5 μm (N = 9,000) was achieved without coating. The migration time was reproducible (17.1 ± 0.22 sec), with a signal height of 0.07 V. SDS was added to the sample to increase LDL negative charge, thereby preventing particle aggregation and interactions between LDL particles and channel

J.M. Ramsey and A. van den Berg (eds.), Micro Total Analysis Systems 2001, 513–514.

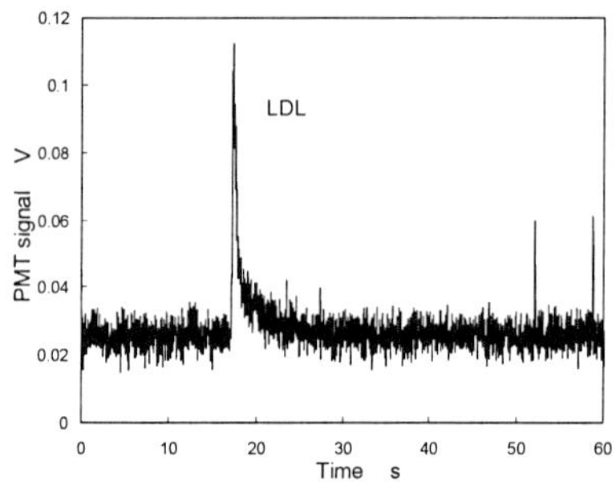

Figure 1: Electropherogram of LDL(NBD) diluted 3:5 with 40 mM Tricine buffer. Peak area= 0.07 ± 0.02 V·s.

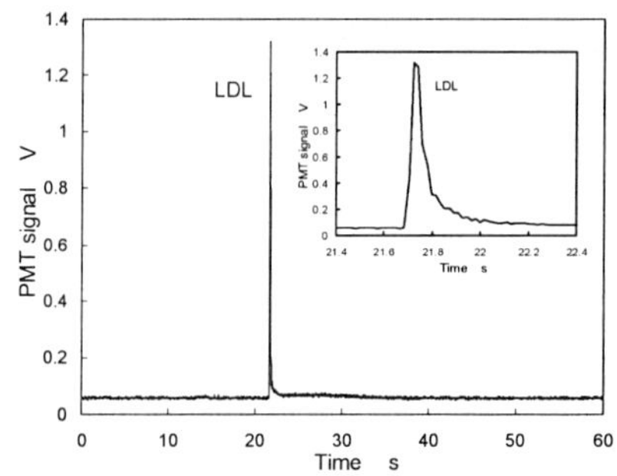

Figure 2: Electropherogram of SDS-LDL(NBD). Same sample as in Fig.1 with 0.3 mM SDS. Peak area= 0.08± 0.02 V·s.

walls. By adding 0.3 mM SDS to the sample (and not to the running buffer), the signal increased to 1.4 V, and the migration time was shifted from 17 to 21 s (Fig. 2). Although SDS is known to denature proteins and break up protein particles, this peak is probably due to intact, SDS-bound LDL particles stained with NBD-ceramide (SDS-LDL(NBD)), and not SDS micelles or apolipoproteins. A number of observations support this conclusion: (1) Control experiments showed that SDS micelles can be detected using NBD-ceramide; however, the SDS concentration used here is much lower than the critical micelle concentration (8.2 mM) and no signal was observed. (2) As reported in the literature [3], 0.3 mM SDS is not enough to delipidate LDL particles. (3) Apolipoproteins could not be detected by LIF, since NBD-ceramide emits only in a lipophilic environment. (4) The areas of the peaks in Figs. 1 and 2 are quite similar, suggesting that the recovery of the particles was the same with and without SDS present, and particle integrity was preserved. In our experiments, peaks achieved in the presence of SDS reached efficiencies up to 0.05 μm (N= 928,000, 2.2×10^7 plates/m), and elution times were reproducible (21.4 ± 0.28 s). This efficiency is very high, and may be due to SDS-coated LDL particles having greater surface uniformity than uncoated LDL.

4. Conclusions

These preliminary results demonstrate fast, efficient, non-oxidized LDL analysis on a chip without the need for dynamic or permanent coatings. Addition of a small amount of SDS to the sample increases efficiency by 2 orders of magnitude without destroying particle structure, thereby fulfilling an important criterion for the development of a diagnostic chip for oxidized LDL.

References

1. J. Stocks and N.E. Miller, *Capillary electrophoresis to monitor the oxidative modification of low density lipoproteins*, J. of Lipid Research, 39, 1305-1309, 1998.
2. G. Schmitz, C. Möllers and V. Richter, *Analytical capillary isotachophoresis of human serum lipoproteins*, Electrophoresis, 18, 1807-1813, 1997.
3. R.D. Macfarlane et al., *Development of a lipoprotein profile using capillary electrophoresis and mass spectrometry*, Electrophoresis, 18, 1796-1806, 1997.

AN INTEGRATED, STACKED MICROLABORATORY FOR BIOLOGICAL AGENT DETECTION

Joon Mo Yang, Janice Bell, Ying Huang, Marcus Tirado, Donald Thomas, Anita H. Forster, Robert W. Haigis, Paul D. Swanson, R. Bruce Wallace, Bob Martinsons and Michael Krihak[1]

Nanogen, Inc., 10398 Pacific Center Court, San Diego, CA 92121, USA

[1]Current Affiliation: DARPA MTO, 3701 N. Fairfax Drive, Arlington, VA22203, USA

Abstract

An integrated, stacked microlaboratory for detecting biological agents through automated electric-field-driven immunoassays and DNA hybridization assays were developed. Versatility of the stacked microlaboratory was demonstrated by various automated assays.

Keywords: Stacked structure, integrated fluidic system, electric-field-driven DNA and immunoassays

1. Introduction

We present here an integrated, stacked mirolaboratory for detecting biological agents, which is capable of performing electric-field-driven immunoassays and DNA hybridization assays starting from the collection of intact bacteria. The developed stacked microlaboratory can be potentially used in a wide range of applications including detection of bacteria and biowarfare agents and genetic identification.

2. Stacked microlaboratory

The stacked microlaboratory was fabricated by orderly laminating several different functional layers, including a patterned polyimide layer (Kapton®) with a flip-chip bonded CMOS chip, a pressure sensitive acrylic adhesive (PSA) layer with a fluidic cutout, an optically transparent polymethyl methacrylate (PMMA) film, a PSA layer with a via, a patterned polyimide layer with a flip-chip bonded silicon chip, a PSA layer with a fluidic cutout, and a glass cover plate layer (Figure 1(a)). Two semiconductor chips were flip-chip bonded to the patterned polyimide layers. One was silicon chip with 16 large and 9 small platinum electrodes (DEP chip), and the other was a CMOS biochip, which had an array of 40 x 40 electrodes (1600-site chip). The fabricated two-level stacked structure was 3" x 3" and 0.109" thick (Figure 1(b)). 18 step electrodes and the DEP chip were located in the upper level, and 6 step electrodes and the 1600-site chip in the lower level. Fluidic connection between the upper level and the lower levels was attained by a centrally located via. The electrode patterns at the opposite edges of the 3" x 3" layer permitted electrical contact to an external control circuit. Pathways for fluidic delivery and removal (fluidic inlet/outlet) were located on the bottom of the stacked structure.

J.M. Ramsey and A. van den Berg (eds.), Micro Total Analysis Systems 2001, 515–516.

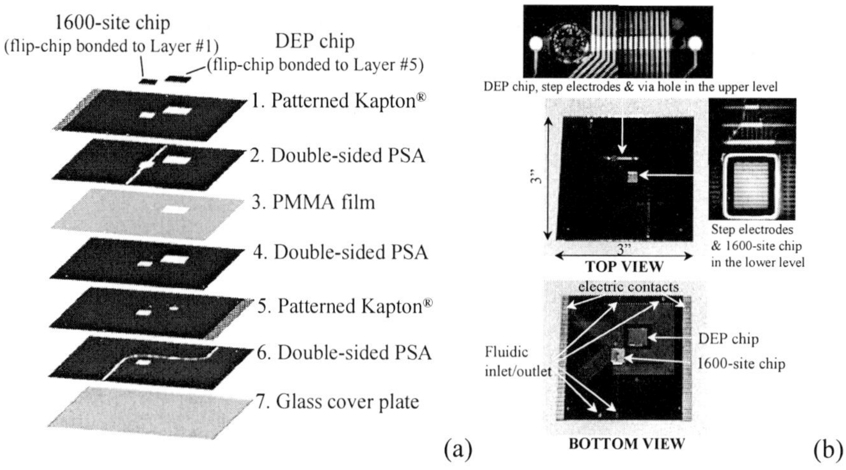

Figure 1. Stacked microlaboratory: (a) Fabrication; (b) Photographs.

3. System integration

Besides the stacked structure, a module for strand displacement amplification (SDA) and other supporting modules were fabricated in parallel and integrated into the system. The fluidic deliveries in the stacked structure were controlled by a syringe pump and several solenoid valves.

4. Demonstration of assay protocols

Texas Red labeled DNA was electrophoretically transported along the step electrodes and through the via. *Escherichia coli* bacteria and Alexa-labeled protein toxin staphylococcal enterotoxin B were detected by electric-field-driven immunoassays on a single DEP chip with specific-to-nonspecific signal ratios of 4.2:1 and 3.0:1, respectively. An electric-field-driven immunoassay for Alexa-labeled SEB was demonstrated on the 1600-site CMOS chip with a specific-to-nonspecific signal ratio of 3.1:1. Furthermore, a comprehensive protocol was performed for DNA analysis of *E.coli* bacteria, which included the following steps: (a) dielectrophoretic collection of *E.coli* bacteria, (2) cell lysis and DNA amplification of the Shiga-like toxin gene (SLT1), and (3) electric-field-driven DNA hybridization assay [1].

Acknowledgement

This work was supported by a DARPA MicroFlumes program (#N66001-98-C-8625).

References

1. J. M. Yang, J. Bell, Y. Huang, M. Tirado, D. Thomas, A. H. Forster, R. W. Haigis, P. D. Swanson, R. B. Wallace, B. Martinsons and M. Krihak, *An integrated, stacked microlaboratory for biological agent detection with DNA and immunoassays*, submitted to Biosensors and Bioelectronics (2001).

A HANTZSCH SYNTHESIS OF 2-AMINOTHIAZOLES PERFORMED IN A MICROREACTOR SYSTEM

E. Garcia-Egido, S. Y. F. Wong and B. H. Warrington
GlaxoSmithKline Pharmaceuticals, New Frontiers Science Park, Harlow, CM19 5AW, UK
(Eduardo_2_Garcia-Egido@sbphrd.com)

Abstract

A Hantzsch synthesis of 2-aminothiazoles was carried out in a heated glass micro-channel reactor under EOF control. The results demonstrate the chip's capability to support the generic synthesis of a related series at higher yield than the macro scale.

Keywords: EOF, micro channel, synthesis, screening, combinatorial libraries

1. Introduction

Microchannel-based chemistry is being perceived widely as a potentially enabling technology for high speed, ultra-high throughput chemical synthesis [eg. 1]. The greatly reduced scale is compatible with modern miniaturised screening techniques. Wittig [2] and Suzuki products [3] have been prepared in a device similar to that described here.

2. Objectives

Previous microchannel-based syntheses were carried out at room temperature. In this work a modified reactor allows greater scope as higher temperatures can be used.

3. Experimental

Reactions were performed in a microreactor in which a borosilicate bottom plate was etched with a T-design channel and then sealed over with a borosilicate top plate. A computer-controlled power supply was used to mobilise reagents and products using EOF. The optimum voltage applied over a 30-minute period was 400V. The reaction solvent used was NMP. The chip was heated at 70 °C using a T-shaped Peltier heater, aligned with the channels and applied to the lower plate. The reaction products were identified and yields quantified by LC-MS using an UV detector and previously prepared calibration curves for starting materials and final products.

4. Results and discussion

Six 2-amino-thiazoles were synthesized from ring-substituted 2-bromo-acetophenones and 1-substituted-2-thioureas applying the experimental conditions described (1).

J.M. Ramsey and A. van den Berg (eds.), Micro Total Analysis Systems 2001, 517–518.

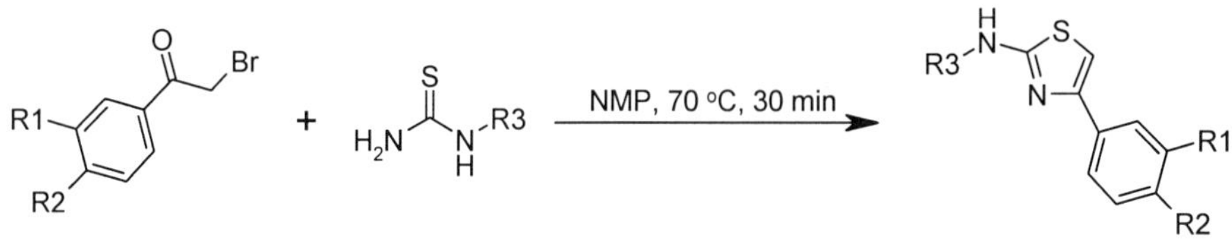

(1)

Comparable/better yields were obtained for reactions performed in microreactor (400V) relative to macro batch reactions carried out using the same conditions (Table 1).

Table 1. Batch vs Microreactor

Reaction No	R_1	R_2	R_3	Batch Yield	Microreactor Yield
1	H	H	Acetyl	44%	63%
2	H	OMe	Acetyl	53%	58%
3	H	Me	Acetyl	59%	77%
4	NO_2	H	Acetyl	96%	99%
5	Br	H	Acetyl	83%	93%
6	H	H	Phenylethyl	99%	99%

5. Conclusions

The results of this and earlier reports [2,3] indicate that microchannel-based technologies could be used to support the generic synthesis of a broad range of compounds. This work demonstrates that a heated device is also possible. Future intentions are to explore a broader range of chemistry under pressure flow (as well as EOF) and the use of polymer supported reagents.

Acknowledgements

The authors thank Prof. S. Haswell for some of the software and hardware used. E.G.-E. thanks the EU for a Postdoctoral Marie Curie fellowship (HMPI-CT-1999-00066).

References

1. Salimi-Moosavi, H.; Tang, T.; Harrison, D. J., *Electroosmotic Pumping of Organic Solvents and Reagents in Microfabricated Reactor Chips*, J. Am. Chem. Soc. 119, pp. 8716-8717, (1997).
2. Skelton, V.; Greenway, G. M.; Haswell, S. J.; Styring, P.; Morgan, D. O.; Warrington, B. H.; Wong, S. Y. F., *The generation of concentration gradients using electroosmotic flow in micro reactors allowing stereoselective chemical synthesis*, Analyst, 126, pp. 11-13, (2001).
3. Greenway, G. M. et al. *The use of a novel microreactor for high throughput continuous flow organic synthesis*, Sens. Actuators B, 63, pp.153-158, (2000).

THE DEVELOPMENT AND EVALUATION OF A MICROFLUIDIC DEVICE FOR THE DETERMINATION OF ORGANIC-AQUEOUS PARTITION COEFFICIENTS TO SUPPORT AGROCHEMICAL AND DRUG DISCOVERY

Brian Law[1], Nicola Colclough[1], Dave Temesi[1], Eric Clarke[2], Julie Deacon[3], Bina Mackenzie[3] and Chris Turner[3].

1. AstraZeneca Pharmaceuticals, Alderley Park, Macclesfield, Cheshire, SK10 4TG, UK.
2. Syngenta, Jealott's Hill International Research Centre, Bracknell, Berkshire, RG42 6EY, UK.
3. CRL, Dawley Rd., Hayes, Middlesex, UB3 1HH, UK.

Abstract

A prototype multi-channel microfluidic device has been developed and tested for the determination of organic-aqueous partition coefficients.

Keywords: Organic-aqueous partition coefficients, drug discovery, agrochemical discovery

1. Introduction

Organic-aqueous partition coefficients (LogP) are important parameters in agrochemical and drug discovery. They are predictive of membrane permeability or tissue distribution and in particular, octanol-aqueous values are frequently used in structure activity relationships to describe biological activity.

The method of choice for this measurement is the 'shake flask' method with analysis by HPLC with UV or MS detection after phase separation. This process is both tedious and time consuming and requires significant sample volumes. We describe here a new method under development involving contact between aqueous and organic 'slugs' moving in a capillary, which is capable of generating accurate LogP data on less than 50μl of sample.

Experimental

The prototype device consists of 8 high precision Kloehn 50300 pumps in parallel, with a 320 μm id capillary 1.5 m long connected to each pump. The capillary is used to aspirate aqueous samples and organic solvent to form a chain of liquid 'slugs' deliminated by air gaps. Shuttling the samples back and forwards brings about partitioning. The capillary can be connected to a standard HPLC UV detector fitted with a 35 nl capillary flow cell for absorbance measurements on the aqueous sample.

A typical analysis at 100:1 volume ratio involves aspirating 18 μl of aqueous sample, followed by 0.36 μl of octanol followed by a further 18 μl of aqueous. The samples are separated from the system solvent by precise air gaps. The slug chain is

J.M. Ramsey and A. van den Berg (eds.), Micro Total Analysis Systems 2001, 519–520.

shuttled back and forth over a 34cm length of capillary (>1.5 x aqueous slug length), to allow partitioning between the aqueous and organic phases and effective circulation and mixing within slugs. Phase integrity is maintained aiding analysis when the slug train is transported into the UV detector. Determination of the concentration in a pre and post-partitioning aqueous sample allows calculation of the partition coefficient.

Results and Discussion

Initial experiments with capillaries of 200 to 700 μm ID indicated 320 μm ID to be optimal in terms shuttling speed, sample volume, time to reach equilibrium and having a stable system without break up of the 'slugs'. As the slugs are shuttled, re-circulation within them reduces the time to reach equilibrium. It was found that forty shuttles (taking 14 min) are sufficient for the system to reach equilibrium, with no difference in LogP values measured using a higher number of shuttles. Further improvements in speed appear possible.

A set of model compounds were tested and results compared with those obtained by standard LogP methods (Table 1). These data show good accuracy and precision. Furthermore, additional tests have also shown the system works well for partitioning into dodecane and additional solvents will also be investigated.

Table 1. Comparison of LogP (octanol-pH 7 buffer) data generated using standard methods at Syngenta with that generated using the new microfluidic device at CRL

	Shake flask	New method		
Compound	LogP	Mean LogP	Std. Dev.	n
Methylphenylsulphone	0.42	0.45	-	1
Nitrobenzylalcohol	1.20	1.25	0.07	5
Nitroacetophenone	1.43	1.48	0.08	9
Nitroanisole	1.97	1.99	0.10	8
Nitroethylbenzoate	2.40	2.08	0.05	5

Conclusions

We have developed a prototype multi-channel microfluidic system for the determination of organic-aqueous partition coefficients. The current system involves manual switching between samples and solvents, although this process is readily automatable. The results obtained for a range of compounds with this initial prototype are in good agreement with those from conventional methods and literature data. The LogP range covered is expected to be similar to that for the shake flask methodology. We intend to further develop this instrument, focusing on speed and throughput as well as evaluating a mass spectrometric detector to improve selectivity and accuracy.

PMMA/PDMS BASED MICROFLUIDIC SYSTEM WITH OPTICAL DETECTION FOR TOTAL HEAVY METALS CONCENTRATION ASSESSMENT

M. Chudy, W. Wróblewski, A. Dybko, Z. Brzózka

Department of Analytical Chemistry, Warsaw University of Technology, Poland

Keywords: micro system, PDMS and plastics machining, PAR, heavy metals determination, optical detection.

Introduction

4-(2-pyridylazo)-resorcinol (PAR) is a highly sensitive photometric reagent for wide variety of metal ions [1]. Heavy metal ions as: Cd, Cu, Hg, Pb, Zn, in aqueous solutions (pH 5 to 6), form with PAR chelates slightly soluble in water. There are a lot of very complicated spectrophotometric methods based on the use of PAR published in the literature. Then the selective determination of metal ion is obtained by extraction under precisely defined condition (pH, extracting agent, organic solvent etc.) The non-selective complexation of counted metal ions can be used for their total level control. It is much easier, faster and is especially important during classification of various natural water samples. Combining extraction free, group reaction with simple and cheap microfluidic PMMA/PDMS system fabrication, useful method for sample classification and further sample analysis can be developed.

Experimental

The "Y"- shape microstructure with a mixing was made in PMMA. The channels were made by fine engraving and sealed with the PDMS (Sylgard 184, Dow Corning) plate (Fig. 1), which was prepared according to the procedure given in the literature [2, 3].

Fig. 1. PMMA/PDMS microfluidic system with mixing area.

J.M. Ramsey and A. van den Berg (eds.), Micro Total Analysis Systems 2001, 521–522.

Solutions of metal ions mixtures and buffered PAR solution contained also Triton X100 were introduced to the chip simultaneously with help of the peristaltic pump. The complexation took place in the mixing area of the chip. The solubility of formed complexes was better when micellar environment was created by the addition of non-ionic surfactant (Triton X100) [4]. Total flow rate of the reagents in the system was not higher than 100 μL/min. The detection was performed in a special flow cuvette and spectra measurements were carried out using diode array spectrophotometer.

Results

For each of chosen metal ion the M^{2+}-PAR complex forming conditions were checked. Next the synthetic mixtures of Cd, Cu, Hg, Pb, Zn ions were prepared as to obtain total molar metal ions concentration from 10^{-6} to 10^{-4} and measured after on-chip reaction with buffered (CH_3COOH/CH_3COONa aq.) PAR solution (Fig.2).

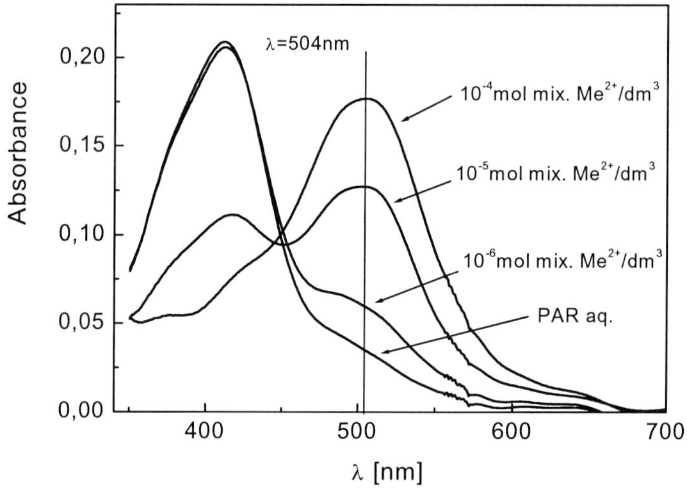

Fig. 2. Spectra of mixtures of Cd^{2+}, Cu^{2+}, Hg^{2+}, Pb^{2+}, Zn^{2+}-PAR complexes (pH 5.5)

Further measurements are planed to perform at λ=525nm after integration on one chip the fibre optic "U"-shape flow cell and all channels structure. The ultra sensitive spectrophotometer (Control Development CDI2MPP, USA) will be used for this purpose.

References

1. K.L. Cheng, K. Ueno, T. Imamura, Handbook of Organic Analytical Reagents, CRC Press, Inc. Boca Raton, Florida 1982.
2. D. C. Duffy, Cooper McDonald, O. J. A. Schueller, G. M. Whitesides, Anal. Chem. 70, 1998, p. 4974-4984.
3. B. Jo, J. Mootrhy, C.J. Beebe, Micro Total Analysis Systems 2000, p. 335-338, 2000 Kluwer Academic Publishers, The Netherlands.
4. P. Paradkar, R. Williams, Anal. Chem. 66, 1994, p. 2752-2756.

APPLICATION OF MICROSEPARATION ARRAYS TO THE DETECTION OF BIOTOXINS IN AEROSOL BACKGROUNDS

J. Fruetel, R. Renzi, R. Crocker, V. VanderNoot, J. Stamps, I. Shokair and D. Yee
Sandia National Laboratories, P.O. Box 969, Livermore, CA 94551-0969

Abstract
We are investigating the use of parallel separation channels that utilize orthogonal separation techniques as a platform for a biosensor to accurately identify protein biotoxins. Two separation techniques were initially chosen for proof of concept demonstration—capillary zone electrophoresis (CZE) and capillary gel electrophoresis (CGE)—and have been implemented in our compact "μChemLab™" device. This device integrates automated high-voltage control of microchip-based separations with laser-induced fluorescence (LIF) detection and on-board data analysis. The effectiveness of this two-channel approach was evaluated using biotoxin-spiked aerosol samples.

Keywords: microseparation arrays, protein analysis, CZE, CGE, LIF

1. Introduction

Highly accurate detection of aerosolized protein biotoxins is of interest for the identification of possible bio-terrorist incidents. The protein complexity of aerosol samples is not known, although total protein content of collector fluids suggest concentrations on the order of 1 μg/mL may be expected (unpublished results). We propose to utilize our two-channel μChemLab™ device to analyze aerosol samples spiked with protein biotoxins to assess the feasibility of detecting low nanomolar concentrations of biotoxins in the presence of realistic backgrounds.

2. Experimental

Analyses were carried out using an in-house built portable device (7 x 8 x 4") which contained: a single 2 x 3" fused silica wafer wet-etched with two 10-cm long channels; a fluidics cartridge containing 8 reservoirs; an LIF optics module utilizing two 392 nm diode lasers for excitation and a PMT for signal collection; a HV control board; and an on-board microprocessor. Separation buffers were 10 mM phytic acid + 2 mM lauryl sulfobetaine, pH 9.5 for the CZE analyses and Beckman 14-200 sieving gel for the CGE separations. The fluid reservoirs incorporated a two-chamber design for the CZE analyses [1] and a single chamber for CGE. Protein toxins (ricin, SEB, ovalbumin and a botulinum toxin surrogate) were labeled with 1 mM fluorescamine (10 second reaction time) then syringe injected into the device. Indoor and outdoor aerosol samples were collected in Livermore, California.

3. Results and discussion

The robustness of the separation methods was tested to determine run-to-run variability and to establish a database for toxin identification. CZE separations could be performed for up to ~35 runs with good reproducibility in elution time (<4% RSD over

J.M. Ramsey and A. van den Berg (eds.), Micro Total Analysis Systems 2001, 523–524.

three days for the internal standard, n = 56). The CZE elution times for the toxins, corrected to the internal standard, varied <2 %. CGE separations were initially much less robust and a database was not constructed. Reproducibility of the separations depended significantly on how well the injection plug was defined during injection as observed by fluorescence imaging and current monitoring.

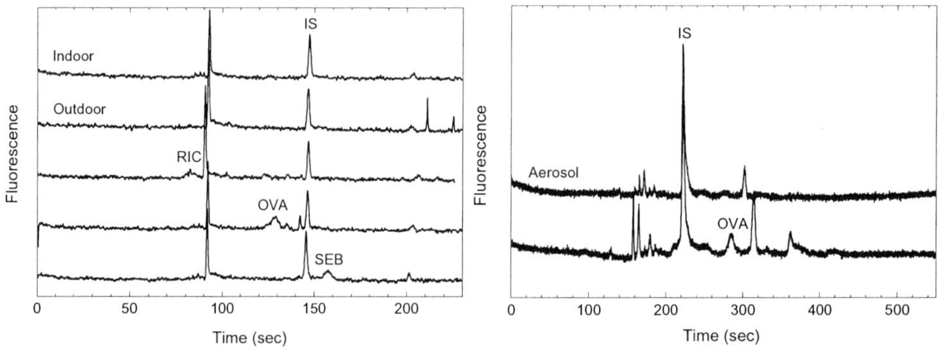

Figure 1. CZE and CGE analyses of aerosol samples alone and spiked with biotoxins.

The impact of the presence of aerosol background species on the identification of toxin peaks by the on-board data analysis software was tested by spiking aerosol samples with toxin standards. CZE and CGE analyses of toxin-spiked samples showed minimal interference of the toxin peaks with aerosol species at ~10^{-8} M toxin (Figure 1). Initial studies of 13 toxin-spiked (or blank) samples indicated that 10 of the samples (77%) were correctly identified using only data from the CZE channel. Missed peaks were due to shifts in elution time and overlap with interferants. Improvements in separation stability and peak recognition algorithms and incorporation of data from a second orthogonal analysis (CGE) are expected to give much improved peak identification rates.

5. Conclusions

Initial studies of aerosol backgrounds with this system suggest few species fluorescently-tag that interfere with toxin identification. A more detailed study of the variations in aerosol content is planned to better predict and understand aerosol effects, and to provide a guide for methods development for these types of analytical devices.

Acknowledgements

Sandia is operated for the United States Department of Energy under contract DE-AC04-94AL85000. Funding for this project is acknowledged from the Department of Energy Chemical and Biological National Security Program.

References

1. S.R. Wallenborg, C.G. Bailey and P.H. Paul, *On-Chip Separation of Explosive Compounds—Divided Reservoirs to Improve Reproducibility and Minimize Buffer Depletion*, Proc. of µTAS 2000, Kluwer Academic Publishers, Dordrecht, The Netherlands, pp. 355-358 (2000).

MICROCHIP-BASED SIMULTANEOUS ON-LINE MONITORING OF CR(III) AND CR(VI) USING HIGHLY EFFICIENT CHEMILUMINESCENCE DETECTION

Dong Jin Kim[1], Won Hui Cho[1], Kyung Won Ro[2], and Jong Hoon Hahn[2*]
[1]School of Environmental Science and Engineering, Pohang University of Science and Technology, South Korea
[2]Department of Chemistry, Pohang University of Science and Technology, South Korea

Abstract

We have developed a micromachined continuous flow analysis (CFA) device for the simultaneous detection of Cr(III) and Cr(VI) with chemiluminescence (CL). By selective injection of the reductant, we can determine the total Cr and Cr(III) quantitatively. Linearity for Cr(III) and Cr(VI) has been established over concentration ranges of 50 ~ 500 ppb and 10 ~ 1000 ppb, and detection limits (S/N = 3) are 10 ppb and 1.3 ppb respectively.

Keywords: on-line monitoring, chromium, chemiluminescence detection

1. Introduction

In aqueous solution chromium exists essentially in Cr(III) and Cr(VI). Because of their difference in toxicity and bioavailability, the facile determination of both chromium species has been highly desirable. This is the first report of development of a microchip system for simultaneous, continuous monitoring of Cr(III) and Cr(VI).

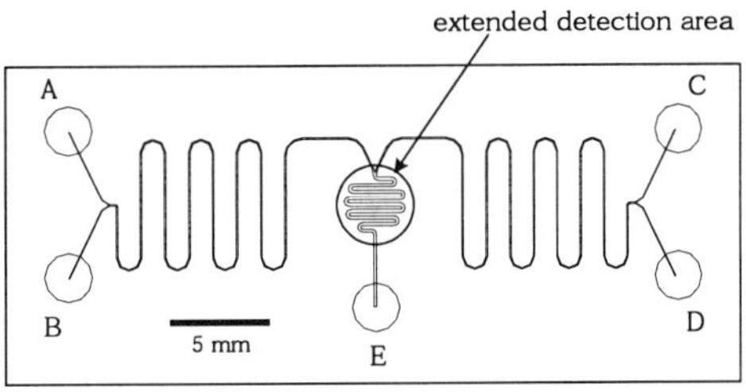

Figure 1. Layout of the CL detection microchip fabricated in PDMS [1]. Reservoirs: (A) sample; (B) reductant (15 mM (pH 3.0) or DI water; (C) 6.8 mM luminol in 20 mM boric acid (pH 11.5); (D) 20 mM hydrogen peroxide in 20 mM boric acid (pH 11.5); (E) waste. Channel width is 50 μm at the starting point, 100 μm at the serpentine mixing channel, and 200 μm at the detection channel. All channel depths are 250 μm.

2. Simultaneous Determination of Cr(III) and Cr(VI)

The chip (Figure 1) has four inlets: two for chemiluminescence (CL) generation reagents, luminol and hydrogen peroxide, one for sample solution containing both Cr(III) and Cr(VI), and the last one for DI water (when we determine Cr(III)) or reductant solution (when we determine total Cr) (Table 1). In order to monitor

J.M. Ramsey and A. van den Berg (eds.), Micro Total Analysis Systems 2001, 525–526.

Cr(III) and Cr(VI) continuously, improve the portability of the system, and meet the flow rate conditions of the microchip, two ultra-low-flow battery-powered peristaltic pump are used to provide reagents and sample solution to the four inlets.

3. High-Performance Chemiluminescence Detection

Figure 2 indicates the pulsation in flow from the pump, which is characteristic of the pumping system. Averaging the peak intensities of CL gives a very reproducible value corresponding to a concentration of a sample. The 33 mm long detection channel beginning at the confluent point of CL reagents and chromium species is compactly folded into a circular shape, the area of which is matched to the photocathode active area of a compact, battery-powerd PMT so that the detection efficiency can be maximized. Excellent linearities and detection limits for Cr(III) and Cr(VI), which satisfy the standards of the Environmental Protection Agency and are much better than previous results [2], have been accomplished. The linear dynamic ranges: 50 ~ 500 ppb ($r^2 = 0.988$) for Cr(III) and 10 ~ 1,000 ppb ($r^2 = 0.998$). The detection limits (S/N = 3): 10 ppb for Cr(III) and 1.3 ppb for Cr(VI).

Table 1. Selection of inlet reagents for Cr(III) and total Cr determination

Inlet	Cr(III)	Total Cr
A	Sample	Sample
B	**DI water**	**Reductant**[a]
C	Luminol	Luminol[b]
D	H_2O_2	H_2O_2[b]

[a] $Cr_2O_7^{2-} + 3K_2SO_3 + 8H^+ \rightarrow 2Cr^{3+} + 6K^+ + 3SO_4^{2-} + 4H_2O$

[b] luminol + $2H_2O_2 + 2OH^- + Cr^{3+}$(catalyst) $\rightarrow$ 3-aminophthalate + $N_2 + 4H_2O + h\nu$

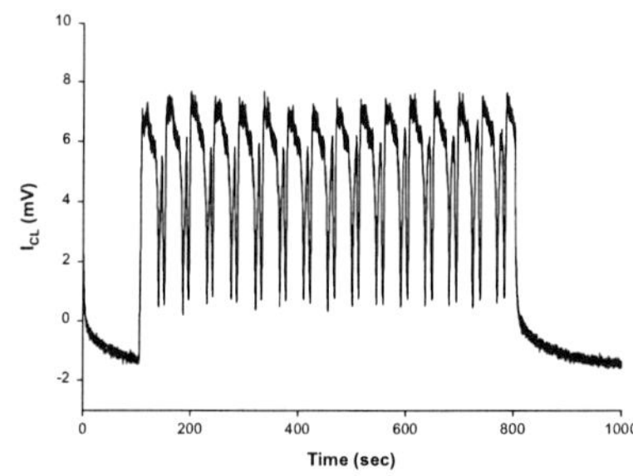

Figure 2. Consistent emission peak shape of 50 ppb Cr(VI). Flow rate is 7.32 μL/min, and detection volume is 1.64 μL.

5. Conclusions

We have developed a microchip system for simultaneous, continuous monitoring of Cr(III) and Cr(VI) with chemiluminescence detection. This approach is expected to facilitate the in situ monitoring of Cr(III) and Cr(VI).

References

1. D. C. Duffy, J. C. McDonald, O. J. A. Schueller, G. M. Whitesides, "Rapid Prototyping of Microfluidic Systems in Poly(dimethylsiloxane)", *Anal. Chem.*, **70**, (1998), 4974-4984.
2. Y. Xu, F. G. Bessoth, Jan C. T. Eijkel, A. Manz, "On-Line Monitoring of Chromium(III) Using a Fast Micromachined Mixer/Reactor and Chemiluminescence Detection", *Analyst*, **125**, (2000), 677-688.

MICRO-FLUIDIC DEVICE FOR DETECTION AND IDENTIFICATION OF AROMATIC VOC BY OPTICAL METHOD

Tsutomu Horiuchi, Yuko Ueno, Osamu Niwa
NTT Lifestyle and Environmental Technology Laboratories,
NTT Corporation, Atsugi, Kanagawa 243-0198, Japan

Abstract

We fabricated a micro-fluidic device that enables us to identify and quantify atmospheric levels of aromatic VOCs by using absorption spectroscopy as the detection method. The device consists of concentration- and detection-cells. We successfully measured sub-ppm levels of toluene gas with this device.

Keywords: VOC, micro-fluidic device, BTX gas, optical detection, environment

1. Introduction

Airborne benzene, toluene, and xylenes (BTX) are aromatic volatile organic compounds (VOCs) of great social and environmental significance. Since they are toxic even at ppb concentrations and their levels of toxicity are different, quantitative identification of VOCs in the air is important. A highly sensitive maintenance-free detection device is needed for practical application of on-line monitoring in the field. We fabricated a micro-fluidic device that enables us to identify and quantify atmospheric levels of aromatic VOCs by using absorption spectroscopy as the detection method.

2. Experimental

The device consists of concentration- and detection-cells (fig. 1a and fig. 1b , respectively). Both cells were fabricated from Pyrex glass plates using a micro-machining technique. The trenches for gas flow lines were fabricated by a dicing method. We used silicon nitride (SiN) film as an adhesive layer for Pyrex-to-Pyrex anodic bonding [1]. The concentration-cell is composed of an adsorbent (SiO_2 powder, fig. 1c) to concentrate VOC gases and a thin-film platinum heater to desorb the concentrated gas thermally. A protruding structure was formed as an adsorbent stopper in the trench by controlling the dicing blade height precisely, which prevented the adsorbent flowing out (fig. 1c). The desorbed gas was introduced into the detection cell. The method we used to fabricate the detection-cell was almost the same. UV/VIS multimode optical fibers were fixed at both ends of the trench in the detection-cell (fig. 1b).

The other ends of the two fibers were connected to a D_2 lamp and a spectrometer, respectively, to measure the absorption spectrum of the desorbed gas. We used the device to detect toluene gas, a typical example of a VOC.

J.M. Ramsey and A. van den Berg (eds.), Micro Total Analysis Systems 2001, 527–528.

3. Results and discussion

We optimized the trench structure to improve the sensitivity. A wide trench was formed above the heater to increase the amount of adsorbent. A narrow trench was connected to the wide trench. We did not put a heater under the narrow trench in order to make the cold trap of desorbed gas effective.

We could measure the absorption spectrum of 10-ppm toluene gas concentrated for 30 min (fig. 2). The concentration-cell could amplify the detection signal of the toluene gas by two orders of magnitude. We successfully measured sub-ppm levels of toluene gas with this device.

4. Conclusions

We fabricated a highly sensitive micro-fluidic device to identify and quantify atmospheric levels of aromatic VOCs. The device is expected to find practical application in on-line monitoring in the field due to its simple operation and easy maintenance.

References

1. A. Berthold, L. Nicola, P. M. Sarro, M. J. Vellekoop, Transducers'99 pp1324-1327, (1999).

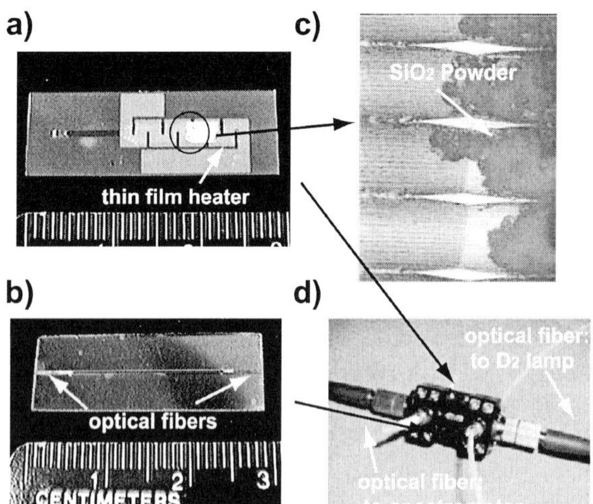

Figure 1. Micro-fluidic device for VOC detection. Concentration-cell (a), detection-cell (b), protruding structure for adsorbent stopper (c), and set up for absorption measurement (d).

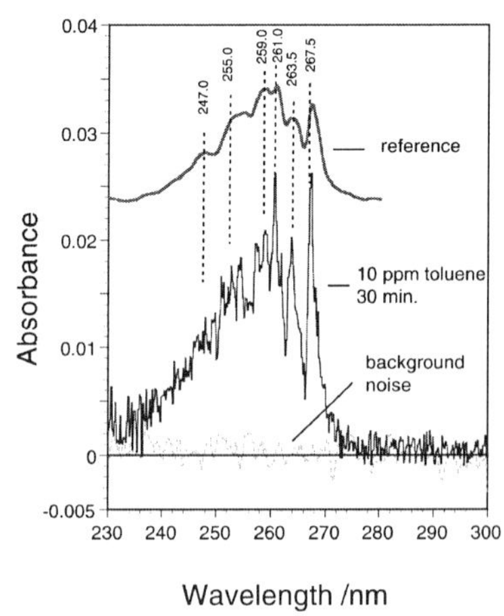

Figure 2. Absorption spectrum of 10-ppm toluene gas concentrated for 30 min.

INTEGRATED MICRO-ANALYZERS WITH FROZEN PLUG VALVES

H.W. Jannasch[1], P.R. McGill[1], M. Zdeblick[2] and Jon Erickson[1]

[1]Monterey Bay Aquarium Research Institute, Moss Landing, CA 95039
[2]Aspire Technology, a Sole Proprietorship in Portola Valley, CA 94028

Abstract. A miniature batch analyzer has been designed that is totally self-contained, able to operate in situ and is pressure tolerant. The analyzer uses frozen-plug micro-valves to digitally control fluid flow, and is capable of operating at any depth within the ocean.

Keywords: microanalysis, micro valves, environmental analysis

1. Introduction. μTAS systems are of interest to oceanographers for in situ monitoring of nutrients and trace elements. Chemical analyzers deployed in the ocean need to be totally self-contained, able to operate in situ and be pressure tolerant. We present a new miniature batch analyzer that is designed for full ocean depth applications. The analyzer uses frozen-plug micro-valves to digitally control fluid flow, and is capable of operating at any depth within the ocean.

2. Frozen plug valves. Most valves used in μTAS systems are either non-sealing (fluidic diodes) or use gas/liquid interfaces that are not suited for in situ operation at high ambient pressures. We show that fluid flow can be digitally controlled in multiple channels using a thermoelectric cooling device (TEC) and miniature heaters. In our instrument, a single TEC is used to freeze eight ~250μm diameter by 1.6 mm channels (Fig. 1). Individual platinum resistive heaters and temperature sense resistors are applied onto glass by photolithography, and sandwiched to a PVC manifold containing the flow channels. To open a valve, the heater under that valve melts the ice plug when desired. A PWM controller operates the freeze/thaw cycle of each valve. The valves can operate up to 2 Hz with the TEC waste heat

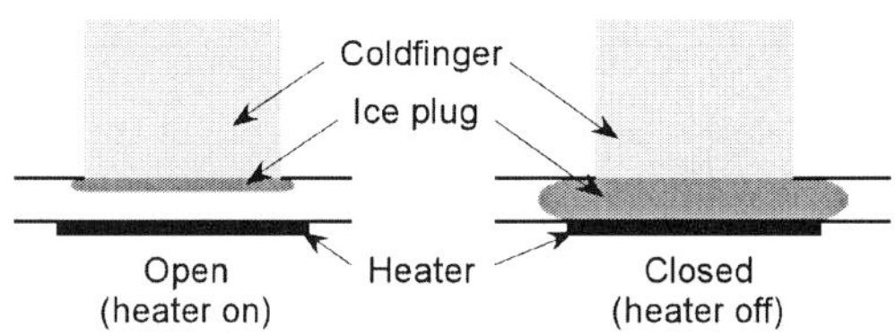

Figure 1: The open and closed state of a single frozen-plug microvalve. A TEC freezes a plug of liquid in the flow channel, which can be thawed by photolithographed resistive heater. Our valves are 250 mm diameter by 1.6 mm long. The valve has no moving parts.

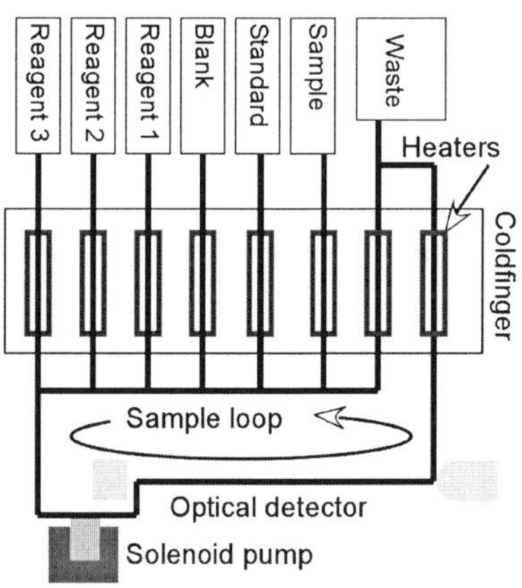

Figure 2: Schematic of the analyzer with a single TEC-cooled coldfinger and eight individually addressed heaters.

J.M. Ramsey and A. van den Berg (eds.), Micro Total Analysis Systems 2001, 529–530.

Figure 3: The integrated chemical analyzer with frozen-plug valves, TEC, heaters, optical detector and pump.

rejected to 6°C water (average seawater temperature).

3. The chemical analyzer. The integrated chemical analyzer incorporates one set of eight valves, a single solenoid pump, and an LED/photodiode optical detector in a 3x3x5 cm package. A PVC flow manifold is configured as a batch analyzer with 6 inputs (sample, standard, blank and three reagents) and a waste exit (Fig. 2). The entire analyzer is shown in Fig. 3. Direct control of the valves, pump and detector permits total sample analysis, on-board standard curves, and sample dilution as necessary.

4. Results and conclusion. Analysis of injecting dye solution demonstrates that >95% mixing takes place within the manifold with fluids pumped around the loop three times. Nitrite in seawater was measured by reaction with sulfanilamide and NED to form a colored azo dye. A five-point standard curve of nitrite analyzed by on-board dilution of a concentrated standard solution has an r^2 of 0.9913 (Fig. 4). Replacing the current PVC manifold with an etched silicon wafer should permit an increase in valve freeze/thaw frequency, an order of magnitude decrease in internal loop volume, and a significant reduction in power requirements.

Acknowledgement. This work has been generously supported by the David and Lucile Packard Foundation.

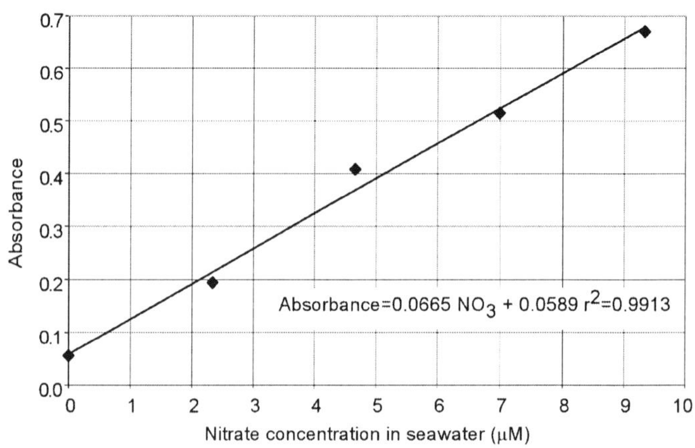

Figure 4: A standard curve measuring nitrite in seawater. The curve was produced by diluting a concentrated standard solution with low-nutrient seawater within the sample loop.

COMPACT WATER-QUALITY MONITORS WITH MICRO-MACHINED FLOW-CELLS

Ryo Miyake[1], Hideo Enoki[1], Sadao Mori[1], Kouichi Tsuzuki[1], Toshihiro Onose[2], Katsutoshi Yamada[2], Tamio Ishihara[2], Kouji Saitou[2], Ikuo Uematsu[2] and Masao Fukunaga[2]
[1]Mechanical Eng. Research Lab., Hitachi, Ltd., [2]Instruments, Hitachi, Ltd.

Abstract
Compact monitors with micro-machined flow-cells for water-quality multi-point monitoring and control system are described. A micro-mixer and an optical flow-cell were integrated in one chip. Today, two types of monitors are available.

Key words: Water quality, online monitoring, micro-mixer, flow injection analysis

Introduction
Public concern with the quality of drinking water is growing in Japan as a result of the accidental pollution with infectious bacteria such as O-157. Frequent and precise control of water quality is now required. To meet this requirement, we have developed a multi-point monitoring and control system that uses the public network (fig. 1)[1]. Key components of this system are compact water-quality monitors connected to a telecommunication network.

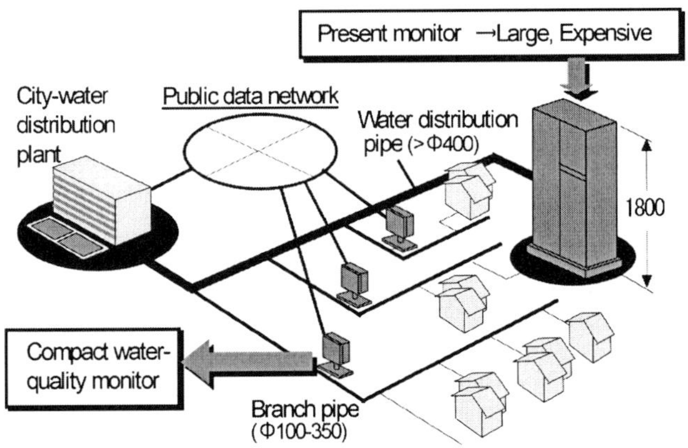

Figure 1. Concept of mult-point monitoring and control system

Compact water-quality monitors with micro-machined flow-cell
Legally, the following parameters must be checked everyday: residual chlorine, turbidity, and chromaticity as the index of water quality. Flow injection analysis (FIA) can measure these parameters accurately. A micro-mixer and an optical flow-cell for FIA were integrated in one chip (fig. 2)[2]. A micro-manifold as a liquid interface was developed. A commercial monitor (Type AN-530) is shown in fig. 3.

Field-test result and summary
A long-term field test was carried out under outside working conditions. The concen-

J.M. Ramsey and A. van den Berg (eds.), Micro Total Analysis Systems 2001, 531–532.

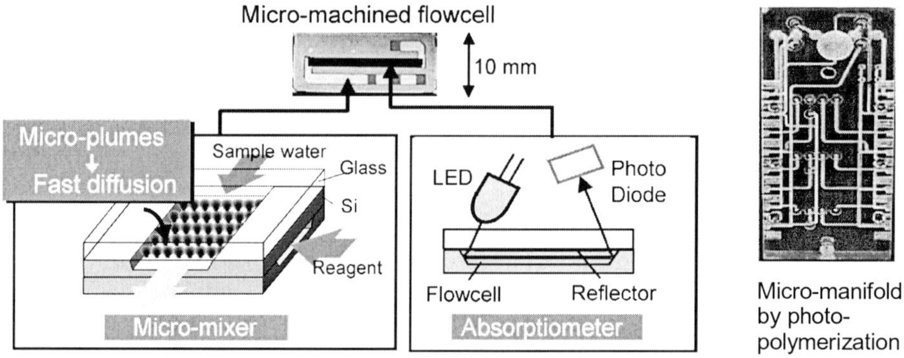

Figure 2. Micro-machined flow-cell with micro-mixers

tration of residual chlorine and turbidity measured by the monitor at a reservoir were compared with those at a city-water distribution plant. The residual chlorine of the two set of data agree well, except for some delay and deviations caused by the distance between two point (fig. 4). The advantages of the new monitor are listed in Table 1.

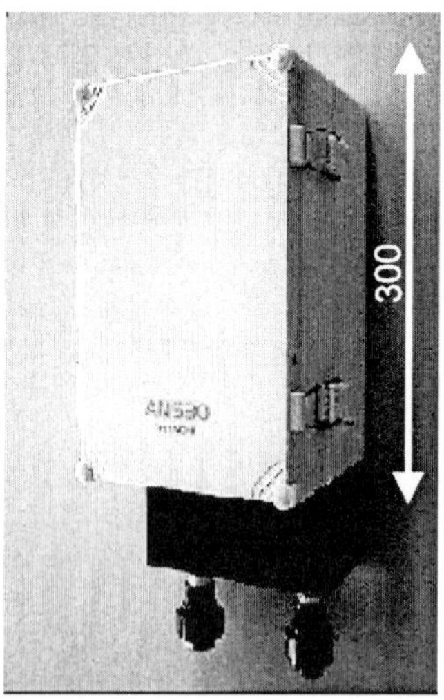

Figure 3. Type AN-530

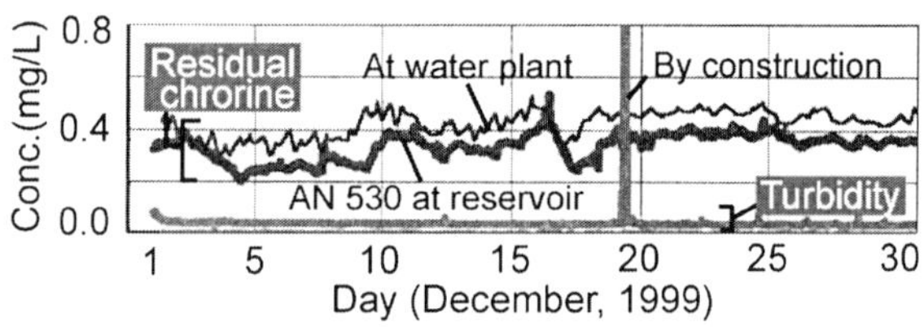

Figure 4. Field-test data by AN 530

Table 1. Comparison with present monitor

Item	Present monitor	AN-530	Advantages
System volume(m^3)	1.4	0.01(1/128)	Easy to install
Area for setting (m^2)	0.8	0.04(1/21)	
Weight(kg)	400	7(1/57)	Reduced setup cost
Sample-water volume (m^3/month)	15	0.06(1/250)	Reduced maintenance cost
Power consumption (VA)	400	12(1/33)	

References

1. K. Yamada, K. Saito, et al, *Water quality monitoring with compact on-line analyzer*, Proc. of 5th Int. Symposium on water supply technology, 2000, pp161-169

2. R. Miyake, et al, *A highly sensitve and small flow-type chemical analysis system with integrated absorptiometric micro-flowcell*, Proc. of MEMS' 97, 1997, pp102-107

INTEGRATION OF CHEMICAL PROCESSING ON A MICROCHIP

Manabu Tokeshi[1], Tomoko Minagawa[2], Kenji Uchiyama[1], Akihide Hibara[2], Kiichi Sato[2], Hideaki Hisamoto[2] and Takehiko Kitamori[1,2]
[1]Integrated Chemistry Project, Kanagawa Academy of Science and Technology, KSP East-307, 3-2-1 Sakado, Takatsu, Kawasaki, Kanagawa 213-0012, Japan
[2]Department of Applied Chemistry, The University of Tokyo, 7-3-1 Hongo, Bunkyo, Tokyo 113-8656, Japan
(tokeshi@pop12.odn.ne.jp)

Abstract

The new methodology for integration of chemical processing on a microchip was proposed. We have demonstrated the integration of whole chemical processing necessary for Co(II) wet analysis on a microchip by using the combination of several unit chemical operations, continuous multi-phase flow and newly designed microchips. Our system corresponds to the integration of about 40 unit operations.

Keywords: Unit chemical operation, chemical processing, multi-phase flow

1. Introduction

From the analogy of the development of electronic integrated circuits (IC), we consider that the introduction of the concept of unit chemical operation leads to the integration of multi-step analytical procedures. In the previous studies, we have demonstrated the integration of fundamental unit operation, i.e., mixing and reaction, solvent extraction, phase separation, solid-phase extraction and heating. In the present work, we demonstrated that the integration of chemical processing including a chelating reaction of divalent cobalt ion with 2-nitroso-1-naphthol (NN), solvent extraction of the cobalt chelates, and decomposition/removal of coexistence metal chelates, on a microchip by the combination of several unit chemical operations. This result corresponds to the integration of about 40 unit operations.

2. Experimental

In order to realize chemical processing in a microchannel, we fabricated a newly designed microchip having the guided structures at the bottom wall of microchannel [1]. The schematic cross sectional view is shown in Figure 1. Using this chip, it is very easy to stabilize an organic/aqueous interface and to separate two phases in a microchannel. Thus, multi-phase flow network in the microchannels can be formed easily using this chip. These are very important things to realize the integration

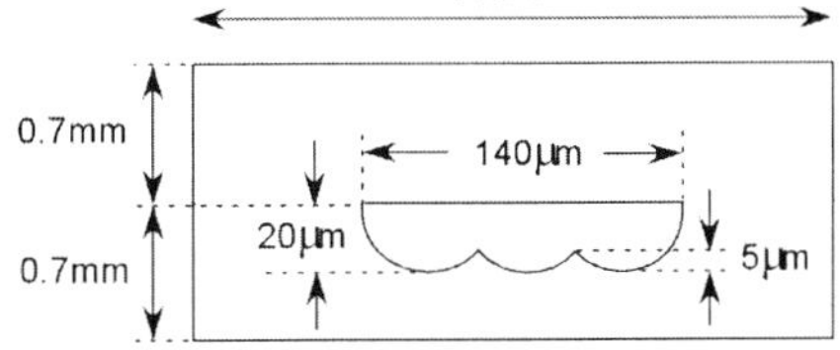

Figure 1. Cross sectional view of the guide structures.

J.M. Ramsey and A. van den Berg (eds.), Micro Total Analysis Systems 2001, 533–534.

of chemical processing including a chemical reaction, solvent extraction, and post-extraction clean up. The chip consists of two areas. One is chelating reaction and solvent extraction area, and the other is decomposition and removal area (Figure 2). The behavior of reaction, extraction and decomposition/removal was monitored by a thermal lens microscope (TLM).

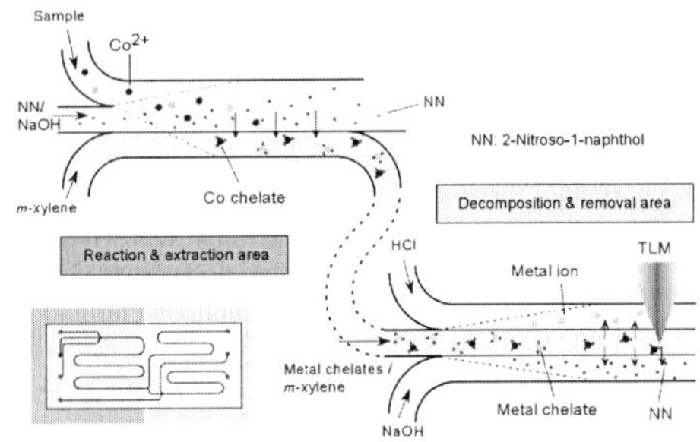

Figure 2. Schematic illustration of chemical processing in microchannels.

3. Results and discussion

Figure 3 shows the results of admixture sample containing Co(II) and Cu(II) ions. At the former area, the chelating reaction of both metals with NN and extraction of these metal chelates proceed as the microchannel length increases. On the other hand, at the latter area, the copper chelates were decomposed and removed by HCl and NaOH as the microchannel length increases. Thus, the TLM signal at the downstream was contributed only by Co chelates. From the results at 3 mm downstream from the confluence of HCl and NaOH, we obtained the calibration curve, which was shown in good linearity. The analyzing time for the sample of this system (below 1 min) is faster than that of the conventional system (ca.2 h).

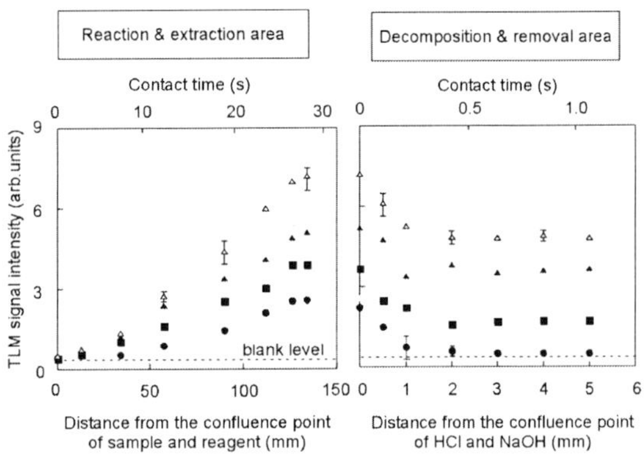

Figure 3. Position-dependent TLM signal in *m*-xylene along the x-axis in a microchannel. The samples were (Δ) 1.5×10^{-7} M, (▲) 1.0×10^{-7} M, (■) 5.0×10^{-8} M and (●) 0 M of Co(II) concentration. All samples was included in 1×10^{-6} M of Cu(II) concentration.

4. Conclusions

The chemical processing realized here corresponds to the integration of eight unit chemical operations, i.e., two-phase formation, mixing and reaction, extraction, phase separation, three-phase formation, decomposition of coexistence, removal of metal ions and removal of reagents, on a microchip. The methodology demonstrated here is expected to apply the development of more complicated system.

Acknowledgements

This research was supported by The SHISEIDO Grant for Science and Technology Research 2000. We acknowledge Miss Nao Aoyama for fabrication of the microchips.

References

1. E.Zubritsky, Multiphase flow on chip, Anal. Chem., 73, 2001, p. 188A.

Developing a Routine Coating Method for Multichannel Flow Networks on a Chip using Pyrolized Poly(dimethylsiloxane)

M.Youssouf Badal, Margaret Wong, Nghia Chiem, Hossein Salimi-Moosavi, D. Jed Harrison
Dept. of Chemistry, University of Alberta, Edmonton, AB, T6G 2G2, Canada

Abstract
Thermally pyrolized poly(dimethylsiloxane) coatings prevent surface adsorption during capillary electrophoresis (CE) separation of proteins.

Keywords: CE, PDMS, protein separation, electroosmotic mobility

1. Introduction/Experimental

Coating of microfluidic capillary electrophoresis (CE) chips is more challenging than coating conventional capillaries, due to the increased difficulty of ensuring uniform coating within a multitude of intersecting channels. The thermal polymerization of moderate viscosity silicone oils (100 centistokes) such as poly(dimethylsiloxane) (PDMS) can provide an easy to apply and satisfactory surface coating for the analysis of compounds that adsorb on a silica surface. The oil is delivered into the channels in 1% CH_2Cl_2 solution. The hydrophobic character of PDMS results in adsorption of hydrophobic proteins, so that PDMS alone is not the optimum choice and a neutral surfactant must be added. The additional adsorption of charged surfactants such as CTAB and SDS on the PDMS provides a means to both control EOF and separate proteins of different pI values [1,2].

2. Results and Discussion

The PDMS coating suppresses electroosmotic flow (EOF) to about 50% of the value seen for an uncoated capillary and adding 0.01% Brij 35 (83.3 μM) further reduces EOF to 25 % [1]. Figure 1 shows the separation of different proteins with significantly different pI values using PDMS coated capillaries. Figure 2 shows the peak observed for fluorescein labeled, goat anti-human Fab fragment separation on a chip. Figure 3 shows the electroosmotic mobility, μ_{eo}, observed across an SDS concentration range of 0 to 0.08 % (2.8 mM), for three different pH values at an ionic strength of 50 mM (phosphate buffer) in the presence and absence of Brij 35.

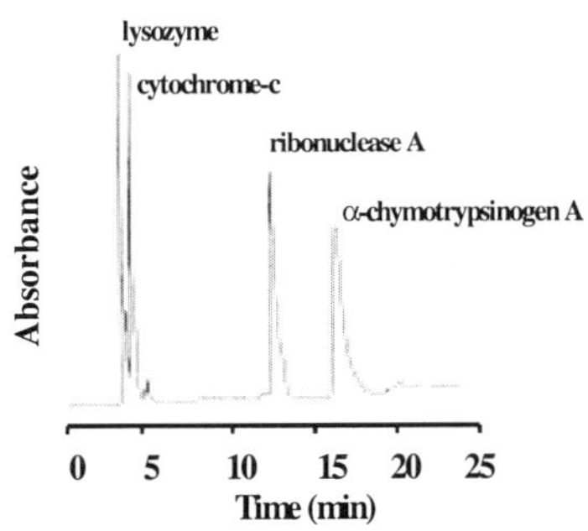

Figure 1: Separation of a protein mixture in a PDMS coated capillary, 0.020 M phosphate buffer (pH 7) with 0.01% Brij 35.

J.M. Ramsey and A. van den Berg (eds.), Micro Total Analysis Systems 2001, 535–536.

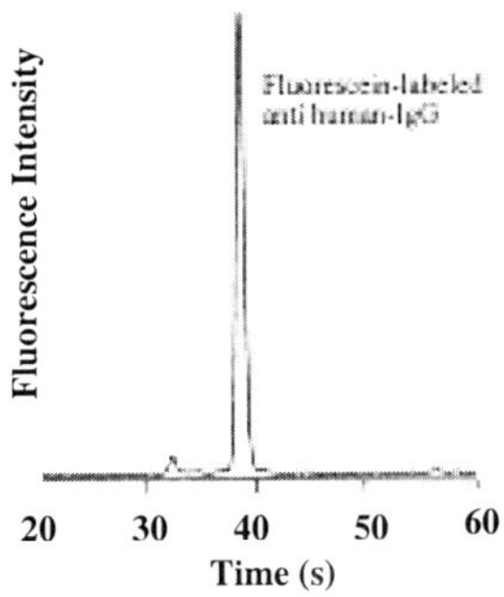

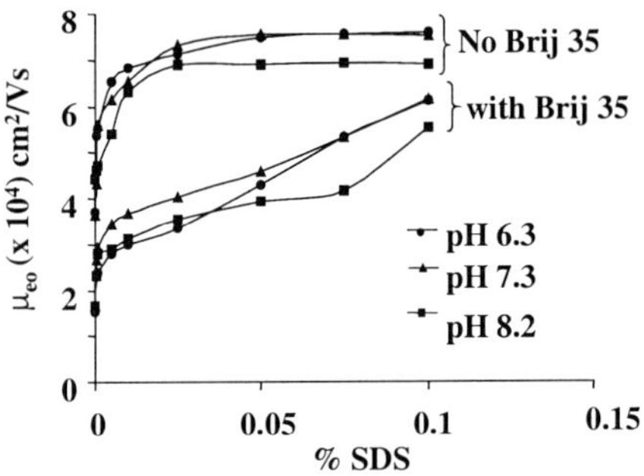

Figure 2: Separation of Fab fragment on a PDMS coated chip.

Figure 3: Electroosmotic mobility, μ_{eo}, vs % SDS

Most of the functional range for SDS control of μ_{eo} is below the critical micelle concentration, so that added SDS can be used to control EOF without necessarily impacting the activity of proteins. This approach could be useful for assays within microfluidic devices, in which EOF is required to mobilize samples and reagents for reaction prior to performing a separation. Figure 4a shows the variation in μ_{eo} with CTAB concentration around the flow reversal range. In Figure 4b, the mobility continued to increase to values in the range of –3.7 to –5.0 x 10^{-4} cm^2/Vs, as [CTAB] increased to 0.1 % (2.74 mM). Brij 35 appears to play a less significant role in the presence of CTAB than it does with SDS.

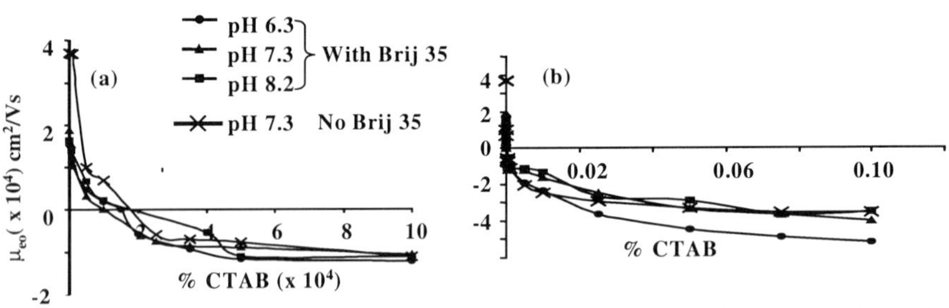

Figure 4: Electroosmotic mobility, μ_{eo}, vs % CTAB. (a) Near flow reversal and (b) across the entire range.

3. Conclusion

Coating capillary surfaces with pyrolized silicone oils provides a convenient means of preparing a hydrophobic PDMS surface in fused silica and microchip CE devices. The addition of a neutral and charged surfactant to the running buffer creates a surface that can be used effectively for separation of proteins and EOF control.

Reference

1. M. Y. Badal, M. Wong, N. Chiem, H. Salimi-Moosavi, D. J. Harrison, submitted to *J. Chromatogr.* A, 2001.
2. US Patent Number 5,074,982, Novoney et al, December 24, 1991.

SAMPLE CONCENTRATION AND SEPARATION ON MICROCHIPS

B. Scott Broyles, Stephen C. Jacobson, and J. Michael Ramsey
Oak Ridge National Laboratory, Oak Ridge, Tennessee 37831-6142 USA

Abstract

Microfabricated devices integrating solid phase extraction and chromatographic separation with solvent programming were demonstrated. Preconcentration and separation were performed on channels 5 μm deep and 25 μm wide coated with a C18 phase, and elution was achieved under isocratic, step, or linear gradient conditions. For the solid phase extraction, signal enhancement factors of 400 over a standard injection of 1.0 s were observed for a 320 s injection.

Keywords: Solid phase extraction, solvent programming, electrochromatography

1. Introduction

Sample concentration and separation are necessary in instances where trace analysis of complex mixtures is desired, and the detection sensitivity is too low to reliably detect and quantify the analytes. In such cases, routinely used methods in liquid phase analysis are solid phase extraction (SPE) and liquid chromatography. In this work microchip solid phase extraction[1] and open channel electrochromatography (OCEC) with solvent programming[2] are coupled together for the analysis of polycyclic aromatic hydrocarbons (PAHs).

2. Experimental

Quartz microchips containing separation and solvent programming elements (Figure 1) were prepared and treated by selectively coating the analysis and sample waste channels with an octadecyltrimethoxysilane/toluene solution.

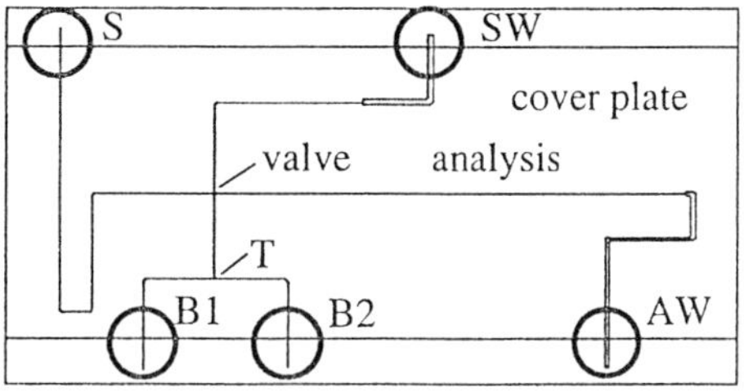

Figure 1. Chip schematic. Reservoirs: (B1) Buffer with 20% acetonitrile, (B2) Buffer with 60% acetonitrile, (S) Sample, (SW) Sample waste, (AW) Analysis waste. (not to scale).

The gated injection method was used to load the sample onto the 5 μm deep, 25 μm wide, 30 mm long, and C18 coated analysis channel (Figure 1). Computer controlled power supplies allowed precise mixing of the buffer solutions from B1 and B2 containing 20% and 60% acetonitrile, respectively. To perform SPE and OCEC, the sample was concentrated on the analysis channel for the desired time, e.g., 160 s, at low

J.M. Ramsey and A. van den Berg (eds.), Micro Total Analysis Systems 2001, 537–538.

organic concentration and separated at high organic concentration under isocratic or gradient conditions.

3. Results and discussion

Figure 2 shows a comparison of a standard 1 s injection of 900 nM pyrene eluted in high organic concentration with a 320 s injection under SPE conditions. The concentration increased linearly with time, and an enhancement factor up to 400 was observed for the 320 s injection. Solid phase extraction was then coupled to electrochromatographic separations. In Figure 3, four PAHs were concentrated for 160 s and separated using a step gradient. Detection limits of 3.1, 1.0, 8.1, and 17 nM were obtained for anthracene, pyrene, 1,2-benzofluorene, and benzo(a)pyrene, respectively.

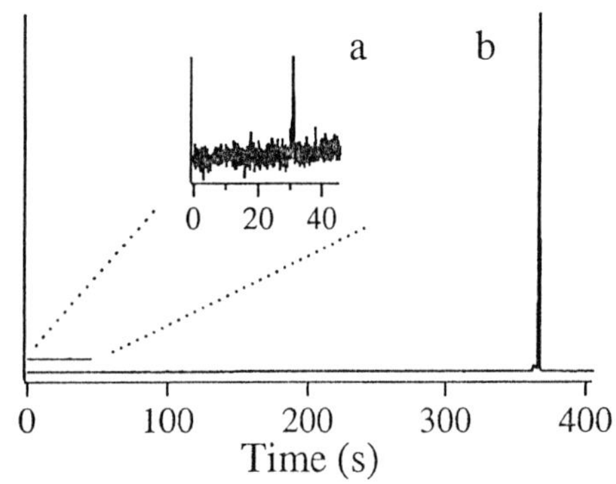

Figure 2. 900 nM pyrene injections. (a) 1 s injection and elution with 56% acetonitrile without concentration. (b) 320 s concentration and elution. (a) is offset from (b) for clarity. Y axis is detector response (V).

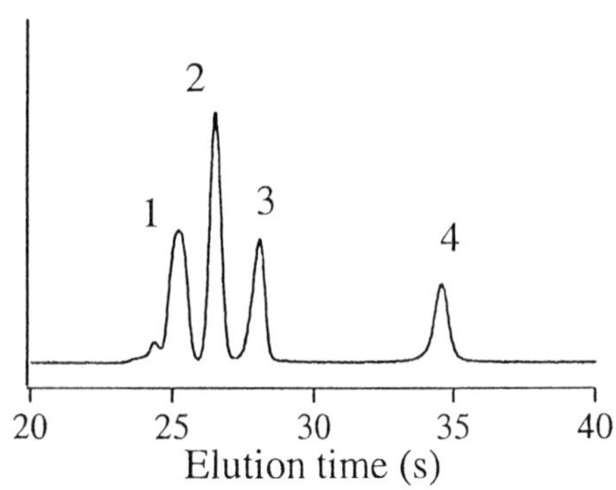

Figure 3. Separation using a step gradient that started at 52% acetonitrile, held for 10 s, and switched to 56% acetonitrile for the remaining elution time. 160 s concentration time. 1) Anthracene- 2.8 μM; 2) Pyrene- 0.9 μM; 3) 1,2 Benzofluorene- 5.8 μM; 4) Benzo(a)pyrene- 5.0 μM. Y axis is detector response (V).

Acknowledgements

Oak Ridge National Laboratory is managed and operated by UT-Battelle, LLC, for the U.S. Department of Energy under contract DE-AC05-00OR22725.

References

1. Kutter, J. P.; Jacobson, S. C.; Ramsey, J. M. *J. Microcolumn Sep.* **2000,** *12*, 93-97.
2. Kutter, J. P.; Jacobson, S. C.; Matsubara, N.; Ramsey, J. M. *Anal. Chem.* **1998,** *70*, 3291-3297.

DIELECTROPHORETIC SEPARATION AND ACCUMULATION OF (BIO)PARTICLES IN MICRO-FABRICATED CONTINUOUS FLOW SYSTEMS

M. Dürr [1], T. Schnelle [2], T. Müller [3] and M. Stelzle [1]

[1]NMI Natural and Medical Sciences Institute, Markwiesenstrasse 55, 72770 Reutlingen, Germany (duerr@nmi.de)
[2]Evotec OAI, Schnackenburgallee 114, 22525 Hamburg
[3]Humboldt University, department of biology, Invalidenstr. 42, 10115 Berlin

Abstract

Automated handling of cells and other bio-particles is becoming increasingly important for lab-on-a-chip devices. This requires sophisticated sample preparation steps like particle accumulation and separation prior to the analysis. This work presents the physical principle, design and experimental results of a novel device (fig.1) for particle accumulation and separation based on dielectrophoretic actuation in a micro flow cell.

Keywords: Dielectrophoresis, Particle accumulation and separation, Lab-on-a-chip

1. Theory

When the device is operated, fluid carrying the particles flows along the channel. Radio frequency voltage is applied to the electrode pairs generating an inhomogeneous electric field which in turn induces dipole moments in the particles (fig. 1). As a result, the particles are repelled from the environment of the electrode pairs (negative dielectrophoresis).

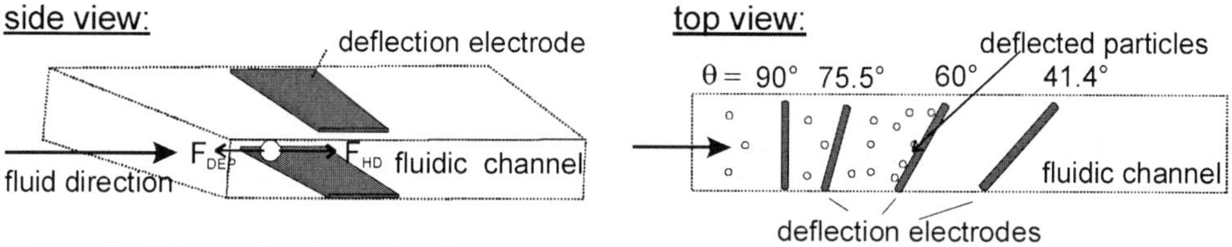

Figure 1. Schematic view of a flow system with deflection electrodes located at the top and at the bottom of the fluidic channel, respectively.

Particles moving along with the fluid experience both dielectrophoretic (F_{DEP}) and hydrodynamic (F_{HD}) forces. The microelectrode bars are oriented at different angels θ with respect to the fluid direction. Typical dimensions of the complete device are 2 mm channel width, 30 mm channel length and 25 μm channel height. Microelectrode width is 5-100 μm.

J.M. Ramsey and A. van den Berg (eds.), Micro Total Analysis Systems 2001, 539–540.

Whether or not a particle penetrates the dielectrophoretic barrier depends on the relative strength of dielectrophoretic

$$F_{DEP} = \frac{27}{32}\pi^2\varepsilon_m \operatorname{Re}(f_{CM})\cdot R^3 \cdot \frac{U^2}{h^3} \qquad (1)$$

and drag force component $\quad F_{HD} = 6\pi \cdot \eta \cdot R \cdot v \qquad (2)$

exerted on particles moving in the center of the channel (fig. 1). Particles are deflected if $F_{DEP} > F_{HD} \cdot \sin\theta$, i.e. the velocity of all deflected particles is $v < v_{\max}$ with

$$v_{\max} = \frac{9\pi\varepsilon_l \operatorname{Re}[f_{CM}]}{64\eta} \cdot \frac{R^2 \cdot U^2}{h^3 \cdot \sin\theta}. \qquad (3)$$

2. Experimental and results

The relevant parameters describing the operation of this system are (1) the particle radius *R*, (2) the applied electrical Potential *U*, (3) the dielectric properties of the particle – liquid system (4) the viscosity, (5) the particle speed *v* in the channel, (6) the deflection angle θ, and (7) the channel height *h*.

Particles of different sizes can therefore be separated by variation of θ (fig. 2). Thus, we were able to concentrate latex particles in a size ranging from 20 µm down to 500 nm using the same device. Fluid throughput can be as high as 150 µl/h.

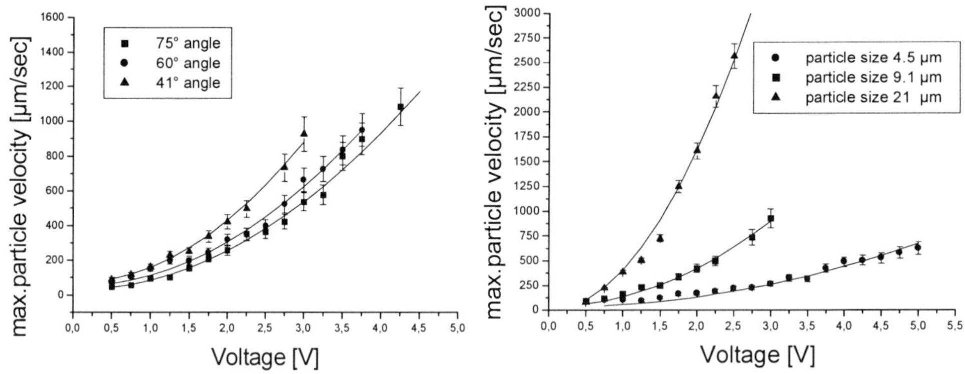

Figure 2. left Deflection of 9.1 µm Latex particles at different angles of the electrodes. **right** Deflection of particles with different sizes at an electrode bar with 41° angle. Fit curves of equation 3 (solid lines) are in very good agreement with experimental values.

References

1. T. Schnelle, T. Müller, G. Gradl, S. Shirley and G. Fuhr, *Dielectrophoretic manipulation of suspended submicron particles,* Electrophoresis 21, 2000, p. 66-73
2. T. Müller, G. Gradl, S. Howitz, S. Shirley, T. Schnelle and G. Fuhr, *A 3-D microelectrode system for handling and caging single cells and particles*, Biosensors & Bioelectronics 14, 1999, p. 247-256

DEVELOPMENT OF A MICRO SYSTEM FOR CIRCULAR CHROMATOGRAPHY USING WAVELET TRANSFORM DETECTION

J.C.T.Eijkel[1], C.Dalton[2], C.J.Hayden[2], J.A.Drysdale[2], Y.C.Kwok[1], A.Manz[1]
[1]AstraZeneca SmithKline Beecham Centre for Anal. Science, Imperial College, London, UK
[2]Institute of Bioelectronics and Molecular Microsystems, University of Wales, Bangor, UK

Abstract
Two components of a circular chromatography system are demonstrated: a magnetohydrodynamic micropump and wavelet transform detection.

Keywords: Magnetohydrodynamic pump, circular chromatography, wavelet transform

1. Introduction

When chromatography is performed in a circular channel with detection at some point(s) along the channel, it is possible to vary the analysis time at will. The plate number will be proportional to analysis time, enabling the generation of extremely high plate numbers for difficult separations but also the reduction of analysis time for simple separations. A pumping method is needed that is suitable for circular channels. Also, a sensitive detection method is needed. Both elements are presented here.

2. Magnetohydrodynamic micropump

When perpendicular magnetic and electrical fields are applied to an electrolyte solution, a Lorentz force on the electrolyte ions results, propelling the solution. Lemoff and Lee constructed the first practicable circular magnetohydrodynamic pump applying ac fields to reduce electrolysis [1]. The present device is based on the same principle but has a much smaller channel depth, necessitating the application of a strong magnetic field and a high current density at high frequency (10-100 kHz). The device (Fig.1) consists of a glass/gold/laminate/glass sandwich with a 30 μm deep and 400 μm wide circular channel of 1 cm radius as well as injection channels defined in the gold. The magnetic field is generated by five toroidal ferrite electromagnets fitted around half the channel circumference (Fig.2). The current is sent between the gold channel walls. A study of pumping characteristics of, and chromatography with the device is ongoing.

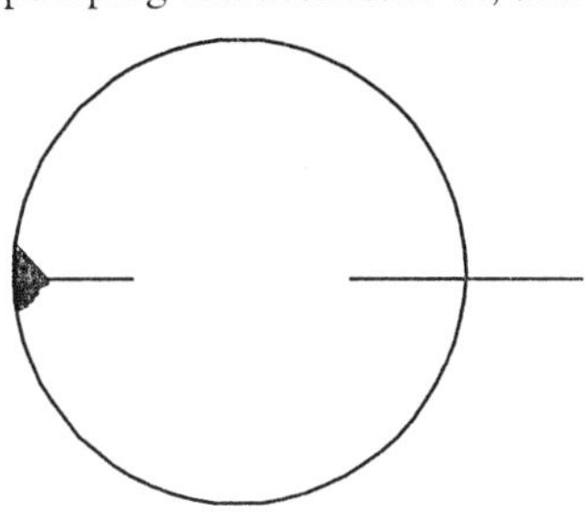

Figure 1. Channel layout of the circular micropump with straight injection channels

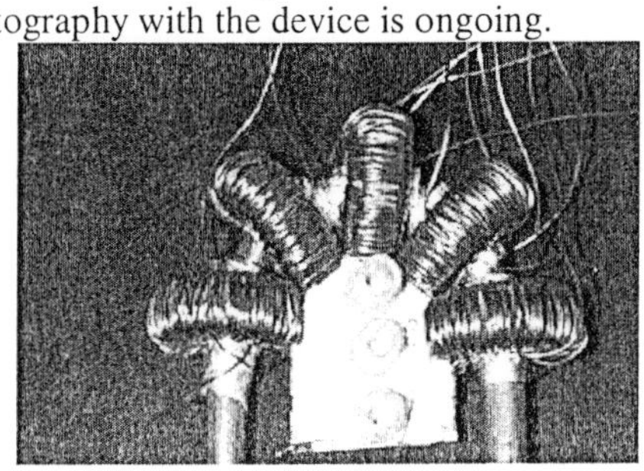

Figure 2. The micropump slotted in the toroidal electromagnets

J.M. Ramsey and A. van den Berg (eds.), Micro Total Analysis Systems 2001, 541–542.

3. Wavelet transform detection

In a set-up where the separation channel is periodically illuminated, moving bands of fluorescent analyte generate a periodic emission signal (the Shah convolute of the detection signal) that can be transformed into frequency (and hence velocity) information by Fourier transformation [2]. This method increases detection sensitivity. For the last step, wavelet transformation offers a good alternative since it produces both frequency and time information. Two examples using data from electrophoretic separations in straight channels will demonstrate this. Fig.3a shows the Shah convolution detection signal generated for a separation of an analyte mixture, and Fig.3b its wavelet transform. Here the wavelet transformation enables not only frequency and hence component resolution of three analytes (y-axis), but it also shows how the fastest analyte leaves the channel first (x-axis). Fig.4 shows a conventional detection of the same sample. Fig.5 shows the detection signal and wavelet analysis obtained for a separation of two analytes [2]. A frequency drift is observed that went undetected in Fourier transform. In circular chromatography, an inhomogeneous coating of the stationary phase could easily cause periodical drifts, and their detection will be important to obtain maximal peak resolution in the frequency domain.

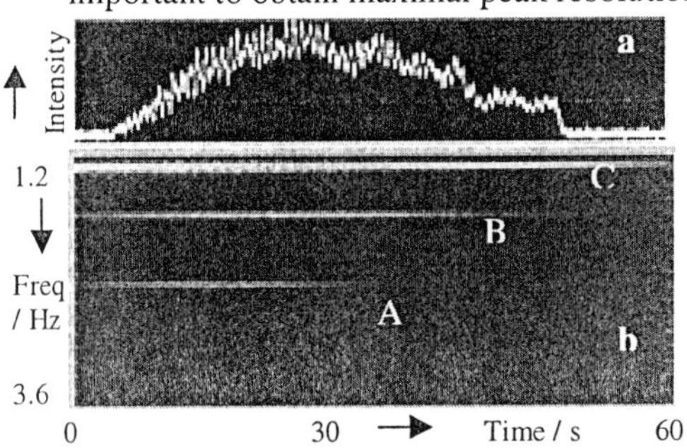

Figure 3. Demonstration of wavelet detection. (a) the convoluted detection signal; (b) the wavelet transform

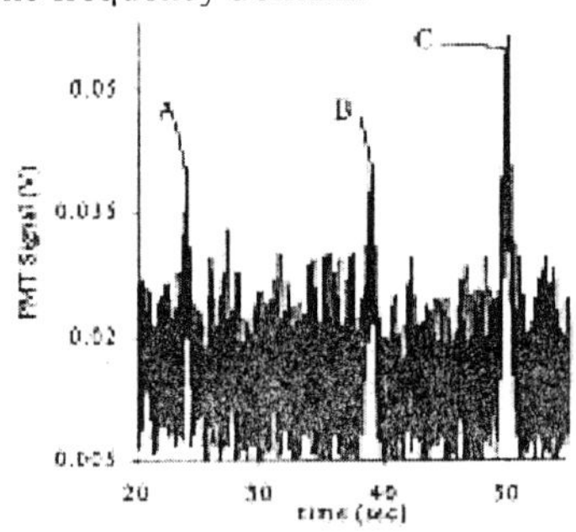

Figure 4. Electropherogram of the same sample containing analytes A, B and C

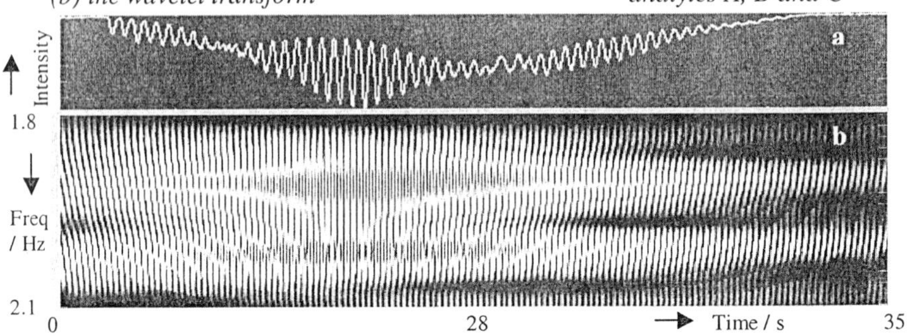

Figure 5. Demonstration of wavelet detection. (a) convoluted detection signal of an electrophoretic separation of two analytes; (b) the wavelet transform

Acknowledgement

The work presented here was partly sponsored by the EPSRC UK, the Department of Trade and Industry (DTI) and the Lab-on-a-Chip consortium (LOC).

References

1. A.V.Lemoff and A.P.Lee, Sensors and Actuators B 63 (2000), 178.
2. H.J.Crabtree, M.U.Kopp and A.Manz, Anal.Chem.71 (1999), 2130.

A MICROMACHINED ELECTRICAL FIELD-FLOW FRACTIONATION (μ-ElFFF) SYSTEM WITH INTEGRATED ELECTRICAL CONDUCTIVITY DETECTION AND COMPLIMENTARY RESONANCE LIGHT SCATTERING DETECTION SYSTEMS

Mason Graff and A. Bruno Frazier
School of Electrical and Computer Engineering, Georgia Institute of Technology, Atlanta, Georgia 30332-0250

Abstract

In this work, a μ-ElFFF system is demonstrated with complimentary on-column electrical conductivity and resonance light scattering (RLS) detection systems. High speed serial fractionations are demonstrated in the μ-ElFFF system using anti-biotin coated metallic nanospheres of various sizes. The fractionations are visualized in real-time using RLS and correlated to the integrated conductivity detectors.
Keywords: Resonance light scattering, Electrical Field-Flow Fractionation, conductivity detector

1. Introduction

Electrical field-flow fractionation (ElFFF) is a separation technique which utilizes an electric field perpendicular to the direction of flow to separate particles based on size and electric charge. ElFFF is an ideal method for sample purification / concentration and can be connected serially with downstream analytical tools to create integrated, multi-compartment analysis systems. In previous work, μ-ElFFF systems with on-chip electrical detectors have been shown to be capable of efficiently separating several types of particles and molecules [1]. Integration of an optically based signal generation and detection method allows real-time visualization of developing separations as well as correlation of the optical and electrical detector signals.

2. Separation of Particles

A μ-ElFFF system (Figure 1) was used to separate anti-biotin coated gold nanospheres. The gold particles ranged from 40-100 nm in diameter. Three sizes of microchannels were used for the separations. The breadths of the channels were 1-, 5- or 10 mm, and each had a height and length of approximately 25 μm and 61.5 mm respectively. An electric field (76-120 kV/m, corresponding to applied voltages of 1.5 – 2.5 volts) maintained between the top and bottom surfaces of the channels was used as the driving force behind the separations. For the integrated, on-column conductivity-based electrical detection system, an AC voltage of 2.475 V_{rms} @ 2 kHz was applied and the resultant AC_{rms} current was monitored using a multi-meter. Separation of the 40 nm and 60 nm particles required less than 1.0 minutes to obtain a resolution of greater than 1. The μ-ElFFF system was used for continuous serial particle separations for more than 100 fractionation runs.

J.M. Ramsey and A. van den Berg (eds.), Micro Total Analysis Systems 2001, 543–544.

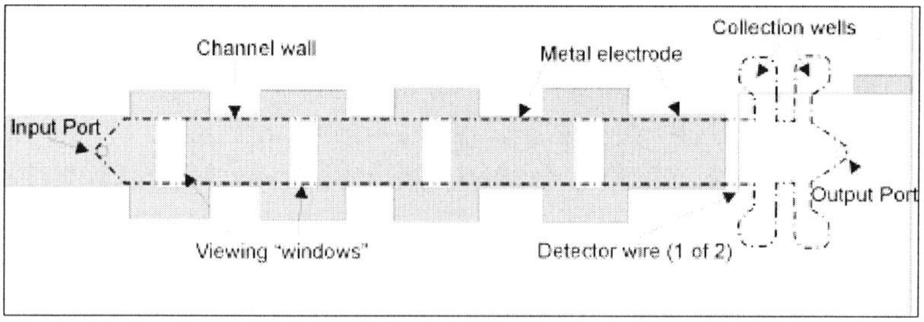

Figure 1. Overview of a μ-ElFFF device (5 mm wide channel) with integrated electrical detector and made for use with RLS detection of particles.

3. Visualization of Particles

Resonance light scattering has been shown to be capable of detecting particles in the 10-1000 nm (diameter) range [2]. Visualization of particles was accomplished using a microscope (modified for RLS detection) with a CCD camera attached that enabled digital image capture and VCR recording of separations (Fig. 2). The separations were viewed at 100X, and correlated with the on-chip conductivity detector's response, showing that we are able to separate different particles with resolutions >1 (Fig. 3).

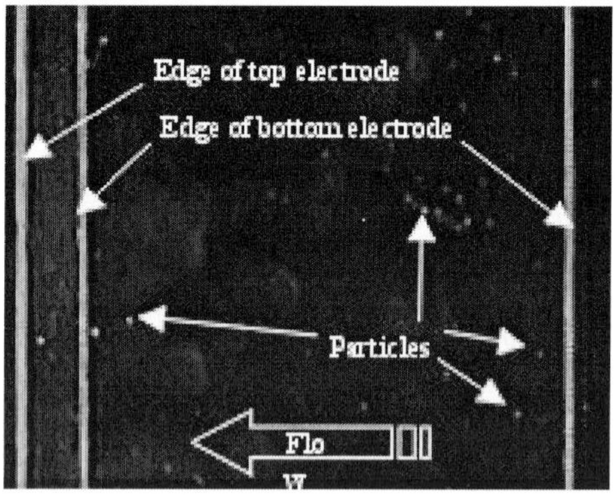

Figure 1. RLS image of particles flowing through a μ-ElFFF device as seen through a "window" in the fractionation electrodes.

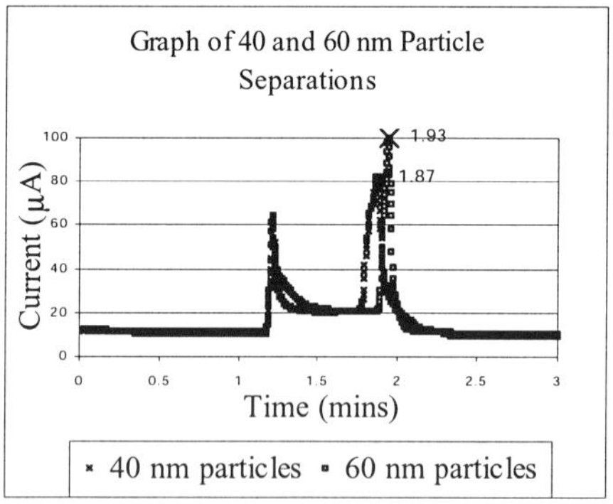

Figure 3. Graph showing plots of 40- and 60-nm particle separations. The first peak is a "void" peak.

References

1. B.K. Gale, K.D. Caldwell, and A.B. Frazier, Geometrical Scaling Effects in Electrical Field Flow Fractionation. 1. Theoretical Analysis, Analytical Chemistry, (73) 2345-2352 (2001).
2. J. Yguerabide and E. E. Yguerabide, Light-Scattering Submicroscopic Particles as Highly Fluorescent Analogs and Their Use as Tracer Labels in Clinical and Biological Applications, Analytical Biochemistry, pp. 137-156, (1998).

INTEGRATION OF A SOLID STATE MICROPUMP AND A SUB-MICROMETRE PARTICLE ANALYSER / SEPARATOR

Nicolas G. Green
Dept Electronics and Electrical Engineering, University of Glasgow, Glasgow, Scotland, UK
(n.green@elec.gla.ac.uk)

Abstract

Microelectrode structures fabricated into microfluidic devices and energised with alternating potentials produce forces on samples of fluids and particles passed through the device. A particle analyser using dielectrophoretic field flow fractionation was designed and fabricated for application to sub-micrometre particles. A solid state micropump based on electric-field-induced fluid flow was also fabricated and the fluid rates measured.

Keywords: dielectrophoretic separation, electrohydrodynamic pumping, sub-micrometre particle analysis

1. Background

Suspensions of particles in fluid experience electric-field-induced forces arising from the interaction of the field and induced charges. The force resulting from the interaction of a non-uniform field and the dipole moment induced in particles, produces a particle movement referred to as Dielectrophoresis (DEP) [1]. The dielectrophoretic force is frequency dependent, a function of the conductivity and permittivity of the suspending fluid and the particle, and unique to a given particle type. DEP has been used for characterisation and separation of a variety of particle types and has recently been applied to the sub-micrometre scale [1]. A successful approach for particle separation using dielectrophoresis is DEP field flow fractionation [2]. In this method, long arrays of microelectrodes are fabricated on planar surfaces, and placed horizontally with the sample of particles suspended above. Different particle types experience a differing DEP force and levitate to different heights in equilibrium with gravity. A parabolic fluid flow is then passed across the electrode array and the particle types experience different fluid velocities and separate spatially. This method has been demonstrated on large particles such as cells [2]. In this work, dielectrophoretic field flow fractionators are designed and fabricated for the separation of sub-micrometre particles.

The non-uniform electric fields generated by the planar microelectrodes also produce an electrical force on gradients of permittivity and conductivity in the fluid. Gradients are typically generated by non-uniform heating of the fluid either by the field or from an external source [3]. Novel techniques based on different physical mechanisms have been developed by the author and co-workers [4]. The potential of the new technique for the purpose of pumping fluid using asymmetric designs of electrodes has also been

J.M. Ramsey and A. van den Berg (eds.), Micro Total Analysis Systems 2001, 545–546.

discussed [5]. This micropump is solid state, with no moving parts and uses low voltage power supplies, ideal for integration into a microsystem to provide the fluid flow required in DEP field flow fractionation.

2. Experimental fabrication

The dielectrophoretic device and the micropump were fabricated using standard photolithography on planar glass surfaces. Signal generators operating in the range 10Hz-20MHz with potentials up to 20 Volts were used to energise the microelectrodes. Sub-micrometre particles in the range 100-500 nanometres, suspended in aqueous solutions were used as test particles.

3. Results and discussion

Preliminary measurements of the AC Electroosmotic fluid flow on the surface of plate electrodes are shown in Figure 1. These high surface velocities produce an average bulk motion an order of magnitude less. This flow rate is suitable for the operation of a DEP separator.

AC Electroosmosis is also generated by the microelectrodes in the separator. The fluid flow in this case restricts the operating frequency range of the separator to 10kHz-20MHz.

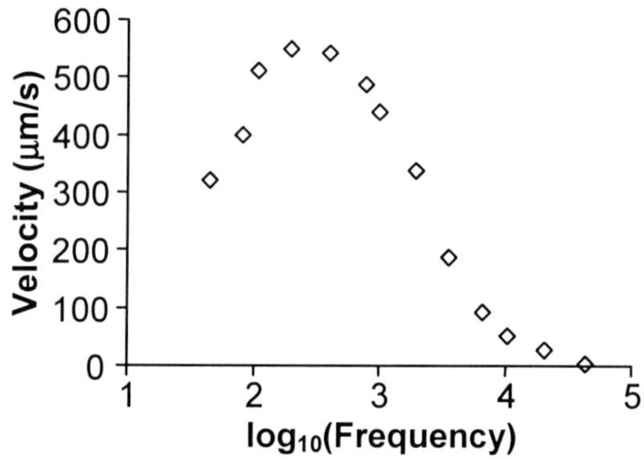

Figure 1. Fluid velocity 5μm from electrode edge for 5Volts pk-pk in KCl of 2.1mS/m.

Acknowledgements

N. G. Green is a Royal Academy of Engineering Post Doctoral Research Fellow at the University of Glasgow. The author would also like to thank Prof H Morgan and Dr Antonio Ramos for discussions and Ms M Robertson for technical assistance.

References

1. H. Morgan, M. P. Hughes and N.G. Green, Separation of sub-micron bio-particles by dielectrophoresis, *Biophysical Journal* 77 pp.516-525 (1999).
2. J. Yang, Y. Huang, X-B. Wang, F. F. Becker and P. R. C. Gascoyne, Cell Separation on Microfabricated Electrodes Using Dielectrophoretic/Gravitational Field Flow Fractionation, *Analytical Chemistry* 71 pp.911-918 (1999).
3. M. Koch, A. Evans and A Brunnschweiler, *Microfluidic Technology and Applications*, Research Studies Press, UK, (2001).
4. N.G. Green, A. Ramos, A. Gonzalez, H. Morgan and A. Castellanos, Fluid flow induced by nonuniform AC electric fields in electrolytes on microelectrodes. I. Experimental measurements, *Physical Review E*, 61 pp.4011-4018 (2000).
5. A. Ajdari, Pumping liquids using asymmetric electrode arrays, *Physical Review E* 61 pp.R45-R48 (2000).

FAST CHIRAL ON-CHIP SEPARATIONS WITH AMPEROMETRIC DETECTION

Maria A. Schwarz and Peter C. Hauser*
The University of Basel, Department of Chemistry, Spitalstrasse 51, 4056 Basel, Switzerland

Abstract

Despite the high separation efficiencies of microfabricated electrophoresis devices, enantiomeric separations require the use of auxiliary agents. By employing cyclodextrins, chiral separations of neurotransmitters and related compounds of clinical importance could be carried out in an unprecendent short time of two minutes.

Keywords: Chiral separation, neurotransmitters, amperometric detection

1. Introduction

The importance of chiral analysis of pharmaceuticals and other biochemically active substances is well recognised as generally only the L-enantiomers show a physiological effect and a large number of samples have to be processed in clinical analysis and pharmaceutical industry. Capillary electrophoresis has been applied successfully. A drastic increase in sample throughput should be possible by carrying out chiral separations in the microchip format.

2. Experimental

Detection was carried out by amperometry, using a simplified cell based on the electrophoretic ground electrode acting as pseudo-reference and counter [1]. Electrochemical detection has the advantages of not requiring analyte derivatization and of relative simplicity. The special arrangement developed allows further simplification in view of the ultimate miniaturization of the overall assembly. In particular the elimination of conventional, liquid junction based reference electrodes is desirable. No compromises in terms of performance had to be made and detection limits in the order of 10^{-7} M were achieved.

3. Results and Discussion

Chiral separations are not possible by electrophoretic separation alone as the isomers show largely identical mobilities. To this end, cyclodextrins are often used in HPLC and capillary electrophoresis as enantiomer-selective complexing agents [2]. As seen in Fig. 1 as an example, it was found possible to separate enantiomers in approximately 2 minutes

J.M. Ramsey and A. van den Berg (eds.), Micro Total Analysis Systems 2001, 547–548.

on a micromachined device with the aid of a cyclodextrin derivative as buffer additive. The species analysed are metabolites of catecholamine neurotransmitters, which are of clinical importance in the diagnosis of Parkinson's disease and related illnesses.

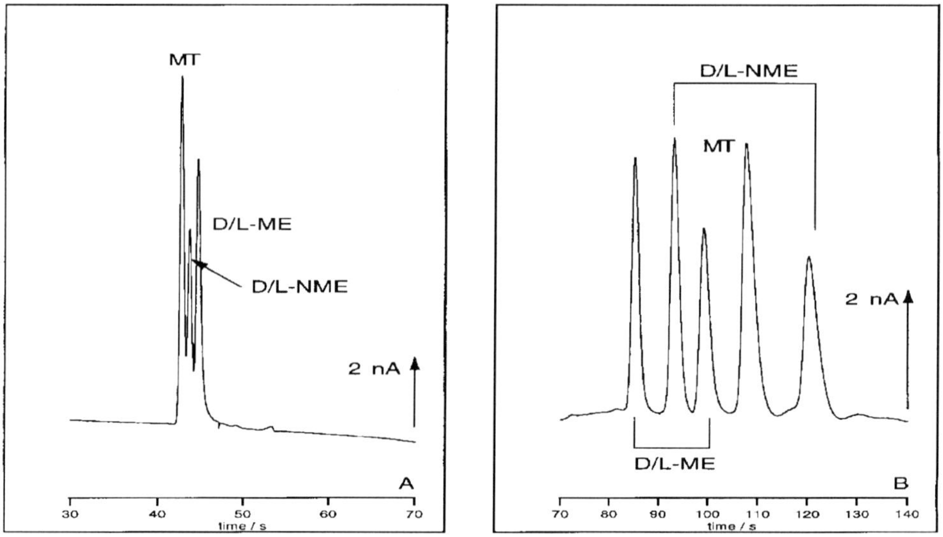

Figure 1 Separation of D/L-metanephrine (ME), D/L-normetanephrine (NME) and methoxytyramine (MT, not chiral) without additive (A) and in the prescence of sulfated cyclodextrin (B). Detection on a Au-electrode.

Acknowledgements

This work was partially funded by the Swiss National Science Foundation under grant number 2100-055403.98.

References

1. M. A. Schwarz, B. Galliker, K. Fluri, T. Kappes, P. C. Hauser, Simplified amperometric detection for a microfabricated electrophoretic separation device, The Analyst, 126, 2001, p. 147-151.
2. S. Fanali, Enantioselective determination by capillary electrophoresis with cyclodextrins as chiral selectors, J. Chrom. A, 875, 2000, p. 89-122.

LIQUID-LIQUID TWO-PHASE CROSSING FLOWS IN GLASS MICROCHIPS BY UTILIZING OCTADECYLSILANE MODIFICATION OF MICROCHANNELS

Akihide Hibara[1], Masaki Nonaka[1], Hideaki Hisamoto[1], Kenji Uchiyama[2], Yoshikuni Kikutani[2], Manabu Tokeshi[2], Takehiko Kitamori[1]
[1]Department of Applied Chemistry, School of Engineering, University of Tokyo, Japan.
[2]Integrated Chemistry Project, Kanagawa Academy of Science and Technology, Japan.

Abstract

A novel liquid-liquid phase separation method in a microchip utilizing surface modification was demonstrated. Microchannels were fabricated on two glass substrates and the bottom plate was modified by octadecylsilane (ODS) group. The ODS modified channel was used for organic flow while microchannel with bare glass surface was used for aqueous flow. Stable liquid-liquid two-phase crossing flows were observed.

Keywords: Multi-phase flow, liquid-liquid interface, surface modification

1. Introduction

In chemistry, biochemistry and chemical engineering, a separation process is doubtlessly one of the most important technologies. By utilizing micro technology, separation methods such as electrophoresis and electrochromatography have been reported by many researchers. Although these batch methods are very effective for analyses, continuous separation methods are preferred for sample split and sequential chemical operations. As an on-flow separation method, we have reported solvent extraction in microchips [1,2]. While multi-phase confluence technique has been investigated well in the previous reports, phase separation in microchips have not been established. In this presentation, we will demonstrate a novel phase separation method utilizing surface modification of microchannels. By using the

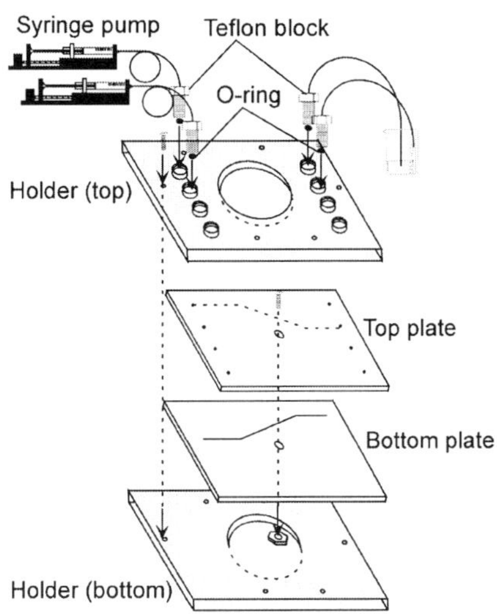

Figure 1. A schematic illustration of a sealing and connection method. The top and bottom plates are pressed by a screw and holders.

J.M. Ramsey and A. van den Berg (eds.), Micro Total Analysis Systems 2001, 549–550.

method, two immiscible phases can interflow and be separated easily. Furthermore, crossing of two straight flows can be realized.

2. Experimental

Microchannels were fabricated on two pyrex glass substrates (top and bottom) by a photolithographic wet-etching method. After the fabrication, the bottom plate was modified by hydrophobic octadecylsilane (ODS) group. To protect the ODS group, the top and bottom plates were not thermally bonded but sealed by using only mechanical pressure (Fig.1). An aluminum holder and a pair of screw and nut were used. The experimental setup around the glass microchip is illustrated in Fig 1. The crossing point of the microchannels was observed by a microscope .

3. Results and discussions

The ODS modified channel (bottom) was used for the organic flow while microchannel with bare glass surface was used for the aqueous flow. A schematic illustraton of microchannels is shown in Fig.2. Each microchannel had 200-μm-width and 50-μm-depth. First, nitrobenzene and water was flowed in the microchannels. Stable crossing of the straight two-phase flows are observed . To confirm the clear phase separation and to analyze hydrodynamic properties of the flow, flow paths of fluorescent particles in aqueous flow were visualized. As shown in Fig.3, the aqueous phase flows as a laminar flow and does not contaminate the organic phase. By controlling the surface energy difference between two microchannels, we can realize stable phase confluence and a clear phase separation. The present technique will be a powerful tool for sequential chemical operations in environmental analyses and microscale combinatorial chemistries in microchips.

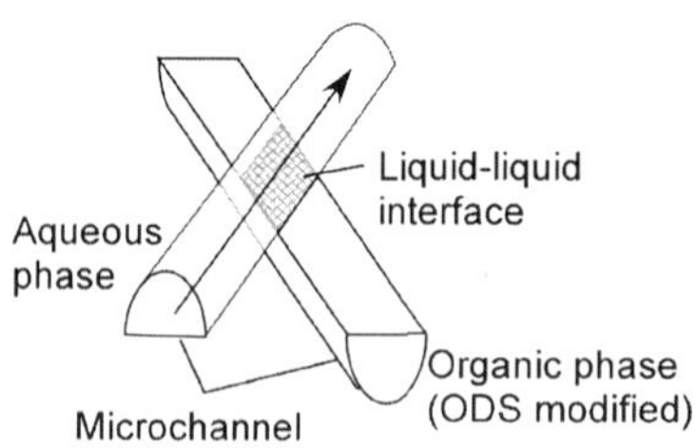

Figure 2. A schematic illustration of two-phase crossing flows of aqueous and organic phases in microchannels.

Figure 3. A Fluorescent image of probe particles dispersed in the aqueous phase.

References

1. M. Tokeshi, T. Minagawa, T. Kitamori, *Integration of a Microextraction System on a Glass Chip: Ion-Pair Solvent Extraction on Fe(II) with 4,7-Diphenyl-1,10-phenanthrolinedisulfonic Acid and Tri-n-octylmethylammonium Chloride*, Anal.Chem. **72,** 2000, p.1711-1714.
2. A. Hibara, M. Tokeshi, K. Uchiyama, H. Hisamoto, T. Kitamori, *Integrated Multilayer Flow System on a Microchip*, Anal. Sci. **17**, 2001, p.89-93.

MULTIPLE SEPARATIONS AT NANOLITER SCALE USING GRADIENT ELUTION

Gérald Jesson and Per Andersson
Gyros AB, Uppsala Science Park, Uppsala, SE 751 83 SWEDEN
Gerald.jesson@gyrosmicro.com

Abstract

A system that runs gradient elution profiles through nanoliter-volume micro-structures within a CD, using a flow-through dispenser [1] and centrifugal force to control liquid flow is demonstrated.

Keywords: dispenser, gradient elution, centrifugal force, CD.

1. Introduction

μ-TAS should enable parallel separation processes, but few have been demonstrated, possibly due to the complexity of the connections and feeding mechanisms required to link the macro and micro-world. Even fewer examples of single gradient elution separations at the nanoliter scale on planar chip substrates have been shown [2]. Here, we present a system for continuous dispensing of gradient elution profiles from the macro-world into the micro-world of a CD containing multiple micro-structures.

2. Experimental

The system configuration is depicted in figure 1.

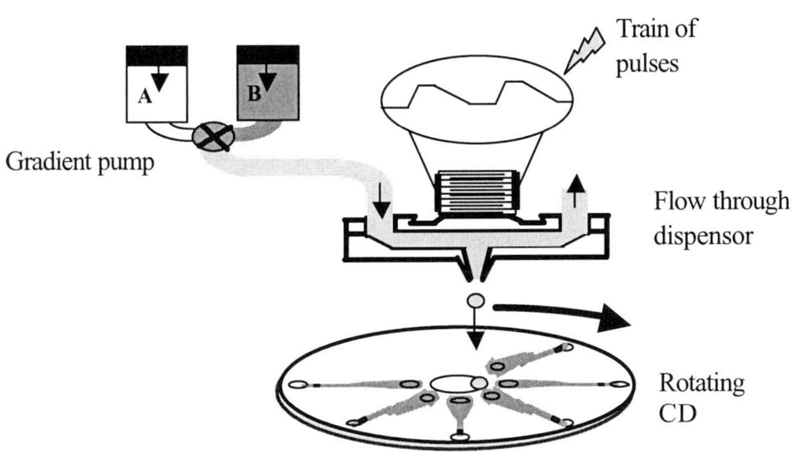

Figure 1. Schematic of the experimental system

J.M. Ramsey and A. van den Berg (eds.), Micro Total Analysis Systems 2001, 551–552.

3. Results and discussion

Each time the CD completes a revolution, a controller triggers a train of pulses to actuate the dispenser. Each droplet is dispensed to an individual CD structure and, by using a triggered train of pulses, several structures are fed simultaneously. The flow entering into the structures is controlled by the angular velocity of the CD and the triggering frequency (e.g. 1 trigger every 2 revolutions). Synchronization between dispensation and the position of the CD inlet structures is important. As shown in figure 2a, the physical properties of the dispensed liquid vary with the gradient profile. This leads to misalignment, but is compensated by adapting the dispensing pulse amplitude as illustrated in figure 2b. Another way to compensate for this is to adjust the pulse triggering.

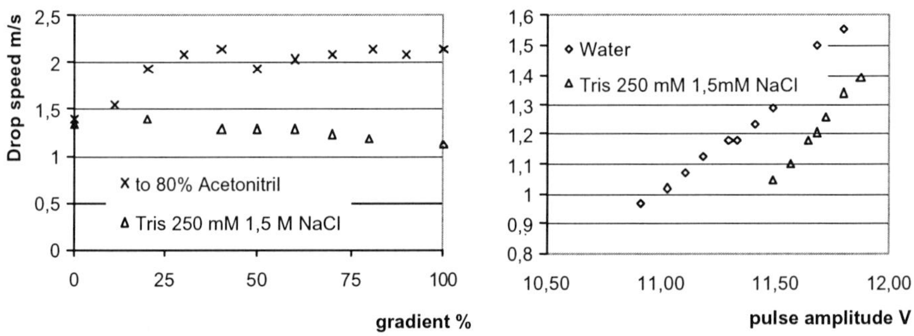

Figure 2a. Variation of the droplets speed in function of the gradient (flow-through)
2b. Variation of the droplets speed in function of the pulse amplitude

4. Conclusions

In this work, we have shown a system that acts as an interface between the macro and micro-world permitting micro-structures to be fed simultaneously with a gradient elution profile. This system has the potential to produce hundreds of simultaneous gradient elution separations in a single run.

Acknowledgements

The authors would like to thanks Jan Sjöberg and Tobias Söderman for their valuable contribution.

References

1. T. Laurell, L. Wallman and J. Nilsson, *Design and development of a silicon microfabricated flow-through cell for on-line picoliter sample handling*, J. Micromech. Microeng. 1999, 9, pp. 369.

2. Jörg P. Kutter, Stephen C. Jacobson, Norio Matsubara, and J. Michael Ramsey, *Solvent Programmed Microchip Open Channel Electrochromatography*, Anal. Chem. 1998, 70, pp. 3291.

DESIGN OF MOVING ELECTRIC FIELD DRIVEN CAPILLARY ELECTROPHORESIS CHIPS

Yu-Cheng Lin[1] and Ching-Yi Wu[2]
[1]1 University Rd., Department of Engineering Science, National Cheng Kung University, Tainan, Taiwan, yuclin@mail.ncku.edu.tw
[2]Electronics Research and Service Organization, ITRI, Taiwan

Abstract

This study reports on detailed designs for low voltage driven CE chips using moving electric fields for DNA separation based upon experimental DNA separation results. A reasonable reduction in the applied voltage is from one-fifth to one-third with respect to the original applied voltage.

Keywords: Moving electric field, Microfluidics, Capillary electrophoresis

1. Introduction

High-applied voltages are needed to generate the necessary high electric field strength for separating biological samples [1,2]. This study reports on detailed designs for low voltage driven CE chips using moving electric fields for DNA separation based upon experimental DNA separation results. Partitioning the separation channel into many smaller separation zones using electrode pairs can lower the applied voltage but maintain electric field strength to separate the samples. A schematic view of the arrayed-electrode design and the mechanism for moving the electric field for driving and separating DNA fragments is shown in Figure 1.

2. Chip design

The design parameters for the separation subdivision, as illustrated in Figure 2, include the electrode width, W_e, the interval between two successive electrodes, W_{sp}, and the span, S, of the separation zone between the powered electrode pair. The total movement distance for these DNA fragments with different migration velocities in one Δt in the *Nth* electrode pair equals the span S.

3. Results and discussion

A finite element simulation for this particular case was conducted to analyze the electric field distribution and uniformity. Figure 3 shows the electric field distribution generated by 50-μm electrodes with a 1000 μm interval inside a 20-μm high channel. The electric field becomes uniform within 2.5% variation of the peak electric field at the region about 35 μm away from the electrode. Therefore, this arrayed-electrode design can generate the necessary uniform electric field to drive and separate DNA fragments.

J.M. Ramsey and A. van den Berg (eds.), Micro Total Analysis Systems 2001, 553–554.

Figure 4 shows how the migration velocities of the DNA fragments affects the applied driven voltage under the assumption that the migration velocities of the DNA fragments are proportional to the applied electric field. Under the experimental results and the calculation, a reasonable reduction in the applied voltage is from one-fifth to one-third with respect to the original applied voltage.

References

1. S.C. Jaconson, R. Hergenroder, L.B. Koutny, R.J. Warmack, and J.M. Ramsey, "Effects of Injection Schemes and Column Geometry on the Performance of Microchip Electrophoresis Devices," Anal. Chem., 66, pp. 1107-1113, 1994.
2. Y.-C. Lin, H.-C. Ho, C.-K. Tseng, and S.-Q. Hou, "A Poly-methylmethacrylate Electrophoresis Microchip with Sample Preconcentrator," Journal of Micromechanics and Microengineering, 11, pp. 189 - 194, 2001.

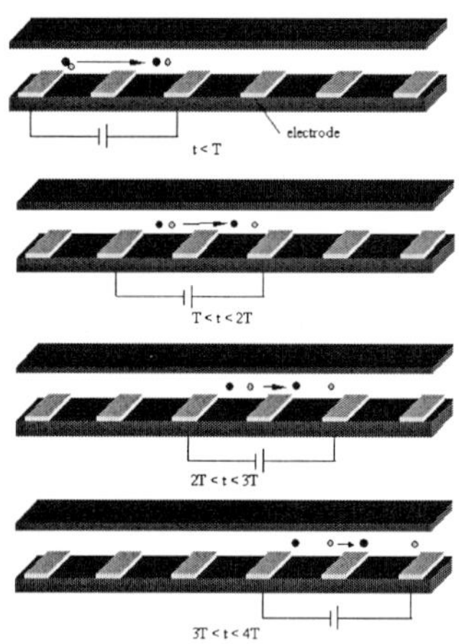

Figure 1: Schematic view of an arrayed-electrode design.

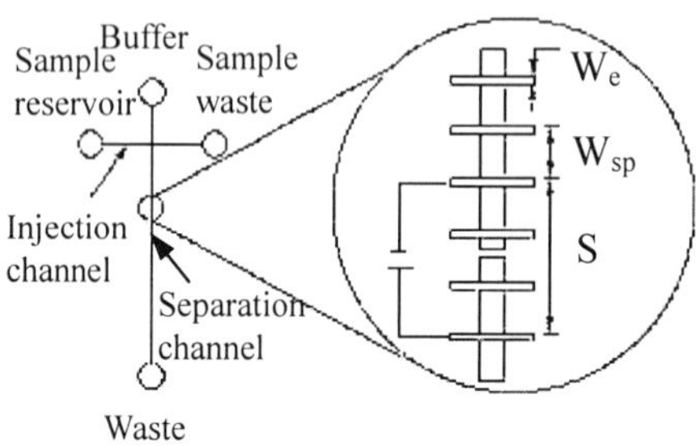

Figure 2: Schematic layout of the subdivision of the separation channel with design parameters.

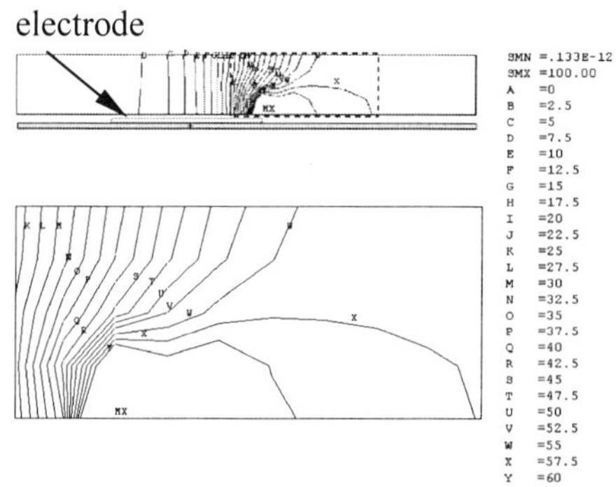

Figure 3: Simulation of the electric field distribution for a cross section of the microchip and close-up of the electric field distribution in the dashed line region.

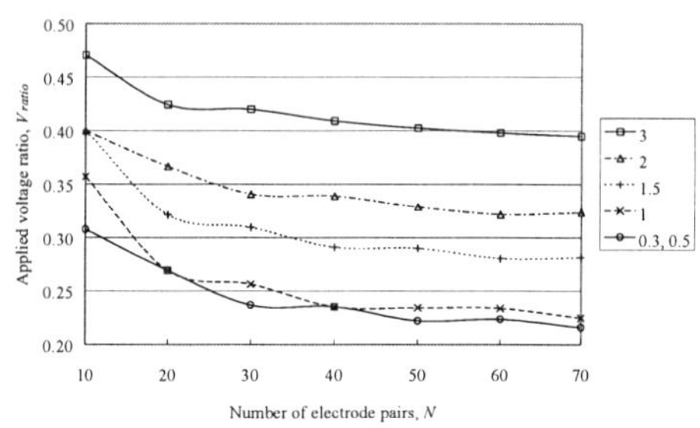

Figure 4: Applied-voltage ratio versus the number of electrode pairs with different migration velocities.

MONOLITHIC SILICA COLUMNS FOR MICRO-HPLC

Norio Ishizuka[1], Hiroyoshi Minakuchi[1], Kazuki Nakanishi[2], Kazuyuki Hirao[2], Masanori Motokawa[3], Hiroshi Kobayashi[3] and Nobuo Tanaka[3]

[1]Kyoto Monotech. Co., 376-5-206 Tsukiyama-cho, Kuze, Minami-ku, Kyoto 601-8203, Japan (E-mail: norio@k-monotech.com)

[2]Department of Material Chemistry, Graduate School of Engineering, Kyoto University, Yoshida, Sakyo-ku, Kyoto 606-8501, Japan

[3]Department of Polymer Science and Engineering, Kyoto Institute of Technology, Matsugasaki, Sakyo-ku, Kyoto 606-8585, Japan

Abstract

Monolithic silica columns having continuous gel skeletons and through-pores were prepared by a sol-gel method in fused-silica capillaries, and evaluated in reversed-phase liquid chromatography. The monolithic silica columns showed performance beyond the limit of conventional particle-packed columns in HPLC.

Keywords: Monolithic silica, sol-gel process, HPLC, capillary, hybrid

1. Introduction

For nearly two decades, columns packed with spherical particles of 5 μm have been most widely used for practical HPLC. The reduction of particle size leads to an increase in the column efficiency in exchange for an increase in the pressure drop where arises an instrumental limitation for the accelerated separation. One of the ways of circumventing this limitation is to fabricate a column made of a single piece of a porous solid consisting of co-continuous small skeletons and large through-pores, which can provide high efficiency and low pressure drop respectively. We previously reported that a much higher column efficiency could be obtained at higher mobile phase velocity by using a monolithic silica column prepared via a sol-gel process rather than a column packed with particles [1, 2]. The preparation of monolithic silica in capillary is advantageous in that small-sized and long monolithic silica can be prepared without clad. Here we report the results describing the chromatographic properties of monolithic silica prepared in fused-silica capillaries.

2. Experimental

Monolithic silica column, MS, was prepared in fused-silica capillary (50 μm ID) according to the method described previously [2, 3]. In the preparation of MS(Hybrid) column, the mixture of tetramethoxysilane and methyltrimethoxysilane in a volume ratio of 3:1 was used as a starting material. Chromatography measurement was carried out at an ambient temperature. Commercial column packed with 5 μm ODS-silica particles, Inertsil ODS 3 was obtained from GL Science.

J.M. Ramsey and A. van den Berg (eds.), Micro Total Analysis Systems 2001, 555–556.

3. Results and discussion

Figure 1 shows that the uniform-sized silica skeletons of about 2 µm and the through-pores of up to about 8 µm were distributed evenly over the cross section of the 50 µmID capillary. The column efficiency of MS per unit length was lower than that of the column packed with 5 µm particles despite the small skeleton size, as shown in Figure 2. It can be attributed to the contribution of slow mass transfer in the mobile phase, which is caused by the presence of large through-pores. MS(Hybrid) showed better performance, due to the smaller through-pores than that of MS. The pressure drop of MS, however, was about 1/25–1/30 that of conventional column. Thus it was possible to generate more than 100,000 theoretical plates at very low pressure.

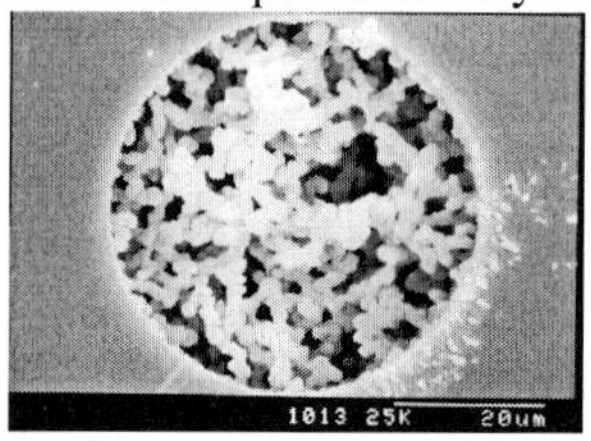

Figure 1. SEM photograph of MS in capillary. Capillary diameter: 50 µm.

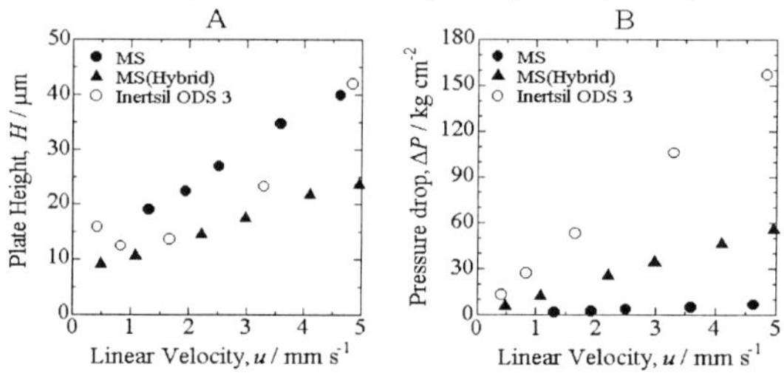

Mobile phase: 80 % methanol
Solute: hexylbenzene
Column size: MS 50 µmID x 130 cm, MS(Hybrid) 50 µmID x 25 cm
Inertsil ODS 3 4.6 mmID x 15 cm
The pressures were normalized to the column length of 15 cm.

Figure 2. A: Van Deemter plots, B: Plots of column back pressure against linear velocity.

4. Conclusions

The preparation of monolithic silica in capillary has made the fabrication process simpler and the preparation of a long column feasible for generating large numbers of theoretical plates with current instrumentation in HPLC.

References

1. K. Nakanishi, *J. Porous. Mater.*, 4 (1997) 67-112.
2. H. Minakuchi, K. Nakanishi, N. Soga, N. Ishizuka, N. Tanaka, *J. Chromatogr. A*, 762 (1997) 135-146.
3. N. Ishizuka, H. Kobayashi, H. Minakuchi, K. Nakanishi, K. Hirao, K. Hosoya, T. Ikegami, N. Tanaka, *J. Chromatogr. A*, submitted.

IN-SITU PREPARATION OF PHOTOPOLYMERIZED SOL-GEL MONOLITHS FOR CAPILLARY ELECTROCHROMATOGRAPHY ON A CHIP

Keisuke Morishima, Maria T. Dulay, Bryson D. Benett, Aaron R. Wheeler, Joselito Qurino and Richard N. Zare
Department of Chemistry, Stanford University, Stanford, CA 94305-5080, USA

Abstract

We have developed a simple and fast procedure for the *in-situ* photopolymerization of a sol-gel monolith having a porous structure on a chip. A solution of methacryloxypropyltrimethoxysilane mixed with an acid catalyst, water, toluene, and a photoinitiator was irradiated at 365nm for 10sec~1min in a microchip to prepare a photopolymerized sol-gel monoliths under an inverted microscope with a mercury lamp.

Keywords: capillary electrochromatography, sol-gel, photopolymerization

1. Introduction

Capillary electrochromatography (CEC) has been regarded as a very promising analytical separation technique that combines the efficiency of capillary zone electrophoresis with the selectivity of liquid chromatography. Although CEC has been applied in many different areas, packed column preparation on a microfluidic chip remain challenges of this technique. One disadvantage of packed columns on a chip is the fabrication of packed particle columns and porous frits of controlled pore sizes, lengths, and high mechanical stabilities. In response to frit fabrication problems, capillary columns controlled by electric field in a chip and monolithic capillaries are being developed as variants of packed capillary columns.[1,2] *In-situ* preparation method of photopolymerized sol-gel (PSG) has many advantages: (1) control of the pore size; (2) control over the placement and length of the PSG segment; (3) high mechanical strength; (4) avoidance of high temperatures that lead to cracking; and (5) rapid photopolymerization at a local area on a microchip. We describe here the *in-situ* one-step preparation method of a PSG having a porous structure on a chip, and the ability of a PSG column in a liquid stream by application of an electric field.[3]

2. Experimental

A solution of methacryloxypropyltrimethoxysilane in the presence of an acid catalyst, water, toluene, and a photoinitiator was irradiated at 365 nm for 10sec~1min to construct a

J.M. Ramsey and A. van den Berg (eds.), Micro Total Analysis Systems 2001, 557–558.

porous monolithic sol-gel column on a chip by a one-step, in-situ process under an inverted microscope with a mercury lamp. Then unreacted solution was removed by rinsing with various solvents inside a microchip. A sol-gel colloidal precursor of the PSG is formed in solution through a series of hydrolysis and condensation reactions. [3] Figure 1 shows SEM image of cross sectional of a PSG photopolymer formed in a microchip and microscopic image of microchannel.

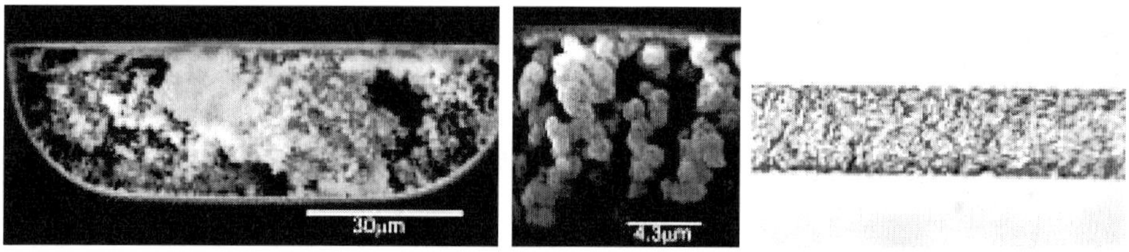

Figure 1. SEM image (left and middle) of cross sectional of a PSG formed with 80%(v/v) toluene in a chip (width 110μm, depth 30μm). Left: whole cross section. Middle: close view of PSG monoliths. Microscopic image (right) of top view of microchannel filled with PSG monoliths.

3. Conclusion

We have developed a simple and fast procedure for the *in-situ* photopolymerization of a sol-gel monolith on a chip. The resulting photopolymer is well-suited for the separation of a variety of neutral compounds in the reversed-phase mode both in voltage-driven systems.[3]

Acknowledgements

This work was supported by Beckman-Coulter, Inc, Stanford Nanofabrication Facility and Sandia National Laboratories.

References

1. K. Morishima, D. W. Arnold, A. R. Wheeler, D. J. Rakestraw, R. N. Zare, "Novel Separation Method on a Chip using Capillary Electrophoresis in Combination with Dielectrophoresis", Proc. Micro Total Analysis Systems, pp.269-272, Enschede, Netherlands, (2000).
2. M. T. Dulay, R. P. Kulkarni, R.N. Zare, "Preparation and Characterization of Monolithic Porous Capillary Columns Loaded with Chromatographic Particles", *Anal. Chem.* **70**, 5103- 5107 (1998).
3. M. T. Dulay, J. Qurino, B. D. Bennett, M. Kato, R.N. Zare, "Photopolymerized Sol-Gel Monoliths for Capillary Electrochromatography", *Anal. Chem.*, in press (2001).

CAPILLARY ELECTROCHROMATOGRAPHY ON MICROCHIPS WITH *IN SITU* FABRICATED PARTICLES

N. A. Penner, B. E. Slentz, and F. Regnier
1393 Brown Bldg, Purdue University, West Lafayette, IN 47907, USA
(fregnier@purdue.edu)

Abstract

Microchips with *in situ* fabricated particles and interconnecting capillaries with varying dimensions were used as CEC columns for peptide separations giving an efficiency of over 600 000 plates/meter.

Keywords: CEC, microchip, peptide separation, COMOSS

1. Introduction

CEC is a hybrid between CE and microcolumn HPLC that combines advantages of both techniques. CEC has the great advantage of circumventing the need for a mechanical pumps and interfacing them to chips. In this work, microfabricated chips with interconnecting capillaries were used as CEC columns. These columns have solved the problems of packing and frits by fabricating particles and channels *in situ*. These collocated monolithic support structures (COMOSS) have high level of channel homogeneity, frequent trans-channel coupling, and total control of channel dimensions and geometry [1].

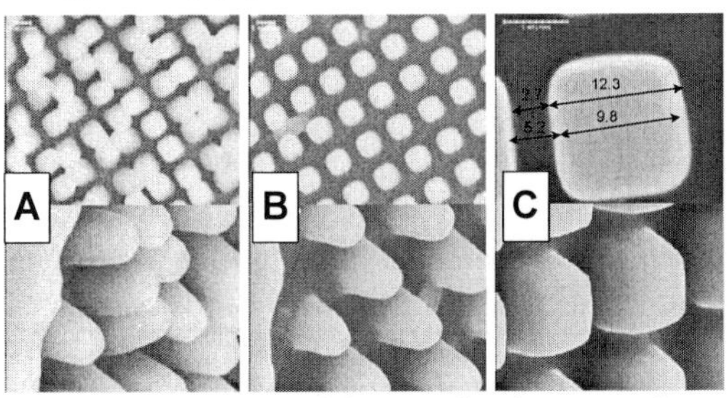

Fig.1. SEM of microfabricated columns with 10μm deep channels. Theoretical dimensions: monoliths: A&B – 5×5μm, C – 10×10μm; channels: A – 3 μm; B&C – 5μm.

2. Results and discussion

The column was designed using the same architectural specifications previously described for COMOSS systems [2]. Microchips were prepared by replicating positive photoresist masters in PDMS. The SEM images of columns are presented in Fig.1.

The surface of PDMS was modified after exposure to an oxygen plasma to get reproducible electroosmotic flow by the following methods:

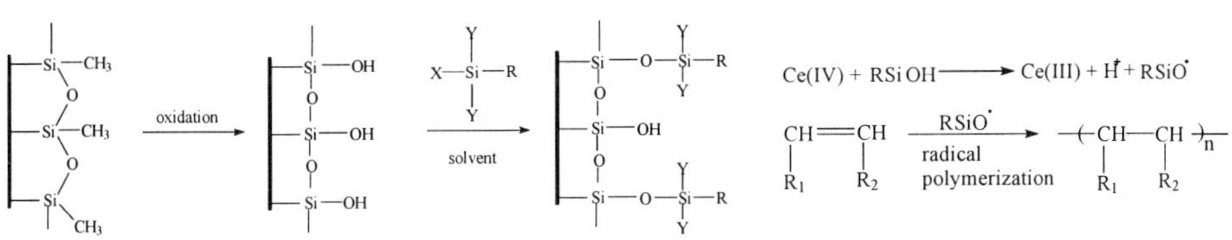

Vinylsulfonic acid, acrylic acid, 2-acrylamido-2-methylpropylsulfonic acid (AMPS), 4-styrenesulfonic acid, and stearyl methacrylate was polymerized with Ce(IV) as a catalyst that resulted in a coating that is stable for a long period of time.

J.M. Ramsey and A. van den Berg (eds.), Micro Total Analysis Systems 2001, 559–560.

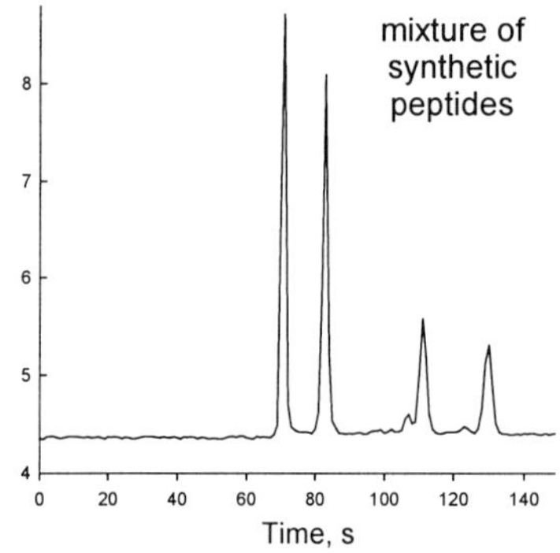

We also created a C_{18}-AMPS stationary phase where AMPS served for production of stable EOF and C_{18} served for selectivity of separation by a two step synthesis (1–Ce(IV)-grafting AMPS, 2–modification with octadecylsilane). The resulting stationary phase gave the most selective and efficient separation of both simple and complex mixtures of peptides (Fig.2&3).

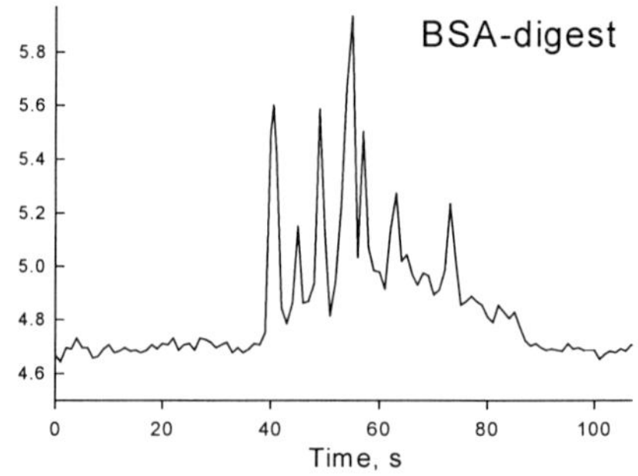

Fig.2&3 Separation of FITC labeled peptides on C_{18}-AMPS; 1mM carbonate buffer (pH 9.0), 1000V.

The reproducible modification of channels allowed us to compare behavior and characteristics of separation columns with different dimensions. The efficiency of separation strongly depends on geometry of resulting channels (Fig.4). The decrease of size of monolithic structures from 10μm (C) to 5μm (B and A) leads to an increase of efficiency of chromatographic peaks. The large drop in efficiency for column C at voltage > 3000V is the result of relatively low trans-channel coupling in comparison with A and B. It is clear from Fig.1 that decreasing channel widths from 5 to 3μm (B to A) results in collapsing monolith structures and therefore in nonhomogeneity of column packing, thus, resulting in reduction of efficiency of separation. Monolithic dimensions of 5×5μm, with channels 5(width)×10(depth)μm was found to give highest efficiency of a peptide separation (620 000 plates/meter).

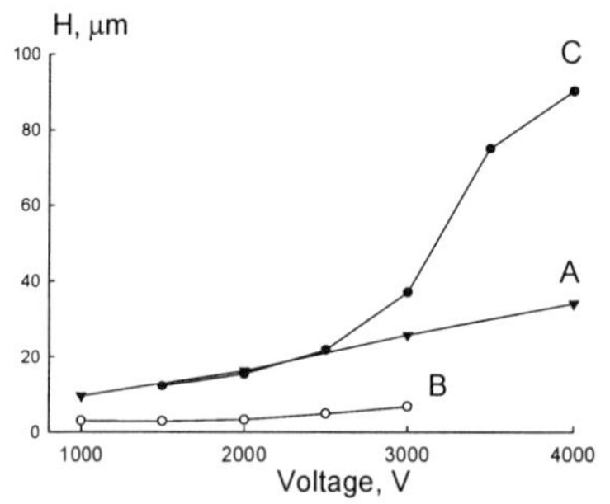

Fig.4 Plot of plate height (H) vs. voltage for microfabricated columns (C_{18}-AMPS modified) of varying dimensions.

4. Conclusions

Microfabricated COMOSS PDMS chips with a modified surface give a reproducible separation of complex peptide mixtures for a long period of time with an efficiency up to 620 000 plates/meter and RSD for retention times below 2.6%. The efficiency of separation strongly depends on geometry of resulting channels.

References

1. B.He, J.Ji, F.Regnier, J.Chromatogr.A. 853 (1999) 257.
2. B.He, N.Tait, F.Regnier, Anal.Chem. 70 (1998) 3790.

PDMS MICROCHIP FOR PRECOLUMN REACTION AND MICELLAR ELECTROKINETIC CHROMATOGRAPHY OF BIOGENIC AMINES

Kyung Won Ro, Kwanseop Lim, and Jong Hoon Hahn*
Department of Chemistry, Pohang University of Science and Technology,
San 31 Hyoja-Dong, Pohang, 790-784, South Korea
(email: hahn@postech.ac.kr)

Abstract

We have developed a microchip to perform fast and efficient precolumn reactions and micellar electrokinetic chromatography (MEKC) of biogenic amines sequentially in poly(dimethylsiloxane) (PDMS). The relative mixing and reaction efficiencies of three different structures of precolumn reactors are investigated. The separation efficiency in a PDMS chip is comparable to that in a glass chip.

Keywords: precolumn reaction, biogenic amines, PDMS microchip

1. Introduction

Fabrication of microfluidic devices in PDMS by replica molding method provides faster, less expensive routes than photolithography and etching methods in glass and silicone [1]. We have fabricated PDMS microchips for the fluorescence derivatization and the separation by MEKC of biogenic amines in successive, and have investigated the efficiencies of precolumn reaction and separation in PDMS chips.

2. Microchip Fabrication

The channel manifold in a PDMS substrate is fabricated using replica molding against a patterned, thick photoresist, SU-8, generated by photolithographic techniques. Three different structures of precolumn reactors of the same volume (4.5 nL) are designed for the comparison of relative mixing and reaction efficiencies. The schematic of the channel structures of precolumn reaction and separation microchips (PRSMs) is shown in figure 1. The dimension of separation channel is 38 mm long, 40 ì m wide and 16 ì m deep. The channels are covered by another PDMS plate and the two PDMS plates are bonded strongly through surface oxidation by corona discharge generated from a Tesla coil.

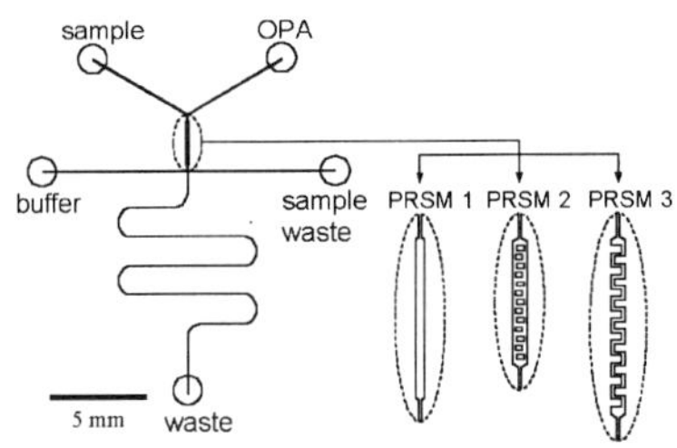

Figure 1. Layout of three precolumn reaction and separation microchips (PRSMs) fabricated in PDMS. The insets show details of the construction of precolumn reactors.

J.M. Ramsey and A. van den Berg (eds.), Micro Total Analysis Systems 2001, 561–562.

3. Results and Discussion

The mixing and reaction efficiencies of three precolumn reactors are investigated by the mixing test of fluorescein and buffer as shown in figure 2. The most efficient mixing is shown in the reaction chamber of PRSM3 and it is mainly ascribed to its structure with a long, narrow channel that facilitates diffusive mixing.

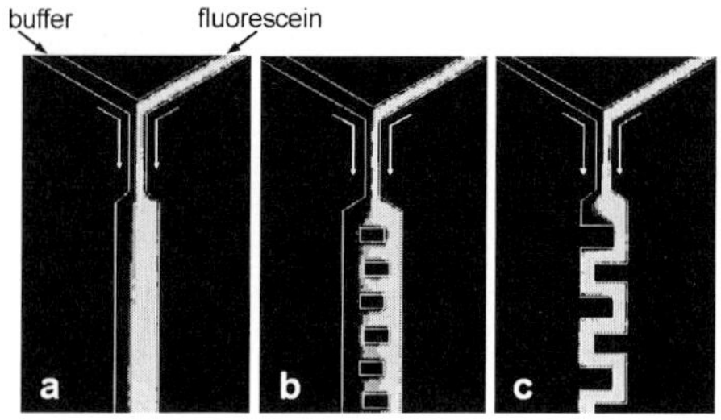

Figure 2. CCD images of fluorescein (white) being mixed with buffer (black) in the reaction chambers of (a) PRSM 1, (b) PRSM 2 and (c) PRSM 3.

The microchips are operated under a continuous reaction mode with gated injections to introduce the reaction product into the separation channel [2]. The biogenic amines, histamine (His), tyramine (Tyr), putrescine (Put) and tryptamine (Trp), are derivatized with o-phthaldialdehyde (OPA) in the reaction chamber, separated by MEKC, and detected by laser-induced fluorescence with He-Cd laser. The optimized separation is carried out by using 25 mM sodium borate buffer (pH 10), 25 mM SDS, 5% (v/v) methanol.

The separation efficiencies in PDMS and glass microchips are compared in the same MEKC running buffer as shown in figure 3, and they have not great difference and are in the range of 17,000 ~ 21,000 theoretical plate numbers. Excellent reproducibility in the migration times (% RSD = 0.2 ~ 0.6) and the relative peak areas (% RSD = 0.7 ~ 1.6) of each amine are realized in a PDMS microchip. The detection limits (S/N=3) of the amines range from 0.05 ppm for His to 0.5 ppm for Trp.

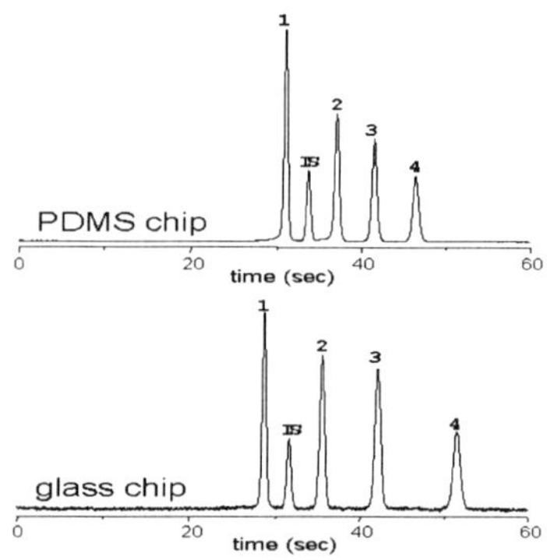

Figure 3. MEKC separations of OPA-biogenic amines in microchips (device: PRSM 3). Peaks: 1, His; 2, Tyr; 3, Put; 4, Trp; IS, internal standard (DNS-OH).

4. Conclusions

In this report, the integration of precolumn reactor with PDMS microchip CE devices has been accomplished. PDMS is an attractive alternative to glass or quartz for the fabrication of microfluidic systems using aqueous media. Integrated microchip-CE systems appear to have a good potential for performing fast analysis of biological, chemical and environmental samples.

References

1. D. C. Duffy, J. C. McDonald, O. J. A. Schueller, G. M. Whitesides, "Rapid Prototyping of Microfluidic Systems in Poly(dimethylsiloxane)", *Anal. Chem.*, **70**, (1998), 4974-4984.
2. S. C. Jacobson, R. Hergenröder, A. W. Moore, Jr., J. M. Ramsey, "Precolumn Reactions with Electrophoretic Analysis Integrated on a Microchip", *Anal. Chem.*, **66**, (1994), 4127-4132.

A MICROFABRICATED THERMO- ELECTRICAL FIELD FLOW FRACTIONATION SYSTEM

Himanshu J Sant and Bruce K Gale*,
Institute for Micromanufacturing, Biomedical Engineering,
Louisiana Tech University, Ruston, LA 71270,USA.
*bkgale@coes.latech.edu

Abstract:
This work introduces a microfabricated Thermo-Electrical Field Flow Fractionation System (μTEFFF) with potential use as an integrated sample preparation system in a μ–TAS. The paper addresses design, modeling, and fabrication of the device.

Keywords: Sample Preparation, Field Flow Fractionation, Separation

1. Introduction

Reports [1], [2] on various miniaturized field flow fractionation (FFF) subtypes have shown their potential as a sample preparation system in the μ-TAS. This paper introduces another microfabricated FFF subtype, a combination of thermal and electrical FFF, which was first described on macroscale in 1991 [3]. This device finds application in preliminary separations prior to sensing or analysis step in stand-alone or μ-TAS based biological assays and protein adsorption studies.

2. Theory

In μ-TEFFF, combined thermal and electrical fields in the direction perpendicular to the flow of sample are used to induce the separation. This transverse field forces particles towards the accumulation wall (Figure 1) and the parabolic velocity profile in the channel causes particles to elute at different times depending upon their protrusion into the channel. Combining the two fields in a μ-TEFFF device enhances selectivity as separation can be based on two criteria, electrophoretic mobility [1] and thermal diffusion [2].

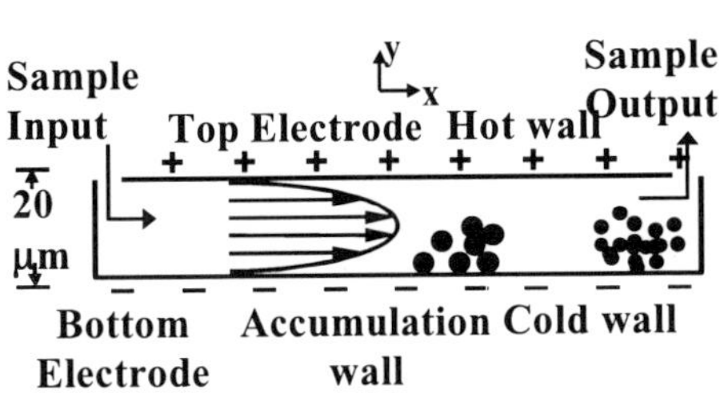

Figure 1 TEFFF system showing input and output ports, application of electric field, thermal field, parabolic flow profile, and relative channel dimensions.

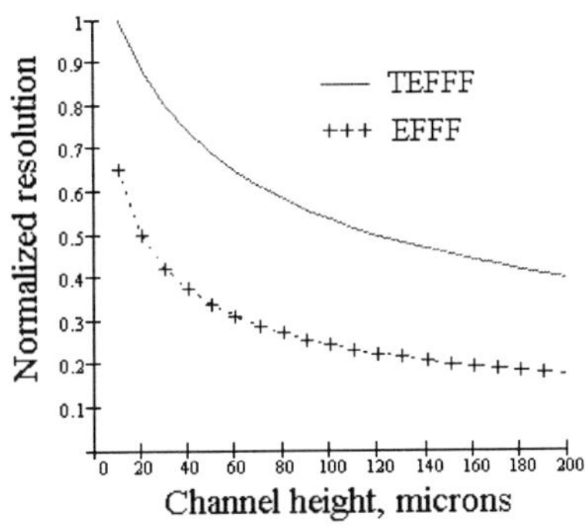

Figure 2 Comparison of normalized resolution for TEFFF and EFFF systems.

J.M. Ramsey and A. van den Berg (eds.), Micro Total Analysis Systems 2001, 563–564.

3. Modeling

Scaling models for TEFFF were prepared to study the effect of miniaturization on performance of the device and to compare this instrument with other FFF systems. The modeling results show improvement in resolution (and other performance measurement criteria) for TEFFF system with miniaturization and it also demonstrates better resolution when compared with electrical-FFF. (Figure 2) Heat transfer calculations were also carried out to decide the operating conditions of the device.

4. Fabrication and Results

The fabrication of the μ-TEFFF system (Figure 3) begins by etching the input and output ports of the channel using ICP, followed by sputter deposition of titanium heater. After spin coating a thin SU-8 insulation layer, titanium and gold electrodes (for electrical field) are deposited. SU-8 serves as the channel walls for the system. A glass slide with titanium and gold electrodes is bonded to the silicon wafer to enclose the channels. After fabrication of μ-TEFFF, integration with various microfluidic components and data collection system was done to complete the experimental setup.

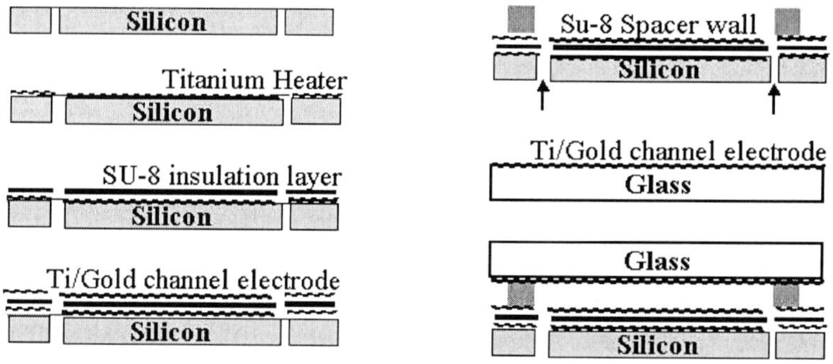

Figure 3 Fabrication flow chart for μ-TEFFF system

The initial characterization of the system has been done using standard polystyrene samples. Experiments were carried out to measure the improvement in the performance characteristics such as resolution and analysis time with encouraging results.

Acknowledgments

Funding for the project was provided by NSF, the United Engineering Foundation, and the Institute for Micromanufacturing at Louisiana Tech University.

References

1. B.K. Gale, K.D. Caldwell, and A.B. Frazier, "A Micromachined Electrical Field-Flow Fractionation System," *IEEE Tran. Biomedical Engineering*, 45, pp. 1459-1469, 1998.
2. T. L. Edwards, B. K. Gale, and A. B. Frazier, "A Micromachined Thermal Field-Flow Fractionation System," in *Proc. of Transducers '99*, 1999, pp. 742-745.
3. G. Liu and J.C. Giddings, "Separation of Particles in Nonaqueous Suspensions by Thermal-Electrical Field-Flow Fractionation," *Anal. Chem.*, 63, pp. 296-299, 1991.

DIELECTROPHORETIC SEPARATION AND TRANSPORTATION OF CELLS AND BIOPARTICLES ON MICROFABRIACTED CHIPS

Junquan Xu[1], Lei Wu[1], Mingxian Huang[1], Weiping Yang[1], Jing Cheng[1,2] and Xiao-Bo Wang[1]
[1]AVIVA Biosciences Corporation, San Diego, CA 92121
[2] Department of Biological Sciences and Biotechnology, Tsinghua University, Beijing, 100084, China

Abstract
Dielectrophoresis (DEP) is the motion of polarized particles in response to non-uniform AC electric fields. This paper describes the application of DEP for separating and transporting cells and bioparticles on microfabricated electrode arrays.

Keywords: dielectrophoresis, cell separation, sample preparation

1. Introduction
Dielectrophoresis has been explored in microchip configurations for a number of cell separation and manipulation applications [1, 2]. Here we report the use of new DEP electrode arrays for cell separation and transportation. Development of such DEP devices is significant in achieving microchip based, integrated bioanalytical systems.

2. Experimental
Thin film microelectrodes of different geometries were fabricated on glass substrates using standard photolithography. A fluidic chamber with a thickness of 500 μm and volume 100 μL was constructed by gluing a molded, plastic housing onto the glass chip. Teflon tubing was glued onto the holes on the two ends of the plastic housing to allow the sample introduction and removal. Human peripheral blood was diluted by 20 to 100 fold in an isotonic (250 mM sucrose) or hypotonic (60 mM sucrose) buffer with final electrical conductivity between 10 and 50 mS/m. For cell separation and transportation, voltage signals of 2 MHz and 300 kHz were applied, respectively.

3. Results and Discussion
For blood cell separation, diluted blood samples were introduced into the fluidic chamber comprising a sinusoidal electrode array. Various electrical field conditions were tested and it was found that at 2 – 10 MHz, DEP forces generated by the microelectrodes were able to capture and retain white blood cells on the electrode edges whilst red blood cells did not experience large attractive forces so they were carried away from the chamber by the flow forces (Fig. 1). Differential DEP responses exhibited by white and red blood cells resulted from the difference in their dielectric properties, which was further increased by suspending the cells in a hypotonic medium.

J.M. Ramsey and A. van den Berg (eds.), Micro Total Analysis Systems 2001, 565–566.

At a flow rate of 5 μl/min, the chip was able to capture several thousands white blood cells on a chip.

For molecule separation, microbeads of 1 – 20 μm with surface-immobilized, specific biomolecules were used for binding the molecules of interest from a solution. The microbeads were caused to subject to positive or negative DEP forces, depending on their dielectric properties (Fig. 2). Separation of DNA and protein molecules was achieved with conductive carbon beads and insulating polystyrene beads.

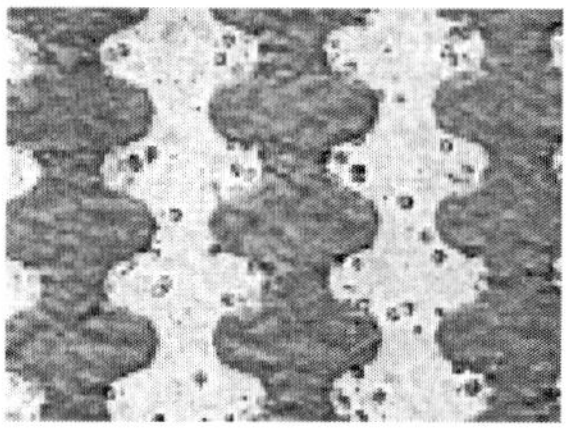

Figure 1. White blood cells were retained by sinusoidal electrodes when a diluted blood sample was flown over the chip.

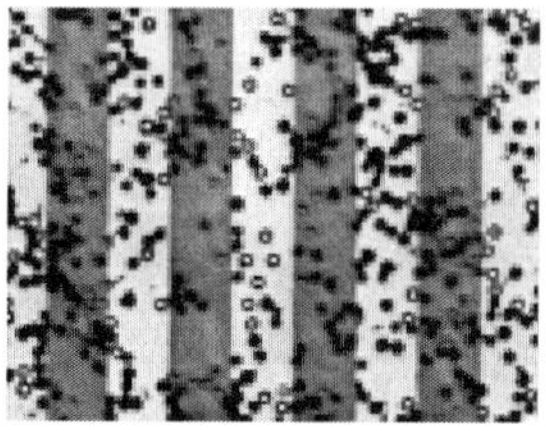

Figure 2.A A mixture of carbon beads and polystyrene beads distributed over a parallel electrode array.

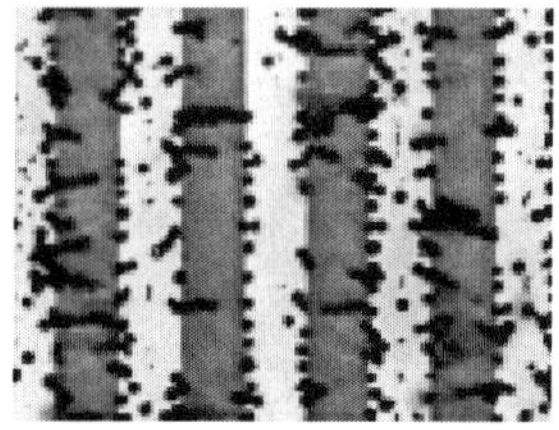

Figure 2.B Application of AC voltages caused carbon beads being collected along electrode edges and polystyrene beads being levitated.

For cell transportation, a DEP chip was developed comprising three interconnected, and electrically-independent branches, which were capable of producing traveling electric fields. A fluidic chamber was constructed with the DEP chip and a guide-electrode chip facing each other (Fig. 3). Blood cells were introduced into the chamber and were transported and switched along the three branches (Fig. 4). Cell kinetic responses depended on the amplitude and frequency of the applied signals.

Figure 3. Transportation chip (bottom) and the guide electrode chip (top) separated by 80 μm.

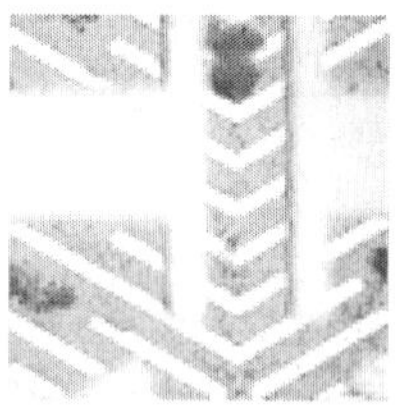

Figure 4.A Transportation of blood cells from top channel on the chip

Figure 4.B The blood cells were switched at the three-way switching location.

Figure 4.C Transportation of blood cells to the left channel on the chip.

Conclusions

Dielectrophoresis provides an effective tool for on-chip manipulating cells and bioparticles.

Reference

1. Y. Huang, K. L. Ewalt, M. Tirado, et al *Electric manipulation of bioparticles and marcromolecules on microfabricated electrodes*, Anal. Chem. 70, pp. 1549-1559, (2001).
2. X-B. Wang, J. Yang, Y. Huang, J. Vykoukal, F.F. Becker and P.R.C. Gascoyne, *Cell separation by DEP field-flow-fractionation*, Anal. Chem., 72, pp. 832-839, (2000).

A MULTI-LAYERED CAPILLARY CHIP FOR HIGH-THROUGHPUT ANALYSIS

Hironobu Yamakawa,[1] Yasuhiko Sasaki,[1] Akira Koide,[1] and Ryo Miyake[1]
Mechanical Engineering Research Laboratory, Hitachi, Ltd.
(e-mail address: yamakawa@merl.hitachi.co.jp)

Abstract

We fabricated a multi-layered chip consisting of four capillary chips that are optically optimised and excited with one beam. Sample fluorescent light from each capillary was detected cleanly, so the multi-layered chip is suitable for high-throughput analysis of plural samples.

Keywords: Micro-chip capillary, thick resist, optical devices, layered chip

1. Introduction

An analysis system with higher throughput (i.e. improved sequencing performance) is required to meet the demands brought by the expanding applications of genetic engineering. Accordingly, we have developed a multi-layered capillary chip with less complicated surrounding devices and lower background noise for high-resolution analysis.

2. Theory

Fig.1 shows a conceptual image of the developed detection system, which has a with multi-layered capillary chip for electrophoretic analysis. The excitation beam goes through the plural capillaries from the bottom of the chip to the top. Fluorescence from sampleflow in the capillaries makes a right angle with excitation beam so the background noise is reduced. Fluorescence from the sampleflow is detected from the side of the chip. Note that this detection system has no complex devices such as optical filters for cutting out the excitation beam.

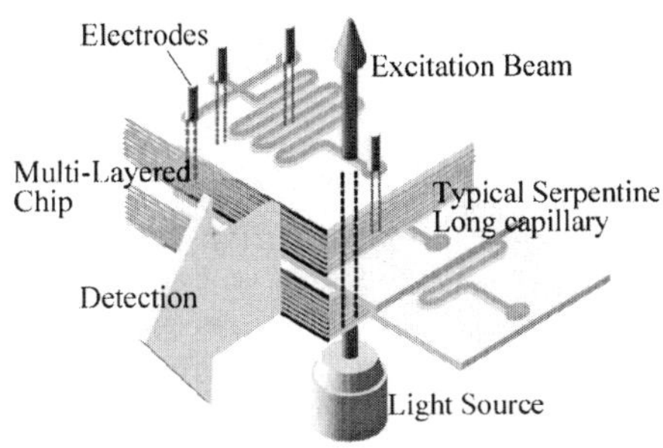

Fig.1: Concept of multi-layered capillary chip and its detection method.

The manufactured test system has the following characteristics. To avoid optical scatter, the cross-section of the capillary has a rectangular shape and the surface of the capillary wall is smooth. Generally, glass channels, which are widely used as capillaries for

J.M. Ramsey and A. van den Berg (eds.), Micro Total Analysis Systems 2001, 567–568.

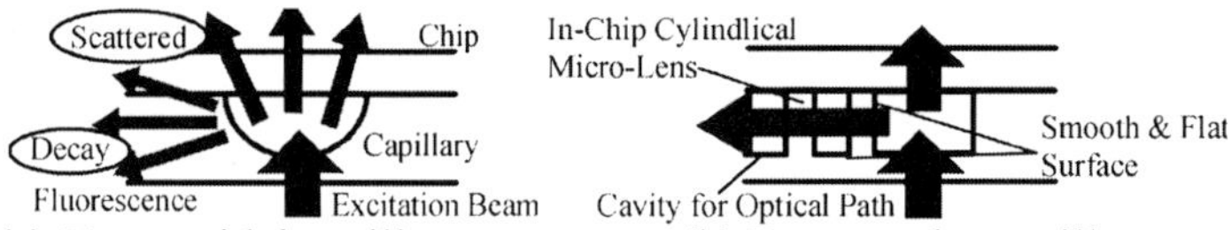

(a) Trapezoidal capillary (b) Rectangular capillary

Fig.2: Optical property necessary for multi-layered capillary chip.

micro-chips, have a trapezoid cross-section and the surface of their walls is rough, which often generates scattered light as shown in Fig.2(a). The rectangular cross-section improves optical performance as shown in Fig.2 (b). It is well known that a forming die using a thick resist (SU-8) has a vertical sidewall. We thus manufactured the chip by moulding elastomer material (PDMS) from such a die [1]. In addition, the sidewall of the capillary must be smooth. We adjusted the fabrication condition (heat and exposure time) for resist SU-8 so that the forming die had a smooth sidewall. Moreover cavities for better optical alignment the optical path and micro-lens in the path were fabricated.

3. Experimental and discussions

We evaluated the optical performance of the multi-layered capillary. Fig.3 shows the manufactured multi-layered chip with a detecting section. Four capillaries and optical lenses are layered so an excitation laser beam can go through them. The laser beam excites the fluorescent dye in the four capillaries and the fluorescence is recorded by CCD camera. Fig.4 shows a fluorescence image from a sample liquid driven by electrophoresis. This image shows that the detection system with a multi-layered capillary chip is one of the best solutions for simultaneous detection of plural samples.

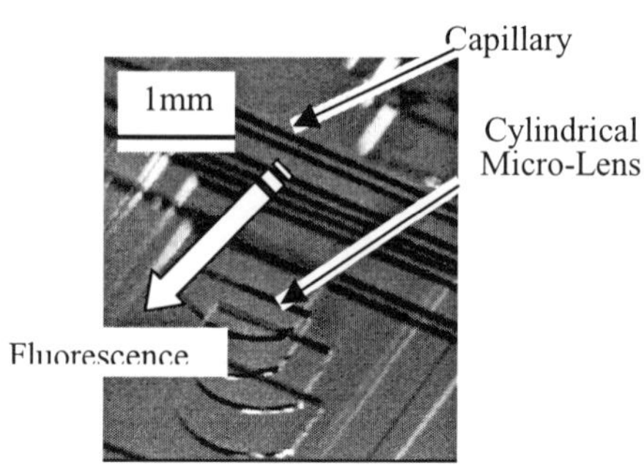

Fig.3: Photograph of a multi-layered chip.

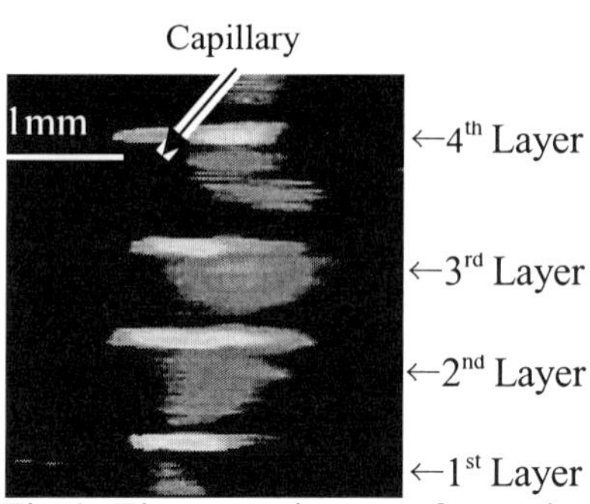

Fig.4: Fluoresce image of sample-flow in observation area of multi-layered capillary chip.

References

1. H. Jo, L. M. V. Lerbeerghe, K. M. Motsegood, D. V. Beebe, Three-Dimensional Micro-Capillary Fabrication in Polydimethylsiloxane (PDMS) Elastomer, J. of Microelectromechanical Systems, Vol. 9, No. 1, Mar. 2000, pp.76-81

A SIGNAL PROCESS METHOD FOR DETECTION OF MICRO-CHANNEL ELECTROPHORESIS

Ji-Feng Chen, Qing-Hui Jin, Jian-Long Zhao, Yuan-Sen Xu
Shanghai Institute of Metallurgy, Chinese Academy of Sciences
865 Changning Road, Shanghai 200050, PR China

Abstract:

This paper presents a signal process method for detection of micro-channel electrophoresis. It was developed and optimized using a laser induced fluorescence (LIF) based detection system. In this detection system, signal was sampled and processed through a novel signal process module. We have used this system to DNA segments separation. The result shows that this signal process method can provide good signal-to-noise ratios.

Keywords: micro-channel electrophoresis, DNA segments, LIF

1. Introduction

Recently, two groups have been using different strategies to complete the sequencing of the human genome. Recent developments have shown that speed and throughput are the main problems in DNA sequencing. Of all the alternative technologies, electrophoresis in multiplexed capillary tubes or micro-channel appears to be most ready for real applications in high-speed high-throughput production environments. In the multiplexed capillary tubes or micro-channel electrophoresis, laser induced fluorescence (LIF) detection has been the most universal method.

2. Experimental

Figure 1 shows the optical diagram of the detection module. In this module, the fluorescence emitted from detection sample was collected by the objective lens and passed through the dichroic beam filter to the process part. The analog signal from the photo-multiplier was processed by the electric signal processing part, which contains low pass filters, amplifiers and other circuit parts. Then it was digitized by A/D sampling card to the computer. The experimental results showed that noises were effectively decreased and signal was greatly amplified when signal passed these processing part. The signal-to-noise ratio was also significantly increased.

3. Results and discussion

In our experiments, the samples are 3 DNA segments which lengths are 12, 23, 117 mer with concentrations of 1.4×10^{-6}M, 6.0×10^{-7}M, 8.0×10^{-7}M.

Figure2 shows the comparison of the PMT output and the processed signal. The signal obtained from the PMT is a small analog voltage signal. The highest voltage of the three peaks is about 50mV. When this signal is processed, the highest voltage is about 5V. The magnitude of amplified signal is 100 times larger than the PMT output.

J.M. Ramsey and A. van den Berg (eds.), Micro Total Analysis Systems 2001, 569–570.

The signal-to-noise ratio (SNR) of the lowest peak in PMT output is about 2~3. The SNR of the lowest peak in amplified signal is 6~8. The ripple of the unprocessed signal is about one tenth of the lowest peak signal. After the filter, the ripple is reduced to about one hundredth of the lowest peak signal. The signal process module greatly amplified the signal and effectively increased the SNR.

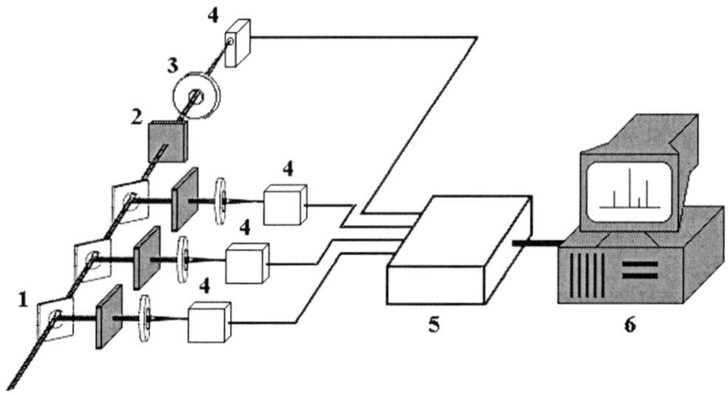

Figure 1. Schematic diagram of the detection module

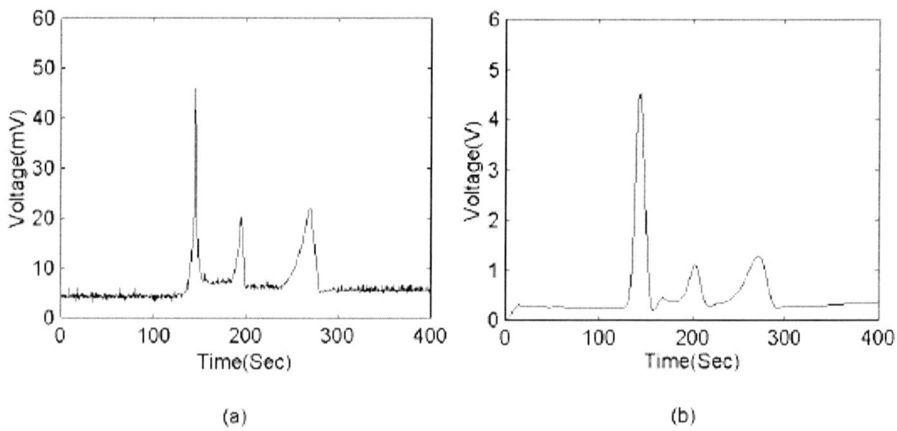

Figure 2. Comparison of the PMT output and the processed signal

4. Conclusions

In this paper, a significant progress has been described, based on the development of the signal process. The signal process method was very effective to amplify and filter the signal and increase the signal-to-noise ratio.

5. Acknowledgments

This research was supported by the Special Funds for Major State Basic Research of China (G1999033107) and Shanghai Science and Technology Development Fund (99QD14030). The authors thank the group of DNA chip in Shanghai Institute of Metallurgy, Chinese Academy of Sciences, for assistance in the experiments.

DIFFUSION IMMUNOASSAY IN POLYACRYLAMIDE HYDROGELS

Anson Hatch and Paul Yager
Department of Bioengineering, Box 352141,
University of Washington, Seattle, WA 98195, USA

Abstract

A modified format for conducting a diffusion immunoassay is described. Polyacrylamide hydrogels containing antibody are used to conduct the assay under non-flow conditions. Fluorescein-labeled biotin and biotin-specific antibody were used as a model system. This assay format shrinks the device area required, reduces the required reagent volumes, and greatly simplifies fluid handling.

Keywords: Immunoassay, DIA

1. Introduction

Microfluidic immunoassay systems such as the diffusion immunoassay (DIA) or capillary electrophoresis could become the basis of portable instruments for rapidly measuring biological analytes and therapeutic drugs [1,2]. Previous DIA work relied on laminar flow in a T-Sensor to appose two solutions. Here, static (non-flow) conditions are described for implementing a DIA (fig. 1). Static conditions reduce the required reagent volumes, shrink the area required, and greatly simplify fluid handling.

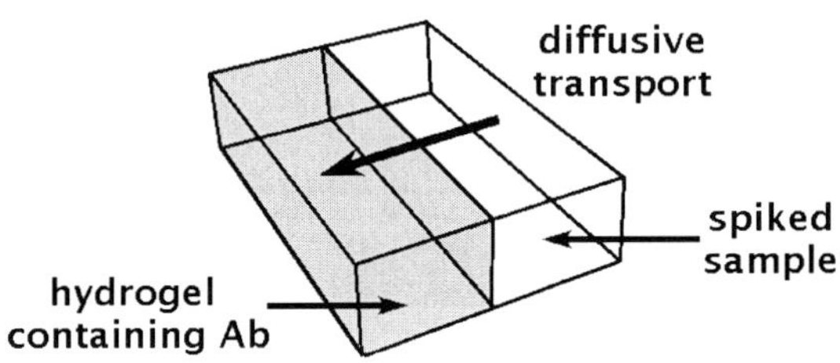

Figure 1. Diffusion immunoassay using a hydrogel. Antibody binding dramatically reduces diffusion of labeled analyte into the hydrogel, which is measured in a competitive assay to determine the concentration of sample.

2. Theory

Immobilizing antibody in a hydrogel allows sample loading by simple means such as capillary action or spotting. A sample containing the analyte of interest is spiked with labeled analyte. The spiked sample is placed in contact with the hydrogel that contains analyte-specific antibody. Labeled and unlabeled analyte are then allowed to diffuse for a short time (seconds to minutes) into the hydrogel and compete for available antibody binding sites. The concentration of analyte influences diffusion of the label as binding

J.M. Ramsey and A. van den Berg (eds.), Micro Total Analysis Systems 2001, 571–572.

to antibody slows diffusion of the label through the gel. Analyte concentrations can be determined by measurement of the intensity of label across the hydrogel.

3. Experimental

Biotin-specific polyclonal antibody was immobilized in acrylamide hydrogels during polymerization. Hydrogels were generated with either photoinitiated polymerization or lithography techniques or standard reaction chemistry initiated by ammonium persulfate. Gels were cast between two glass coverslips separated by a 100 μm spacer layer. A 7.5% gel concentration with a 37.5:1 ratio of acrylamide monomer:crosslinker was used. Fluorescein-biotin concentrations were 20 nM and biotin specific antibody concentrations were approximately 500 nM.

4. Results and discussion

The analyte accumulated at the edge of hydrogels containing antibody (Fig. 2A). The analyte diffused freely into hydrogels without antibody. A hydrogel area of 500 μm by 500 μm was more than sufficient to conduct the assay. The required device area was reduced by a factor of 50 compared to flow conditions tested previously. Model simulations of experimental conditions predict the observed behavior (Fig. 2B).

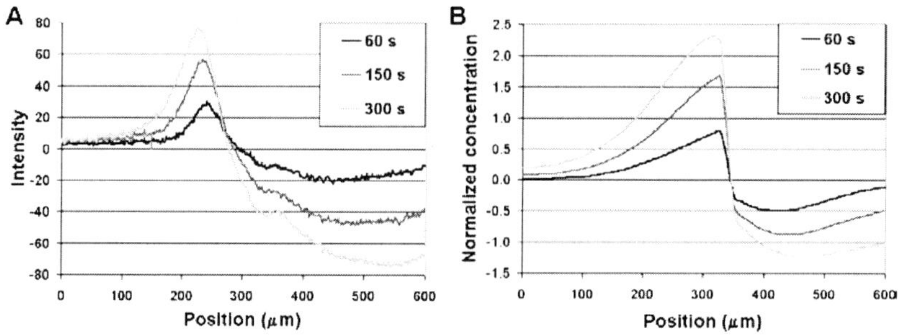

Figure 2. Changes in concentration of analyte (initially on the right) over time. A) experimental B) model results. Accumulation of sample is observed at the hydrogel edge while it is depleted in the sample.

5. Conclusions

It is possible to conduct a diffusion immunoassay under non-flow conditions using acrylamide hydrogels. Great savings in device size and simplification of fluid handling was realized. This assay format may be useful for developing portable instrumentation to rapidly detect analytes below the nM level.

References

1. A. Hatch, *et al.* "A rapid diffusion immunoassay in a T-sensor", *Nat. Biotech.*, **19** (5): 461-465 May 2001.
2. D. Schmalzing, S. Buonocore, C. Piggee, "Capillary electrophoresis-based immunoassays", Electrophoresis, **21** (18): 3919-3930, December 2000.

INTEGRATED BACK-SIDE PHOTODIODES AS PART OF A SINGLE SUBSTRATE INTEGRATED CHEMILUMINESCENT DETECTION SYSTEM

A.M. Jorgensen, W. Rong, O. Geschke, H. Klank, P. Telleman, J.P. Kutter
MIC-Mikroelektronik Centret, DTU, Building 345east, 2800 Lyngby, Denmark
(amj@mic.dtu.dk)

Abstract

Back-side contacted photodiodes detecting chemiluminescence in a microsystem are presented. These photodiodes absorb radiation far from their active regions and have a number of advantages compared to traditional photodiodes when integrated in micro total analysis systems (μTAS).

Keywords: Integration, Chemiluminescence, Back-side Photodiode

1. Introduction

Advanced micro total analysis systems (μTAS) depend heavily on integrating functionality. One critical step is to integrate fluidics and optical detection. The system investigated here combines a passive fluidic network and photodiodes. The photodiodes are back-side contacted, resulting in metal contacts on only one side, which facilitates hermetic sealing of the fluidic channels on the front-side.

2. Theory

Back-side contacted photodiodes are planar silicon photodiodes. A key feature is that the incident radiation is absorbed on the opposite side of the wafer. This absorption leads to generation of electron/hole pairs. The minority carriers will diffuse in the bulk material until they either recombine or reach the depletion region where they are swept out by the built-in field.

3. Experimental

To have a relatively simple tool to evaluate the functionality and performance of back-side photodiodes with different fluidic layouts prior to monolithic integration, we employed a hybrid solution consisting of two parts, a silicon chip containing the backside photodiodes and a polymer chip containing the fluidic system (**Fig. 1**). The fluidic system is realized by laser-micromachining a PMMA slide (2cm by 4cm) with a CO_2 laser. The silicon chip is of dimensions 1cm by 2cm and the back-side planar photodiodes are located in an array underneath the detection area. Details of the photodiode fabrication are presented elsewhere (Jorgensen *et al.* in Proceedings of SPIE Microelectronic and MEMS Technology, 2001, Edinburgh, Scotland, UK).

J.M. Ramsey and A. van den Berg (eds.), Micro Total Analysis Systems 2001, 573–574.

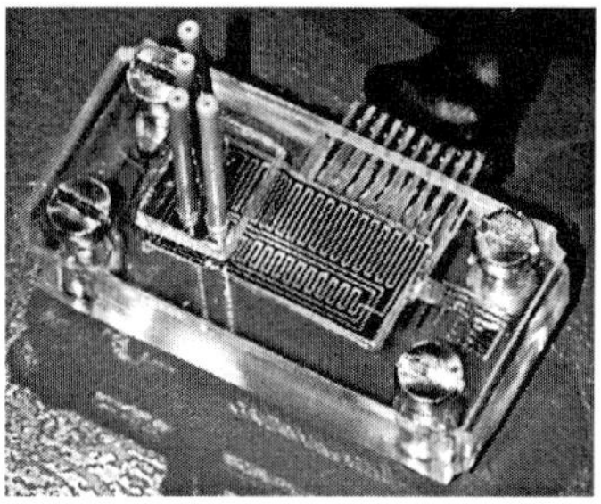

Fig. 1 Photograph showing the whole system assembled with a holder.

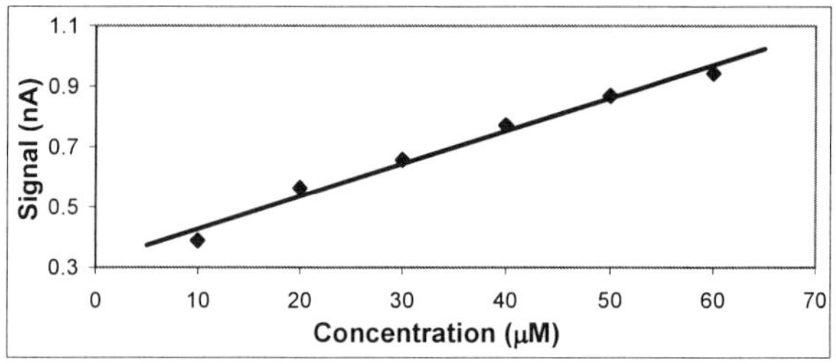

Fig. 2 Calibration curve for H_2O_2. A plugsize of 47μl was used. The flowrates were 30μL/min for all solutions.

4. Results and discussion

During the experiment syringe pumps were used for pumping the liquids. The luminol and ferricyanide solutions were pumped at constant flowrates of 30μL/min. Hydrogen peroxide plugs of 47μl, of various concentrations, were transported by a phosphate buffer at 30μL/min. The eight photodiodes were coupled in parallel and a Keithley 2400 source/measure unit carried out the actual measurements. A calibration curve (**Fig. 2**) was obtained for hydrogen peroxide concentrations of nominal 10μM to 60μM. Three samples were measured at each concentration and the reproducibility was about 2%. The system performs nicely and yields very reproducible measurements over the concentration range examined.

5. Conclusion

Back-side contacted photodiodes have been shown to perform well in a chemiluminescence detection system. Yielding very reproducible measurements over a range of concentrations. Having proved the functionality of back-side photodiodes the next step is monolithic integration where we have realised structures with microchannels, hermetic sealing (**Fig. 3**) and working photodiodes (**Fig. 4**).

Fig. 3 Front side of the monolithically integrated detector system. The cover lid is anodically bonded to ensure a hermetic seal.

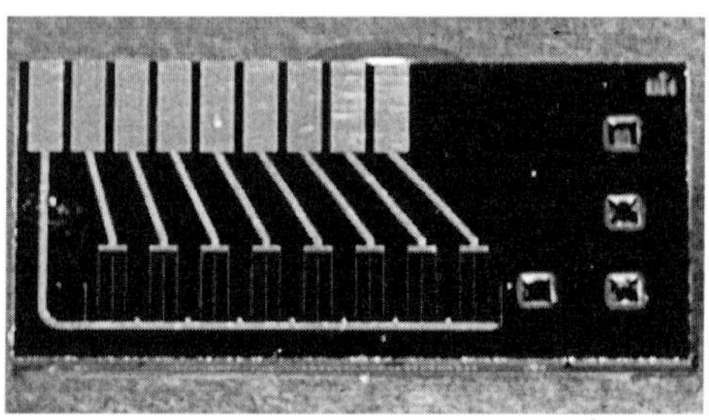

Fig. 4 Backside of the monolithically integrated detector system showing the 8 photodiode array and the fluidic via connections

REAL-TIME FLUORESCENCE DETECTION OF DNA IN 5 μl CAPILLARY CHANNELS FOR MINIMAL RESIDUAL DISEASE QUANTIFICATION USING THE ACAPELLA-5K HIGH-THROUGHPUT AUTOMATED ANALYSIS SYSTEM

Tobias P. Mann[1], Mark R. Holl[1], Mohan S. Saini[1], Charles H. Fisher[1], William H. Pence[2], Stephen E. Moody[2], David L. Cunningham[2], Daniel E. Sabath[3], Deirdre R. Meldrum[1]

[1] Department of Electrical Engineering Genomation Laboratory, University of Washington, Seattle, Washington, USA

[2]Orca Photonic Systems, Inc., Redmond, Washington, USA

[3]School of Medicine, University of Washington, Seattle, Washington, USA

Abstract

This paper describes a system for quantification of template DNA in a sample using fluorescence detection of DNA amplification during a real-time PCR reaction in a 5 μl glass capillary channel. The capillary system allows aliquots to be drawn from small samples efficiently and with no cross contamination when many samples are processed in an automated high-throughput sample preparation system, the Acapella-5K.

Keywords: Real-time PCR, fluorescence detection, minimal residual disease, microfluidics

1. Introduction

Accurate analysis of minute quantities of target human genomic DNA (1-10,000 copies) in the presence of background DNA requires exacting attention to sample handling. Our microchannel capillary-based methods enable the precise transfer of submicroliter sample fractions from standard storage and preparation disposables without loss or contamination of the source archival DNA. At the heart of this research is a "smart assay" schema whereby the high-throughput pipelined serial architecture of the ACAPELLA-5K (A5K) capillary-based processing system [1] is leveraged in concert with MRD (minimal residual disease) workstation hardware (in development) to implement an "auto-focus" feedback-control loop. An important advantage of the A5K architecture for rare event detection is directly related to the microfluidic nature of the system and its high-throughput architecture. This architecture allows efficient sample segmentation by micro-volume capillary-based processing as well as repeated quantitations for improved accuracy. Additional scientific needs addressed by the hardware for microfluidic high-throughput real time PCR quantitation [3] are: (1) design of specific primers and probes using existing chemistries and development of new methods, (2) optimization of the reagent compositions used in real time PCR reactions. Thermal cycling is performed using Peltier modules, custom capillary heater interface platens, and control algorithms implemented in LabVIEW software.

J.M. Ramsey and A. van den Berg (eds.), Micro Total Analysis Systems 2001, 575–576.

2. Method

The real-time PCR quantitation thermal cycling conditions used were: 95 °C initial denaturation of DNA and activation of AmpliTaq Gold DNA polymerase, 10 min; 40 cycles of denaturation, 95 °C, 5 s; annealing, 50 °C, 10 s; extension, 72 °C, 45 s. The reaction volume was 2 μl and was carried out in glass capillaries (OD, 0.86 mm; ID, 0.34 mm; length, 55 mm).

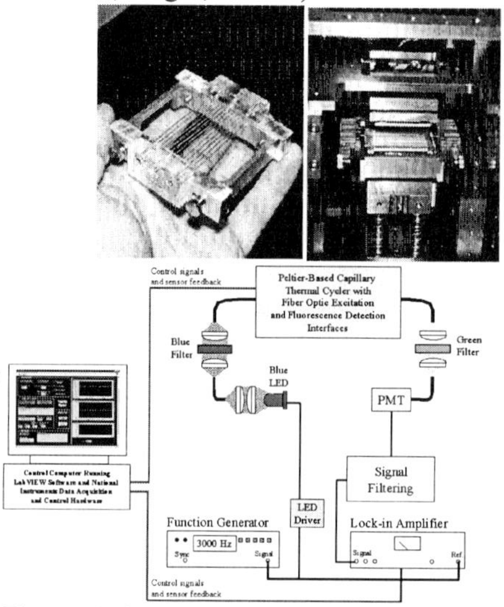

Figure 1: Optical Analyzer Testbed: Experimental cassette with optical and fluidic interface seals, cassette in thermal cycler, and functional layout of the MRD research workstation.

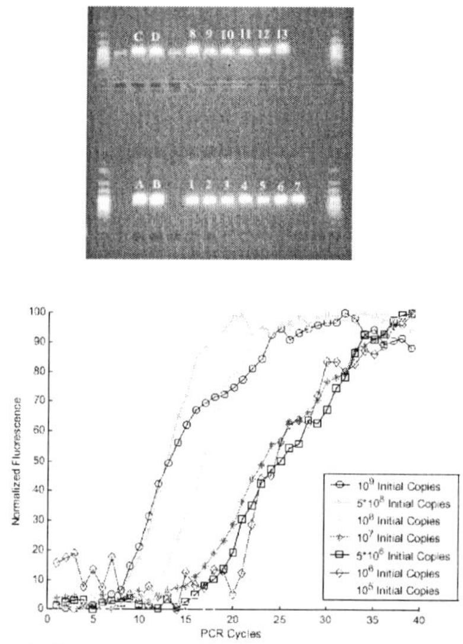

Figure 2: Robust PCR validation (lanes 1-13 on workstation; A-D are controls thermocycled on a commercial air thermocycler), and initial copy count estimation within one order of magnitude.

3. Conclusions

We have demonstrated the ability to reliably estimate the amplification rate of PCR reactions in real time with a ±5% accuracy and to estimate sample initial copy count within 1 order of magnitude. Based on these results, we have also developed a conceptual design for a workstation that functions as an add-on module within the A5K automated sample processing system architecture.

Acknowledgements

This research was supported by NIH Grant #1 R21 CA84691.

References

1. D. R. Meldrum, H. T. Evensen, W. H. Pence, S. E. Moody, D. L. Cunningham, and P. J. Wiktor, Genome Res, vol. 10, pp. 95-104, 2000.
2. J. Peccoud, and C. Jacob, Gene Quantification., Ed. F. Ferré, Boston, Birkhauser, pp. 111-128, 1997.
3. C. Wittwer, and K. Ririe, et al. (1997). Fluorescence Monitoring of Rapid Cycle PCR Quantification. Gene Quantification., Ed. F. Ferré, Boston, Birkhauser, pp. 129-144, 1997.

MAGNETO-HYDRODYNAMICALLY (MHD) DIRECTED FLOW IN MICROFLUIDIC NETWORKS

Richard B.M. Schasfoort, Regina Lüttge and Albert van den Berg.
University of Twente, P.O.Box 217, 7500 AE Enschede, The Netherlands

Abstract

Magneto-hydrodynamic flow is applied to generate rotation of liquid around a wire electrode. An interesting approach is to apply the effect in a microfabricated device. A buffer solution is pumped by EOF in a magnetic field through the channel network. The Lorentz-force causes the flow to deviate into one of the side channels. Thus, an external magnetic flow-switch for use in EOF driven flow systems has been realized.

Key words: magnetic mixing, magneto-hydrodynamic (MHD), rotational flow, electrokinetic pumping

1. Introduction

Magneto-hydrodynamic flow is a principle to generate macroscopic liquid flow [1]. Using coupled AC-electrical and magnetic fields the principle has been used by Lemoff and Lee to pump a liquid in a setup with current generating electrodes perpendicular to the flow-channel [2]. Here we present two new phenomena based upon the MHD effect.

2. Experimental

With a simple setup of a planar and a wire electrode we generated an electric field in a liquid-containing microvial by applying a potential difference of 9V between the electrodes. In a following experiment devices were fabricated by Micronit using powder-blasting. The technological process is described according to Wensink [3]. A 4-compartment holder with electrodes was connected to a computer controlled power supply (IBIS 411, EOF control unit, IBIS Technologies). In both experiments a permanent NeFeB-magnet (1.3 Tesla, diameter: 5 mm, Edmond Scientific Instr.), "Uniform Dyed Microspheres" (diameter 2.60 μm, type PS (425,480), yellow fluorescent, Bangs Laboratories, Inc.) and PBS (Sigma) were used in the experiments. A fluorescent inverted microscope (Olympus) connected to a video image system was used to view the flow profiles.

3. Results and discussion

The first experiment leads to highly non-uniform electrical current lines that in combination with a permanent magnet placed below the microvial generate a rotation of the liquid around the wire electrode. This rotation is caused by the Lorentz-force. Reversing the direction of the magnetic field results in a reversal of the rotating fluid flow which confirms control of magneto-hydrodynamic mixing. The rotation velocity is proportional to the current and salt concentration of the liquid. The velocity vector of cat- and anions will be influenced by the magnetic field. In the second experiment liquid

J.M. Ramsey and A. van den Berg (eds.), Micro Total Analysis Systems 2001, 577–578.

is pumped electrokinetically in a fluid manifold as shown in Figure 1 by applying EOF between compartment position 4 and 2. Due to the Lorentz force generated by the permanent magnet placed below the central interconnections and the electric current of $\pm$ 200 μA at 200 Volts the magneto-hydrodynamic flow bends into the left-hand electrically floating channel. Applying this principle an external, magnetic flow-switch for use in electrokinetically driven flow systems has been realized.

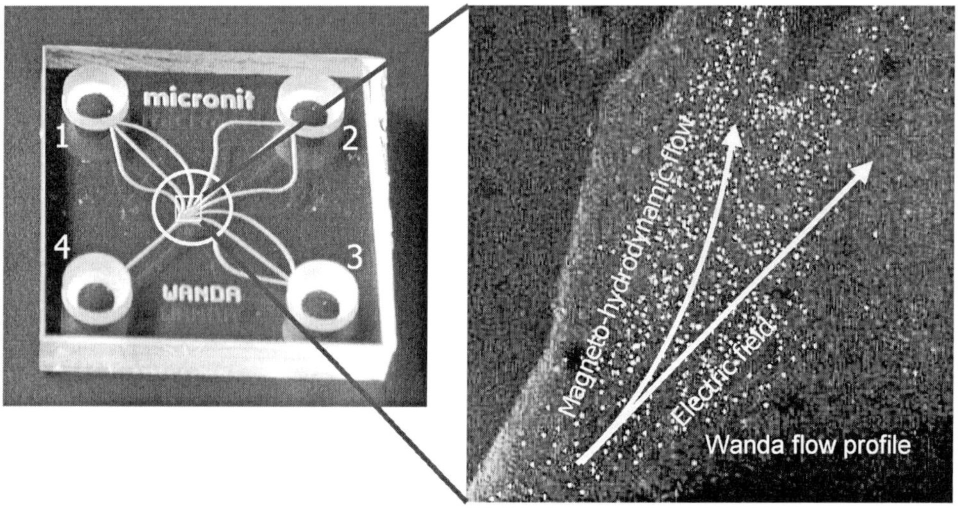

Figure 1. Powder-blasted "WANDA" device. Channel diameter = 30 μm x 100 μm. The ring in the middle is the location of the magnet area under the device (left). The square is the vision area of the microscope. Visualised liquid flow (right).

4. Conclusion

Two new MHD-based devices have been realized for liquid mixing or switching. The mixer based on the centered wire electrode design enables the controlled generation of liquid vortices in a microvial, whereas the flow-switch realized in the WANDA device may be used for collection of biochemical species or even magnetic separation.

References

1. G.S. Dulikravich, S.R. Lynn, *Unified Electro-Magneto-Fluid Dynamics (EMFD): Introductory Concepts*, International Journal of Non-Linear Mechanics, (1997), vol. 32, no. 5, 913-922 (10)
2. A.V. Lemoff, A.P. Lee, *An AC magneto hydrodynamic micropump*, Sensors and Actuators B 63 (2000) 178-185
3. H. Wensink, J.W. Berenschot, H.V. Jansen, M.C. Elwenspoek, *High Resolution Powder Blast Micromachining*, 13th Int. Workshop on Micro Electro Mechanical Systems (MEMS 2000), Miyazaki, Japan, (2000), 769-774

A NOVEL QUARTZ CRYSTAL MICROBALANCE-BASED PLANAR SENSOR DESIGN FOR HIGHLY SENSITIVE NON-LABELING BIOMOLECULE DETECTION

Hiroyuki Sota[1], Hiroshi Yoshimine[1], Masanori Gotoh[1], Yukio Hasegawa[1] and Yoshio Okahata[2]
[1]Tokyo R&D, Amersham Pharmacia Biotech, Japan
[2]Department of Biomolecular Engineering, Tokyo Institute of Technology, Japan
(Correspondence: hiroyuki.sota@jp.apbiotech.com)

Abstract

In this work, we present a novel versatile planar sensor chip design (QCM chip) for a μl-scale on-line biosensor. The resulting 27 MHz sensor signal exhibited a very low noise level over a range of flow rates up to 100 μl/min, promising advanced biosensor performance with high effective sensitivity.

Keywords: Biosensor, Quartz Crystal Microbalance, QCM Chip, Non-labeling Detection, Real-time Monitoring

1. Introduction

To extend the capabilities of microfluidic devices aimed at state-of-the-art analysis of biological samples, a simple but sensitive liquid-phase sensor is desired. From this technical viewpoint, a biosensor based upon quartz crystal microbalance (QCM) technology has been recognized as one of the most promising sensors. Its advantages include high theoretical sensitivity, simple and low-cost instrumentation as well as real-time non-labeling detection. However, realization of such a sensor that achieves both high practical sensitivity and reproducibility has long been stymied mainly due to the design challenges imposed by an aqueous environment. Both the physical properties of the liquid and the contrivances necessary to prevent electrical short currents increase the electro-mechanical load on the oscillator causing the oscillation quality (represented as a parameter, Q-value) to deteriorate. Here, we present a novel versatile planar sensor chip design (QCM chip) and supporting electronic and fluidic systems for a μl-scale on-line sensor in order to overcome the hurdles to practicality.

2. Experimental

The key to avoiding deterioration of oscillation under a liquid load is that the sensor crystal's mounting should minimize the stress applied both by the fixative and the hydraulic pressure. Figure 1 shows our new planar QCM sensor chip, in which a rectangular AT-cut quartz oscillator (4.0 x 1.6 mm x 60 μm; fundamental frequency: 27 MHz) is flexibly fixed and sealed along the whole of its circumference to a solid support by an elastic sealant. This chip design not only obviates the need for direct handling of

J.M. Ramsey and A. van den Berg (eds.), Micro Total Analysis Systems 2001, 579–580.

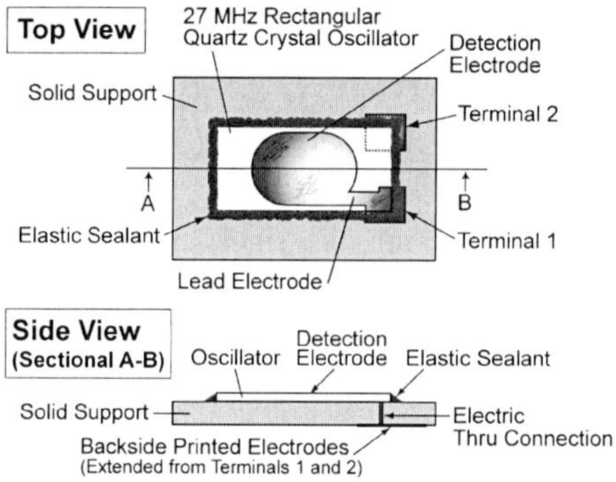

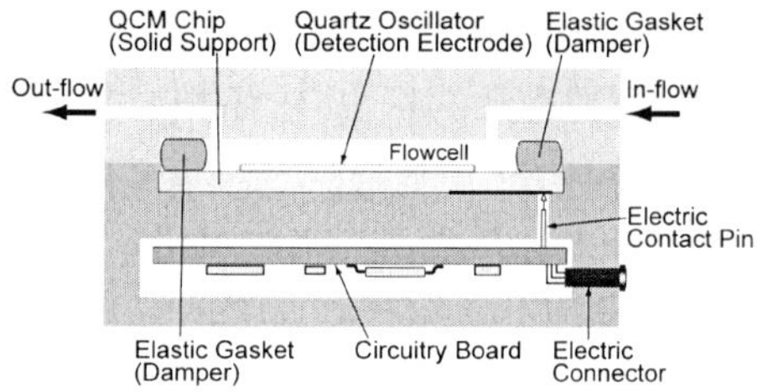

Figure 1 (left): Schematic View of the QCM chip.

Figure 2 (above): Sectional View of the flowcell with the QCM chip installed.

the delicate quartz crystal but also enables the oscillator to be held indirectly by sandwiching both faces of the solid substrate elastically into the flowcell (Figure 2).

4. Results and discussion

The QCM chip was tested for noise level under aqueous flow. Figure 3 shows a comparison of signal stability between an oscillator employing our new mounting design and the conventional one in which a bare 27 MHz oscillator is sandwiched between two O-rings in a flowcell. In the absence of flow, the QCM sensor achieved an excellent reduction in the noise level down to ±0.05 Hz with the synergistic electronic and fluidic improvements. Moreover, substantially the same low noise level was retained over a range of the flow rates up to 100 μl/min, while noise levels in the conventional design increased with the flow rate. We attribute this flow rate-independent noise level of the new QCM design to its mounting, a non-adhesive but distributive contact with the substrate, which efficiently prevents quality (Q-value) loss due to bulk deformation upon exposure of the crystal to increased hydraulic pressure.

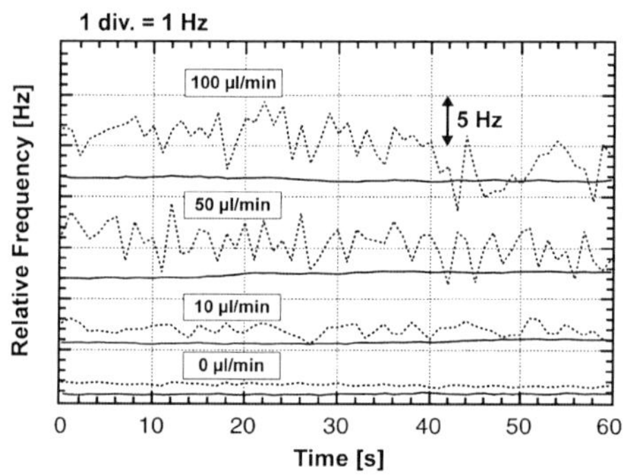

Figure 3: Noise level comparison for the two different on-line sensor systems at various flow rates. (Solid line): surface-mount QCM chip, (Dotted line): conventional mounting (O-ring mediated).

5. Conclusions

This advanced planar sensor design coupled with a simple but sophisticated system holds great promise for integration into various microfluidic systems (μTAS).

Acknowledgements

This work was partly supported by the Ministry of Agriculture, Forestry and Fishery of Japan (MAFF).

ON-LINE FTIR-MICROSCOPY FOR MECHANISM STUDIES OF NOVEL NITRATION ROUTES IN MICROREACTORS

T. Tuercke, W. Schweikert, D. Boškovic, H. Krause, S. Loebbecke
Fraunhofer-Institut Chemische Technologie ICT
P.O. Box 12 40, 76318 Pfinztal, Germany

Abstract

The on-line analysis of strong exothermic nitrations in flat silicon microreactors by means of FTIR microscopy is described. Intermediates and products can be IR spectroscopically identified, quantified and localized. The reaction progress in microreactors can thus be monitored on-line.

Keywords: FTIR microscopy, on-line analysis, nitration, urea, intermediates

1. Introduction

Many microreactors are based on fluidic structures made of silicon. Due to the transparency of silicon in the mid-infrared spectral range, FTIR spectroscopy is a suitable method to analyse chemical processes within these microfluidic structures.
Nitrations are among the basic reactions used in chemical synthesis. By applying microreactors nitrations can be carried out in a thermally controlled way. Furthermore, the combination of microreactors with fast on-line analytical techniques can be applied to investigate new synthetic routes for nitration processes, like the nitration of dialkyl substituted ureas used as precursors for pharmaceuticals and energetic materials.

2. Experimental

The setup of the FTIR spectroscopic monitoring of microreaction processes is schematically shown in Fig. 1. The flat silicon microreactor is mounted on a positionable stage and connected with the reactants supply. The key advantage of applying a microscopic technique is the high spatial resolution (< 10µm) which ensures that educts and products of the nitration process can be identified and quantified at different zones inside the microreactor. By focusing the IR beam consecutively on different reaction zones the progress of the reaction can be IR spectroscopically monitored.

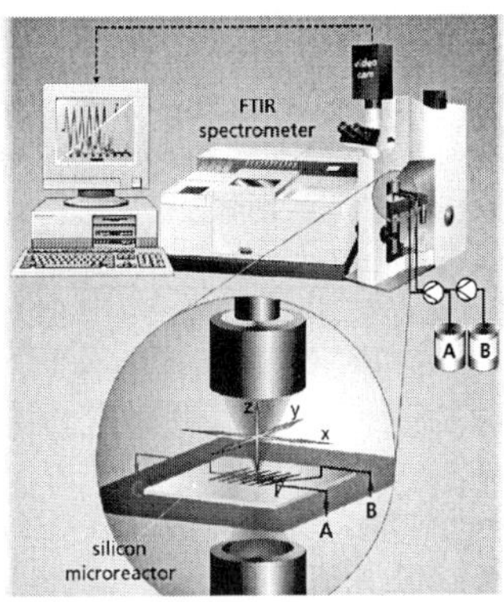

Figure 1. Experimental setup

* Author to whom correspondence should be addressed (sl@ict.fhg.de).

J.M. Ramsey and A. van den Berg (eds.), Micro Total Analysis Systems 2001, 581–582.

3. Exemplary results

A new approach to a quantitative synthesis of N,N'-diethyl-N,N'-dinitro-urea (III) was investigated by using the much cheaper N,N'-diethyl-thiourea (I) as starting material. By performing this nitration in macroscopic batch reactors an insufficient complex mixture of products is obtained giving no evidence for the reaction pathway.

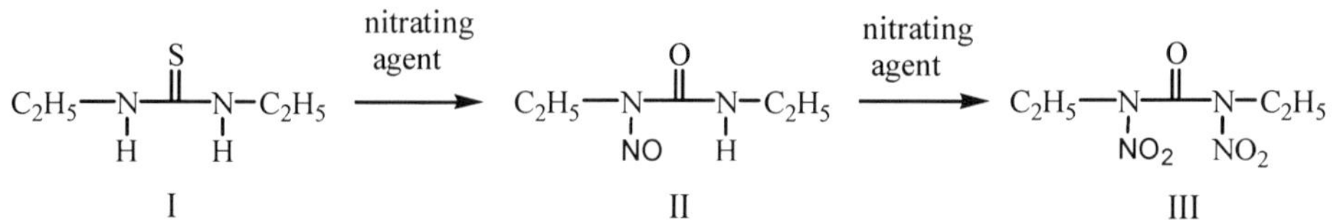

Figure 2. Nitration of N,N'-diethyl-thiourea to N,N'-diethyl-N,N'-dinitro-urea

By applying microreactors in combination with FTIR microscopy the individual reaction steps could be identified (Fig. 2). The first step is the oxidation of the thiourea to N,N'-diethyl-urea accompanied by a nitrosation step forming the mono-substituted nitroso derivative (II) which is finally nitrated to the dinitro-substituted target product (III). The oxidation occurs instantaneously so that no characteristic absorption bands of the starting material (thiourea) can be detected at the reactor entrance. However, the progress of the nitrosation can be clearly monitored and quantified along the microfluidic reaction channels (Fig. 3).

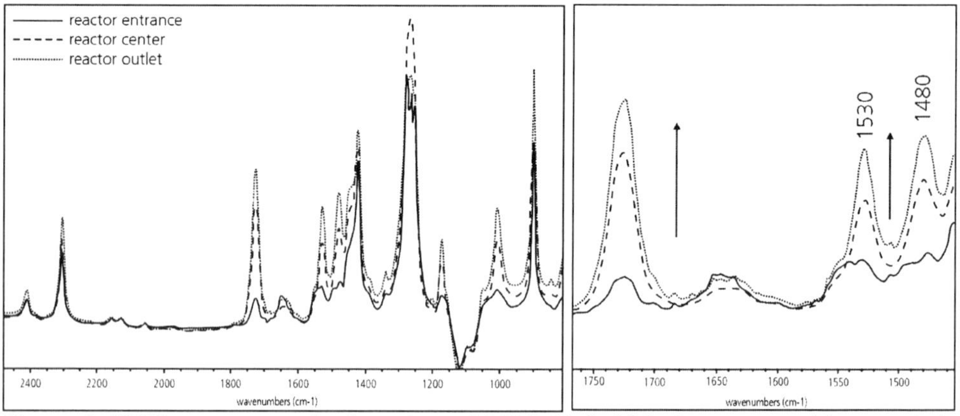

Figure 3. FTIR spectroscopic detection of N,N'-diethyl-N-nitroso-urea (characteristic absorption bands of the nitroso product (II) are located at 1722, 1530, 1480, and 1170 cm^{-1} respectively, increasing continually due to reaction progress)

4. Conclusions

The combination of microfluidic silicon reactors with on-line FTIR spectroscopic analysis is a suitable tool for process analysis and process development. The results obtained can now be applied to an optimized process engineering of the macroscopic synthesis of N,N'-diethyl-N,N'-dinitro-urea.

PROTEIN MICROARRAYS FABRICATED BY MICROCONTACT PRINTING

H. D. Inerowicz[1] F. E. Regnier[1], S. W. Howell[2], R. Reifenberger[2]
[1]Department of Chemistry, Purdue University, West Lafayette, IN 47907
[2] Department of Physics, Purdue University,West Lafayette, IN 47907
(fregnier@purdue.edu)

Abstract
Immunoassays were performed using protein microarrays fabricated by microcontact printing (μCP) techniques. The developed immunoassays were studied by fluorescence microscopy (FM) and scanning probe microscopy (SPM).

Keywords: protein microarrays, microcontact printing, immunoassay

1. Introduction

Microarray technology is evolving from genomics to proteomics. Protein arrays will be useful in miniaturized biosensors, clinical immunological assays, and a series of basic research applications ranging from mapping the proteome to examining protein: protein interactions. But unlike DNA microarrays, the technology for generating protein microarrays is still developing. The micropatterning of proteins by μCP on solid substrates is of great interest today for this reason [1]. The low cost of fabrication compared to lithography [2], as well as simplicity of transferring proteins without loss of biological activity, makes this method very attractive.

2. Theory

Immunoassays are possible because of the targeted interaction between specific proteins. Thus, by patterning a substrate, various proteins can be distinguished by morphological and optical modifications of targeted regions of the substrate.

3. Experimental

Polydimethysiloxane (PDMS) elastomer based stamps, after appropriate surface treatment, have been found to be an excellent material for the directed transferring of proteins from solution to a micro-pattern on a substrate. PDMS stamps are manufactured by curing Sylgrad 184 (Dow Corning, Midland) on a positive photoresist master. Immunoassay reagents included rabbit IgG, mouse IgG, human IgG, donkey anti-rabbit IgG FTIC, goat anti-mouse IgG FTIC and goat anti-human IgG FTIC, all of which were obtained from Pierce, Rockford, IL. Glass substrates were cleaned and silanized with chlorodimethyloctadecylsilane from toluene solution. PDMS stamps were treated in a plasma cleaner before equilibrating with a protein solution in PBS buffer at a pH=7.4. The micropatterned designs on the stamps were either strips of 2 μm in width or 2 to 10 μm squares separated by channels of the same width. Selected antigens were stamped on a silanized glass substrate to form patterned immunoassay platforms . The unfunctionalized areas on the platforms were exposed to a solution of a

J.M. Ramsey and A. van den Berg (eds.), Micro Total Analysis Systems 2001, 583–584.

second antigen. To perform an immunoassay, the platforms were then incubated in a solution of fluorescein isothiocyanate (FTIC) tagged antibodies (see Figure 1).

Results

Figures 2 and 3 show representative immunoassays using FM and SPM. Figure 2a shows the FM detection of anti-mouse IgG (bright squares) and Figure 2b shows the detection of antihuman IgG (bright surrounding regions). Figure 3a displays a false-colored topographic image acquired by contact SPM. For the case of anti-mouse IgG, the bright square region of the substrate was found to be on average 5 nm higher than the surrounding human IgG regions. Similarly, our SPM studies indicate that for anti-human IgG detection, the region surrounding the mouse IgG squares is on average 4 nm higher (see Figure 3b). The AFM images also indicate that regions of antigen bonding are not uniform. The data also show little cross-reaction between mouse and human IgG patterns. Patterns consisting of three different proteins were also obtained by crossed stamping of PDMS strips (results not shown).

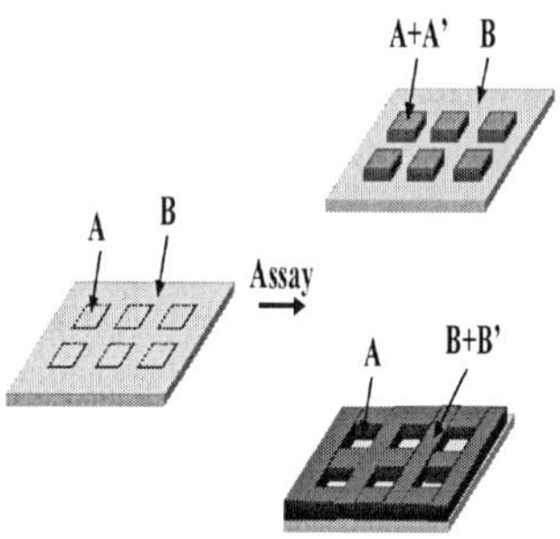

Figure 1. A schematic of the immunoassay.

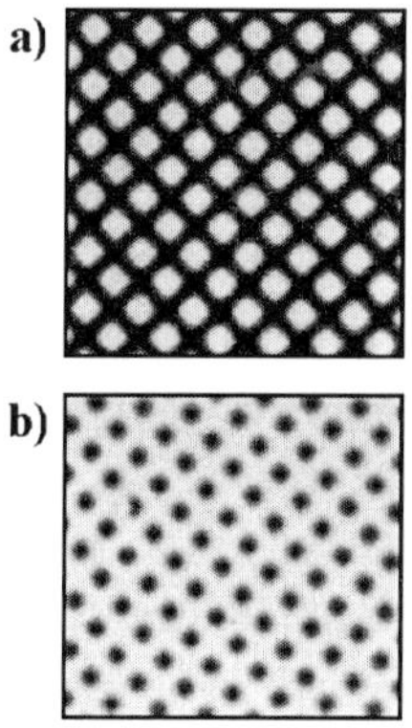

Figure 2. Fluorescence image of a) anti-mouse IgG; b) anti-human IgG assays.

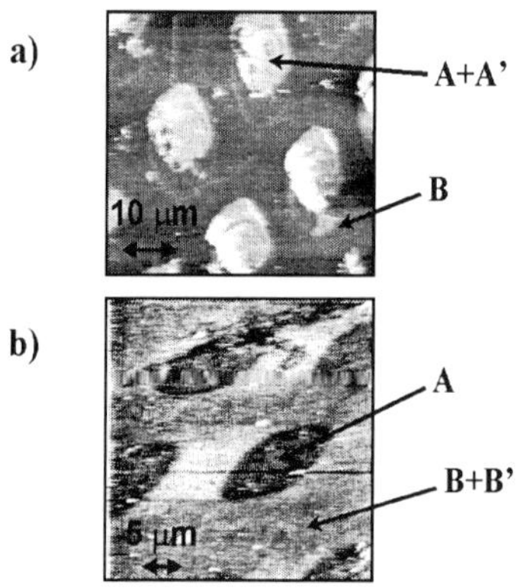

Figure 3. SPM images of a) anti-mouse IgG; b) anti-human IgG assays.

Conclusions

In summary, we have demonstrated a simple, inexpensive and effective method to fabricate protein microarrays using μCP techniques. This methodology will be useful in the development of biosensors based on protein microarrays.

Acknowledgment

The authors acknowledge B.E.Slentz for help in preparation of PDMS stamps.

References

1. A. Bernard, J.P. Renault, B. Michel, H.R. Bosshard and E. Delamarche, Adv. Materials, 12, 1067 (2000).
2. V.W. Jones, J.R. Kenseth, M.D. Porter, C.L. Mosher and E. Henderson, Anal. Chem. 70, 1233, (1998).

SURFACE MODIFICATION AND HYBRIDIZATION ON A THERMAL GRADIENT DNA CHIP

T. Kajiyama[1,2], T. Sakazume[1,2], Y. Miyahara[2], S. Surrey[3], D.J. Graves[4], P. Wilding[5], L.J. Kricka[5], P. Fortina[1]
[1]Dept Pediatrics, The Children's Hospital of Philadelphia, Philadelphia, PA 19104, USA
[2]Hitachi, Ltd, Instruments, Hitachinaka-shi, Ibaraki-ken, 312-8504 Japan
[3]Dept Medicine, Jefferson Medical College, Philadelphia, PA 19107, USA
[4]Dept Chemical Engineering, and [5]Dept Pathology and Laboratory Medicine, University of Pennsylvania, Philadelphia, PA 19104, USA

Abstract
We developed a method for attaching oligonucleotide (ON) probes to the silicon nitride surface of a thermal gradient DNA chip. We verified that aminosilane and polylysine were effective in generating reactive surface amino groups, and phenyl-diisothiocyanate (PDC) was effective to link amino modified ON to the chip surface. We used the chip to detect single-base changes by allele-specific ON hybridization.

Keywords: DNA chip, silicon nitride, surface chemistry, Si-island, ASOH

1. Introduction

Silicon chips containing 100 islands, each of which can be maintained at distinct temperatures, have been manufactured for use in DNA array hybridization studies [1]. The 500 x 500 μm silicon islands are created under an SiO_2/Si_3N_4 membrane (10 μm thick) by anisotropic etching. Heaters and temperature sensors are created with a diffusion layer in each island. We investigated different surface chemistries and potential diagnostic applications of such devices.

2. Surface treatment and attachment of oligonucleotides

Three methods were studied to generate reactive amino groups on the silicon nitride (Si_3N_4) chip surface based on treatment with (i) dendrimer (Figure 1A); (ii) polylysine (Figure 1B); (iii) (aminoethylamino) propyl-trimethoxysilane (Figure 1C). Dendrimer and aminosilane treatments differ from polylysine in that they create "spacer" groups between the surface and the amino moieties. Amino-modified oligonucleotide (AON) probes (~20-25 n) were attached following surface treatment with PDC. AON then were spotted with a robotic device [2]. Free PDC groups were blocked with glycine. Attachment results for AON probes to the various surfaces as a function of probe concentration (0.5-50 μM) (Figure 2) show strong signals for deposition of probes at concentrations >25 μM (S/B 130 – 190). Dendrimer-coated silicon nitride surfaces produced spots that spread from their original deposition site (Figure 2A), hence we focused on use of polylysine and aminosilane-coated surfaces (Figure 2B, C). In addition aminosilane-coated surfaces did not require a 2-week "curing" period (Figure

J.M. Ramsey and A. van den Berg (eds.), Micro Total Analysis Systems 2001, 585–586.

2D), as did the polylysine and dendrimer; hence we decided to use silane-coated surfaces for hybridization studies.

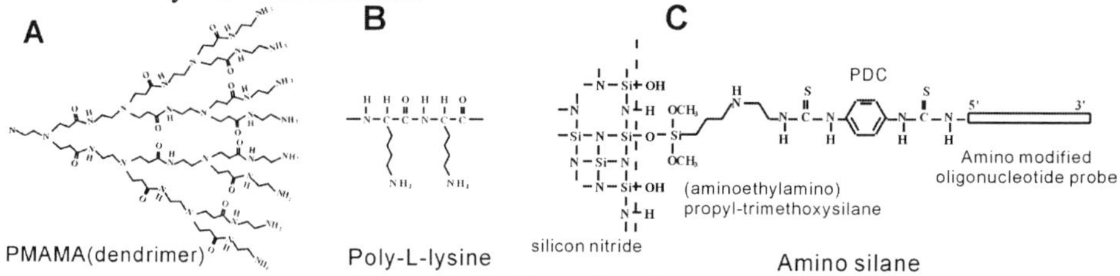

Figure 1. Surface chemistry

3. Detection of single-base changes by allele-specific oligonucleotide hybridization

We tested for a single-base change in the Factor VII gene (R503G) by first attaching AON probes corresponding to normal and mutant alleles at separate array registers. Three different dye-tagged PCR targets representing the three different genotypes (N/N, N/M and M/M) were generated from genomic DNA. PCR products were heat denatured and annealed to the chips. Results of target hybridization (Figure 3) demonstrate the distinction among the three genotypes. Current experiments involve varying island temperatures to define optimal conditions for assessing simultaneously single-base changes in a variety of genes.

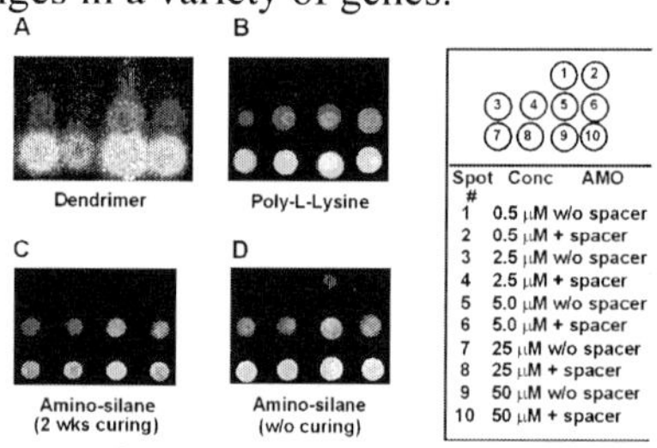

Figure 2. Attachment results

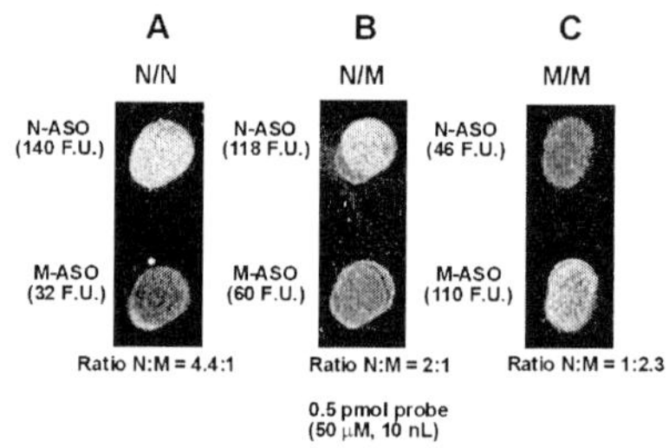

Figure 3. Hybridization results

4. Conclusions

Aminosilane, polylysine and dendrimer were effective for generating reactive amino groups on the silicon nitride (Si_3N_4) chip surface. Aminosilane was preferred because it did not require curing. We verified that the DNA chip was effective for detection of single-base changes by allele-specific oligonucleotide hybridization.

Acknowledgements

We thank Dr. Yokoyama of Hitachi Ltd. and Dr. Terada, Mr. Kunimoto, Mr. Morimoto and Mr. Tanabe of Hitachi ULSI Systems Co. for their useful discussions.

References

1. T. Kajiyama, K. Murakawa, Y. Miyahara, *Thermal gradient chip*, Proc. Micro Total Analysis Systems 2000, pp. 505-508, (2000)
2. D. Graves, H-J. Su, S. McKenzie, S. Surrey, P. Fortina, *System for preparing microhybridization arrays on glass slides*, Analytical Chemistry 270, pp. 5085-5092, (1998)

PHOTOCHEMICAL SELECTIVE SURFACE MODIFICATION USING MICROMIRROR ARRAY (MMA) AND PROTEIN PATTERNING

Dong-Sik Shin[1], Kook-Nyung Lee[2], Woo-Jae Chung[1], Ki-Hoon Jang[1], Yong-Kweon Kim[2], and Yoon-Sik Lee[1]

[1]School of Chemical Engineering, Seoul National University, Kwanak PO Box 34, Seoul 151-742

[2]School of Electrical Engineering & Computer Science, Seoul National University, Kwanak PO Box 34, Seoul 151-742

Email : yslee@snu.ac.kr

Abstract

The array of immobilized biomolecules has emerged as a tool for biochip fabrication. With a digital micromirror array (MMA), maskless photolithography process can be achieved. We describe the method of protein patterning using MMA system.

Keywords: Protein patterning, maskless photolithography, micromirror array

1. Introduction

With recent developments in microfabrication processes and surface modification, on-site analysis, which is a powerful method for diagnosis of diseases, can be achieved by biochips of a protein array. Here, streptavidin was chosen as a model protein, since biotin-streptavidin chemistry is being used extensively for protein immobilization.

2. Maskless photolithography

If MMA system is used as a virtual photomask, maskless photolithography process can be achieved so that patterned DNA probes [1] or protein ligands can be synthesized on a substrate by selective illumination to the chip surface (Fig. 1). In this paper, we utilized nitroveratryloxycarbonyl (NVOC) group as a photolabile protecting group for protein patterning.

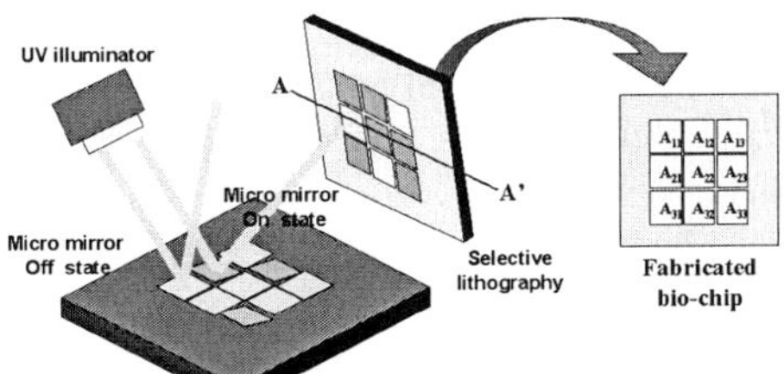

Figure 1. Schematic view of photochemical surface modification using MMA.

J.M. Ramsey and A. van den Berg (eds.), Micro Total Analysis Systems 2001, 587–588.

3. Experimental

The glass slide was aminosilanized with 3-aminopropyltriethoxysilane (APTS). Then, NVOC-6-aminocaproic acid was coupled to the aminated glass. The resulting NVOC-protected surface was selectively cleaved by using UV light (365 nm) and MMA (mirror size: 50×50μm^2). The condition of complete cleavage of NVOC group was optimized by checking the irradiation time under constant UV intensity (30 mW/cm^2). After selective NVOC cleavage, biotin was coupled to the surface. The slide was then immersed in a phosphate buffer (pH 7.4) solution of FITC conjugated streptavidin.

4. Results and discussion

The complete NVOC-cleavage occurred after 6 min (Fig. 2), so we set the UV irradiating time as 10 min. The protein-patterned image was observed by a confocal fluorescence microscopy (Fig. 3).

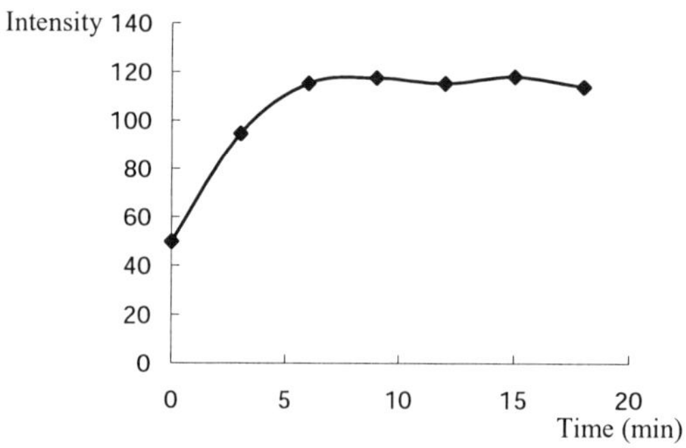

Figure 2. Fluorescence intensity measurement as UV irradiation time of NVOC group goes on.

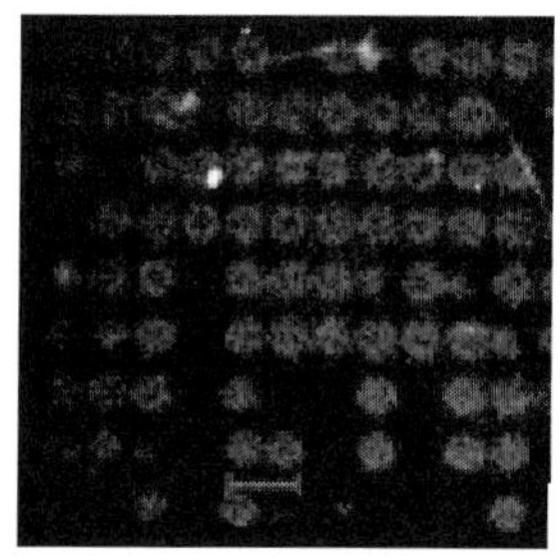

Figure 3. The image of patterning FITC conjugated streptavidin.

5. Conclusions

NVOC-protected surface was selectively irradiated by UV illuminator using MMA. We could see that projected micromirror patterns on the glass substrate corresponded to that of FITC fluorescence image. And FITC labeled streptavidin was successfully patterned.

Acknowledgement

This work was supported by the Nano Bioelectronics and Systems Research Center of Seoul National University, which is an ERC supported by the Korean Science and Engineering Foundation (KOSEF).

Reference

1. S. Singh-Gasson, R. D. Green, Y. Yue., C. Nelson, F. Blattner, M. R. Sussman, and F. Cerrina, "Maskless fabrication of light-directed oligonucleotide microarrays using a digital micromirror array", *Nature Biotechnology*, Vol. 17, October 1999, p. 974-978.

A MICRO REACTOR DEVICE FOR THE UGI FOUR COMPONENT CONDENSATION (4CC) REACTION

Victoria Skelton[1], Stephen J. Haswell[1], Peter Styring[2], Brian Warrington[3] and Stephanie Wong[3]

[1]Department of Chemistry, Faculty of Science and Environment, University of Hull, Cottingham Road, Hull, UK, HU6 7RX, [2]Department of Chemical and Process Engineering, The University of Sheffield, Mappin Street, Sheffield, UK, S1 3JD, [3]GlaxoSmithKline, New frontiers Science Park (North), third Avenue, Harlow, Essex, UK, CM19 5AW (email: v.skelton@hull.ac.uk)

Abstract

This paper describes a micro reactor system, in which the solution phase Ugi four-component condensation (4CC) reaction was investigated. The micro reactor is the first of its kind to perform the Ugi reaction on-chip using electro osmotic flow (EOF) producing the reaction intermediate *in-situ*. Currently, the reaction intermediate (an imine) can be synthesised successfully to a maximum yield of 94.3±4.4% (n=12) with a conversion to the final product of 60%. However, research is continuing into the final stage of the reaction to improve the reactions reproducibility.

Keywords: Synthesis, Chemistry, Micro reactors

Introduction

The main research themes into micro reactors have concentrated on combinatorial and fine chemical synthesis[1]. Micro reactors offer these chemistries a variety of advantages in particular rapid process optimisation of a single or multi step reaction, small reagent volumes, large surface to volume ratios, reaction selectivity, scale up rather than scale out and the isolation of pure reaction products. In this paper, the development of a micro reactor for the Ugi 4CC reaction between an aldehyde, primary amine, an acid and a cyclohexylisocyanide is described using EOF as the mobilisation mechanism is described. The Ugi reaction was first reported towards the end of the 1950's allowing the synthesis of biologically active compounds. Traditionally, the Ugi is a one-pot synthesis between the above-mentioned compounds, which traditionally has been performed at 0°C due to its highly exothermic nature. In this paper, the authors demonstrate that all the reaction components can be added sequentially increasing reaction selectivity.

Experimental

The micro reactor, Figure 1 was prepared using photolithography and wet etching techniques resulting in a final channel geometry of 200 μm wide and 100 μm deep with an overall outer dimension of 20 mm x 20 mm square and 25 mm deep[2]. Using the micro reactor outlined above, a solution of cyclohexylisocyanide in methanol (0.03M, 50 μL) was added to reservoir A and methyl 4-formylbenzoate in methanol (0.075M, 50

J.M. Ramsey and A. van den Berg (eds.), Micro Total Analysis Systems 2001, 589–590.

μL) was added to reservoir B. Benzyl amine methanol (0.03M, 50 μL) was added to reservoir C and 2-nitrobenzoic acid in methanol (0.03M, 50 μL) was added to reservoir D. Finally, 30 μL of methanol was added to reservoir E, which was the collection reservoir and common ground for the micro reactor. A high voltage power supply developed by Kingfield Electronics (Sheffield, UK) was used to apply an optimum external voltage at reservoirs A to D of 400 V. The micro reactor was left to run for a period of 20 minutes after which samples were collected for analysis off-chip using High Performance Liquid Chromatography (HPLC) and Gas Chromatography coupled with a Mass Spectrometer (GC-MS).

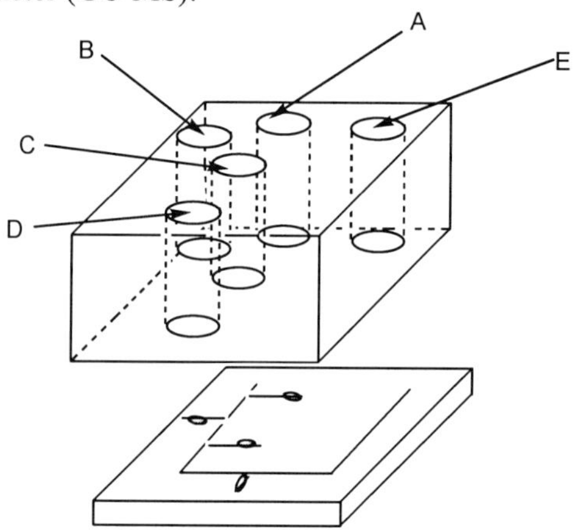

Figure 1. Schematic of the micro reactor used for the Ugi 4CC reaction

Results and discussion

For the initial stages of the reaction, the formation of the imine intermediate, the yields calculated were very high. The average yield obtained for the imine intermediate was 94.3 ± 4.4% (M^+ = 252, Rt = 13.4 minutes by GC-MS). This can be readily isolated on-chip before the final stage of the reaction occurs increasing the reaction selectivity. Currently, the reaction intermediate can be converted to the final product anywhere in the region of 15 % to 60% at either room or elevated temperatures. However, research is currently continuing in to the improvement including reproducibility as well as yield for the latter stage of the reaction.

References

1. S. J. Haswell, R. J. Middleton, B. O'Sullivan, V. Skelton, P. Watts and P. Sytring, *Chem. Commun.*, 2001, 391

2. T. McCreedy, *Anal Chim Acta*, 2001, **427**, 39-43.

DUAL-PROTEIN MICRO ARRAYS DEPOSITED BY μ-STAMPS AND μ-WELLS

[1]F. G. Tseng, [2]H. M. Huang, [3]C. Y. Huang, [4]S. C. Lin, and [5]C. C. Chieng
[1,3,4,5]ESS, [2]LS Dept., National Tsing Hua University, Taiwan,

Abstract

This paper reports a novel protein stamping mean, employing μ-stamps and protein wells to simultaneously transfer various proteins onto an array. Testing results have shown that each stamped protein sample can be clearly identified with uniform deposition, within 7% size and 5% gray level variation. This method can be applied for high-throughput disease diagnosis and drug screening.

Key words: Protein mciroarrays, Micro stamps, Protein diagnosis, drug screening

1. Introduction

The development and application of protein arrays have become an important trend in modern biomedical diagnoses because of micro-array's ability of massive parallel process. The conventional way to produce micro arrays, widely employed for DNA spotting, utilized a needle array and computerized robot system to select and spot numerous bio reagents. This arraying system has the drawbacks of high cost, large size variation (about 20%), and serial and long spotting process which may malfunction proteins from dry-out. Other methods, including photolithography, inkjet printing, spray printing, and stamping (for references please refer to [1, 2]), either have limitations on selection of sample types or numbers, or have complexity on fabrication or operation process, not mature enough for large protein-array application.

2. Design concept and fabrication

In this paper, a novel μ-stamping system for protein arrays is utilized as shown in Fig. 1 [1]. In contrast to other patterning methods, such as lithography and inkjet printing, the μ-stamping is a gentle mechanical-contact process and would not degrade protein samples. When contacting with the wells containing various protein samples, the hydrophilic stamp surfaces pick up and hold protein solutions on a desired array (Fig. 1(a)) simultaneously. The protein samples are then transferred to the bio-reaction surface by mechanical contact (Fig. 1(b) and (c)). Because of the compliant silicone used in the micro stamp, uniform deposition is achieved. Fig. 2 shows the fabricated micro stamps and micro wells, respectively [1, 2].

3. Testing results and conclusions

The possibility for multiple-protein transference by this stamping method is explored in this paper. Two proteins, 10 μg/μl rabbit and mouse IgG (Sigma), were filled into two compartments of the protein wells, respectively. Micro stamps with surface area 350μm square and 100 μm space in a 4 by 4 array were used to pick up proteins from the corresponding well positions, and spot them onto a APTS-BS3 [2] coated slide in seconds, then let dry for 0.5 hour. After horse serum protein was used for blocking for 0.5 hour, Cy3-conjugated anti-rabbit IgG and Cy5-conjugated anti-mouse IgG (Pharmacia) were pipetted simultaneously, to label the antigen proteins on the specific areas. Fig. 3 shows the results of bound protein arrays with different scanning scheme, (a) excitation of both Cy-3 and Cy-5, (b) excitation of Cy-3 only, (c)

J.M. Ramsey and A. van den Berg (eds.), Micro Total Analysis Systems 2001, 591–592.

excitation of Cy-5 only (by GenePix 4000B, Axon instruments), demonstrating very clean and specific binding of the stamped proteins. The spot size and gray level for different spot positions are also shown in Fig. 4 (a) and (b), respectively. Very uniform deposition, size variation less than 7% and gray level variation less than 5%, were obtained by this micro-stamping method. No cross contamination was observed from the staining process through the results of Fig. 3(b) and (c).

References

[1] F. G. Tseng, et al., Transducers'01, pp. 330-333, June (2001).
[2] F. G. Tseng, et al., Fresenius J. Anal. Chem., will appear, Sep. (2001).

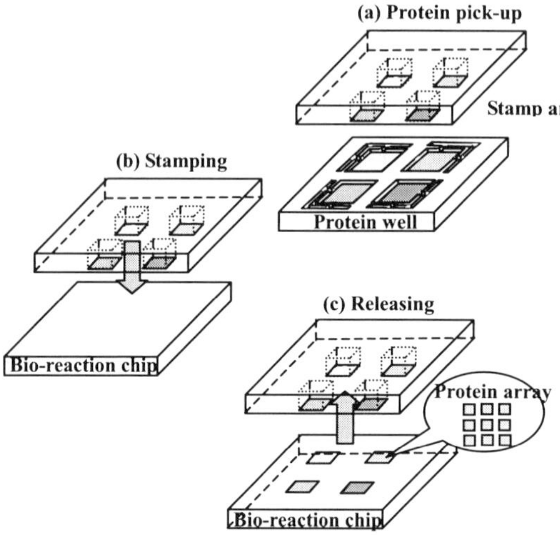

Fig.1 Concept of protein array printing by μ-stamps and μ-protein wells

Fig. 2 Micro stamps and protein wells

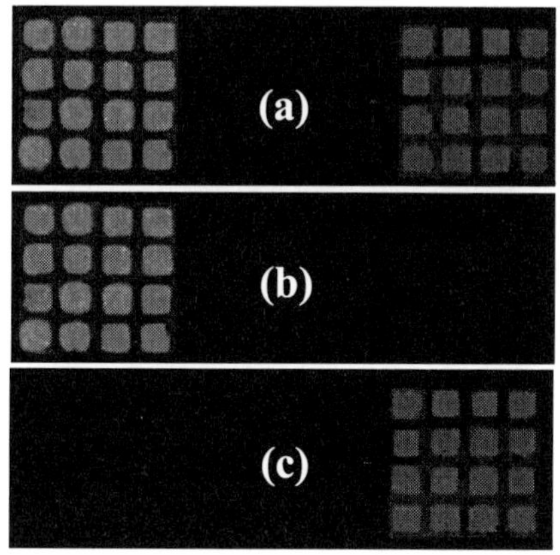

Fig. 3 Fluorescent stamped protein arrays

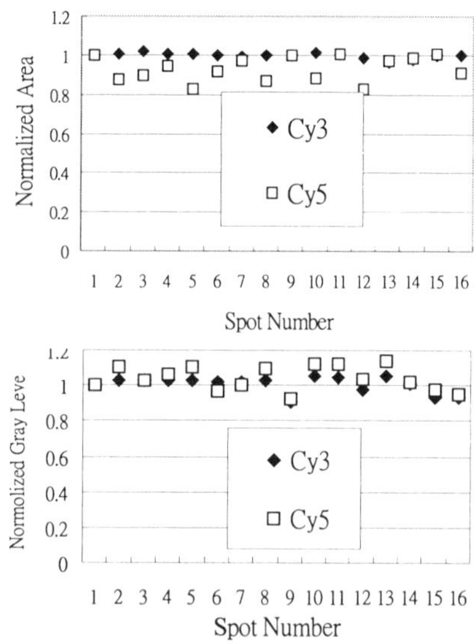

Fig. 4 Normalized (a) area size, (b) gray level for different protein spots.

PATTERNING POLYMER-SUPPORTED LIPID BILAYERS

Kevin C. Weng[1,3] and Curtis W. Frank[2,3]
[1]Department of Materials Science and Engineering, Stanford University
[2]Department of Chemical Engineering, Stanford University, Stanford, CA 94305
[3]Center on Polymer Interfaces and Macromolecular Assemblies (CPIMA)

Abstract

This project aims to micro-pattern polymer-supported lipid bilayers by corralling the supramolecular architectures with chemical/physical barriers. To achieve this goal, random lipo/glyco-copolymer was synthesized and characterized for interfacial properties. The copolymer mixed with free and fluorescently labeled lipids was transferred onto patterned photoresist on a planar glass surface by the Langmuir-Blodgett technique. Fluorescence microscopy, FRAP and AFM are employed to examine the resulting patterns as well as two-dimensional mobility of the supported membrane systems.

Keywords: Patterning, polymer-supported lipid bilayers, microarrays

1. Introduction

Planar supported lipid bilayers have been investigated not only for their ideal role as model biological membranes in biophysical studies but also for their great potential in applications as biosensors and lab-on-a-chip devices. Several polymer-supported lipid bilayer systems have been developed and characterized in our group [1]. One major direction that has evolved is to impose micro-patterns on the membrane systems by exploiting the surface interactions and lateral mobility of the lipids. In this research, we aim to present a simple method to pattern the polymer-supported lipid bilayers. With the sequence of steps by which the microarrays are created, each element can be tuned to desired chemical composition or biofunctionalization. The new approach should lead to versatile new functions for the artificial membrane systems.

2. Theory - Architecture of the Patterned Artificial Membrane

The polymer inserted between the solid surface and lipid bilayer is intended to provide a soft cushion that both prevents denaturation of membrane-incorporated proteins and stabilizes the bilayer structure. Random D-glucose-2-propenoate/N,N-di(octadecyl)acrylamide-copolymer (lipo/glyco-copolymer, home-synthesized) mixed with L-α-phosphatidylcholine (Egg-PC, Avanti Polar Lipids, Alabaster, AL) can physisorb on hydrophilic glass surface or be chemically grafted on the benzophonone functionalized surface by photochemical reaction. To corral the polymer-supported lipid bilayers, good physical/chemical barriers that provide physical constraint on lateral interdiffusion and resistance to interfacial film transfer are needed. Standard photoresist emerged to be an ideal material for this purpose. This positive photoresist is relatively hydrophobic compared with the silanol-abundant glass surface and its patterning technique has long been applied in the electronics industry. An array is therefore

J.M. Ramsey and A. van den Berg (eds.), Micro Total Analysis Systems 2001, 593–594.

created with each element being a complete and tunable system consisting of lipo/glyco-copolymer supported lipid bilayer.

3. Experimental

Glass slides were treated with Piranha solution and baked at 400°C for 4 hours to ensure oxidation. The static water contact angle was shown to be negligible. Mylar masks with 150μm features were designed by commercial drawing software and printed by a high resolution laser printer. Shipley 3612 or SPR220-7 positive photoresist was spin-coated onto the slides and exposed to 365nm and 405nm UV light, respectively, by contact mode. The patterned barrier of approximately 1.65μm of Shipley 3612 or 7μm of SPR220-7 (dry thickness) was formed after development and post-baking. Langmuir film of lipo/glyco-copolymer mixed with free lipids and fluorescent probe (NBD-DPPE, Avanti Polar Lipids, Alabaster, AL) was pre-organized at the air/water interface using a KSV 5000 minitrough (KSV Instruments, Helsinki, Finland) and then transferred onto the substrates at surface pressure of 27mN/m. The substrate was pulled out from subphase at the speed of 0.5mm/min. Epifluorescence images were taken by Nikon Eclipse ε800 microscope.

4. Results and discussion

Patterns of the transferred film that conform to the photoresist on surface were observed by epifluorescence microscopy. The edges are rough perhaps due to polymer entanglement and transfer conditions. Additional results from the following experiments will be presented at the time of the conference. The film on the benzophenone-functionalized substrate will be treated by photochemical reaction to form the chemically attached proximal layer of the lipids. Small unilamellar vesicles (SUVs) of various lipids are prepared by standard extrusion process. Vesicle fusion will be employed to complete the bilayers. Fluorescence microscopy, FRAP and AFM are the main tools to examine the patterned polymer-supported membrane systems.

5. Conclusion

Patterned photoresist provides a means to confine planar polymer-supported membrane systems in an array format. Patterning of polymer-supported proximal layer of lipids established a good foundation for subsequent modification to create spatially-addressable microarrays of polymer-supported lipid bilayers. Detailed construction and investigation of the systems will be presented at the conference.

Acknowledgments

Center on Polymer Interfaces and Macromolecular Assemblies (CPIMA), Stanford University

References

1. W.W. Shen, S.G. Boxer, W. Knoll and C.W. Frank, *Polymer-Supported Lipid Bilayers on Benzophenone-Modified Substrates,* Biomacromolecules, Vol.2(#1), pp. 70-79, Spring 2001

EVALUATION OF SILANOL CONCENTRATION ON QUARTZ GLASS SURFACE FOR EOF STABILITY OF CE CHIP

Takahiro Nishimoto, Hirohisa Abe, Yoichi Fujiyama, Masaki Kanai, Masakazu Akechi, Hiroaki Nakanishi
Technology Research Laboratory, Shimadzu Corporation, Kyoto, Japan
(e-mail: nisimoto@shimadzu.co.jp)

Abstract
The effects of heat treatment during fabrication process and NaOH treatment after fabrication on surface silanol (Si-OH) concentration profile are discussed. FTIR-ATR (Fourier transform Infrared Spectroscopy - Attenuated Total Reflection) method is applied to measure the surface OH concentration of materials. The result revealed that FTIR-ATR can be a simple solution for estimation of OH concentration on the glass surface.

Keywords: quartz glass, silanol concentration, FTIR-ATR, electroosmotic flow

1. Introduction
Quartz glass is the most suitable material for electrophoresis devices because it has high UV transmittance, very high purity and excellent chemical properties. We have reported the characteristics of CE microchips fabricated on quartz glass substrates [1][2]. The next subject of the quartz glass chip for CE is how to get the reproducibility of the analysis. It is needless to say that not only geometrical reproducibility but also chemical condition reproducibility of channel surface have to be realized.

2. Simulation
In the fabrication sequence of CE chips made of glass, some steps contains heat treatment that may change the OH concentration profile in the substrates. The change of OH concentration profile by heat treatment due to re-distribution and out diffusion is calculated using Fick's low shown below.

$$\partial C / \partial t = D \times \partial^2 C / \partial x^2$$

Here, C:concentration, t:time, x: distance. The distribution coefficient, D, was determined by another experiment as 2×10^{-9}cm^2/s at the temperature of 1100°C and agreed approximately with that in reference [3]. The calculated profiles before and after heat treatment are shown in figure 1(by courtesy of Shin-Etsu Chemical Co.).

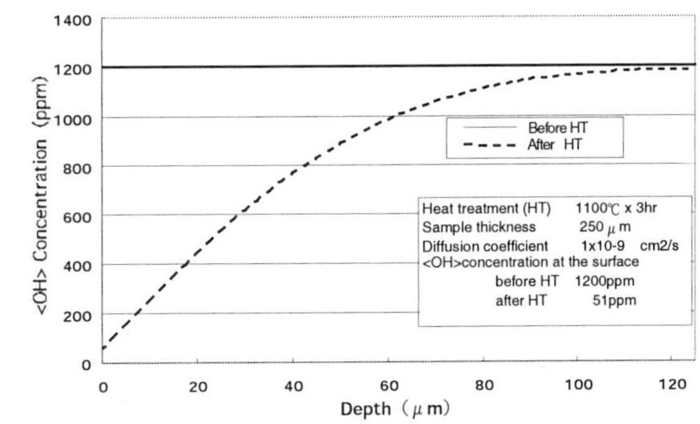

Figure1. Calculated OH concentration profiles of before and after heat treatment.

J.M. Ramsey and A. van den Berg (eds.), Micro Total Analysis Systems 2001, 595–596.

3. Experimental

Heat treatment(HT): Quartz glass substrates, in which hydroxyl (OH^-) have concentration of 1200 ppm, were heat treated in the furnace (1100°C×3hr : as same as annealing condition during fabrication). After treatment, glass samples were attached to Germanium ATR prism and IR absorption by surface OH were measured as shown in figure 2. *NaOH treatment(NaOH):* Other type of quartz glass substrates, in which hydroxyl have concentration of about 50 ppm were dipped into 0.1M NaOH solution (3min or 30min).

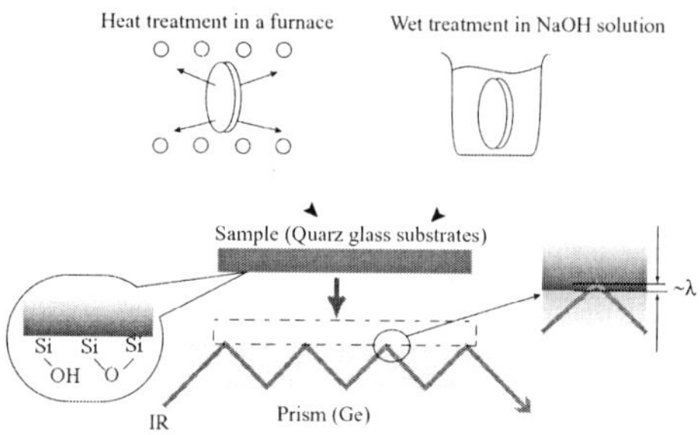

Figure 2 Schematic figure of sanple treatment and FTIR-ATR measurement.

4. Results and discussion

Figure 3 shows the ATR absorption spectrum obtained by subtraction of spectra: <before HT> - <after HT >. At the wavenumber of ~3600cm^{-1}, OH concentration reduction is observed. Figure 4 is the spectra which are obtained by subtractionof spectra: <after NaOH> - <no treatment>. In any subtraction, two spectra have been scaled to make the 1100 cm^{-1} Si-O band equal in height in order to compensate the difference of light penetration depth. This results showed that FTIR-ATR method is effective for evaluating the surface OH concentration. Furthermore, it is possible to find out the optimized treatment condition for CE chips by using the method.

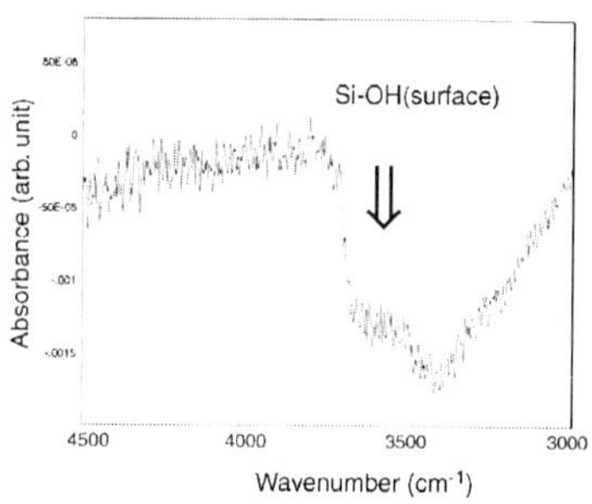

Figure 3. Subtracted FTIR-ATR spectrum(<before HT> - <after HT>)

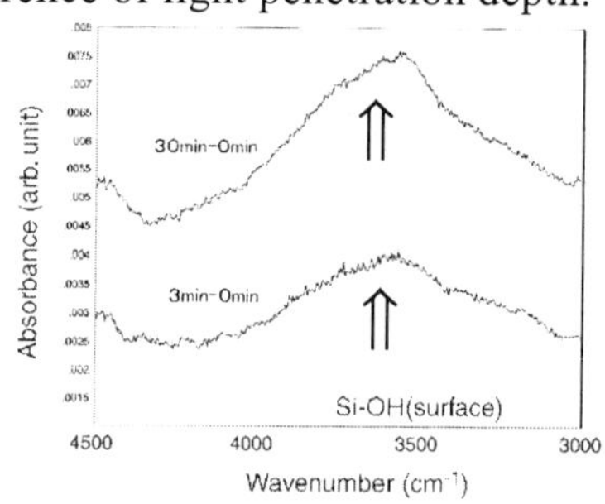

Figure 4. Subtracted FTIR-ATR spectra. <after NaOH> - <no treatment>

References

1. T. Nishimoto, H. Nakanishi, H. Abe, A. Arai, R. Nakamura, A. Yotsumoto, S. Shoji, Microfabricated Chips for Capillary Electrophoresis on Quartz Glass Substrates Using a Bonding with Hydrofluoric Acid, proceeding of μ-TAS'98, Banff, Canada(1998)311-314
2. T. Nishimoto, Y. Fujiyama, H. Abe, M. Kanai, H. Nakanishi, A. Arai, Microfabricated CE Chips with Optical Slit for UV Absorption Detection, proceeding of μ-TAS2000, Enschede, Netherland(2000)395-398
3. P.B.McGinnis and J.E.Shelby, J. Non-Cryst. Solids, 179 p.185(1994)

ELECTROCHROMATOGRAPHY ON-CHIP WITH LOW DEAD-VOLUME LATERAL CAPILLARY CONNECTIONS

D. Solignac[1], A. Sayah[1], S. Heimgartner[1], L. Szekely[2], R. Freitag[2], and M. A. M. Gijs[1]

[1]Institute of Microsystems, Swiss Federal Institute of Technology Lausanne, CH-1015 Lausanne EPFL, Switzerland

[2]Center of Biotechnology, Swiss Federal Institute of Technology Lausanne, CH-1015 Lausanne EPFL, Switzerland

Abstract

We report on a technology for the production of microfuidic chips with lateral low dead-volume connections to external capillary tubes. The chips are integrated in a commercial CE-system to perform electrochromatography experiments.

Keywords : Electrochromatography, microchip, CEC, powder blasting, low dead-volume connector

1. Introduction

Capillary electrochromatography (CEC) is a hybrid bio-analytic separation technique, which combines electro-osmotic pumping, as used in electrophoresis devices, with chromatographic separation. The technique allows the separation of uncharged molecules by specific affinity interactions with a stationary phase. We demonstrate CEC *on-chip*, thanks to the development of novel sol-gel stationary phases within the separation channel and low dead-volume lateral chip connections, by which our chip is connected to a commercial CE system.

2. Microchip fabrication and capillary connections

The CEC chip is composed of a single 40 cm long channel (figure 1) formed by lithographic processing of a pyrex wafer and bonding with a second wafer [1]. We connect the inlet and the outlet of the channel to fused silica capillaries, allowing the integration of our chip within a standard CE system, thereby benefiting from an automated sample injection, transport by EOF, UV detection and data handling.

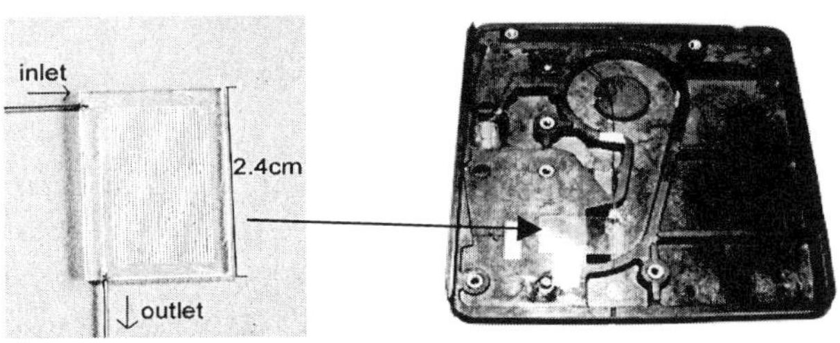

Figure 1 : CEC chip in pyrex glass and its insertion in a commercial cartridge.

J.M. Ramsey and A. van den Berg (eds.), Micro Total Analysis Systems 2001, 597–598.

The fabrication of the low dead-volume (~ 0.1 nl) fluidic connectors involves a powder blasting/gluing technology. A special design of the metallic mask for powder blasting [2] and a careful tuning of the angle of the powder jet allows to realize cavities which fit exactly around the capillaries. Figure 2a shows one wafer with the powder blasted cavity aligned with an HF etched channel. Two structured wafers are aligned and fusion bonded, giving a vertical profile at the end of the channel (see figure 2b) for insertion and gluing of a capillary tube (figure 2c).

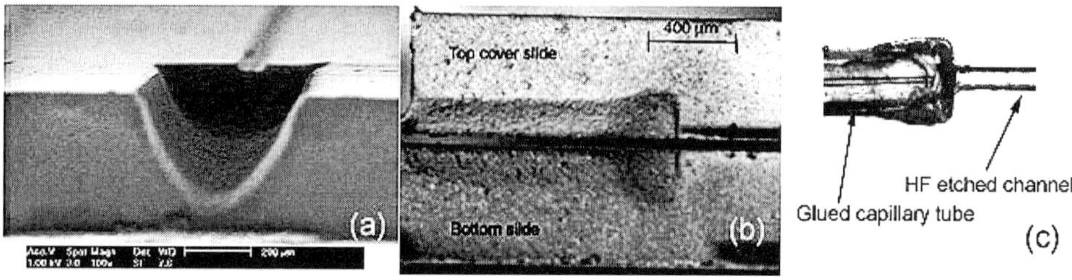

Figure 2: (a) Scanning Electron Microscopy graph of the powder blasted cavity. (b) Optical photograph showing the section of the cavity after assembly of the two wafers. (c) Optical photograph of a capillary tube glued with epoxy into the cavity.

3. CEC separation on-chip

A self-containing open tubular stationary phase using sol-gel chemistry [3] is developed and adapted for the CEC microchip. Separation experiments are carried out using a Beckman P/ACE System 5500 with diode array detection. Only the microchip channel is coated by the reverse stationary phase and the detection is performed on-chip at 280 nm. A mixture of three uncharged poly-aromatic compounds is separated in less than 7 minutes in the coated channel [4], proving the feasibility of our approach.

4. Conclusion

We have introduced a new way of connecting capillary tubes to microfluidic chips using the powder blasting technique and applied this technology to demonstrate on-chip CEC separations. For the future, this capillary-to-chip connection technique offers a new option for coupling fluidic chips to analytic instrumentation.

References

1. See, for example: C. S. Effenhauser, A. Manz, H. M. Widmer, Anal. Chem. 65, pp. 2637-2642 (1993), and literature cited therein.
2. E. Belloy, I. Zalunardo, A. Sayah, and M. A. M. Gijs, Proceed. SPIE's 2000 Symposium and Educational Program on Micromachining and Microfabrication, Sept. 18-20, Vol. 4174, pp.467-476, Santa-Clara, USA, 2000.
3. S. Constantin, R. Freitag, J. Chromatogr. A 887, pp. 253-263 (2000).
4. S. Constantin, R. Freitag, D. Solignac, A. Sayah and M. A. M. Gijs, Sens. & Actuators, in press (2001).

CONSECUTIVE MICROCONTACT PRINTING – LIGANDS FOR ASYMMETRIC CATALYSIS IN SILICON CHANNELS

Helene Andersson[1], Christina Jönsson[2], Christina Moberg[2], Göran Stemme[1]
[1]Department of Signals, Sensors and Systems, [2]Department of Chemistry, Organic Chemistry, Royal Institute of Technology, S-10044 Stockholm, Sweden

Abstract

Consecutive microcontact printing (μCP) has been developed to enable multiple functionalization of silicon surfaces, such as the immobilization of chiral ligands. The technique involves two subsequent printing steps using unstructured PDMS stamps. The pattern is already defined on the substrate, consisting of etched channels. Hence, no precise alignment is needed between the two printing steps. A carboxylic acid group-containing reagent was initially printed onto the silicon oxide surface and transformed to an anhydride. In the second printing step an ester bond was formed with the hydroxy-functionalized ligand. The formed molecular layers were evaluated by contact angle measurements, SEM and electron spectroscopy for chemical analysis (ESCA), indicating that the consecutive microcontact printing was successful. The channels were then sealed with a low temperature bonding technique using an adhesive PDMS film, which does not destroy the printed ligand.

Keywords: microcontact printing, chiral ligand, catalysis, solid phase screening

1. Introduction

High throughput screening for the discovery of new chemical compounds and processes is currently attracting much attention. The purpose of the screening is either to optimize the selectivity and/or reactivity of a chemical reaction or to optimize the properties of a product. The efforts to combine organic chemistry and micromachining are increasing but, although potentially very useful, catalysis-on-a-chip is not yet very well explored. A microfluidic device for high throughput screening would reduce the catalyst and reagent consumption and facilitate the separation and control of the end product since the catalyst or reagent is immobilized on the chip surface.

A possible chip design for performing combinatorial assays with immobilized catalysts would be an array of channels or reaction chambers with common or separate inlets and outlets, where the catalysts are attached on the internal walls. To enable screening of several catalysts on the same chip a convenient and effective technique for applying different catalysts in the different elements constituting the array is needed. Consecutive microcontact printing has therefore been developed to enable multiple functionalization of structured silicon surfaces, such as the immobilization of chiral ligands [1].

2. Chemistry Design

Asymmetric catalysis is a powerful technique used *e.g.* by the pharmaceutical industry in the search for new active compounds for pharmaceuticals. We have chosen to attach a chiral ligand (a pybox (2.6-bis(oxazolyl)pyridine) derivative) capable of performing asymmetric catalysis on the silicon surface. A carboxylic acid group-containing reagent (4-(chlorosulfonyl)benzoic acid) was initially printed onto the silicon oxide surface and transformed to an anhydride. In the second printing step an ester bond was formed with the

J.M. Ramsey and A. van den Berg (eds.), Micro Total Analysis Systems 2001, 599–600.

hydroxy-functionalized ligand, see scheme 1. The details about the printing procedure are described in [1].

4. Results and Discussion

The consecutive microcontact printing technique was first evaluated on unstructured silicon surfaces. Contact angle measurements were performed on each molecular layer to verify the µCP resulting in the following averaged values: silicon oxide 22°, anhydride 33°, pybox 96.5°. The results show that the surface characteristics changed after each surface treatment and the values correspond well with the hydrophobicity of the outermost atom layers.

The chemical composition of the surface layer was studied with ESCA. In the analysis, wide spectra were first run to detect elements present in the surface layer. This was followed by detailed spectra for each element, and then the relative surface composition was obtained from the quantification of C, O, Si and N. The ESCA results show that the carbon content increases with the number of layers and the presence of nitrogen from the immobilized ligand. In addition, measurements of the anhydride layer clearly show a peak at the shift expected for the carbon atom in the anhydride-bond.

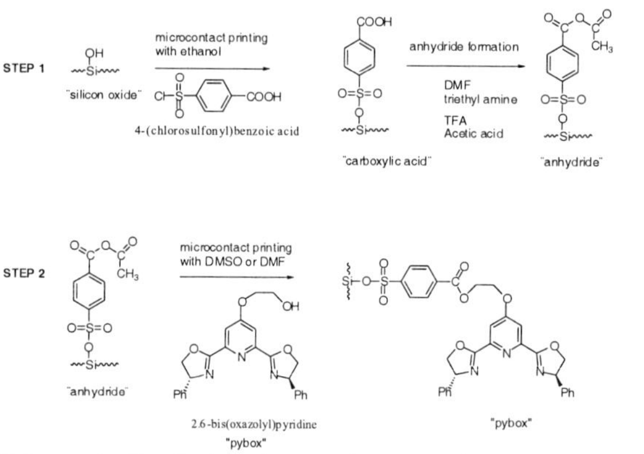

Scheme 1. Synthetic scheme of consecutive microcontact printing of 4-(chlorosulfonyl)benzoic acid and a 2,6-bis(oxazolyl)pyridine.

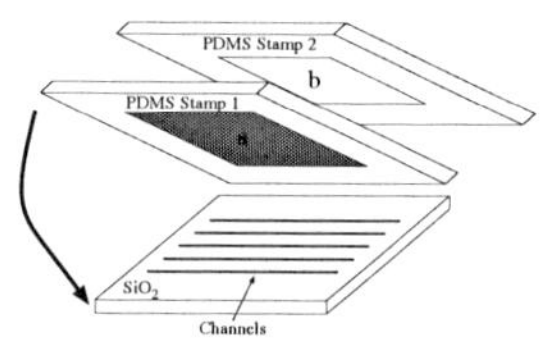

Fig. 1. A schematic of the principle of consecutive microcontact printing in channels.

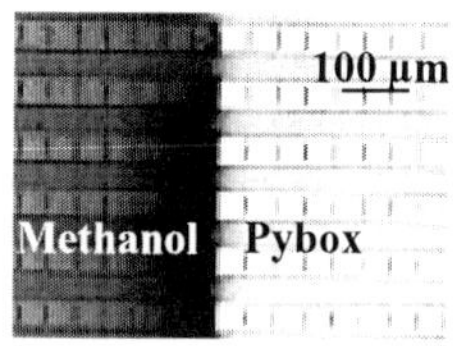

Fig. 2. A photo showing a wetting test where methanol was stopped by the printed pybox.

Finally, the consecutive microcontact printing technique was applied to immobilize the ligand in etched channels (50 µm wide and 50 µm deep), see figure 1. The printing was evaluated by wetting tests since it was not possible to perform ESCA measurements in the channels. In figure 2 a photo shows that methanol fills the array of channels until the beginning of the printed patch of pybox indicating that the ligand was successfully immobilized. Chiral ligands for asymmetric catalysis other than pybox are also currently immobilized using the consecutive µCP technique.

5. Conclusions

In this study microcontact printing is used in a novel manner. It enables a convenient method for performing complex surface modification of etched structures, which is a frequently appearing problem in biochemical microfluidic systems.

References

[1] H. Andersson, C. Jönsson, C. Moberg, G. Stemme, *Sensors and Actuators B*, 79, 78-84.

A MICRO BEAD AFFINITY CHROMATOGRAPHY FOR HEPATITIS C VIRUS (HCV) RNA REPLICASE DETECTION USING RNA APTAMER IN BLOOD

Suhyung, Cho[1,2], Sang-Ho Lee[4], Yong-Kweon Kim[4], Byung-Gee Kim[1,2,3]
[1]Interdisciplinary Program for Biochemical Engineering and Biotechnology,
[2]Institute for Molecular Biology and Genetics, [3]School of Chemical Engineering,
[4] School of Electrical Engineering and Computer Science,
Seoul National University, Korea

Abstract

Identification of *Hepatitis C virus (HCV)* RNA replicase with RNA aptamer immobilized on magnetic bead was performed from patient's blood and analyzed by MALDI-TOF mass spectrometer.

Keywords : Aptamer, bead affinity chromatography, microanalysis, MALDI-TOF

1. Introduction

Detection of pathogens and diagnosis of specific disease are executed in affinity chromatography system through antibody or aptamer. In this study, we detected specifically RNA replicase of *HCV* using aptamer bead in microfluidic system [1].

2. Theory

Affinity chromatography is widely used as a means of separation and purification with specific properties. We have coupled the technologies of protein affinity chromatography with sub-picomole protein identification by MALDI- TOF MS

3. Experimental

In the microfabrication of microfluidic device for micro reactor, liquid injection holes were formed by anisotropic wet etching, and microchannels were etched by a silicon deep RIE process as the desired depth (100~200μm). Main chamber size was 2.4×3 mm^2. Here, biotinylated aptamer was immobilized on a magnetic streptavidin bead (Fig. 1) and used total blood volume was 40μl blood. After loading of blood, several washing steps were performed and we used 8M urea as buffer to elute the captured protein.

J.M. Ramsey and A. van den Berg (eds.), Micro Total Analysis Systems 2001, 601–602.

4. Results and discussion

For the analysis of the eluted sample, protein was digested by trypsin for 12hour. As a result, we detected that the eluted protein was *HCV* RNA replicase through " *Profound* " database search (Fig. 2). Experimental peptide mass coverage of the protein was 54% and Z score was 2.43.

5. Conclusion

We analyzed the RNA replicase of *HCV* through micro bead affinity chromatography reactor with immobilized RNA aptamer.

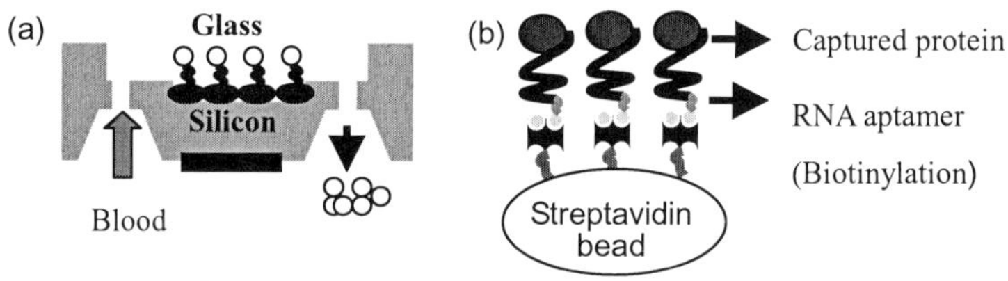

Figure 1.(a) Microbead affinity chromatography reactor. In this chamber, magnetic bead was fixed using bottom magnet. (b) Scheme of aptamer immobilized magnetic bead

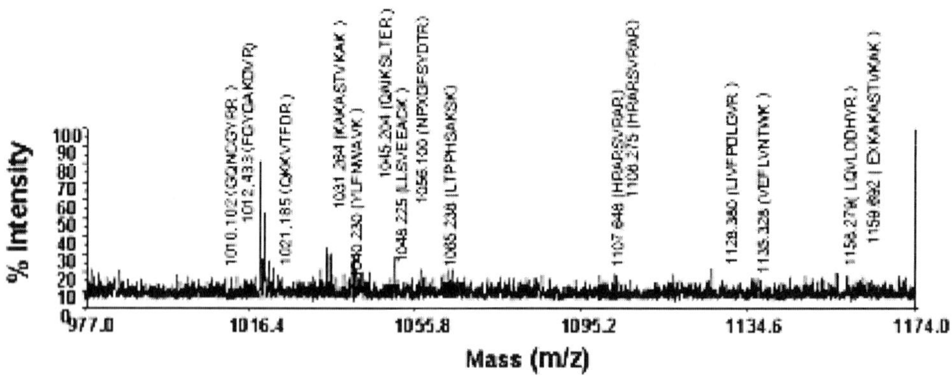

Figure 2. Mass spectrum of the HCV RNA replicase which eluted from patient blood.

Acknowledgments

This research was supported by the Intelligent Microsystems Program (IMP) that is the project of the " 21C Frontier R&D program "

References

1. Timonthy S. Romig, Carol Bell, Daniel W. Drolet, *J. Chrom.*, Aptamer affinity chromatography:combinatorial chemistry applied to protein purification, 731, pp. 275-284, (1999)

LASER MODIFICATION OF CHANNELS TO REDUCE BAND BROADENING OR TO INCREASE MIXING

Timothy J. Johnson, David Ross, Michael Gaitan, Laurie E. Locascio
National Institute of Standards and Technology, 100 Bureau Drive, Stop 8394, Gaithersburg, MD 20899-8394, USA

Abstract

An excimer-laser system was used to spatially alter the charge and/or geometry of imprinted, polymer channels. Subablation pulses were used to increase the charge on a wall of a 90°-turn without changing the physical morphology of the turn and resulted in a 40 % reduction in plate height. For rapid reagent mixing, slanted wells were etched near a T-channel junction. The wells caused lateral transport and mixing of two confluent streams. For an electroosmotic flowrate of 1.07 cm/s, the streams were 82.1 % mixed after just 0.038 cm compared with 2.2 cm for diffusional mixing.

Keywords: Excimer Laser, Polymer Channels, Band Broadening, Mixing

1. Introduction

The reduction of band broadening is an important research topic for μTAS devices and has been well documented for electrokinetic flow [1]. Attempts to correct for it have mainly focused on geometrical changes to the channel (e.g., channel restrictions or increasing the radius of curvature) [2-3]. Here we report an alternative approach, the use of an excimer laser to spatially increase the surface charge, and therefore the local electroosmotic (EO) flowrate, on the outside wall of a 90°-turn in PMMA.

The ability to rapidly mix incoming microstreams is another relevant research topic. In these laminar flow systems, flow-splitting techniques are used to reduce the diffusional equilibration time [4]. Recently He et al. [5] developed a picoliter mixer that utilized flow splitting, but also incorporated lateral transport to increase the mixing rate. Here, we describe a simple mixer design fabricated by photoetching a series of alternating wells near a T-channel junction in polycarbonate (PC) and show that lateral transport occurs to mix two streams under high EO flowrates.

2. Experimental

For the band broadening experiments, PMMA was used for the channel substrate and lid. The microchannel imprinting and laser modification parameters are discussed elsewhere [6]. The laser light modified a linear region, 500 μm, of the outer wall of the turn [6]. Uncaging a fluorescein dye in the channel and using fluorescence microscopy allowed for the detection and quantification of band broadening.

For the mixing experiments PC was imprinted with a T-channel. The laser light etched a series of slanted wells near the channel junction, as shown in Figure 1a. Substrates were sealed with poly(ethylene terephthalate glycol). Captured fluorescence images of two incoming streams allowed for the quantification of mixing, Figures 1b,c.

J.M. Ramsey and A. van den Berg (eds.), Micro Total Analysis Systems 2001, 603–604.

3. Results and Discussion

To quantify band broadening, the plate heights for nonmodified and laser-modified, 90°-turns were compared to a straight channel and to theoretical limits. It was determined that the modified turn, relative to a nonmodified turn, had a decrease in plate height of 40 % ± 4 % [6].

To quantify the extent of mixing, the deviation in the fluoresence profile across the channel from the profile of a perfectly mixed solution is calculated at the exit of the mixer. It can be seen from Figure 2 that the etched mixer causes a significant increase in the rate of mixing. At an EO flowrate of 1.07 cm/s, two streams were 82.1 % mixed after 0.038 cm compared to 2.2 cm for diffusional mixing.

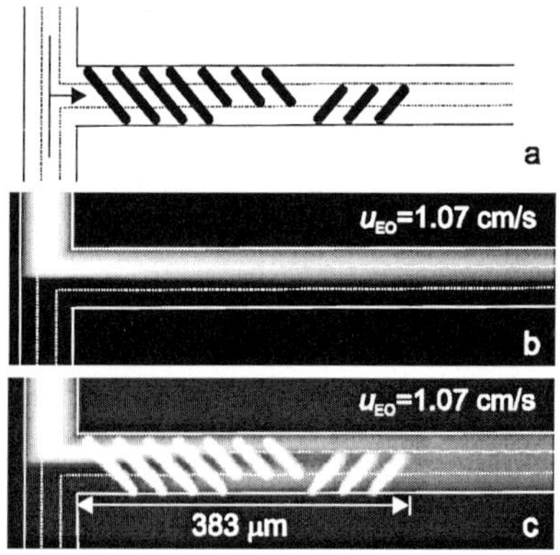

Figure 1. (a) Schematic of a T-channel with etched wells. Fluorescence microscopy images for (b) no mixer and (c) mixer.

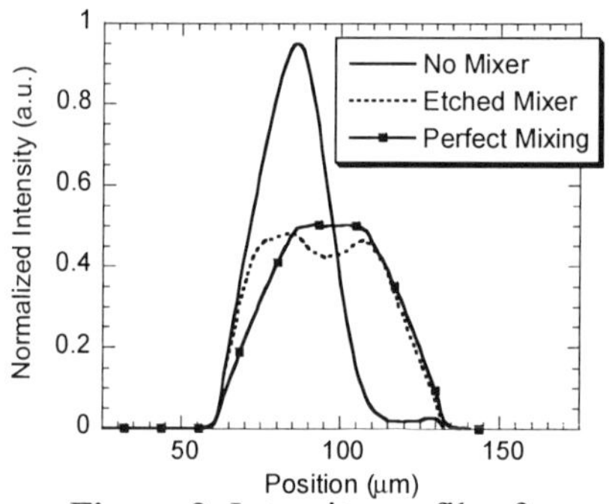

Figure 2. Intensity profiles for Figs. 1b, 1c, and perfect mixing.

4. Conclusions

An excimer laser was used to alter the surface charge and/or geometry of imprinted polymer channels. This post-modification technique has many potential applications for μTAS devices, and has been successfully applied to the reduction of band broadening and to rapid reagent mixing.

References

1. C.T. Culbertson, S.C. Jacobson, and J.M. Ramsey, *Dispersion sources for compact geometries on microchips*, Anal. Chem., **70**, pp. 3781-3789, (1998).
2. B.M. Paegel, L.D. Hutt, P.C. Simpson, and R.A. Mathies, *Turn geometry for minimizing band broadening in microfabricated capillary electrophoresis channels*, Anal. Chem., **72**, pp. 3030-3037, (2000).
3. C.T. Culbertson, S.C. Jacobson, and J.M. Ramsey, *Microchip devices for high-efficiency separations*, Anal. Chem., **72**, pp. 5814-5819, (2000).
4. F.G. Bessoth, A.J. deMello, and A. Manz, *Microstructure for efficient continuous flow mixing*, Anal. Commun., **36**, pp. 213-215, (1999).
5. B. He, B.J. Burke, X. Zhang, R. Zhang, and F.E. Regnier, *A picoliter-volume mixer for microfluidic analytical systems*, Anal. Chem., **73**, pp. 1942-1947, (2001).
6. T.J. Johnson, D. Ross, M. Gaitan, and L.E. Locascio, *Laser modification of preformed polymer microchannels*, Anal. Chem., ASAP Article, (2001).

A LASER-POLYMERIZED THIN FILM SILICA SURFACE MODIFICATION FOR SUPPRESSION OF CELL ADHESION AND ELECTROOSMOTIC FLOW IN MICROCHANNELS

Brian J. Kirby[1], Aaron R. Wheeler[2], Timothy J. Shepodd[1], Julia A. Fruetel[1], Ernest F. Hasselbrink[1], Richard N. Zare[2]
[1] Sandia National Laboratories, PO Box 969, Livermore, CA 94551-0969
[2] Chemistry Department, Stanford University, Stanford, CA

Abstract

A laser-polymerized thin film for silica surface modification is presented. This technique enables photopatterned surface charge modification consistent with optical tweezer techniques. This charge modification may be applied for chip-based cell techniques and multi-dimensional separation techniques that require nonuniform zeta potential on a microfluidic chip.

Keywords: surface modification, zeta potential, cell adhesion

1. Introduction

Silica surface charge interferes with a number of microfluidic chip-based techniques (*e.g.*, capillary isoelectric focusing (cIEF)) by inducing unwanted electroosmotic flow (EOF); this surface charge can also interfere with chip-based cell techniques by causing cells to electrostatically adhere to cell walls. Two-dimensional separations may require selectively eliminating zeta potential (ζ) on one part of a microfluidic chip while ζ is retained in other sections. Use of optical tweezers for cell manipulation on-chip [1] further requires that surface modifications be nonabsorbent and nonscattering at the laser wavelength.

2. Surface Modification Mechanism

An aqueous mixture of 3-(trimethoxysilyl)-propyl acrylate (Aldrich) and acetic acid is applied to pretreat the silica surface (Figure 1), covalently bonding the acrylate intermediary to exposed surface charge sites. This intermediary affects ζ only slightly, and hence does not inhibit electroosmotic flow. Following this pre-treatment, frequency-tripled Nd:YAG laser emission (355 nm) is used to locally polymerize acrylamide to the intermediary. Flushing the solution leaves a photopatterned acrylamide thin film.

3. Measurements

Streaming potential was used to measure ζ in silica capillaries with and without surface modifications. For 1 mM phosphate buffer solutions (K+ cation, pH 7.3 +/- 0.2), ζ was measured at –83 +/-3 mV, -75 +/- 3 mV, and 0 +/- 1 mV for untreated, pretreated, and polymerized surfaces, respectively; the photopatterned polyacrylamide thin film

J.M. Ramsey and A. van den Berg (eds.), Micro Total Analysis Systems 2001, 605–606.

creates an uncharged (ζ=0) surface but the pretreat step does not significantly affect ζ. A solution containing murine bone marrow mast cells (d=10 μm), glucose, and phosphate buffer was brought into contact with untreated and polymerized silica surfaces on glass cover slides and a wet-etched glass chip (30 x 150 μm channels). Cell adhesion was inferred from images after flushing with solution. On cover slides, cells appear post-flush only if adhered to the surface (Figure 2). In microchannels, adhered cells appear the same regardless of flow rate, while cells in the bulk appear only as a faint background in the flowing image. Figure 3 shows cell solutions in microchannels with and without flow (flowrates as high as 20 cm/s were used). The polyacrylamide thin film prevents mast cell adhesion.

References

1. A.R. Wheeler, K. Morishima, D.W. Arnold, A.B. Rossi, and R.N. Zare, Single Organelle Analysis with Integrated Chip Electrophoresis and Optical Tweezers, in *Micro Total Analysis Systems 2000*, A. van den Berg et al. (eds.), (2000).

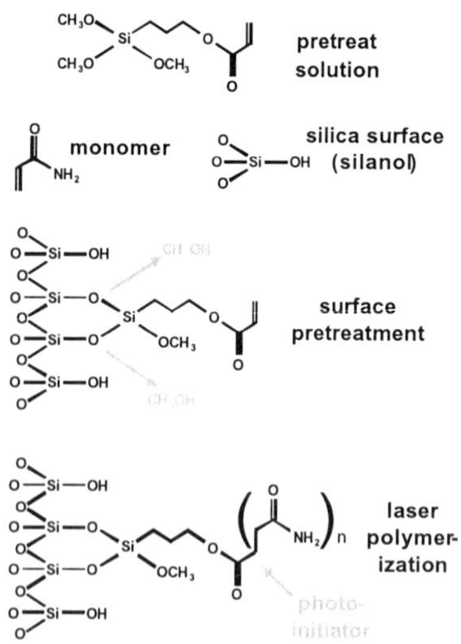

Figure 1. Surface pretreatment and laser-polymerization steps.

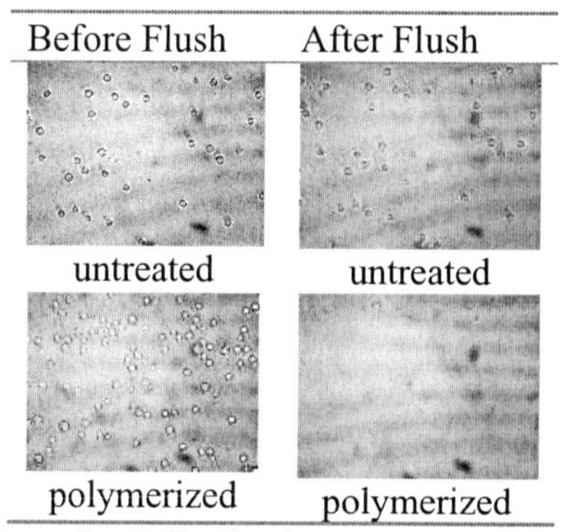

Figure 2. Cell adhesion on cover slides.

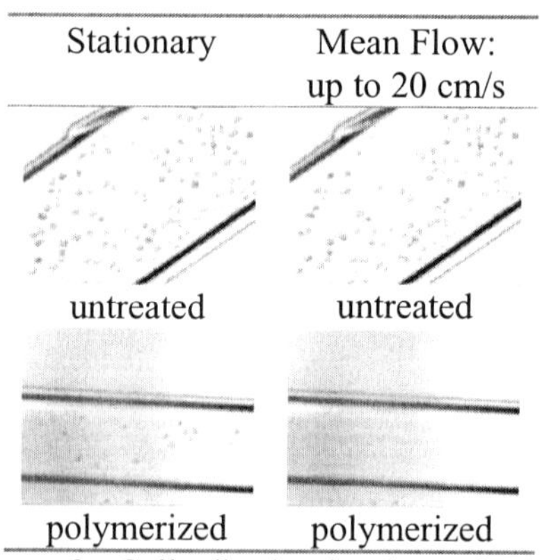

Figure 3. Cell adhesion in glass microchannels.

CHEMICAL FOUNDATIONS FOR PLASTIC-BASED MICRODEVICES: CHEMICALLY MODIFIED POLY(METHYL METHACRYLATE) AND POLY(CARBONATE) SUBSTRATES USED IN THE FABRICATION OF MICROANALYTICAL DEVICES

Robin L. McCarley*, Alyssa C. Henry, Alison Smith, Bikas Vaidya, Michelle Galloway, Yun Wang and Steven A. Soper,
Department of Chemistry, Louisiana State University
Baton Rouge, Louisiana 70803
tunnel@lsu.edu

Abstract

Polymer surfaces are modified so as to control the properties of microanalytical systems constructed from such polymers using hot embossing fabrication techniques. We demonstrate here the modification of poly(methyl metacrylate), PMMA, and poly(carbonate), PC, surfaces using solution- and gas-phase methods.

Keywords: Surface Modification, Interfacial Characterization, Plastics

1. Introduction

Polymeric substrates for the construction of microanalytical devices have become important due to the ease of which such materials can be formed into microdevices using molding, embossing, and micromachining techniques [1]. We discuss here the surface modification of a variety of polymers currently being investigated for use in the construction of microanalytical devices, such as capillary electrophoresis "chips" and biomolecule arrays. Poly(methyl methacrylate), PMMA, and poly(carbonate), PC, surfaces can be derivatized [2, 3] to yield surfaces that can be made hydrophobic, contain organic and biological moieties, or that are coated with conducting metal films.

2. Experimental

All chemicals were used as received from Aldrich. Solvents used were of spectroscopic or HPLC grade. PMMA and PC sheets were purchased from Goodfellow and MSC Industrial Suppliers. PMMA and PC sheets were derivatized and then subsequently evaluated using procedures published earlier [2,3,4].

3. Results and Discussion

Poly(methyl methacrylate), PMMA, surfaces can be derivatized to yield amine-termininated PMMA, AT-PMMA [2,3]. In addition, AT-PMMA can be further functionalized such that a variety of ionizable, hydrophilic groups are present, thus

J.M. Ramsey and A. van den Berg (eds.), Micro Total Analysis Systems 2001, 607–608.

leading to control over the electroosmotic flow, EOF, in PMMA microchannels, Figure 1.

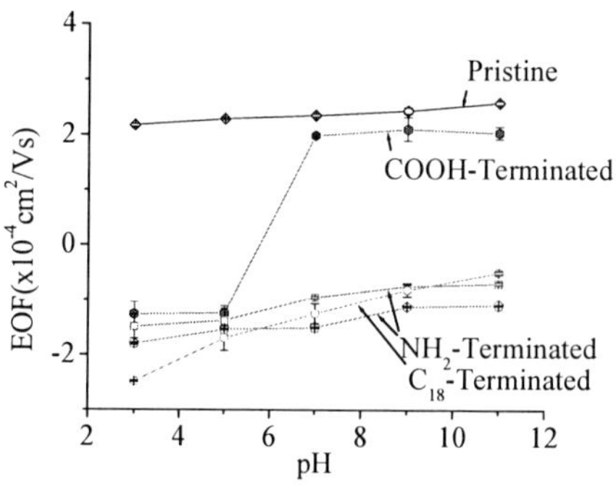

Figure 1. EOF profiles for a variety of chemically modified PMMA surfaces.

Patterns (arrays, etc.) of a variety of materials on amine-terminated PMMA surfaces can be achieved by protection of the amine groups with photo-labile groups, removal of such groups with light, and subsequent attachment of the material of choice.

Similar protocols can be used to modify the surface of poly(carbonate), PC. In particular, sulfonation of PC is readily achieved through the use of solution and gas-phase methods.

5. Conclusions

PMMA and PC surfaces can be chemically modified using covalent attachment protocols to yield a variety of surface moieties. These surface functional groups can be used to control the EOF in microfabricated separation devices or act as a foundation onto which metal films or biomolecules can be attached.

Acknowledgements
The financial support of NIH and NSF is gratefully acknowledged.

References

1. S.A. Soper, S.M. Ford, S. Qi, R.L. McCarley, K. Kelly, and M.C. Murphy, *Microelectromechanical Systems Fabricated in Polymeric Materials: Applications in Chemistry and Life Sciences*, Anal. Chem., *72*, 643A-651A (2000).

2. A.C. Henry, T.J. Tutt, M. Galloway, Y. Davidson, C.S. McWhorter, S.A. Soper, and R.L. McCarley, *Chemical Modification of PMMA Used in the Construction of Microanalytical Devices*, Anal. Chem., *72*, 5331-5337 (2000).

3. B. Karandikar, J. Puschett, and K. Matyjaszewzki, Polymer Prep., *30,* 250-251 (1989).

4. Xiaohua Huang, Manuel. J. Gordon and Richard N. Zare, *Current-Monitoring Method for Measuring the Electroosmotic Flow Rate in Capillary Zone Electrophosis*, Anal. Chem., *60*, 1838-1840 (1988).

5. M. Galloway and S.A. Soper, unpublished results, (2001).

EXPERIMENTAL AND NUMERICAL STUDY OF ELECTROKINETIC AND PRESSURE DRIVEN FLOWS IN STRAIGHT AND CURVED MICRO-CHANNELS

Y. Chen[1], Z. Chen[1], G.-P. Li[2], M. Bachman[2], and D. Banerjee[3]
[1]Department of Biomedical Engineering and Beckman Laser Institute, University of California, Irvine, CA, USA.
[2]Department of Electrical and Computer Engineering, University of California, Irvine, CA, USA.
[3]COVENTOR, 7 Corporate Park, Suite 260, Irvine, CA 92606, USA.
debban@coventor.com

Keywords: Optical Doppler Tomography (ODT), Intralipid, electro-osmotic flow.

Abstract

Electrokinetic and pressure driven flows were studied in straight and curved PDMS microchannels of nominal cross-sections of 50 μm × 30 μm and 100 μm × 30 μm. Aqueous solution of Intralipid™ (Manufacturer: Pharmacia AB) was used as the test liquid. Velocity profiles at various cross sections were measured experimentally using the technique of Optical Doppler Tomography (ODT). ODT uses a low coherence light source to perform tomographic cross-sectional velocity scans. Flow rate and average velocity measurements were performed for various pressure drops and different applied potentials. Effects of surface properties on flow rates were studied using surface treated and non-surface treated microchannel walls. Detailed numerical analyses of flow behavior is presented and compared with the experimental data. Analytical formulations were derived for flow in a curved microchannel to understand the flow profile obtained from experiments and numerical simulations.

Figures 1 and 2 are obtained from simulation results using Coventorware (Coventor, USA) for flow in straight and curved microchannels. Figures 3 and 4 are obtained from tomography experiments conducted for pressure driven flow of Intralipid in straight and curved microchannels. The experimental results and numerical calculations are found to be in good agreement. Figures 5 and 6 are obtained from simulation of electrosmotic flow using Coventorware in a straight channel using 20 mM aqueous solution of KH_3PO_4. From the simulations results it is observed that the velocity profile in electrosomotic flow is relatively flat compared to pressure driven flow. This result is qualitatively confirmed by ODT scans which show a plateau in velocity for the central region of the channel. Oxygen plasma surface treatments to the PDMS were found to increase the degree of electro-osmotic flow by fourfold. The velocity magnitudes and profiles are consistent for simulation predictions and experimental measurements. The simulation results show that the average velocity is 1.1 mm/s which is consistent with the experimental result of 1.09 mm/s for an electrosmotic

J.M. Ramsey and A. van den Berg (eds.), Micro Total Analysis Systems 2001, 609–610.

mobility of 3.3 cm^2/Vs. Consistent results were reported by [1] while comparing experimental data with numerical simulations for electrosmotic and pressure driven flow in microchannels.

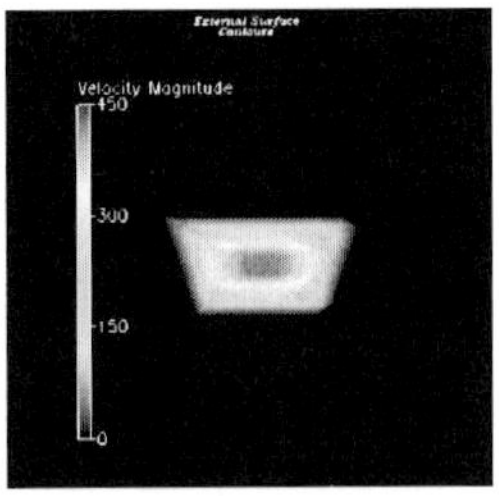

Figure 1. Pressure driven flow simulations for a straight channel (30 μm × 50 μm).

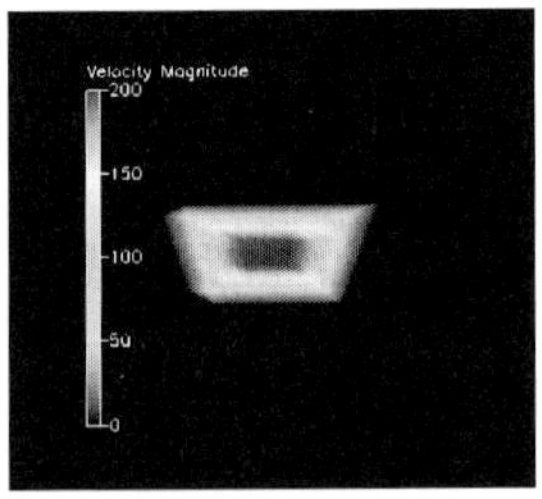

Figure 2. Pressure driven flow simulations - curved channel (30μm×50μm, 5mm radius).

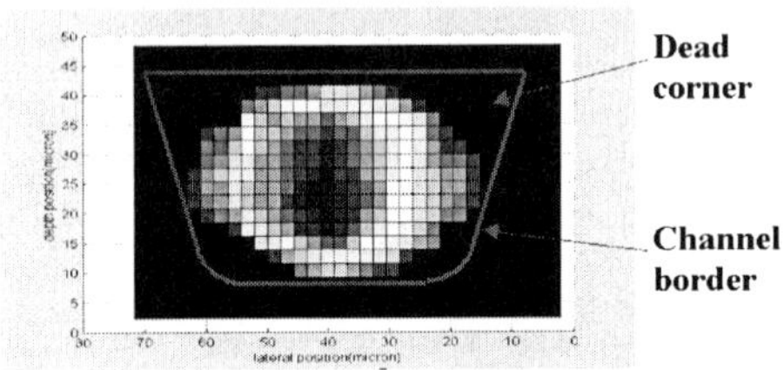

Figure 3. Pressure driven flow experiment for straight channel (30 μm × 50 μm) using ODT. Parabolic velocity profile is observed.

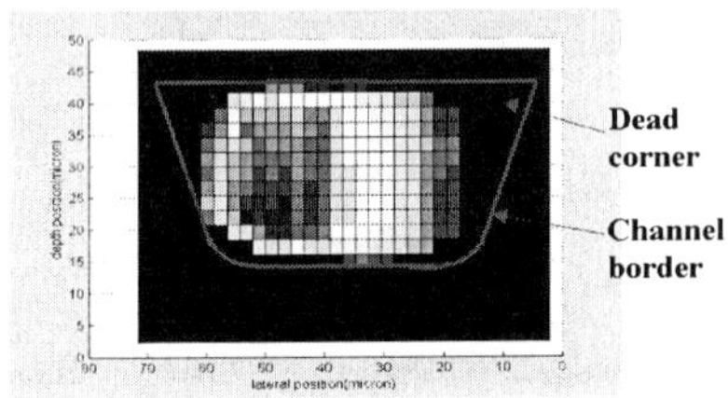

Figure 4. Pressure driven flow experiment for curved channel (30μm×50μm, 5mm radius). Note: higher velocity at inside edge.

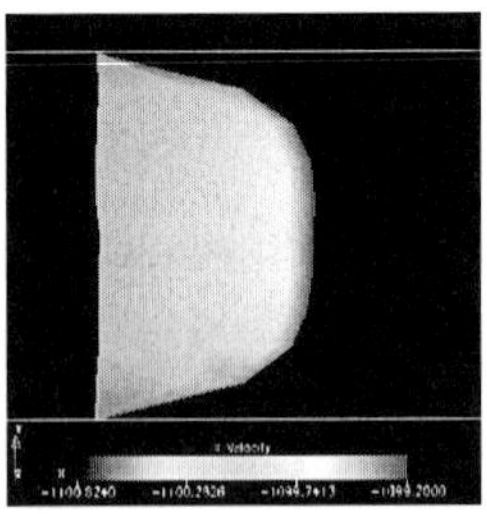

Figure 5. Side view of velocity field for electrosomotic flow simulations in a straight microchannel (30 μm × 100 μm).

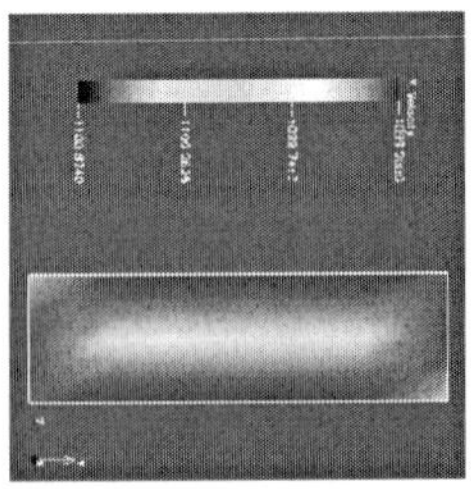

Figure 6. Cross sectional view of velocity field for electrosomotic flow simulations in a straight microchannel (30 μm × 100 μm).

References

1. Lutteri G-L., Dodge A., Boer G., Lichtenberg J., Verporte E., de Rooij N., "*Consequences of Opposing Electrokinetically and Pressure-Induced Flows in Microchannels of Varying Geometries*", Micro Total Analyses Systems 2000, pp. 351-354, Kluwer Academic Publishers.

DESIGN RULE EVALUATION FOR MICRO-SCALE FLOW RESTRICTIONS: COMPARING EXPERIMENT AND THEORY

Jian-Bin Bao,[1] Cameron Skinner,[1] N. Chiem,[2] Fahima Ouchen,[1] Paul C. H. Li,[1] D. Jed Harrison[1]

[1]Dept. of Chemistry, University of Alberta, Edmonton, AB T6G 2G2, Canada

[2]Conventor Inc., San Mateo, CA 94404, USA

Abstract

Control of the direction of fluid flow within a network of flow paths in a microfluidic device is paramount to its proper operation. Pressure driven flow requires either active valving for control, or passive control through variation of the flow resistance of each intersecting channel by adjusting its geometry. In order to achieve rational design of passive flow control, an accurate model describing fluid flow in micron-scale devices must be available. In this report we compare the predictions of the Navier-Stokes equations with experimental results for three-port devices with flow restrictors. The good agreement seen indicates that flow restrictors with 1-μm dimensions can be designed using solutions to the Navier-Stokes equations.

Keywords: Pressure, Navier-Stokes equations, Fluid mechanics

1. Introduction

While solutions to the Navier-Stokes equations are quite accurate for dimensions above 10 μm, there is uncertainty as to their accuracy below 2 μm. A single research group has studied such device dimensions, their data indicating drastic deviation from predicted friction factors for a channel 0.8 μm deep [1]. The geometric form factor for Poisseuille flow in a shallow slot-like rectangle is in the range of 0.001, versus 0.14 for a square cross-section [2], and the validity of the form function at such small dimensions is uncertain. Thus, an evaluation of the Navier-Stokes equations for use as a design tool for multichannel devices with dimensions on the order of a micron is needed.

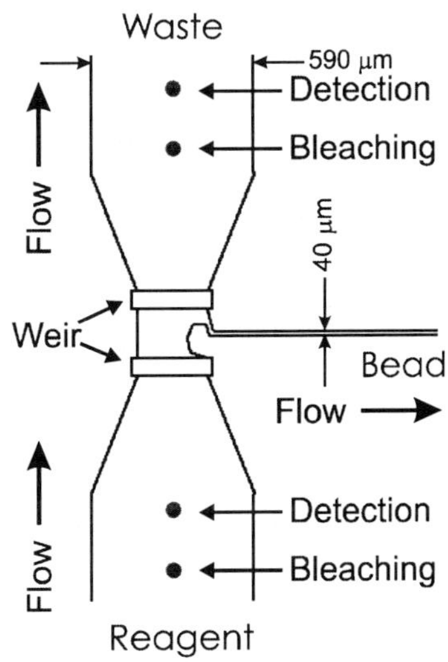

Figure 1. Three-port device (laser spots are not in scale): channels 10 µm deep, weir 1 or 3 µm deep

2. Experimental

Fluorescein photobleaching was performed setting up a time of flight measurement of average linear flow rate, along the path from Reagent to

J.M. Ramsey and A. van den Berg (eds.), Micro Total Analysis Systems 2001, 611–612.

Waste reservoirs (Fig. 1). The first laser beam bleached a spot using a chopper to create a square wave pattern of 0.19 μM fluorescein. Velocities were measured many *entrance lengths* downstream as a function of pressure, with a fluorescence detector. The proportion of flow into the bead channel versus across the second weir was evaluated experimentally and using theoretical expressions.

3. Results and discussion

Velocities determined up- and down-stream of the flow restricting weirs in Figure 1 are shown in Figure 2 for two weir depths. The cross-sections at the measurement points were the same, so the ratio of velocities is also the ratio of volume flow rates, Q_W%. The use of a 1-μm weir was observed to give 14% leakage along the bead channel, while a 3-μm deep weir presents less flow resistance, so that about 96% of the flow traversed the second weir. Analytical solutions to the Navier-Stokes equations were employed to describe rectangular channels, while FlumeCAD software was used to evaluate the form factor in the case of the bead channel, which is narrow enough to require evaluation as a D-shape, rather than a rectangle. The measured and calculated values of Q_W% agree within ±2%.

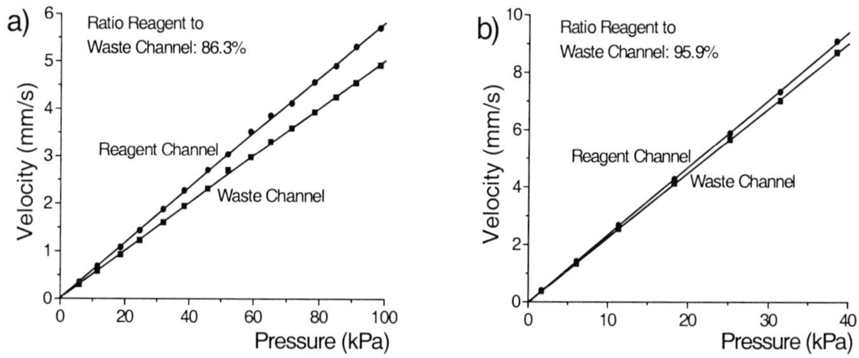

Figure 2. Experimental flow rates for Waste and Reagent channels. Ratios of the velocities give Q_W%. a) 1-μm deep weir; b) 3-μm deep weir

4. Conclusions

Our results demonstrate that solutions to the Navier-Stokes equations provide an accurate means for predicting pressure driven flow within dimensions of 1 μm or larger. Flow restrictions with micron dimensions can be accurately designed with these equations. Confidence in the theory allows for the establishment of design rules with absolute predictive power for the micron scale.

References

1. Pfahler, J.; Harley, J.; Bau, H.; Zemel, J. *Sens. Actuators* **A21-23**, 431 (1990)
2. Longwell, P. A. *Mechanics of Fluid Flow*, McGraw-Hill, New York, 1966

OPTIMIZATION OF FIELD AMPLIFIED SAMPLE STACKING ON A MICROCHIP

Rajiv Bharadwaj[1], Juan G. Santiago[2]
[1] Department of Chemical Engineering, Stanford University
[2] Department of Mechanical Engineering, Stanford University

Abstract

An experimental scheme has been developed to study field amplified sample stacking at a single fluid-fluid interface on a microchip. Using a simple cross-channel configuration a high-gradient initial step-function in the buffer concentration field was established which causes an unsteady sample concentration process. The concentration process is wave-like in character and is analyzed using epifluorescence imaging. Experiments at varying fields suggest the existence of an optimum electric field for a given concentration ratio.

Keywords: Sensitivity, sample stacking, optimization

1. Introduction

Sensitivity to low analyte concentrations is a crucial challenge towards the development of robust miniaturized bioanalytical devices. Field amplified sample stacking is a promising method of achieving high sensitivity. Although conceptually simple, the transport phenomena associated with sample stacking in microchip-based separation systems is in general a complex coupling of convective-diffusion, electrostatics, and electrokinetics.

2. Theory

Figure 1 depicts the stacking process in a simple capillary system without electroosmotic flow (EOF). A first-order analysis of stacking, which neglects diffusion, suggests that the stacked-to-initial concentration ratio of the analyte can be approximated as $C_f / C_i = E_S / E_B = \gamma$. In presence of EOF, the gradient in buffer

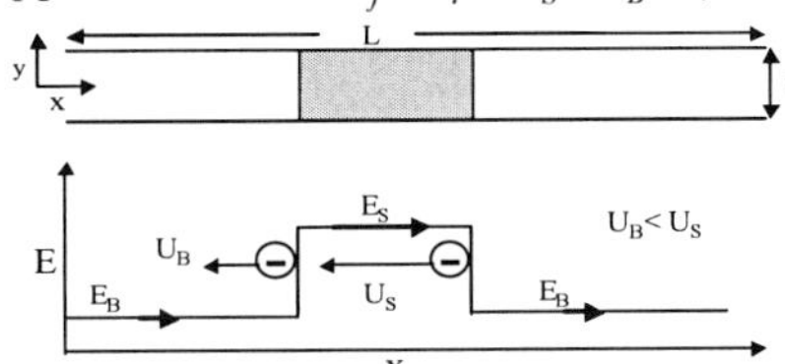

Figure 1: The sample buffer is lower concentration than the background buffer. The sample region has a lower electrical conductivity and hence a higher electric field. The analyte ions, represented as circles, experience higher velocity inside the sample region than in the background buffer region and stack at the rear sample plug boundary.

concentration leads to gradients in both the electric field and the electroosmotic wall mobility along the axis of the channel. Application of Gouy-Chapman-Stern model [1] of the electrical double layer suggests that the sample-zone-to-background EOF velocity ratio will scale as $U_{EOF,S} / U_{EOF,B} \propto \gamma \sqrt{\gamma}$. This mismatch causes dispersion of the stacked sample and hence limits the maximum concentration enhancement [2]. Thus

J.M. Ramsey and A. van den Berg (eds.), Micro Total Analysis Systems 2001, 613–614.

increase in γ increases stacking but also increases the dispersion and therefore optimization of system parameters is essential for efficient stacking.

3. Experimental

In order to experimentally study the dispersive processes associated with stacking, we have chosen to study the sample stacking flow field associated with a single step change in conductivities (Figure 2).

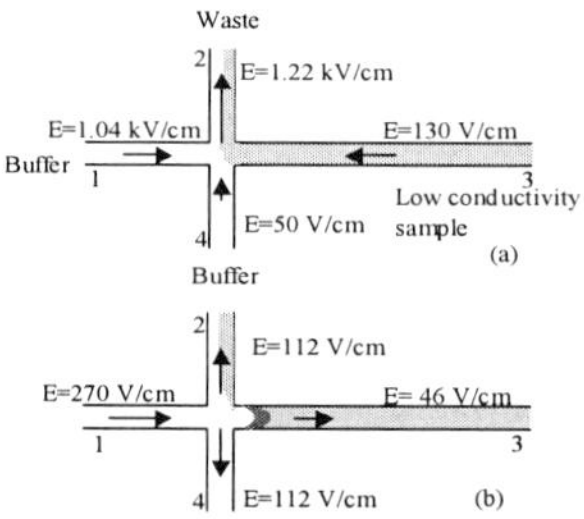

Figure 2: a) Electrokinetic injection scheme for generating a well-defined interface between the sample and the background buffer using electrokinetic injection; b) Conditions for driving the high conductivity buffer into the stacking channel. In all cases the sample was 25 μm Fluorescein dye in a 5mM Borate buffer and background buffer was 25 mM Borate Buffer, pH = 9, leading to $\gamma = 5$. The channel was 50 μm wide and 20 μm deep with an isotropic etch profile.

4. Results and Discussion

Figure 3 shows the instantaneous CCD images of the concentration profiles during stacking. EOF mismatch induced pressure gradient causes the curvature of the analyte band. In this initial study we performed experiments at a range of electric fields(5 –100 V/cm). Figure 4 shows the measured (column averaged) axial concentration profiles in the microchannel as a function of time for two representative electric fields. It is observed that the stacking is higher at higher field. Our data suggests that, at low electric fields the electromigrational flux is balanced by the diffusive fluxes which counter stacking. At higher electric fields, the electromigrational flux is more important relative to the dispersion effects of diffusion and pressure driven convection. In all cases, the maximum achievable concentration is much less than that predicted by simple one-dimensional theory. Also, at high electric fields, the peak concentration is progressively less sensitive to electric field.

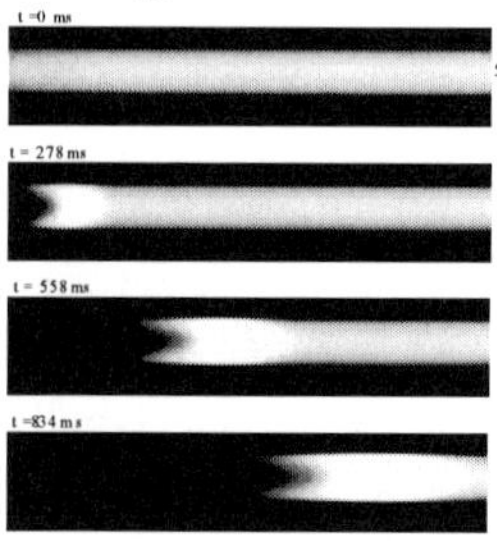

Figure 3

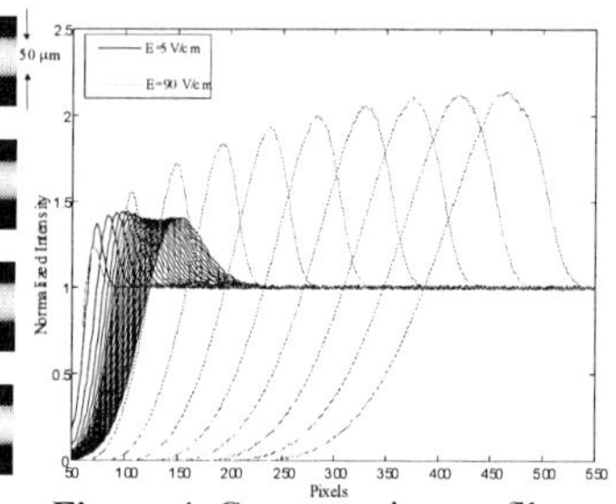

Figure 4: Concentration profiles

5. Conclusions

We have developed a robust experimental approach to study stacking at a single buffer/buffer interface on a microchip. Clear evidence of EOF mismatch induced pressure gradients is presented and data was obtained for a range electric fields. Experiments indicate that at low electric fields the concentration increase is simply limited by diffusion. At higher fields dispersion due to both diffusion and EOF-mismatch induced convection is important. We are currently performing a detailed parametric experimental and theoretical study for this process.

6. References:

1. R.J. Hunter, Zeta Potential in Colloid Science Principles and Applications, Academic Press, San Diego, 1988.

2. D.S. Burgi, R.L. Chien, Optimization in sample stacking for high-performance capillary electrophoresis, Anal. Chemistry 63, 1991, p. 2042-2047.

SOLUTE DISPERSION BY ELECTROOSMOTIC FLOW IN NONUNIFORM MICROFLUIDIC CHANNELS

M.T. Blom[1], E. F. Hasselbrink[2], H. Wensink[1], A. van den Berg[1]
[1] University of Twente, MESA+ Research Institute, Enschede, The Netherlands, P.O. Box 217, 7500 AE Enschede, The Netherlands, m.t.blom@el.utwente.nl
[2] Sandia National Laboratories, PO Box 969, MS 9951, Livermore, CA 94550

Abstract

In this work the effect of nonuniform channel shape on the dispersion of a solute by electroosmotic flow in microfluidic systems was studied using fluorescence imaging.

Keywords: dispersion, electroosmotic flow, band broadening

1. Introduction

The phenomenon of dispersion is a central issue determining the resolution of several types of liquid phase chromatography, and is an important component of overall transport through electrokinetic flow systems. The severe dispersion of solute as it travels through a U-turn, for example, has attracted recent attention for this reason [1]. In "ideal" electroosmotic flow through uniform, straight channels (in the limit of thin double layers), solute dispersion is simply due to axial diffusion. However, in straight microfabricated channels, channel nonuniformities may arise due to surface roughness (e.g., due to etching or powderblasting fabrication methods), or imperfections (e.g., etch pits, waviness due to poor resist adhesion, or imperfect bonding of channel lids). Other nonuniformities may be intentional, e.g., engineered post arrays for capillary electrochromatography. In this work, we measure dispersion in several types of nonuniform channels, and report their geometric Taylor-Aris dispersion coefficients α_o. The dispersion coefficient relates the effective diffusion coefficient to the binary diffusion coefficient D_{12} as a function of the Peclet number: $D_{eff} = D_{12}(1 + \alpha_o Pe^2)$ [2,3].

2. Measurements

Measurements are performed in glass channels using a custom-built fluorescence imaging apparatus. The two types of perturbations studied are (a) well-defined perturbations (square-wave and sawtooth) of various wavelengths and amplitudes (Fig. 1), and (b) fine-scale random roughness introduced by a powderblasting channel fabrication method. Samples of Rhodamine-110 dye (in a 50:50 solution of 20mM phosphate buffer and acetonitrile) are injected into several parallel channels of different waviness, as shown in Fig. 2. As the samples move

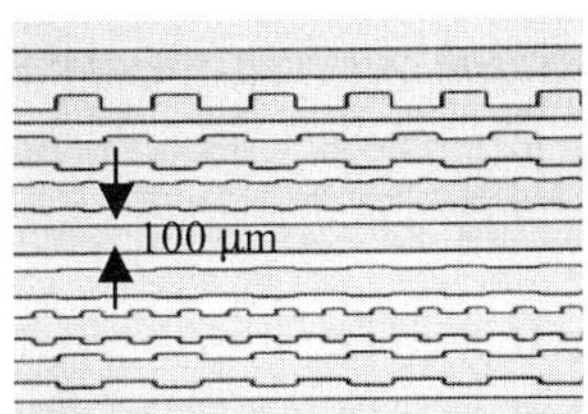

Fig. 1: Square-wave channels having perturbations of various amplitudes (0, 3 or 10 μm) and wavelengths (100 or 200 μm).

J.M. Ramsey and A. van den Berg (eds.), Micro Total Analysis Systems 2001, 615–616.

through the channels, they assume a Gaussian profile of width $\sigma^2 = 2D_{eff}t + \sigma_o^2$, where t is time and σ_o is the initial plug width. Gaussian fits to profiles in a single channel give values of σ^2 at various times; plotting this versus time gives a straight line of slope $2D_{eff}$.

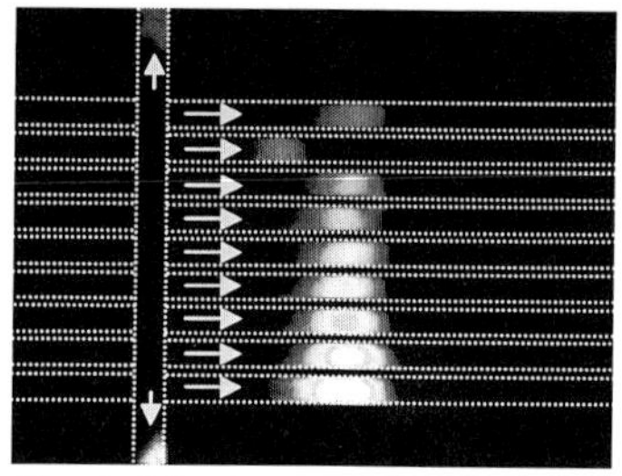

Fig. 2: Injection part of the test chip showing injected plugs of fluorescent dye.

3. Results and discussion

Measurements of D_{12} (with no flow, Pe = 0) were repeatable to within 2%. Performing the experiment at several Peclet numbers (varied by increasing the applied electric field), provides D_{eff} versus Pe^2, which gives the dispersion coefficient α_o for a given geometry. Results for several geometries are shown in Fig. 3. Comparing α_o for 3 and 10 μm perturbations, we find that they differ by a factor of about 50 for both wavelengths that were studied. Thus the effect of increasing amplitude is more than quadratic in this case. In the most dispersive channel, at Pe=1000 the effective diffusivity would be more than 25 times the binary diffusion coefficient.

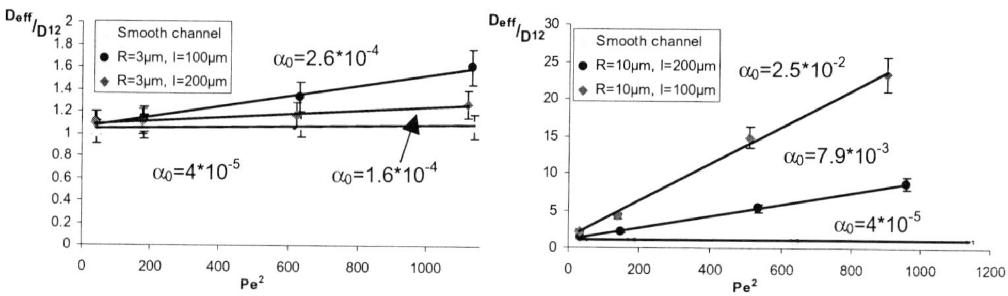

Fig. 3: Graphs showing the measured ratio of D_{eff} over D_{12} at various Pe^2 numbers for square-wave channels from Fig. 1 with amplitudes R = 3 or 10 μm and wavelengths l = 100 or 200 μm.

4. Conclusions

Dispersion coefficients in different geometries have been measured and compared.

Acknowledgements

The authors wish to thank Sandia National Labs for providing the imaging facilities and the Dutch Technology Foundation STW for financial support.

References

1. *S. K. Griffiths, R. H. Nilson, Low-dispersion turns and junctions for microchannel systems, Anal. Chem. (2001), no. 73, p. 272-278.*
2. *G. Taylor, Dispersion of soluble matter in solvent flowing slowly through a tube, Proc. Roy. Soc. A219 (1953), p. 186-203.*
3. *R. Aris, On the dispersion of a solute in a fluid flowing through a tube, Proc. Roy. Soc. A235 (1956), p. 67-77.*

FLUID MIXING DESIGN: COMPARISON OF SIMPLE VERSUS COMPLEX MODELING METHODS

Siew Bang Cheng[1], Cameron D. Skinner[1], Walter Allegretto[2], D. Jed Harrison[1]

[1]Department of Chemistry, [2]Department of Mathematical Sciences
University of Alberta, Edmonton, AB, Canada, T6G 2G2

Abstract

This paper reports on three modeling stages used: a simple lumped element model in which each channel is represented by discrete resistors and the results are calculated using PSPICE; a sheet resistance model, in which each channel is represented by a network of resistors and PSPICE or a home written program is used; a finite element model is solved using Coventor s FlumeCAD software. A detailed evaluation comparing these models with experimental results will be presented.

Keywords: Modeling, Electroosmotic Flow, PSPICE, Simulation

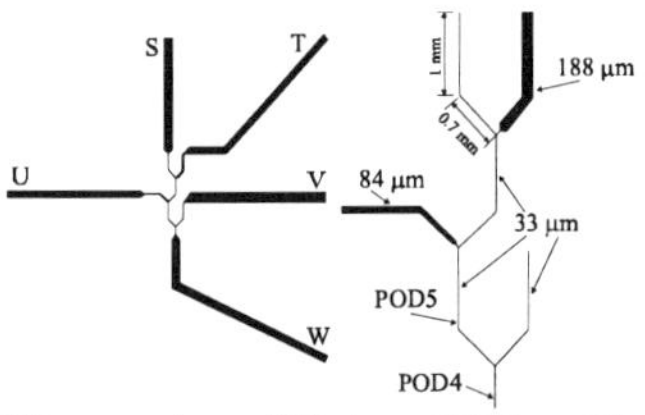

Figure 1a: Diluter D64 with channel widths indicated in the exploded view. S, T, U and W represent reservoirs with sample present in S, and the detection is indicated as PODs.

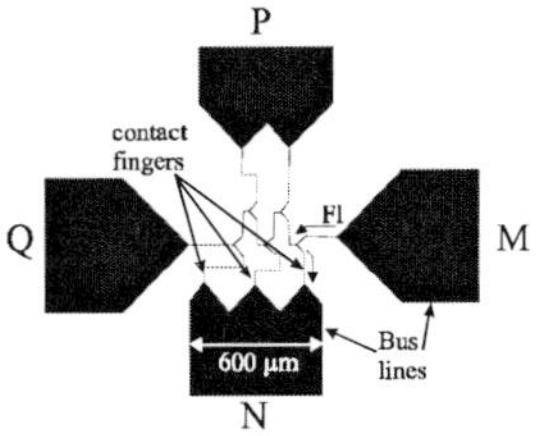

Figure 1b: Expanded view of D32. The four wide channels (bus lines) extend ~1 cm to reservoirs labeled M, N, P and Q. Arrows indicated the observed flow direction of fluorescein (Fl) sample in M when driven bv electrokinetic forces.

1. Introduction

The increasing complexity of microfluidic flow paths has begun to require modeling of the design to minimize development time.[1-3] For electrokinetically pumped devices, treatment of the electrical impedance of the design provides a first order approximation of the device performance.[1,3] However, the relative low conductivity of flow channels leads to voltage gradients at junctions that can affect performance. Simple lumped element models will not describe the non-uniform current densities that arise at intersections, and so may not describe the overall performance. More sophisticated modeling is required.

2. Results and Discussion

The two designs in Figure 1, D32 and D64, mix sample with reagent streams in a series of steps, progressively performing a 32 or 55 times dilution. Equal potentials are applied to all reservoirs, except waste, which is held at ground. D64 uses a sequential series of dilutions created by the differing channel cross-sections. Modeling of this design using simple resistors to represent each channel (lumped element approach) with Microsims PSPICE agreed with 2-D modeling performed using Coventor s FlumeCAD, in terms of dilution factors at each junction. In each case the electrical elements alone were considered, and the current was taken to be proportional to the EOF. Experimental tests gave dilution factors that agreed with modeling. The smaller dilution steps are in good agreement with the values predicted by the models, however, the results for the 1:4.8

J.M. Ramsey and A. van den Berg (eds.), Micro Total Analysis Systems 2001, 617–618.

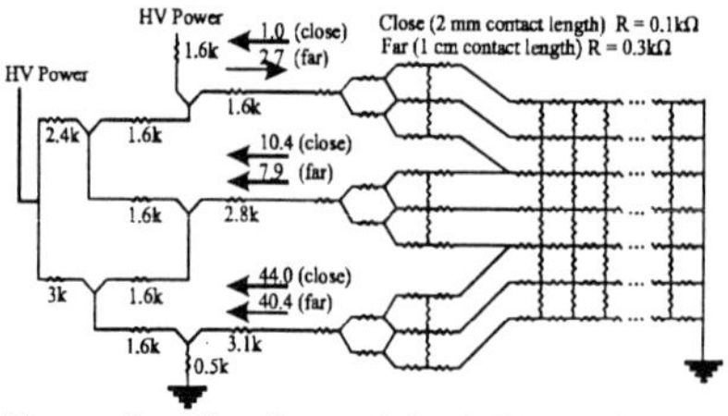

Figure 2: Circuit model of D32 with the wide channel connected to reservoir N represented as a network of resistors. Resistor values of 0.3 kΩ versus 0.1 kΩ were used to represent 1 cm and 2 mm contact channel length, respectively. The arrows show the direction of calculated current flow, with the current magnitude indicated above the arrow.

mixing stage are in poor agreement. This is consistent with the need to consider leakage effects at intersections as mixing ratios increase, and supports the need for multi-step serial diluter designs.[2,4]

The D32S design was modeled using a lumped element network. During design, the wide contact arms were considered as electrical bus lines , delivering equal potential via the fingers to each of the small channels. The calculated overall dilution factor was 32, compared to 4.5 measured experimentally. Flow was observed to go into the first finger of the bus line contact, rather than having buffer flow out of this channel as predicted by the lumped element model. A sheet resistance model, shown in Figure 3, was used to explain this effect. The model used arbitrary assignments for each resistive element, but the results clearly showed that current could flow into the bus line in some fingers, and out in others. The design was then modeled using FlumeCAD and the results are illustrated in a 3-D plot in Figure 4. The elevation corresponds to potential, and the indicated lines represent the fluid network. It is clear that the potential drops towards the fingers of the bus line , in agreement with the sheet resistance model. The FlumeCAD model predicted a dilution factor of 4.1, in reasonable agreement with the value of 4.5 determined experimentally.

3. Conclusions

Flow networks can be described by a simple lumped element model whenever the electrical impedance is the dominant factor, and leakage effects and other fluid mechanics terms do not play a role. When the potential drop (or current density lines) across a channel become important a sheet resistance model can be used. A finite element approach is the same as the sheet resistance model, with each resistor corresponding to a node, unless additional physics is present in the model to include other contributions to the fluid mechanics besides the electrical terms. Either model appears able to describe the electrokinetic contributions to the flow when a lumped element model is insufficient, and so assist in device modeling and design. Importantly, experiment and theory can agree quantitatively.

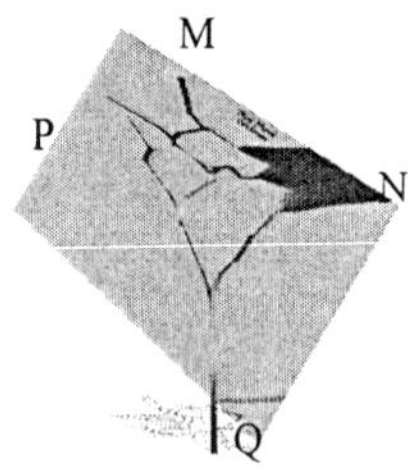

Figure 3: FlumeCAD simulation results showing the potential gradient plot. The voltages are shown cascading down from the inlet M to the outlet Q. The channel connecting the first intersection to the wide channel has a downward slope indicating the direction of voltage drop across the channel segment.

References

1. Seiler, L.; Fan, Z.H.; Fluri, K.; Harrison, D.J. *Anal. Chem.* **1994**, *66*, 3485-3491
2. Cheng, S.B.; Skinner, C.D.; Harrison, D.J. *Proc. μTAS Workshop*, **1998**, 157-160
3. Bousse, L.; Kopf-Sill, A.; Parce, J.W. *Transducer 97*, 499-502
4. Jacobson, S.C.; McKnight, T.E.; Ramsey, J.M. *Anal. Chem.*, **1999**, *77*, 4455-4459

SHAPE EFFECT ON FLUID FILLING FOR MICROFABRICATED RESERVOIR

F. G. Tseng[1], I. D. Yang[1], K. H. Lin[2], Y. T. Tseng[1], and C. C. Chieng[1]*

[1]ESS/[2]PME Department, National Tsing Hua University, Hsinchu, Taiwan, 30043

Abstract

This study demonstrates the shape effect on filling process of protein solution/water into micro-wells for microfabricated microchips. Two approaches are employed (1) high speed camera system is set up to record the liquid slug moving in and out of the micro-well and (2) numerical simulations by solving first principle equations provide the details of the flow process.

Keywords: filling process, surface tension, micro-fabrication, micro-reservoir

1. Introduction

Since surface tension is the major driving force of fluids for microchannel flow and the contact angle relating solid-liquid-gas gives the directions of this force, the well shape profile will contribute not only force directions but also force magnitude. Previous investigations have reported that the success of filling strongly depends on the hydrophilic nature of wall surface [1,2], this work reports that the wall shape profile enhances the fluid driving and helps the reservoir fluid-filling, specifically for OVA (ovalbumin) protein solutions.

2. Results and Discussions

Flow Systems consist of microchannels and microwells corresponding to wells of (I) no inclined wall and (II) 45° inclined wall. (Figs.1a and 1b) The computations are performed by grid systems of 22,400 and 27,000 nodes. Same methodologies as Refs 1 & 2 are used in numerical simulations, i.e. homogenous model in cooperation with Volume-of-Fluid (VOF) methodology and Continuum Surface Force (CSF) model. For System I, a large air bubble is entrained in the well and full filling is failed although the wall surface is hydrophilic. (Fig. 2a) The inclined wall surface of System II directs the liquid to contact well surface and make the filling successful (Fig.2b). Figures 3a and 3b display the high-speed camera pictures of filling processes for two micro-fabricated flow systems simulating to Flow Systems I and II. The cross section of the fabricated channel was formed as a V-groove by an isotropic etching in KOH. Over etching in KOH was performed so that an inclined (rounded) plane interconnecting channels and wells were obtained to ease the fluid flowing into the reservoir. Hydrophilic surface inside the flow device was also achieved. Figure 3a

J.M. Ramsey and A. van den Berg (eds.), Micro Total Analysis Systems 2001, 619–620.

shows that the protein solution only reaches part of well surface and bottom part of the well is occupied by the air bubble, and the complete filling can be smoothly achieved if the well shape is modified with rounded entrance (Fig.3b) even the height of the well has a 100 μm increase. Both numerical simulations and experiments reveal that the modification of well shape is very effective to enhance full protein filling.

References

1. K.T. Ma, F.G. Tseng, and C.C.Chieng, AIAA-2001-1014 (2001)
2. F.G. Tseng, et. al. IEEE Transducers 01, pp.1518-1521, Munich, Germany (2001)

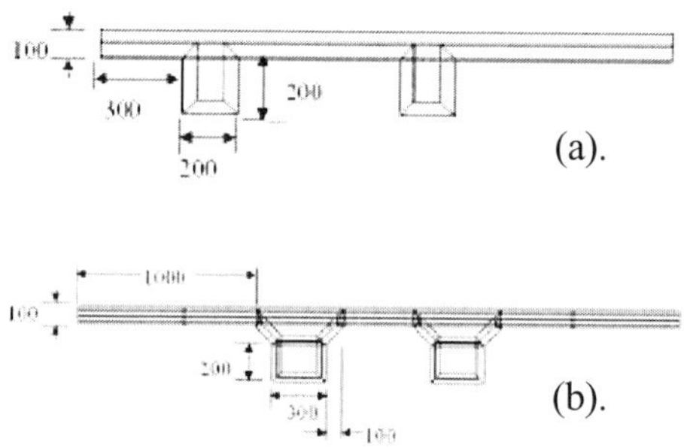

Fig. 1: Configurations of Flow Systems I and II

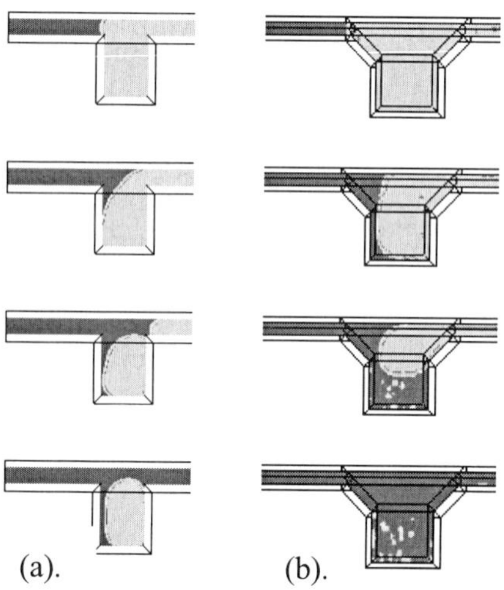

Fig. 2: Flow Filling Processes Vin=0.5 m/s, Re= 27.53, protein (dark), air (gray)

Fluid: OVA solution 1mg/ml
Protein solution velocity: about 0.2m/s
Re=0.0023, dimension unit: μm

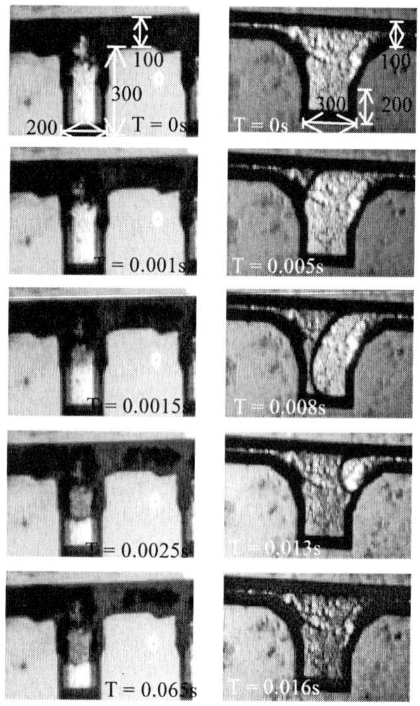

Fig 3a (left), Fig.3b (right): Protein filling processes through Flow Systems I (straight well) and II (well of rounded entrance)

AN ELECTROKINETIC ACTIVE MICROMIXER

Jin-Woo Choi, Chien-Chong Hong, and Chong H. Ahn
MicroSystems and BioMEMS Lab
University of Cincinnati, Cincinnati, OH 45221-0030, USA

Abstract

In this work, a novel active micromixer using electrohydrodynamic (EHD) convection phenomena [1] has been demonstrated, simulated, and characterized. Two different liquid samples injected through the realized electrokinetic micromixer have been successfully mixed after passing the mixer. Numerical simulation has also proved the concept of the micromixer. Since the mixer includes no moving parts, reliable and robust mixing in a microchannel can be rapidly achieved. The novel active micromixer realized in this work can be widely applicable for micro total analysis systems (μ-TAS) and biochips.

Keywords: Active miromixer, electrohydrodynamic, microfluidic

1. Introduction

Mixing of the liquid samples is frequently required to increase reaction probability so as to improve the detection and analysis capability of micro total analysis systems for detection of biological molecules or for analyzing DNA in microfluidic systems. There are, however, some difficulties in realizing reliable micro mixing devices, because the fluid in microchannels shows as a laminar flow characteristic in most cases due to low Reynolds number. In the previous work [2], we introduced a new active micro mixer using electrohydrodynamic convection phenomena. Figure 1 describes schematic illustration of the active micromixer.

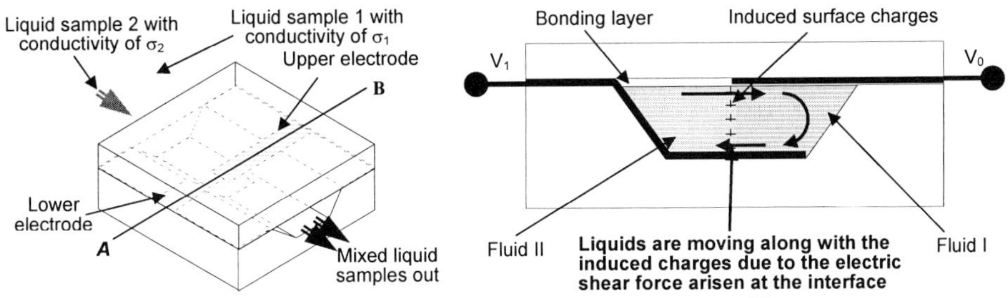

Figure 1. An electrokinetic active micromixer: (a) schematic illustration and (b) cross sectional view along A-B showing the convection and active mixing mechanism.

2. Simulation and Experiments

According to the electromagnetic theory, surface charges are induced and accumulated on the boundary of a dielectric material (which is a liquid sample in this case). When a non-uniform external electric field is applied over the surface charges, the charges will be moved with liquids due to a shear stress generated at the interface layer between the liquids to be mixed. These phenomena can continuously occurs and

J.M. Ramsey and A. van den Berg (eds.), Micro Total Analysis Systems 2001, 621–622.

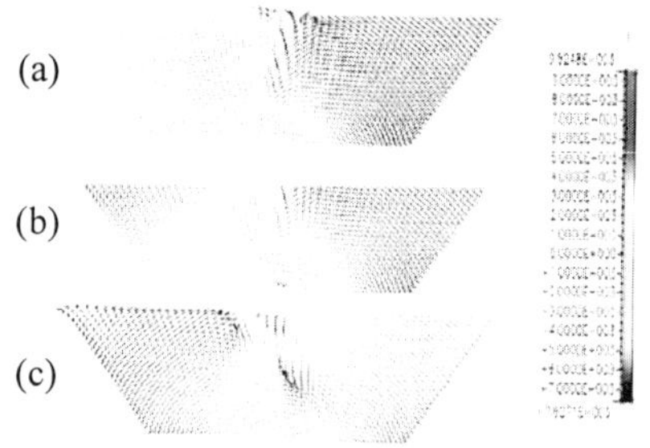

Figure 2. Numerical simulation for velocity vector after applying electric fields: (a) t = 1 ms; (b) t = 10 ms, and (c) t = 100 ms.

thus the convection of the liquid samples will continue until the liquid samples get fully mixed to eliminate the interfacial shear stress. To prove the idea, we have simulated the active micromixer using CFD-ACE+ and the results are shown in Figures 2 and 3 for the different time frames after applying voltage into electrodes. The electric potential in the fluids is also re-distributed as the fluids get mixed, which shows a firm evidence of the mixing concept (Figure 3).

To demonstrate the proposed mixing concepts, two different liquid samples were chosen: DI water (low conductivity) and KCl solution (1 mM, high conductivity). A dye was added to the KCl solution for optical monitoring. Two liquid samples were injected through the fabricated device as shown in Figure 4 (a). By applying electric field to the electrodes, the flowing liquid samples were fully mixed after passing the electrodes, due to the electric shear force generated at the interface between the liquid samples. Figure 4 (c) obviously shows the functionality of the realized active micromixer, clearly demonstrating two separate liquid streams before reaching the electrodes and only one liquid stream after passing the mixing zone.

(a) At t = 1 ms (b) At t = 100 ms

Figure 3. Electric potential distribution in the microchannel: (a) t = 1 ms and (b) t = 100 ms.

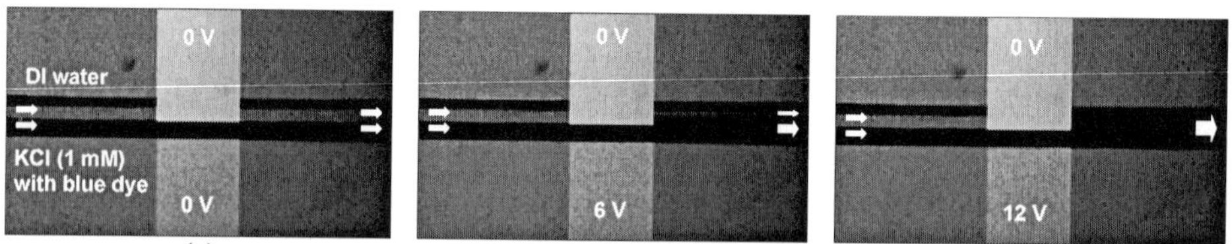

(a) (b) (c)

Figure 4. Mixing test results of the fabricated micro mixer: mixing between DI water and 1 mM of KCl solution.

3. Conclusions

A novel concept of the electrokinetic active micromixer has been proved, realized, and characterized in this work. With the demonstrated device, active mixing control of electrolytes and microparticles can be envisaged toward microfluidic-based biochemical analysis systems and biochip applications.

References

[1] H. A. Haus and J. R. Melcher, *Electromagnetic Fields and Energy*, Prentice Hall, 1989.

[2] J.-W. Choi and C. H. Ahn, "An Active Micro Mixer Using Electrohydrodynamic Convection," *Tech. Digest of Solid-State Sensor and Actuator Workshop 2000*, Hilton Head Island, SC, 2000, pp. 52-55.

CHARACTERIZATION OF FLOWS IN LASER-MACHINED POLYMERIC MICROCHANNELS

G. Goranovic[1], H. Klank[1], C. Westergaard[2], O. Geschke[1], P. Telleman[1], J.P. Kutter[1]

[1]MIC - Mikroelektronik Centret, DTU, Build. 345 East, 2800 Kgs. Lyngby, Denmark;

[2]DANTEC, 2740 Skovlunde, Denmark

Abstract:

We have examined pressure-driven and electroosmotically-driven flows in laser-machined microchannels having characteristic Gaussian-like cross-sections and various degrees of surface roughness.

Keywords: polymeric microchannels, laser micromachining, microflow simulations

Introduction:

Polymer-based microfluidic devices are becoming increasingly exploited in various bio/chemical applications because of their inexpensive and disposable nature, as well as their chemical resistance and compatibility with biological materials. We have examined in closer detail the influence of some characteristic properties (cross-section and surface features) of laser-machined polymeric microchannels on flows inside these channels.

Fabrication:

As an alternative to imprinting a silicon template into a polymer (hot-embossing method), in this study we have applied an even faster method of micromachining channels in polymers using an infrared CO_2 laser, a technique suitable for exceptionally fast prototyping. Several materials have been tested, and the most useful polymer for the purpose was found to be PMMA. The two main characteristics of the structures produced by this method are a Gaussian-like channel profile (Fig.1) and a certain degree of surface roughness (Fig. 2).

Figure 1. SEM picture of a bonded laser-machined microchannel.

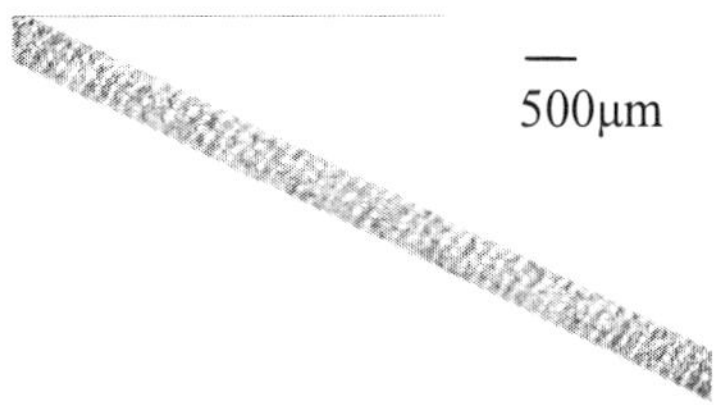

Figure 2. Perpendicular ripples in a laser-machined microchannel.

J.M. Ramsey and A. van den Berg (eds.), Micro Total Analysis Systems 2001, 623–624.

Results and discussion:

We have simulated microflows in several channel profiles with three types of surface roughness patterns (ripples perpendicular to the flow, parallel to the flow, and a check-board pattern). Typical dimensions of the features (width, height) were 10-30 μm.

In pressure-driven flows, apart from producing different flow resistances (Fig. 3), the ripples at the bottom were found to affect the flows in shallow channels more than in deeper channels because the former have bigger hydraulic areas. Surface roughness contributes also to the velocity profile, which changes from a parabolic (in the smooth part of the channel) into a Gaussian-like shape (in the rippled part), as confirmed by microPIV (Particle Image Velocimetry) experiments (Fig. 4).

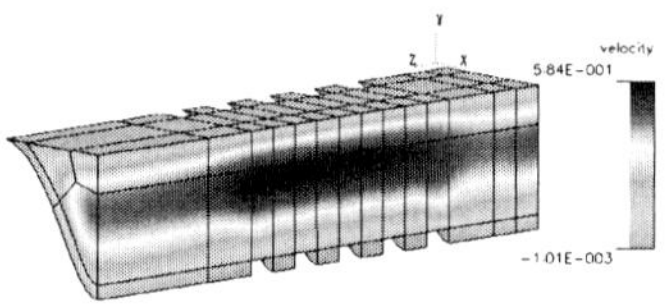

Figure 3. Simulated velocity field in a Gaussian channel with perpendicular ripples (symmetry plane cut).

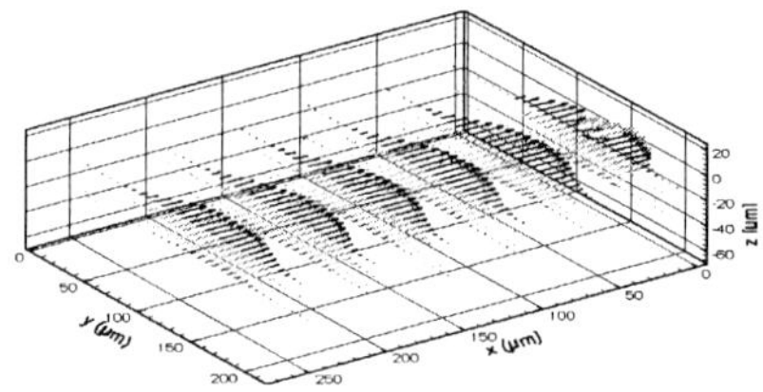

Figure 4. Gaussian-like velocity profiles in a rippled plastic channel at several planes determined by PIV measurements.

In the case of electroosmotic flow the surface roughness produces an unequal distribution of surface charges resulting in variations of the zeta potential and thus in the velocities within the double layer (Fig. 5), that lead eventually to an increased sample dispersion. Additional dispersion is produced when using a different polymer material for the channel covers (e.g. PDMS) (Fig. 6). Indeed, these results are supported by recent experimental measurements in plastic microchannels[1].

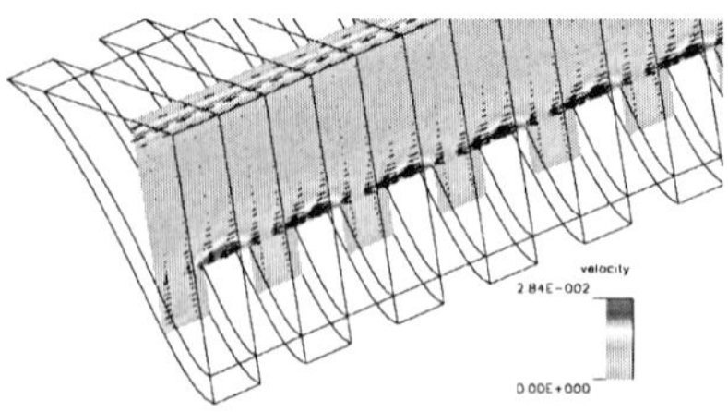

Figure 5. Velocity variations at alternating ripples due to a difference in the ζ-potential in a plane parallel to the symmetry plane.

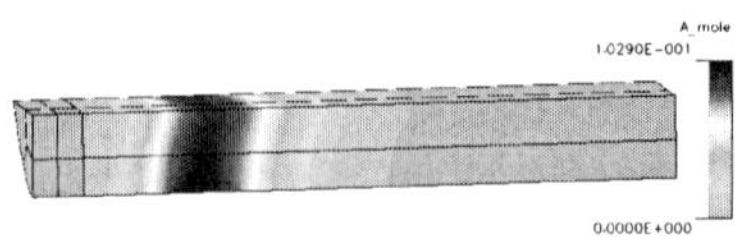

Figure 6. Simulated dispersion of a sample in a smooth, hybrid plastic channel.

References

1. Imaging of electroosmotic flows in plastic microchannels, D. Ross, T.J. Johnson, L.E. Locascio, Analytical Chemistry, 2001; *73*(11); 2509-2515

A FLUOROGENIC ASSAY USING PRESSURE DRIVEN FLOW ON A MICROCHIP

Matthew Kerby and Ring-Ling Chien
Caliper Technologies Corporation
605 Fairchild Drive, Mountain View, CA 94043, USA

Abstract

A fluorogenic enzyme assay was conducted on microchips using pressure driven flow. Real-time on-chip dilutions were performed in both low viscosity buffer and an enzyme solution containing high viscosity solution. Flow regulation of mixed viscosity fluids requires a pressure control on each arm of the chip contributing to the overall flow. An 8-channel pressure controller was used to regulate the pressure above all wells, thereby controlling the dilution ratios of buffer, substrate and enzyme with different viscosity. Well pressures maintained a constant concentration of enzyme in the detector channel while adjusting the flow contribution of substrate and buffer.

Keywords: Enzyme assay, hydrodynamic flow, on-chip dilution

1. Introduction

Enzyme assays in microchip format have been demonstrated by several groups [1-3]. Miniaturization of the assay shows significant advantages in terms of reagent saving, throughput, reproducibility, and data quality. Most papers published in this subject utilize electrokinetic mechanism for fluid transport and control in microfluidic channels. The purpose of this poster is to demonstrate the use of hydrodynamic flow control to perform an enzyme assay involving on-chip dilution with mixed viscosity solutions.

2. Experimental

A schematic diagram of the chip design optimized for hydrodynamic flow is shown in Figure 1. A multiport controller was constructed that provides basic control and detection capabilities needed for microfluidic microchips. The controller has 8 independent peristaltic pumps that could provide either positive or negative pressure with respect to ambient. The output of the controller is then connected to a chip manifold that interfaces with the microchip. The chip interface is mounted on a microscope stage of a modified Nikon TE300 inverted microscope for epi-fluorescence detection. The multiport module is typically run using a script containing the order, duration and magnitude of each function such as the pressure settings.

3. Results and Discussion

A fluorogenic assay for human T-cell phosphatase (TCPTP) was chosen in our experiment. The TCPTP enzyme catalyzes the removal of a phosphate group from DiFMUP to produce the fluorogenic product DiFMU. Enzyme solutions with viscosities of water (1cP) and 50% glycerol (6cP), a common cryoprotectant, were

J.M. Ramsey and A. van den Berg (eds.), Micro Total Analysis Systems 2001, 625–626.

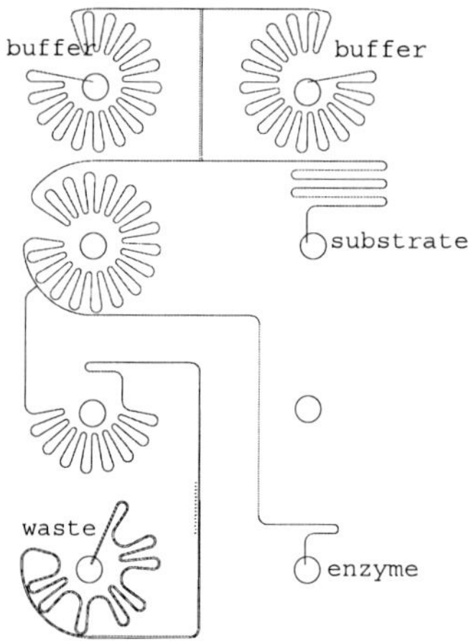

Figure 1. Chip design

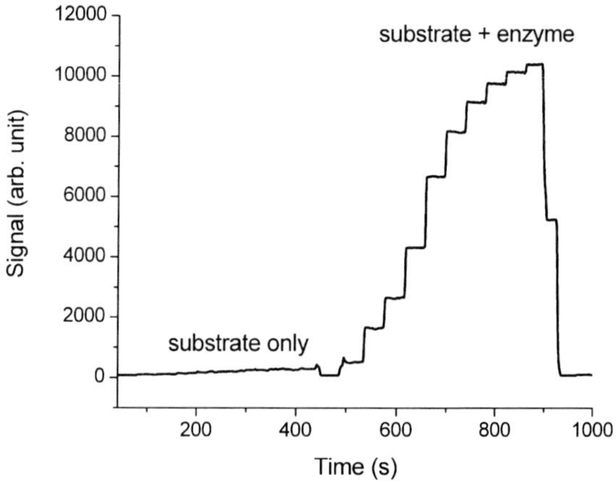

Figure 2 On-chip dilution

assayed using on-chip dilution. The DiFMUP substrate signal without enzyme was obtained first by filling the enzyme well with plain buffer. After running the multiple dilution script, the buffer is removed, replaced with the active TCPTP enzyme solution, and the same dilution script is executed again. Figure 2 shows the raw data of both the signal from enzymatic reaction and the signal from substrate only. Similar experiments were also performed for enzyme in high viscosity buffer. To validate the chip experiment, we also repeated the enzyme assay with the traditional method using a microtiter plate. Excellent agreement was obtained between these experiments.

4. Conclusions

We have designed a system that allows flexible and selective transportation of fluids within microfluidic channels using either electrokinetic or hydrodynamic forces. In addition to controlling dilution with buffers that have the same viscosity, we also showed good experimental results for buffers with different viscosity.

Acknowledgements

We wish to thank Derek Louch for designing the chip, Roger Deltroff for the flow solver program. We also like to thank Andrea Chow, Theo Nikiforov and J. Wallace Parce for advice and technical discussion.

References

1. A. G. Hadd, D. E. Raymond, J. W. Halliwel, S. C. Jacobson, and J. M. Ramsey, *Microchip device for performing enzyme assay*, Anal. Chem. **69**:3407-3412 (1997).
2. L. Bouss, C. B. Cohen, T. Nikiforov, A. Chow, A. Kopf-Sill, R. Dubrow, and J. W. Parce, *Electrokinetically controlled microfluidic analysis systems*, Annu. Rev. Biophys. Biomol. Struct. **29**:155-181 (2000).
3. C. B. Cohen, E. Chin-Dixon, S. Jeong, and T. Nikiforov, *A microchip based enzyme assay for protein kinase A*, Anal. Biochem. **273**:89-97 (1999).

OPTIMIZATION AND REALIZATION OF ELECTRO-OSMOTICALLY DRIVEN MICROSYNTHESIS SYSTEMS

R.E. Oosterbroek[1], M.H. Goedbloed, A. van den Berg
MESA[+] Research Institute, University of Twente, P.O. Box 217,
7500 AE Enschede, The Netherlands, [1]R.E.Oosterbroek@el.utwente.nl

Abstract

The fabrication, optimization and testing of an electro-osmotically driven microsynthesis system for Wittig reactions are described. For this purpose numerical models have been developed for injection / mixing and system design and optimization.

Keywords: microsynthesis, electro-osmosis, modeling, microfluidics

1. Introduction

In drug discovery, where process parameters need to be optimized, microchips allow a high level of automated data analysis. Essential in these systems is a good predictable flow control. This means a reduction of the influence of reservoir height on the generated flow rate and avoiding back flow of reaction products. Furthermore fast, uniform mixing is wanted of the reactants.

2. Design and fabrication

The chips are fabricated in glass, using a powder blasting process and consist of 3 inlet channels and one outlet branch, connected to the main channel. To ensure a well-defined electrode configuration and easy cleaning, the field Pt electrodes were deposited at the walls of the entrance and exit holes by using a shadow mask. From a previous presented design it is known that back-flow is a big problem for doing accurate synthesis. Hence additional resistance structures (shallow channels) were integrated. Figure 1 shows a picture of the complete chip and the holder for electrical and fluid connections.

3. Flow injection & mixing analysis

The injection and diffusive mixing were analyzed with fluorescent dyes. Figure 2a shows a picture of a typical profile of an injection and subsequent diffusion of fluorescent dye in the main stream. Figure 2b shows the simulation results (CFDRC) which resembles the measurements. In order to speed up the mixing, further simulations were performed to obtain a better vertically layered flow.

J.M. Ramsey and A. van den Berg (eds.), Micro Total Analysis Systems 2001, 627–628.

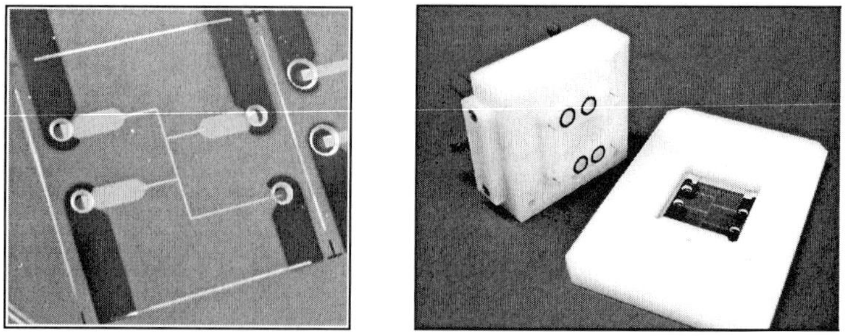

Figure 1 A). Picture of the improved microsynthesis chip: 3 inlet channels, 1 outlet. Pt electrodes are integrated for electro-osmotic flow control, which are connected with the spring-loaded pins of the connection block shown in figure B).

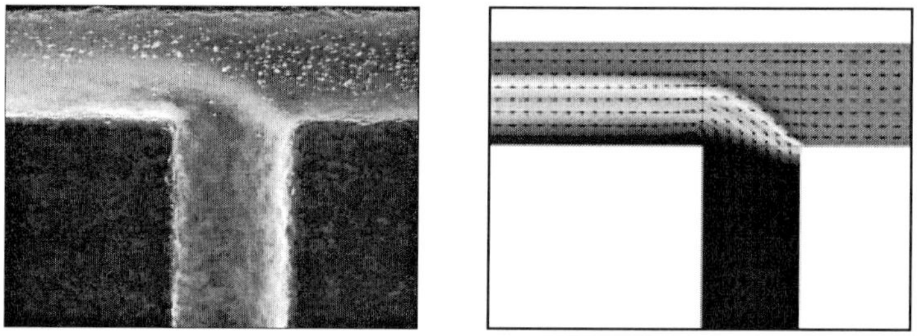

Figure 2 A) Measured injection / mixing profile at one of the channel junctions and B) the simulated profile.

4. System analysis

A bondgraph system model, implemented in he software package 20-SIM was developed to predict the voltages needed to drive the 3 entrance ports. Voltages, currents, pressures, flows and EOF-velocity and Poiseuille velocities are computed in each channel. Especially the latter values are very important for designing EOF systems since a net positive oriented flow can be generated whereas pressure-driven back-flow (Poiseuille flow) of e.g. reaction products might occur.

5. Conclusions

A microsynthesis chip has successfully been designed, fabricated and tested to reduce back-flow. Geometric and system modeling tools have been used to analyze the injection / mixing and system behavior. The bondgraph system model showed to be a very powerful tool to describe EOF in complex channel networks with constant or varying osmotic mobilities.

Acknowledgements

We like to thank Steve Haswell and his group for the cooperation and Avantium B.V. for their financial support.

COMPUTER MODELING AS A DESIGN TOOL FOR MICROFLUIDIC STRUCTURES

I.R.Perch-Nielsen[1*], G. Goranovic[1], H. Klank[1], U.D.Larsen[2], J.P.Kutter[1], P.Telleman[1], and A.Wolff[1].

[1]Mikroelektronik Centret, DTU, 2800 Lyngby, Denmark, [*]irp@mic.dtu.dk

[2]Chempaq, Lyngby, Denmark

Keywords: coaxial sheathing, microfluidics, simulations, computer modelling

Introduction

As the field of microfluidics is maturing, increasingly intricate microsystems are fabricated. This complexity is making the conventional approach of intuitive design of microfluidic structures more difficult or even impossible. In many cases numerous rounds of design, fabrication and testing are required to arrive at microstructures with optimal performance. Computer modeling can significantly shorten this process by suggesting the optimal design on beforehand. The behavior of individual components as well as the interplay between integrated components can be predicted. Here we present the optimization by computer modeling of a microfluidic structure for single step coaxial flow. Experimental characterization of the fluid behavior in the microfluidic structures is presented and compared to predictions from the model.

Simulation and fabrication

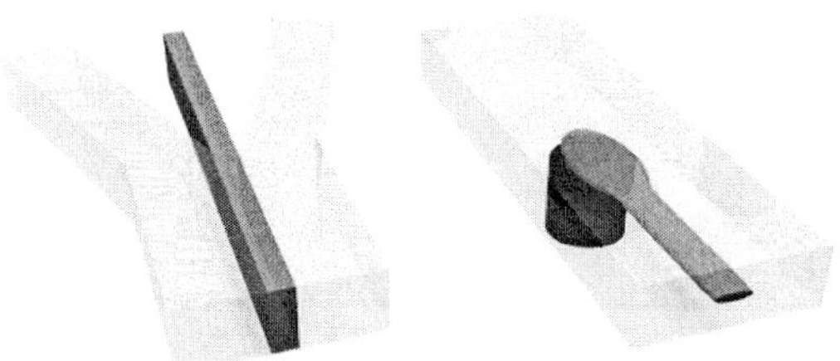

Figure 1: Schematic representation of different sheathing strategies: 2D sheathing (Left) and 3D or coaxial sheathing (Right). The 'smoking chimney' structure on the right achieves single step coaxial flow by injecting the sample stream perpendicular to the carrier stream.

In previous work [1,2,3] we were confronted with the need for a simple monolithically integrated component for single-step coaxial flow to replace 2D sheathing (fig. 1). The idea for the chimney structure was inspired by simulations, before it was actually fabricated. In this study simulations were performed with the CFD-ACE+ 6.2 software (CFD Research Corporation, Huntsville Alabama), a multiphysics package based on the Finite-Volume Method. The mesh-independence test runs were made within the range between 15000 and 48000 cells. Although operated in the laminar flow regime, a rather fine mesh was needed to account for the more complex geometry of the design that was fabricated and experimentally examined. Several of the designs were realized in silicon in a three-step process of reactive ion etching (RIE) in a plasma of SF_6 and O_2. In the first step, the front etching of the 50 μm deep basin (lowest part of the chimney) and the inner part of the circular tube was made using the first mask. In the second step, the 100 μm deep channels were completed (second mask). Finally, the hole for the sample inlet was made in the 250 μm deep etching from the backside of the wafer (third mask). To complete the structure, a glass lid was anodically bonded on top of the etched geometry.

J.M. Ramsey and A. van den Berg (eds.), Micro Total Analysis Systems 2001, 629–630.

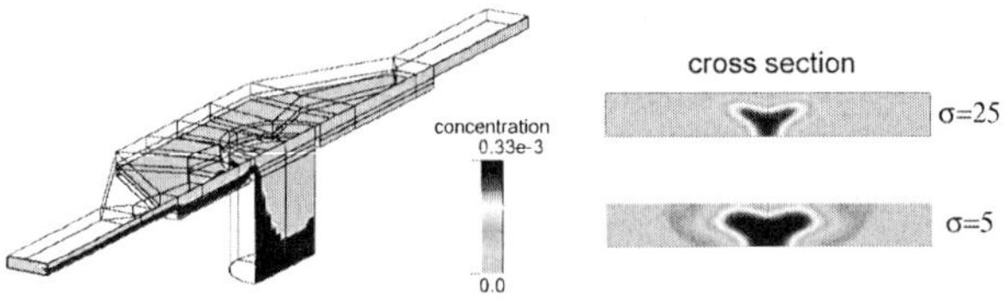

Figure 2: Computer modelling of the chimney. A perspective view of the chimney, and a cross section of the sample fluid concentration after the flow has developed downstream.

Experimental

The behavior of fluids in the microstructures was tested by adding dyes to the different liquid streams (fig. 3). As fluorescein is transparent it was possible to map fluid layer thicknesses by measuring fluorescein signal intensities, i.e. concentrations. The controlling parameter σ is defined as the ratio between the sheathing and sample flow.

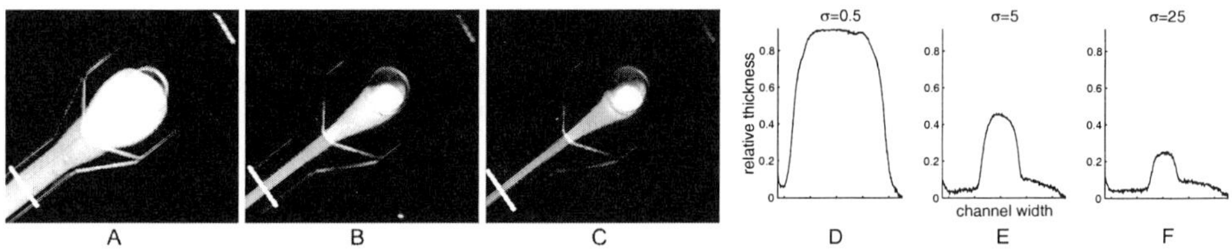

Figure 3 A-C: The chimney seen from above with fluorescein flowing through. The σ ratio is set to 0.5, 5 and 25 respectively. **D-F:** The intensity measured along the white lines in fig. A-C. Increasing the flow ratio results in an increased focusing of the stream not only from the sides (reduced peak width) but also from the top and bottom (reduced peak height). These results are in agreement with simulation predictions.

Conclusion

The simulations of the chimney structure have predicted the focusing effect, which was then experimentally confirmed. Generally, higher sheathing velocity and smaller chimney diameter produce stronger effects with higher uniform velocity distributions within the sample stream. From our results it is clear that computer modeling is a valuable tool for the design of microfluidic components where reasonable comparisons between model and experiment can be achieved. Nevertheless, both the development of computer modeling tools for microsystems and equipment for the quantitative characterization of flows in microstructures are still in their infancy and will require further optimization to allow for routine application. We are currently adapting particle image velocimetry (PIV) to microfluidic structures to further enhance the evaluation of simulation results.

References

[1] W. Rong, J.P. Kutter, et. al.. Sixth World Congress on Biosensors, May, 2000, San Diego, p 284.

[2] A. Wolff, U.D. Larsen, P. Friis, and P. Telleman, Proceedings of EuroSensors XIV, 2000, pp. 133-134

[3] P. Telleman, U.D. Larsen, A. Wolff, and J.P. Kutter, Proceedings of SPIE symposium on Micromaching and Microfabrication, 2000, pp. 30-39.

MOLECULAR TRANSPORT IN MULTIPHASE FLOW SYSTEM ON MICROCHIP

Mariana Surmeian[1], Hideaki Hisamoto[1], Akihide Hibara[1], Maxim N.Slyadnev[2], Kenji Uchiyama[2], Takehiko Kitamori[1,2]
[1] Department of Applied Chemistry, School of Engineering, the University of Tokyo, Hongo, Bunkyo, Tokyo 113-8656, Japan
[2]Integrated Chemistry Project, Kanagawa Academy of Science and Technology, Sakado, Takatsu, Kawasaki 213-0012, Japan
surmeian@icl.t.u-tokyo.ac.jp

Abstract

We have investigated the integration of three-layer flow system (aqueous – organic – aqueous) on a microchip and the molecular transport process from the donor aqueous phase to the acceptor aqueous phase inside the micro channel. As model analyte, we have used Methyl Red.

In order to explain the experimental results a numerical simulation of the liquid/liquid/liquid extraction processes in the micro channel has been performed. We have presumed the liquid/liquid/liquid extraction process as a mass transfer on stationary phases according the first Fick's law and first order transport kinetics at each interface. The calculated profiles fit well experimental results.

Key words: Microchips, Thermal lens microscopy, Interface, Molecular transport.

1. Introduction

Recent microchip technology provides wide variety of applications in chemistry. Many of those are involving analytical chemistry using capillary electrophoresis. In contrast to those, we have been demonstrated the fundamental chemical operations using microchips. In our previous studies, formation of liquid-liquid interface inside the micro channel played an important role for effective molecular transport between two phases [1].

The present paper presents the integration of the three-layer flow system (aqueous – organic – aqueous) on a microchip [2] and the molecular transport process from the aqueous donor phase to the acceptor aqueous phase in continuous flow using Methyl Red as an analyte example. In this case, thickness of organic layer sandwiched by two aqueous layers formed inside the micro channel is expected to be μm order (60μm). Usually this kind of three-layer flow cannot be demonstrated in macro scale, because the mixing of two aqueous phases occurs due to the gravity.

2. Experimental

We have used a micro channel, 190 μm wide and 22 μm deep and 30 mm length. Two aqueous phases: 1) donor phase: methyl red (0.8×10^{-5} – 1.6×10^{-5} M in 0.1M HCl solution) and 2) acceptor phase: 0.1M HCl solution and an organic phase: cyclohexane, were introduced from three inlets of the microchip to form three layer flow under continuous flow conditions.

J.M. Ramsey and A. van den Berg (eds.), Micro Total Analysis Systems 2001, 631–632.

As detection, a thermal lens microscope was utilized to monitor the molecular transport process inside the microchip.

3. Results and Discussion

A stable three laminar flow of aqueous/organic solvent/aqueous was performed in a micro channel. In this case, thickness of the organic phase sandwiched by the two aqueous phases was ca. 60 μm, and it can be considered as a thin liquid organic membrane (figure 1). During the continuous flow, analyte was rapidly extracted along the micro channel from the donor phase into the organic phase, and then released into the acceptor phase. After few seconds, the equilibrium has been achieved and the concentration in the donor and acceptor phases become equals (figure 2).

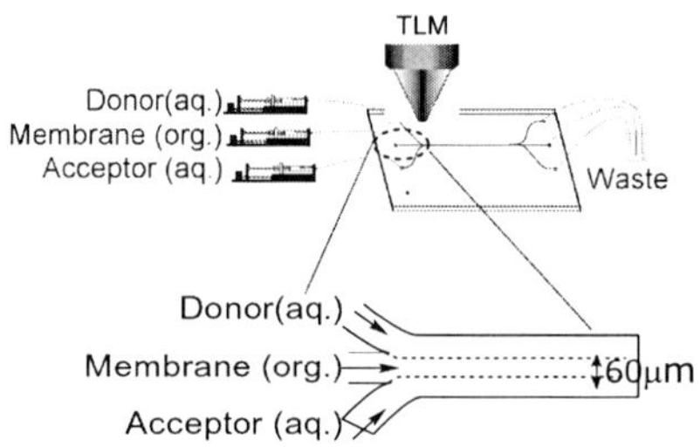

Figure 1. Liquid membrane inside the microchannel and the molecular transport measurement

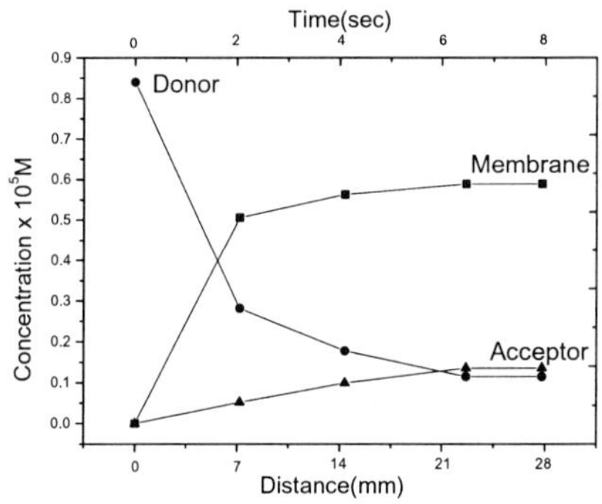

Figure 2. The concentration course of the model compound along the micro channel in the three layers system

The mathematical simulation model we have used, presuming the liquid/liquid/liquid extraction as a mass transfer process according the first Fick's law and at each interface a first order transport kinetic, fit well experimental data.

4. Conclusions

We have successfully developed for the first time, a three-layer flow, water / organic membrane flow / water, system in a micro channel on a chip to study the molecular transport in micro space.

This kind of experiment is possible only in micro space and is expected to be applied for the rapid liquid membrane transport that makes it possible of eliminating slow diffusion process which is a typical time consuming factor of conventional liquid membrane transport.

References

1. M.Tokeshi, T.Minagawa, T. Kitamori, Anal.Chem., 72, 1711-1714, (2000).
2. A.Hibara, M.Tokeshi, K.Uchiyama, H.Hisamoto, T.Kitamori, Anal.Sci., 17, 89-93, (2001).

HYDRODYNAMIC ADDRESSING FLOW CELL: DESIGN CONSIDERATIONS AND CALCULATIONS

Mattias Tidare[1], Anders Olsson[2], Håkan Roos[1] and Ulf Lindberg[3]
[1]Biacore AB, Rapsgatan 7, SE-754 50 Uppsala, Sweden
[2]Acreo AB, Electrum 236, SE-164 40 Kista, Sweden
[3] Solid State Electronics, P.O. Box 534, SE-751 21 Uppsala, Sweden

Abstract

This paper presents experimental and CFD (Computational Fluid Dynamic) design optimization of the flow cell for hydrodynamic addressing (HA), a concept that was presented earlier this year [1, 2].

Keywords: SPR, Hydrodynamic addressing, Flow cell, CFD

1. Introduction

In contrast to a conventional flow cell having one inlet and outlet, the HA flow cell has two inlets and one outlet, see Figure 1. Simultaneously injected fluids will flow along defined 'lanes' in the flow cell. When characterizing biomolecular interactions with modified sensor surfaces this is advantageous [1]. With a two-inlets flow cell it is possible to modify each side of the sensor surface separately, leaving the middle sensing spot unmodified as a reference, as illustrated in Figure 1.

Figure 1. Coupling of ligand to the sensing surface and sample injection

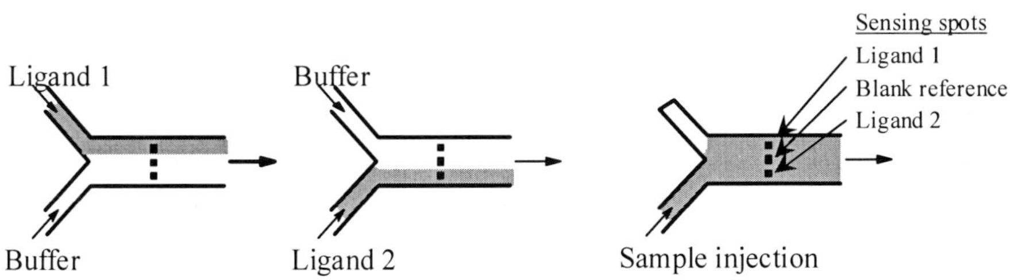

Immobilization of ligand 1 on the upper part of the flowcell.

Immobilization of ligand 2 on the lower part of the flowcell.

Sample injection over a surface with two ligands and a blank reference.

2. Theory

When designing a robust HA flow cell, several parameters have to be considered. For instance, in the immobilisation step is it crucial that the borderline between the fluids is straight at the sensing spots. Therefore these spots have to be placed far away from the intersection of the fluids. On the other hand, during the sample injection step the sample will be depleted as it moves along the sensor surface and consequently the sensing spots should be placed close to the above-mentioned intersection. A suitable

J.M. Ramsey and A. van den Berg (eds.), Micro Total Analysis Systems 2001, 633–634.

compromise has to be found between these conflicting demands. Also, for referencing purposes it is vital that the sample reaches all sensing spots simultaneously.

3. Experimental

Ten different flow cells where fabricated using milling in transparent plastics. The flow of one colored and one transparent fluid was monitored with a video camera.

4. Results and discussion

To reduce expensive and time consuming prototyping the experiments were complemented with numerical simulations using the commercial CFD software ANSYS/Flotran. The relevance of the simulations where tested qualitatively by comparing the shape of the borderline of some prototypes and corresponding simulations. This revealed that in most cases 3D models was necessary due to the complicated flow pattern through the flow cell. The conflicting demands are shown in Figure 2. The detection spots in figure 2A are far from the inlets resulting in too large analyte depletion. 2B: Shorter distances to the detection spots but curved borderlines. 2C: Straight borderline but larger time differences between detection spots.

Figure 2. CFD simulated streamlines with the flow from left to right. The plot is taken 10 μm above the active surface. The the dots represent10 ms time intervals. The three rectangles mark where the detection takes place (including mechanical tolerances). The vertical lines in flow cell C outlines an area of different depth.

5. Conclusions

The simulations give accurate results and are a useful complementary tool for the design of a robust flow cells. Experiments and simulations show similar results and the simulations makes it possible to select a robust design. The use of CFD made it also made possible to study the flow profile at any point. Futhermore, we can evaluate the three important aspects discussed above: borderline straightness, depletion length, and sensing simultaneity.

References

1. R. Stålberg, S. Sjölander, H. Roos, H. Sjödin, M. Tidare, M. Malmqvist *Method and Device for Laminar Flow on a Sensing Surface*, Patent, WO9936766.
2. M. Tidare, H. Roos and U. Lindberg, *Hydrodynamic Addressing; a Flowcell for Improved Signal Quality in Bioanalytical Systems*, Transducers'01, pp. 1154-1157, (2001).

RHEOLOGICAL SHEAR EXPERIMENTS IN MICRON SCALE GEOMETRIES

K. Johan A. Westin, Chang-Hwan Choi and Kenneth S. Breuer
Div. of Engineering, Brown University, Box D. Providence RI 02912, USA
(johan_westin@brown.edu)

Abstract

An experimental procedure designed for rheological testing of shear-dependent viscosities using small sample volumes is described. Excellent results are obtained with simple (Newtonian) fluids at flow rates smaller than 1 nl/s using a channel with 0.9 μm gap height. However, anomalous behavior is observed with aqueous (ionic) solutions, stressing the importance of chemical and/or electrokinetical surface effects.

Keywords: Microchannels, rheology, surface effects, non-Newtonian fluids

1. Introduction

The lack of knowledge of material properties is a severe limiting factor when designing biofluidic and microfluidic systems. Many biological fluids are non-Newtonian, and the material properties can be expected to alter when the geometrical scale of the device become comparable to the molecular scale. An increased flow rate due to slip effects near the walls can be expected ([1],[2]). Moreover, the rheological testing needs to be accomplished using small test volumes but with high accuracy. In the present study an apparatus and procedure for testing shear rate dependent viscosities has been built based on pressure-driven flow through micro channels.

2. Experimental set-up

The microchannels are fabricated using standard micromachining techniques, with thermally grown silicon dioxide to define the channel height. The channel geometry is etched in the oxide layer, followed by a KOH etch to define inlet/outlet ports through the handle wafer. Finally a capping silicon wafer is bonded to close the channel. The principle for the measurements is to track the liquid free surface in a precision bore hole (ϕ2 mm) using a laser distance meter (figure 1a). In the present set-up the flow is driven by pressurized nitrogen.

3. Results and discussion

The channel heights of the presently used microchannels are not known with sufficient accuracy, and so a calibration liquid (Silicon oil, DC 200, 0.65 cSt) is used for this purpose. In figure 1b, the calibration is made from a single measurement at 1MPa (the solid circle), giving a channel height of 0.9 μm. The solid lines in figure 1b correspond to theoretically-predicted pressure drop based on the above channel height

J.M. Ramsey and A. van den Berg (eds.), Micro Total Analysis Systems 2001, 635–636.

and the relative (macroscopic) viscosity ratio between silicon oil and ethanol. The results are consistent also at flow rates less than 1 nl/s.

Since most biological materials are dissolved in aqueous solutions, it is required that reliable data can be obtained with water as test fluid. The present results, obtained with a 0.1 M saline solution, show a higher pressure drop than predictions based on a Newtonian viscous flow. This may be due to electrokinetic effects (see e.g. [3]). However, there are also time-dependent variations, showing an initial transient followed by a slower decrease of the flow rate (figure 2). Many possible causes for this have been explored and rejected and the result is currently unexplained.

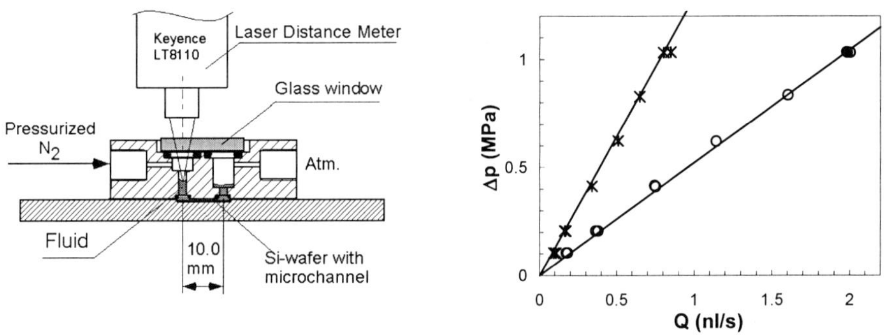

Figure 1. (a) Schematic of experimental set-up. (b) Pressure drop versus flow rate using (o) Silicon oil (0.65 cSt) and (*) Ethanol. Channel height 0.9 μm. See text for details.

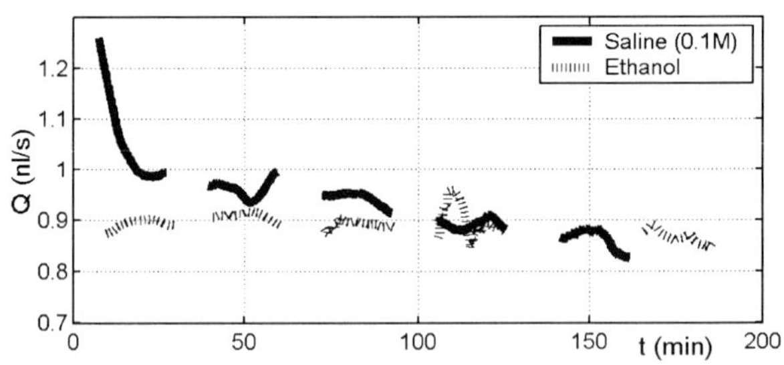

Figure 2. Time dependence of measured flow rates (Saline 0.1M and Ethanol).

Acknowledgements

We wish to thank B. Caswell, P. Richardson and M. Sibulkin for useful comments.

References

1. Y. Cohen, PhD thesis, Univ. of Delaware, Newark, (1981).
2. B. Caswell, K. Breuer and P. Richardson, Proceedings, *3rd Pacific Rim Conf. On Rheology, Vancouver*, July 8-13, (2001).
3. L. Ren, W. Qu and D. Li, *Interfacial electrokinetic effects on liquid flow in microchannels,* Int. J. Heat and Mass Transf. 44, 3125-3134, (2001).

Miniaturisation – What's in it for chemistry

Stephen J. Haswell
Department of Chemistry, University of Hull, Hull HU6 7RX, UK

Abstract

Research over the past 2-3 years has clearly demonstrated that a range of liquid based chemical synthesis can be carried out in so-called micro reactors [1-8]. Based on such work, this paper will reflect on the real chemical advantages such methodology offers the chemist and how these relate to currently accepted techniques.

Keywords: Micro reactor, chemical synthesis, Wittig reaction

In very simplistic terms, a chemical reaction proceeds when one molecule comes into close proximity with another such that a favorable exchange of electron energy occurs. In general, this is a bimolecular event where component A reacts with component B to form a quantity of product C or a combination of products. Whilst the energy exchange in such a process is extremely fast, its initiation and control at the right time and place (often in multi step reactions), underpins the practical nature of reaction chemistry. In particular, the rate of a reaction is governed by the equilibrium between the reactants and a product(s), where conditions such as concentration and temperature are manipulated to favor product yield. In addition, product selectivity is crucial for many chemical reactions/intermediates and side reaction mechanisms need to be minimized *via* the manipulation of the equilibrium conditions and/or time constants.

Traditionally laboratory scale reactions rely heavily upon already established experimental techniques/methods that mix many milliliters of reactants together in a common solvent at a suitable concentration (i.e. dictated by the amount (weight) required) for a given period of time. Reaction processes may include the addition of a catalyst, variation in temperature and the addition of specific reagents, which favor a desired geometric molecular interaction. In practical terms to achieve the required reaction and generate the desired product a whole range of techniques such as stirring, refluxing, distillation, crystallization and chromatographic separations are used. In an ideal chemical environment, the chemist would like the ability to select and control (in spatial and temporal terms) a chemical reaction, such that they can synthesize the product in a preferred quantity and location. In addition the opportunity to generate large amounts of quality data relating to reaction mechanisms would clearly be of value in its own right for generating physical organic information on chemical processes, a luxury not widely available to the synthetic chemist at present.

So what are the real attributes miniaturization can offer reaction chemistry that cannot be realized using current methodology, as simply making a reactor smaller is not reason enough. Perhaps the most important properties of micro reactors which makes them unique is the presence of diffusive mixing under laminar flow conditions and their

J.M. Ramsey and A. van den Berg (eds.), Micro Total Analysis Systems 2001, 637–639.

efficient heat dissipation property arising from their low mass reaction volume compared to the very large heat sink properties of the effective container.

Diffusion can be considered as the last stage of any mixing process, which traditionally uses mechanical stirring to overcome the mass transfer difficulties of getting two reactant molecules sufficiently close enough to react. Thus mixing with out mass in a micro system allows the rapid generation of products as reactants can be positioned in close proximity to each other for efficient diffusion to occur. In practical terms, reaction products can be produced in a matter of seconds/minutes compared with laboratory scale taking hours or even days. However, the fundamental properties of the chemistry performed in a micro reactor are no different to that performed in the laboratory but the real chemical advantage lies with the fact that a larger quantity of reactions can be carried in a given unit of time. Thus for the first time, statistical design of experiments and optimization models have been successfully applied in a robust way to well established chemical reactions. The important message here is that we have not fundamentally altered the underlying chemistry so the realization of kinetic date etc for a reaction is transferable to bulk systems given that the mass diffusion issues can be addressed. The capacity however for massive scale synthesis offers a richness of chemical information not attainable through conventional methodology.

In chemical terms the diffusive reaction environment within a micro reactor leads to the generation of localized concentration and thermal gradiants which will have a direct bearing on the equilibrium distribution of both the starting materials and products. Whilst the ability to manipulate the equilibrium of a reaction is not a new concept in chemistry, the ease and selectivity with which it can be achieved is a unique feature of the micro reactor environment. This has already be illustrated using the well known Wittig reaction (Scheme 1) for which in this example 2-nitrobenzyltriphenylphosphonium bromide is reacted with methyl 4-formylbenzoate, in dry degassed methanol, to generate the resultant mixture of *cis*- and *trans*-nitro stilbene esters [4].

Scheme 1. The Wittig Reaction.

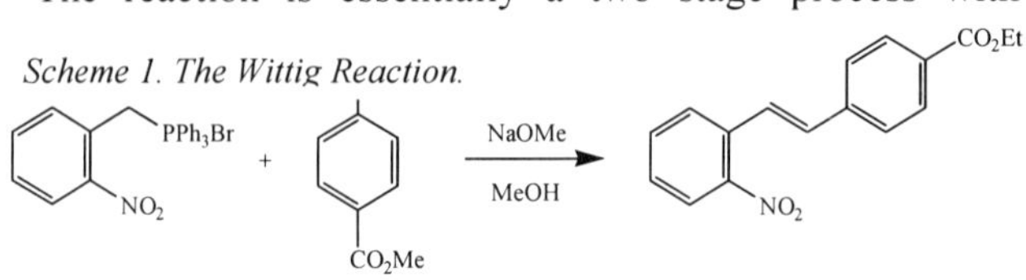

The reaction is essentially a two stage process with the triphenylphosphonium bromide initially reacting with the sodium methoxide in a fast pre-equilibrium step to produce the ylide intermediate which then follows by a 2nd order reaction with the methyl 4-formylbenzoate to produce the *cis* or *trans* sterio isomers of the corresponding ester. In addition nitrotoluene is produce as a side reaction with triphenylphosphine oxide and sodium bromide also being generated as reaction products. Using a micro reactor of the type shown in Figure 1 we are able to spatially and temporally control both the formation and concentration of the ylide at step one and likewise the conditions for producing the isomeric products. In practical terms simply adjusting the voltages applied to the reagent reservoirs (i.e. controlling flow rate) in this electroosmotically pumped system results in a direct correlation between the ratios of *cis* and *trans* products produced. This level of chemical control is not possible using traditional one pot methodology. Furthermore, it is a relatively simple process to change the methyl 4-formylbenzoate for alternative aldehydes or ketones to allow rapid analog chemistry to be performed [3].

As indicated above reactions such as the Wittig can be carried out under electroosmotic flow conditions. The interesting aspect of using electrokinetic based methodology, from a chemical point of view, is the opportunity to use *in situ* the electrophoretic mobility of reagents, reactants and intermediates to also influence chemical equilibrium. In the example above, the mobility of the base is influenced by electophoretic forces, induced by the different applied voltages in the adjacent reservoirs, rather than by the rate of diffusion. In real terms, the methoxide molecules are mobilized under electrophoretic forces moving freely towards the 2-nitrobenzyltriphenylphosphonium bromide held in methanol producing a much faster reaction rate than possible with diffusive mixing. This represents unprecedented control of chemical mixing and illustrates one of the clear strength of using micro reactors in reaction chemistry. There are many more such examples e.g. the removal of water from a reaction to effectively drive equilibrium to product formation in the enamine reaction [9].

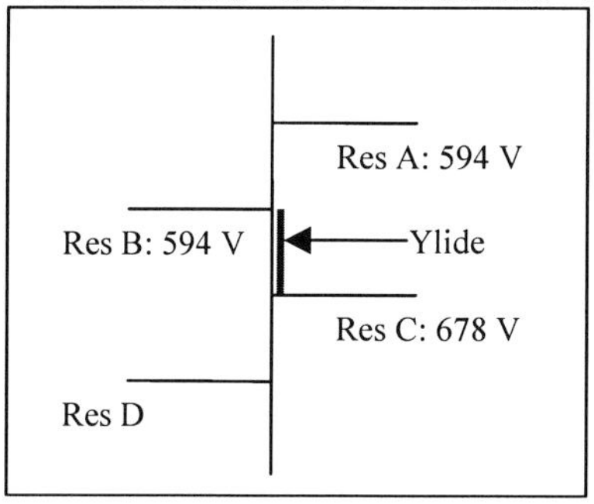

Figure 1 Schematic of a micro reactor for the Wittig reaction: A contains 2-nitrobenzyltriphenylphosphonium bromide, B sodium methoxide ,C methyl 4-formylbenzoate with the product being collected in reservoir D. The voltages applied to Pt electrodes in each reservoir are shown relative to D which is ground.

In conclusion there now appeasers to be clear evidence that performing chemical reaction in a micro reactor environment offers unique chemical advantages in terms of controlling product selectivity, yield and rate of production. Clearly the question of how do we generate sufficient quantities of material for a chemical production.

References

1. S.J. Haswell, R.J. Middleton, B. O'Sullivan, V. Skelton, P. Watts and P. Styring, Chem. Commun., , 391 (2001).
2. G.M. Greenway, S.J. Haswell, P. Styring, D.O. Morgan, V. Skelton, B. Warrington, S. Wong, Analyst, 126, 7, (2001).
3. V. Skelton, G.M. Greenway, S.J. Haswell, P. Styring, D.O. Morgan, B.H. Warrington, S. Wong, Analyst, 126, 11 (2001).
4. W. Ehrfeld, V. Hessel and H. Lowe, Microreactors, Wiley-VCH, (2000).
5. M. Sands, S.J. Haswell, S.M. Kelly, V. Skelton, D.O. Morgan, P. Styring and B. Warrington, Lab. On a Chip, in press (2001).

CONTINUOUS MULTI CHEMICAL PROCESSING IN 2D AND 3D MICROCHANNEL NETWORKS ON CHIP

Takehiko Kitamori[1,2]
[1]Department of Applied Chemistry, School of Engineering,
The University of Tokyo, 7-3-1 Hongo, Bunkyo-ku, Tokyo 113-8656, Japan
[2]Integrated Chemistry Project, Kanagawa Academy of Science and Technology
KSP East 307, 3-2-1 Sakado, Takatsu, Kawasaki, Kanagawa 213-0012, Japan
(kitamori@icl.t.u-tokyo.ac.jp)

Abstract

We have established three methods for enabling integration of general chemical systems onto microchips. One method is continuous flow chemical processing for realizing multistep chemical processing in liquid microspace. Another method is thermal lens microscopy for ultrasensitive determination of non-fluorescent molecules. The third method is construction of 3D microchannel networks. Appling these methods, any kind of chemical system, including analytical and synthetic, can, in principle, be designed and integrated onto microchips. These methodologies and some examples of micro-analytical, synthetic, and combinatorial synthetic systems will be presented.

Keywords: general chemical system, analysis, synthesis, combinatorial chemistry, 3D microchannel network, thermal lens microscope, micro unit operation

Introduction

Volume reduction and high throughput are the most prominent benefits that reduction in space affords. It is well known that in analytical systems, sample preparation and separation steps are dramatically improved by the innovative reduction of time, waste, and space enabled by miniaturization. Recently, the same demands are also arising from a wide range of chemical technologies and sciences, such as chemical synthesis, microchemical plants, combinatorial synthesis, etc.

Although the principle creating these size effects, like diffusion in microspace, is well understood, chemical operations and processing such as mixing, reaction, separation, and detection have still been difficult to integrate onto microchips. Our research group has developed original methodologies to overcome these difficulties. These methods are the continuous flow chemical processing[1)] (CFCP) using the micro unit operations (MUOs), the thermal lens microscope[2)] (TLM), and the multiphase flow system[3)] in 2D and 3D microchannel networks in a glass microchip[4)]. These techniques are for reaction and process control, for ultrasensitive detection, and for integration of complicated chemical processing, respectively. These methodologies are summarized in Fig. 1.

Micro unit operations and continuous flow chemical processing

J.M. Ramsey and A. van den Berg (eds.), Micro Total Analysis Systems 2001, 640–642.

The first method is designing of micro chemical processes into a microchip. We have developed the microchannel structures for mixing, reaction, extraction, phase separation, phase confluence, cell cultivation, and so on without any mechanical

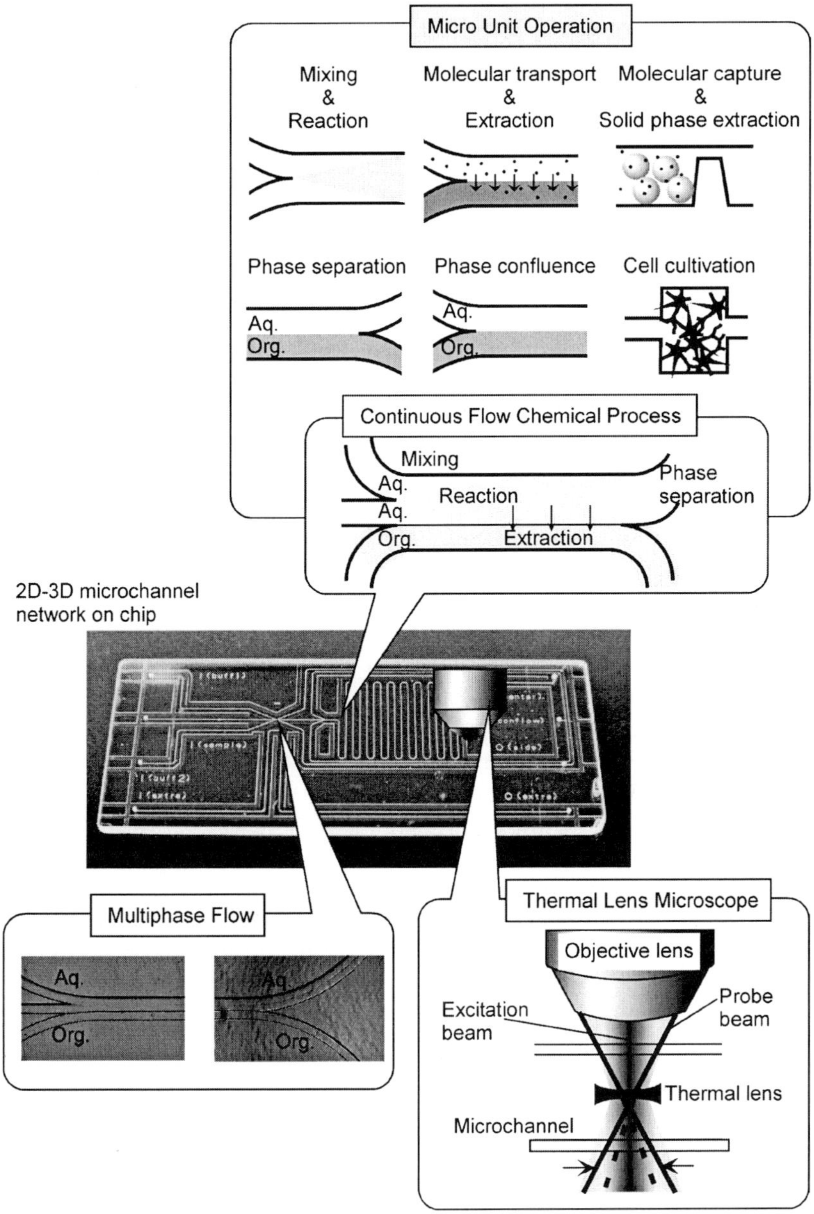

Figure 1. The methods for integration of a chemical system onto a microchip.

operation and electrokinetic forces. By combination of these MUOs we can construct CFCP like a chemical plant on a microchip. Chemical reaction and molecular transportation in the multiphase flow CFCP are controlled by flow velocity and chemical potential.

Thermal lens microscope

An ultrasensitive determination method for non-fluorescent molecules at single molecule level is also indispensable for micro chemical systems. The absolute amount of analyte of 1 nM concentration in 1-μm cubic space (1 fL) is as small as 1 ymol (10^{-24}mol). Our thermal lens microscope (TLM) satisfies these demands[2]. We are now trying to reduce the dimensions of TLM to palm-top size and this goal will be realized in the near future.

Multiphase flow system in 2D and 3D microchannel network

Stable multiphase flow[3], aqueous/organic/aqueous flow for example, was achieved by using micro laminar flow and guide structure in microchannels. This technique is very important to realize CFCP in microchips. Furthermore, 3D piping with two-level crossings of microchannels is necessary for integration of complicated systems, such as the one for combinatorial chemistry[4].

Examples of practical on-chip micro chemical systems

We have presented the concept of integration of general chemical systems onto microchips along with some practical examples of integrated chemical devices for environmental analysis[5], immunoassay chip[6], multi-ion sensing device[7], organic synthesis chip[8], and micro combinatorial chip[4]. For example, the immunoassay chip was applied to a clinical test for cancer diagnosis. We proved its performance to be 100 times faster (50hrs to 30 min) and 100 times more sensitive (1ng/ml to 10pg/ml) to conventional test as well as sufficiently accurate for practical diagnosis[9].

These examples demonstrate that our method can be applied to various kind of chemical processes not only for analysis but for synthesis, as well.

References

1. M. Tokeshi et al., submitted
2. M. Tokeshi, M. Uchida, A, Hibara, T. Sawada and T. Kitamori, Anal. Chem., 73, 2112-2116 (2001)
3. A. Hibara, M. Tokeshi, K. Uchiyama, H. Hisamoto and T. Kitamori, Anal. Sci., 17, 89-93 (2001)
4. Y. Kikutani et al. in preparation
5. M. Tokeshi, T. Minagawa and T. Kitamori, Anal. Chem., 72, 1711-1714 (2001)
6. K. Sato, M. Tokeshi, T. Odake, H. Kimura, T. Ooi, M. Nakao and T. Kitamori, Anal. Chem., 72, 1144-1147 (2001)
7. H. Hisamoto et al, submitted
8. H. Hisamoto et al. submitted
9. K. Sato, M. Tokeshi, H. Kimura and T. Kitamori, Anal. Chem. 73, 1213-1218 (2001)

DESIGN OF A TOOLBOX FOR FABRICATION OF ANALYTICAL MICROFLUIDIC SYSTEMS USING POROUS POLYMER MONOLITHS

Frantisek Svec [1,2], Cong Yu [1], Thomas Rohr [2] and Jean M. J. Fréchet [1,2]
[1] Department of Chemistry, University of California, Berkeley, CA 94720-1460,
[2] E.O. Lawrence Berkeley National Laboratory, Berkeley, CA 94720, USA

Abstract
Two building blocks required for the construction of complex microfluidic systems, namely a preconcentration unit and a mixer, have been designed, prepared, and evaluated. These elements consisting of porous polymer monoliths were prepared *in situ* via a UV initiated polymerization. The function of the monolithic concentrator was demonstrated with the enrichment of Coumarin 519, a hydrophobic tetrapeptide, and green fluorescent protein. An increase in concentration by a factor as high as 10^3 was achieved. Monolithic mixer with a length of only 5 mm and an internal volume of 45 nL enabled efficient mixing of two streams in a broad range of flow velocities.

Keywords: Polymer, monolith, microfluidic, toolbox, mixer, concentrator

1. Introduction

One of the persisting problems of microfluidic chip technologies is almost exclusive use of open channel architectures. Hence, the surface to volume ratio characterizing these channels is rather small and the channel walls can handle only very small amounts of compounds. The issue of surface area in the macroscopic devices can be solved by packing them with porous particles that also enable the introduction of specific chemistries. However, only a few reports have dealt with attempts to enhance the limited surface area in channels of a microchip. Examples of these efforts are fabrication of arrays of ordered tetragonal posts [1], porous silicate membrane [2], and a cavity containing octadecyl silica beads [3] Another problem is the lack of efficient mixing within the channels since it is difficult to achieve high Reynolds numbers required for turbulent flow. The simplest solution to the mixing problem is the use of a T-piece [4]. The other option is dividing both phases into a larger number of small parallel channels, mixing the streams in each of these channels, and recombining the sub-streams again into a single channel [5,6]. A wave-like architecture with multiple intersecting channels enables mixing in picoliter volume [7]. This report describes the simple and straightforward preparation of porous polymer monoliths and their use in microfluidic devices.

2. Experimental

The channel was vinylized using a 30 % solution of 3-(trimethoxysilyl)propyl methacrylate in acetone and allowed to react at room temperature in dark for 24 hours. The channel was washed with acetone and dried using a stream of nitrogen. The porous polymer monoliths were prepared from specific polymerization mixtures that were first

J.M. Ramsey and A. van den Berg (eds.), Micro Total Analysis Systems 2001, 643–645.

purged with nitrogen for 10 min. The microchip channel was then filled with the mixture and the holes sealed with a tape. The surface of the chip was covered with a mask leaving an open window allowing a specific section of the channel to be exposed to the UV light for 3 h in a reactor equipped with two 365 nm 8W UV tubes with an overall intensity of 1150 μW/cm^2 at a distance of 7.6 cm.

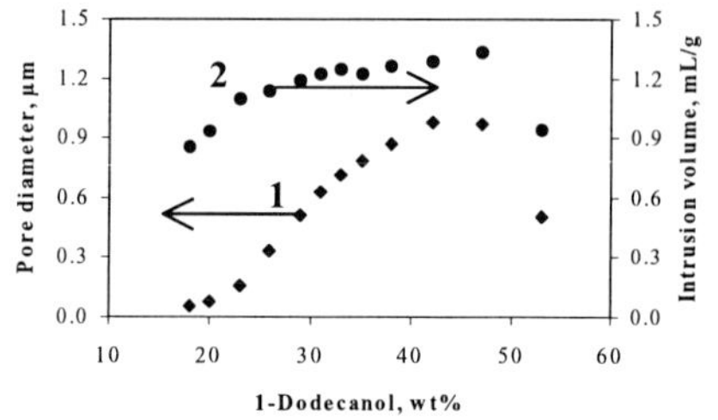

Figure 1. Effect of dodecanol on modal pore diameter (1) and pore volume (2) of monoliths prepared from mixtures of 24% 2-hydroxyethyl methacrylate, 16% ethylene dimethacrylate, x% dodecanol, and (60-x)% cyclohexanol using UV polymerization initiated by dimethoxy-acetophenone.

3. Results and discussion

Porous polymer monoliths. Monolithic porous polymers are easily prepared in situ by photoinitiated polymerization [8]. Photolithographic technique using a mask allows free radical polymerization to proceed only in the irradiated zones. Monolithic with desired surface chemistries is easily prepared by copolymerization of suitable monomers. The porous properties and consequently the flow resistance of the monolithic device are controlled by addition of a suitable porogenic mixture (Figure 1) The low flow resistance of the monolithic polymer enables high flow rates of up to 10 μL/min that is equivalent to a linear flow velocity of 50 mm/s.

Solid-phase extraction. The broad array of chemistries that are obtainable by simple polymerization mixtures is exemplified on monoliths containing polymerized ionizable [2-(methacryloyloxy)ethyl]trimethyl ammonium chloride and hydrophobic butyl methacrylate functionalities. The function of the monolithic concentration device was first tested using very dilute 10 nmol/L solutions of Coumarin 519. The adsorption was carried out at a flow rate of 3 μL/min (15 mm/s) and run time for 0.2 mL sample is 67 min. The flow rate in the desorption step has a significant effect on the extent of enrichment. For example, while a concentration enhancement of 337 times with the hydrophobic concentrator was achieved at a flow rate of 3 μL/min, a much higher enrichment of 1650 was obtained at a lower flow rate of 115 nL/min. The overall capacity determined by frontal analysis is 1.93 x 10^{-10} mol or 8.3 x 10^{-3} mol/L. However, a more realistic application is enrichment of peptides and proteins. Fig-

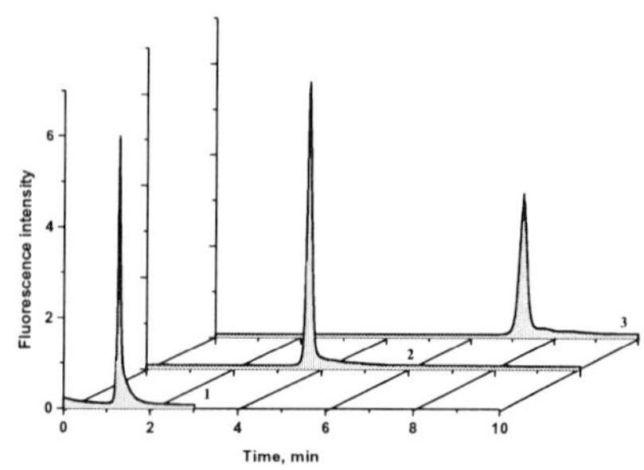

Figure 2. Elution of green fluorescent protein (loading 18.5 nmol/L, 200 μL, 3 μL/min) using 1:1 CH_3CN-water at a flow rate of 3 μL/min (curve 1), 1.03 (2), and 0.53 μL/min (3).

ure 2 shows peak of green fluorescent protein for which an increase in concentration by a factor as high as 10^3 was easily achieved.

Monolithic mixer. The performance of the monolithic mixers was tested by pumping aqueous solutions of two fluorescent dyes at various flow rates through a specifically designed microchip. The best results were achieved with a monolithic mixer containing very large irregular pores. Figure 3 shows the mixing performance of porous monoliths prepared from polymerization mixtures that included 60% (1) to 95% (6) of porogen. The plots clearly demonstrate the increased efficiency of 6, which is considerably better than that of all the other monoliths. A mixing length of only 5 mm (internal volume 45 nL) is sufficient for mixing using a broad range of flow velocities.

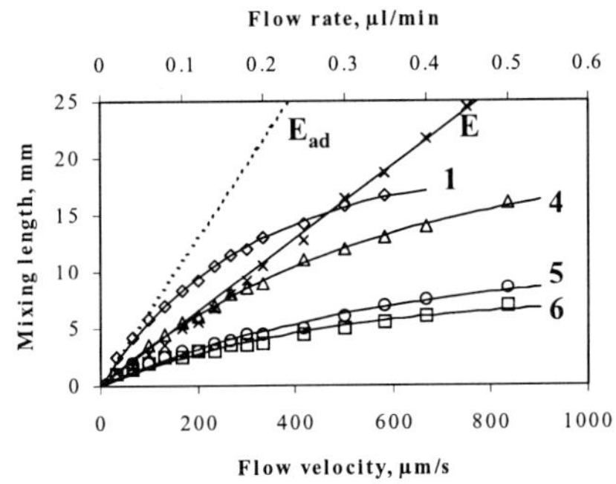

Figure 3. Length of the flow path required to achieve complete mixing in an empty channel (**E**), and channels containing porous polymer monoliths varying in pore size and pore volume. $\mathbf{E}_{ad}$ represents the length calculated for an empty channel with a cross section adjusted to 50 %.

5. Conclusions

Porous polymer monoliths with tailored both porous properties and surface chemistry prepared by photolithographic technique in situ are well suited for the preparation of various building blocks required for the construction of complex microfluidic devices.

Acknowledgments

This work was supported by the Office of Nonproliferation Research and Engineering of the U.S. Department of Energy under Contract No. DE-AC03-76SF00098.

References

1. B. He, N. Tait and F.E. Regnier, *Anal. Chem.* 70, 3790 (1998).
2. J. Khandurina, S.C. Jacobson, L.C. Waters, R.S. Foote and Ramsey J.M. *Anal. Chem.*, 71, 1815 (1999).
3. R.D. Oleschuk, L.L. Shultz-Lockyear, Y. Ning and D.J. Harrison *Anal. Chem.*, 72, 585 (2000).
4. S.C. Jacobson, T.E. McKnight and J.M. Ramsey, *Anal. Chem.*, 71, 4455 (1999).
5. F.G. Bessoth, A.J. DeMello and A. Manz *Anal. Commun.*, 36, 213 (1999).
6. O.T. Guenat, D. Ghiglione, W.E. Morf and N.F. DeRooij *Sens. Actuat. B-Chem.*, 72, 273 (2001).
7. B. He, B.J. Burke, X. Zhang, R. Zhang and F.E. Regnier, *Anal. Chem.* 73, 1942 (2001).
8. C. Yu, F. Svec and J.M.J. Fréchet *Electrophoresis* 21, 120 (2000).

HDC-CHIP: AN INTEGRATED MICROMACHINED SEPARATION SYSTEM FOR POLYMERS AND PARTICLES

E. Chmela[1], M.T. Blom[2], J.G.E. Gardeniers[2], A. van den Berg[2], R. Tijssen[1]
[1] Department of Chemical Engineering, University of Amsterdam, The Netherlands
[2] MESA+, University of Twente, The Netherlands

Abstract
Development of a novel on-chip HDC system for analytical separation of (bio)polymers and particles is presented, consisting of an extremely shallow separation channel with integrated injection, fabricated using silicon and glass microtechnology. Successful sample injections and separation tests are shown.

Keywords: on-chip hydrodynamic chromatography, open channel, size separation, fluidics

1. Introduction

Size characterisation of biopolymers, synthetic polymers and particles is one of the main issues in today's analytical chemistry. Here, Hydrodynamic Chromatography (HDC) offers fast and efficient analysis, however, classical instrumentation partly limits its performance. Therefore, development of an on-chip HDC system is presented.

2. Theory

In narrow conduits (effective size ≤ 1 μm) with laminary flowing liquid (fig. 1) larger molecules (size range from 0.002 to 0.2 of the conduit size) are transported faster than smaller ones as they cannot fully access slow-flow regions near the conduit walls [1]. In HDC this is used for analytical separation in similar applications as traditional Size Exclusion Chromatography (SEC). Unlike the latter, HDC has no stationary phase and therefore is faster and more efficient. However, part of its performance is lost by imperfect instrumentation when performed classically in packed columns or open capillaries [2]. The former

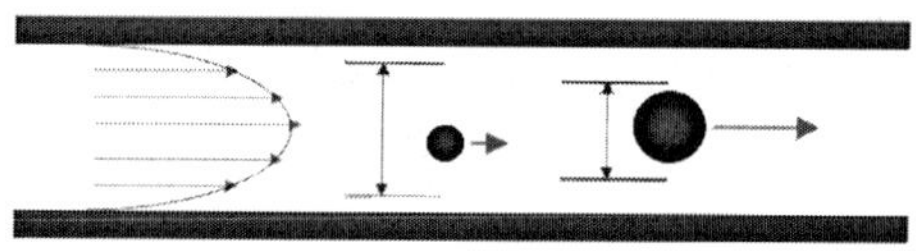

Figure 1: Principle of HDC separation. Larger analytes cannot sample low velocities near the channel wall and therefore move faster.

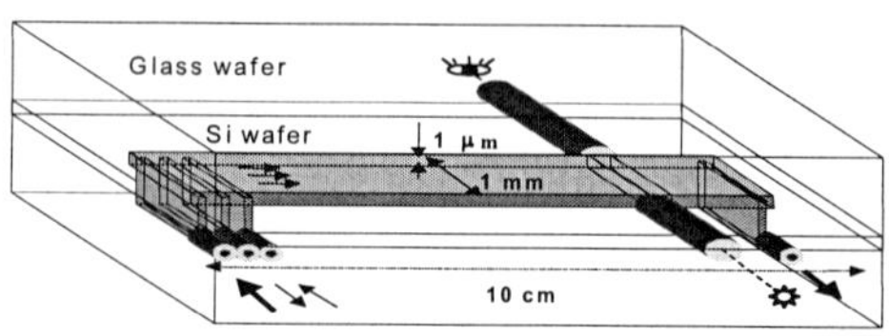

Figure 2: HDC-Chip layout (not in scale). Slits on the left form the solvent inlet and the injector, a future detector is depicted by an optical fibre at the end of the separation channel.

J.M. Ramsey and A. van den Berg (eds.), Micro Total Analysis Systems 2001, 646–648.

introduces extra-column peak broadening, the latter suffers from insensitive detection.
An integrated on-chip HDC system (fig. 2) should both preserve the efficiency and provide more material for detection by using a thin but wide separation channel. Another advantages are negligible solvent and sample consumption or easier temperature control.

3. System Design and Technology

The injector and separation channel are etched into a Si wafer and covered with thermally bonded Pyrex glass (fig. 2). Channel cross-section uniformity and well defined depth around 1 μm are crucial parameters. Its length is limited by the 4" Si wafer size to 8 cm, yet it should provide an impressive number of 100,000 theoretical plates. The channel width, which determines sensitivity of future integrated detectors is restricted by working pressures (1-10 bar) to about 1 mm. A microviscodetector is being developed simultaneously [3] and will be integrated downstream the channel, possibly together with an optical microdetector. The injector (fig. 3), designed using Computer Fluid Dynamics (CFD) simulations (fig. 4), consists of a small part of the separation channel, a resistance channel and inlet and outlet slits, connected through holes in top glass to a valve flow-actuating system and pressurised vessels with the eluent. Common action of the valves and of the properly chosen resistances of the microchannels allow reproducible injection of a rectangular sample plug.

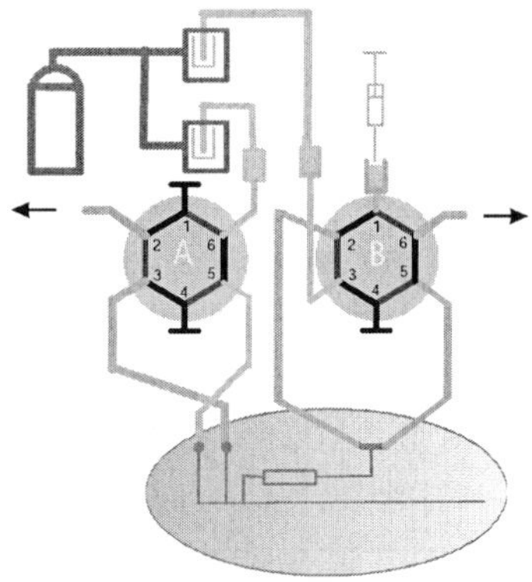

Figure 3: Experimental setup. The HDC-Chip is clamped to a macro flow-actuating system, solvent being pressed in from by Helium pressurised vessels.

4. Experimental

Injection test and separations were performed on a prototype chip with 80 mm x 0.5 mm (or 1 mm) x 0.001 mm (L x w x d) separation channel and 0.3 mm injection length. A fluorescence imaging microscope was used for the detection. Fluorescein (0.05 mg/ml) alone or mixed with 20 nm and 100 nm fluorescent polystyrene nanoparticles (0.2 - 0.5 mg/ml) in 10 mM phosphate buffer (pH 7.0) served as samples. The same buffer

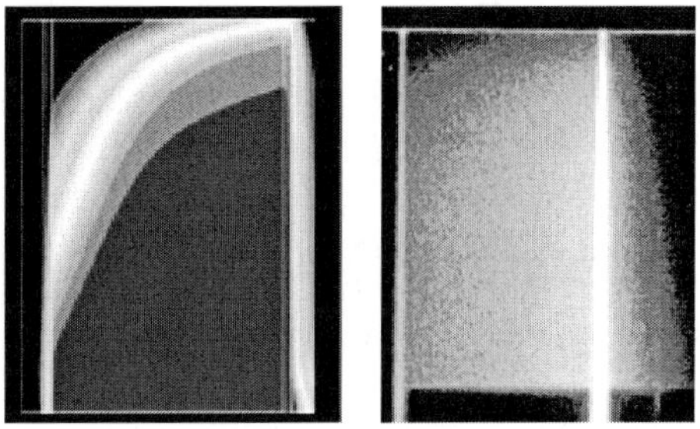

Figure 4: Injection. 3D-CFD (left) and experiment (right). Filling stage, after 3 seconds. Injection length is 0.3 mm channel width 0.5 mm.

was used as the eluent.

5. Results and Discussion

The injection confirms our preliminary CFD studies (fig. 4), separations (fig. 5) achieved within 2 minutes (working pressure 4 bar) indicate the high performance of the device. Unlike in 0.5 mm wide channels, the zones in 1 mm wide channels show some bending, likely due to a deformed top glass. This translates into increased channel thickness and thus enhanced flow speed in the middle of the channel. Particles retention corresponds to HDC theory, while excessive retardation of fluorescein suggests wall interaction and thus potential of the device also for μLC.

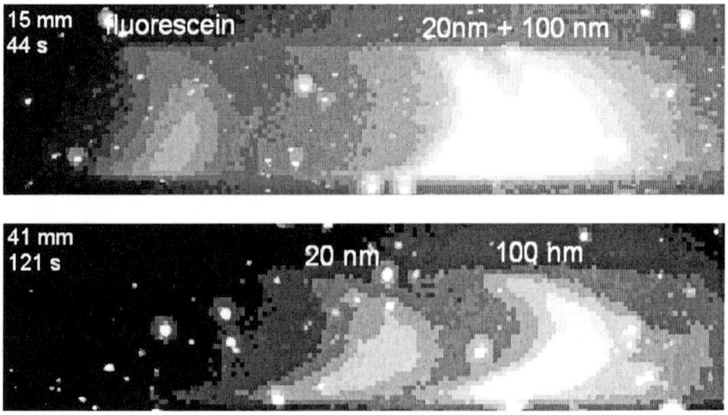

Figure 5: Separation (in 1 mm wide channel) of fluorescein (0.005 mg/ml), 20 nm (0.25 mg/ml) and 100 nm (0.5 mg/ml) fluorescent polystyrene particles in 10 mM phosphate buffer (pH 7.0). Working pressure 4 bar. Fluorescein is most retarded (above), further downstream also particles get completely resolved (bellow). Distance and time passed form the injection are indicated on the left.

6. Conclusions

A prototype Hydrodynamic Chromatography chip with integrated injection and 1 μm deep separation channel has been designed and created. Fast and efficient separation of nanoparticles is demonstrated.

Acknowledgements

This project is funded by the Dutch Technology Foundation STW, project nr. AAC4556.

References

1. E.A. DiMarzio, C.M. Guttman, *Separation by flow,* Macromolecules 3, 1970, p. 131-146.
2. R. Tijssen, J. Bos, M.E. van Kreveld, *Hydrodynamic chromatography of macromolecules in open microcapillary tubes*, Anal. Chem. 58, 1986, p. 3036-3044.
3. F.H.J. van der Heyden, M.T. Blom, J.G.E. Gardeniers, E. Chmela, M. Elwenspoek, R. Tijssen, A. van den Berg, *A micro viscosity detector for a planar hydrodynamic chromatography (HDC) system,* Proceedings μTAS 2000, 2000, p. 595-598.

RAPID SEPARATION OF PEPTIDES AND AMINO ACIDS IN GLASS MICROCHIPS BY REVERSED-PHASE ELECTROCHROMATOGRAPHY

Anup K. Singh, Daniel J. Throckmorton, and Timothy J. Shepodd
Chemical and Radiation Detection Laboratories
Sandia National Laboratories, CA 94551, USA
(aksingh@sandia.gov)

Abstract

We have developed microfabricated glass chips for reversed-phase separation followed by laser-induced fluorescence detection of peptides and amino acids. Hydrophobic acrylate-based porous polymer monoliths were cast in the channels by photopolymerization. Use of UV-light for initiation of polymerization allows for selective patterning of stationary phase in the channels for optimal design of injection and detection regions. Charged functionalities such as either sulfonic acids or quaternary amines were incorporated during polymerization for generation of electroosmotic flow. These monoliths can be cast *in situ* in less than 10 minutes, are very reproducible with respect to separation characteristics, and allow easy manipulation of separation parameters such as charge, hydrophobicity, and pore size. Moreover, the solvent used to cast the polymer is chosen so as to allow electroosmotic flow to flush the channels thereby obviating any need for application of high pressures for conditioning. Rapid (50 seconds) and efficient (up to 600,000 plates/m) separations of peptides and amino acids were achieved in the microchip.

Keywords: electrochromatography, CEC, reversed-phase, peptide separation

1. Introduction

With the completion of human genome sequencing, the next revolution lies in developing techniques for separation and analysis of thousands of proteins and peptides present in cells. Microfabricated systems are attracting significant attention in the area of proteomics because of their portability, speed of analysis, potential for multiplexing and high throughput, and ability to analyze minute sample volumes. Chromatography, because of its outstanding separation power and versatility, is the most widely used separation technique and its adaptation to microfabricated devices holds substantial promise for rapid analysis of complex biological samples. The major roadblock in performing chromatography in chips is requirement of placing stationary phase in channels. Unlike packed capillary or HPLC columns, uniform packing of channels in a chip with particles and creation of retaining frits are very troublesome. To circumvent these problems, chips whose channels contain cast-in-place uv-initiated polymers have been developed and employed for separation of biological molecules.

J.M. Ramsey and A. van den Berg (eds.), Micro Total Analysis Systems 2001, 649–651.

3. Experimental

Glass chips were microfabricated by using standard lithography followed by wet etching and thermal bonding. The standard chips contained tee- or offset-junctions, where the length of the main channel was 6.5 cm, the injection arms were 1 cm each, and the distance from the junction to the waste reservoir was 5 cm. The channels were 20-40 μm deep and 90-130 μm wide. Polymer monoliths were prepared as described elsewhere [1]. A mask made out of thin blackboard cardboard placed directly underneath the chip was used for photopatterning. Peptides and amino acids were labeled off-chip with NDA in presence of KCN [1].

4. Results and discussion

Figure 1(a) shows a cartoon of a glass chip used for separations. The inset of Figure 1(a) shows the scanning electron micrograph (SEM) of the negatively-charged porous polymer monolith containing lauryl groups cast in a channel by uv-initiation. The mean pore diameter by porosimetry is ~1 μm and surface area by BET analysis is 1-3 m^2/g [2]. Figure 1(b) shows the interface between open and packed channel, the sharpness of interface indicating the ability to control the patterning process. We have made chips where polymer was cast between injection intersection to a point a few mm away from the waste reservoir. This was done to keep both the injection path and detection window devoid of polymer, allowing for faster injection and more sensitive detection.

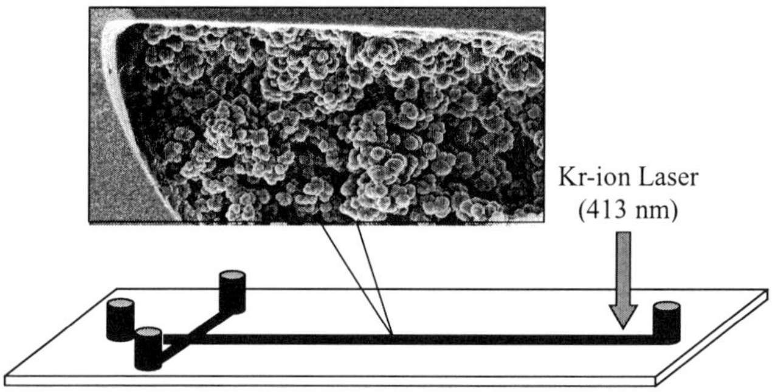

Figure 1(a): Sketch of a glass microchip used for separation. The isotropically-etched channels are 20-40 micron deep, ~100 micron wide, and the length from T-junction to detector is 5 cm. The inset shows scanning electron micrograph of porous polymer monolith cast in the channels by photoinititaion using a UV-lamp.

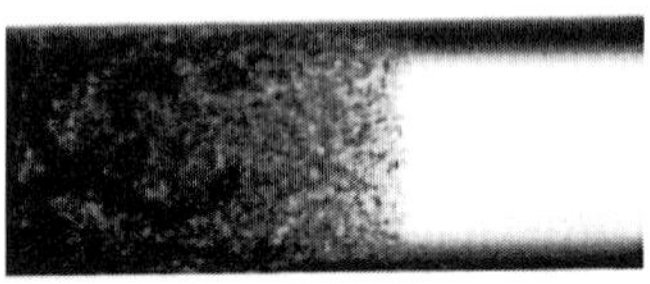

Figure 1(b): Polymer-air interface in a ~100 μm wide channel

Figure 2 (inset) illustrates the isocratic separation of five bioactive peptides in less than 50 seconds at a field-strength of ~700 V/cm. The peptides 4 and 5, Ile-Angiotensin III and Angiotensin III, have the same pI (8.75) and the same number of amino acids, only the last amino acid is different (Ile in Angiotensin III and Phe in Ile-Angiotensin III). Despite the small difference in properties, these two peaks are almost baseline resolved even at a high speed of separation proving the high resolving power of chip electrochromatography. Figure 2 also shows the effect of electric-field strength on the speed and resolution of separation. Increase in applied voltage greatly increases the separation speed but at the cost of efficiency. Peptide peaks 1, 2, and 3 attain the maxima in efficiency at 2 kV (with efficiency of peak 3 being 609,000 plates/m) while peaks 4 and 5 have the maximum efficiency at 1 kV . Figure 3 shows baseline separation of 5 amino acids labeled with NDA (at 10^{-7} to 10^{-8} M concentration). Upon labeling with NDA, R is neutral while S, G, F, and W carry one negative charge each and hence, separation of the last 4 amino acids is predominantly due to reversed-phase interaction.

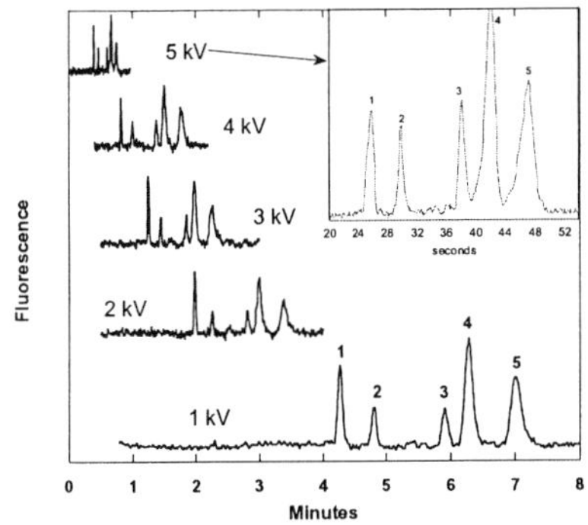

Figure 2: Separation of 5 bioactive peptides labeled with NDA in the microchip. The buffer is acetonitrile: 25 mM borate (30:70) containing 10 mM octane sulfonate. The peptides are (1) papain inhibitor, (2) proctolin, (3) Opioid fragment 90-95, (4) Ile-angiotensin III, and (5) angiotensin III. The 1-4 kV runs were performed in a chip with 20 micron-deep channel, the 5 kV run was performed in a chip with 40 micron-deep channel.

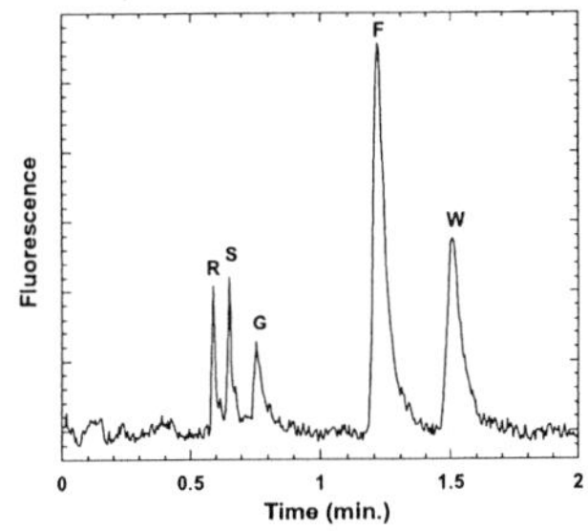

Figure 3: Separation of 5 amino acids. Acetonitrile:20 mM phosphate (10:90), 770 V/cm.

5. Conclusions

We have demonstrated the speed and resolving power of chip-based electrochromatography for separation of peptides and amino acids using chips containing photo-patterned acrylate polymers.

Acknowledgements

This work was funded by the LDRD program at Sandia National Laboratories. Sandia is operated for the U. S. Department of Energy under contract DE-AC04-94AL85000.

References

1. R. Shediac, S.M. Ngola, D.J. Throckmorton, D.S. Anex, T.J. Shepodd, A.K. Singh. Reversed-phase electrochromatography of amino acids and peptides using porous polymer monoliths, *J. Chrom. A*, 2001, *925*, 251-263
2. S.M. Ngola, Y. Fintschenko, W-Y. Choi, T.J. Shepodd. Conduct-as-cast polymer monoliths as separation media for capillary electrochromatography, *Anal. Chem.*, 2001, *73*, 849-856.

CONFORMATIONAL STUDY OF A LONG DNA IN MICRO- AND NANOFABRICATED DEVICES

Masanori Ueda [1,2] , Yuzuru Takamura [3] , Yasuhiro Horiike [3] , Tomonari Dotera [4] and Yoshinobu Baba [1,2]

[1] Department of Medicinal Chemistry, Faculty of Pharmaceutical Sciences, The University of Tokushima, Tokushima 770-8505, Japan

[2] CREST, Japan Science and Technology Corporation, Chiba292-0812, Japan

[3]Department of Materials Science, School of Engineering, The University of Tokyo, Tokyo 113-8656, Japan

[4]Saitama Study Center, The University of the Air, Saitama 331-0851, Japan

(e-mail: uedam@fc.ph.tokushima-u.ac.jp)

Abstract

We studied statics and dynamics of a long DNA molecule confined in micro- and nanofabricated channels. Based on the investigation, we aim to develop micro- and nanofabricated devices for analysis of chromosomal DNA from an individual cell.

Keywords: single molecule analysis, nanochip, entropy, confinement, DNA

1. Introduction

When a large flexible polymer is confined in the smaller space than its Flory radius, the confinement energy should be taken account of in the statics and the dynamics of the polymer conformation. The confinement energy depends on the geometry of the confinement. In this study, using nanofabrication technology, we investigated the confinement effect on the conformation of a long DNA molecule.

2. Theory

Considering a real chain trapped in a thin tube of diameter $D << R_F$ (R_F is Flory radius), the confinement energy should be represented by the following form

$$F_{conf} = TN\left(\frac{a}{D}\right)^{5/3} \qquad (1)$$

where a is an effective length of the polymer, N is a number of segments. Note that D is still larger than a [1].

The above formula doesn't include details in the geometry of the thin tube. Recently, T. Dotera has calculated the entropy changes based on curvature effect of a confined polymer [2]. In the simplest case, where a polymer is confined in a thin tube with curvature radius R and diameter D, the entropy gain should be represented as following

$$\Delta\sigma = N\frac{a^{5/3}D^{1/3}}{16}\left(\frac{1}{R}\right)^2 \qquad (2)$$

J.M. Ramsey and A. van den Berg (eds.), Micro Total Analysis Systems 2001, 652–654.

The entropy change calculated here is that in an equilibrium state, however, the result implies that the similar geometrical effect exists even in the dynamics of a single molecule under external fields.

3. Experimental

The microcapillary was fabricated on a quartz plate with a 2 cm x 2 cm area. The fabrication process was followed by the previously described method [3]

Figure 1 shows SEM image of microcapillary. Both depth and width of the channel are ca.0.5 μm. The channel has straight parts and curved parts. The curvature radius R is 5 μm.

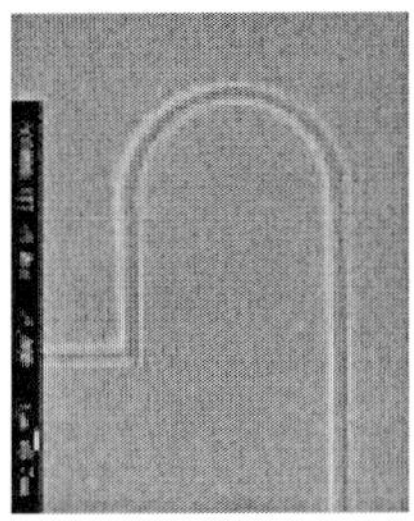

Figure 1. SEM image of microcapillary

4. Results and discussion

Figure 2 shows fluorescence images of T4 DNA migrating in the microchannel. T4 DNA molecules are migrating in the channel with the larger size than the Flory radius R_F (ca.3 μm) in a free solution. The photographs were taken with the interval of 0.5 s.

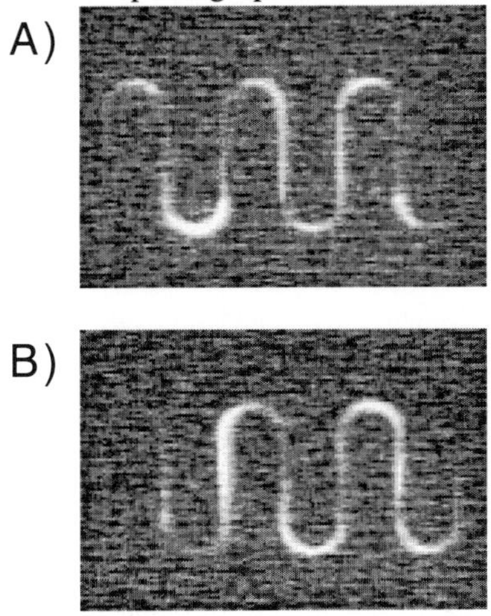

Figure 2. Fluorescence images of DNA in microcapillary. Time interval between A) and B) is 0.5 s. DNA molecules were hydrodynamicaly driven from left to right.

Curvature effect in DNA migration should be investigated from both points of view of migrating rate and migrating size of DNA. Now, we are investigating the cross-correlation between migrating rate and migration size in straight and curved region.

Acknowledgements
The present work was partially supported by the Grant of Core Research for Evolutional Science and Technology (CREST) from Japan Science and Technology Corporation (JST), the Grant from New Energy and Industrial Technology Development Organisation (NEDO) of the Ministry of Economy, Trade and Industry, Japan, a Grant in-Aid for Scientific Research from the Ministry of Health, Labour, and Welfare, Japan, and a Grant-in-Aid for Scientific Research from the Ministry of Education, Science and Technology, Japan.

References

1. P-G. de Gennes, *Scaling Concepts in Polymer Physics,* Cornell University Press, Ithaca and London, (1979).
2. T. Dotera, submitted for publication.
3. A. Oki *et al.*, *Electrophoresis*, vol.22, pp341-347, (2001).

NOVEL INJECTION METHODS FOR MINIATURISED GAS CHROMATOGRAPHY

O.P.Naji[1], F.G.Bessoth[2], A.Manz[1]

[1] Analytical Centre, Chemistry Department, Imperial College, London, UK
[2] CASECT Ltd., Exhibition Road, London SW7 2AY

Abstract

This paper describes two novel injection methods that can provide gas sample injection on chip without the use of integrated valves. The first successful demonstration of the use of an electrical plasma as an injection technique for gas chromatography is presented.

Keywords: gas chromatography, injection, plasma

1. Introduction

Injection in micro GCs for gaseous samples usually involves the integration of solenoid or membrane valves on chip [1] thus increasing the complexity of the fabrication process. Two novel injection methods are presented that illustrate successful formation of small gaseous sample plugs that do not require the use of integrated valves.

2. Experimental

2.1 Double-plasma (DP) injector/detector

The DP setup comprises two plasma generators and a separation column (Fig.1). Plasma 1 (the "injector") is situated at the start of the column, and plasma 2 (the "detector" [1][2]) at the end of the column. Both plasma sources comprise pairs of co-axial opposing Tungsten wire electrodes (50µm diameter, 2mm separation) inserted into 250µm ID capillaries. Connection to the column (HP INNOWAX, 250µm ID, 30 m) is made by butting both capillary faces together and enclosing the joint with PTFE sleeving to ensure a small dead volume. The gas sample is generated by injecting the analytes (Chloroform and Dichloromethane: 0.3 µl, 1:1 ratio) into a glass vessel (0.5L) containing Helium at a pressure of 20psi. The voltages applied[2] were typically in the range of 1 to 2 KV with ballast resistors of 3 to 5 MΩ in series. During conduction,

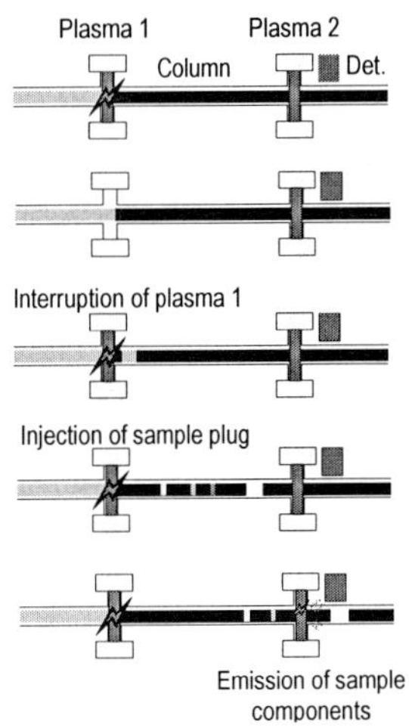

Fig. 1: *Schematic of plasma injection process.*

[1] Strictly speaking this plasma is being used as the excitation source for optical emission spectroscopy and is therefore only a component of the detection system.
[2] The actual voltages across the plasmas during conduction are significantly lower and have a complex dependency [3]. Electrically, they also typically exhibit non-Ohmic behaviour.

J.M. Ramsey and A. van den Berg (eds.), Micro Total Analysis Systems 2001, 655–657.

the electrical resistance of the plasmas is relatively small and the current drawn is therefore dominated by the ballast, thus in the range of ≈ 0.2 to 0.6 mA. Spectral measurements were made using an Ocean Optics S2000 spectrometer with a 600 lines/mm grating, giving a resolution of ≈ 1 nm over a spectral range of 200 to 800 nm. Emission lines characteristic of CH, Cl and H [4] are observed concurrently. A column temperature of 40°C was maintained during these experiments.

2.2 Injection cross

A schematic of the injection channel system employing the use of external valves is illustrated in Fig. 3. The injection part consists of a simple channel intersection, with one channel introducing the carrier gas, one for the sample inlet, one for the waste outlet (opposite the sample inlet) and one leading to the separation column. In this thermally bonded, glass-glass device, all channels are 25 μm deep and 250 μm wide and the length of the separation channel is 2 m. In sampling mode, the gaseous analyte is drawn through the injection intersection by vacuum. In order to minimize diffusion of the sample into the adjacent channels, a small carrier gas stream flows simultaneously from both the carrier gas inlet as well as the column side into the waste exit. For the injection of a sample plug, the valves of sample and waste ports are closed and pressure applied from the carrier gas port, separating off a sample plug that is transported through the separation stage. Micro pressure regulator, valves and pumps are controlled by a Scorpion micro-controller.

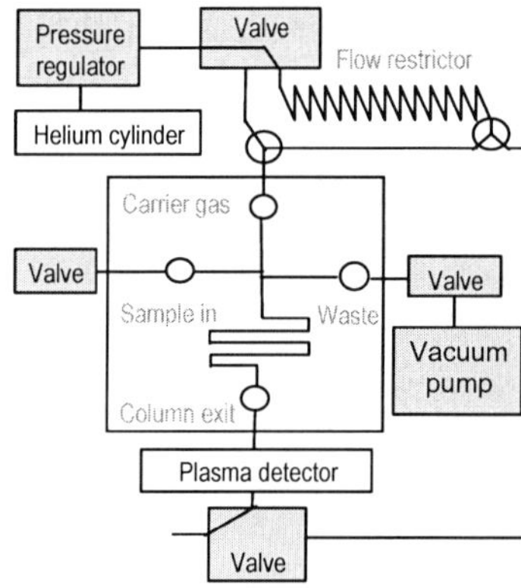

Fig.2: *Schematic (simplified) of the injection cross principle*

3. Results and Discussion

3.1 Double-plasma (DP) injector/detector

In the DP experiments sample is continuously introduced into the system, via the first plasma, into the column. This plasma effects continuous modification (fragmentation / ionization) of the sample, resulting in a reduction in intensity of most emission lines observed at the detector. Interruption of plasma 1 for a short time interval leads to the introduction of a short plug of unmodified sample into the separation column. The plug volume is a product of the sampling flow rate (~40μl/s) and the time of the plasma interruption (~500ms). The sample components subsequently separate and are detected. The first published gas chromatogram using this novel injection technique is illustrated in Fig. 3. It is worth noting that in order to define a time zero, in fact both plasmas are interrupted simultaneously, the resultant decrease in all emission is evident. Under these conditions the retention times for CH_2Cl_2 and $CHCl_3$ are approximately 75s and 105s respectively. It is important to note that by using emission spectroscopy, the chromatogram is obtained at a number of wavelengths simultaneously. In order to approximate the efficiency of the fragmentation by the 1st plasma we can compare the intensity observed for the maximum peak height with the value prior to and following injection. Applying this to the most efficient of the fragmentation processes (which appears to be associated with the CH bond in CH_2Cl_2), yields an efficiency of ~60%. It is noted that the emission at 431 nm (CH) for $CHCl_3$ is significantly less efficient, whilst, for example, the emission due to atomic Cl is comparable for both analytes. A more detailed and comprehensive presentation of results will be described in a proceeding publication.

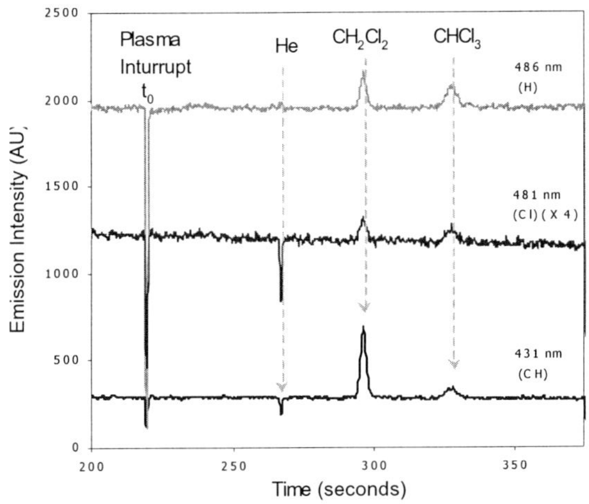

Fig.3: *Gas chromatogram obtained using "plasma injector".*

3.2 Injection cross

Preliminary results obtained using the injection cross are illustrated in Fig. 4, in which 3 successive injections of Nitrogen (500ms duration) have been applied. Detection was performed with a dc plasma in a fused silica capillary off-chip. Whilst some perturbation to the emission spectra is evident at time of switching (t_0), a clear and reproducible sample plug is detected ~20s after each injection.

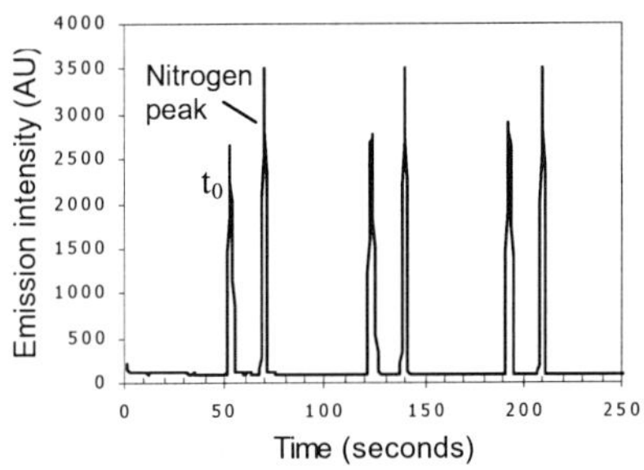

Fig.4: *Three repeat injections of nitrogen using the injection cross on chip. Detection at 359nm(N).*

4. Conclusions:

Two techniques have been developed that show great potential for micro-injection on chip without the need for complex fabrication. In particular, the use of plasmas as both the injection source and detector illustrates their importance in the field of μ-TAS.

References:

1. S.C. Terry, J.H. Herman, and J.B. Angell, *A gas chromatographic air analyzer fabricated on a silicon wafer,* IEEE Trans. El. Dev., ED-26, 12, pp.1880-1886, (1979).
2. J.C.T. Eijkel, H. Stoeri, and A. Manz, *A molecular emission detector on chip employing direct current microplasma,* Anal. Chem. 71, pp. 2600-2606, (1999).
3. Y.P. Raizer, *Gas discharge physics*, Springer, Berlin, (1991).
4. R.W.B. Pearse & A.G. Gaydon, *The identification of molecular spectra*, Chapman & Hall Ltd., London, (1950).

EXPANDING THE CAPABILITIES AND APPLICATIONS OF GAS PHASE MINIATURE CHEMICAL ANALYSIS SYSTEMS (μChemLab™)

Greg Frye-Mason, Rich Kottenstette, Curt Mowry, Cathy Morgan, Ron Manginell, Pat Lewis, Carolyn Matzke, George Dulleck, Larry Anderson, and Doug Adkins
Microanalytical Systems Dept., MS-0892, PO Box 5800
Sandia National Laboratories, Albuquerque, New Mexico, 87185-0892

Abstract

Sandia National Laboratories has developed both gas and liquid phase chemical analysis systems. An autonomous hand-held system has been fabricated and demonstrated for sensitive and selective detection of gas phase chemical warfare agents. Recent efforts have focused on maturing this technology and extending its application to other analytical needs. We have now fabricated an on-chip packed gas chromatography column and demonstrated its ability to separate gases such as methane, ethane, ethylene, and acetylene. We have also demonstrated the use of a thermally isolated membrane to rapidly modify and pyrolize fatty acids to form volatile fatty acid methyl esters (FAMEs). This new tool is useful in analyzing biological samples. Finally, we have used rapid temperature ramping of our on-chip open tubular columns to enable separation of low volatility analytes such as explosives and FAMEs. These new capabilities significantly extend the applicability of Sandia's μChemLab™ technology.

Keywords: preconcentrator, gas chromatography (GC), temperature ramp

1. Introduction

A schematic of the gas phase analysis system is shown in Figure 1. Sensitive and selective analysis is accomplished using three components: (1) a microfabricated sample preconcentrator that selectively collects the analyte of interest and generates a rapid and concentrated chemical pulse, (2) a deep-etched open tubular gas chromatograph column that provides rapid separations, and (3) a chemically selective detector based on an array of polymer coated surface acoustic wave (SAW) devices [1]. The system includes two parallel analysis channels to provide enhanced chemical discrimination and versatility. We have fabricated and demonstrated an autonomous hand-held system that can detect trace concentrations of semi-volatile chemical warfare

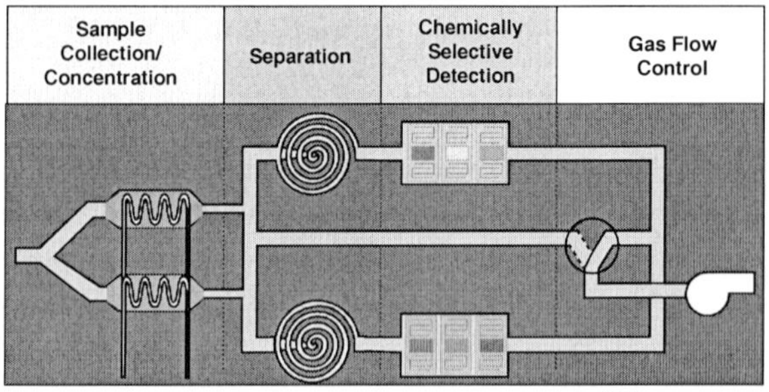

Fig. 1: Schematic of gas phase μChemLab™ system.

J.M. Ramsey and A. van den Berg (eds.), Micro Total Analysis Systems 2001, 658–660.

agents, even in the presence of large chemical background interferents (e.g., water vapor, volatile solvents, diesel) [1]. With this successful demonstration as a backdrop, this paper focuses on recent advances to expand the capabilities of our gas phase μChemLab^{TM} technology. This expansion is driven by applications with diverse analyte sets ranging from gases to very low volatility compounds.

2. Separation of Gases

We have fabricated miniature packed gas chromatography columns (Figure 2) by using deep reactive ion etching to form 300x300 μm channels. Posts at one end contain the packing material. Different porous packing materials have been utilized including polymers and carbon. Figure 3 shows a separation using a 20-cm long carbon-packed column (using a conventional inlet and flame ionization detector (FID)). These packed GC columns give effective separation of gases and highly volatile organic compounds.

Fig. 2: Scanning electron micrographs of packed microGC column components.

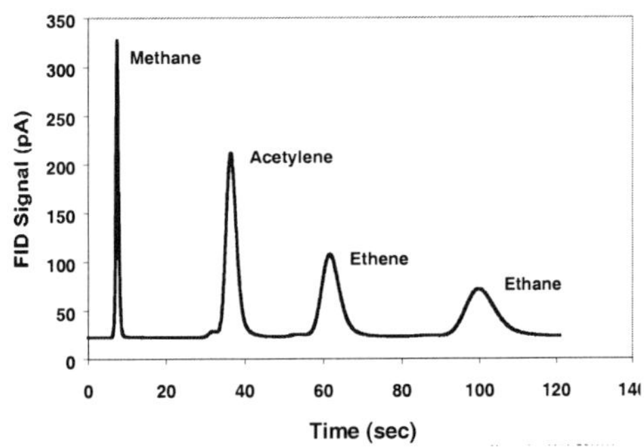

Fig. 3: Separation of gases using a 20-cm carbon packed microGC column.

3. Chemical Modification and Pyrolysis

Rapid screening of biological species can be achieved based on their unique fatty acid profiles [2]. For this application, we have developed a microfabricated pyrolysis stage that derivatizes the fatty acids to their more volatile Fatty Acid Methyl Esters (FAMEs) and volatilizes these products into the GC column. The pyrolysis stage consists of a thermally isolated membrane with an on-chip heater. A liquid sample containing fatty acids and a methylating reagent are added to the membrane. The membrane is heated rapidly (in ~10 msec after solvent evaporation) performing the methylation and volatilization. Figure 4 shows an example of the generation of FAME profiles from two fatty acid mixtures (analyzed by conventional GC with mass spectrometric detection). This new capability expands the ability of the gas phase μChemLab^{TM} system to analyze a variety of species in liquid and solid samples.

4. Temperature Ramp Chromatography

Another new capability, on-chip heaters that provide temperature ramp chromatography, expands the μChemLab^{TM} to lower volatility analytes. This capability has been used to separate very low volatility explosives (e.g., TNT) using a temperature ramp from 100 to 150°C. Figure 5 shows the broad range of volatilities that can be separated with this technique. Here, FAMEs with carbon lengths from 8 to 17 carbons are separated in a single run. The boiling points range from 194 to ~430°C, illustrating the power of temperature ramp chromatography for our system.

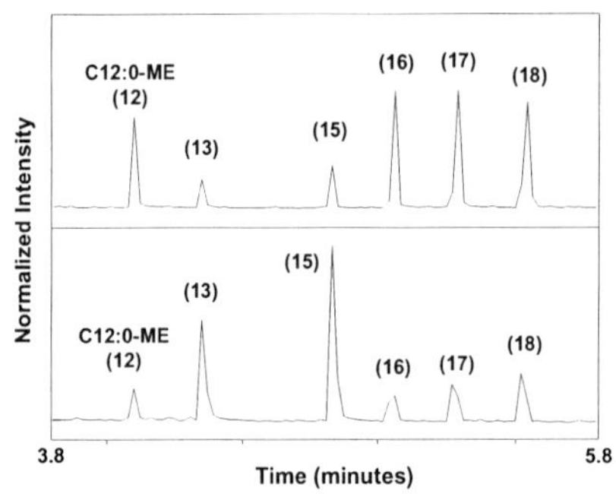

Fig. 4: Fatty acid methyl esters produced using micropyrolysis stage.

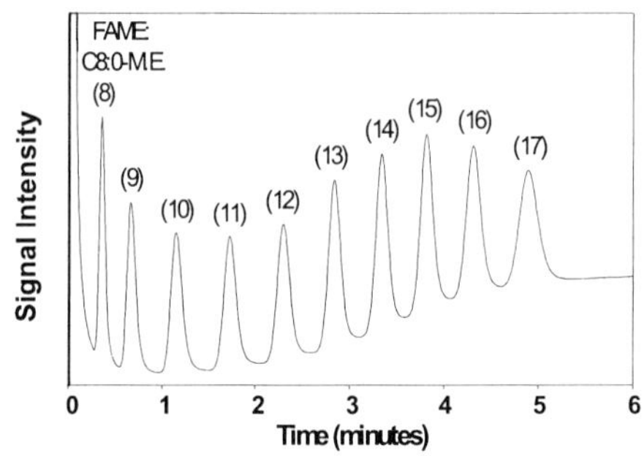

Fig. 5: Separation of fatty acid methyl esters C8:0-ME through C17:0-ME using temperature ramp (60-150°C, 20°C/sec).

5. Conclusions

The advances described above are allowing us to take a technology that was initially designed to selectively detect semivolatile chemical warfare agents and modify it for diverse analytes ranging from gases or volatile solvents to low volatility explosives to fatty acids in liquid samples.

Acknowledgements

Sandia National Laboratories is a multiprogram laboratory operated by Sandia Corporation, a Lockheed Martin Company, for the United States Department of Energy under Contract DE-AC04-94AL85000.

References

1. G. C. Frye-Mason, et al., *Hand-held miniature chemical analysis system (μChemLab^{TM}) for detection of trace concentrations of gas phase analytes*, Proc. Micro Total Analysis Systems 2000, Kluwer Academic Publisher, Dordrecht, The Netherlands, pp. 229-232, (2000).
2. C. Morgan, C. Mowry, R. Manginell, G. Frye Mason, R. Kottenstette, and P. Lewis, *Rapid Identification of Bacteria with Miniaturized Pyrolysis/GC Analysis,* Proceedings SPIE – International Society for Optical Engineering, Vol. 4205, pp. 199-206, (2001).

DEVICE LEVEL INTEGRATION TO FORM A PARALLEL MICROFLUIDIC REACTOR SYSTEM

David J. Quiram,[1] James F. Ryley[2], James W. Ashmead[2], Richard D. Bryson[2], Daniel J. Kraus[2], Patrick L. Mills[2], Russell E. Mitchell[3], Mark D. Wetzel[2], Martin A. Schmidt[1], Klavs F. Jensen[1]

[1]Massachusetts Institute of Technology, Cambridge, MA 02139

[2]E.I. du Pont de Nemours and Company, Wilmington, DE 19880

[3]ENSER Corporation, Cinnaminson, NJ, 08077

Abstract

This work demonstrates the integration of multiple microreactors operating in parallel with microvalves, micro-pressure sensors, and electronic circuitry to form a miniature reactor system. This system is equivalent to conventional laboratory reactor systems for catalyst testing but in one tenth of the space.

Keywords: microreactor, microreaction technology, microfluidic reactor, packaging

1. Introduction

Microfabrication technology presents the opportunity to create highly instrumented, micro-scale chemical reactors that bring with them the potential for new analytical capabilities, point-of-use synthesis, and highly parallel screening methods. The challenges of fabricating these devices have been well-addressed during the past decade, and microreactors have been successfully tested for a number of commercially relevant chemistries[1-2]. However, the use of microelectro-mechanical system (MEMS) reaction devices, such as the MIT thin-film microreactor, shown in Figure 1, also gives rise to a spectrum of new engineering challenges with respect to reactor system integration and scale-up schemes.

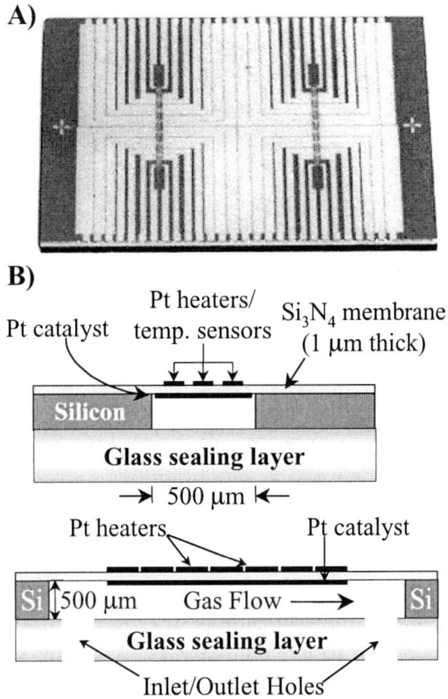

Figure 1. A) Two channel microreactor used in scale-up system B) Channel cross-sections

2. System Design

The microreactor system developed is similar to conventional laboratory reactor

J.M. Ramsey and A. van den Berg (eds.), Micro Total Analysis Systems 2001, 661–663.

systems that are in use at DuPont and other companies. It consists of instrumentation to produce a mixed feed gas, a gas distribution system to the reactors, the reaction devices, and the product gas analysis system. For this project, only the product gas analysis system is not included in the miniaturization.

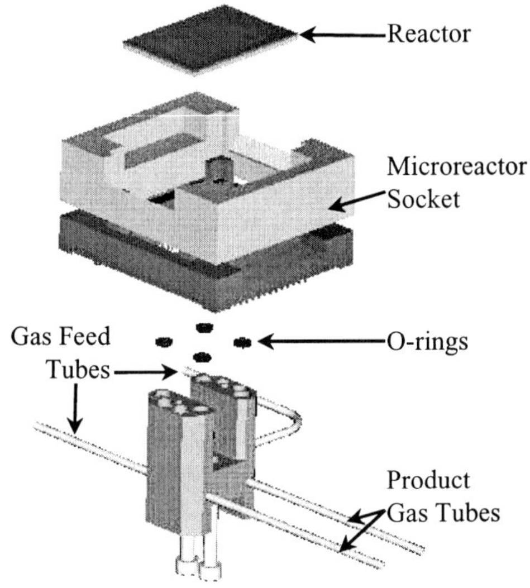

Figure 2. Microreactor packaging assembly with the Texas Instruments' DieMate™ Socket.

Packaging microfluidic devices has traditionally been problematic since it is difficult to address the need to make electrical connections and fluidic connections simultaneously. In order to avoid the expense of developing a microreactor packaging scheme, a commercially available Known Good Die socket was used as the starting point. Texas Instruments' DieMate socket (Figure 2) met the requirements since it provided a quick and simple method for electrical connections. Futhermore, with some modifications the socket could also be adapted for providing fluidic connections. This was done using a stainless steel manifold that is inserted in the socket, under the microreactor die (see Figure 2).

System level intregration between the packaged microreactor and other devices and circuits was achieved by mounting the DieMate™ socket on PC boards. In addition, MEMS shut-off and proportional valves from Redwood Microsystems were also mounted on PC boards to control gas flow in the system. As an example a microreactor board is shown at left in Figure 3. These devices along with their control circuitry were then integrated together in a 6U CompactPCI chassis manufactured by Kaparel, shown at right in Figure 3. Electrical connections between the boards were made through a standard telephony backplane and custom-built rear I/O PC boards.

The microreactor system is comprised of a temperature controller card that regulates temperature for auxiliary heaters in the system; a gas mixing board that mixes the feed gas for the microreactors; two microreactor boards that each contain two independent microreactors; and two heater circuit boards that provide power to the microreactor thin film heaters. A National Instruments embedded real time controller is used to provide closed-loop control and monitor system alarms at a cycle time of 500 Hz. A host PC, running LabVIEW® 6i, is used as the human machine interface at the supervisory control and data acquisition level.

Figure 3. Reactor board with DieMate™ socket, two Redwood Microsystems gas flow manifolds, and device electronics (left); and the completed microreactor system (right).

3. Application Example

The microreactor system has been tested with the partial oxidation of methane to form carbon monoxide and hydrogen (syngas) along with carbon dioxide and water. 90% oxygen conversion was obtained with 20% selectivity to CO and 5% selectivity to H_2 at ~700°C. The temperature limits and the design of the thin-film microreactor prevented achieving higher temperatures and higher selectivities toward syngas. However, this reaction did demonstrate the operability of the system.

4. Summary

A highly integrated, gas-phase microfluidic reaction system for high temperature partial oxidation reactions has been demonstrated. The thin-film heaters and temperature sensors required considerable support electronics and novel methods for real-time control. Further design work could extend the concepts developed here to accommodate entirely liquid phase or gas-liquid phase reactions. With longer time constants and lower temperature requirements these systems should be simpler to design.

Acknowledgements

The authors thank the DARPA MicroFlumes Program for financial support under contract F30602-97-2-0100. DJQ thanks the National Science Foundation for his graduate fellowship.

References

1. K.F. Jensen, *Microreaction engineering -- is smaller better?* Chemical Engineering Science **56**, pp. 293-303 (2001).
2. W. Ehrfeld, V. Hessel, and H. Löwe, *Microreactors: New Technology for Modern Chemistry*, Wiley-VCH, New York, (2000).

MONOLITHIC VERSUS MODULAR INTEGRATION OF A MICRO-FIA SYSTEM FOR AMMONIUM DETERMINATION

T.T. Veenstra, R.M. Tiggelaar, R.G.P. Sanders, J.W. Berenschot, J.G.E. Gardeniers[1], J.M. Wissink[2], R. Mateman[2], M.C. Elwenspoek, A. van den Berg

MESA+ Research Institute, University of Twente, Enschede, The Netherlands

[1] Kymata Netherlands, Enschede, The Netherlands

[2] 3T BV, Enschede, The Netherlands

(t.t.veenstra@el.utwente.nl)

Abstract

Two differently constructed μ-FIA ammonium detection systems have been realized and tested. A monolithic system (Full-Lab-On-A-Chip) shows reproducible measurements for 10 and 20 mM ammonium samples. A modular version of this system could measure ammonium concentrations down to 1.67 mM.

Keywords: Lab-On-A-Chip, FIA-system, Ammonium

1. Introduction

Micro analysis systems can be setup in two ways. The first approach is the integration of all necessary components on one chip. The second way is to fabricate all components separately and define a certain interconnection method.

Both system approaches have their specific advantages. Monolithic integrated micro analysis systems can be designed optimally in terms of interfacing all involved components. This results in small dead volumes in the whole system, which minimizes dispersion effects and hence allows fast analysis. A modular built up version of the same system has the major advantage that a failing component can be replaced easily.

In this paper we present two realized microsystems for the determination of the concentration of ammonium. The first system is a monolithic Full-Lab-On-A-Chip system. The second system has the same functionality as the monolithic system, but is built up in a modular way. The components are coupled by means of the MATAS technology which in a way is analogous to a PCB type of integration [1].

2. Theory

The working principle of both systems is based on the reaction mechanism of Berthelot for the detection of ammonium [2]. This reaction is implemented in a FIA system. The Berthelot reaction requires two reagents to be added to the ammonium sample. The available ammonium in the sample will be converted to indophenol blue (a blue dye). The concentration of the formed dye is measured in a light absorption cell and is directly related to the initial ammonium concentration.

J.M. Ramsey and A. van den Berg (eds.), Micro Total Analysis Systems 2001, 664–666.

Both systems consist of four pumps, two mixers, a reaction chamber and an optical detection cell. The physical layout of the monolithic system is shown in figure 1a. A photograph of the realized monolithic system is shown in figure 1b.

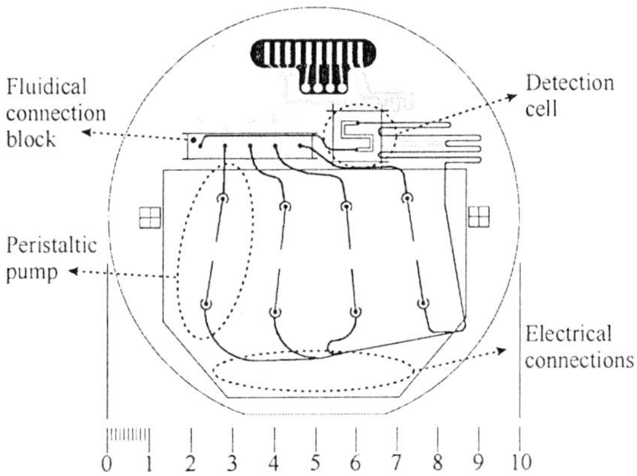

a)

b)

Figure 1: The monolithic analysis system: a) layout b) photograph

The construction principle of the modular system is shown in figure 2a. The incorporated components are made with the aid of micromachining techniques. Subsequently, the realized components are glued in PCB module housings. These housings are made using standard machining techniques (e.g. CNC controlled milling). The realized modules are soldered into a contact board. This contact board consists of two parts, which are glued on top of each other. The upper part (PCB) takes care of the electrical connections, whereas the lower part comprises the fluidical connections between the different modules. A photograph of the system is shown in figure 2b.

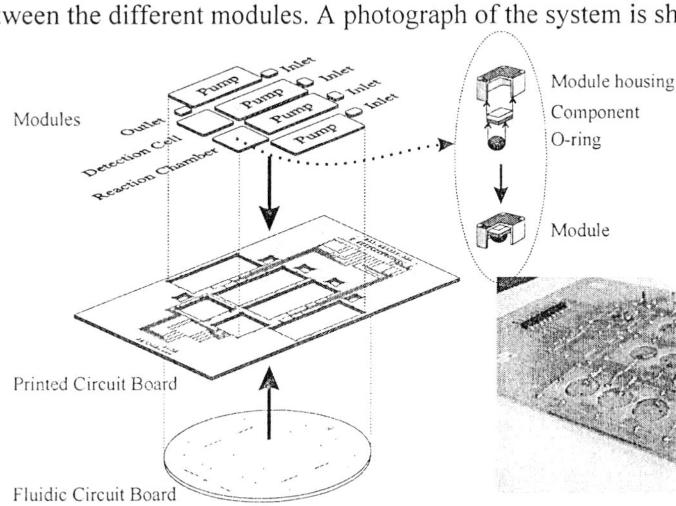

Figure 2: The modular analysis system: a) layout b) photograph

3. Experimental

Performed preliminary concentration measurements with both systems are shown in figure 3. For the monolithic system, concentrations for 10 and 20 mM were reproducibly detected. The modular built up system is capable of detecting concentrations down to 1.67 mM.

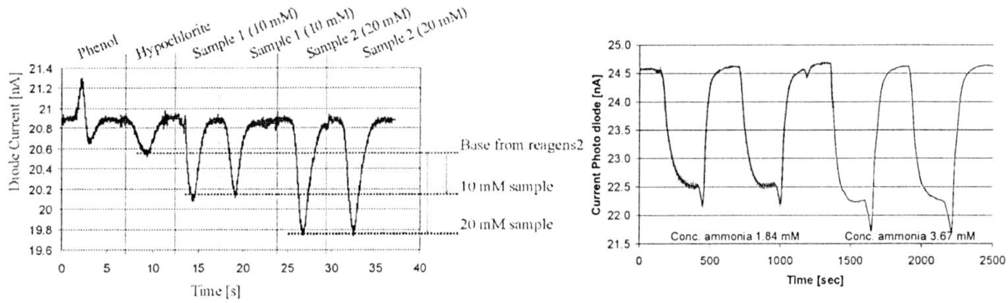

a) b)

Figure 3: Measurement readings of the: a) monolithic system b) modular system

4. Conclusions and Discussion

Both a monolithic and a modular built-up analysis system for the measurement of ammonium (using Berthelot's reaction mechanism to convert ammonium into indophenol blue) were realized and tested.

Though preliminary measurements prove the working principle and indicate the possibility of reproducible measurements, the detected ammonium concentrations are not yet in the desired concentration range (down to 6 μM). However, the detection range can be adapted to the desired range by decreasing the flow rate in the system. This will result in a higher degree of completion of the Berthelot reaction, thus changing the detection limits of the system towards lower concentrations. Also a modified Berthelot reaction can be applied, resulting in a faster analysis.

Though the realized systems have been designed for the Berthelot reaction for the detection of ammonium, the system control parameters (e.g. temperature, residence time) can be changed easily to accommodate any other two-step chemical reaction.

Acknowledgements

The Dutch Technology Foundation (STW) is acknowledged for funding this research.

References

1. *J. Wissink, A. Prak, M. Wehrmeijer, R. Mateman, Novel Low Cost Modular Assembly Technologyfor uTAS using PCB-technology, Proceedings Volume 2 MICRO.tec 2000 - VDE World Microtechnologies Congress, pp. 51-56, September 25-27, 2000, EXPO Hannover, Germany*
2. *P.L.Searle, The Berthelot or indophenol reaction and its use in the analytical chemistry of nitrogen - A review, Analyst 109, pp549-568, 1984*

ON-CHIP INTEGRATION OF MULTI ION SENSING SYSTEM BASED ON MICROFLUIDIC CONTROL

Hideaki Hisamoto[1], Takayuki Horiuchi[1], Kenji Uchiyama[2], Manabu Tokeshi[2], Akihide Hibara[1], and Takehiko Kitamori[1, 2]

[1] Department of Applied Chemistry, Graduate School of Engineering, The University of Tokyo, 7-3-1 Hongo, Bunkyo-ku, Tokyo 113-8656, Japan

[2] Integrated Chemistry Project, Kanagawa Academy of Science and Technology, 3-2-1 Sakado, Takatsu-ku, Kawasaki, Kanagawa 213-0012, Japan

(e-mai: hisamoto@appchem.t.u-tokyo.ac.jp)

Abstract

Multi ion-sensing device using single microchip was successfully performed by microfluidic control. While conventional ion sensors enable single ion sensing for single device, our system enables multi ion sensing for single device. The key technologies to realize the system are segmented flow injection of plural organic phases containing different ion-selective neutral ionophores and subsequent formation of organic/aqueous two layer flow. The integrated microfluidic system propose here provides significant advantages compared with conventional ion sensors using solvent polymeric membrane.

Keywords: Ion pair extraction, Microfluidic control, Multi Ion sensing, Segmented flow injection

1. Introduction

We have proposed exploitation of the characteristics provided by a liquid microspace in order to develop widely applicable chemical systems using microchips. Among the characteristics of a liquid microspace, the short molecular diffusion distance and large specific interfacial area are quite advantageous for effective liquid/liquid extraction [1-3].

The microfabricated device also allows us to exploit characteristics of microfluidics such as formation of laminar multi-phase flow. When the organic phase and aqueous phase are introduced into microchannel, stable laminar flow forms and continues to the end of the microchannel [4].

Very recently, we reported the on-chip integration of neutral ionophore-based ion pair extraction as an advanced ion-sensing scheme using microfabricated devices [5]. However, further exploitation of characteristics provided by microfluidics can be expected to realize a novel multi-ion sensing system, which is not easily demonstrated by using conventional ion sensor technology.

In order to perform multi-ion sensing using a single microchip, we propose here a new methodology involving neutral ionophore-based ion pair extraction, segmented flow injection of plural organic phases, and formation of laminar two-phase flow with an aqueous sample solution.

J.M. Ramsey and A. van den Berg (eds.), Micro Total Analysis Systems 2001, 667–669.

2. Operation principle of the multi-ion sensing device

Figure 1 shows the operation principle of the multi-ion sensing using a microchip. Different organic phases containing the same lipophilic pH indicator dyes, but different ionophores are introduced sequentially into the microchannel by on-off switching of syringe pumps. In this case, organic phase without ionophore is introduced in between them, in order to avoid contamination. Aqueous sample solution containing different ions is introduced from the other inlet to form laminar two-phase flow with the segmented organic phase. The selective ion pair extraction reaction proceeds during flow, thus, different ions can be selectively extracted into different organic phase segments depending on the selectivity of neutral ionophores contained in the respective organic phase segments. Downstream in the flow, the ion pair extraction reaction becomes equilibrated, thus, downstream detection of the color change of organic phase allows sequential and selective multi-ion sensing in the single aqueous sample solution containing multiple ions. Here we demonstrate successful multi-ion sensing of potassium and sodium ions using potassium and sodium selective neutral ionophores.

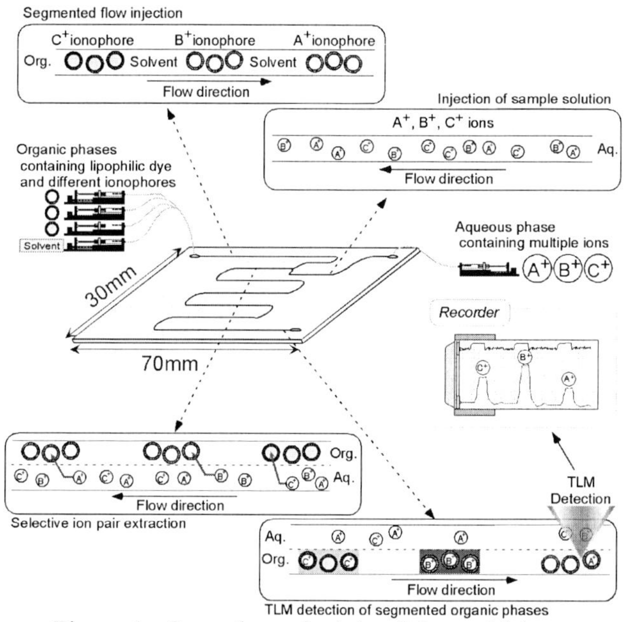

Figure 1. Operation principle of the multi-ion sensing device

3. Experimental

Glass microchip possessing wet-etched microchannel of 150μm wide and 45 μm deep was used. In order to realize the concept, we used KD-A3 as lipophilic pH indicator dye and valinomycin (for K^+) and DD16C5 (for Na^+) as neutral ionophores. The butylacetate containing KD-A3 and Valinomycin (5×10^{-6} M each), that without ionophore, and that containing KD-A3 and DD16C5 were sequentially introduced into one inlet of microchannel by on-off switching of syringe pumps to form segmented flow of organic phase. For aqueous phase, 0.05M Tris-HCl buffer solution (pH 10) of alkaline metal chloride salts were used. Thermal lens microscope (TLM) detection were performed at the organic phase, 166mm down stream from the confluence point.

4. Results and discussion

As shown in Figure 2, sequential determination of potassium and sodium ions was successfully performed. Minimized volume of single organic phase segment to obtain equilibrium response without dilution by cross dispersion of two organic phase segments, was ca. 500 nl in our system. While several milli grams of expensive reagents are required for the preparation of ion sensor membrane, our system requires only 1.7 ng and 2.8 ng for the amount of dye molecule and ionophore molecule, respectively, in one organic phase segment. It means more than 10^6 sequential measurements can be achieved in our system without any signal degradation, which has been a typical problem of conventional ion sensors.

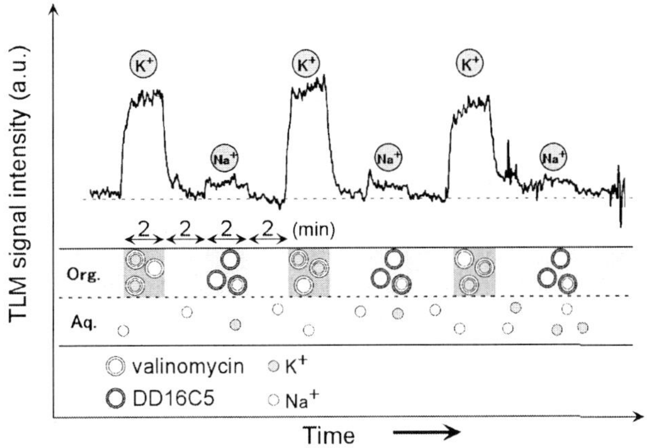

Figure 2. Multi ion-sensing of potassium and sodium ions

5. Conclusions

We have demonstrated the sequential ion-sensing system employing segmented flow injection of organic phase and formation of two-layer flow inside the microchannel. The proposed system allowed determination of multiple ions in one sample solution containing the ions. The methodology proposed here is expected to be applicable to the multi-ion analyses required in clinical and environmental chemistry.

Acknowledgements

We gratefully acknowledge Prof. Koji Suzuki, Department of Applied Chemistry, Keio University, for supplying the lipophilic pH indicator dye molecule (KD-A3) and the highly sodium-selective neutral ionophore (DD16C5). This work was partially supported by Grants for Scientific Research from the Ministry of Education, Science, and Culture, Japan.

References

1. M. Tokeshi, T. Minagawa, T. Kitamori, *Anal. Chem.*, **72**, 1711-1714, (2000).
2. K. Sato, M. Tokeshi, T. Sawada, T. Kitamori, *Anal. Sci.*, **16**, 455-456. (2000).
3. M. Tokeshi, T. Minagawa, T. Kitamori, *J. Chromatogr. A*, **894**, 19-23 (2000).
4. A. Hibara, M. Tokeshi, K. Uchiyama, H. Hisamoto, T. Kitamori, *Anal. Sci.* **17**, 89-93 (2001).
5. H. Hisamoto, T. Horiuchi, M. Tokeshi, A. Hibara, T. Kitamori, *Anal. Chem.* **73**, 1382-1386 (2001).

FULLY AUTOMATED SAMPLE PREPARATION FOR PATHOGEN DETECTION PERFORMED IN A MICROFLUIDIC CASSETTE

M. T. Taylor, P. Belgrader, R. Joshi, G. A. Kintz, M. A. Northrup
Cepheid, 1190 Borregas Ave., Sunnyvale, CA
(taylor@cepheid.com)

Keywords: Microfluidic processing, automated sample preparation, DNA analysis

Abstract
Sample preparation steps including bacterial spore concentration by filtering, washing to remove PCR inhibitors, ultrasonic lysis, pumping of elution containing DNA, mixing with PCR-mix, thermal cycling and real-time PCR were performed automatically in a microfluidic cassette.

1. Introduction
Molecular testing for pathogens based on nucleic acid identification requires several sample preparation steps to release and isolate the DNA for polymerase chain reaction (PCR) analysis. These labor-intensive procedures can include spore/cell concentration, centrifugation, filtration, washing, lysis, pipetting, mixing, vortexing, and handling. We have reduced a complex multi-step procedure for spore detection, developed a fully automatic processing system, and integrated all sample preparation and PCR functions into a miniature cassette as part of a micro total analysis system (μTAS).

2. Theory and Experimental
All microfluidic processing was performed in a disposable plastic cassette (Fig. 1a & b) mounted into the cassette-station of an instrument (Fig. 1c) with inputs for valving, pumping, sonication, mixing and PCR analysis. The cassette contains membrane valves and diaphragm pumps, which are driven by pressure/vacuum nozzles built into the instrument cassette-station. The station also contains an ultrasonic transducer for spore/cell lysis and an I-CORE module from a Smart Cycler (Cepheid, Sunnyvale, CA) for PCR, both of which interface with the cassette.

As the sample is processed and becomes more purified it is moved into clean regions. To achieve this, 9 valves and 3 pumps operated sequentially for peristaltic action [1] are built into the cassette. The first 100-μL pump is for sample filtering, washing and extracting eluate. The second 60-μL pump is used for mixing the eluate with master-mix and filling the reaction tube. The third 25-μL pump empties the reaction tube so the PCR product is available for post-PCR analysis, (i.e., electrophoresis, mass spectrometry, etc.). Pumps and valves consist of an elastic silicone rubber membrane in-between two opposing spherical cavities formed in the plastic layers. Membrane theory [2] was implemented to predict the relationship between the pressure P and the

J.M. Ramsey and A. van den Berg (eds.), Micro Total Analysis Systems 2001, 670–672.

spherical radius R of a membrane with base radius r_0, thickness t_0, and modulus E (Fig. 2). Figure 2b shows the equation used to determine the geometry of the cavity that confines the stroke of the membrane. Pumping required less than 20 kPa to produce a repeatable volume of liquid for every cycle (Fig 2c). This insures complete valve sealing with operation of the pumps and valves at 140 and 280 kPa, respectively.

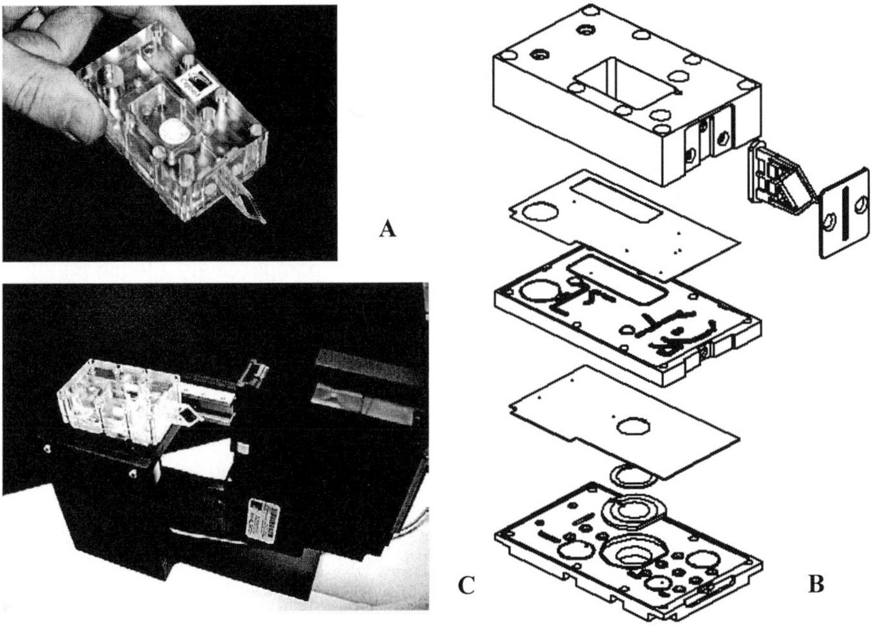

Figure 1: (A) Miniature sample preparation cassette (38 x 64 x 25 mm). (B) Exploded view of cassette showing gaskets between plastic layers which contain valves, pumps, and channels. Also shown are the 100-μL PCR reaction tube and filter-sonication region. (C) Cassette is shown being loaded into cassette-station which contains pressure/vacuum nozzles, ultrasonic transducer, and Smart Cycler I-CORE (Cepheid).

3. Results and Discussion

Spore samples (10^4 of *Bacillus subtilis* in 1-mL) were introduced via pipette and then pumped through a 0.22-μm filter (Millipore Corp, Bedford, MA). This was followed by a wash with DI water (2-mL). Spores trapped on the filter were sonicated for 15 s in the presence of 30 mg of 106-μm glass beads (Sigma Chem., St. Louis, MO) to aid in spore disruption. The elution (30-μL) from the disrupted spore solution was pumped into a chamber containing PCR master-mix (70-μL) and toggled in and out using the 60-μL pump for mixing. The mixture was then finally pumped into the 100-μL reaction tube for real-time PCR analysis. Quantitative real-time PCR, known to be an indicator of the amount of spore genomic DNA, was used to detect the presence or absence of spores in the test solution (Fig. 3). The spore solution produced a significant positive signal compared to no signal using a control solution with no spores.

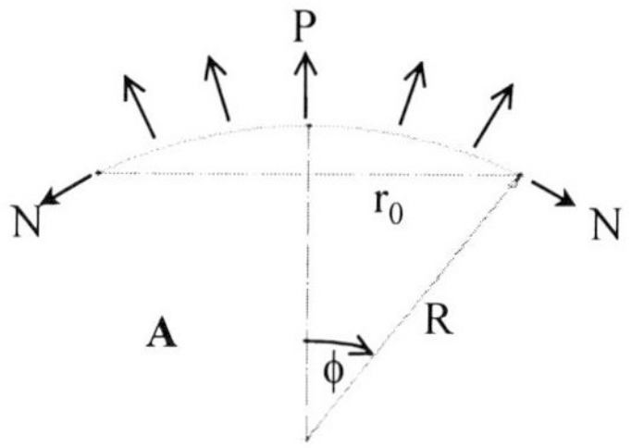

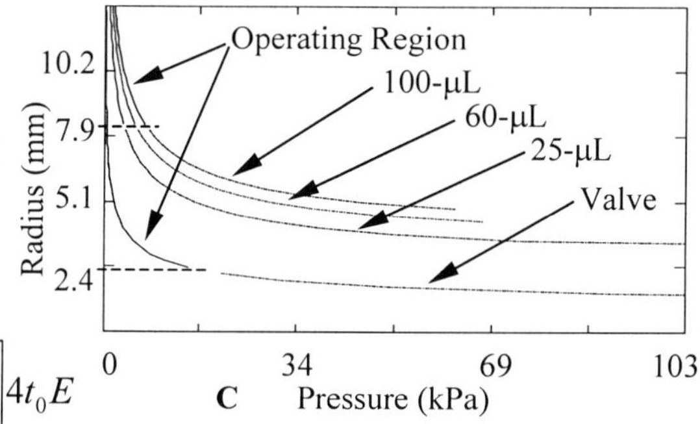

B $$P = \left[\frac{1}{r_0} \sin^{-1}\left(\frac{r_0}{R} \right) - \frac{1}{R} \right] 4 t_0 E$$

Figure 2: (A) Membrane under pressure P with spherical radius R, N normal force/unit length. (B) Equation from membrane theory. (C) Radius as a function of pressure for valves (cavity 2.4 mm R), and for 25-, 60-, and 100-µL pumps (cavity 7.9 mm R).

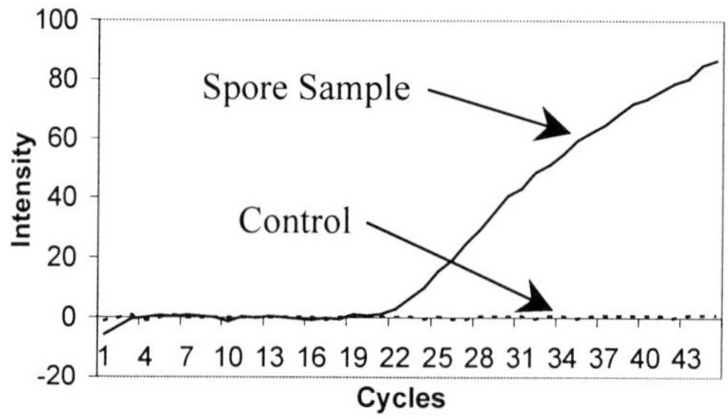

Figure 3: Real-time PCR analysis indicating the quantitative detection of spore DNA from the processed spore sample. No signal was obtained from the negative control.

4. Conclusions

These results show the promise of a fully automated, miniaturized, pathogen detection system. We are continuing development of these portable systems for lab, field, and hospital applications.

Acknowledgement

The authors would like to acknowledge the support of DARPA's Advanced Pathogen Detection Program (N65236-98-1-5414).

References

1. M. T. Taylor, F. Raisi, P. Belgrader, F. Pourahmadi, A. E. Herr, G. A. Kintz, and M. A. Northrup, *Microfluidic Bioanalysis Cartridge with Interchangeable Microchannel Separation Components*, Transducers'01/Eurosensors XV, 11th International Conference on Solid-State Sensors and Actuators, Munich, 2001, Springer, pp. 1214-1217 (2001).

2. C. Ugural and S. K. Fenster, *Advanced Strength and Applied Elasticity*, 3rd Ed. Prentice Hall, 1995.

Authors Index

A

B

C

D

E

I

J

K

L

M

N

O

P

Q

R

S

T

U

V

W

X

Y

Z

Subject Index

F

G

H

I

J

Q

R

S

T

U

V

W

XYZ